Solid Mechanics
in Engineering

Solid Mechanics in Engineering

RAYMOND PARNES

Department of Solid Mechanics
Materials and Systems
Tel-Aviv University

JOHN WILEY & SONS, LTD

British Library Cataloguing in Publication Data

A catalogue record for this book is available from the British Library

ISBN 978-0-471-49300-6

For

SOUAD

and our children

DAVID, DANNA and NICOLE

ERRATA

Solid Mechanics in Engineering

Raymond Parnes   0 471 49300 7

Please note that the figure numbering within Section 14.8 is incorrect. The first figure in this section **should be** Figure 14.8.1 and **not** 14.8.35. Figure 14.8.1 should be 14.8.2 and all subsequent figures are similarly incorrect up to the end of the section, where Figure 14.8.15 becomes 14.8.16.

The publishers very much regret this error and hope that this clarification will help to avoid any confusion

Contents

◇ indicates sections that may be omitted on a first reading without loss of continuity.

◊ indicates sections that may be omitted on a first reading without loss of continuity.

◊ indicates sections that may be omitted on a first reading without loss of continuity.

◊ indicates sections that may be omitted on a first reading without loss of continuity.

PART C   Energy methods and virtual work

14   Basic energy theorems, principles of virtual work and their applications to structural mechanics   537

Preface

This book is intended to serve as a textbook, typically covered in two semesters, primarily for undergraduate students in the fields of aeronautical, civil and mechanical engineering. Nevertheless several chapters—particularly the latter—can be incorporated in a first-year graduate program. The book is based on class notes for courses in solid mechanics and mechanics of materials that I have taught over the past 30 years in the United States, Europe and Israel. The reader is assumed to be familiar with the basic ideas of mechanics, principally statics.

In general, as an overview, the book presents the material with an emphasis on theoretical concepts. However, although fundamental concepts are emphasised, technological, practical and design applications are illustrated throughout by a large number of examples.

From a perusal of the Table of Contents, it might appear that the book is similar to other textbooks on the subject. Indeed, much of the subject matter is rather 'classical', as it treats a subject that has been part of engineering curricula for many decades, if not longer. The question then arises: Why another book on this subject? I believe that the approach of this book to the subject is quite different from most others and that its rational approach and level is such that it will appeal to instructors who wish to emphasise the fundamental ideas of solid mechanics.

One might say that the text is a compromise between the approach as found, for example, on the European continent (where the most general theory is first presented leading later to simplified and approximate 'strength of material' results) and the approach of engineering schools in the United States, where an ad hoc treatment is often preferred. In an ad hoc approach, many of the fundamental and unifying ideas are overlooked and neglected in most textbooks on the subject.

To avoid this pitfall, here the basic concepts are first developed in Part A of the book, starting with an introductory chapter (Chapter 1) where it is stressed that solid mechanics is based on three fundamental ideas: namely (a) the laws of mechanics, (b) the kinematic (i.e. geometric) equations describing the deformation of a solid and (c) the basic equations (i.e. constitutive equations) that describe the general material behaviour of a solid. These fundamental ideas are illustrated in the first introductory chapter by some simple one-dimensional examples. The next few chapters (Chapters 2–4) are devoted to a careful development of the three basic concepts (that were introduced in Chapter 1) of solid mechanics: stress, as a measure of the intensity of internal force (Chapter 2); strain, as a measure of the intensity of deformation (Chapter 3) and material behaviour, on a phenomenological level, as described by various types of constitutive equations (Chapter 4). In this last chapter in particular, the reader is given a general global view of materials that are classified into several types. The concepts of 'micro' and 'macro' scales, necessary for a modern approach to an understanding of materials, are introduced. These features serve to give the student a better idea of the overall picture of the behaviour of solids. In Chapter 5, the results of the previous chapters are summarised and the general approach to the analysis of specific problems is presented.

It should be pointed out that several features of Part A do not appear, to the author's knowledge, in most other textbooks dedicated for courses in this subject. To cite but one example: the equality of the conjugate shear stress is shown to be equally valid for bodies not in a state of equilibrium. (Too often, based on the usual approach using static equilibrium, students are erroneously led to believe that this property holds true only for a body in equilibrium. As a consequence, the symmetric property of the tensor is then believed to be valid only for a state of equilibrium.)

The fundamental ideas that were developed in Part A are then applied in Part B: Chapters 6–11 cover the basic applications to simple structural elements encountered in practice: axial behaviour, torsion, flexure and buckling. More advanced topics (such as general torsion, unsymmetric bending of beams, etc.), which are usually covered in a second semester, are treated in Chapters 12 and 13 and in the chapters of Part C where the concepts of energy and virtual work are carefully developed. Each topic is illustrated by means of numerous illustrative examples. As a particular feature, believing that it is not sufficient to only show the solution, many examples are immediately followed by extensive comments in order to provide an interpretation of the solution and encourage the student to develop physical insight. Such comments are often accompanied by graphs to illustrate the effects of the governing parameters.

In developing the relations for the behaviour of various elements, the derivations follow either from plausible physical assumptions or from direct conclusions based on the deformation pattern. At each stage, it is emphasised which solutions are 'exact' within the theory and which are 'approximate'. To illustrate this point, I cite but three examples: (1) After having developed the simple expressions for the axial stress and elongation of a prismatic rod, the expressions are not blithely applied to rods of varying cross-sections. Instead, the reader is shown *quantitatively* when such an approximation is permissible. (2) After having developed the expression for the flexural stress in beams, it is *shown* (using the equations of equilibrium derived in Chapter 2) that this expression is 'exact' only for pure bending or when the moment varies linearly along the beam. (3) While discussing the deflections of relatively stiff beams using linearised Euler–Bernoulli beam theory, a clear upper bound to the error is derived; the student is not expected to merely accept, with no quantitative explanation, that the beam must be sufficiently stiff. These examples serve to illustrate the underlying philosophy of the book, namely that the reader must be provided with rigorous analytical explanations and is not expected to accept 'hand waving' explanations.

Another goal of this text has been to eliminate unnecessary errors, simplifications or misconceptions that often arise in introductory courses – errors that are to be later undone, as the student pursues more advanced studies. Thus, I have attempted to write a text so that students 'get it right the first time'. From the above comments, it is clear that as the principal aim of the treatment is to provide the student with a broad and fundamental understanding of basic principles, the book attempts to present the reader with several unifying ideas. In this respect, the book provides the student with a thorough preparation for more advanced studies in the field.

A word about the mathematical level: although not requiring a knowledge of 'higher mathematics', it is assumed that the student has a good preparation in differential and integral calculus, differential equations and linear algebra (i.e., a reasonable knowledge of mathematics). I have not attempted to avoid mathematics where it is appropriate and necessary, particularly in the latter chapters of the book. (Too often, students question as to why they are subjected as undergraduates to mathematics courses when little use is made of what they have studied.) In fact, I view a course in solid mechanics as an excellent opportunity to expose students to the application of mathematics to engineering problems, to reinforce their mathematical studies and thus enhance their analytical abilities.

Although, as mentioned above, the text has been prepared for a two-semester course, instructors may wish to skip various sections at their discretion. I have indicated by means of a symbol (◇) certain sections and subsections that may be omitted on a first reading without loss of continuity.

Over 600 problems, of varying degrees of difficulty, are included in the text. Most of these are not numerical in nature, since I believe students should be encouraged to first work out the solutions algebraically. Those problems (not necessarily more difficult) that require a deeper understanding of the subject or a more sophisticated approach are indicated by an asterisk (*). Answers to about half of the problems are provided. Since any modern engineering curriculum provides students with a reasonable facility with computers, computer-related problems can be found throughout the text. These problems are generally not of an artificial nature; rather they *require* the use of a computer (e.g., for the solution of transcendental or quartic algebraic equations). Students are therefore encouraged to write algorithms using, at their discretion, FORTRAN or other software such as MAPLE, MATHEMATICA or MATLAB.

Finally, I acknowledge my debt to my teachers who had taught me while I was an undergraduate and graduate student at Columbia University; in particular, to Raymond Mindlin and Mario Salvadori from whom I learned to understand the beauty of mechanics and applied mathematics and who taught me much of what I know today.

I wish to thank my department colleagues Leslie Banks-Sills, Shmuel Ryvkin and Leonid Slepyan, who read parts of the manuscript during the preparation of this text, and for their discussions, comments and suggestions, as well as Dan Givoli of the Technion, Israel Institute of Technology. A word of thanks is also due to the Department of Engineering of the University of Cambridge (UK) where this book was completed while I was on sabbatical leave, unencumbered by usual university obligations.

I would appreciate receiving comments—both positive and negative—and suggestions for further improvements from readers of this text.

Raymond Parnes

Basic concepts

1

Introductory concepts
of solid mechanics

1.1 Introduction

Solid mechanics is a branch of mechanics that has many applications. In ancient times, solid mechanics was of interest primarily for the construction of structures and buildings. The pragmatic knowledge of this subject was based on empirical rules that were accumulated, based on both the successes (and failures) in the construction of previous structures. Starting with Galileo and Newton, attempts were made to determine rational laws governing the general behaviour of solids. Great progress was made in the understanding of the behaviour of solid bodies during the eighteenth and nineteenth centuries, notably by Bernoulli, Euler, Coulomb, Navier, Poisson, Cauchy and others. The study of mechanics has continued into the present century, and well-developed theories and principles have been elaborated. An understanding of the basic laws of solid mechanics is of particular importance in mechanical, aeronautical and civil engineering. With the advent of modern materials, it has been necessary to develop more refined theories to ultimately achieve the most efficient design of the relevant structures.

The study of solid mechanics has as its goal, the determination of the deformation and internal forces existing in a body when subjected to external loads. Solid mechanics is based on the following:

(a) *Physical laws that describe the behaviour of solids in accordance with experimental data obtained in a laboratory.* The laws must thus accord with the general behaviour as found in the real world.
(b) *Mathematical deductions to express these laws, based on simplifying assumptions.* Such assumptions must often be made to render the solutions tractable. In other words, one wishes to model a problem in the simplest fashion, provided it leads to solutions that adequately describe the actual behaviour of the body.

Let us consider a typical problem that, in its simplicity, reveals several aspects of our study. We consider a plank AB of length L, which rests at two ends. We wish to know if it can support a person whose weight is $P = Mg$, standing at a distance a from the left end, as shown in Fig. (1.1.1a). From our previous study of rigid-body mechanics, we know that the supports at A and B can be idealised as 'simple supports'. Furthermore, we may represent the force exerted by the person on the plank by a concentrated force P. Thus we replace the actual problem by the

3

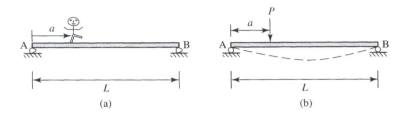

Figure 1.1.1

idealised model shown in Fig. (1.1.1b). From rigid-body mechanics, we can then find the reactions at A and B, which clearly depend on the force P.

Now, since the plank is not rigid, it will deform and assume a curved shape, as shown in the figure, when subjected to the load P. From our study of solid mechanics, we should want to determine (a) the maximum load that the plank can support, (b) the displacement of all points of the plank, i.e., the shape of the deformed curve and (c) the internal forces existing within the plank. Clearly, these quantities depend upon the geometry of the plank, namely, the length L, the position a of P and the geometry of the cross-section and the material. In this case, since we assume that the plank is made of wood, the behaviour depends also on the type of wood. However, if the plank were made of steel or some other material, the characteristics of the material would obviously have to enter into our calculations.

In particular, to establish the maximum force that the plank can bear, we must first determine not only the internal forces that exist but also the *intensity* of the internal forces. The intensity of internal forces will lead us to the concept of *stress*. On the other hand, to determine the deformation, we will require a measure of the *intensity of deformation*; this will lead us to the concept of *strain*.

Now, the analysis of this simple problem must evidently be expressed in terms of mathematical equations. Thus, it should be clear from this discussion that we will encounter three types of equations, namely

(a) Equations of mechanics (in this case, equations of equilibrium) that are written in terms of forces and/or 'stresses'.
(b) Kinematic equations, i.e., equations describing the geometry and deformation of the body. Clearly, these are written in terms of the displacement of points of the body and involve 'strain'.
(c) Equations that describe the general mechanical behaviour of the material and which are characteristic of given materials. These equations will involve the material properties of the body. Such equations are called, in general, **constitutive equations**.

Thus, based on the discussion of this simple problem, we remark that three basic concepts exist in the study of solid mechanics: stress, strain and the intrinsic behaviour of a material, as described by its constitutive equations.

In the first part of this book, we shall consider these concepts in detail and establish them on a firm mathematical basis in order to enable us to apply them to relevant engineering problems.

In the next sections of this chapter, we first consider some idealisations that are used in solid mechanics and introduce and elaborate on the above concepts via some simple one-dimensional problems.

1.2   Forces, loads and reactions – idealisations

In studying the behaviour of deformable bodies, we must consider how loads are applied. Furthermore, in considering the application of loads and their reactions, it is usually necessary to make certain idealisations. The extent and type of idealisations are, of course, dependent on the degree of refinement of the analysis that we require.

In general, all forces and loads are represented by means of vectors. We recall that in rigid body mechanics, a force can be represented by a 'sliding vector'; i.e., one can write equations by considering a force anywhere along its line of action. However, in studying deformable bodies, this is no longer true: *one must stipulate the actual point of application of the force.* To show this, consider two bodies [Figs. (1.2.1a and b)] subjected to three forces F_1, F_2 and F_3. If these are rigid bodies then, in both cases, the equations of equilibrium, $\sum F = 0$ and $\sum M = 0$, are the same. Since, by definition, a rigid body does not deform, we do not consider any internal deformation. However, if the bodies are deformable bodies, then clearly, the body in Fig. (1.2.1a) is in compression while that in Fig. (1.2.1b) is under tension. It is therefore evident that the two bodies will behave quite differently; the first will tend to become smaller, while the second will become larger. Thus we see that in solid mechanics it is necessary to prescribe not only the line of action of a force but also the point of application of the force.

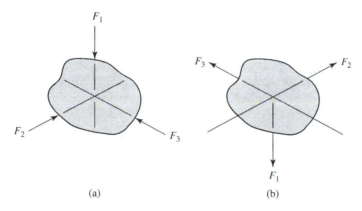

(a)   (b)   Figure 1.2.1

(a) Types of loads

In solid mechanics, a body may be subjected to two types of applied loads: contact forces and body forces.

Contact forces are forces that are applied to the body, usually on its external surface, by direct contact [Fig. (1.2.2a)]. *Body forces* are forces that act upon a

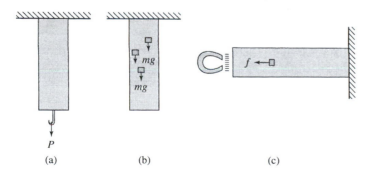

(a)   (b)   (c)   Figure 1.2.2

body through action at a distance. Such forces are assumed to act on the particles of the body and may thus be either constant or may vary throughout the body. Gravitational force is an example of a body force: for each particle of mass m, the particle 'feels' the gravitational attraction [Fig. (1.2.2b)]. A magnetic force acting on, say, an iron bar, is another example of a body force [Fig. (1.2.2c)]. It is noted that body forces have units of Newton/metre^3 (N/m^3).

(b) Representation of forces and loads

In our previous discussion, we represented all forces by means of vectors. This clearly is an idealisation of a concentrated force acting at a point. Such a case can never, in fact, exist in nature, for we know that a force can only be applied over some small but finite area; it is for this reason that we state that a concentrated force is but an idealisation. We therefore consider the following representations.

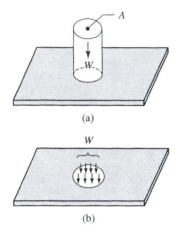

(a)

(b)

Figure 1.2.3

(i) *Distributed loads.* A load acting over a finite area is called a **distributed load**. For example, consider a cylindrical block, whose cross-section is A and whose weight is W, resting on a plate [Fig. (1.2.3a)]. For simplicity, we assume that the weight of the cylinder is evenly distributed over the cross-section. The distributed load that acts upon the plate is then represented by W/A [Fig. (1.2.3b)]. Note that the units of this distributed load are N/m^2.

(ii) *Concentrated force or point load.* Consider a distributed load, as described above, acting over a small area ΔA about some point x_p, y_p [Fig. (1.2.4a)].

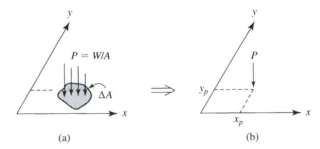

(a)   (b)

Figure 1.2.4

Let $p = W/\Delta A$ denote this distributed load which, in effect, is a pressure. We now consider the case where the pressure $p(x, y)$ increases indefinitely, i.e., $p \to \infty$, while $\Delta A \to 0$. The resultant force is then defined as

$$P(x_p, y_p) = \lim_{\substack{\Delta A \to 0 \\ p \to \infty}} \iint\limits_{\Delta A} p(x, y)\, dA. \qquad (1.2.1)$$

Thus, Eq. (1.2.1) is to be taken as the mathematical definition of an idealised concentrated load P acting at x_p, y_p [Fig. (1.2.4b)]. Note that the unit of P is the Newton. Therefore, when we represent a concentrated load by means of a vector, we implicitly use the idealisation defined mathematically by Eq. (1.2.1). We shall find that this idealisation is generally acceptable in our study of solid mechanics.

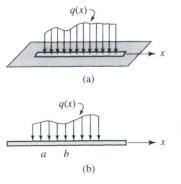

Figure 1.2.5

(iii) *Line loads.* If a distributed load is acting over a relatively thin strip, it is called a **line load** and is a function of a single coordinate, say, x [Fig. (1.2.5)]. We note that the line load may be constant or may be a function, which we denote by $q(x)$. Note too that the units of q are N/m. Thus we observe that, at any point, there exists a load of intensity $q(x)$ given in N/m. The resultant force R

of the load between any two points, $x = a$ and $x = b$, is then given by

$$R = \int_a^b q(x)\,dx. \qquad (1.2.2)$$

Note that the resultant force is represented by the area under the load function $q(x)$. Line loads are commonly encountered in the study of beams.

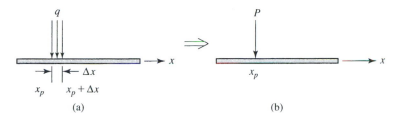

(a)   (b)

Figure 1.2.6

Concentrated loads acting on beams are idealisations, defined in a manner analogous to Eq. (1.2.1). Thus, consider a line load $q(x)$ acting over a small distance Δx between two points x_p and $x_p + \Delta x$ [Fig. (1.2.6a)]. As before, let us now consider the case where the intensity $q(x)$ of this distributed load increases indefinitely, i.e., $q(x) \to \infty$, while $\Delta x \to 0$. The resultant force is then defined as

$$P(x_p) = \lim_{\substack{\Delta x \to 0 \\ q \to \infty}} \int_{x_p}^{x_p + \Delta x} q(x)\,dx. \qquad (1.2.3)$$

Thus, Eq. (1.2.3) is the one-dimensional definition of an idealised concentrated load P acting at $x = x_p$ [Fig. (1.2.6b)].

As we have observed, the concept of the concentrated force P acting at a point, as defined either by Eq. (1.2.1) or by Eq. (1.2.3), is an artificial one and is physically unrealistic. However, it is a useful concept that simplifies considerably the solution of many problems. Moreover, because this concept is an artificial one, it can often lead to solutions that contain spurious discontinuities at the point of application of P, which contradict actual physical behaviour. Upon realising that the basic idealisation is indeed artificial, we are then usually willing to disregard these spurious results.

(c) Reactions and constraints – idealisations

A body is usually supported in such a way that at certain points no motion can take place; constraints are therefore said to exist at these points. For example, in the beam shown in Fig. (1.2.7a), a constraint against translation in the x- and y-directions exists at point A, while at point B there is a constraint against motion in the y-direction. These constraints are idealised by a 'pin' at A and a 'roller' at B. The forces that provide these constraints (i.e., which prevent the motion) are called **reactions**: note that to each constraint there corresponds a component of the reaction. Thus at point A, there exist reactive components R_{Ax} and R_{Ay} while at point B there exists only a single component, R_{By} [Fig. (1.2.7b)]. We refer to point A as a (frictionless) pin support and to point B as a roller support. Collectively, the beam is said to be 'simply supported'. The above are in fact idealisations, in the sense that we are referring to points in the beam at which concentrated reactive components (i.e. concentrated forces) are acting.

We note that a 'pin' can exist within a structure composed of several elements. For example, consider the structure shown in Fig. (1.2.8a) consisting of two elements

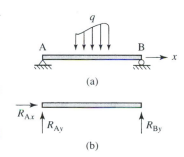

(a)

(b)

Figure 1.2.7

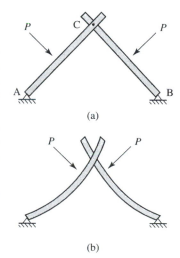

(a)

(b)

Figure 1.2.8

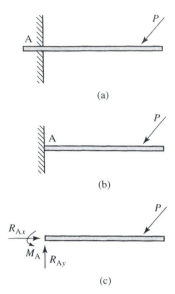

Figure 1.2.9

AC and BC supported by pins at points A and B. The two elements are connected by a pin at point C. The role of this pin at point C is to provide a constraint against relative translation of the two elements; that is, AC and BC cannot move apart. Note, however, that if the two elements are not rigid, then there can indeed be relative rotations of these two elements [Fig. (1.2.8b)]. Thus we observe that a pin does not provide any constraint against rotation. [A pin existing within components of a structure is also referred to as a (frictionless) 'hinge'. Thus a hinge cannot transmit a moment from one component of a structure to another.]

Now consider a beam shown in Fig. (1.2.9a), embedded in a rigid support at point A. The idealised model then is as shown in Fig. (1.2.9b). Such a support provides constraints against both translation and rotation at the point. In addition to the reactive components R_{Ax} and R_{Ay}, the reaction providing the constraint against rotation is a moment M_A as shown in Fig. (1.2.9c). Such a beam is called a **cantilevered beam** with a fixed end at A.

From the above discussion, we therefore find it particularly useful to define a reaction, in general, as follows: *a reaction at a point is the force that is required to satisfy a given corresponding prescribed constraint (i.e., to prevent a prescribed motion) of a body or structure at the point.*

1.3 Intensity of internal forces – average stresses

We introduce here the concept of stress as a measure of the intensity of internal forces acting within a body via a simple problem.

Consider a structure, shown in Fig. (1.3.1a), consisting of a beam, BC, pinned at points B and C. A wire, CD, of cross-sectional area A, supports the pin at C and is attached to point D. A vertical force $P = 18,000\,\text{N} = 18\,\text{kN}$ acts at point C as shown.

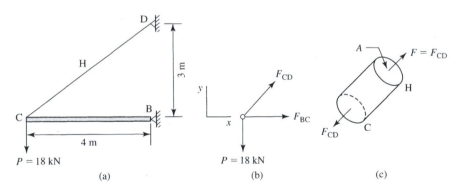

Figure 1.3.1

From statics, we note that both the element and the wire are subjected to forces only at their ends; therefore, being two-force bodies, it necessarily follows that the lines of action of the resultant end forces must fall along the longitudinal axis of both BC and CD.

Isolating the pin at C, we therefore have, from statics [Fig. (1.3.1b)],

$$\uparrow + \sum F_y = 0 \rightarrow\rightarrow 3/5 F_{CD} = 18 \rightarrow\rightarrow F_{CD} = 30\,\text{kN},$$

$$\overset{\rightarrow}{+} \sum F_x = 0 \rightarrow\rightarrow 4/5 F_{CD} + F_{BC} = 0 \rightarrow\rightarrow F_{BC} = -24\,\text{kN},$$

where F_{CD} and F_{BC} are the axial forces in the respective elements. Note that the wire CD is therefore under tension and the element BC is in compression.

Let us now investigate the internal force in the wire. To this end, we imagine that we 'cut' the wire at some arbitrary point H [Fig. (1.3.1c)]. From equilibrium we conclude immediately that there exists a normal force $F = F_{CD} = 30$ kN acting on any cross-section. Let us assume that this force is uniformly distributed over the cross-sectional area A. Then the intensity of the force per unit area is F/A. We denote this quantity by $\overline{\sigma}$ and will refer to it as the **average normal stress**; thus

$$\overline{\sigma} = \frac{F}{A}. \tag{1.3.1}$$

Note that the units of σ in the SI system are N/m^2. This unit is defined as a Pascal (Pa); i.e., 1 N/m^2 = 1 Pa. In engineering practice, one often deals with quantities that are given in thousands or millions of Pascals; thus we often use multiples of Pascals, defined as follows:

$$1 \text{ N/m}^2 = 1 \text{ Pa}$$
$$10^3 \text{ N/m}^2 = 1 \text{ kPa (kilo-Pascal)}$$
$$10^6 \text{ N/m}^2 = 1 \text{ MPa (mega-Pascal)}$$
$$10^9 \text{ N/m}^2 = 1 \text{ GPa (giga-Pascal)}.$$

Now, let us say that the diameter of the wire is 15 mm. Then $A = \pi r^2 = 56.25\pi$ mm$^2 = 1.77 \times 10^{-4}$ m^2. Hence

$$\overline{\sigma} = \frac{30.0 \times 10^3 \text{ N}}{1.77 \times 10^{-4} \text{ m}^2} = 169.5 \times 10^6 \text{ Pa} = 169.5 \text{ MPa}.$$

We wish to determine whether the wire can withstand this stress. Now it is evident that various materials are 'stronger' than others. In more precise terms, the maximum stress that a particular material can withstand is a characteristic of the material. Let us say that CD is made of low-carbon steel. For such steel, the maximum ultimate stress that the material can withstand in tension is, based on laboratory tests, $\sigma_{ult} = 400$ MPa. (The ultimate normal stress, σ_{ult}, of various steels can be found, e.g., in tables giving these properties.) Clearly, since in this case, $\overline{\sigma} < 400$ MPa, the steel wire can sustain the load $P = 18$ kN. However, in designing the structure, one usually wishes to provide for a 'safety factor'. This may be done by introducing a maximum *allowable stress*, σ_{allow}, for the material. We therefore define the safety factor, S.F., as[†]

$$\text{S.F.} = \frac{\sigma_{ult}}{\sigma_{allow}};$$

thus

$$\sigma_{allow} = \frac{\sigma_{ult}}{\text{S.F}}.$$

If we choose, for example, a safety factor of 2.0, it follows that $\sigma_{allow} = 400/2.0 = 200$ MPa. Hence, we may state that since $\overline{\sigma} = 169.5 < 200$ MPa for the problem at hand, the structure is 'safe' according to the prescribed safety factor of 2.0.

Let us now assume instead that CD is made of aluminium with $\sigma_{ult} = 160$ MPa and S.F. = 1.6 such that $\sigma_{allow} = 100$ MPa. We therefore conclude that the wire cannot

---

[†] The allowable stresses, σ_{allow}, for a given engineering structure, are usually given according to engineering specifications and depend on the type of structure to be designed, on conditions of loadings, variations of material properties, etc.

sustain the applied load and will fail since $\overline{\sigma} > \sigma_{\text{allow}}$. Therefore a thicker wire is required. In order to satisfy the condition $\overline{\sigma} < \sigma_{\text{allow}}$, we require a cross-sectional area

$$A = \frac{F}{\sigma_{\text{allow}}} = \frac{30 \text{ kN}}{100 \text{ MPa}} = \frac{30 \times 10^3}{100 \times 10^6} = 0.30 \times 10^{-3} \text{ m}^2 = 300 \text{ mm}^2,$$

and therefore an aluminium wire of radius $r = \sqrt{A/\pi} = 9.77$, or of diameter $d = 19.54$ mm. We would then use in practice a wire of 20-mm diameter.

The preceding analysis is a primitive example of engineering design.

Now there exists a second type of stress: shearing stress. To introduce this, let us first consider two smooth plates connected by a rivet (whose cross-section is A) and subjected to two forces P as shown in Fig. (1.3.2a). Clearly the rivet is holding the plates together. Let us 'cut' the body along the plane BC (we imagine that the rivet has been cut along this plane) and isolate the lower portion as a free body. It is important to realise that whenever we make such an imaginary cut in a body, the internal forces of the whole body must then be considered as external forces acting on the isolated free body.

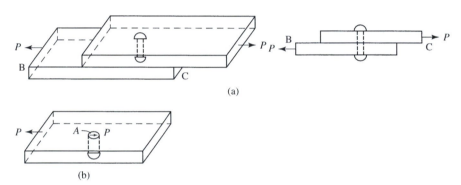

Figure 1.3.2

Equilibrium conditions for this lower portion require that a force P act along the surface of the cross-section [Fig. (1.3.2b)]. We observe that the force P here is acting tangentially in the cross-sectional area. As before, we assume that this force is uniformly distributed over the area A; we denote the average intensity of this force by τ; i.e.,

$$\overline{\tau} = \frac{P}{A}. \tag{1.3.2}$$

We refer to this quantity, acting tangentially to the area A, as the average **shear stress** in the rivet and note too that the SI unit is Pa.

Thus, we conclude that there can exist two types of stresses, a normal stress and a tangential shear stress. Both are measures of the intensity of force per unit area. In the above two cases, there existed but one type of stress on each area. We shall soon discover that both normal and shear stress may exist simultaneously on a given area.

It should be emphasised here that we have only obtained the average stresses based on the assumption, in both cases, that the internal force on an area is uniformly distributed. This is sufficient to introduce the basic idea of stress as intensity of force per unit area. However, we will find in our later discussion that in many bodies, internal forces are not always distributed uniformly over an area and that it will be therefore necessary to define the stress at a point.

1.4   Intensity of a normal force acting over an area – refinement of the concept: normal stress at a point

Consider a rod subjected to an axial force P such that the normal force on each section, lying in the y–z plane, is $F = P$ [Fig. (1.4.1a)]. As in the previous discussion, the average normal stress is $\overline{\sigma} = F/A$, based on the assumption that the force F is uniformly distributed over the area A. We now wish to refine this concept; i.e., we no longer will assume the uniform distribution but instead assume that the stress varies over the cross-sectional area, i.e., $\sigma = \sigma(y, z)$. Let us now consider that the total area to be the sum of small elemental areas ΔA, such that on each of these areas an incremental part of the total force ΔF is acting [Fig. (1.4.1b)]: clearly, the incremental force ΔF is then given by

$$\Delta F \simeq \sigma(y, z)\Delta A, \qquad (1.4.1a)$$

where the incremental area ΔA surrounds the point O [Fig. (1.4.1b)]. (Note that the symbol $\simeq$ has been used here; although σ varies only slightly over ΔA, it cannot be assumed to be constant over the ΔA.) Hence

$$\sigma(y, z) \simeq \frac{\Delta F}{\Delta A}. \qquad (1.4.1b)$$

Clearly, as $\Delta A \to 0$, $\Delta F \to 0$. However, taking the limit as $\Delta A \to 0$, we have

$$\sigma(y, z) = \lim_{\Delta A \to 0} \frac{\Delta F}{\Delta A} = \frac{dF}{dA}, \qquad (1.4.1c)$$

which represents the normal stress component at the point $O(y, z)$.

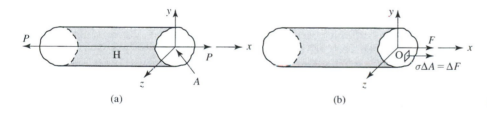

(a)   (b)   **Figure 1.4.1**

Thus we see, according to Eq. (1.4.1c), that the normal stress at a point can now be defined as the limiting case of incremental normal force divided by incremental area at the point. In the next chapter, we will generalise this concept to forces that are not necessarily normal to the plane.

Now, the total resultant force F acting on the cross-section is evidently given by $F = \sum \Delta F$ and therefore, as $\Delta A \to 0$ and the number of incremental areas approaches infinity, from Eq. (1.4.1a) we obtain, in the limit,

$$F = \int dF = \iint_A \sigma \, dA. \qquad (1.4.2)$$

We note that the internal force F, as given by Eq. (1.4.2), represents a resultant force due to a summation of all the σ stresses acting over incremental areas. It is in this sense that internal forces are often referred to as **internal stress resultants**.

We observe, in passing, that if σ is constant, i.e., not dependent on the particular element dA, then from Eq. (1.4.2),

$$F = \sigma \iint\limits_{A} dA = \sigma A, \tag{1.4.3}$$

and we recover the expression $\sigma = F/A$. Thus we see that, only in this particular case, is the average normal stress, $\overline{\sigma}$, equal to the actual normal stress existing at each point of the cross-section.

1.5   Average stresses on an oblique plane

We consider again a rod of cross-section A that is in equilibrium when subjected to forces P acting along the x-axis [Fig. (1.4.1a)]. In our previous treatments, the normal stress was found on a cross-section lying in the z-plane perpendicular to the applied load. However, we are not limited to making a cut only along planes of a cross-section; indeed we may cut the bar along any plane passing through an arbitrary point H and isolate the two parts as a free body. Let us therefore imagine, for example, that we cut the bar at some arbitrary point, H, by means of a plane whose unit normal $\boldsymbol{n}$ lies in the $x-y$ plane and is inclined with respect to the x-axis by a given angle θ as shown in Fig. (1.5.1). We then isolate the left segment. Let A' denote the area of the oblique plane of the cut.

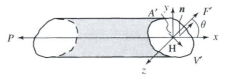

Figure 1.5.1

Then, since any arbitrary part of the bar is in equilibrium, a force P clearly must be acting on the cut in order to satisfy equilibrium in the x-direction. We now resolve this force into two components: a component F' acting normal to the plane of the cut and a component V' acting tangential to the plane as shown in Fig. (1.5.1).

Now, from equilibrium

$$\sum F_x = 0 \rightarrow\rightarrow F' \cos\theta + V' \sin\theta = P, \tag{1.5.1a}$$

$$\sum F_y = 0 \rightarrow\rightarrow F' \sin\theta - V' \cos\theta = 0. \tag{1.5.1b}$$

Equations (1.5.1) represent two equations in two unknowns, whose solution is readily given by

$$F' = P \cos\theta, \tag{1.5.2a}$$

$$V' = P \sin\theta. \tag{1.5.2b}$$

We thus find that on this oblique plane the equilibrium conditions require that both a normal force F' and a tangential force V' act on the plane of the cut. The average intensity of these forces is then, by Eqs. (1.3.1) and (1.3.2),

$$\overline{\sigma} = \frac{F'}{A'} = \frac{P \cos\theta}{A/\cos\theta} = \frac{P}{A} \cos^2\theta, \tag{1.5.3a}$$

$$\overline{\tau} = \frac{V'}{A'} = \frac{P \sin\theta}{A/\cos\theta} = \frac{P}{A} \sin\theta \cos\theta = \frac{P}{2A} \sin 2\theta. \tag{1.5.3b}$$

We therefore observe that there exist on the plane of the cut both normal stresses and tangential stresses. In particular, if $\theta = 45°$, $\overline{\sigma} = P/(2A)$ and $\overline{\tau} = P/(2A)$. Note too that if $\theta = 0$, then $\overline{\sigma} = P/A$ and $\overline{\tau} = 0$; that is, we recover the average stresses existing on a cross-section.

1.6   Variation of internal forces and stresses with position

In our previous discussion, forces and stresses acting on cross-sections were found to be independent of the location of the cross-section. (For example, in the cases examined in Sections 3 and 5, point H along the longitudinal axis was arbitrary.) This clearly is not the general case; one often wishes to determine how the internal forces vary with the location of a cross-section. To this end, consider a triangular plate ABC of mass density ρ (having a weight W) and constant thickness t, hanging from a support BC and subjected to gravity g acting downward [Fig. (1.6.1a)]. We wish to determine the internal forces acting at any cross-section DE, located at a distance y from the apex A due to its own weight. Note that the total weight W of the plate is given by $W = \rho g b t h / 2$.

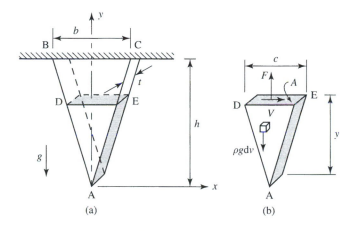

(a)   (b)   Figure 1.6.1

If we isolate the lower portion [Fig. (1.6.1b)], we note that a gravitational force $\rho g \, dv$ acts on each element of incremental volume dv of this portion of the plate. (Note that the gravitational forces are in fact body forces as described previously in Section 2.) Having made the cut along DE, we note too that in general there may exist a resultant force F and shear force V acting along the plane of the cut.

From equilibrium, $\sum F_x = 0 \rightarrow \rightarrow V = 0$ and $\sum F_y = 0 \rightarrow \rightarrow F - \iint_V \int \rho g \, dv = 0$ where the triple integral is over the volume of the segment ADE. Since ρg is constant, noting that the width c of DE is $c = by/h$ and that the volume of the lower segment is $tcy/2$, we have

$$F(y) = \rho g \frac{tcy}{2} = \rho g \frac{bt}{2h} y^2 = W (y/h)^2 . \qquad (1.6.1a)$$

The area of the cross-section at any location y is given by $A(y) = ct = bty/h$, and hence the average normal stress acting on any cross-section is

$$\overline{\sigma}(y) = \frac{F(y)}{A(y)} = \frac{\rho g}{2} y. \qquad (1.6.1b)$$

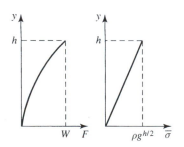

The internal force $F(y)$ and average stress $\overline{\sigma}_y$ are thus seen to vary parabolically and linearly, respectively, with y [Fig. (1.6.2)].

Figure 1.6.2

1.7   Strain as a measure of intensity of deformation

Having defined stress as a measure of the intensity of internal forces in a body, we now consider the second concept of importance in solid mechanics; namely, strain as a measure of the intensity of deformation. This is a purely geometric (or kinematic) concept, not intrinsically related to forces. It is important because one wishes to know how a mechanical system deforms for, in reality, no body is perfectly rigid.

To illustrate this concept, consider two rods, A and B, resting freely on a table: rod A is 10 cm long and rod B is 100 cm long, as shown in Figs. (1.7.1a and b), respectively. Assume that in each case, the rods undergo the same elongation, which we denote by δ. Then it is evident that if, for example, $\delta = 1$ cm, this elongation is far more significant for the short bar (whose length has changed from 10 to 11 cm) than for the long bar (whose length has changed from 100 to 101 cm). In other words, the 'intensity' of the deformation of rod A is much greater than that of rod B. We wish to have a measure to this 'intensity'. Clearly a useful quantity is δ/L, where L is the original length.

Figure 1.7.1

(a)   (b)

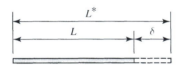

Figure 1.7.2

Thus, assume that a rod, originally of length L, deforms to a length L^* [Fig. (1.7.2)]. Then $\delta = L^* - L$. We define the ratio

$$\bar{\epsilon} = \frac{\delta}{L} = \frac{L^* - L}{L} \tag{1.7.1}$$

as the average strain or the average **engineering strain**. Note that the strain ϵ is a non-dimensional quantity.

In most problems encountered in engineering practice, one deals with materials that are relatively stiff (e.g., steel, aluminium, or other metals). Hence the strain that one encounters in such materials is often very small, i.e., of the order $O(10^{-3})$. For example, if a steel bar has length $L = 60$ cm $= 0.6$ m and elongates by $\delta = 1.5$ mm $= 1.5 \times 10^{-3}$ m, the resulting average strain is, according to Eq. (1.7.1),

$$\bar{\epsilon} = \frac{1.5 \times 10^{-3}}{0.6} = 2.5 \times 10^{-3}.$$

In this discussion, we have implicitly assumed that the bar deforms uniformly along its length. However, let us say that it deforms due to heating, which is not uniform along the longitudinal axis. The average strain then does not provide an indication of the deformation at any point along the bar. We must therefore consider the deformation more carefully and find the strain at each cross-section of the bar.

Let x denote the longitudinal axis of an undeformed rod [Fig. (1.7.3a)]. Let us assume the horizontal displacement u of any cross-section located at the coordinate x is known, that is, $u = u(x)$. Now, the rod may be considered to be composed of a series of elements, each of length Δx. If $u(x)$ denotes the displacement of the cross-section originally at x, then $u(x + \Delta x) = u(x) + \Delta u$ is the displacement of the right side of the deformed element [Fig. (1.7.3b)]. The average strain *of the*

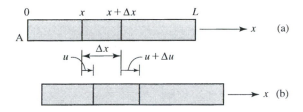

Figure 1.7.3

element, given by Eq. (1.7.1), is then

$$\bar{\epsilon} = \frac{[u(x) + \Delta u] - u(x)}{\Delta x} = \frac{\Delta u}{\Delta x}. \tag{1.7.2}$$

Taking the limit, we have

$$\epsilon(x) = \lim_{\Delta x \to 0} \frac{\Delta u(x)}{\Delta x}$$

or

$$\epsilon(x) = \frac{du(x)}{dx}. \tag{1.7.3}$$

Thus we have obtained the strain $\epsilon \equiv \epsilon_x$ in the x-direction, which exists at any cross-section. It is of importance to observe that the strain ϵ is dependent on the *relative* displacements of points in the rod.

Note that from Eq. (1.7.3), $du(x) = \epsilon_x \cdot dx$. Hence integrating this expression, we have

$$\int_{u(0)}^{u(L)} du = \int_0^L \epsilon(x) \, dx, \tag{1.7.4a}$$

that is,

$$u_L - u_0 = \int_0^L \epsilon(x) \, dx. \tag{1.7.4b}$$

Since the total elongation $\delta \equiv u_L - u_0$, we thus have an explicit expression for the elongation in terms of the strain at any point, namely

$$\delta = \int_0^L \epsilon(x) \, dx. \tag{1.7.4c}$$

Note that if $\epsilon(x)$ is constant, then $\delta = \epsilon L$.

Example 1.1: A rod, originally of length L, is heated non-uniformly [Fig. (1.7.4)]. The increase in temperature, ΔT, is given by $\Delta T = kx(L - x)$, where k is a constant. The temperature increase at the mid-point is known to be 50°C. Determine the elongation δ of the bar if α is the coefficient of thermal expansion.

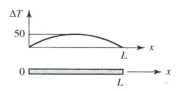

Figure 1.7.4

Solution: Since $\Delta T(x = L/2) = 50°C$, $k = 200/L^2$. Consequently, $\Delta T = \frac{200}{L^2} x(L - x)$. Noting that the strain $\epsilon(x) = \alpha \Delta T(x)$, from Eq. (1.7.4c),

$$\delta = \alpha k \int_0^L x(L - x)\,\mathrm{d}x = \frac{200\alpha}{L^2} \int_0^L x(L - x)\,\mathrm{d}x = \frac{100\alpha}{3} L.$$

Let us assume that the length of the bar is $L = 100$ cm and that it is made of steel, for which $\alpha = 11.7 \times 10^{-6}[1/°C]$. (Note that the units of α are $[1/°C]$. The elongation is then $\delta = 3.9 \times 10^{-2}$ cm $= 0.39$ mm. $\square$

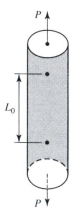

Figure 1.8.1

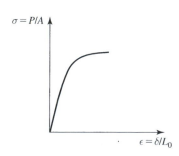

Figure 1.8.2

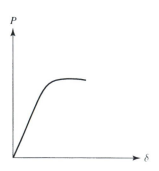

Figure 1.8.3

1.8 Mechanical behaviour of materials

As we have previously noted, it is evident that if a solid body is subjected to forces, it will respond differently depending on the material of which it is made. For example, steel, aluminium and wood behave differently. We thus must find a way of describing the general mechanical behaviour of these materials and express the behaviour in mathematical terms. The equations that describe the mechanical behaviour of a material are referred to, in general, as **constitutive equations** and are based on experimental evidence. These equations, which must describe the real behaviour of materials as they exist in nature, are established from experiments performed in a laboratory. The simplest test that one can perform on a given material is a standard tension test, which we now describe.

Let us consider a specimen of cross-sectional area A to which we apply a slowly increasing axial load P. Let us assume that we initially inscribe two points on the rod a distance L_0 apart [Fig. (1.8.1)]. (The length L_0 is commonly called a **gauge length**.) Now, as we slowly increase the force P from zero, at each value of loading, we can measure the distance L between these two points and thus we obtain the elongation $\delta = L - L_0$ for each value of P. Clearly, δ and P are related and we can plot a P–δ curve as shown in Fig. (1.8.2). This curve, however, is not of much use, for it depends on the dimensions of the specimen being tested and not intrinsically on the material itself; i.e., it depends on (a) the cross-section of the rod, A, and (b) the original gauge length, L_0. Therefore, let us calculate

$$\overline{\sigma} = \frac{P}{A} \quad \text{and} \quad \overline{\epsilon} = \frac{\delta}{L_0} \tag{1.8.1}$$

and plot the $\overline{\sigma}$–$\overline{\epsilon}$ curve as shown in Fig. (1.8.3). This curve is called the standard **stress–strain curve**. (At this point we drop the notations $\overline{\sigma}$ and $\overline{\epsilon}$ and will refer to the stress σ and strain ϵ with the clear understanding that these are average values.) We observe that the σ–ϵ curve, being a curve of force per unit area versus elongation per unit length, represents the behaviour of the material itself, since it is independent of the geometry of the rod.

Note that in describing the test, it was mentioned that the load is slowly applied, and thus we implicitly assume that the curve is independent of the rate of loading. This is true of many (but not all) materials, e.g., metals such as steel and aluminium, provided the loading rate is sufficiently small. We therefore assume that the behaviour is 'rate-independent'.

We now consider the σ–ϵ curve obtained from a standard test on a typical ductile material: for example 'low-carbon steel', whose σ–ϵ curve is shown schematically in Fig. (1.8.4). (This steel, which is of common use, consists mainly of iron and relatively small quantities of carbon.)

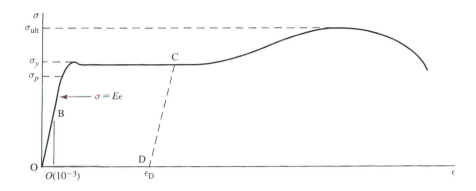

Figure 1.8.4

The behaviour may be described as follows: as the load is increased from zero, the stress–strain relation is essentially linear, up to a certain stress level, σ_p, called the **proportional limit**. Thus the proportional limit is defined as the maximum stress for which a linear stress–strain relation exists. As the load increases, the stress reaches a value σ_y at which, with no further increase in the load, the metal deforms continuously; that is, the steel suddenly 'yields'. The stress σ_y is therefore called the **yield point** of the material. A typical value of the strain ϵ at which yielding occurs in steel is 10^{-3}. This yielding usually continues over a wide range of strain. (Note that the curve is parallel to the ϵ axis during yielding and may be so until the strain reaches a value of the order of 0.2.) At this point, it is necessary to increase the load to cause the rod to elongate further; hence the stress increases. This behaviour is called **strain-hardening**. The stress thus increases over this range and reaches a maximum value σ_{ult}, called the **ultimate stress** or more precisely the **nominal ultimate stress**. At this point, the stress drops off sharply while the rod continues to elongate until rupture takes place. For reasons that would become apparent in Chapter 4, the entire σ–ϵ curve obtained during the test is referred to as the **nominal stress–strain curve**.

Let us now return to consider the initial behaviour with $\sigma \leq \sigma_p$, i.e., where the stress σ does not exceed the proportional limit. It was observed that the σ–ϵ relation in this range is linear. Consequently, this initial part of the curve can be represented by the equation of a straight line:

$$\sigma = E\epsilon, \tag{1.8.2}$$

where the constant E, representing the slope of the line, is called the **Young's Modulus**. Note that the units of this modulus are given in $N/m^2(1\ N/m^2 = 1\ Pa)$. A typical value for steel is $E = 200 \times 10^9\ Pa = 200\ GPa$. We observe that E is a measure of the stiffness of the material. Thus, the larger the value of E, the greater the stress required to deform the material.

It should be mentioned here that the relation $\sigma = E\epsilon$ (valid for $\sigma \leq \sigma_p$) is a typical example, perhaps the simplest, of what is called, in general, a constitutive equation. Indeed not all materials can be described by such a simple constitutive equation.

Now, let us consider again that we start to load the material from point O to some arbitrary point B such that $\sigma = \sigma_B \leq \sigma_p$ and that we then remove the load; i.e., we 'unload' the material. We find, upon loading and unloading repeatedly, that the behaviour follows the same σ–ϵ curve provided that we remain below σ_p. Indeed, if we load the material within this range and then remove the load completely, the strain, after removal of the load, is $\epsilon = 0$. Thus the body recovers its initial length. The material is therefore said to exhibit elastic behaviour in this range. As a result, the constant E is often called the **Modulus of elasticity**, a terminology that seems to be preferred in engineering practice. In fact, the material is said here to be *linearly elastic*. (It should be pointed out that although the linear elastic range is in the range $\sigma \leq \sigma_p$, there may also exist a nonlinear elastic region where the $\sigma-\epsilon$ relation is no longer linear. We leave this for a later study.)

Now, let us assume that we load the rod from point O until we reach point C on the $\sigma-\epsilon$ curve of Fig. (1.8.4). At this point, we then remove the load. We would find that as we remove the load, the behaviour follows a line CD, which is parallel to the original straight line portion of the $\sigma-\epsilon$ curve. Thus, when we have completely removed the load, there is a permanent strain ϵ_D in the material, represented by point D. The material thus is said to behave as a *plastic material*, if σ exceeds σ_p.

The above is a description of what takes place during a typical tension test on steel. At this stage, we have not attempted to provide explanations for the behaviour; we have only given a description of the phenomena. A more elaborate explanation for the behaviour will be given at a later stage, but the more fundamental study of the phenomena in terms of the structure of the material requires an understanding at the atomic or crystal level. Such studies belong to the field of materials science. At this stage, we are content to provide a phenomenological description of the behaviour of materials to pursue our study of solid mechanics.

1.9 Summary

In this chapter, we have introduced the three basic concepts that are required in the study of solid mechanics: stress, strain and the constitutive equations. These concepts were introduced by means of simple one-dimensional problems. In our future study, it will be necessary to generalise these basic concepts to enable us to treat typical problems of solid mechanics as encountered in engineering practice.

PROBLEMS

In all the following problems, assume that the stresses are identical to the average stresses.

Throughout this book, problems that require a deeper understanding of the subject or a more sophisticated approach have been indicated by an asterisk (*).

1.1: Two cylindrical rods AB and BC, welded together, as shown in Fig. (1P.1), are subjected to a force 30 kN at B and an unknown force P at C. Determine (a) the force P such that the same normal stress exists in each segment of the rod and (b) the force P such that the tensile stress in BC is equal in magnitude to the compressive stress in AB. Indicate whether the P is a tension or compression force in each case.

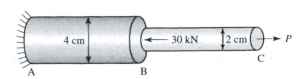

Figure 1P.1

1.2:   The shearing stress at failure of a steel plate is given as $\tau = 90$ MPa. (a) Determine the force P required to punch a 20-mm diameter hole if the thickness of the plate is $t = 4$ mm, as shown in Fig. (1P.2); (b) What is the average normal stress $\bar{\sigma}$ in the punch when subjected to this force?

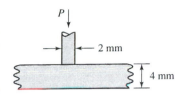

Figure 1P.2

1.3:   A rectangular block of brass ($E = 100$ GPa) and allowable stress $\sigma = 120$ MPa, whose cross-section is 40 mm × 60 mm, supports a compressive load P. Determine the maximum force P that may be applied if the block is not to shorten by more than 0.05%.

1.4:   An aluminium control rod of circular cross-section is to be designed to lengthen by 2 mm when a tensile force $P = 40,000$ N is applied. If the allowable stress is $\sigma = 20$ MPa and $E = 70$ GPa, determine (a) the smallest permissible diameter D and (b) the shortest length of the rod.

1.5:   The frame shown in Fig. (1P.5) consists of three pin-connected 3-cm diameter rods. Determine the average normal stress in rods AB and AC if the force $P = 60$ kN.

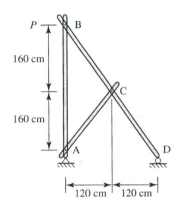

Figure 1P.5

1.6:*   A truss, consisting of two pin-connected rods AB and AC (each made of the same material whose density is ρ (N/m³) and each having the same cross-sectional area A), supports a force P as shown in Fig. (1P.6). The truss is to be designed such that $|\sigma|$, the maximum allowable normal stresses (in absolute value), are the same in each member.

(a) Show that under these conditions, the weight W of the truss is

$$W = \frac{\rho P L}{\sigma}\left[\frac{1}{\tan \beta} + \frac{2}{\sin 2\beta}\right].$$

(b) Assuming that P is much greater than the total weight W of the truss, determine the angle β for which the truss has minimum weight, thus yielding an optimal design. What is the minimum weight W expressed non-dimensionally as $\frac{W}{\rho P L/\sigma}$?

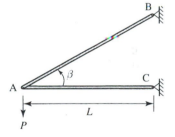

Figure 1P.6

1.7:   Two tubular rods are connected, as shown in Fig. (1P.7), by means of an adhesive whose allowable shear stress is $\tau = 4$ MPa. Determine the permissible axial force P that the connection can carry.

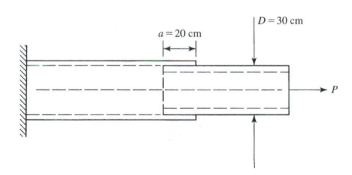

Figure 1P.7

1.8:   A steel rod of length $L = 2$ m, fixed at A, is heated by a linearly varying temperature $\delta T(x)$, as shown in Fig. (1P.8). The coefficient of thermal expansion is $\alpha = 11.7 \times 10^{-6} [1/°C]$.

(a) What is the strain $\epsilon(x)$ at any point x of the rod? (b) Determine the displacement $u(x)$ at any cross-section of the rod; (c) What is the total elongation ΔL of the rod? (d) Determine the average strain $\bar{\epsilon}$ in the rod.

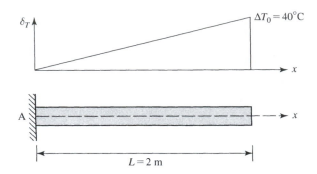

Figure 1P.8

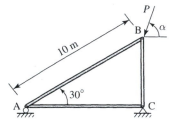

Figure 1P.9

The following problems are to be solved using a computer.

1.9:   A three-bar truss, each of whose members has a cross-sectional area $A = 2.0\,\text{cm}^2$, is subjected to a load $P = 30\,\text{kN}$ that is inclined with respect to the x-axis by a varying angle α, as shown in Fig. (1P.9). Plot the normal stress in AB and BC as a function of α for $0 \le \alpha \le 180°$.

1.10:   For the truss of Problem 1.6, plot the weight of the truss in non-dimensional terms, i.e. $\frac{W}{\rho P L / \sigma}$, as a function of β for $0 < \beta < 90°$.

2

Internal forces and stress

2.1 Introduction

We have seen that stress as a measure of the intensity of internal forces is a fundamental concept in solid mechanics. In the previous chapter, the idea of stress was introduced by means of some simple one-dimensional cases. However, the concept of stress is more complex. As will become evident from the treatment below, the quantities found previously are only components of stress. In order to develop more fully the concept of stress, it is necessary to consider first the three-dimensional case which, in general, exists in reality. From this more general case, we consider the simpler two- and one-dimensional cases.

Since the stress in a body is dependent on the existing internal forces, we first consider and analyse these forces.

2.2 Internal force resultants

Consider a body, located in an x, y, z coordinate system, under a set of external forces $F_1, F_2, \ldots, F_n$ [Fig. (2.2.1)]. According to Newton's laws of mechanics, the body must satisfy two basic principles: the principle of linear momentum and the principle of angular momentum. If the body is in equilibrium then these principles reduce to the vector equations $\sum F = 0$ and $\sum M = 0$. Moreover, since all

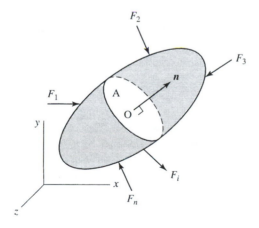

Figure 2.2.1

points in the body are in a state of equilibrium, it is clear that these laws must be satisfied for any arbitrary portion of the body. Let us therefore imagine that we cut the body by means of a plane (whose area is A) that passes through some point O and that we isolate the two portions of the body. First, we recognise that there are

21

an infinite number of planes passing through the point O. To identify any specific plane, we construct a unit vector, normal to the plane, which we denote by n. The vector n thus specifies the orientation of the plane in the x, y, z coordinate system. We shall hereafter refer to this plane as the n-plane. Upon isolating the two portions of the body, it is important to recognise that the top part of the body exerts forces on the lower portion and that the lower portion exerts (equal and opposite) forces on the upper portion.

Having isolated the two portions of the body, we examine, for example, the lower portion as shown in Fig. (2.2.2) and consider it as a 'free body'. In addition to the known external forces F_i, which act on the original exterior surface of the body, we must also represent the forces acting on the plane of the cut that the upper portion exerts on the lower portion. We wish to determine these unknown internal forces using the laws of mechanics. These forces acting on the plane of the body will, in general, be distributed over the cut in some arbitrary way. At this stage, we are not interested in determining the distribution of forces but wish merely to determine the resultant effect. Now, whatever the distribution of these internal forces, it is known that any force system can always be represented by a single resultant and a moment; we denote these by the vectors F^* and M^*, respectively, shown as acting on the n-plane in Fig. (2.2.2).[†] Note that while the vectors F^* and M^* represent the resultant internal force system, this force system must be considered as part of 'external forces' when acting on the entire cut section of area A of the isolated free body.

If this free body is in equilibrium, then

$$\sum F|_{\text{ext}} + F^* = 0, \qquad (2.2.1a)$$

$$\sum M|_{\text{ext}} + M^* = 0, \qquad (2.2.1b)$$

where $F|_{\text{ext}}$ and $M|_{\text{ext}}$ represent the forces and moments due to the externally applied force system. Clearly, in principle, the two unknown vectors F^* and M^*, which are required to maintain the isolated portion of the body in equilibrium, can be found from these two vector equations. We now define two other mutually perpendicular directions by means of unit vectors s and t, both of which lie in the plane of the cut [Fig. (2.2.3)]. Since n is perpendicular to the n-plane, the three vectors n, s and t are said to form an orthogonal triad. We may then resolve the internal resultant F^* and M^* into scalar components as follows:

$$F^* = Fn + V_s s + V_t t, \qquad (2.2.2a)$$

$$M^* = Tn + M_s s + M_t t. \qquad (2.2.2b)$$

The component F appearing above is referred to as a **normal force component** or more briefly, the **normal force** acting on the n-plane. The components V_s and V_t, which act tangentially to the n-plane, are called the **shear forces** in the s- and t-directions, respectively [Fig. (2.2.4a)].

The component of M^* in the n-direction, T, represents the moment about the normal n-axis; i.e., $T \equiv M_n$. It is therefore called the **torsional moment** (or the **torque**) since it tends to twist the body. On the other hand, the components M_s and

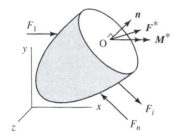

Figure 2.2.2

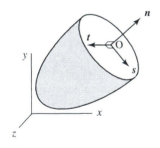

Figure 2.2.3

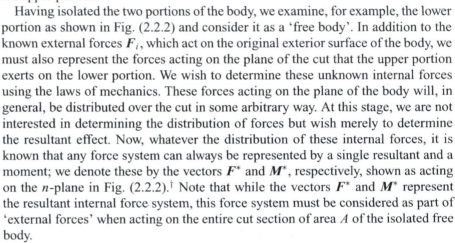

[†] To distinguish the moment vector from the force vector, the moment vector is drawn with a double arrow.

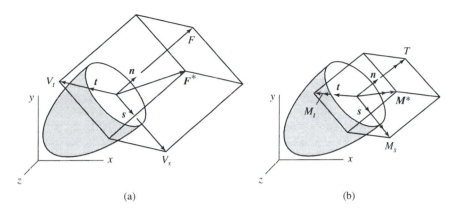

Figure 2.2.4

M_t are called **bending moments** about the s- and t-directions since they tend to bend the body about the s- and t-axes, respectively [Fig. (2.2.4b)].

We emphasise here that the above quantities represent resultant forces and moments acting on the n-plane and that, at this stage, no consideration is made as to their distribution over the plane of the cut.

Since F^* and M^* represent the resultant effect of the upper portion on the lower portion of the body, it is clear, according to Newton's Third Law, that the lower portion also exerts an effect that is equal and opposite on the upper portion of the body [Fig. (2.2.5)].

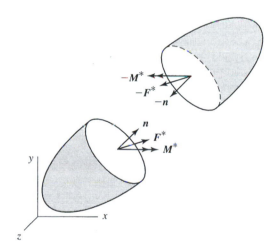

Figure 2.2.5

We illustrate these ideas in the following examples.

Example 2.1: A body, consisting of a pipe lying in the x–z plane, is subjected to a vertical force $P = 250$ N as shown in Fig. (2.2.6). Determine the components of the resultant forces and moments at the sections a–a and b–b.

Solution: In the following calculations, positive forces indicate that the force is acting in the positive coordinate direction; positive moments act about positive axes according to the right-hand rule. (For the present, this will suffice. However, we will find, in future treatments, that it is necessary to adopt a different sign convention appropriate to solid mechanics.)

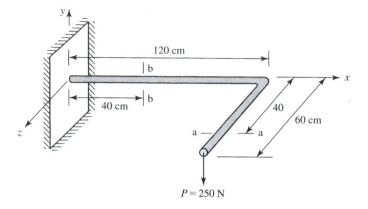

Figure 2.2.6

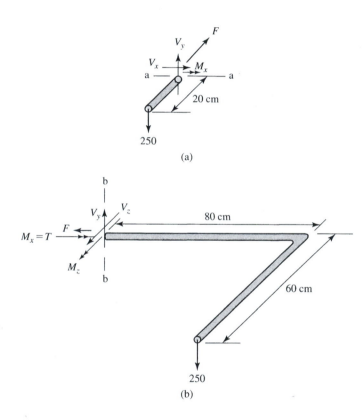

Figure 2.2.7

(a) *Section* a–a: To determine the forces at a–a, we consider the isolated body shown in Fig. (2.2.7a). Note that the shear forces at this section are V_x and V_y.[†]

$$\sum F_x = 0 \to\to V_x = 0,$$

$$+\uparrow \sum F_y = 0 \to\to V_y - 250 = 0 \to\to V_y = 250\,\text{N},$$

$$\sum F_z = 0 \to\to F = 0.$$

[†] For simplicity, only non-zero moments are shown in the figure.

Letting $(\sum M_x)_a$ denote the sum of the moments about an x-axis passing through a–a, etc. we have

$$-\overset{\curvearrowleft}{} + \left(\sum M_x\right)_a = 0 \to \to (20)(250) + M_x = 0 \to \to M_x = -5000 \text{ N-cm},$$

$$\left(\sum M_y\right)_a = 0 \to \to M_y = 0, \qquad \left(\sum M_z\right)_a = 0 \to \to M_z \equiv T = 0.$$

(b) *Section b–b:* Making a cut at b–b and isolating the free body [Fig. (2.2.7b)], we have

$$\sum F_x = 0 \to \to F = 0,$$

$$+\uparrow \sum F_y = 0 \to \to V_y - 250 = 0 \to \to V_y = 250 \text{ N},$$

$$\sum F_z = 0 \to \to V_z = 0.$$

$$-\overset{\curvearrowleft}{} + \sum (M_x)_b = 0 \to \to (60)(250) + M_x = 0 \to \to M_x \equiv T = -15,000 \text{ N-cm}.$$

Note that the minus sign indicates that T acts in the opposite sense to that shown in Fig. (2.2.7b).

$$\left(\sum M_y\right)_b = 0 \to \to M_y = 0$$

$$+\left(\sum M_z\right)_b = 0 \to \to M_z - (80)(250) = 0 \to \to M_z = 20,000 \text{ N-cm}. \qquad \square$$

Example 2.2: A beam ABC consists of two elements, fixed at A, pinned at point B and simply supported at point C, as shown in Fig. (2.2.8). Determine the internal force resultants occurring at points D and E, due to the uniform line load w (N/m) acting between B and C.

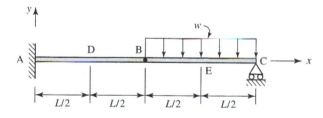

Figure 2.2.8

Solution: We first note that this is a two-dimensional problem. Therefore, since all forces F act in the x–y plane, the only moments that can exist are moments about the z-axis, which we denote below by M, i.e., $M \equiv M_z$.

Considering the body ABC, we note that there exist four unknowns R_{Ax}, R_{Ay}, M_A and R_C [Fig. (2.2.9a)]. The equilibrium equations for this body are

$$\sum F_x = 0 \to \to R_{Ax} = 0, \qquad (2.2.3a)$$

$$+\uparrow \sum F_y = 0 \to \to R_{Ay} + R_C - wL = 0, \qquad (2.2.3b)$$

$$\overset{\curvearrowleft}{} + \left(\sum M_z\right)_A = 0 \to \to -M_A + 2LR_C - wL(3L/2) = 0. \quad (2.2.3c)$$

Having immediately found $R_{Ax} = 0$, we observe that we are left with two remaining equations [Eqs. (2.2.3b) and (2.2.3c)] and three unknowns (R_{Ay}, R_C and M_A);

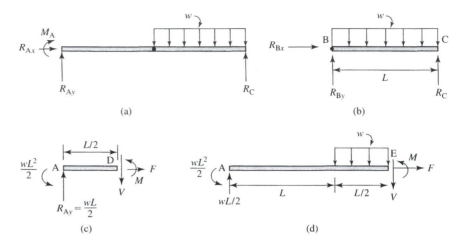

Figure 2.2.9

we therefore require a third equation. However, a pin exists at point B. Therefore $M_B = 0$ since (as the pin, by definition, does not provide a constraint against rotation at B) no moment can be transmitted by the pin from one part of the beam to the other. This provides us with the additional equation: we thus make a 'cut' at B and isolate BC as a free body [Fig. (2.2.9b)]. Note that although we must now show the forces R_{Bx} and R_{By} (as external forces), which represent the effect of the segment AB on BC, these do not appear in the moment equation if taken about point B. Thus, for member BC, we have

$$\curvearrowright^{\prime} + \left(\sum M_z\right)_B = 0 \rightarrow\rightarrow R_C L - (wL)(L/2) = 0 \rightarrow\rightarrow R_C = wL/2.$$

Substituting this in Eqs. (2.2.3b) and (2.2.3c), we have

$$R_{Ay} = -R_C + wL = wL/2 \quad \text{and} \quad M_A = 2R_C L - 3wL^2/2 = -wL^2/2.$$

Note that the minus sign appearing in M_A indicates that it is acting in a direction opposite to the assumed direction shown in Fig. (2.2.9a).

Having found the external reactions, we may now find the internal force resultants at D and E:

At D: Isolating A–D [Fig. (2.2.9c)],

$$\sum F_x = 0 \rightarrow\rightarrow F = 0,$$

$$+ \uparrow \sum F_y = 0 \rightarrow\rightarrow wL/2 - V = 0 \rightarrow\rightarrow V = wL/2,$$

$$\curvearrowright^{\prime} + \left(\sum M_z\right)_D = 0 \rightarrow\rightarrow M + wL^2/2 - (wL/2)(L/2) = 0$$

$$\rightarrow\rightarrow M = -wL^2/4.$$

At E: Treating AE as an isolated body [Fig. (2.2.9d)],

$$\sum F_x = 0 \rightarrow\rightarrow F = 0,$$

$$+ \uparrow \sum F_y = 0 \rightarrow\rightarrow wL/2 - wL/2 - V = 0 \rightarrow\rightarrow V = 0,$$

$$\curvearrowright^{\prime} + \left(\sum M_z\right)_E = 0 \rightarrow\rightarrow M + wL^2/2 - (wL/2)(3L/2)$$

$$+ (wL/2)(L/4) = 0 \rightarrow\rightarrow M = wL^2/8.$$

It should be noted that the same results at section E could be obtained by considering equilibrium of EC as a free body. The reader is urged to verify this.   □

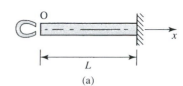

(a)

Example 2.3: A magnet is attached at the free end of an iron bar whose cross-sectional area is A, as shown in Fig. (2.2.10a). The attraction force f is found to decay exponentially, that is,

$$f(x) = c e^{-x/L} \,(\text{N/m}^3)$$

as shown in Fig (2.2.10b), where c is a constant. Determine the normal force $F(x)$ that exists at any cross-section.

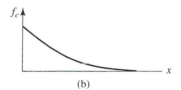

(b)

Solution: We note that the magnet exerts body forces on each element of the rod. Let us make a cut at some arbitrary cross-section located at x [Fig. (2.2.10c)]. Then, since at any cross-section ξ, $f(\xi) = c e^{-\xi/L}$, from $\sum F_x = 0$, the total force on the cross-section is given by

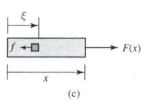

(c)

$$F(x) = \int_0^x f(\xi) A \, d\xi = cA \int_0^x e^{-\xi/L} \, d\xi = -cALe^{-\xi/L}\Big|_0^x = cAL\left(1 - e^{-x/L}\right).$$

(2.2.4)

The variation of $F(x)$ is shown in Fig. (2.2.10d).   □

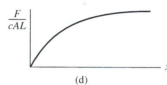

(d)

Figure 2.2.10

2.3 State of stress at a point: traction

It is evident that the internal force system F^* and M^* shown in Fig. (2.2.2) may be considered to be composed of small increments ΔF^* and ΔM^*, each acting over a small area ΔA surrounding any point O of the n-plane [Fig. (2.3.1)]. We now examine the area surrounding this point.

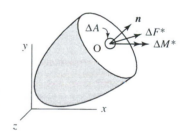

Figure 2.3.1

(a) Traction

Let us confine our attention to point O and the increment of area ΔA surrounding it. Now if we shrink the area ΔA to zero, it is clear that both $\Delta F^* \to 0$ and $\Delta M^* \to 0$. We now define the following ratio:

$$T_n = \lim_{\Delta A \to 0} \frac{\Delta F^*}{\Delta A}.$$

(2.3.1)

Furthermore, we assume that[†]

$$\lim_{\Delta A \to 0} \frac{\Delta M^*}{\Delta A} = 0.$$

(2.3.2)

This assumption is based on experimental evidence and, in general, is found to be valid for solids encountered in engineering practice.

We first observe that T_n, which acts in the same direction as ΔF^*, is a vector. The quantity T_n is called the **traction**. We observe, too, that according to its definition, the traction T_n represents an intensity of force per unit area and that it acts in some arbitrary direction with respect to the n-plane. In the SI system, this quantity is given in units of N/m^2, which, as we noted in Chapter 1, is defined as a Pascal (Pa).

[†] As a result of this assumption, we eliminate the existence of what are known in solid mechanics as 'stress couples' in the body.

Having noted that the vectors $\Delta \boldsymbol{F}^*$ and $\boldsymbol{T}_n$ act in some arbitrary direction with respect to the n-plane, we resolve them into their scalar components in the n-, s- and t-directions defined by the unit vectors $\boldsymbol{n}$, $\boldsymbol{s}$ and $\boldsymbol{t}$, respectively, as described in the previous section.

Following Eq. (2.2.2a), the scalar components of $\Delta \boldsymbol{F}^*$ in the n-, s- and t-directions are denoted by ΔF, ΔV_s and ΔV_t, respectively. The quantity ΔF thus represents an increment of normal force while ΔV_s and ΔV_t, which act tangentially to the n-plane at point O, represent increments of the shear forces in the s- and t-directions, respectively. Thus we have

$$\Delta \boldsymbol{F}^* = \Delta F \boldsymbol{n} + \Delta V_s \boldsymbol{s} + \Delta V_t \boldsymbol{t}. \tag{2.3.3}$$

Substituting this in Eq. (2.3.1), we may therefore write

$$\boldsymbol{T}_n = \lim_{\Delta A \to 0} \left[\frac{\Delta F}{\Delta A} \boldsymbol{n} + \frac{\Delta V_s}{\Delta A} \boldsymbol{s} + \frac{\Delta V_t}{\Delta A} \boldsymbol{t} \right]. \tag{2.3.4}$$

Now we denote the limits of these ratios as follows:

$$\sigma_n = \lim_{\Delta A \to 0} \frac{\Delta F}{\Delta A}, \qquad \tau_{ns} = \lim_{\Delta A \to 0} \frac{\Delta V_s}{\Delta A}, \qquad \tau_{nt} = \lim_{\Delta A \to 0} \frac{\Delta V_t}{\Delta A}. \tag{2.3.5}$$

Hence

$$\boldsymbol{T}_n = \sigma_n \boldsymbol{n} + \tau_{ns} \boldsymbol{s} + \tau_{nt} \boldsymbol{t}. \tag{2.3.6}$$

The traction $\boldsymbol{T}_n$, shown in Fig. (2.3.2), which represents the intensity of the force acting on an n-plane per unit area, thus has been resolved into components normal and tangential to the n-plane.

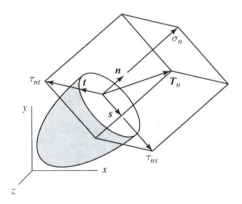

Figure 2.3.2

The following definitions and remarks are now in order:

- It is important to emphasise that the traction $\boldsymbol{T}_n$ depends on the particular n-plane passing through the point O. In general, a different traction exists on each n-plane.
- The scalar quantities, σ_n, τ_{ns} and τ_{nt}, are stress components, which represent intensity of force per unit area. Clearly, they have units of Pascals.
- σ_n acts on a plane whose normal is $\boldsymbol{n}$ and acts in the direction of n. We shall refer to σ_n as the **normal component**.
- τ_{ns} and τ_{nt} act tangentially to the n-plane in the s- and t-directions, respectively. Thus τ_{ns} and τ_{nt} are called the **shear components** acting on the n-plane. The

first subscript in the τ terms refers to the plane over which it is acting while the second subscript refers to the direction in which it acts.

■ Note that it is not necessary to prescribe two subscripts to the normal component σ_n since it is understood that this component acts in the same n-direction, which is normal to the n-plane. Indeed, the symbol σ_n has been introduced to distinguish it from the shear stress components τ_{nt} and τ_{ns}. However, at times, it is more appropriate to use a different notation: namely, $\sigma_n \equiv \tau_{nn}$. Thus, when using the letter τ to indicate a normal component, it is then also necessary to use two subscripts.

We now examine these quantities when referred to a Cartesian coordinate system x, y, z (with unit vectors i, j, k, respectively).

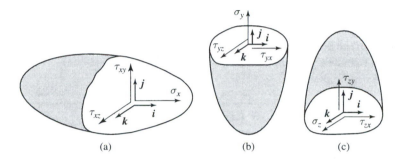

Figure 2.3.3

Let us consider the particular case where n lies in the x-direction, and let s and t lie in the y- and z-directions, respectively. Thus, here $n \to i, s \to j, t \to k$ [Fig. (2.3.3a)]. The traction is said to be acting on the x-plane; we denote this traction by T_x. Hence, we have

$$T_x = \sigma_x i + \tau_{xy} j + \tau_{xz} k$$

or, in the alternative notation,

$$T_x = \tau_{xx} i + \tau_{xy} j + \tau_{xz} k. \tag{2.3.7a}$$

Similarly, for the y-plane, let $n \to j, s \to i, t \to k$ [Fig. (2.3.3b)] so that

$$T_y = \tau_{yx} i + \sigma_y j + \tau_{yz} k$$

or

$$T_y = \tau_{yx} i + \tau_{yy} j + \tau_{yz} k. \tag{2.3.7b}$$

For the z-plane, we let $n \to k, s \to i, t \to j$ [Fig. (2.3.3c)]. Then

$$T_z = \tau_{zx} i + \tau_{zy} j + \sigma_z k$$

or

$$T_z = \tau_{zx} i + \tau_{zy} j + \tau_{zz} k. \tag{2.3.7c}$$

In passing, we observe from Eqs. (2.3.7) that the various stress components are given by the following scalar products; namely,

$$\sigma_x = T_x \cdot i, \qquad \sigma_y = T_y \cdot j, \qquad \sigma_z = T_z \cdot k \tag{2.3.8a}$$

$$\tau_{xy} = T_x \cdot j, \qquad \tau_{xz} = T_x \cdot k, \qquad \text{(etc.)}. \tag{2.3.8b}$$

Thus, in general, we may write

$$\sigma_n = \boldsymbol{T}_n \cdot \boldsymbol{n} \qquad (2.3.9a)$$

and

$$\tau_{ns} = \boldsymbol{T}_n \cdot \boldsymbol{s}, \qquad (2.3.9b)$$

where $\boldsymbol{n}$ and $\boldsymbol{s}$ are unit vectors ($|\boldsymbol{n}| = |\boldsymbol{s}| = 1$) given by $\boldsymbol{n} = n_x \boldsymbol{i} + n_y \boldsymbol{j} + n_z \boldsymbol{k}$ and $\boldsymbol{s} = s_x \boldsymbol{i} + s_y \boldsymbol{j} + s_z \boldsymbol{k}$.

(b) Sign convention

At this stage of our treatment, it is necessary to adopt a sign convention.

- We first define a positive and negative plane with respect to a coordinate system as follows:

  A positive (negative) plane is one for which the outward normal is acting in the positive (negative) coordinate direction. [Thus the planes shown in Figs. (2.3.3) are all positive planes.]

  At times it is convenient to use a different terminology: we refer to the plane as a 'face'; thus one refers to the 'positive x-face' instead of the 'positive x-plane', the two terms being synonymous.

- Positive and negative components are defined as follows:

  A positive component acts on a positive face in a positive coordinate direction; or
  A positive component acts on a negative face in a negative coordinate direction.

  Therefore, in accordance with this convention,

  A negative component acts on a positive face in a negative coordinate direction; or
  A negative component acts on a negative face in a positive coordinate direction.

It is convenient to represent the above components acting at a point, which appear in Figs. (2.3.3), by means of a single figure. This representative figure is shown in Fig. (2.3.4) where all the stress components are acting. It is important to note that this figure is but a pictorial representation that permits one to show the components acting on the positive and negative planes by means of a single figure. *The element as shown in Fig. (2.3.4), therefore, is not meant to necessarily represent a physical element of the body.*

We observe that all the stress components shown in Fig. (2.3.4) are positive components in accordance with the above sign convention. According to our definition and sign convention, we also observe from this figure that a positive σ_n component, $\sigma_n > 0$, indicates tension while a negative σ_n component, $\sigma_n < 0$, indicates compression.

(c) The stress tensor

Using the symbol $\tau_{nn} \equiv \sigma_n$, we have seen that there exist nine components at a point; these are shown in the array

$$\begin{pmatrix} \tau_{xx} & \tau_{xy} & \tau_{xz} \\ \tau_{yx} & \tau_{yy} & \tau_{yz} \\ \tau_{zx} & \tau_{zy} & \tau_{zz} \end{pmatrix}.$$

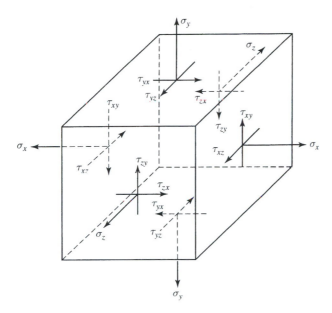

Figure 2.3.4

These components define the stress at a point; the components are referred to as **stress components**. Therefore, in order to define the state of stress existing at a point, it is necessary to specify nine scalar components. (It is interesting to note the analogy with a vector in three-dimensional space: in order to specify a vector, it is necessary to specify three scalar components.)

The array of these nine stress components is called the **stress tensor**, which we sometimes denote by the symbol τ. We shall see that it is not simply because it is represented by an array that we call τ a stress tensor, but rather because the components obey certain specific laws and τ possesses certain specific properties. These laws and properties will be found in our subsequent treatment. We mention, however, that the stress tensor is said to be a **second-rank tensor**, since two subscripts are required to specify its scalar components.[†]

At this point, we should also note that in order to specify the traction (vector) T_n, it is necessary to know three components of stress acting on the n-plane. Therefore, it follows that if the traction on three different (orthogonal) planes passing through a point are known, then all the components of stress are known.

(d) Equality of the conjugate shear stresses

We have observed above that, in principle, it is necessary to specify nine stress components in order to define the state of stress existing at a point. While this is true, we shall find that three of these are not independent.

Now, in general, the stress components will vary from point to point. That is, $\tau_{xx} = \tau_{xx}(x, y, z)$, $\tau_{xy} = \tau_{xy}(x, y, z)$, etc. or symbolically, $\tau = \tau(x, y, z)$. Let us isolate an infinitesimal element $\Delta x \, \Delta y \, \Delta z$, having density ρ, surrounding a point O (which is located at its centre) through which x-, y- and z-axes are assumed

---

[†] The terminology *tensor* is perhaps new to the reader. Indeed, at this stage we do not attempt to justify the use of this term but simply accept it as a name. As we shall see, in order for a mathematical quantity to be called a tensor, it must obey specific laws. We shall find that the stress tensor obeys these laws and it is for this reason that it is called a tensor.

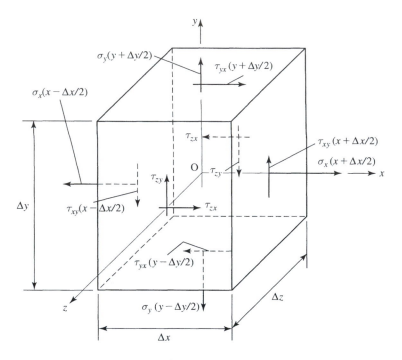

Figure 2.3.5

to pass [Fig. (2.3.5)]. Since these are axes of symmetry, they therefore are also principal axes of the element. Now, let us assume that stress components are acting on each face of the element as shown in the figure. (Note that all components shown in the figure are positive according to our sign convention. For simplicity, we have not shown stress components acting in the z-direction.) Since the element is infinitesimal, we may also assume that the components are acting at the centre of each face. It is important to note that Fig. (2.3.5) represents a real physical element of the body (as opposed to Fig. (2.3.4), which was merely a convenient pictorial representation). However, although the element is small, we may not assume that the stress components on opposite faces are the same. Thus, for example, the stresses on the right (positive) face of the element, which is located at the coordinate $x + \Delta x/2$, must be assumed to be different than those on the left (negative) face, which is located at the coordinate $x - \Delta x/2$; that is,

$$\sigma_x(x \pm \Delta x/2, y, z) = \sigma_x(x, y, z) \pm \Delta\sigma_x,$$

$$\tau_{xy}(x \pm \Delta x/2, y, z) = \tau_{xy}(x, y, z) \pm \Delta\tau_{xy} \qquad \text{(etc.).}$$

Similarly,

$$\sigma_y(x, y \pm \Delta y/2, z) = \sigma_y(x, y, z) \pm \Delta\sigma_y,$$

$$\tau_{yx}(x, y \pm \Delta y/2, z) = \tau_{yx}(x, y) \pm \Delta\tau_{yx},$$

where $y \pm \Delta y/2$ represent the y-coordinate of the top and bottom plane, respectively.

In addition to the stress components, we must also assume that a body force $\boldsymbol{B}$ acts through point O. These forces (which were defined in Chapter 1, and have units of N/m^3) act at various points of the body. Denoting the components of $\boldsymbol{B}$ by

B_x, B_y and B_z, we have

$$\boldsymbol{B} = B_x \boldsymbol{i} + B_y \boldsymbol{j} + B_z \boldsymbol{k}. \tag{2.3.10}$$

Now, this element must satisfy the principle of angular momentum. In particular, let us consider the equation of angular momentum about the z-axis:

$$\sum M_z = \dot{H}_z, \tag{2.3.11a}$$

where $\dot{H}_z \equiv \frac{\mathrm{d}H_z}{\mathrm{d}t}$, the rate of change in angular momentum about the z-axis is given by[†]

$$\dot{H}_z = I_{zz}\ddot{\theta}_z + (I_{xx} - I_{yy})\dot{\theta}_x\dot{\theta}_y. \tag{2.3.11b}$$

In the above equation, I_{xx}, I_{yy} and I_{zz} represent the mass moments of inertia about the x-, y- and z-principal axes, respectively; $\dot{\theta}_x$, $\dot{\theta}_y$ represent the angular velocities and $\ddot{\theta}_z$ denotes the angular acceleration about the z-axis. For simplicity, let us consider the case where the element does not rotate about the x- or y-axes. We then have, from Eqs. (2.3.11a) and (2.3.11b),

$$\sum M_z = I_{zz}\ddot{\theta}_z. \tag{2.3.11c}$$

We note further that in the expression for M_z, components acting in the z-direction will not appear in Eq. (2.3.11c); it is for this reason that they were omitted from Fig. (2.3.5).

We recall that the moment of inertia I_{zz} appearing above can be written as

$$I_{zz} = \rho \Delta x \Delta y \Delta z k_z^2, \tag{2.3.12}$$

where k_z is the radius of gyration of the element about the z-axis. (The expressions for I_{xx} and I_{yy} are similar with k_z replaced by k_x and k_y, respectively). It is important to note that k_z is of the order of $\Delta\ell$ (where $\Delta\ell$, a characteristic dimension of the element, is an infinitesimal of the same order as Δx, Δy or Δz) .

Taking moments about the z-axis passing through point O (and observing that all the stress components σ_x, σ_y, τ_{zx} and τ_{zy}, as well as the body forces pass through this axis and therefore do not contribute to M_z), we find, from Eq. (2.3.11c),

$$\left\{ \left[\tau_{xy}\left(x + \frac{\Delta x}{2}, y, z\right) \right] \Delta y \Delta z \right\} \frac{\Delta x}{2} + \left\{ \left[\tau_{xy}\left(x - \frac{\Delta x}{2}, y, z\right) \right] \Delta y \Delta z \right\} \frac{\Delta x}{2}$$

$$- \left\{ \left[\tau_{yx}\left(x, y + \frac{\Delta y}{2}, z\right) \right] \Delta x \Delta z \right\} \frac{\Delta y}{2}$$

$$- \left\{ \left[\tau_{yx}\left(x, y - \frac{\Delta y}{2}, z\right) \right] \Delta x \Delta z \right\} \frac{\Delta y}{2} = I_{zz}\ddot{\theta}_z. \tag{2.3.13}$$

Now,[‡]

$$\tau_{xy}\left(x \pm \frac{\Delta x}{2}, y, z\right) \simeq \tau_{xy} \pm \frac{\partial \tau_{xy}}{\partial x}\frac{\Delta x}{2}, \tag{2.3.14a}$$

$$\tau_{yx}\left(x, y \pm \frac{\Delta y}{2}, z\right) \simeq \tau_{yx} \pm \frac{\partial \tau_{yx}}{\partial y}\frac{\Delta y}{2}. \tag{2.3.14b}$$

[†] See, for example, Beer and Johnston, *Vector Mechanics for Engineers.*
[‡] We assume here that all stress components vary 'smoothly' with x, y and z.

Substituting Eqs. (2.3.12) and (2.3.14) in Eq. (2.3.13),

$$[\tau_{xy} - \tau_{yx}]\Delta x \Delta y \Delta z \simeq \rho \Delta x \Delta y \Delta z k_z^2 \ddot{\theta}_z \tag{2.3.15a}$$

and dividing through by $\Delta x \Delta y \Delta z$:

$$[\tau_{xy} - \tau_{yx}] \simeq \rho k_z^2 \ddot{\theta}_z. \tag{2.3.15b}$$

Taking the limit as $\Delta x \to 0$, $\Delta y \to 0$ and $\Delta z \to 0$, and recalling that k_z^2 is of order $\Delta \ell^2$, it follows that

$$\lim_{\Delta \ell \to 0} k_z = 0. \tag{2.3.15c}$$

We therefore obtain in the limit, $\tau_{xy} - \tau_{yx} = 0$; that is,

$$\tau_{xy} = \tau_{yx}. \tag{2.3.16a}$$

It is important to understand that by taking the limit as $\Delta x \to 0$, $\Delta y \to 0$ and $\Delta z \to 0$ (and implicitly $\Delta \ell \to 0$), we have established that the property $\tau_{xy} = \tau_{yx}$ exists *at a point*.

Similarly, taking moments about the x- and y-axes and proceeding in the same manner, we obtain

$$\tau_{yz} = \tau_{zy} \tag{2.3.16b}$$

and

$$\tau_{zx} = \tau_{xz}, \tag{2.3.16c}$$

respectively.

The equalities given by Eqs. (2.3.16) are referred to as the equality of the 'conjugate shear stresses' at a *point*. Thus we have found that the shear stress components acting at a point in perpendicular directions on any two mutually perpendicular planes are always equal. These equalities at a point are shown in Fig. (2.3.6). Again, it emphasised that Fig. (2.3.6) is merely a pictorial representation, which permits the representation of the shear stresses existing at the various planes passing through point O and does not represent a physical element.

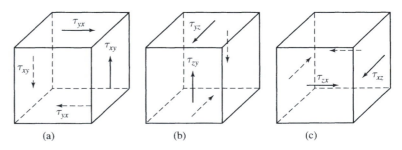

Figure 2.3.6 (a)   (b)   (c)

As a result of the equality of the three conjugate shear stresses, the stress tensor at a point contains only six *independent* components, namely

$$\begin{pmatrix} \tau_{xx} & \tau_{xy} & \tau_{xz} \\ \tau_{yx} = \tau_{xy} & \tau_{yy} & \tau_{yz} \\ \tau_{zx} = \tau_{xz} & \tau_{zy} = \tau_{yz} & \tau_{zz} \end{pmatrix}.$$

We observe immediately that the stress tensor is *symmetric*. Thus we refer to the stress at a point as being represented by a *second-rank symmetric tensor*.

It is important to emphasise that equilibrium is *not* a requirement for the equality of the conjugate shear stresses; indeed the equality of the conjugate shear stresses is valid for bodies with angular accelerations. It should be pointed out, however, that the equality of the conjugate shear stresses follows also from the assumption given by Eq. (2.3.2). In fact, if this assumption were not valid, we would find that $\tau_{xy} \neq \tau_{yx}$, etc. and nine independent components would remain.

Now, there exist many bodies where the stress state at a point is such that all the stress components that act in a particular direction vanish. For example, such a case exists in a plate that lies in the x–y plane and which is subjected to forces that lie only in this plane [Fig. (2.3.7)]; in this case, all stress components acting in the z-direction will vanish and the array representing the resulting two-dimensional stress tensor contains only four non-zero components. We refer to the body as being in a state of 'plane stress'. The stress tensor, in this two-dimensional case, is written as

$$\begin{pmatrix} \tau_{xx} & \tau_{xy} \\ \tau_{yx} = \tau_{xy} & \tau_{yy} \end{pmatrix}.$$

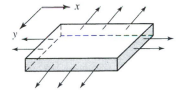

Figure 2.3.7

We thus observe that for this two-dimensional case, there exist only three independent stress components: τ_{xx}, τ_{yy} and $\tau_{xy} = \tau_{yx}$. We shall analyse such a state of stress in greater detail in a subsequent section.

Having shown that the principle of angular momentum leads to the equality of the conjugate shear stresses at a point, it remains for us to satisfy the principle of linear momentum for any element in a body.

2.4 Stress equations of motion and equilibrium

Consider a body of mass density ρ in a x, y, z coordinate system as shown in Fig. (2.4.1). In general, such a body may be subjected to external forces $\boldsymbol{F}$ as well as body forces $\boldsymbol{B}$ that act at various points of the body and have components B_x, B_y and B_z, as defined in Eq. (2.3.10).

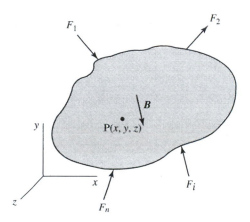

Figure 2.4.1

Due to the forces $\boldsymbol{F}$ and $\boldsymbol{B}$, it is clear that the various points of the body will displace and internal stresses will exist within the body. Let us consider a point $P(x, y, z)$ located in the body. From the previous sections, we have established that there exist nine stress components, six of which are independent ($\sigma_x, \sigma_y, \sigma_z, \tau_{xy} = \tau_{yx}, \tau_{yz} = \tau_{zy}, \tau_{zx} = \tau_{xz}$).

We first denote the displacements by

$$\boldsymbol{u} = u\boldsymbol{i} + v\boldsymbol{j} + w\boldsymbol{k}. \qquad (2.4.1a)$$

Now, as we observed previously, the stress components may vary from point to point, i.e., in general, they are functions of x, y and z. The variation of the stress state throughout the body is often referred to as the **stress field**. We shall assume that the stress field is continuous; i.e., there exist no discontinuities in the stresses and that all partial derivatives with respect to the coordinates exist.

Let us consider an element $\Delta x \, \Delta y \, \Delta z$ at the general point P(x, y, z) where, here, we have taken the point P to be at the corner of the element as shown in Fig. (2.4.2). Now, according to the principle of linear momentum, $\sum \boldsymbol{F} = m\ddot{\boldsymbol{u}}$, where m, the mass of the element, is given by $\rho \Delta x \, \Delta y \, \Delta z$. Isolating this element as a free body, the above stresses are considered as 'external forces' acting upon it. Applying the principle of linear momentum in the x-direction, the stress components acting in the x-direction as well as the body force B_x are as shown in Fig. (2.4.2). As before, we may assume that the stresses act at the centre of each face.

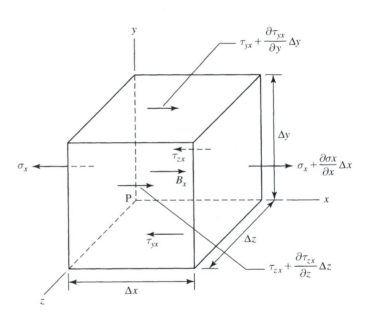

Figure 2.4.2

From linear momentum in the x-direction, i.e., $\sum F_x = m\ddot{u}$, we have

$$-\sigma_x \Delta y \Delta z + (\sigma_x + \Delta\sigma_x)\Delta y \Delta z - \tau_{yx}\Delta x \Delta z + (\tau_{yx} + \Delta\tau_{yx})\Delta x \Delta z$$
$$- \tau_{zx}\Delta x \Delta y + (\tau_{zx} + \Delta\tau_{zx})\Delta x \Delta y + B_x\Delta x \Delta y \Delta z = \rho \Delta x \Delta y \Delta z \, \ddot{u} \quad (2.4.2a)$$

or

$$\Delta\sigma_x \Delta y \Delta z + \Delta\tau_{yx}\Delta x \Delta z + \Delta\tau_{zx}\Delta x \Delta y + B_x\Delta x \Delta y \Delta z = \rho \Delta x \Delta y \Delta z \, \ddot{u}. \qquad (2.4.2b)$$

Upon dividing by $\Delta x \, \Delta y \, \Delta z$, we obtain

$$\frac{\Delta\sigma_x}{\Delta x} + \frac{\Delta\tau_{yx}}{\Delta y} + \frac{\Delta\tau_{zx}}{\Delta z} + B_x = \rho\ddot{u} \qquad (2.4.2c)$$

Taking the limits as $\Delta x \to 0$, $\Delta y \to 0$ and $\Delta z \to 0$, and recalling that in the limit, by definition, the above ratios are partial derivatives, we obtain

$$\frac{\partial \sigma_x}{\partial x} + \frac{\partial \tau_{yx}}{\partial y} + \frac{\partial \tau_{zx}}{\partial z} + B_x = \rho \ddot{u}. \qquad (2.4.3a)$$

Similarly, $\sum F_y = m\ddot{v}$ and $\sum F_z = m\ddot{w}$, yield respectively,

$$\frac{\partial \tau_{xy}}{\partial x} + \frac{\partial \sigma_y}{\partial y} + \frac{\partial \tau_{zy}}{\partial z} + B_y = \rho \ddot{v}, \qquad (2.4.3b)$$

$$\frac{\partial \tau_{xz}}{\partial x} + \frac{\partial \tau_{yz}}{\partial y} + \frac{\partial \sigma_z}{\partial z} + B_z = \rho \ddot{w}. \qquad (2.4.3c)$$

If the body is in equilibrium, $\ddot{u} = \ddot{v} = \ddot{w} = 0$, and hence

$$\frac{\partial \sigma_x}{\partial x} + \frac{\partial \tau_{yx}}{\partial y} + \frac{\partial \tau_{zx}}{\partial z} + B_x = 0, \qquad (2.4.4a)$$

$$\frac{\partial \tau_{xy}}{\partial x} + \frac{\partial \sigma_y}{\partial y} + \frac{\partial \tau_{zy}}{\partial z} + B_y = 0, \qquad (2.4.4b)$$

$$\frac{\partial \tau_{xz}}{\partial x} + \frac{\partial \tau_{yz}}{\partial y} + \frac{\partial \sigma_z}{\partial z} + B_z = 0. \qquad (2.4.4c)$$

It is important to note that the above equations are valid for any body, i.e., they do not depend upon a particular material. In fact, these equation are valid for fluids as well as for solid bodies.

Although Eqs. (2.4.3)–(2.4.4), known as the equations of motion and equilibrium, respectively, were derived in terms of a Cartesian coordinate system, it should be mentioned that similar equations (although of different form) exist for other coordinate systems.

Equations (2.4.3) and (2.4.4) demonstrate that the stress components may not vary arbitrarily from point to point within a body. They must vary in a prescribed manner such that they satisfy Eqs. (2.4.3) for an accelerating body, or Eqs. (2.4.4), if the body is in equilibrium; otherwise they will violate the principles of linear momentum or equilibrium.

Example 2.4: The stress field of a body is given as

$$\sigma_x = ax^4, \qquad \sigma_y = 6ax^2 y^2, \qquad \sigma_z = cxyz^2,$$
$$\tau_{xy} = bx^3 y, \qquad \tau_{yz} = cx^2 yz, \qquad \tau_{zx} = cx^3 y,$$

where a, b, c are constants (whose units are Pa/m^6). In addition, the body forces are known to be zero throughout the body.

Under what conditions do these stresses represent a state of equilibrium at *all* points of the body?

Solution: In order for the body to be in equilibrium, the above stress field must satisfy Eqs. (2.4.4). Substituting in these equations, with $\mathbf{B} = 0$,

$$4ax^3 + bx^3 = 0 \qquad \text{(i)}$$
$$3bx^2 y + 12ax^2 y + cx^2 y = 0 \qquad \text{(ii)}$$
$$3cx^2 y + cx^2 z + 2cxyz = 0 \qquad \text{(iii)}$$

We first observe that in Eq. (iii), the constant c is the coefficient to terms of various *different* powers of the coordinates x, y and z. Therefore, equilibrium in the z-direction can be satisfied at all points P(x, y, z) of the body only if $c = 0$. The remaining two equations, (i) and (ii), reduce to

$$(4a + b)x^3 = 0 \quad \text{and} \quad 3(4a + b)x^2y = 0,$$

respectively, and are satisfied at *all* points P(x, y, z) only if $b = -4a$.   □

2.5   Relations between stress components and internal force resultants

For simplicity, let us consider a rod of cross-sectional area A whose outward normal is in the x-direction as shown in Fig. (2.5.1). We denote here the normal force acting on this plane by F and the shear forces in the y- and z-directions by V_y and V_z, respectively.

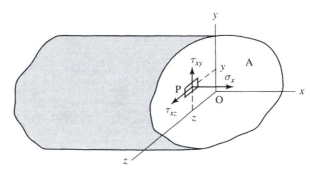

Figure 2.5.1

The components of the traction at any point P(x, y) on this plane are given by [cf. Eq. (2.3.5)]

$$\sigma_x = \lim_{\Delta A \to 0} \frac{\Delta F}{\Delta A}, \tag{2.5.1a}$$

$$\tau_{xy} = \lim_{\Delta A \to 0} \frac{\Delta V_y}{\Delta A}, \tag{2.5.1b}$$

$$\tau_{xz} = \lim_{\Delta A \to 0} \frac{\Delta V_z}{\Delta A}. \tag{2.5.1c}$$

Therefore, acting on an infinitesimal area ΔA about the point P, we have

$$\Delta F \simeq \sigma_x \Delta A, \tag{2.5.2a}$$

$$\Delta V_y \simeq \tau_{xy} \Delta A, \tag{2.5.2b}$$

$$\Delta V_z \simeq \tau_{xz} \Delta A. \tag{2.5.2c}$$

Now the resultant of these forces on the entire plane is, clearly, the sum of all these incremental forces over the area, i.e.,

$$F = \sum \Delta F, \tag{2.5.3a}$$

$$V_y = \sum \Delta V_y, \tag{2.5.3b}$$

$$V_z = \sum \Delta V_z. \tag{2.5.3c}$$

Substituting Eqs. (2.5.2) and taking the limit as the incremental areas shrink to zero, we obtain

$$F = \iint_A \sigma_x(y, z)\, dA, \tag{2.5.4a}$$

$$V_y = \iint_A \tau_{xy}(y, z)\, dA, \tag{2.5.4b}$$

$$V_z = \iint_A \tau_{xz}(y, z)\, dA. \tag{2.5.4c}$$

For the case where $\sigma_x = \text{const.}$, $F = \sigma_x \iint_A dA = \sigma_x A$. Similarly, if τ_{xy} and τ_{xz} are constant over the area, we obtain $V_y = \tau_{xy} A$ and $V_z = \tau_{xz} A$; hence

$$\sigma_x = \frac{F}{A}, \tag{2.5.5a}$$

$$\tau_{xy} = \frac{V_y}{A}, \tag{2.5.5b}$$

$$\tau_{xz} = \frac{V_z}{A}. \tag{2.5.5c}$$

The first expression, $\sigma_x = F/A$, is often found to be true for a prismatic bar subjected to a system of applied axial forces such that a resultant normal force F acts on the section. However, we shall find that this result depends on the line of action of F. (We shall study this case in greater detail in Chapter 6.)

On the other hand, as we shall see in Chapter 8, the expressions $\tau_{xy} = V_y/A$ and $\tau_{xz} = V_z/A$ cannot, in general, represent the true stress components at all points in a section. These expressions merely yield some *average* shear stress component on the section, as found, for example, in Chapter 1.

Consider now the moments resulting from the stresses acting on the section. We note from Fig. (2.5.1) that the stresses acting on the element ΔA produce incremental moments about the y- and z-axes, respectively:[†]

$$\Delta M_y = z\sigma_x(y, z)\Delta A, \tag{2.5.6a}$$

$$\Delta M_z = -y\sigma_x(y, z)\Delta A. \tag{2.5.6b}$$

Hence, as before, upon taking the sum and the limiting case as $\Delta A \to 0$, we obtain

$$M_y = \iint_A z\sigma_x(y, z)\, dA, \tag{2.5.7a}$$

$$M_z = -\iint_A y\sigma_x(y, z)\, dA. \tag{2.5.7b}$$

The moments M_y and M_z about axes that lie in the plane of the cross-section are called **bending moments** since they tend to bend a straight rod into a curved shape.

[†] The signs of the moments are according to the right-hand rule.

It is worthwhile to observe that if $\sigma_x = $ const., then

$$M_y = \sigma_x \iint_A z \, dA, \tag{2.5.8a}$$

$$M_z = -\sigma_x \iint_A y \, dA. \tag{2.5.8b}$$

If, in particular, point O is the centroid of the cross-section [Fig. (2.5.1)], then $M_y = M_z = 0$; i.e., the moments about y- and z-centroidal axes due to $\sigma_x = $ const. acting over a section are zero.

Consider now the incremental moment about the longitudinal x-axis caused by the stress components τ_{xy} and τ_{xz} (note that σ_x can produce no moment about this axis since it acts parallel to the x-axis):

$$\Delta M_x = y(\tau_{xz}\Delta A) - z(\tau_{xy}\Delta A), \tag{2.5.9a}$$

which yields a total moment

$$M_x = \iint_A (y\tau_{xz} - z\tau_{xy}) \, dA. \tag{2.5.9b}$$

The moment M_x about the longitudinal x-axis of a rod is called the **torsional moment** since it tends to twist the rod, and is usually denoted by $T \equiv M_x$.

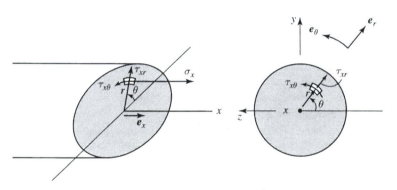

Figure 2.5.2

In the above, we have discussed the state of stress in a Cartesian x, y, z system. Now, if the geometry of a body is defined in terms of another coordinate system, for example, a cylindrical coordinate system, it is clearly more reasonable to express the stress components in this more natural system rather than in a Cartesian system. For example, if we have a circular rod, we would use a polar system defined by coordinates r, θ, x [Fig. (2.5.2)]. In this case, it is customary to define the unit vectors in this coordinate system by the symbols $\mathbf{e}_r, \mathbf{e}_\theta, \mathbf{e}_x$, respectively. The resulting traction at any point on the x-plane is then

$$\mathbf{T}_x = \sigma_x \mathbf{e}_x + \tau_{xr}\mathbf{e}_r + \tau_{x\theta}\mathbf{e}_\theta. \tag{2.5.10}$$

These stress components are shown acting on the x-plane in Fig. (2.5.2). It is worthwhile to mention that the equality of the conjugate shear stresses remains equally valid irrespective of the coordinate system used. Thus, since the r-, θ- and x-directions are mutually perpendicular, we have, for this system,

$$\tau_{xr} = \tau_{rx}, \qquad \tau_{r\theta} = \tau_{\theta r}, \qquad \tau_{x\theta} = \tau_{\theta x}. \tag{2.5.11}$$

It should be emphasised that, at all points, the stress $\tau_{x\theta}$ acts in a circumferential direction in this coordinate system.

The expressions for the moments given by Eqs. (2.5.6)–(2.5.9), were obtained for a Cartesian x, y, z coordinate system. Expressions for the moments M_y and M_z in terms of the stresses $(\sigma_x, \tau_{xr}, \tau_{x\theta})$ acting on the x-plane are

$$M_y = -\iint\limits_A (r \cos\theta)\sigma_x \, dA, \qquad (2.5.12a)$$

$$M_z = -\iint\limits_A (r \sin\theta)\sigma_x \, dA, \qquad (2.5.12b)$$

where θ is measured from the negative z-axis as shown in Fig. (2.5.2).[†]

On the other hand, the torsional moment $T \equiv M_x$ is given by

$$T = \iint\limits_A r\tau_{x\theta}(r, \theta) \, dA. \qquad (2.5.12c)$$

This last expression will be found particularly useful in treating the problem of torsion of rods having circular cross-sections.

Example 2.5: Stress components acting on a rectangular cross-section as shown in Fig. (2.5.3) are given by

$$\sigma_x = \alpha y^2 z, \qquad \tau_{xy} = \beta(d^2 - y^2), \qquad \tau_{xz} = \gamma(d^2 - y^2)(b^2 - z^2),$$

where α, β and γ are coefficients. Determine the resultant internal forces and moments acting on the section.

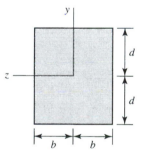

Figure 2.5.3

Solution: From Eqs. (2.5.4)

$$F = \iint\limits_A \sigma_x(y, z) \, dA = \alpha \int_{-d}^{d} y^2 \, dy \int_{b}^{b} z \, dz = \alpha y^3/3|_{-d}^{d} \cdot z^2/2|_{-b}^{b} = 0$$

$$V_y = \iint\limits_A \tau_{xy}(y, z) \, dA = 2\beta b \int_{-d}^{d} (d^2 - y^2) \, dy = 2b\beta[d^2 y - y^3/3]|_{-d}^{d} = \frac{8\beta bd^3}{3}.$$

Similarly,

$$V_z = \iint\limits_A \tau_{xz}(y, z) \, dA = \frac{64\gamma}{9} b^3 d^3.$$

From Eqs. (2.5.7),

$$M_y = \iint\limits_A z\sigma_x(y, z) \, dA = \alpha \int_{-d}^{d} y^2 \, dy \int_{-b}^{b} z^2 \, dz = \frac{\alpha y^3}{3}\bigg|_{-d}^{d} \cdot \frac{z^3}{3}\bigg|_{-b}^{b} = \frac{4\alpha}{9} b^3 d^3$$

$$M_z = -\iint\limits_A y\sigma_x(y, z) \, dA = -\alpha \int_{-d}^{d} y^3 \, dy \int_{-b}^{b} z \, dz = 0.$$

[†] See previous footnote, p. 39.

By direct substitution in Eq. (2.5.9b), we find

$$M_x = \iint\limits_A (y\tau_{xz} - z\tau_{xy})\,dA = 0.$$

□

Example 2.6:   Stress components acting on a circular cross-section of radius R and area A as shown in Fig. (2.5.2) are given by

$$\sigma_x = \frac{\sigma_0 r^2}{R^2}, \qquad \tau_{xr} = \frac{\tau_0 r(R-r)}{R^2}, \qquad \tau_{x\theta} = \tau_0 \frac{r}{R},$$

where σ_0 and τ_0 are constant stress values.

Determine the resulting normal force F, the bending moments M_x, M_y and the torsional moment $T \equiv M_x$ acting on this cross-section.

Solution:

$$F = \iint\limits_A \sigma_x(y, z)\,dA = \frac{\sigma_0}{R^2} 2\pi \int_0^R r^3\,dr = \frac{\pi \sigma_0 R^2}{2} = \frac{\sigma_0 A}{2}$$

$$M_y = -\iint\limits_A (r\cos\theta)\sigma_x\,dA = -\frac{\sigma_0}{R^2} \int_0^R r^4\,dr \int_0^{2\pi} \cos\theta\,d\theta = 0$$

since $\int_0^{2\pi} \cos\theta\,d\theta = 0$.

Similarly,

$$M_z = -\iint\limits_A (r\sin\theta)\sigma_x\,dA = 0.$$

$$T = \iint\limits_A r\tau_{x\theta}(r, \theta)\,dA = \frac{\tau_0}{R} 2\pi \int_0^R r^3\,dr = \tau_0 \frac{\pi R^3}{2}.$$

□

2.6   Stress transformation laws for plane stress

(a) Derivation

We recall that there exists a two-dimensional state of stress in which all stress components in a particular direction, say z, vanish. In this case the stress tensor at a point was seen to be represented by the array

$$\begin{pmatrix} \tau_{xx} & \tau_{xy} \\ \tau_{yx} = \tau_{xy} & \tau_{yy} \end{pmatrix}.$$

Now consider an element at a point P. Let us assume further that all stress components do not vary with the z-coordinate. We recall that such a two-dimensional state of stress is called **plane stress**. Since the z-dependency has been eliminated, we may represent the state of stress by means of a two-dimensional figure of an infinitesimal element as shown in Fig. (2.6.1). (We may think of this element as having unit thickness in the z-direction.) In this figure, we have drawn all stresses

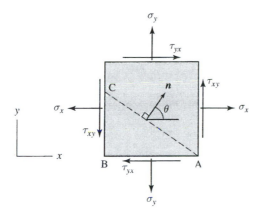

Figure 2.6.1

as positive according to our adopted sign convention. Let us assume that the three independent stress components are given with respect to the x, y coordinates; i.e., we assume that the stresses acting on both the x- and y-faces are known. In addition, body forces B may be assumed to be acting on the element. We now wish to determine the stress components that exist on any other arbitrary n-face whose normal lies in the x–y plane.

To this end, we define the n-plane, by means of a normal n, which makes an angle θ (positive counter-clockwise) with respect to the positive x-axis. We further define, as before, the unit vector t as being tangential to the plane; i.e., n and t are mutually perpendicular. (Note that the angle between t and the positive x-axis then is always $\theta + \pi/2$ according to our definition.) Since we wish to determine the stress components existing on this plane, we therefore 'cut' the element along this plane and isolate it as a free body [Fig. (2.6.2)].

Note that the isolated portion is now a triangular element ABC. Let Δx, Δy and Δs denote the infinitesimal lengths of AB, BC and AC, respectively. Clearly, having made the 'cut' there must exist unknown normal and shear stresses acting on the n-plane. We denote these by σ_n and τ_{nt}, respectively.

Now, this element must satisfy the (vector) equation of linear momentum

$$\sum F = m\ddot{u}, \tag{2.6.1}$$

where the mass of the element is

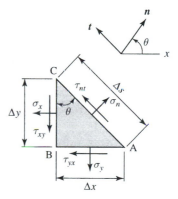

Figure 2.6.2

$$m = \frac{1}{2}\rho\,\Delta x\,\Delta y = \frac{1}{2}\rho\,\Delta s^2 \sin\theta\cos\theta = \frac{1}{4}\rho\,\Delta s^2 \sin 2\theta \tag{2.6.2}$$

and ρ is the mass density.

Instead of resolving the above linear momentum equation in x- and y-components, it is more convenient here to resolve the vector equation in the n- and t-directions; thus we must satisfy the scalar equations

$$\sum F_n = m\ddot{u}_n \tag{2.6.3a}$$

and

$$\sum F_t = m\ddot{u}_t, \tag{2.6.3b}$$

where $\ddot{u}_n$ and $\ddot{u}_t$ are the accelerations in the n- and t-directions, respectively.

From Eq. (2.6.3a), we obtain

$$\sigma_n \Delta s - (\sigma_x \Delta y)\cos\theta - (\sigma_y \Delta x)\sin\theta - (\tau_{xy}\Delta y)\sin\theta$$

$$- (\tau_{yx}\Delta x)\cos\theta + B_n \frac{\Delta s^2}{4}\sin 2\theta = \frac{\rho}{4}\Delta s^2 \sin 2\theta \ddot{u}_n, \qquad (2.6.4a)$$

where $\boldsymbol{B}$ is represented in terms of its components as $\boldsymbol{B} = B_n \boldsymbol{n} + B_t \boldsymbol{t}$. But

$$\Delta x = \Delta s \sin\theta, \qquad \Delta y = \Delta s \cos\theta. \qquad (2.6.4b)$$

Dividing Eq. (2.6.4a) through by Δs and substituting Eqs. (2.6.4b) in it, we find

$$\sigma_n - \sigma_x \cos^2\theta - \sigma_y \sin^2\theta - 2\tau_{xy}\sin\theta\cos\theta = \frac{1}{4}\Delta s \sin 2\theta \, (\rho \ddot{u}_n - B_n). \qquad (2.6.5)$$

Upon taking the limit as $\Delta s \to 0$, the right-hand side goes to zero, and it follows that

$$\sigma_n = \sigma_x \cos^2\theta + \sigma_y \sin^2\theta + 2\tau_{xy}\sin\theta\cos\theta. \qquad (2.6.6a)$$

Recalling the identities,

$$\cos^2\theta = \frac{1}{2}(1 + \cos 2\theta), \qquad \sin^2\theta = \frac{1}{2}(1 - \cos 2\theta), \qquad \sin 2\theta = 2\sin\theta\cos\theta,$$

we obtain an alternate form of Eq. (2.6.6a), namely

$$\sigma_n = \frac{\sigma_x}{2}(1 + \cos 2\theta) + \frac{\sigma_y}{2}(1 - \cos 2\theta) + \tau_{xy}\sin 2\theta$$

or

$$\sigma_n = \frac{\sigma_x + \sigma_y}{2} + \frac{\sigma_x - \sigma_y}{2}\cos 2\theta + \tau_{xy}\sin 2\theta. \qquad (2.6.6b)$$

Similarly, from $\sum F_t = m\ddot{u}_t$,

$$\tau_{nt}\Delta s + (\sigma_x \Delta y)\sin\theta - (\sigma_y \Delta x)\cos\theta - \tau_{xy}(\cos^2\theta - \sin^2\theta)\Delta s$$

$$= \frac{1}{4}\Delta s^2 \sin 2\theta(\rho \ddot{u}_t - B_t). \qquad (2.6.7)$$

Again, dividing through by Δs, using Eqs. (2.6.4b), and taking the limit as $\Delta s \to 0$, we obtain

$$\tau_{nt} = \tau_{xy}(\cos^2\theta - \sin^2\theta) - (\sigma_x - \sigma_y)\cos\theta\sin\theta. \qquad (2.6.8a)$$

Upon noting that $\cos^2\theta - \sin^2\theta = \cos 2\theta$, we obtain an alternative form for τ_{nt}, namely

$$\tau_{nt} = \tau_{xy}\cos 2\theta - \frac{\sigma_x - \sigma_y}{2}\sin 2\theta. \qquad (2.6.8b)$$

Equations (2.6.6a) and (2.6.8a) or alternatively Eqs. (2.6.6b) and (2.6.8b) are called the **transformation laws** for plane stress. These expressions thus permit us to determine the stress components that exist on an arbitrary plane passing through a point in terms of the stresses existing on the x- and y-coordinate planes. One may also think of these laws as prescribing the stress components in any coordinate system (here the n–t system) in terms of the scalar quantities in the x–y system. Thus they transform the scalar quantities in one coordinate system to another coordinate system. It is for this reason that they are referred to as 'transformation laws'. In physical terms, the two-dimensional state of stress at a point, $(\sigma_x, \sigma_y, \tau_{xy})$, is equivalent to the state of stress defined by the components $(\sigma_n, \sigma_t, \tau_{nt})$. This equivalence is shown pictorially in Fig. (2.6.3).

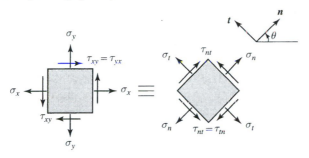

Figure 2.6.3

It should be emphasised that the derivation of the above transformation laws does *not require an equilibrium state* and thus these expressions are also valid at all points of a body undergoing accelerations. It is also important to emphasise that these laws are true *for any specific point* of a body; indeed one can only refer to the state of stress *at a point*. This feature should be clearly evident since, in the process of deriving these laws, it was necessary to take the limit $\Delta s \to 0$.

We shall find that the transformation laws possess certain interesting properties; these will be investigated in the next section.

Example 2.7: A body is subjected to forces such that σ_y is the only non-zero stress component at a point. The remaining stress components $\sigma_x = \tau_{xy} = 0$ [Fig. (2.6.4a)]. Determine the stress components that exist on planes whose normals are oriented by 45° and 135° with respect to the *x*-plane.

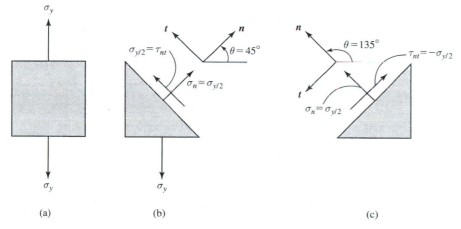

(a)   (b)   (c)   Figure 2.6.4

Solution: On the 45° *n*-plane: $\theta = 45°$, $\sin 2\theta = 1$, $\cos 2\theta = 0$. Therefore, from Eqs. (2.6.6b) and (2.6.8b),

$$\sigma_n = \frac{\sigma_y}{2}, \qquad \tau_{nt} = \frac{\sigma_y}{2} \quad \text{[Fig. (2.6.4b)]}.$$

On the 135° *n*-plane: $\theta = 135°$, $\sin 2\theta = -1$, $\cos 2\theta = 0$. Therefore

$$\sigma_n = \frac{\sigma_y}{2}, \qquad \tau_{nt} = -\frac{\sigma_y}{2} \quad \text{[Fig. (2.6.4c)]}.$$

It is important to note the directions of the shear stresses on the two given planes. □

Further examples illustrating the transformation laws are deferred to a subsequent section.

(b) Remarks on the transformation laws
(stress as a tensor; invariants of a tensor)

Any quantity for which its (two-dimensional) scalar components transform from one coordinate system to another, according to Eqs. (2.6.6) and (2.6.8), is called a **two-dimensional symmetric tensor** of rank 2. Here, in particular, the tensor is a stress tensor. However, there are other quantities, for example, moments and products of inertia (viz., I_{xx}, I_{yy}, $-I_{xy}$), which transform according to the same laws. Therefore, one may state that the moment and products of inertia are also scalar components of a second-rank symmetric tensor. Thus, *by definition, a tensor is a mathematical quantity that transforms according to certain laws.*

Tensors, as governed by their transformations laws, possess several properties. We develop these properties (in two dimensions) for the second-rank symmetric stress tensor.

Recall that Eq. (2.6.6b) represents the normal stress component σ_n acting on the n-plane whose normal $\boldsymbol{n}$ is inclined at an angle θ with respect to the x-axis. We also recall that the unit vector $\boldsymbol{t}$ was then defined as being inclined at an angle $\theta + \pi/2$ with respect to the x-axis. Consequently, the normal stress component σ_t is given by

$$\sigma_t = \frac{\sigma_x + \sigma_y}{2} + \frac{\sigma_x - \sigma_y}{2} \cos 2(\theta + \pi/2) + \tau_{xy} \sin 2(\theta + \pi/2) \quad (2.6.9a)$$

or

$$\sigma_t = \frac{\sigma_x + \sigma_y}{2} - \frac{\sigma_x - \sigma_y}{2} \cos 2\theta - \tau_{xy} \sin 2\theta. \quad (2.6.9b)$$

Adding Eqs. (2.6.6b) and (2.6.9b) we obtain immediately

$$\sigma_n + \sigma_t = \sigma_x + \sigma_y = I_{\sigma_1} \quad \text{(constant)}. \quad (2.6.10a)$$

From Eq. (2.6.10a) we observe that, *for any given point, the sum of the normal stresses in any two orthogonal directions is a constant.* While the three-dimensional case is beyond the scope of our treatment, we state here that for this case,

$$\sigma_x + \sigma_y + \sigma_z = I_{\sigma_1}; \quad (2.6.10b)$$

i.e., the sum of the normal stresses in any three orthogonal directions is a constant.

Similarly, from Eqs. (2.6.6b), (2.6.8b) and (2.6.9b), we find, after some simple algebraic manipulations,

$$\sigma_n \sigma_t - \tau_{nt}^2 = \sigma_x \sigma_y - \tau_{xy}^2 = I_{\sigma_2} \quad \text{(constant)}. \quad (2.6.11)$$

The constants I_{σ_1} and I_{σ_2} appearing in Eqs. (2.6.10) and (2.6.11) are called **invariants**. These equations demonstrate that the two-dimensional symmetric stress tensor possesses two invariant quantities that are *true for any set of mutually perpendicular stress components*, irrespective of their orientation in space. (In the three-dimensional case, which again is beyond the scope of our treatment, we obtain three invariants.) These invariant properties are significant characteristic properties of tensors.

(c) Transformation law of a vector: the vector as a tensor

In our previous development, we found that the symmetric (two-dimensional) array of the stress components transformed according to a law given by Eqs. (2.6.6) and (2.6.8). This law was stated as being the transformation law for a symmetric tensor of rank 2. We further observed that there exist, in this case, two invariants that remain valid irrespective of the orientation of the coordinates x, y.

We now digress from our study of solid mechanics to demonstrate that the concepts of transformation laws and invariants, as found in the investigation of the stress tensor, are concepts that have been encountered previously, namely, for a vector.

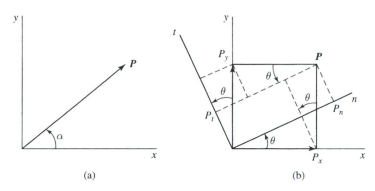

(a)   (b)   **Figure 2.6.5**

To this end, we consider a two-dimensional vector P lying in an x–y plane. Now, a vector is defined by two quantities: (a) its magnitude $|P|$ and (b) its orientation; for example, the angle α with respect to the x-axis (in the two-dimensional case) [Fig. (2.6.5a)]. On the other hand, this vector may be defined, instead, by its two scalar components P_x and P_y as shown in Fig. (2.6.5b). We further note that the square of the magnitude of P is given by

$$|P|^2 = P_x^2 + P_y^2. \tag{2.6.12}$$

Now, instead of defining the vector P by means of its x- and y-components, it is clear that the vector P may also be defined by means of any other two mutually orthogonal components. Let us therefore construct another set of orthogonal coordinates, n and t, oriented with respect to the x–y system by an angle θ as shown in Fig. (2.6.5b). We denote the scalar components in this n–t system by P_n and P_t, respectively. Then, clearly,

$$P_n = P_x \cos\theta + P_y \sin\theta, \tag{2.6.13a}$$

$$P_t = -P_x \sin\theta + P_y \cos\theta. \tag{2.6.13b}$$

Equations (2.6.13) are, in fact, transformation laws that transform the scalar components of a vector in a single coordinate system, the x–y system, into the components of another coordinate system, the n–t system. They are the analogues to the transformation laws given by Eqs. (2.6.6) and (2.6.8) for the second-rank symmetric tensor.

Let us consider the quantity $(P_n^2 + P_t^2)$. Substituting Eqs. (2.6.13) we have

$$\begin{aligned}
P_n^2 + P_t^2 &= \left(P_x^2 \cos^2\theta + P_y^2 \sin^2\theta + 2P_x P_y \sin\theta, \cos\theta\right) \\
&\quad + \left(P_x^2 \sin^2\theta + P_y^2 \cos^2\theta - 2P_x P_y \sin\theta \cos\theta\right), \\
&= P_x^2 + P_y^2
\end{aligned}$$

that is,

$$P_n^2 + P_t^2 = P_x^2 + P_y^2 = |\boldsymbol{P}|^2 = I. \tag{2.6.14}$$

This last equation expresses the *invariant quality of the vector*; i.e., the sum of the squares of *any two orthogonal components of the two-dimensional vector equals a constant*. Here, this constant has a simple physical interpretation: it represents the square of the magnitude of the vector. Thus we observe that a vector possesses a single invariant. In our examination of the two-dimensional symmetric stress tensor, we noted that there exist two invariants, I_{σ_1} and I_{σ_2}. Thus the invariant represented by Eq. (2.6.14) represents the analogue to Eqs. (2.6.10) and (2.6.11).

Indeed, in mathematics, a vector is referred to as a *first-rank tensor*. (Note that only a single subscript is required to specify its components.) Thus, quantities such as a vector, or stress are called tensors since they obey specific transformation laws such that certain invariant properties are maintained for all coordinate systems.

2.7 Principal stresses and stationary shear stress values

(a) Principal stresses: stationary values of σ_n

We have seen in the previous section that if a (two-dimensional) state of stress, σ_x, σ_y and $\tau_{xy} = \tau_{yx}$, is known at a point, then the normal stress σ_n and shear stress τ_{nt} for any n-plane passing through this point can be obtained by means of the transformation laws. We note that $\sigma_n = \sigma_n(\theta)$ and $\tau_{nt} = \tau_{nt}(\theta)$. Now it is obvious that, as they depend on θ, these stress components have maximum and minimum values, i.e. stationary values [Fig. (2.7.1)]. We first investigate the stationary values of σ_n.

Treating σ_n as a function of θ, the necessary condition for stationary values is

$$\frac{\mathrm{d}\sigma_n}{\mathrm{d}\theta} = 0. \tag{2.7.1}$$

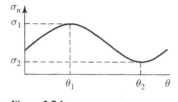

Figure 2.7.1

From Eq. (2.6.6b),

$$\frac{\mathrm{d}\sigma_n}{\mathrm{d}\theta} = -(\sigma_x - \sigma_y)\sin 2\theta + 2\tau_{xy}\cos 2\theta. \tag{2.7.2}$$

Setting this derivative to zero, we obtain

$$\tan 2\theta = \frac{\tau_{xy}}{\frac{\sigma_x - \sigma_y}{2}}. \tag{2.7.3}$$

Now Eq. (2.7.3) possesses two relevant roots, θ_1 and θ_2, which define two planes on which the maximum and minimum stresses σ_n act [Fig. (2.7.1)]. These maximum and minimum values of σ_n are called collectively the **principal stresses**, and the planes upon which they act are referred to as the **principal planes**.

Before examining these roots in detail, we first observe, by comparing Eqs. (2.6.8b) and (2.7.2), that

$$\frac{\mathrm{d}\sigma_n}{\mathrm{d}\theta} = 2\tau_{nt}. \tag{2.7.4}$$

Hence we immediately conclude that $\tau_{nt} = 0$ on the plane for which $\frac{\mathrm{d}\sigma_n}{\mathrm{d}\theta} = 0$. Thus we have established that on a principal plane, $\tau_{nt} = 0$; that is, *no shear stress component exists on a principal plane*.

We now turn our attention to a more thorough examination of Eq. (2.7.3). We have remarked that this equation possesses two roots, θ_1 and θ_2. Clearly, since σ_n is a function of θ, these roots correspond to the maximum and minimum values of σ_n [Fig. (2.7.1)] . We shall hereafter associate σ_1 and σ_2 with the maximum and minimum values of σ_n, respectively (algebraically, $\sigma_2 \leq \sigma_1$) acting on the corresponding principal planes defined by θ_1 and θ_2.

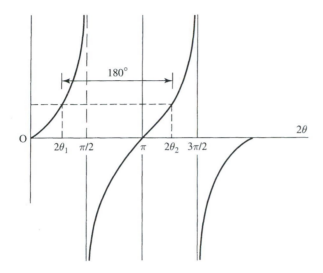

Figure 2.7.2

Upon plotting Eq. (2.7.3) as a function of 2θ [Fig. (2.7.2)], one observes that a relation exists between the two roots, namely $2\theta_2 = 2\theta_1 \pm \pi$; that is,

$$\theta_2 = \theta_1 \pm \pi/2. \qquad (2.7.5)$$

Thus we immediately conclude that *the two principal planes are mutually perpendicular.*

Now, having defined θ_1 to be the plane of maximum σ_n, it is necessary to identify definitely the θ_1 root. Indeed, from Fig. (2.7.2), it is certainly not clear which root corresponds to θ_1 and which to θ_2. However, since the maximum value σ_1 occurs at $\theta = \theta_1$, it necessarily follows that

$$\left. \frac{d^2\sigma_n}{d\theta^2} \right|_{\theta=\theta_1} < 0 \qquad (2.7.6)$$

for $\theta = \theta_1$. From Eq. (2.7.2), we find

$$\frac{d^2\sigma_n}{d\theta^2} = -2[(\sigma_x - \sigma_y)\cos 2\theta + 2\tau_{xy}\sin 2\theta]$$

$$= -4\tau_{xy}\sin 2\theta \left[\frac{\sigma_x - \sigma_y}{2\tau_{xy}} \left(\frac{1}{\tan 2\theta} \right) + 1 \right]. \qquad (2.7.7a)$$

Hence, making use of Eq. (2.7.3), we find

$$\frac{d^2\sigma_n}{d\theta^2} = -4\tau_{xy}\sin 2\theta \left(\frac{1}{\tan^2 2\theta} + 1 \right). \qquad (2.7.7b)$$

Noting that the term in parenthesis is always positive, we observe that the sign of the second derivative depends on the sign of the product $\tau_{xy}\sin 2\theta$. Hence it follows that $\frac{d^2\sigma_n}{d\theta^2} < 0$ at $\theta = \theta_1$ (a) if $\tau_{xy} > 0$ and $0 < 2\theta_1 < \pi$ or (b) if $\tau_{xy} < 0$ and $-\pi < 2\theta_1 < 0$.

We thus have established a *criterion* to identify the θ_1 principal plane upon which the *maximum* principal stress acts; namely

$$\text{If } \tau_{xy} > 0, \quad \text{then } 0 < \theta_1 < \pi/2$$
$$\text{If } \tau_{xy} < 0, \quad \text{then } -\pi/2 < \theta_1 < 0.$$

Having established this criterion, it is useful to examine further $\tan 2\theta$ as a function of 2θ [as shown in Fig. (2.7.3)]. If the quantity $\frac{\tau_{xy}}{\sigma_x - \sigma_y} > 0$, then the root θ_1 according to our criterion will correspond to, say, point A_1 if $\tau_{xy} > 0$; if $\frac{\tau_{xy}}{\sigma_x - \sigma_y} > 0$ and $\tau_{xy} < 0$, then the root θ_1 will correspond to point B_1. On the other hand, if $\frac{\tau_{xy}}{\sigma_x - \sigma_y} < 0$ and $\tau_{xy} > 0$, then the root θ_1 will correspond to point C_1 while if this quantity is negative and $\tau_{xy} < 0$, then θ_1 will correspond to point D_1.

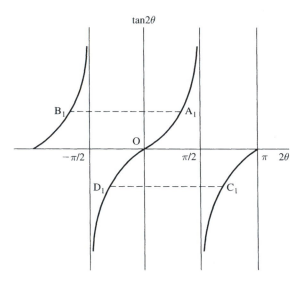

Figure 2.7.3

Now, we recall that the two principal planes defined by θ_1 and θ_2 are mutually perpendicular. Thus, once the angle θ_1 is known, the θ_2 plane, upon which the minimum principal stress σ_2 acts, is given by Eq. (2.7.5). For consistency we shall use

$$\theta_2 = \theta_1 + \pi/2. \tag{2.7.8}$$

Having found the roots θ_1 and θ_2, the principal stresses σ_1 and σ_2 may be given by Eq. (2.6.6b). To determine $\sin 2\theta_1$ and $\cos 2\theta_1$, we may use the trigonometric identities

$$\sin 2\theta = \frac{\tan 2\theta}{\sqrt{1 + \tan^2 2\theta}} = \frac{\tau_{xy}}{\sqrt{[(\sigma_x - \sigma_y)/2]^2 + \tau_{xy}^2}} \tag{2.7.9a}$$

and

$$\cos 2\theta = \frac{1}{\sqrt{1 + \tan^2 2\theta}} = \frac{(\sigma_x - \sigma_y)/2}{\sqrt{[(\sigma_x - \sigma_y)/2]^2 + \tau_{xy}^2}}. \tag{2.7.9b}$$

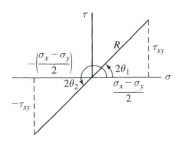

Figure 2.7.4

On the other hand, we may turn to Fig. (2.7.4) where σ_n is taken as the abscissa and τ_{nt} as the ordinate. We note from this figure that $\tan 2\theta_1$ is in agreement with

Eq. (2.7.3). Denoting the hypotenuse by R, we have

$$R = \sqrt{\left(\frac{\sigma_x - \sigma_y}{2}\right)^2 + \tau_{xy}^2}. \tag{2.7.10a}$$

Then clearly,

$$\cos 2\theta_1 = \frac{\sigma_x - \sigma_y}{2R}, \qquad \sin 2\theta_1 = \frac{\tau_{xy}}{R}. \tag{2.7.10b}$$

Similarly, since $\theta_2 = \theta_1 + \pi/2$,

$$\cos 2\theta_2 = -\frac{\sigma_x - \sigma_y}{2R}, \qquad \sin 2\theta_2 = -\frac{\tau_{xy}}{R}. \tag{2.7.10c}$$

Substituting this in Eq. (2.6.6b),

$$\sigma_1 = \frac{\sigma_x + \sigma_y}{2} + \frac{1}{R}\left(\frac{\sigma_x - \sigma_y}{2}\right)^2 + \frac{\tau_{xy}^2}{R} = \frac{\sigma_x + \sigma_y}{2} + \frac{1}{R}\left[\left(\frac{\sigma_x - \sigma_y}{2}\right)^2 + \tau_{xy}^2\right]$$

or

$$\sigma_1 = \frac{\sigma_x + \sigma_y}{2} + \sqrt{\left(\frac{\sigma_x - \sigma_y}{2}\right)^2 + \tau_{xy}^2}. \tag{2.7.11a}$$

Similarly,

$$\sigma_2 = \frac{\sigma_x + \sigma_y}{2} - \sqrt{\left(\frac{\sigma_x - \sigma_y}{2}\right)^2 + \tau_{xy}^2}. \tag{2.7.11b}$$

In passing, it is worthwhile noting, from Eqs. (2.7.11), that

$$\sigma_1 + \sigma_2 = \sigma_x + \sigma_y, \tag{2.7.12}$$

i.e., we observe again that the sum of the normal stresses on any two orthogonal planes is an invariant at a point.

(b) Maximum and minimum shear stress components

The planes of stationary shear stress are determined from τ_{nt}, given by Eq. (2.6.8b), in a similar manner. As in the preceding analysis, the necessary condition for stationary values of τ_{nt} is

$$\frac{d\tau_{nt}}{d\theta} = 0. \tag{2.7.13}$$

Noting that

$$\frac{d\tau_{nt}}{d\theta} = -2\tau_{xy}\sin 2\theta - (\sigma_x - \sigma_y)\cos 2\theta, \tag{2.7.14}$$

we find that the planes of stationary shear stresses are given by the roots of

$$\tan 2\theta = -\frac{\sigma_x - \sigma_y}{2\tau_{xy}}. \tag{2.7.15}$$

As before, this equation possesses two relevant roots, which we denote here by θ_{s1} and θ_{s2}, and where again,

$$\theta_{s2} = \theta_{s1} \pm \pi/2. \tag{2.7.16}$$

Now we observe that the right-hand side of Eq. (2.7.15) is the negative reciprocal of the right-hand side of Eq. (2.7.3).

Recalling that for any angle ϕ, $\tan(\phi + \pi/2) \cdot \tan \phi = -1$, we conclude that

$$\theta_{s1} = \theta_1 - \pi/4 \tag{2.7.17a}$$

and by Eq. (2.7.16),

$$\theta_{s2} = \theta_1 + \pi/4, \tag{2.7.17b}$$

where we have chosen the positive sign. (It is noted that the positive and negative signs appearing in Eq. (2.7.16) define the same physical planes given by θ_{s2}; however, one will represent a positive face and the other, a negative face.)

The maximum and minimum values of τ_{nt} are then obtained by substituting the appropriate values for θ_{s1} and θ_{s2} in Eq. (2.6.8b). The trigonometric quantities appearing in this equation can be obtained by a construction shown in Fig. (2.7.5). Noting that again, the hypotenuse R is given by

$$R = \sqrt{\left(\frac{\sigma_x - \sigma_y}{2}\right)^2 + \tau_{xy}^2},$$

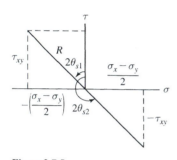

Figure 2.7.5

we have

$$\sin 2\theta_{s1} = -\frac{\sigma_x - \sigma_y}{2R}, \tag{2.7.18a}$$

$$\cos 2\theta_{s1} = \frac{\tau_{xy}}{R}. \tag{2.7.18b}$$

Similarly

$$\sin 2\theta_{s2} = \frac{\sigma_x - \sigma_y}{2R}, \tag{2.7.18c}$$

$$\cos 2\theta_{s2} = -\frac{\tau_{xy}}{R}. \tag{2.7.18d}$$

Upon substituting in Eq. (2.6.8b), we find, after simple algebraic manipulations,

$$\tau_{max} = \sqrt{\left(\frac{\sigma_x - \sigma_y}{2}\right)^2 + \tau_{xy}^2}. \tag{2.7.19a}$$

and

$$\tau_{min} = -\sqrt{\left(\frac{\sigma_x - \sigma_y}{2}\right)^2 + \tau_{xy}^2}. \tag{2.7.19b}$$

Using Eqs. (2.7.11), we readily observe from Eq. (2.7.19a), that

$$\tau_{max} = \frac{\sigma_1 - \sigma_2}{2}. \tag{2.7.20a}$$

Since, in our two-dimensional analysis, we have assumed that $\sigma_z = 0$, it should be mentioned that this last expression is valid provided $\sigma_1 > 0$ and $\sigma_2 < 0$.

Although, as we have previously stated, a three-dimensional analysis is beyond the scope of our present treatment, we mention here that in such cases, there exist three principal stresses, $\sigma_3 \leq \sigma_2 \leq \sigma_1$, and that

$$\tau_{max} = \frac{\sigma_1 - \sigma_3}{2}. \qquad (2.7.20b)$$

(c) Summary of results

It is worthwhile to summarise several of the basic results obtained relating to principal planes and stresses and to stationary shear stresses.

- Principal planes are mutually perpendicular.
- The shear stress $\tau_{nt} = 0$ on a principal plane.
- $\tau_{max} = \frac{\sigma_{max} - \sigma_{min}}{2}$
- Planes of stationary shear stress are oriented at $45°$ with respect to the principal planes.

These results are summarised pictorially in Fig. (2.7.6) where n_1, n_2 denote the unit normal to the respective principal planes, and n_{s1}, n_{s2} denote the unit normal to the planes of stationary shear.

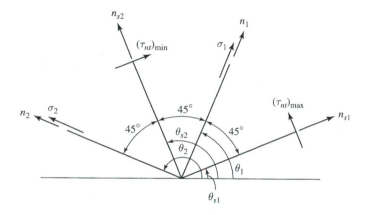

Figure 2.7.6

Finally, it is observed that we have *not* found that the normal stress σ_n acting on a plane of stationary shear stress is zero. It is left as an exercise to show that on such planes, $\sigma_n = \frac{\sigma_1 + \sigma_2}{2}$.

(d) Parametric representation of the state of stress: the Mohr circle

In the preceding section, the transformation law for the normal stress σ_n [Eq. (2.6.6b)]

$$\sigma_n = \frac{\sigma_x + \sigma_y}{2} + \frac{\sigma_x - \sigma_y}{2} \cos 2\theta + \tau_{xy} \sin 2\theta$$

was established. Let us rewrite this as

$$\sigma_n - \frac{\sigma_x + \sigma_y}{2} = \frac{\sigma_x - \sigma_y}{2} \cos 2\theta + \tau_{xy} \sin 2\theta. \qquad (2.7.21a)$$

We also recall the transformation law for τ_{nt}, namely Eq. (2.6.8b), which we repeat here:

$$\tau_{nt} = \tau_{xy} \cos 2\theta - \frac{\sigma_x - \sigma_y}{2} \sin 2\theta. \qquad (2.7.21b)$$

We now take the square of each of Eqs. (2.7.21) and add them; we obtain

$$\left[\sigma_n - \frac{\sigma_x + \sigma_y}{2}\right]^2 + \tau_{nt}^2 = \left(\frac{\sigma_x - \sigma_y}{2}\right)^2 + \tau_{xy}^2. \tag{2.7.22}$$

However we recognise the right-hand side of Eq. (2.7.22) as being R^2, where R, defined by Eq. (2.7.10a), is a known quantity; i.e.,

$$R = \sqrt{\left(\frac{\sigma_x - \sigma_y}{2}\right)^2 + \tau_{xy}^2}. \tag{2.7.23}$$

Thus, we have

$$\left[\sigma_n - \frac{\sigma_x + \sigma_y}{2}\right]^2 + \tau_{nt}^2 = R^2. \tag{2.7.24}$$

Now since σ_x, σ_y and τ_{xy} are given, Eq. (2.7.24) has the form

$$(\sigma_n - a)^2 + \tau_{nt}^2 = R^2, \tag{2.7.25}$$

where $a = \frac{\sigma_x + \sigma_y}{2}$ is known.

Equation (2.7.25) clearly has the same form as $(x - a)^2 + y^2 = R^2$, which is the equation of a circle with centre at $x = a$, $y = 0$ and radius R in an x–y plane. Hence, if we construct a σ_n–τ_{nt} plane (with σ_n as abscissa and τ_{nt} as ordinate) and plot Eq. (2.7.25) in this plane [Fig. (2.7.7)],[†] we recognise that it represents a circle whose centre is at $[(\sigma_x + \sigma_y)/2, 0]$ and whose radius R is given by Eq. (2.7.23).

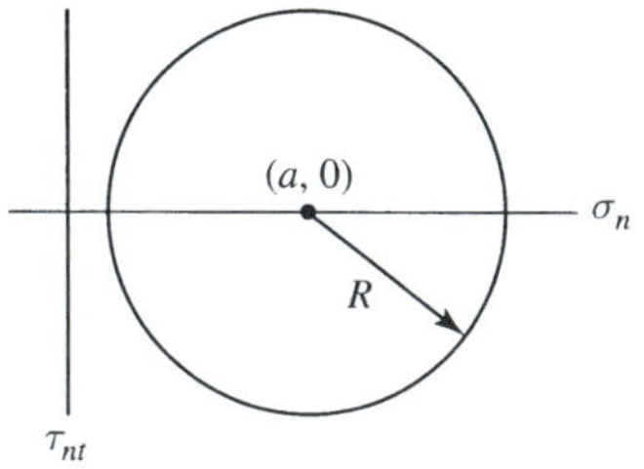

This feature of the transformation law was first observed by Mohr and the circle is called a **Mohr circle**. We thus recognise that the Mohr circle is but a parametric representation of the transformation laws (2θ being the parameter), and that the coordinates of each point on the circle represent the normal stress σ_n and shear stress τ_{nt} acting on the various planes passing through a point.

In order to determine the Mohr circle, we recall that, in general, three quantities are required to define any circle: either three points on the circle, or two points lying on the circle and the coordinates of the centre. Having established that the centre O of the Mohr circle $[(\sigma_x + \sigma_y)/2, 0]$ lies on the abscissa, and that each point represents the stress components on a different plane of the body, it is sufficient to know only two points: namely, (a) the point representing the stress components existing on the x-plane (σ_x and τ_{xy}) and (b) the point representing the components on the y-plane (σ_y and τ_{yx}).

Thus, we may construct the Mohr circle as follows [Fig. (2.7.8)]:

- Define the σ_n–τ_{nt} space with positive τ_{nt} in the downward direction. (Note that positive downward is an arbitrary choice.)
- Plot the stresses acting on the x-plane; we denote this point by P. Note that on the x-face, $\sigma_n = \sigma_x$, $\tau_{nt} = \tau_{xy}$ [see Fig. (2.7.9a)].
- Plot the stresses acting on the y-plane; we denote this point by Q. Note that on the y-face, $\sigma_n = \sigma_y$, $\tau_{nt} = -\tau_{yx} = -\tau_{xy}$ [see Fig. (2.7.9b)]. Note too that point Q is diametrically opposite to point P on the Mohr circle.

Figure 2.7.7

[†] Note that positive τ_{nt} has been taken downward. The reason for this choice will soon become apparent.

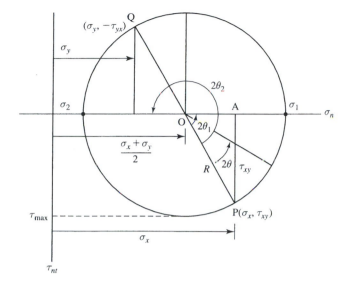

Figure 2.7.8

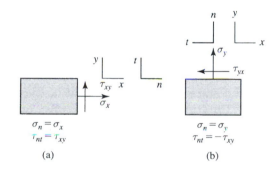

Figure 2.7.9

- Construct the line connecting points P and Q. This line PQ then intersects the σ_n axis at point O, the centre of the circle.
- Draw a circle with radius $R = OP = OQ$ about its centre, point O.

By constructing the Mohr circle in this manner, we observe readily that the coordinates of the centre are $[(\sigma_x + \sigma_y)/2, 0]$. Furthermore one establishes immediately that the circle will intersect the σ_n-axis at two points; namely

$$\sigma_1 = \frac{\sigma_x + \sigma_y}{2} + R, \qquad \sigma_2 = \frac{\sigma_x + \sigma_y}{2} - R, \qquad (2.7.26)$$

which as we observe, represent the principal stresses σ_1 and σ_2 as found previously [see Eqs. (2.7.11)].

We recall now that the transformation laws, as given by Eqs. (2.6.6b) and (2.6.8b), contain trigonometric terms whose argument is 2θ, where positive θ is measured counter-clockwise with respect to the x-axis. Hence, to determine the point representing the stress (σ_n, τ_{nt}) existing on any arbitrary n-plane whose normal $\boldsymbol{n}$ is inclined with respect to the x-axis at angle θ, we measure 2θ counter-clockwise with respect to the line OP (since we recall that P represents the x-plane). (The motivation for defining τ_{nt} positive as downward in the σ_n–τ_{nt} space now becomes apparent; in both the physical x–y space and the σ_n–τ_{nt} space, positive θ is counter-clockwise.)

We may verify from the Mohr circle that the principal plane on which σ_1 acts is given by

$$\tan 2\theta_1 = \frac{\text{AP}}{\text{OA}} = \frac{\tau_{xy}}{\frac{\sigma_x - \sigma_y}{2}},$$

which agrees with Eq. (2.7.3).

Other properties of the state of stress at a point, which were previously established analytically, may be readily observed from the Mohr circle; namely

- The principal planes are orthogonal.
- Planes of maximum and minimum shear stress are oriented at 45° with respect to the principal planes.
- $\tau_{max} = R$, $\tau_{min} = -R$ or $\tau_{max} = \frac{\sigma_1 - \sigma_2}{2}$.

The essential feature of the Mohr circle is that it gives us a complete pictorial representation of the state of stress existing at a point. Although it is useful in this sense, as developed here, it should nevertheless only be considered essentially as a parametric representation of the analytical two-dimensional transformation laws derived previously.

In all the following examples, we first solve the problems analytically and then verify the solutions via the Mohr circle.

The reader should pay special attention to the related solutions of Examples 2.8, 2.9 and 2.10.

Example 2.8: Given the state of stress $\sigma_x = 14$ MPa, $\sigma_y = -10$ MPa, $\tau_{xy} = 5$ MPa [Fig. (2.7.10a)]. Determine the principal stresses and the corresponding principal planes.

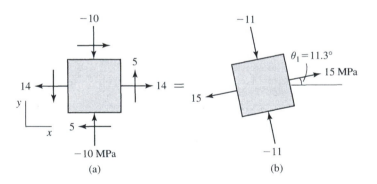

Figure 2.7.10

(a)   (b)

Solution: Using Eqs. (2.7.11),

$$\sigma_{1,2} = \frac{14 + (-10)}{2} \pm \sqrt{\left(\frac{14 - (-10)}{2}\right)^2 + 5^2} = 2 \pm 13.$$

Therefore, $\sigma_1 = 15$ MPa, $\sigma_2 = -11$ MPa.

From Eq. (2.7.3), the principal planes are determined according to the roots of the equation

$$\tan 2\theta = \frac{\tau_{xy}}{\frac{\sigma_x - \sigma_y}{2}} = \frac{5}{12} = 0.416.$$

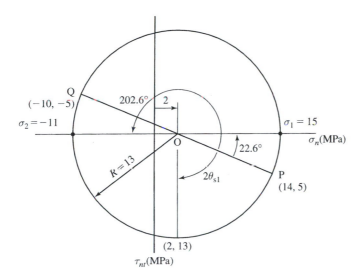

Figure 2.7.11

According to the established criterion [following Eqs. (2.7.7)], since $\tau_{xy} > 0$, $2\theta_1 = 22.6 \to\to \theta_1 = 11.3°$. [This corresponds to point A_1 of Fig. (2.7.3).] Therefore, by Eq. (2.7.5), $\theta_2 = \theta_1 + \pi/2 = 101.3°$.

The corresponding Mohr circle is shown in Fig. (2.7.11) and the principal stresses and planes are shown in Fig. (2.7.10b).

From the Mohr circle, we note too that $\tau_{max} = \frac{\sigma_1 - \sigma_2}{2} = 13$ MPa and that $\theta_{s1} = \theta_1 - \pi/4 = -33.7°$. □

Example 2.9: Given the state of stress $\sigma_x = 14$ MPa, $\sigma_y = -10$ MPa, $\tau_{xy} = -5$ MPa [Fig. (2.7.12a)]. Determine the principal stresses and the corresponding principal planes.

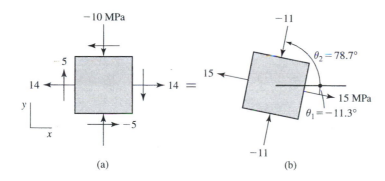

(a)　　　　(b)　　　　**Figure 2.7.12**

Solution: Using Eqs. (2.7.11), the principal stresses are, as before, $\sigma_1 = 15$ MPa, $\sigma_2 = -11$ MPa. Moreover, from Eq. (2.7.3),

$$\tan 2\theta = \frac{\tau_{xy}}{\frac{\sigma_x - \sigma_y}{2}} = -\frac{5}{12} = -0.416.$$

According to the established criterion, since $\tau_{xy} < 0$, $2\theta_1 = -22.6 \to\to \theta_1 = -11.3°$. [This corresponds to point D_1 of Fig. (2.7.3).] Therefore, by Eq. (2.7.5), $\theta_2 = \theta_1 + \pi/2 = 78.7°$.

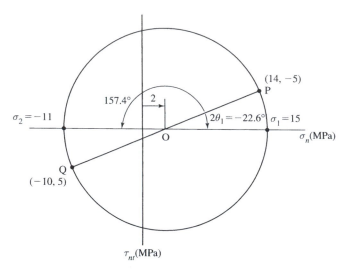

Figure 2.7.13

The corresponding Mohr circle is shown in Fig. (2.7.13) and the principal stresses and planes acting on the physical element are shown in Fig. (2.7.12b). ☐

Example 2.10: Given the state of stress $\sigma_x = -14$ MPa, $\sigma_y = +10$ MPa, $\tau_{xy} = -5$ MPa [Fig. (2.7.14a)]. Determine the principal stresses and the corresponding principal planes.

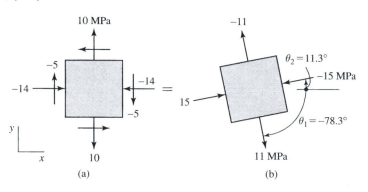

Figure 2.7.14

(a)   (b)

Solution: Using Eqs. (2.7.11),

$$\sigma_{1,2} = \frac{-14 + 10}{2} \pm \sqrt{\left(\frac{-14 - 10}{2}\right)^2 + 5^2} = -2 \pm 13.$$

Therefore, $\sigma_1 = 11$ MPa and $\sigma_2 = -15$ MPa.

From Eq. (2.7.3), the principal planes are determined according to the roots of the equation

$$\tan 2\theta = \frac{\tau_{xy}}{\frac{\sigma_x - \sigma_y}{2}} = -\frac{-5}{12} = 0.416.$$

Since $\tau_{xy} < 0$, $2\theta_1 = -157.4° \rightarrow\rightarrow \theta_1 = -78.7°$. [This corresponds to point B_1 of Fig. (2.7.3).] Therefore, according to Eq. (2.7.5), $\theta_2 = \theta_1 + \pi/2 = 11.3°$.

The corresponding Mohr circle is shown in Fig. (2.7.15) and the principal stresses and planes acting on the physical element are shown in Fig. (2.7.14b).

From the Mohr circle, we note too that $\tau_{max} = \frac{\sigma_1 - \sigma_2}{2} = 13$ MPa and that $\theta_{s1} = \theta_1 - \pi/4 = -123.7°$. ☐

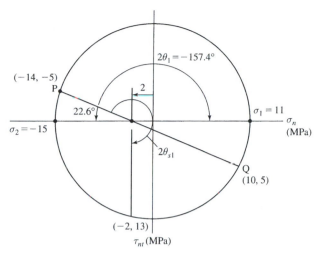

Figure 2.7.15

Example 2.11: Given the state of stress $\sigma_x = \sigma_y = 0$ and $\tau_{xy} = \tau_0 > 0$ [Fig. (2.7.16a)]. Determine the principal stresses and planes.

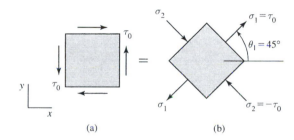

(a)   (b)   Figure 2.7.16

Solution: From Eqs. (2.7.11), the principal stresses are $\sigma_1 = \tau_0$ and $\sigma_2 = -\tau_0$. The principal planes defined by the roots of the equation,

$$\tan 2\theta = \frac{\tau_{xy}}{\frac{\sigma_x - \sigma_y}{2}} = \frac{\tau_0}{0} \to \infty.$$

Hence, since $\tau_0 > 0$, $\theta_1 = 90° \to \to \theta_1 = 45°$ and therefore $\theta_2 = 135°$. The Mohr circle, shown in Fig. (2.7.17), is observed to be a circle of radius $R = \tau_0$ with centre at the origin of the $\sigma_n - \tau_{nt}$ plane.

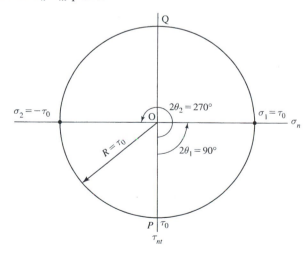

Figure 2.7.17

The state of stress prescribed in this example and shown in Fig. (2.7.16) is called a state of **pure shear**. □

Example 2.12: Given the state of stress $\sigma_x = 0$, $\sigma_y = 10$ kPa, $\tau_{xy} = 0$ [Fig. (2.7.18a)]. Determine σ_n and τ_{nt} on a plane whose normal n is inclined at (a) $\theta = 30°$ and at (b) $\theta = 120°$ with respect to the x-axis.

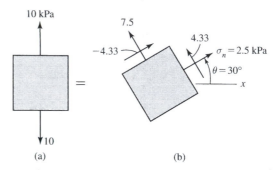

Figure 2.7.18

(a)   (b)

Solution:

(a) From Eq. (2.6.6b),

$$\sigma_n = \frac{\sigma_x + \sigma_y}{2} + \frac{\sigma_x - \sigma_y}{2} \cos 2\theta + \tau_{xy} \sin 2\theta.$$

Substituting the appropriate values and noting that $\cos 60° = 0.5$, $\sin 60° = \sqrt{3}/2$, we find $\sigma_n = 2.5$ kPa.

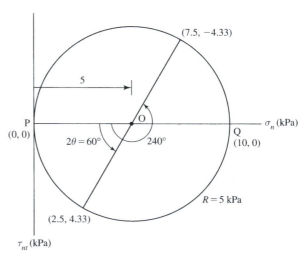

Figure 2.7.19

Similarly, from Eq. (2.6.8b),

$$\tau_{nt} = \tau_{xy} \cos 2\theta - \frac{\sigma_x - \sigma_y}{2} \sin 2\theta = 2.5\sqrt{3} = 4.33 \text{ kPa}.$$

(b) Using $\theta = 120°$, we obtain similarly $\sigma_n = 7.5$ kPa and $\tau_{nt} = -2.5\sqrt{3} = -4.33$ kPa.

The stress components acting on the physical element are shown in Fig. (2.7.18b) and the Mohr circle representation is given in Fig. (2.7.19). □

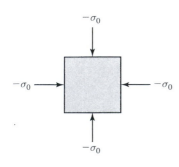

Figure 2.7.20

Example 2.13: Given the state of stress $\sigma_x = \sigma_y = -\sigma_0$ (where $\sigma_0 > 0$) and $\tau_{xy} = 0$ [Fig. (2.7.20)]. Analyse the state of stress.

Solution: According to Eqs. (2.6.6b) and (2.6.8b), we note that $\sigma_n = -\sigma_0$ and $\tau_{nt} = 0$ for *all values of* θ. Furthermore, according to Eq. (2.7.23), the radius of the Mohr circle $R = 0$; hence we note that the Mohr circle, in this case, degenerates to a point located at $(-\sigma_0, 0)$ in the σ_n–τ_{nt} plane [Fig. (2.7.21)].

Such a state of stress is called a **hydrostatic state of stress** at a point. Note that in this case the normal stress is the same for *all* planes and no shear stress exists on any plane passing through this point. $\square$

Figure 2.7.21

$\diamond$2.8 Cartesian components of traction in terms of stress components: traction on the surface of a body[†]

It is often more convenient to express the traction T_n on a given n-plane in terms of its Cartesian components rather than its normal and tangential components, σ_n, τ_{ns} and τ_{nt}, as in Eq. (2.3.6). Although the expressions developed below are valid for any n-plane at an interior point of a body, they are particularly useful in expressing external contact forces acting on the surface S of the body. As shown in Fig. (2.8.1), these contact forces may be either distributed over a given area of S or concentrated forces or couples.[‡] Such distributed forces are thus prescribed in terms of traction vectors T_n where n is the unit normal vector at any point to the surface S. For simplicity, we confine the discussion to a two-dimensional system of plane stress in the x–y plane; the unit normal vector is given by

$$n = \cos\theta\, i + \sin\theta\, j. \qquad (2.8.1)$$

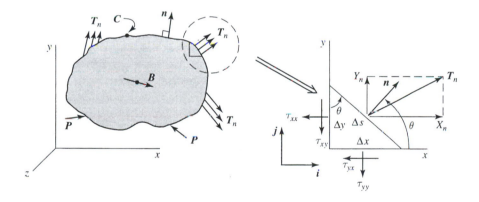

Figure 2.8.1 and **Figure 2.8.2**

Now, for a two-dimensional case, the traction T_n acting on an n-plane can be represented in terms of two perpendicular components in the n- and (say) t-directions; i.e.,

$$T_n = \sigma_n n + \tau_{nt} t. \qquad (2.8.2a)$$

Since any vector can be defined by its scalar components, the traction may instead be given in terms of its scalar components X_n and Y_n in the x- and y-directions, respectively; i.e.,

$$T_n = X_n i + Y_n j. \qquad (2.8.2b)$$

[†] Subject material in this section is optional and, as it is not necessary in the first 13 chapters, may be deferred until a reading of Chapter 14. Throughout this book, the symbol ($\diamond$) has been used before certain sections/subsections that may be omitted on first reading, without loss of continuity.

[‡] We note that a concentrated force represents a particular case of distributed forces in which the intensity of the distributed force per area tends to infinity as the area tends to zero. [See Eqs. (1.2.1) and (1.2.3).]

From the development of Section 6, it is clear that there exists a relation between the components X_n and Y_n and the stress components at all points of the body including points on the surface S. We obtain these relations by proceeding exactly as with the derivation of the stress transformation laws. To this end, consider an element having density ρ near the surface S [Fig. (2.8.2)] and let $\sigma_x \equiv \tau_{xx}$, τ_{yy} and τ_{xy} be the stresses acting on this element. Then from the principle of linear momentum, $\sum F_x = m\ddot{u}$, where $m = \frac{\rho}{4}\Delta s^2 \sin 2\theta$ [see Eq. (2.6.2)] , we have

$$-\tau_{xx}\Delta y - \tau_{yx}\Delta x + X_n \Delta s = \frac{\rho}{4}\Delta s^2 \sin 2\theta\, \ddot{u}. \tag{2.8.3}$$

Dividing through by Δs,

$$-\tau_{xx}\frac{\Delta y}{\Delta s} - \tau_{yx}\frac{\Delta x}{\Delta s} + X_n = \frac{\rho}{4}\Delta s \sin 2\theta\, \ddot{u},$$

noting that $\frac{\Delta y}{\Delta s} = \cos\theta$, $\frac{\Delta x}{\Delta s} = \sin\theta$, and taking the limit as $\Delta s \to 0$, we obtain

$$X_n = \tau_{xx}\cos\theta + \tau_{yx}\sin\theta. \tag{2.8.4a}$$

The component of traction Y_n in the y-direction is obtained similarly from $\sum F_y = m\ddot{v}$:

$$-\tau_{yy}\Delta x - \tau_{xy}\Delta y + Y_n \Delta s = \frac{\rho}{4}\Delta s^2 \sin 2\theta\, \ddot{v}.$$

Dividing through, as before, by Δs and taking the limit as $\Delta s \to 0$, we obtain[†]

$$Y_n = \tau_{yy}\sin\theta + \tau_{xy}\cos\theta. \tag{2.8.4b}$$

We may now rewrite these expressions in a slightly different form that is more appropriate for future developments. Just as we denoted the angle θ as defining the orientation of the unit normal $\boldsymbol{n}$ with respect to the x-axis, we now denote the angle ψ as the orientation with respect to the y-axis [Fig. (2.8.3)]; then

$$\boldsymbol{n} = \cos\theta\, \boldsymbol{i} + \cos\psi\, \boldsymbol{j}, \tag{2.8.5}$$

where $\sin\theta = \cos\psi$. Therefore

$$X_n = \tau_{xx}\cos\theta + \tau_{yx}\cos\psi, \tag{2.8.6a}$$
$$Y_n = \tau_{xy}\cos\theta + \tau_{yy}\cos\psi. \tag{2.8.6b}$$

Letting $\ell_x \equiv \cos\theta$ and $\ell_y \equiv \cos\psi$ be the 'direction cosines' of the vector $\boldsymbol{n}$, we may write

$$\boldsymbol{n} = \ell_x \boldsymbol{i} + \ell_y \boldsymbol{j}, \tag{2.8.7}$$

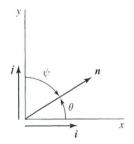

Figure 2.8.3

[†] It is worthwhile to mention here that, while we have established the desired results, namely Eqs. (2.8.4), we could also obtain them in a simpler fashion.
  Treating $\boldsymbol{T}_n$, given by Eq. (2.8.2a), as a vector, its scalar components in the x- and y-directions are, respectively,

$$X_n = \boldsymbol{T}_n \cdot \boldsymbol{i}, \qquad Y_n = \boldsymbol{T}_n \cdot \boldsymbol{j}. \tag{a}$$

Then, since

$$\boldsymbol{n}\cdot\boldsymbol{i} = \cos\theta, \qquad \boldsymbol{n}\cdot\boldsymbol{j} = \sin\theta; \qquad \boldsymbol{t}\cdot\boldsymbol{i} = -\sin\theta, \qquad \boldsymbol{t}\cdot\boldsymbol{j} = \cos\theta, \tag{b–e}$$

substituting the expressions for σ_n and τ_{nt} given by Eqs. (2.6.6a) and (2.6.8a), respectively, into Eq. (2.8.2a), using Eqs. (b to e) and performing the required operations, leads directly to Eqs. (2.8.4).

Hence, using this notation, the components of the traction T_n are rewritten simply as

$$X_n = \ell_x \tau_{xx} + \ell_y \tau_{yx} \tag{2.8.8a}$$

$$Y_n = \ell_x \tau_{xy} + \ell_y \tau_{yy}. \tag{2.8.8b}$$

It is important to observe that if, for a given ℓ_x and ℓ_y, X_n and Y_n are known, then in general, it is not possible to determine *all* the stress components at a point on the surface S.[†] (Indeed it may not be possible to find any of the stress components.) However, if τ_{xx}, τ_{yy} and τ_{xy} are known, then clearly X_n and Y_n are determined.

The expressions of Eqs. (2.8.8) will prove useful in relating the components of traction on the surface of a body in terms of the stress components existing at points on the surface S.

Example 2.14: Consider a rod (of unit width) subjected to given applied tractions T_n as shown in Fig. (2.8.4). What are the known stress components in each sector of the surface boundary S, namely in the sectors AB, BC, CD, DE, etc. (*Note:* q_i here are given in units of Pa.)

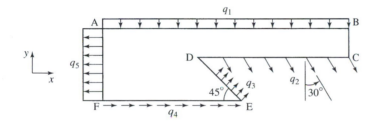

Figure 2.8.4

Solution:

On AB:
$\ell_x = 0$, $\ell_y = 1$; $X_n = 0$, $Y_n = -q_1$ are given.
It follows that $\tau_{yy} = -q_1$, $\tau_{xy} = 0$; τ_{xx} remains unknown.
On BC:
$\ell_x = 1$, $\ell_y = 0$; $X_n = Y_n = 0$ are given.
It follows that $\tau_{xx} = \tau_{yx} = 0$; τ_{yy} remains unknown.
On CD:
$\ell_x = 0$, $\ell_y = -1$; $X_n = 0.5q_2$, $Y_n = -0.5\sqrt{3}q_2$ are given.
It follows that $\tau_{yy} = 0.5\sqrt{3}q_2$; $\tau_{yx} = -0.5q_2$; τ_{xx} remains unknown.
On DE:
$\ell_x = \ell_y = 0.5\sqrt{2}$; $X_n = Y_n = 0.5\sqrt{2}q_3$ are given.
None of the stress components (τ_{xx}, τ_{yy}, τ_{xy}) can be determined.
On EF:
$\ell_x = 0$, $\ell_y = -1$; $X_n = q_4$, $Y_n = 0$ are given.
It follows that $\tau_{yx} = -q_4$, $\tau_{yy} = 0$; τ_{xx} remains unknown.
On AF:
$\ell_x = -1$, $\ell_y = 0$; $X_n = -q_5$, $Y_n = 0$ are given.
It follows that $\tau_{xx} = q_5$, $\tau_{xy} = 0$; τ_{yy} remains unknown.

[†] This is evident since it is clearly impossible to solve for three unknowns (τ_{xx}, τ_{yy}, τ_{xy}) from the two simultaneous equations, Eqs. (2.8.8).

PROBLEMS

Section 2

2.1: A 12-m long rigid bar is suspended by two wires and supports a load of 1200 N, as shown in Fig. (2P.1). What are the components of the internal force system at a cross-section (a) 2 m from each end and (b) at the centre?

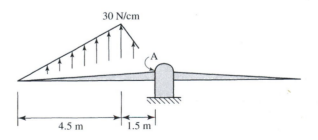

Figure 2P.1

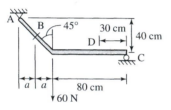

Figure 2P.2

2.2: The bent rod shown in Fig. (2P.2) is simply supported at A and by a roller at C. Find the components of the internal force resultants at cross-sections B and D.

2.3: Member ABCD, shown in Fig. (2P.3a), is welded at A to a rigid plate a–e, which is anchored to the ground by means of two bolts b and c. A force of 600 N is applied as shown at D. (a) Find the normal force, shear force and moment at the cross-section A and C. (b) If the plate is attached to the ground by means of the two bolts as shown in Fig. (2P.3b), determine the forces in each bolt and indicate whether in tension or compression.

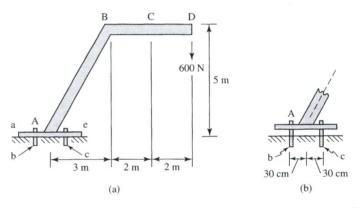

(a)   (b)

Figure 2P.3

2.4: The upward lifting force acting on a helicopter rotor blade is distributed as shown in Fig. (2P.4). Determine the bending couple and shear force acting on the cross-section at A.

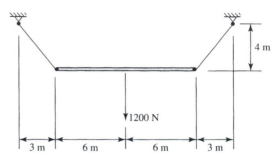

Figure 2P.4

2.5:* The lift force on the wing of an airplane, shown in Fig. (2P.5), is given as

$$q(x) = q_0 \sin(\pi x/2L).$$

Determine the bending couple and shear force acting on the cross-section at A.

2.6: A thin triangular plate having thickness t hangs under its own weight, as shown in Fig. (2P.6). The density of the plate is ρ (N/m^3). Determine, at any cross-section a distance y from the top, the internal force system (consisting of a moment and a normal force acting at the centre of each cross-section).

2.7: A solid cone made of a material whose density is ρ (N/m^3) hangs from a pin at its vertex, as shown in Fig. (2P.7). Determine the normal force acting on a cross-section located at a distance y from the vertex.

2.8:* A magnet is attached to the ends of an iron rod whose cross-section is A, as shown in Fig. (2P.8). The attraction force acting at any distance, x, is given as $f(x) = \frac{c}{\sqrt{1+\alpha x^2/L^2}}$ where c is a constant having dimensions (N/m^3) and α is a non-dimensional constant. (a) Plot $f(x)$ in the range $0 \le x/L \le 1$ for several values of α: $\alpha = 0, 1, 2, 5$. (b) Determine the normal force $F(x)$ at any cross-section, as a function of x. (c) Using a series expansion for the expression for (b) obtained above, show that for $\alpha \to 0$, the normal force approaches $F(x) = cAx$.

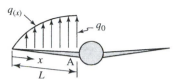

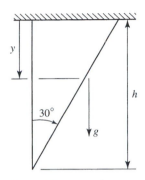

Figure 2P.5

Figure 2P.6

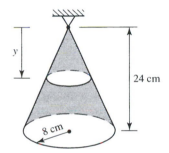

Figure 2P.7

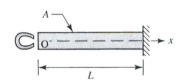

Figure 2P.8

2.9: Express the shear force $V(x)$ and moment $M(x)$ as a function of x for the beams shown in Figs. (2P.9a–j) and sketch the variation with x. (*Note:* Assume, for all cases, positive shear force V and moment M, as shown in the figure).

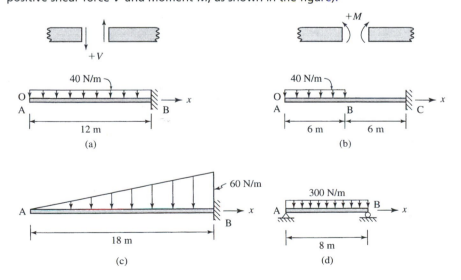

Figure 2P.9

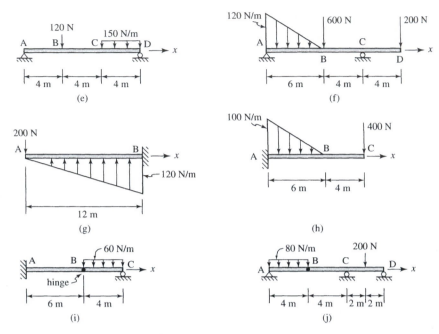

Figure 2P.9 (Continued)

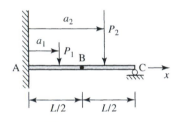

Figure 2P.10

2.10:   Express the shear force $V(x)$ and moment $M(x)$ as a function of x in terms of P_1, P_2, a_1 and a_2 in the two regions $0 \leq x < a_1$ and $a_1 < x \leq a_2$ for the beams shown in Fig. (2P.10).

2.11:   Express the shear force $V(x)$ and moment $M(x)$ as a function of x in terms of w and P for the beam shown in Fig. (2P.11). Sketch the variation with x if $P = wL$.

2.12:   Express the shear force $V(x)$ and moment $M(x)$ within the span AC as a function of x for the beams shown in Figs. (2P.12a–e) and sketch the variation with x.

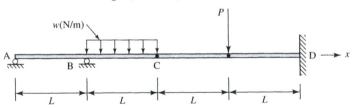

Figure 2P.11

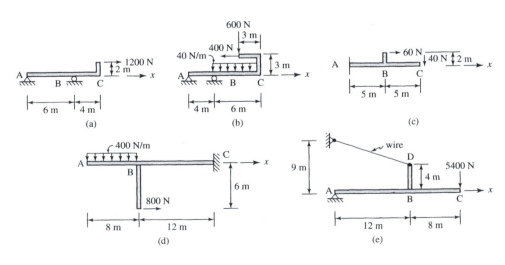

Figure 2P.12

2.13: The rod shown in Fig. (2P.13) is subjected to an eccentric load as shown. Determine the shear force, and bending and torsional moments as a function of x.

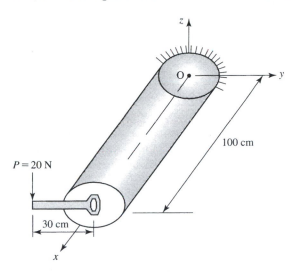

Figure 2P.13

2.14: The bent pipe shown in Fig. (2P.14) is subjected to a force P having components P_x, P_y and P_z as shown. Determine the components of the internal force system at any cross-section. Express the answers in terms of x, y and z where appropriate.

2.15: (a) A thin circular member, AB, lying in the x–y plane, as shown in Fig. (2P.15a), is subjected to two forces P_x and P_y. Determine the resulting internal forces at any cross-section in terms of R and θ. (b) If the circular member is subjected to a bending moment M_0 and a torsional moment T_0 as in Fig. (2P.15b), what is the resulting internal force system at any cross-section?

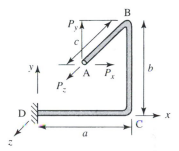

Figure 2P.14

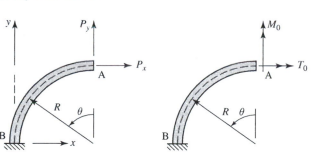

(a)   (b)

Figure 2P.15

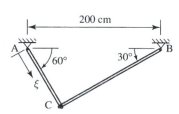

Figure 2P.16

2.16: Two rods, AC and BC, each weighing 10 N/m, lie in a vertical plane and are pinned at each end, as shown in Fig. (2P.16). Determine, as a function of ξ (shown in the figure), the internal force system (normal force, shear force and moment) at any cross-section of the rod AC.

2.17:* A beam of length L is pinned at A and is to be supported by a roller located at point B, as shown in Fig. (2P.17). The beam is subjected to a uniformly distributed load w (N/m). Determine the ratio b/L for which the largest absolute value of the bending moment in the beam is a minimum. What is this value?

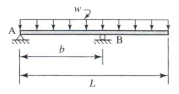

Figure 2P.17

Section 4

2.18:   The following stress field is found to exist in a body:

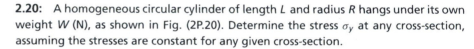

$$\sigma_x = ax^2yz, \qquad \sigma_y = bxyz^3, \qquad \sigma_z = 0,$$
$$\tau_{xy} = 3xy^2z, \qquad \tau_{yz} = cz^2(6y^2 - 5xz^2), \qquad \tau_{zx} = 3xyz^2,$$

where *a, b* and *c* are constants. Assuming no body forces act on the body, for which values of *a, b* and *c* is the body in a state of equilibrium?

2.19:*   At any arbitrary point of a beam subject to zero body forces, the stress components are given as $\sigma_y = \sigma_z = \tau_{yz} = 0$ where the remaining stress components are not zero. Show that the equilibrium equations are satisfied only if σ_x has the form $\sigma_x = a + bx$ where *a* and *b* are functions of *y* and/or *z* or are constants.

Section 5

2.20:   A homogeneous circular cylinder of length *L* and radius *R* hangs under its own weight *W* (N), as shown in Fig. (2P.20). Determine the stress σ_y at any cross-section, assuming the stresses are constant for any given cross-section.

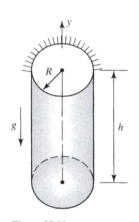

Figure 2P.20

2.21:   The stress distribution on a beam having a triangular shape, as shown in Fig. (2P.21) is given by

$$\sigma_x = Ay + Bz^2, \qquad \tau_{xy} = \tau_{xz} = 0.$$

Determine the normal force *F* and the moments M_y and M_z due to this stress distribution in terms of *a, A* and *B*.

2.22:   A circular cylinder of radius *R* is twisted at its ends by a torsional moment, $T \equiv M_x$, as shown in Fig. (2P.22). (a) The stress distribution at any cross-section is given as $\sigma_x = \tau_{xr} = 0$, $\tau_{x\theta} = kr$, where *r* is the varying radial coordinate and *k* is an undetermined constant. Evaluate *k* and express $\tau_{x\theta}$ in terms of *r, R* and *T*. (b) If the stress distribution due to *T* is given as $\sigma_x = \tau_{xr} = 0$, $\tau_{x\theta} = \tau_0$ (a constant), evaluate τ_0 in terms of *R* and *T*.

2.23:*   The stress distribution on a circular cross-section lying in the *y–z* plane, as shown in Fig. (2P.23), is given by $\sigma_x = 0$, $\tau_{xy} = -200z$ (MPa) and $\tau_{xz} = 200y$ (MPa). Determine the components of the internal force system acting on the cross-section.

2.24:*   The beam shown in Fig. (2P.24) is subjected to a bending moment *M* about the *z*-axis. The stress distribution on a cross-section is given by

$$\sigma_x = \begin{cases} -\sigma_0, & c \leq y \leq h/2 \\ -\sigma_0 y/c, & -c \leq y \leq c \\ \sigma_0, & -h/2 \leq y \leq -c \end{cases}$$

where *c* is a constant, $0 \leq c \leq h/2$, and σ_0 is a given constant stress. (a) Sketch the distribution of σ_x as a function of *y*. (b) Determine *M* in terms of *b, c, h* and σ_0. (c) For a constant σ_0, what is the value of *c* for which the moment *M* is a maximum? What is the value of this maximum moment?

2.25:   A bent bar having a square cross-section ($b \times b$) is subjected to eccentric forces, as shown in Fig. (2P.25). The stress distribution on any cross-section is assumed to be $\sigma_y = B + Cy$, $\tau_{xy} = \tau_{xz} = 0$, where *B* and *C* are constants. (a) Determine *B* and *C* in terms of P, b and e. (b) What is the stress at point d if $e = 0$ and if $e = 4b$? Indicate whether tension or compression in both cases.

2.26:* A member, ABC, with $L = 1$ m, is welded at A to a rigid plate c–d whose dimensions are 50 cm $\times$ 50 cm, as shown in Fig. (2P.26). The plate is anchored to the ground by means of a single bolt (which is anchored in the same plane as ABC). A force of 11,000 N is applied at B and a force of 1000 N is applied at C as shown. (a) Assuming that the pressure exerted between the ground and the plate varies linearly from c to d, determine the maximum pressure. Where does it occur? (b) The load at B is removed such that the member ABC is subjected only to the force $P = 1000$ N at C. Assuming that the linearly varying pressure between the ground and the plate can only be compressive, determine (i) the maximum compressive pressure and (ii) the force exerted by the bolt if $b = 10$ cm.

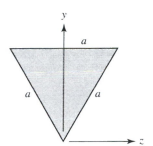

Figure 2P.21

Figure 2P.22

Sections 6

2.27: Verify the expression for the second invariant of (plane) stress given by Eq. (2.6.11).

2.28: At a given point in a body in a state of plane stress with $\tau_{xy} = 0$, the ratio of the invariants, $\frac{I_{\sigma_2}}{I_{\sigma_1}}$, is found to be equal to 300 MPa. If it is known that the stress $\sigma_y = -5\sigma_x$, determine σ_x and $\tau_{nt}|_{max}$.

2.29: On a plane passing through an arbitrary point P, two rectangular Cartesian systems, (x, y) and (n, t), are constructed as shown in Fig. (2P.29). For each of the plane stress cases listed below, (i) determine the required quantities and (ii) sketch the equivalent states of stress (in the two coordinate systems).

(a) $\sigma_x = 200$, $\sigma_y = 400$, $\tau_{xy} = 400$ MPa; $\theta = 30°$. Find σ_n, σ_t, τ_{nt}.
(b) $\sigma_x = -400$, $\sigma_y = 0$, $\tau_{xy} = 300$ MPa; $\theta = -30°$. Find σ_n, σ_t, τ_{nt}.
(c) $\sigma_x = 0$, $\sigma_y = 0$, $\tau_{xy} = 300$ MPa; $\theta = 45°$. Find σ_n, σ_t, τ_{nt}.
(d) $\sigma_x = 1200$, $\sigma_y = 800$, $\tau_{xy} = -800$ kPa; $\theta = 120°$. Find σ_n, σ_t, τ_{nt}.
(e) $\sigma_n = -100$, $\sigma_t = -50$, $\tau_{nt} = 100$ MPa; $\theta = 30°$. Find σ_x, σ_y, τ_{xy}.
(f) $\sigma_x = 200$, $\sigma_y = 100$, $\sigma_n = 50$ kPa; $\theta = 45°$. Find σ_t, τ_{nt}.
(g) $\sigma_n = 100$, $\sigma_y = 200$, $\tau_{nt} = 0$ MPa; $\theta = 60°$. Find σ_x, τ_{xy}, σ_t.
(h) $\sigma_n = 100$, $\sigma_y = -200$, $\tau_{nt} = 0$ MPa; $\theta = 60°$. Find σ_x, τ_{xy}, σ_t.

Figure 2P.23

2.30: Let n, s and t be three directions in a given x–y plane such that the n-direction lies along the x-axis, as shown in Fig. (2P.30). (a) Determine τ_{xy} in terms of σ_n, σ_s and σ_t if $\alpha = 45°$ and (b) Determine τ_{xy} in terms of σ_n, σ_s and σ_t if $\alpha = 60°$.

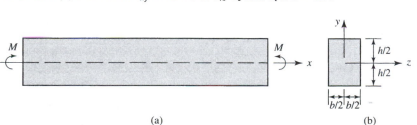

(a)   (b)

Figure 2P.24

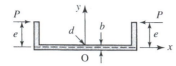

Figure 2P.25

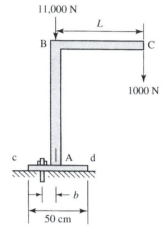

Figure 2P.26

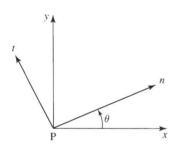

Figure 2P.29

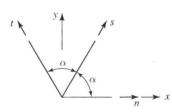

Figure 2P.30

2.31: For each of the following plane stress states at a point lying in the x–y plane of Fig. (2P.29), determine the angle θ of the direction n with respect to the x–axis.
  (a) $\sigma_x = \sigma_y = 100$, $\tau_{xy} = 500$, $\sigma_n = 400$ MPa.
  (b) $\sigma_x = 100$, $\sigma_y = -100$, $\tau_{xy} = 150$, $\sigma_n = 0$ MPa.
  (c) $\sigma_x = -\sigma_y = -\tau_{xy} = c$, ($c$ = constant) $\sigma_n = 0$ MPa.

2.32: A rectangular block is formed by gluing together two wooden wedges, as shown in Fig. (2P.32) where $a = 6$ cm, $b = 8$ cm and $c = 4$ cm. The joint fails if the shear stress in the adhesive exceeds 600 kPa. What is the permitted range of σ_y if a compressive stress $\sigma_x = -300$ kPa and shear stress $\tau_{xy} = 400$ kPa are applied and all other stress components are zero. (*Note:* Assume the stress state in the block is uniform at all points).

2.33: A rectangular block is formed by gluing together two wooden wedges, as shown in Fig. (2P.32) where $a = b = c$. The joint fails if the normal tension stress at the adhesive interface exceeds 400 kPa. What is the maximum permitted value of σ_x if a compressive stress $\sigma_y = -300$ kPa is applied and all other stress components are zero. (*Note:* Assume the stress state in the block is uniform at all points).

2.34: The shear stress distribution on a circular cross-section lying in the y–z plane, as shown in Fig. (2P.34), is given by $\tau_{xy} = C_1 z$ and $\tau_{xz} = C_2 y$. Show that if $C_2 = -C_1$, the resultant shear stress $\tau = \sqrt{\tau_{xy}^2 + \tau_{xz}^2}$ acting on the plane at any point P is directed in the circumferential θ-direction; i.e., $\tau = \tau_{x\theta}$.

Section 7

2.35: For each of the plane stress cases listed below (with $\sigma_z = \tau_{xz} = \tau_{yz} = 0$), (i) determine the principal stresses σ_1 and σ_2, (ii) determine τ_{max}, (iii) determine the principal directions with respect to the x-axis as defined by θ_1 and θ_2, (iv) sketch an element showing principal stresses and directions and (v) sketch the appropriate Mohr circle showing σ_1, σ_2, $2\theta_1$, $2\theta_2$ and τ_{max} on the circle.
  (a) $\sigma_x = 60$, $\sigma_y = 0$, $\tau_{xy} = 40$ MPa.
  (b) $\sigma_x = 200$, $\sigma_y = -200$, $\tau_{xy} = -200$ kPa.
  (c) $\sigma_x = 900$, $\sigma_y = 100$, $\tau_{xy} = 200$ MPa.
  (d) $\sigma_x = 400$, $\sigma_y = 800$, $\tau_{xy} = -600$ kPa.
  (e) $\sigma_x = -200$, $\sigma_y = -100$, $\tau_{xy} = 200$ MPa.
  (f) $\sigma_x = 2000$, $\sigma_y = 500$, $\tau_{xy} = -500$ kPa.
  (g) $\sigma_x = -120$, $\sigma_y = 40$, $\tau_{xy} = -20$ MPa.
  (h) $\sigma_x = 240$, $\sigma_y = 0$, $\tau_{xy} = 120$ MPa.
  (i) $\sigma_x = -200$, $\sigma_y = 100$, $\tau_{xy} = 320$ kPa.
  (j) $\sigma_x = 0$, $\sigma_y = 240$, $\tau_{xy} = 120$ MPa.

2.36: States of plane stress are shown by means of Figs. (2P.36 a–e). For each of the cases listed below, (i) determine the principal stresses and directions, (ii) sketch the equivalent states of stress and (iii) sketch the appropriate Mohr circle.

2.37: Let n, s and t be three unit vectors lying in a plane as shown in Fig. (2P.30). (a) If $\sigma_n = 100$ MPa, $\sigma_s = 50$ MPa and $\sigma_t = 20$ MPa, determine the principal stresses and directions (with respect to the vector n) if $\alpha = 45^0$ and show these by means of a sketch. (b) If $\sigma_n = 100$ MPa, $\sigma_s = -20$ MPa and $\sigma_t = 60$ MPa, determine the principal stresses and directions (with respect to the vector n) if $\alpha = 60^0$ and show these by means of a sketch.

2.38: A circular cylindrical shaft of radius R is subjected to an axial force F and a torsional moment T, as shown in Fig. (2P.38a). The resulting normal and shear stresses at the surface of the shaft, as shown in Fig. (2P.38b), are given respectively by $\sigma_x = F/A$ and $\tau_{x\theta} = TR/J$, where A is the cross-sectional area of the shaft and J is a geometrical property of the cross-section. (a) Determine the principal stresses at the surface. (b) Determine the principal direction with respect to the x-axis if $T = 4FR$ and $J/A = k^2$ (a constant). (c) Repeat parts (a) and (b) for the case where the axial force $F = 0$ and $T \neq 0$.

2.39:* States of plane stress at a point, lying in the x–y plane of Fig. (2P.29), are given as follows:

(a) $\sigma_x = 80$, $\sigma_y = -120$ MPa. If $\sigma_1 = 220$ MPa, determine σ_2, θ_1 and θ_2 (i) if it is known that $\tau_{xy} > 0$ and (ii) if it is known that $\tau_{xy} < 0$.

(b) $\sigma_x = 80$, $\sigma_y = 120$ MPa. If $\sigma_1 = 220$ MPa, determine σ_2, θ_1 and θ_2 (i) if it is known that $\tau_{xy} > 0$ and (ii) if it is known that $\tau_{xy} < 0$.

(c) $\sigma_y = 40$, $\tau_{xy} = -30$ MPa. If $\sigma_1 = 80$ MPa, determine σ_2, θ_1 and θ_2.

(d) $\sigma_y = 40$, $\tau_{xy} = -30$ MPa. If $\sigma_2 = -80$ MPa, determine σ_1, θ_1 and θ_2.

Note: Verify answers via the appropriate Mohr circle.

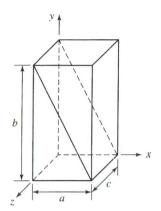

Figure 2P.32

Section 8

2.40: A rectangular plate ABCD of thickness t, lying in the x–y plane as shown in Fig. (2P.40), is subjected to a loading of in-plane surface tractions. The stress field is given as

$$\sigma_x = C_1 \sin(kx), \qquad \sigma_y = C_2 y^2 \sin(kx), \qquad \tau_{xy} = C_3 y \cos(kx); \quad k = \pi/2a,$$

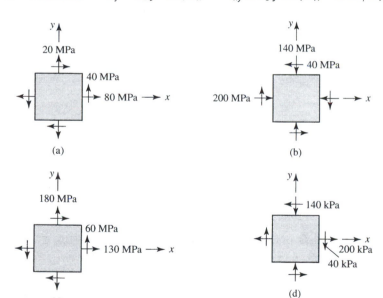

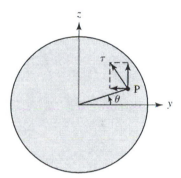

Figure 2P.34

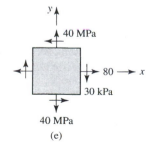

Figure 2P.36

Figure 2P.38

(a)   (b)

where C_1, C_2 and C_3 are constants. (a) For what ratios C_2/C_1 and C_3/C_1 does this stress field represent a state of equilibrium? (b) Determine the Cartesian components of the surface tractions, X_n and Y_n, acting on each of the segments *AB*, *BC*, *CD* and *DA* and show these by means of a sketch. (c) Verify that for the ratios determined in (a), the plate is globally in equilibrium, i.e., the external forces representing the surface tractions over the entire boundary ABCD maintain the plate in equilibrium with respect to both force and moment equilibrium. (Take moments about point *A*.)

2.41: A trapezoidal plate ABCD of thickness t, lying in the x–y plane as shown in Fig. (2P.41), is subjected to a loading of in-plane surface tractions. The stress field is given by

$$\sigma_x = C_1 x^2 y, \qquad \sigma_y = C_2 y^3, \qquad \tau_{xy} = C_3 xy^2,$$

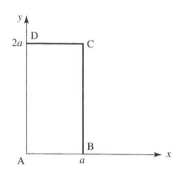

Figure 2P.40

where C_1, C_2 and C_3 are constants having units of N/m^5. Repeat parts (a), (b) and (c) of Problem 2.40 for this case.

2.42: In Section 7, the principal directions and stresses for the case of plane stress were obtained by setting $\frac{d\sigma_n}{d\theta} = 0$. It was then observed that the shear stress, τ_{nt}, vanishes on the principal plane. Alternatively, the principal plane can be *defined* as the plane on which the shear stress vanishes. It then follows that the traction, T_n, acting on this plane is in the principal direction, that is, $T_n = \sigma n$ where σ is a (scalar) constant. (a) Using this alternative definition and the expressions of Eqs. (2.8.8), show that this leads to the following homogeneous equations on the unknowns ℓ_x and ℓ_y:

$$(\tau_{xx} - \sigma)\ell_x + \tau_{xy}\ell_y = 0, \tag{i}$$

$$\tau_{xy}\ell_x + (\tau_{yy} - \sigma)\ell_y = 0, \tag{ii}$$

where ℓ_x and ℓ_y are defined by Eqs. (2.8.7). Show that the condition required for the existence of a solution to Eqs. (i) and (ii) leads to a quadratic equation

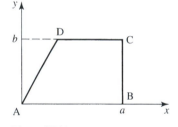

Figure 2P.41

$$\sigma^2 - I_{\sigma_1}\sigma + I_{\sigma_2} = 0 \tag{iii}$$

[where I_{σ_1} and I_{σ_2} are the plane stress invariants; see Eqs. (2.6.10), (2.6.11)] whose roots, σ_1 and σ_2, are the two values of the principal stresses given by Eqs. (2.7.11). (b) Show that the ratio ℓ_y/ℓ_x leads to Eq. (2.7.3), which defines the principal directions. *Note:* In the framework of linear algebra, $\sigma_{1,2}$ are the eigenvalues and ℓ_x and ℓ_y define the eigenvectors *n* of the problem.

The following problems are to be solved using a computer.

2.43: Using the transformation laws for plane stress [Eqs. (2.6.6a) and (2.6.8a)], write a computer program to determine σ_n, σ_t and τ_{nt} for any given state of stress, σ_x, σ_y, τ_{xy} and θ. Check the program by using some of the stress states given in Problem 2.29.

2.44: Given a state of plane stress, σ_x, σ_y and τ_{xy}, write a program to determine the principal stresses σ_1 and σ_2 and the principal directions θ_1 and θ_2. Check the program by using some of the stress states given in Problem 2.35.

2.45: Given a state of plane stress, $\sigma_x \equiv \tau_{xx}$, $\sigma_y \equiv \tau_{yy}$ and τ_{xy}, (a) write a program to find the roots, σ, of Eq. (iii) of Problem 2.42, that is, to determine the principal stresses σ_1 and σ_2; (b) determine the principal directions θ_1 and θ_2. Check the program by using some of the stress states given in Problem 2.35.

2.46: Given the state of plane stress, $\sigma_x = 50$, $\sigma_y = 100$, $\tau_{xy} = 150$ MPa. On what plane (defined by the angle θ of its normal n with respect to the x-axis) is the normal stress $\sigma_n = 225$ MPa? *Note*: The value θ can only be determined numerically.

3

Deformation and strain

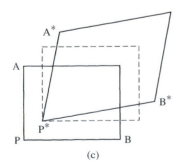

Figure 3.2.1

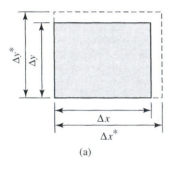

(a)

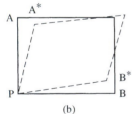

(b)

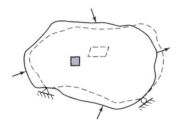

(c)

Figure 3.2.2

3.1 Introduction

Forces, when applied to a body, will evidently cause the body to deform. Since there exists no body in nature that is infinitely stiff, the concept of a rigid body, as used in rigid-body mechanics, is merely an idealisation. In solid mechanics, we are specifically concerned with the study of deformable bodies; in fact, as we have seen in Chapter 1, a primary goal of solid mechanics is to determine the deformation of a body that is subjected to external loads. Consequently, we require a means to describe mathematically the deformation and, in particular, we wish to describe the **intensity of the deformation** of a body. In Chapter 1, this idea, introduced for a simple one-dimensional case, led to the concept of strain. Since bodies usually are not one-dimensional, it is necessary to examine and generalise the concept of strain.

Our goal in this chapter will be to define a measure of the deformation of a body. Now, 'deformation' is essentially described by the changes of geometry of a body. Therefore, in this chapter, we will be concerned only with defining the geometrical changes that occur in a body *irrespective of the cause* of the deformation. The deformation may be caused by external forces or perhaps by changes in temperature of the body, but at this stage of our study, the causes are totally immaterial: we are interested here only in deformation as an intrinsic concept.

3.2 Types of deformation

Consider a body initially at rest. Let us assume, for example, that a set of forces is applied, which causes the body to move. If the body is idealised as a rigid body, the motion in general, will be a combination of translation and rotation such that the distance between any two points in the body remains constant. However, if the body is a deformable body, then in addition to the translation and rotation, the elements of the body will deform as shown in Fig. (3.2.1). In order to describe the deformation of the body, we first examine a small element. To this end, let us consider the simple two-dimensional rectangular element $\Delta x\,\Delta y$ [Fig. (3.2.2a)]. Such an element can undergo two types of deformation:

(a) The element may undergo a change in size: the length Δx changes to Δx^* i.e., $\Delta x \rightarrow \Delta x^*$; similarly, $\Delta y \rightarrow \Delta y^*$. We observe that in this case the element retains its rectangular shape [Fig. (3.2.2a)].

(b) The element may undergo a change in shape without any change in length of Δx or Δy; in this case, the element becomes a parallelogram [Fig. (3.2.2b)]. This distortion of the element, may thus be described by the angle change from

74

its original right angle $\angle$APB to the angle $\angle$A*PB*. We denote the change in angle by γ; thus $\gamma \equiv \angle$APB $- \angle$A*PB*.

In general, however, an element does not undergo only one type of deformation but undergoes simultaneously a change in size as well as a distortion. The total motion of the element may thus be decomposed into (a) rigid-body motion, (b) change in size and (c) distortion as is shown in Fig. (3.2.2c). (The position after only rigid-body motion is shown by the dashed lines in this figure.)

From this discussion, we therefore conclude that two measures of deformation are required: namely (i) elongation (or shortening) of a line element and (ii) changes in angles. We therefore seek a means to describe the deformation mathematically. This description is expressed in terms of a quantity called **strain**.

3.3 Extensional or normal strain

Consider a point P in a body located in an x, y, z coordinate system. Let point Q be a neighbouring point, an infinitesimal distance Δs from P [Fig. (3.3.1a)]. The points P and Q can be defined in the coordinate system by means of the position vectors r_P, r_Q, respectively. The vector $\overrightarrow{PQ}$ is then represented by $\Delta s\, n$, where n is a unit vector that defines the orientation of the infinitesimal line segment PQ [Fig. (3.3.1b)].

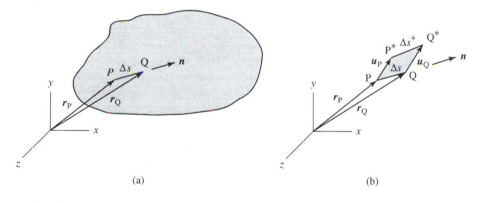

(a)   (b)   **Figure 3.3.1**

Now, let us assume that point P displaces to P* by u_P and Q displaces to Q* by u_Q, such that the distance $|$P*Q*$|$ is Δs^*. We define the **extensional strain** ϵ_n of an infinitesimal line segment at point P, which is *originally* oriented in the n-direction, as

$$\epsilon_n(\text{P}) = \lim_{\substack{\text{Q}\to\text{P} \\ \Delta s \to 0}} \frac{\Delta s^* - \Delta s}{\Delta s}. \tag{3.3.1}$$

From its definition, a positive extensional strain, $\epsilon_n(\text{P}) > 0$, denotes an extension (lengthening) of the segment, while $\epsilon_n(\text{P}) < 0$ denotes a contraction (shortening) of the segment. From its definition, it is clear that ϵ_n is a non-dimensional quantity.

It is important to observe that the deformed segment P*Q* is not necessarily parallel to the segment PQ. Thus, according to our definition, ϵ_n denotes the extensional strain of the line segment that was oriented in the n-direction in its initial or *undeformed state*.

It should be noted that if the orientation of the unit vector n is in the x-, y- or z-direction, then the respective extensional strains for segments lying in these coordinate directions are denoted by ϵ_x, ϵ_y and ϵ_z. From the definition of the extensional strain, the new (deformed) length Δs^* of the original segment is given by

$$\Delta s^* \simeq (1 + \epsilon_n) \, \Delta s \qquad (3.3.2a)$$

and the change in length, $e \equiv ds$, of Δs is

$$e \simeq \epsilon_n \, \Delta s. \qquad (3.3.2b)$$

Having defined the extensional strain ϵ_n at a point, let us consider a line segment AB within a deformable body of *finite* length L, and which is initially oriented in the n-direction as shown in Fig. (3.3.2). Due to deformation of the body, assume that the segment AB deforms to the curve A*B*. Let us also assume that the extensional strain ϵ_n existing at all points along AB is known. We wish to determine the length of the curve A*B* as well as its elongation.

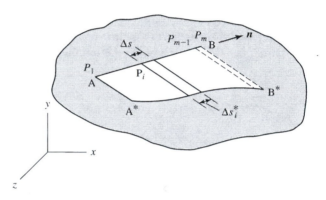

Figure 3.3.2

Now, clearly, we may consider the original line AB to be composed of N number of infinitesimal segments of equal lengths Δs as shown. Let P_1 denote point A and P_m denote point B with P_i denoting some intermediate point. If ϵ_n is known at all points, then a typical segment Δs between P_i and P_{i+1} becomes Δs_i^*. From Eq. (3.3.2a), we then have

$$\Delta s_i^* \simeq [1 + \epsilon_n(P_i)] \, \Delta s. \qquad (3.3.3)$$

[Note that, as opposed to Eq. (3.3.1), Eqs. (3.3.2) and (3.3.3) have been written with a $\simeq$ sign since the expressions are not taken in the limit, i.e., they are written for a small segment in the neighbourhood of a point, but not *at* the point.]

Hence if we divide the segment AB into a large number of segments Δs, we have

$$\Delta s_1^* \simeq [1 + \epsilon_n(P_1)] \, \Delta s$$
$$\Delta s_2^* \simeq [1 + \epsilon_n(P_2)] \, \Delta s$$
$$\vdots$$
$$\Delta s_i^* \simeq [1 + \epsilon_n(P_i)] \, \Delta s$$
$$\vdots$$
$$\Delta s_{m-1}^* \simeq \left[1 + \epsilon_n\left(P_{m_1}\right)\right] \, \Delta s.$$

The deformed length L^* of A^*B^* is then

$$L^* = \sum_{i=1}^{m-1} \Delta s_i^* = \sum_{i=1}^{m-1} [1 + \epsilon_n(P_i)] \, \Delta s. \qquad (3.3.4)$$

Now, if we increase the number of increments infinitely such that $\Delta s \to 0$, then in the limit the summation becomes, by definition,

$$L^* = \int_0^L [1 + \epsilon_n(s)] \, ds = L + \int_0^L \epsilon_n(s) \, ds, \qquad (3.3.5)$$

where $\epsilon_n(s)$ denotes that ϵ_n depends on the parameter s along the line AB. Note that the change in length, ΔL, of AB is given by

$$\Delta L = \int_0^L \epsilon_n(s) \, ds. \qquad (3.3.6)$$

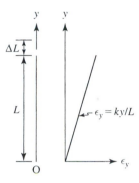

Figure 3.3.3

Example 3.1: A wire, located along the y-axis, is heated in such a way that the strain at any point y is given by $\epsilon_y = ky/L$, where k is a constant [Fig. (3.3.3)]. Determine the change in length of the wire.

Solution: From Eq. (3.3.6),

$$\Delta L = \int_0^L \epsilon_y(y) \, dy = \frac{k}{L} \int_0^L y \, dy = \frac{kL}{2}.$$

Note that the average strain $\bar{\epsilon}_y$, given by $\bar{\epsilon}_y = \Delta L/L = k/2$, is equal to the exact extensional strain only at the point $y = L/2$.   □

Example 3.2: A wire of finite length L, initially lying in the x-direction, is stretched along a rigid track, which is a parabola $y(x) = bx^2$. All points of the wire displace in the y-direction only [Fig. (3.3.4)]. Compute the strain ϵ_x at all points x of the wire.

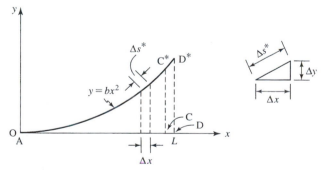

Figure 3.3.4

Solution: We consider the wire in its initial position to be composed of a number of infinitesimal segments, each of length Δx. [Note that here Δx replaces Δs of Eq. (3.3.1).] Then, from Eq. (3.3.1),

$$\epsilon_x = \lim_{\Delta x \to 0} \frac{\Delta s^* - \Delta x}{\Delta x},$$

where Δs^* is the deformed length of the segment of the wire when stretched along the track. From geometry,

$$\Delta s^* = \sqrt{[(\Delta x)^2 + (\Delta y)^2]} = \sqrt{[1 + (\Delta y / \Delta x)^2]}\, \Delta x.$$

Hence

$$\epsilon_x = \lim_{\Delta x \to 0} \{\sqrt{[1 + (\Delta y / \Delta x)^2]} - 1\} = \sqrt{[1 + (dy/dx)^2]} - 1.$$

But $dy/dx = 2bx$. Therefore

$$\epsilon_x(x) = \sqrt{1 + 4b^2 x^2} - 1.$$

Observe that when $x = 0$, $\epsilon_x = 0$ and that the maximum strain occurs at $x = L$. This may be readily observed from Fig. (3.3.4); the greatest deformation occurs in the segment CD, which deforms to C^*D^*.

It is of interest to note that if $bx \ll 1$, i.e., if bx is an infinitesimal quantity, then making use of the binomial theorem, it follows that $\epsilon_x(x) = 2b^2 x^2$, i.e., the strain is a quadratic function of x.   □

3.4   Shear strain

Consider again point P and two neighbouring points Q and R, such that the infinitesimal segments, PQ and PR, are mutually perpendicular as shown in Fig. (3.4.1). Thus $\angle RPQ = \pi/2$. Further let the unit vectors $\boldsymbol{n}$ and $\boldsymbol{t}$ denote the orientation of the line segments PQ and PR, respectively. As before, assume that due to deformation the displacements $P \to P^*$, $Q \to Q^*$ and $R \to R^*$ are given by $\boldsymbol{u}_P$, $\boldsymbol{u}_Q$ and $\boldsymbol{u}_R$, respectively.

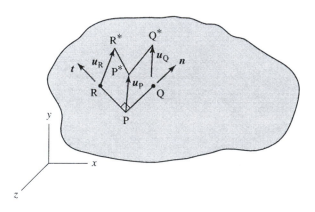

Figure 3.4.1

We define the shear strain γ_{nt} at P as the change in angle between two line segments originally in the orthogonal n- and t-directions; thus

$$\gamma_{nt}(P) = \frac{\pi}{2} - \lim_{\substack{Q \to P \\ R \to P}} \angle R^*P^*Q^*. \qquad (3.4.1)$$

We note that since the shear strain defines the change in angle between two line segments emanating from point P in the n- and t-directions, it is necessary to use two subscripts with γ to define these two directions. Moreover, it should be clear that, in general, the shear strain will be different depending on the orientation of the n- and t-directions.

It is also worthwhile to observe that $\gamma_{nt} > 0$ signifies that the angle between the n- and t-directions *decreases*. Note too that the shear strain component, having units of radians, is a non-dimensional quantity.

Finally, it is important to observe that, based upon its definition, $\gamma_{nt} = \gamma_{tn}$.

Example 3.3: Points A and C of a rectangular plate shown in Fig. (3.4.2) displace to points A* and C* along the x- and y-axes, respectively, so that the rectangle is deformed into a parallelogram. Lines that were initially parallel to the x–y axes remain parallel lines. (a) Compute the shear strain γ_{xy} at all points in the plate. (b) Compute the shear strain γ_{xy} assuming that $\gamma_{xy} \ll 1$.

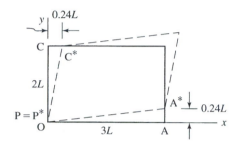

Figure 3.4.2

Solution:

(a) At point P,

$$\gamma_{xy} = \pi/2 - \angle C^*P^*A^* = \angle A^*P^*x + \angle C^*P^*y$$

or

$$\gamma_{xy} = \tan^{-1}(0.24/3) + \tan^{-1}(0.24/2) = 0.0798 + 0.1194 = 0.1993$$

(b) Assuming $\gamma_{xy} \ll 1$, the angles between the sides of the parallelogram and the x- and y-directions are small. Therefore, recalling that for $\alpha \ll 1$, $\sin\alpha \simeq \tan\alpha \simeq \alpha$, we have, at P,

$$\gamma_{xy} = 0.24/3 + 0.24/2 = 0.2000.$$

Note that, using the property of small arguments, the percent error is $(0.1993 - 0.2000)0/0.1993 = -0.0037 = -0.37\%$.

Since all parallel lines were stated to remain parallel after deformation, the angle changes are the same at all points and hence the shear strain γ_{xy} is constant throughout the plate. □

Example 3.4: A plate ABCD lying in the x–y plane [Fig. (3.4.3a)] is deformed such that C → C*, D → D*, etc. Point P displaces in the y-direction to P* by an amount $\Delta = \alpha b$ (where α is a constant) such that the diagonals PA and PB remain straight lines in the deformed state [Fig. (3.4.3b)]. Calculate the shear strains γ_{nt}, where the n- and t-directions lie along the diagonals.

Solution: From geometry, it is evident that the angle ∠BPA of the undeformed plate is a right angle. We denote the angle ∠AP*B by θ. Then, since point P moves in the

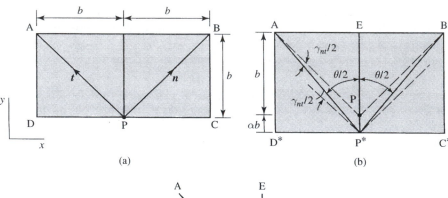

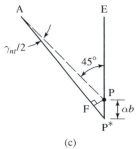

Figure 3.4.3

(c)

y-direction, by symmetry, we observe that the shear strain at P is

$$\frac{\gamma_{nt}}{2} = \frac{\pi}{4} - \frac{\theta}{2} \tag{3.4.2a}$$

and hence

$$\tan \frac{\gamma_{nt}}{2} = \tan \left(\frac{\pi}{4} - \frac{\theta}{2} \right). \tag{3.4.2b}$$

Recalling the trigonometric identity,

$$\tan(x \pm y) = \frac{\tan x \pm \tan y}{1 \mp \tan x \tan y},$$

we find

$$\tan \frac{\gamma_{nt}}{2} = \frac{1 - \tan(\theta/2)}{1 + \tan(\theta/2)}. \tag{3.4.2c}$$

Now, from geometry,

$$\tan \frac{\theta}{2} = \frac{b}{b(1 + \alpha)} = \frac{1}{1 + \alpha}. \tag{3.4.2d}$$

Substituting in Eq. (3.4.2c),

$$\tan \frac{\gamma_{nt}}{2} = \frac{1 - \frac{1}{1+\alpha}}{1 + \frac{1}{1+\alpha}} = \frac{\alpha}{2 + \alpha}, \tag{3.4.3a}$$

and therefore,

$$\gamma_{nt} = 2 \tan^{-1} \frac{\alpha}{2 + \alpha}. \tag{3.4.3b}$$

For example, if $\alpha = 0.02$, $\gamma_{nt} = 0.0198$, while if $\alpha = 0.3$, $\gamma_{nt} = 0.259$.

In the above analysis, we determined the strain γ_{nt} for two values of α: (a) an infinitesimal value, $\alpha = 0.02$, describing a small deformation that results in an infinitesimal strain $\gamma_{nt} = 0.0198$ and (b) a finite value, $\alpha = 0.3$, describing a relatively large deformation. We now re-examine the problem for the first case of infinitesimal strain.

If $\gamma_{nt} \ll 1$, by definition, we observe that the *rotations are small*; in particular, the diagonal AP rotates by an infinitesimal amount. As a result, we note that the angle $\angle AP^*E \simeq 45°$. Moreover, from Fig. (3.4.3b), we observe that the angle $\angle PAP^* = \gamma_{nt}/2$. Let us construct the line segment PF perpendicular to AP^* [Fig. (3.4.3c)]. Examining the triangle FPP^*, $|PF| = \alpha b \sin 45° = \alpha b \sqrt{2}/2$. Then

$$\frac{\gamma_{nt}}{2} = \frac{|PF|}{|AP|} = \frac{\alpha b \sqrt{2}/2}{b\sqrt{2}} = \frac{\alpha}{2} \qquad (3.4.4)$$

and hence $\gamma_{nt} = \alpha$.

If α is small, e.g. $\alpha = 0.02$, the resulting strain, $\gamma_{nt} = 0.02$, is infinitesimal and differs from the exact result given above by 1%, while if α is not infinitesimal, e.g. $\alpha = 0.3$, we obtain a relatively inaccurate value, i.e., $\gamma_{nt} = 0.3$, with an error of 15.6%. □

We thus observe that if the strains and rotations are infinitesimal, we may obtain extremely accurate results using a much simplified analysis. We shall find this conclusion to be true in general.

3.5 Strain–displacement relations

Since strain is a measure of the deformation of a body, it is clear that it depends on the displacements of points within the body. The variation of displacements, $\boldsymbol{u}$, given as a function of the spatial coordinates [e.g., $\boldsymbol{u} = \boldsymbol{u}(x, y, z)$ in a Cartesian coordinate system] is referred to as a **displacement field**. In particular, the strain at points within the body is a result of the *relative* displacements of various points within the body. Although points of a rigid body also may undergo relative displacements, the distance between *any* two points of a rigid body must remain constant. In a deformable body, however, the distance between any two points, in general, does not remain constant and as a result, both extensional and shear strains will exist at the various points of the body. In our treatment below, explicit relations for the strains in terms of the displacement field will be derived. However, to provide a better insight in the analysis of strain resulting from known displacements in a body, we first examine the resulting strains in some simple problems.

For simplicity, we shall investigate two-dimensional cases; i.e., cases where all displacements are in a plane. We consider here the plane of deformation to be the x–y plane, where (a) there exists no displacement component in the z-direction and (b) $\boldsymbol{u} = \boldsymbol{u}(x, y)$. This type of two-dimensional deformation represents a case called **plane strain**.

(a) Some preliminary instructive examples
We examine the strains resulting from given displacements in a body via the following examples.

Example 3.5: Consider a plate ABCD lying in the x–y plane. The sides of the plate are unity and point A is assumed to lie initially at the origin of the

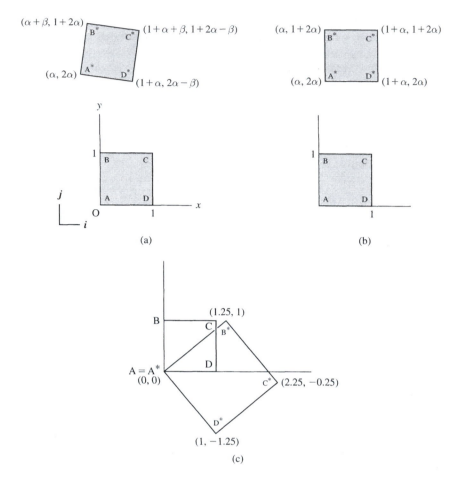

Figure 3.5.1

x, y coordinate system [Fig. (3.5.1a)]. Let the displacements of the plate be $\mathbf{u} = u\mathbf{i} + v\mathbf{j}$, where the components in the x- and y-directions are given by the expressions

$$u(x, y) = \alpha + \beta y, \qquad v(x, y) = 2\alpha - \beta x, \qquad \beta \ll 1. \qquad (3.5.1)$$

Here α and β are positive constants. Note that while β is given as infinitesimal, α is finite. Determine (a) the average extensional strains $\bar{\epsilon}_x$ along AD and $\bar{\epsilon}_y$ along AB and (b) the shear strain at point A.

Solution: We first determine the position of the plate after deformation. The original coordinates of A are $(0, 0)$. Due to the displacement, A $\rightarrow$ A*, and hence the coordinates of A* become $(0 + u_A, 0 + v_A) = (\alpha, 2\alpha)$. Similarly, the coordinates of B$(0, 1)$, which moves to B*, become $(0 + u_B, 1 + v_B) = (\alpha + \beta, 1 + 2\alpha)$. Points C* and D* can be determined similarly. Thus, after deformation the coordinates of the corners of the plate are given by

$$\text{A}^*: (\alpha, 2\alpha); \qquad \text{B}^*: (\alpha + \beta, 1 + 2\alpha);$$

$$\text{C}^*: (1 + \alpha + \beta, 1 + 2\alpha - \beta); \qquad \text{D}^* : (1 + \alpha, 2\alpha - \beta).$$

Since the displacement components u and v vary linearly with the x- and y-coordinates, the straight edges of the plate ABCD remain straight edges in the deformed plate A*B*C*D*, as shown in Fig. (3.5.1a). From the figure it would appear that the displacement pattern, as given, results in a rigid-body motion. We therefore examine the new lengths of the elements. For example, for the edge AB, we obtain

$$|A^*B^*|^2 = [(\alpha + \beta) - \alpha]^2 + [(1 + 2\alpha) - 2\alpha]^2 = 1 + \beta^2$$

and hence

$$|A^*B^*| = \sqrt{1 + \beta^2}. \tag{3.5.2a}$$

Recalling now the binomial expansion

$$\sqrt{1 + x} = 1 + x/2 - x^2/8 + \cdots, \quad x \ll 1,$$

it follows that

$$|A^*B^*| = 1 + \frac{\beta^2}{2} - \frac{\beta^4}{8} + \cdots. \tag{3.5.2b}$$

Since $\beta \ll 1$, we neglect all infinitesimals of the second order. Hence

$$|A^*B^*| \simeq 1. \tag{3.5.3a}$$

Similarly, we find

$$|B^*C^*| \simeq 1. \tag{3.5.3b}$$

$$|C^*D^*| \simeq 1, \tag{3.5.3c}$$

$$|D^*A^*| \simeq 1. \tag{3.5.3d}$$

Thus, according to this 'first-order' analysis, the lengths of the edges of the plate do not change, and hence, according to its definition, we conclude that the average extensional strains $\bar{\epsilon}_x = 0$ (of the segments AD and BC) and $\bar{\epsilon}_y = 0$ (of AB and CD).

We leave it as an exercise to show that at point A, $\gamma_{xy} = 0$ if $\beta \ll 1$.   □

Several features of these results should be emphasised:

- The finite parameter α appearing in this problem defines rigid-body translation. To show this, we set $\beta = 0$ and, for example, let $\alpha = 1.25$. The element then undergoes translation as shown in Fig. (3.5.1b). Observe that such a finite value of α does not result in deformation of the element.
- Consider now a finite value of β, say $\beta = 1.25$. Setting $\alpha = 0$, the element assumes the position as shown in Fig. (3.5.1c). Thus we observe that a finite value of β results both in *rotation and deformation* of the element. We also note from Eq. (3.5.2) that for *finite β* the strains do not vanish since $|A^*B^*| \neq |AB|$, etc.
- On the other hand, if β is an infinitesimal, i.e., $|\beta| \ll 1$, then as we have seen in Fig. (3.5.1a), the rotations are small (to first order in β) and we note, for example from Eq. (3.5.3a), that the strains vanish if all second-order infinitesimals are neglected.
- The strains that were obtained are *average* strains over the length of the sides; i.e., we have not found the strains at a point.

Example 3.6: We consider the same plate ABCD lying in the *x–y* plane as in the previous example. However, the displacement components, *u* and *v*, in the *x-* and *y*-directions are now given respectively by

$$u(x, y) = \alpha + \beta y, \qquad v(x, y) = 2\alpha + \beta x, \quad \beta \ll 1. \tag{3.5.4}$$

Note that the displacement field defined by Eq. (3.5.4) differs from that of Eq. (3.5.1) only by the change of sign of the βx term appearing in the v-displacement component. (We shall find that this sign difference results in a very different displacement pattern from that of the previous example.) Determine (a) the average extensional strains $\bar{\epsilon}_x$ and $\bar{\epsilon}_y$, (b) the average extensional strain along the line AC and (c) the shear strain γ_{xy} at point A.

Solution: As in Example 3.5, the coordinates of the corners of the deformed plate are readily obtained; namely

$$A^*: (\alpha, 2\alpha); \qquad B^*: (\alpha + \beta, 1 + 2\alpha);$$

$$C^*: (1 + \alpha + \beta, 1 + 2\alpha + \beta); \qquad D^*: (1 + \alpha, 2\alpha + \beta).$$

The resulting position of the deformed plate is shown in Fig. (3.5.2). The new lengths of the edges of the plate $A^*B^*C^*D^*$, calculated as in the previous example, are

$$|A^*B^*| \simeq 1, \tag{3.5.5a}$$

$$|B^*C^*| \simeq 1, \tag{3.5.5b}$$

$$|C^*D^*| \simeq 1, \tag{3.5.5c}$$

$$|D^*A^*| \simeq 1, \tag{3.5.5d}$$

where the symbol $\simeq$ is used to indicate that the relation is approximate up to first order in β for $\beta \ll 1$. Therefore, the average strains are $\bar{\epsilon}_x = 0$ and $\bar{\epsilon}_y = 0$ along the respective line segments. Consequently, we conclude that the deformed plate $A^*B^*C^*D^*$ is a parallelogram (and more specifically a rhombus). Now, although the extensional strains vanish for line segments that lie in the x- and y-directions, it is clear from Fig. (3.5.2) that the line segment AC of the undeformed plate changes length and hence the extensional strain of this line segment does not vanish. Moreover, we note that, as opposed to the previous example, the shear strain $\gamma_{xy} \neq 0$ since clearly the right angles no longer remain right angles at point A.

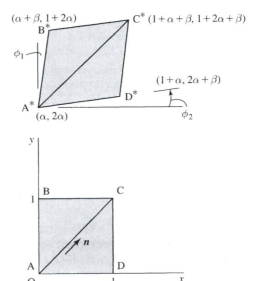

Figure 3.5.2

We first define the orientation of the line segment AC by means of the unit vector n [Fig. (3.5.2a)]. Then, according to the definition given in Eq. (3.3.1), the average strain $\bar{\epsilon}_n$ along AC is given by

$$\bar{\epsilon}_n = \frac{|A^*C^*| - |AC|}{|AC|} = \frac{|A^*C^*| - \sqrt{2}}{\sqrt{2}} \tag{3.5.6a}$$

since $|AC| = \sqrt{2}$.

Now, $|A^*C^*|^2 = [(1 + \alpha + \beta) - \alpha]^2 + [(1 + 2\alpha + \beta) - 2\alpha]^2 = 2(1 + \beta)^2$ and therefore

$$|A^*C^*| = \sqrt{2}(1 + \beta). \tag{3.5.6b}$$

Substituting in Eq. (3.5.6a),

$$\bar{\epsilon}_n = \frac{\sqrt{2}(1 + \beta) - \sqrt{2}}{\sqrt{2}} = \beta. \tag{3.5.7}$$

To calculate γ_{xy} at A, we denote the inclination of A^*B^* and A^*D^* with respect to the y- and x-directions by ϕ_1 and ϕ_2, respectively [Fig. (3.5.2)]. Then, clearly, since the shear strain represents the change in the right angle, $\gamma_{xy} = \phi_1 + \phi_2$. Now, since $\beta \ll 1$, the rotations in this example are infinitesimals. Noting that $\sin\phi_1 \simeq \frac{\beta}{1} = \beta$ and (since for $x \ll 1$, $\sin x \simeq x$) therefore $\phi_1 = \beta$. Similarly, $\phi_2 = \beta$. Therefore $\gamma_{xy} = 2\beta$ at point A. From Fig. (3.5.2), we might anticipate that the shear strain $\gamma_{xy} = 2\beta$ at all points in the plate, although at this stage we cannot prove this assertion. $\square$

We observe from the above two examples, that the expressions are considerably simplified when the strains and rotations at all points of a body are small. In many problems encountered in engineering practice, we find that this is precisely the case. [For example, in Chapter 1, we noted that strains in the elastic range of steel were of the order of $O(10^{-3})$.] We therefore shall derive expressions for the strains in terms of displacements under the above assumption of infinitesimal strains and rotations.

(b) Strain–displacement relations for infinitesimal strains and rotations

For simplicity, we consider a two-dimensional body lying in the x–y plane, as shown in Fig. (3.5.3). Let $P(x, y)$ represent a general point in the body, and let

$$u(x, y) = u(x, y)i + v(x, y)j \tag{3.5.8}$$

denote the displacement of any point P.

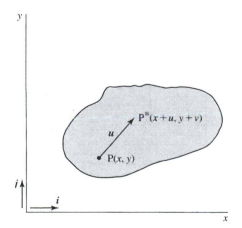

Figure 3.5.3

We analyse the deformation of the body under the following assumptions:

- The displacements $u(x, y)$ vary continuously with the spatial coordinates and possess continuous partial derivatives with respect to these coordinates.
- The strain components at all points P are infinitesimal.
- All elements of the body undergo infinitesimal relative rotations. By 'infinitesimal relative rotations' we mean rotations such that neighbouring points undergo small rotations with respect to each other.

Consider now a rectangular element PQSR having sides Δx and Δy, as shown in Fig. (3.5.4). Due to deformation, P $\rightarrow$ P*, Q $\rightarrow$ Q* and R $\rightarrow$ R*. The coordinates of these points after deformation are given by

$$P^*: (x + u, y + v); \qquad Q^*: [x + \Delta x + (u + \Delta u), y + (v + \Delta v)];$$

$$R^*: [x + (u + \Delta u), y + \Delta y + (v + \Delta v)]$$

Now, from Eq. (3.3.1),

$$\epsilon_x(P) = \lim_{\Delta x \to 0} \frac{|P^*Q^*| - |PQ|}{|PQ|}$$

or

$$\epsilon_x(P) = \lim_{\Delta x \to 0} \frac{|P^*Q^*| - \Delta x}{\Delta x} \qquad (3.5.9a)$$

since $|PQ| = \Delta x$.

We note that

$$|P^*Q^*| = \sqrt{[(\Delta x + \Delta u)^2 + \Delta v^2]} = \sqrt{[(1 + \Delta u/\Delta x)^2 + (\Delta v/\Delta x)^2]}\, \Delta x.$$

Hence, by Eq. (3.5.9a),

$$\epsilon_x = \lim_{\Delta x \to 0} \sqrt{[(1 + \Delta u/\Delta x)^2 + (\Delta v/\Delta x)^2]} - 1$$

or

$$\epsilon_x = \sqrt{[(1 + \partial u/\partial x)^2 + (\partial v/\partial x)^2]} - 1. \qquad (3.5.9b)$$

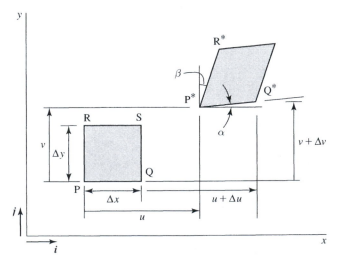

Figure 3.5.4

Similarly, from its definition,

$$\epsilon_y(P) = \lim_{\Delta y \to 0} \frac{|P^*R^*| - \Delta y}{\Delta y}, \qquad (3.5.10a)$$

we obtain

$$\epsilon_y = \sqrt{[(1 + \partial v/\partial y)^2 + (\partial u/\partial y)^2]} - 1. \qquad (3.5.10b)$$

We now examine the order of magnitude of the ratios $\frac{\partial u}{\partial x}, \frac{\partial v}{\partial x}, \frac{\partial u}{\partial y}, \frac{\partial v}{\partial y}$ under the restriction of small strains and small relative rotations.

Let us first consider the rotations $PQ \to P^*Q^*$ and $PR \to P^*R^*$. Denoting the orientations of the deformed segments with respect to the x- and y-directions by α and β, respectively [Fig. (3.5.4)], we note that the segment PQ at point $P(x, y)$ undergoes a rotation given by

$$\tan \alpha = \lim_{\Delta x \to 0} \frac{\Delta v}{\Delta x + \Delta u} = \lim_{\Delta x \to 0} \frac{\Delta v/\Delta x}{\left(1 + \frac{\Delta u}{\Delta x}\right)}$$

or

$$\tan \alpha = \frac{\partial v/\partial x}{1 + \partial u/\partial x}. \qquad (3.5.11a)$$

If the relative rotation α is infinitesimal, then for $\alpha \ll 1$, $\tan \alpha \simeq \alpha$; i.e.,

$$\alpha = \frac{\partial v/\partial x}{1 + \partial u/\partial x}. \qquad (3.5.11b)$$

Let us now *assume* momentarily that $\partial u/\partial x \ll 1$; i.e., $\partial u/\partial x$ is also infinitesimal. It then follows from Eq. (3.5.11b) that $\alpha = \partial v/\partial x$ and therefore $\partial v/\partial x$ must also be an infinitesimal; thus

$$\alpha = \partial v/\partial x \ll 1. \qquad (3.5.12a)$$

Similarly, by examining the rotation of PR $\to$ P*R*, we conclude that if the relative rotation of RP is small and if $\partial v/\partial y \ll 1$, then

$$\beta = \partial u/\partial y \ll 1. \qquad (3.5.12b)$$

Using the property of Eq. (3.5.12a), and *neglecting infinitesimals of second order*, we find, from Eq. (3.5.9b),[†] that

$$\epsilon_x = \sqrt{[1 + \partial u/\partial x]^2} - 1 \qquad (3.5.13a)$$

or

$$\epsilon_x = \left(1 + \frac{\partial u}{\partial x}\right) - 1, \qquad (3.5.13b)$$

[†] Note that $\sqrt{[1 + \partial u/\partial x]^2} = \sqrt{[1 + 2\partial u/\partial x + (\partial u/\partial x)^2]}$. Hence, alternatively, by neglecting the second-order infinitesimal $(\partial u/\partial x)^2$ in addition to $(\partial v/\partial x)^2$ in Eq. (3.5.9b) and making use of the binomial theorem,

$$\sqrt{1 + x} = 1 + x/2 - x^2/8 + \cdots,$$

we obtain Eqs. (3.5.13b) and (3.5.14a).

and hence

$$\epsilon_x = \frac{\partial u}{\partial x}. \tag{3.5.14a}$$

Note that the above development was based on our previous assumption that $\partial u/\partial x$ is an infinitesimal. It then follows that ϵ_x as given by Eq. (3.5.14a) is also infinitesimal. Thus the entire development is consistent with small strain theory.

Similarly, using the property of Eq. (3.5.12b), we find, from Eq. (3.5.10b), that

$$\epsilon_y = \frac{\partial v}{\partial y} \tag{3.5.14b}$$

is the infinitesimal strain in the y-direction.

The shear strain γ_{xy} can be obtained by noting that the change in angle at point P is given by

$$\gamma_{xy} = \frac{\pi}{2} - \lim_{\substack{\Delta x \to 0 \\ \Delta y \to 0}} \angle R^*P^*Q^* = \alpha + \beta.$$

From Eqs. (3.5.12), we have

$$\gamma_{xy} = \frac{\partial v}{\partial x} + \frac{\partial u}{\partial y}. \tag{3.5.15}$$

Thus we have obtained explicit two-dimensional expressions for the infinitesimal strains in terms of the displacements $\boldsymbol{u} = \boldsymbol{u}(x, y)$:

$$\epsilon_x = \frac{\partial u}{\partial x}, \tag{3.5.16a}$$

$$\epsilon_y = \frac{\partial v}{\partial y}, \tag{3.5.16b}$$

$$\gamma_{xy} = \frac{\partial v}{\partial x} + \frac{\partial u}{\partial y}. \tag{3.5.16c}$$

Hence, if the displacement field $\boldsymbol{u}(x, y)$ is known at all points of a body, the strain components are immediately obtained by taking the partial derivatives as given by the above equations.

The following comments are now in order. The derivation of the above expressions for the strain components was based on several assumptions:

(a) Relative rotations of the line segments are small. [This assumption led to Eq. (3.5.11b).]
(b) The partial derivatives $\partial u/\partial x$ and $\partial v/\partial y$ are infinitesimals. [This additional property was necessary to obtain Eqs. (3.5.12), namely that $\partial v/\partial x$ and $\partial u/\partial y$ are infinitesimals.]
(c) Note that as a result of the assumption (b), we conclude that the resulting strains ϵ_x and ϵ_y, given by Eqs. (3.5.16a) and (3.5.16b), are also infinitesimals.

Similarly, from assumption (a), it follows by Eq. (3.5.16c) that γ_{xy} is also infinitesimal.

Thus to summarise, the strain–displacement relations given by Eqs. (3.5.16) are valid for a body whose elements undergo infinitesimal strains and infinitesimal relative rotations. The expressions are said to be within the limitations of *small strain theory*. Moreover, since all these derivatives are of the first order (essentially all second-order infinitesimal terms are neglected), small strain theory is often also referred to as being a *linear theory*.

For simplicity, the above expressions were derived for the two-dimensional case. For a three-dimensional body in an x, y, z coordinate system, with displacements $\boldsymbol{u}$ given by

$$\boldsymbol{u}(x, y, z) = u\boldsymbol{i} + v\boldsymbol{j} + w\boldsymbol{k}, \tag{3.5.17}$$

one finds the following equivalent expressions consistent with small strain theory:

$$\epsilon_x = \frac{\partial u}{\partial x}, \tag{3.5.18a}$$

$$\epsilon_y = \frac{\partial v}{\partial y}, \tag{3.5.18b}$$

$$\epsilon_z = \frac{\partial w}{\partial z} \tag{3.5.18c}$$

and

$$\gamma_{xy} = \frac{\partial v}{\partial x} + \frac{\partial u}{\partial y}, \tag{3.5.18d}$$

$$\gamma_{yz} = \frac{\partial w}{\partial y} + \frac{\partial v}{\partial z}, \tag{3.5.18e}$$

$$\gamma_{zx} = \frac{\partial u}{\partial z} + \frac{\partial w}{\partial x}. \tag{3.5.18f}$$

Example 3.7: A body is deformed such that the displacements at any point P are given by

$$\boldsymbol{u} = \frac{1}{B}[x^3 y\boldsymbol{i} + 3y^4\boldsymbol{j} + (z-4)^2 xy\boldsymbol{k}], \quad 0 \le x, \quad y \le 1,$$

where $B \gg 1$ is a constant. Determine the strain components at all points of the body.

Solution: From Eqs. (3.5.18),

$$\epsilon_x = \frac{\partial u}{\partial x} = \frac{1}{B}3x^2 y, \qquad \epsilon_y = \frac{\partial v}{\partial y} = \frac{1}{B}12y^3, \qquad \epsilon_z = \frac{\partial w}{\partial z} = \frac{2}{B}(z-4)xy,$$

$$\gamma_{xy} = \frac{\partial v}{\partial x} + \frac{\partial u}{\partial y} = \frac{x^3}{B}, \qquad \gamma_{yz} = \frac{\partial w}{\partial y} + \frac{\partial v}{\partial z} = \frac{1}{B}(z-4)^2 x,$$

$$\gamma_{zx} = \frac{\partial u}{\partial z} + \frac{\partial w}{\partial x} = \frac{1}{B}(z-4)^2 y.$$

□

Example 3.8: We reconsider the plate ABCD examined previously in Example 3.5 [Fig. (3.5.1)] of this section, for which the displacement components at each point are, as before,

$$u(x, y) = \alpha + \beta y, \qquad v(x, y) = 2\alpha - \beta x, \quad \beta \ll 1.$$

Determine the strain components at each point $P(x, y, z)$.

Solution: For this two-dimensional problem, Eqs. (3.5.16) yield

$$\epsilon_x = \epsilon_y = \gamma_{xy} = 0$$

at all points. Thus, since the strains are zero everywhere, we conclude that the given displacement field represents a rigid-body motion. □

It is important to observe that in Example 3.5, we were only able to find that *average* values of the strains along the line segments were zero. Hence, we previously could only surmise that the plate ABCD undergoes only rigid-body motion, since it was not possible to prove that the strains vanish everywhere. It is only from the derived expressions of Eq. (3.5.16) which yield the strain at all points, that we have conclusively shown that the plate undergoes only rigid-body motion.

Example 3.9: We reconsider the plate ABCD examined previously in Example 3.6 [Fig. (3.5.2)] of this section, for which the displacements components at each point are, as before,

$$u(x, y) = \alpha + \beta y, \qquad v(x, y) = 2\alpha + \beta x, \qquad \beta \ll 1.$$

Determine the strain components $\epsilon_x, \epsilon_y, \gamma_{xy}$ at any point $P(x, y, z)$.

Solution: For this two-dimensional problem, Eqs. (3.5.16a) and (3.5.16b) yield again $\epsilon_x = \epsilon_y = 0$ at all points. Thus the extensional strains of any line segment parallel to the x- or y-axes vanish everywhere.

However, for the shear strain γ_{xy}, given by Eq. (3.5.16c), we find

$$\gamma_{xy} = \frac{\partial v}{\partial x} + \frac{\partial u}{\partial y} = 2\beta.$$

Thus, as we anticipated in Example 3.6, the shear strain is *constant* throughout the plate. □

There also exist cases where the displacement field $u(x, y, z)$ is unknown but where the strain field $\epsilon(x, y, z)$ is known within a body. In such cases, the strain–displacement relations, Eqs. (3.5.18), can be integrated to yield, together with appropriate boundary conditions on the surfaces, displacements u at all points within the body. We illustrate this for a two-dimensional case in the following example.

Example 3.10: A rectangular plate ABCD, lying in the x–y plane, is deformed to A*B*C*D*, as shown in Fig. (3.5.5), such that lines AB and AD remain straight lines and $\delta/L \ll 1$ and $\delta/h \ll 1$. The extensional strains at any point $P(x, y)$ are

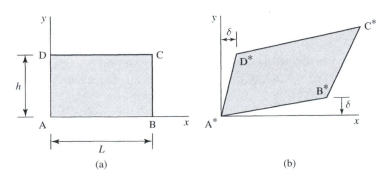

(a)   (b)

Figure 3.5.5

given as

$$\epsilon_x = a\frac{x^2 y}{L^2 h}, \qquad \epsilon_y = a\frac{xy^2}{Lh^2},$$

where $a \ll 1$.

By integration of the strain–displacement relations, (a) determine the displacement field $u(x, y)$; (b) determine the coordinates of points B*, C* and D*; (c) show that the edges B*C* and C*D* remain straight lines; (d) determine the angle between the edges B*C* and C*D* at point C*.

Solution:

(a) Using Eq. (3.5.18a), $\frac{\partial u}{\partial x} = ax^2 y/L^2 h$, integration yields $u(x, y) = ax^3 y/3L^2 h + A(y)$, where $A(y)$ is a function of y only. Since the edge AD remains a straight line, $u(0, y) = \frac{\delta \cdot y}{h}$; it then follows that $A(y) = \frac{\delta \cdot y}{h}$. Hence,

$$u(x, y) = a\frac{x^3 y}{3L^2 h} + \frac{\delta \cdot y}{h}.$$

Similarly, using Eq. (3.5.18b), $\frac{\partial v}{\partial y} = axy^2/Lh^2$, integration yields $v(x, y) = axy^3/3Lh^2 + B(x)$, where $B(x)$ is a function of x only. Noting that $v(x, 0) = \frac{\delta \cdot x}{L}$, it follows that $B(x) = \frac{\delta \cdot x}{L}$ and hence

$$v(x, y) = a\frac{xy^3}{3Lh^2} + \frac{\delta \cdot x}{L}.$$

Therefore

$$\mathbf{u} = \left(a\frac{x^3 y}{3L^2 h} + \frac{\delta \cdot y}{h}\right)\mathbf{i} + \left(a\frac{xy^3}{3Lh^2} + \frac{\delta \cdot x}{L}\right)\mathbf{j}.$$

(b) The coordinates of points B*, C* and D* are then calculated as follows:

$$x_{B^*} = x_B + u(L, 0) = L, \qquad\qquad y_{B^*} = y_B + v(L, 0) = \delta,$$
$$x_{C^*} = x_C + u(L, h) = L + aL/3 + \delta, \qquad y_{C^*} = y_C + v(L, L) = h + ah/3 + \delta,$$
$$x_{D^*} = x_D + u(0, h) = \delta, \qquad\qquad y_{D^*} = y_D + v(0, h) = h.$$

(c) The coordinates of points along B*C* are given by

$$x^* = L + u(L, y) = L + (1/h)(aL/3 + \delta) \cdot y,$$
$$y^* = y + v(L, y) = y + ay^3/3h^2 + \delta.$$

Since x^* is a linear function of y, the edge B*C* is a straight line. Similarly, the coordinates of points along C*D* are

$$x^* = x + u(x, h) = x + ax^3/3L^2 + \delta,$$
$$y^* = h + v(x, h) = h + (1/L)(ah/3 + \delta) \cdot x.$$

Since y^* is a linear function of x, the edge C*D* is a straight line.

(d) Using Eq. (3.5.18c), $\gamma_{xy} = ax^3/3L^2 h + ay^3/3Lh^2 + \delta(1/h + 1/L)$. Hence, after simplification,

$$\gamma_{xy}|_C = \gamma_{xy}(L, h) = \left(\frac{1}{Lh}\right)[a(L^2 + h^2) + (L + h)\delta]$$

Therefore, $\angle D^*C^*B^* = \pi/2 - \gamma_{xy}|_C$.

Note that B*C* is not parallel to A*D* nor is C*D* parallel to A*B*. $\qquad\qquad \square$

3.6 State of strain

As has been seen, the deformation of a body can be described completely by means of the extensional and shear strains at all points of the body.

Consider now the three-dimensional body shown in Fig. (3.6.1) in an x, y, z coordinate system, which undergoes deformation. We may consider this body to be composed of an infinite number of infinitesimal parallelepipeds (e.g. cubes) and may regard the total deformation of the body as the total effect of the deformation of the elemental cubes. Let us therefore examine the deformation of an individual cube, neglecting rigid-body motion (since by definition it does not contribute to the deformation).

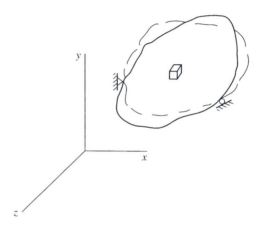

Figure 3.6.1

We note first that the sides originally of lengths Δx, Δy and Δz, respectively, may change lengths, so that the new lengths are $(1 + \epsilon_x)\,\Delta x$, $(1 + \epsilon_y)\,\Delta y$ and $(1 + \epsilon_z)\,\Delta z$, as shown in Fig. (3.6.2a).

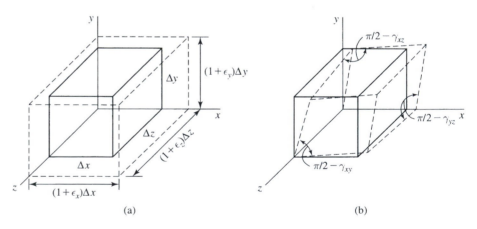

Figure 3.6.2   (a)   (b)

We note too that the elementary rectangular parallelepiped can distort to a general parallelepiped as shown in Fig. (3.6.2b). As is clear from our previous discussion, this distortion can be measured by the changes in angle between the coordinate lines; that is, the angle changes between x and y, y and z, and z and x line segments, which we denote by γ_{xy}, γ_{yz} and γ_{zx}, respectively.

Hence we note that to describe completely the deformation of an element at a point, we require six independent strain quantities: ϵ_x, ϵ_y, ϵ_z, γ_{xy}, γ_{yz} and γ_{zx}, as shown in the array

$$
\begin{pmatrix}
\epsilon_x & \gamma_{xy} & \gamma_{xz} \\
\gamma_{yx} & \epsilon_y & \gamma_{yz} \\
\gamma_{zx} & \gamma_{zy} & v\epsilon_z
\end{pmatrix}.
$$

We observe that this array is a symmetric array since, *by definition,* $\gamma_{xy} = \gamma_{yx}$, $\gamma_{yz} = \gamma_{zy}$, $\gamma_{xz} = \gamma_{zx}$.

Now, as in the case of stress, there exist two-dimensional cases of strain where the displacement in a particular direction, say w in the z-direction, is zero and where the remaining non-zero displacements, u and v, are functions only of x and y. Thus $u = u(x, y)i + v(x, y)j$. As previously mentioned, such a two-dimensional state of strain is called **plane strain**. From Eqs. (3.5.18), it follows that $\epsilon_z = \gamma_{yz} = \gamma_{zx} = 0$, and hence for the case of plane strain, we have

$$
\begin{pmatrix}
\epsilon_x & \gamma_{xy} \\
\gamma_{yx} & \epsilon_y
\end{pmatrix},
$$

where $\gamma_{xy} = \gamma_{yx}$.

The above arrays have the same appearance as the arrays for the symmetric second-rank tensor of Section 3 of Chapter 2. We might therefore be inclined to believe that these arrays also represent second-rank symmetric tensors. However, in order to make this assertion, we must prove that the scalar strain components transform according to the same transformation laws as the scalar stress components. In the following section, we derive the appropriate transformation law for these scalar components and will discover that the above arrays *do not* represent second-rank tensors.

3.7   Two-dimensional transformation law for infinitesimal strain components

Let us assume that the infinitesimal strains ϵ_x, ϵ_y and γ_{xy} are known at any point of a body situated in an x, y coordinate system. We derive here expressions for the strain components in any arbitrary direction (of the x–y plane) in terms of the above known strain components; we call these derived expressions, as in the case of stress, the **transformation laws**.

It is possible to derive these laws by means of two different approaches: a geometric approach and a more formal analytic approach in which we make use of the strain–displacement relations of Section 3.5.

Each approach has its advantages. However, because the geometric approach provides more physical insight into the concept of strain, we first derive the transformation law for the extensional strain from simple geometric considerations.

(a) Geometric derivation

For simplicity we examine a two-dimensional state of strain at a point, i.e., we assume that all points undergo displacements in the x- and y-directions only. Consider therefore an element PAQB with sides $\Delta x \, \Delta y$ in the neighbourhood of a point P, which undergoes displacements such that P $\rightarrow$ P*, A $\rightarrow$ A*, Q $\rightarrow$ Q* and B $\rightarrow$ B*, as shown in Fig. (3.7.1). As we have noted previously, an element may undergo both rigid-body displacements and deformation. However, since we have shown that the strains depend solely on the deformation, we disregard all rigid-body

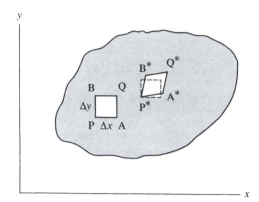

Figure 3.7.1

motion. Therefore, taking P as our reference point, we disregard the displacement $\boldsymbol{u}_P = \overrightarrow{PP^*}$ and analyze the deformation with respect to point P* in the deformed state.

Let us assume that the deformation of the element is known, i.e., the infinitesimal strains ϵ_x, ϵ_y and $\gamma_{xy} = \gamma_{yx}$ *are known at the given point* P.

We now pose the following question: if the above strains are known, what is the strain of a line segment PQ whose initial orientation (before deformation) was at some angle θ with respect to the x-axis [Fig. (3.7.2a)]? (Note that θ is taken positive in the counter-clockwise direction.)

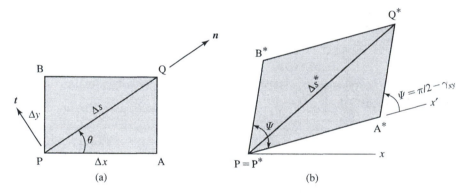

Figure 3.7.2

(a)   (b)

To this end, let the unit vector $\boldsymbol{n}$ define the initial orientation of the line segment PQ, and let the unit vector $\boldsymbol{t}$, lying in the x–y plane, define a direction perpendicular to $\boldsymbol{n}$. Further, let $\Delta s = |PQ|$. In more precise terms, we wish to determine the extensional strain ϵ_n of the segment PQ. In addition, we shall determine the change in angle between the original n- and t-directions due to the deformation of the element. Clearly, this analysis is purely a problem of geometry.

Now since the strains are known, the deformation of the rectangular element of Fig. (3.7.2a) is known and appears as in Fig. (3.7.2b). Note that we have drawn the deformed element as a parallelogram. We may justify this on the following bases: (a) the element PAQB is assumed initially to be infinitesimal and consequently the strain at the adjacent points in the neighbourhood of P must be *approximately* the same as at P, and (b) in the limit, we shall shrink the element to a point by taking $\Delta x \rightarrow 0$ and $\Delta y \rightarrow 0$.[†]

[†] Consequently, while the geometric analysis for the infinitesimal element is, in general, an approximation, it is exact only in the limit (i.e., at a point).

Having denoted the initial length of the line segment PQ by Δs, we let $\Delta s^* = |P^*Q^*|$. Now, according to Eq. (3.3.2a),

$$\Delta s^* \simeq (1 + \epsilon_n) \Delta s \tag{3.7.1a}$$

and

$$|P^*A^*| \simeq (1 + \epsilon_x) \Delta x, \tag{3.7.1b}$$

$$|P^*B^*| \simeq (1 + \epsilon_y) \Delta y. \tag{3.7.1c}$$

Furthermore, for the parallelogram $P^*A^*Q^*B^*$, we observe that $|A^*Q^*| = |P^*B^*|$ and that the angle $\psi \equiv \angle B^*P^*A^* = \pi/2 - \gamma_{xy}$ [Fig. (3.7.2b)].

We now make use of the cosine law for the triangle $P^*A^*Q^*$:

$$(\Delta s^*)^2 \simeq |P^*A^*|^2 + |A^*Q^*|^2 + 2|P^*A^*| \cdot |A^*Q^*| \cos \psi. \tag{3.7.2}$$

Noting that $\cos \psi \simeq \cos(\pi/2 - \gamma_{xy}) = \sin \gamma_{xy} \simeq \gamma_{xy}$ for infinitesimal γ_{xy} and substituting Eqs. (3.7.1) in Eq. (3.7.2) we obtain

$$(1 + \epsilon_n)^2 \Delta s^2 = (1 + \epsilon_x)^2 \Delta x^2 + (1 + \epsilon_y)^2 \Delta y^2 + 2(1 + \epsilon_x)(1 + \epsilon_y)\gamma_{xy} \Delta x \Delta y. \tag{3.7.3a}$$

Expanding

$$\left(1 + 2\epsilon_n + \epsilon_n^2\right) \Delta s^2 \simeq \left(1 + 2\epsilon_x + \epsilon_x^2\right) \Delta x^2 + \left(1 + 2\epsilon_y + \epsilon_y^2\right) \Delta y^2$$
$$+ 2(1 + \epsilon_x + \epsilon_y + \epsilon_x\epsilon_y)\gamma_{xy} \Delta x \Delta y.$$

Neglecting all second-order infinitesimals, we have

$$(1 + 2\epsilon_n) \Delta s^2 \simeq (1 + 2\epsilon_x) \Delta x^2 + (1 + 2\epsilon_y) \Delta y^2 + 2\gamma_{xy} \Delta x \Delta y \tag{3.7.3c}$$

and noting that $\Delta s^2 = \Delta x^2 + \Delta y^2$, we obtain, after dividing through by Δs^2,

$$2\epsilon_n \simeq 2\epsilon_x(\Delta x/\Delta s)^2 + 2\epsilon_y(\Delta y/\Delta s)^2 + 2\gamma_{xy}(\Delta x/\Delta s)(\Delta y/\Delta s). \tag{3.7.3d}$$

Now

$$\cos \theta = \Delta x/\Delta s, \tag{3.7.4a}$$

$$\sin \theta = \Delta y/\Delta s. \tag{3.7.4b}$$

Noting, upon taking the limit, that the approximation $\simeq$ becomes an equality, we finally obtain[†]

$$\epsilon_n = \epsilon_x \cos^2 \theta + \epsilon_y \sin^2 \theta + \gamma_{xy} \sin \theta \cos \theta. \tag{3.7.5a}$$

Using the standard trigonometric identities [see Eq. (2.6.6)], Eq. (3.7.5a) can be written in the alternate form:

$$\epsilon_n = \frac{\epsilon_x + \epsilon_y}{2} + \frac{\epsilon_x - \epsilon_y}{2} \cos 2\theta + \frac{\gamma_{xy}}{2} \sin 2\theta. \tag{3.7.5b}$$

Equations (3.7.5) thus provide a means to obtain the extensional strain ϵ_n in any given n-direction with respect to the x, y coordinate system, provided ϵ_x, ϵ_y and γ_{xy} are known. These equations thus represent the transformation law for the extensional strain from the x, y coordinate system to another coordinate system, oriented with respect to the first by an angle θ.

[†] See previous footnote.

We observe that Eqs. (3.7.5) resemble, in form, the transformation law, Eqs. (2.6.6) for the normal stress σ_n given in Chapter 2. However, a casual comparison reveals that they are somewhat different since a factor of $1/2$ appears in the γ_{xy} term in Eq. (3.7.5b), which does not appear in Eq. (2.6.6b). We shall not pursue this comment, but will return to this remark only after deriving the transformation law for the shear strain term.

Now, although the transformation law for γ_{nt} can be obtained by a similar geometric approach, the derivation is less straightforward since it leads to some rather cumbersome geometry. We therefore resort to a different analytical approach that yields the transformation laws for both extensional and shear strain components.

(b) Analytic derivation of the transformation laws

We consider a body in an x–y plane undergoing deformation where the displacement of any point $P(x, y)$ in the plane [Fig. (3.7.3a)] is given by

$$u(x, y) = ui + vj, \tag{3.7.6}$$

where $u(x, y)$ and $v(x, y)$ describe the displacement field. Then, if all strains and rotations are infinitesimal, the strain–displacement relations are [Eqs. (3.5.16) repeated here]

$$\epsilon_x = \frac{\partial u}{\partial x}, \tag{3.7.7a}$$

$$\epsilon_y = \frac{\partial v}{\partial y}, \tag{3.7.7b}$$

$$\gamma_{xy} = \frac{\partial v}{\partial x} + \frac{\partial u}{\partial y}. \tag{3.7.7c}$$

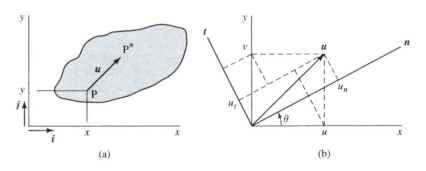

Figure 3.7.3   (a)   (b)

It is useful to note that the extensional strain component in a particular direction is given by the partial derivative of the displacement component in the given direction with respect to the coordinate in the *same* direction. On the other hand, the terms for the shear strain component are expressed in terms of partial derivatives of the displacement components in a given direction with respect to coordinates in the orthogonal direction.

Now, instead of resolving the displacement vector u into components in the x- and y-directions, we recognise that the vector may also be resolved into components in any arbitrary orthogonal directions. To this end, we first define two directions by means of the orthogonal unit vectors n and t, where n is inclined by the angle θ

with respect to the x-axis [Fig. (3.7.3b)]. Then,

$$\mathbf{u} = u_n \mathbf{n} + u_t \mathbf{t}. \tag{3.7.8a}$$

From this figure we note that the components of the vector, u_n and u_t, in the n- and t-directions, respectively, are given by

$$u_n = u \cos \theta + v \sin \theta, \tag{3.7.8b}$$

$$u_t = -u \sin \theta + v \cos \theta. \tag{3.7.8c}$$

It is clear that we may consider the inclined axes as representing a new n, t coordinate system. Hence, we may consider u_n and u_t to be functions of the coordinates n and t; i.e., $u_n = u_n(n, t)$ and $u_t = u_t(n, t)$. Therefore, analogously to Eqs. (3.7.7), we may write

$$\epsilon_n = \frac{\partial u_n}{\partial n}, \tag{3.7.9a}$$

$$\epsilon_t = \frac{\partial u_t}{\partial t}, \tag{3.7.9b}$$

$$\gamma_{nt} = \frac{\partial u_t}{\partial n} + \frac{\partial u_n}{\partial t}. \tag{3.7.9c}$$

Now, there exists also a relation between the x, y coordinate system and the n, t system. Consider, for example, an arbitrary point P, which may be represented by $P(x, y)$ in the x, y coordinate system, or by $P(n, t)$ in the n, t coordinate system [Fig. (3.7.4)]. Then from the figure, we observe that the following relations exist between the coordinate systems (x, y) and (n, t):

$$x = n \cos \theta - t \sin \theta, \tag{3.7.10a}$$

$$y = n \sin \theta + t \cos \theta; \tag{3.7.10b}$$

that is, we consider $x = x(n, t; \theta)$, and $y = y(n, t; \theta)$, where θ is a parameter. From Eqs. (3.7.10), we note that

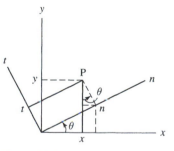

Figure 3.7.4

$$\frac{\partial x}{\partial n} = \cos \theta, \tag{3.7.11a}$$

$$\frac{\partial x}{\partial t} = -\sin \theta, \tag{3.7.11b}$$

$$\frac{\partial y}{\partial n} = \sin \theta, \tag{3.7.11c}$$

$$\frac{\partial y}{\partial t} = \cos \theta. \tag{3.7.11d}$$

Taking partial derivatives of $u_n = u_n[x(n, t), y(n, t)]$, we have, from Eq. (3.7.9a),

$$\epsilon_n = \frac{\partial u_n}{\partial n} = \frac{\partial u_n}{\partial x} \frac{\partial x}{\partial n} + \frac{\partial u_n}{\partial y} \frac{\partial y}{\partial n} \tag{3.7.12a}$$

and, making use of Eqs. (3.7.11a) and (3.7.11c), we find

$$\epsilon_n = \frac{\partial u_n}{\partial x} \cos \theta + \frac{\partial u_n}{\partial y} \sin \theta. \tag{3.7.12b}$$

Substituting from Eq. (3.7.8b), we obtain

$$\epsilon_n = \left(\frac{\partial u}{\partial x} \cos \theta + \frac{\partial v}{\partial x} \sin \theta \right) \cos \theta + \left(\frac{\partial u}{\partial y} \cos \theta + \frac{\partial v}{\partial y} \sin \theta \right) \sin \theta \quad (3.7.12c)$$

or

$$\epsilon_n = \frac{\partial u}{\partial x} \cos^2 \theta + \frac{\partial v}{\partial y} \sin^2 \theta + \left(\frac{\partial v}{\partial x} + \frac{\partial u}{\partial y} \right) \sin \theta \cos \theta \quad (3.7.13)$$

from which, using Eqs. (3.5.16), we can write

$$\epsilon_n = \epsilon_x \cos^2 \theta + \epsilon_y \sin^2 \theta + \gamma_{xy} \sin \theta \cos \theta. \quad (3.7.14)$$

We note that Eq. (3.7.14) is identical to Eq. (3.7.5a), i.e., we have rederived the transformation law for ϵ_n via this more formal, analytic approach. This approach permits us to obtain the transformation law for γ_{nt} (i.e., the change in angle between the n- and t-directions) in a very simple manner.

We first note that

$$\gamma_{nt} = \frac{\partial u_t}{\partial n} + \frac{\partial u_n}{\partial t}. \quad (3.7.15a)$$

Again, considering u_n and u_t to be functions of $[x(n, t), y(n, t)]$, taking the partial derivatives with respect to n and t and making use of Eqs. (3.7.10), we find

$$\frac{\partial u_t}{\partial n} = \frac{\partial u_t}{\partial x} \cos \theta + \frac{\partial u_t}{\partial y} \sin \theta, \qquad \frac{\partial u_n}{\partial t} = -\frac{\partial u_n}{\partial x} \sin \theta + \frac{\partial u_n}{\partial y} \cos \theta. \quad (3.7.15b)$$

Substituting in Eq. (3.7.15a), combining terms and making use of Eqs. (3.7.8), we obtain

$$\gamma_{nt} = \left(\frac{\partial u}{\partial y} + \frac{\partial v}{\partial x} \right) \cos^2 \theta - \left(\frac{\partial v}{\partial x} + \frac{\partial u}{\partial y} \right) \sin^2 \theta + 2 \left(\frac{\partial v}{\partial y} - \frac{\partial u}{\partial x} \right) \sin \theta \cos \theta, \quad (3.7.15c)$$

which, again by Eqs. (3.5.16), we can write as

$$\gamma_{nt} = \gamma_{xy}(\cos^2 \theta - \sin^2 \theta) - 2(\epsilon_x - \epsilon_y) \sin \theta \cos \theta. \quad (3.7.16)$$

(c) The infinitesimal strain tensor – two-dimensional transformation laws

It is useful to collect together the transformation laws for ϵ_n and γ_{nt} as previously derived:

$$\epsilon_n = \epsilon_x \cos^2 \theta + \epsilon_y \sin^2 \theta + \gamma_{xy} \sin \theta \cos \theta, \quad (3.7.17a)$$

$$\gamma_{nt} = \gamma_{xy}(\cos^2 \theta - \sin^2 \theta) - 2(\epsilon_x - \epsilon_y) \sin \theta \cos \theta. \quad (3.7.17b)$$

Using, as before, the standard trigonometric identities, the alternative form of these laws is

$$\epsilon_n = \frac{\epsilon_x + \epsilon_y}{2} + \frac{\epsilon_x - \epsilon_y}{2} \cos 2\theta + \frac{\gamma_{xy}}{2} \sin 2\theta, \quad (3.7.18a)$$

$$\gamma_{nt} = \gamma_{xy} \cos 2\theta - (\epsilon_x - \epsilon_y) \sin 2\theta. \quad (3.7.18b)$$

We note that Eqs. (3.7.17) and (3.7.18) resemble the transformation laws for the stress components [Eqs. (2.6.6) and (2.6.8) respectively] as derived in Chapter 2. They would be identical laws if σ_x and σ_y could be replaced by ϵ_x and ϵ_y and τ_{xy} by γ_{xy}. However, we note that in Eqs. (3.7.17) and (3.7.18) the γ_{xy} terms differ by a factor of $1/2$ with the corresponding τ_{xy} terms of Eqs. (2.6.6) and (2.6.8).

Now, we recall from Chapter 2, that a second-rank symmetric tensor is said to be one for which its scalar components transform (in two dimensions) according to the laws given by Eqs. (2.6.6)–(2.6.8). Indeed, it was precisely because the stress components obey these transformation laws that stress was said to be a second-rank symmetric tensor.

In the case of the array shown in Section 5 of this chapter, we observe that because of the factor $1/2$, the transformation laws for these scalar components do not satisfy the transformation laws for a second-rank symmetric tensor. Consequently, the array of scalar components shown in Section 5 does *not* represent a tensor.

Therefore, let us now define the quantities

$$\epsilon_{xy} = \frac{1}{2}\gamma_{xy}, \quad \epsilon_{yz} = \frac{1}{2}\gamma_{yz}, \quad \epsilon_{zx} = \frac{1}{2}\gamma_{zx}, \quad \epsilon_{nt} = \frac{1}{2}\gamma_{nt}. \quad (3.7.19)$$

Furthermore, as in the case of stress, we introduce a new notation for the extensional strain, namely, $\epsilon_{xx} \equiv \epsilon_x$, $\epsilon_{yy} \equiv \epsilon_y$ and $\epsilon_{nn} \equiv \epsilon_n$.

With this change in notation, and taking into account the definition of ϵ_{xy} etc., we write the following array:

$$\begin{pmatrix} \epsilon_{xx} & \epsilon_{xy} & \epsilon_{xz} \\ \epsilon_{yx} & \epsilon_{yy} & \epsilon_{yz} \\ \epsilon_{zx} & \epsilon_{zy} & \epsilon_{zz} \end{pmatrix}$$

where, now

$$\epsilon_{xx} \equiv \epsilon_x = \frac{\partial u}{\partial x}, \quad (3.7.20a)$$

$$\epsilon_{yy} \equiv \epsilon_y = \frac{\partial v}{\partial y}, \quad (3.7.20b)$$

$$\epsilon_{zz} \equiv \epsilon_z = \frac{\partial w}{\partial z}, \quad (3.7.20c)$$

$$\epsilon_{xy} = \frac{1}{2}\left(\frac{\partial v}{\partial x} + \frac{\partial u}{\partial y}\right), \quad (3.7.20d)$$

$$\epsilon_{yz} = \frac{1}{2}\left(\frac{\partial w}{\partial y} + \frac{\partial v}{\partial z}\right), \quad (3.7.20e)$$

$$\epsilon_{zx} = \frac{1}{2}\left(\frac{\partial u}{\partial z} + \frac{\partial w}{\partial x}\right). \quad (3.7.20f)$$

With these new definitions, the transformation laws of Eqs. (3.7.18) become

$$\epsilon_n = \frac{\epsilon_x + \epsilon_y}{2} + \frac{\epsilon_x - \epsilon_y}{2}\cos 2\theta + \epsilon_{xy}\sin 2\theta, \quad (3.7.21a)$$

$$\epsilon_{nt} = \epsilon_{xy}\cos 2\theta - \frac{\epsilon_x - \epsilon_y}{2}\sin 2\theta. \quad (3.7.21b)$$

Comparing Eqs. (3.7.21) with Eqs. (2.6.6b)–(2.6.8b), we observe that they are now *identical* since there now exists a direct correspondence:

$$\tau_{xx} \longleftrightarrow \epsilon_{xx}, \quad \tau_{yy} \longleftrightarrow \epsilon_{yy}, \quad \tau_{xy} \longleftrightarrow \epsilon_{xy},$$

$$\tau_{nn} \longleftrightarrow \epsilon_{nn}, \quad \tau_{nt} \longleftrightarrow \epsilon_{nt}.$$

Thus, Eqs. (3.7.21) represent the (two-dimensional) transformation laws for the array of scalar components shown above. The strain components of this array, as in the case of stress, are therefore components of a second-rank symmetric tensor. Hence, we have shown that strain at a point is represented by a second-rank tensor; thus we refer to the *strain tensor* at a point.

Several remarks are now in order:

- According to its definition, the shear strain $\epsilon_{xy} = \frac{1}{2}\gamma_{xy}$, for example, represents $\frac{1}{2}$ the angle change between two line segments that lay initially in the x- and y-directions *before deformation* occurred.
- Double subscripts nn appearing in the extensional strain ϵ_{nn} indicate extension of a line segment that was oriented initially in the n-direction *before deformation*. (For convenience, we sometimes will use the notation ϵ_n. Thus, in cases where the symbol ϵ appears with only one subscript, it will denote extensional strain according to the identity, $\epsilon_{nn} \equiv \epsilon_n$.)
- Having established that the strain components at a point are components of a second-rank symmetric tensor, these components have the same properties as the stress tensor. In particular, they satisfy the same invariant properties. Thus, analogously to the case of plane stress [see Eqs. (2.6.10)–(2.6.11)], the two-dimensional strain components satisfy the condition

$$\epsilon_n + \epsilon_t = \epsilon_x + \epsilon_y = I_{\epsilon 1} \quad \text{(constant)} \tag{3.7.22a}$$

and

$$\epsilon_n \epsilon_t - \epsilon_{nt}^2 = \epsilon_x \epsilon_y - \epsilon_{xy}^2 = I_{\epsilon 2} \quad \text{(constant)}, \tag{3.7.22b}$$

where $I_{\epsilon 1}$ and $I_{\epsilon 2}$ are invariants. As in the case of stress, for the three-dimensional strain tensor, there exists a third invariant.

Although we have not treated the case of three-dimensional strain, we mention here, that in that case, the first invariant is then given by

$$\epsilon_x + \epsilon_y + \epsilon_z = I_{\epsilon 1}, \tag{3.7.23}$$

that is, the sum of three extensional strain components *at a point* in any three mutually perpendicular directions is always a constant. In Section 10 of this chapter below, we will find that this invariant lends itself to a physical interpretation.

Hereafter, and throughout the book, *the expression* shear strain *will signify* ϵ_{xy}, ϵ_{nt}, *etc.; that is, it will signify one-half the angle change.*

3.8   Principal strains and principal directions of strain: the Mohr circle for strain

The extensional strain at a point ϵ_n clearly varies with the orientation θ of the n-direction since according to Eq. (3.7.21a), $\epsilon_n = \epsilon_n(\theta)$. Treating ϵ_n as a function of θ, the necessary condition for stationary values is

$$\frac{d\epsilon_n}{d\theta} = 0. \tag{3.8.1}$$

We might therefore proceed with the analysis exactly as with the case of two-dimensional stress in Chapter 2. However, in the previous section, we have observed that the two-dimensional strain components ϵ_x, ϵ_y and $\epsilon_{xy} \equiv \gamma_{xy}/2$ satisfy the same transformation law as the two-dimensional stress components σ_x, σ_y and τ_{xy}. Therefore all expressions derived for the principal directions, principal stresses, etc., of Chapter 2 (Section 7), will be the same, if we replace σ_x by ϵ_x, σ_y by ϵ_y and τ_{xy} by ϵ_{xy}. Thus, it is not necessary to rederive the expressions for the strains; we need only replace the stress terms with the corresponding strain terms. In particular, corresponding to Eq. (2.7.3) we obtain the equation

$$\tan 2\theta = \frac{\epsilon_{xy}}{\frac{\epsilon_x - \epsilon_y}{2}} \tag{3.8.2}$$

whose two relevant roots are denoted by θ_1 and θ_2 with $\theta_2 = \theta_1 + \pi/2$.

However, although the mathematics follow by analogy, it is necessary to make a distinction in interpreting these angles. While θ_1 and θ_2 of Chapter 2 denoted the orientation of the normal to the principal planes, here θ_1 and θ_2 denote the mutually perpendicular *principal directions* of the strain. Thus θ_1 denotes the direction of the line segment (existing at a point) for which the maximum value of ϵ_n occurs and θ_2 denotes the direction of the line segment which has the minimum value of ϵ_n.

Analogously to the case of stress, the shear strain $\epsilon_{nt} = 0$, where n and t lie in the two orthogonal principal (strain) directions. Thus, the right angle existing between the two orthogonal principal directions (in the undeformed state) does not change as a result of deformation; i.e., line segments lying in the body in these two orthogonal directions remain mutually perpendicular after deformation.

The stationary values of ϵ_n are called the **principal strains**. As with stress, we let ϵ_1 and ϵ_2 denote the maximum and minimum algebraic values of strain, respectively. Then corresponding to Eqs. (2.7.11), these are given by

$$\epsilon_1 = \frac{\epsilon_x + \epsilon_y}{2} + \sqrt{\left(\frac{\epsilon_x - \epsilon_y}{2}\right)^2 + \epsilon_{xy}^2} \tag{3.8.3a}$$

and

$$\epsilon_2 = \frac{\epsilon_x + \epsilon_y}{2} - \sqrt{\left(\frac{\epsilon_x - \epsilon_y}{2}\right)^2 + \epsilon_{xy}^2}. \tag{3.8.3b}$$

Thus, for any given values of ϵ_x, ϵ_y and ϵ_{xy}, we find that a line segment oriented at an angle θ_1 with respect to the x-axis will undergo the largest extension, ϵ_1. Furthermore, the smallest extensional strain (which, if $\epsilon_n < 0$, signifies a contraction) will occur for a line segment defined by $\theta_2 = \theta_1 + \pi/2$. This is shown in Fig. (3.8.1a) where the two principal directions are defined by the lines n_1 and n_2. (We note here that the relevant root of θ_1 lies in the quadrant as defined by the criteria established in Chapter 2, for stress: namely here, according to the sign of ϵ_{xy}.)

We note that the angle between the two orthogonal principal directions remains a right angle since the shear strain $\epsilon_{12} = 0$. Thus, line segments PC and PD (which lie originally in the directions n_1 and n_2 respectively) lie, after deformation, in the perpendicular directions n_1^* and n_2^* [Fig. (3.8.1a)]. All other sets of mutually perpendicular line segments at a point will not remain orthogonal after deformation since then, $\epsilon_{nt} \neq 0$. Analogously with the case of stress, the largest and smallest shear strain will occur for two line segments [denoted by n_{s1} and n_{s2} in Fig. (3.8.1b)], which are oriented at 45° with respect to the principal directions. Letting $n_{s1} \rightarrow n_{s1}^*$, $n_{s2} \rightarrow n_{s2}^*$, we note that n_{s1}^* and n_{s2}^*, the directions after deformation, are no longer mutually perpendicular [Fig. (3.8.1b)].

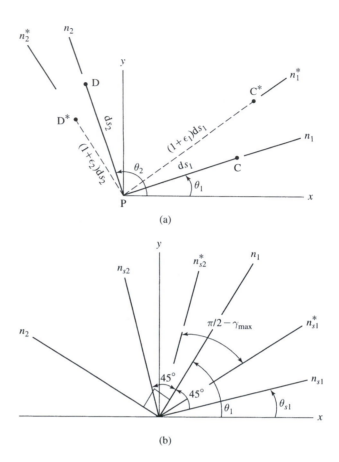

(b)

As in the case of stress, the transformation laws lead to a parametric representation analogous to Eqs. (2.7.25), namely

$$(\epsilon_n - a)^2 + \epsilon_{nt}^2 = R^2 \tag{3.8.4a}$$

with

$$R = \sqrt{\left(\frac{\epsilon_x - \epsilon_y}{2}\right)^2 + \epsilon_{xy}^2} \tag{3.8.4b}$$

and where $a = \frac{\epsilon_x + \epsilon_y}{2}$.

Thus one can construct a Mohr circle for strain with radius R and with the centre of the circle lying on the ϵ_n-axis at coordinates $[(\epsilon_x + \epsilon_y)/2, 0]$ in the ϵ_n–ϵ_{nt} space. The construction follows exactly as with the Mohr circle for stress [Fig. (3.8.2)].

The principal strains ϵ_1 and ϵ_2 are then given by

$$\epsilon_1 = \frac{\epsilon_x + \epsilon_y}{2} + R, \tag{3.8.5a}$$

$$\epsilon_2 = \frac{\epsilon_x + \epsilon_y}{2} - R. \tag{3.8.5b}$$

We illustrate the analysis by means of two typical examples.

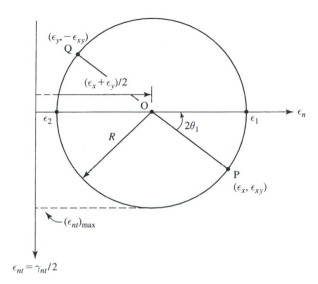

Figure 3.8.2

Example 3.11: A plate is subjected to shear forces such that at a point, the strains are $\epsilon_x = \epsilon_y = 0$ and $\epsilon_{xy} = 2k$ (where $0 < k \ll 1$). Determine (a) the extensional strain ϵ_n for a line segment oriented at $\theta = 30°$ with respect to the x-axis, (b) the shear strain ϵ_{nt} and (c) the principal strains and directions at this point. Draw the Mohr circle representing the state of strain at this point.

Solution:

(a) From Eq. (3.7.21a), the extensional strain in the n-direction is

$$\epsilon_n = \epsilon_{xy} \sin 60° = k\sqrt{3}.$$

Similarly, the strain ϵ_t in the t-direction (with $\theta = 30 + 90 = 120°$) is

$$\epsilon_t = \epsilon_{xy} \sin 240° = -k\sqrt{3}.$$

(b) The shear strain ϵ_{nt}, obtained from Eq. (3.7.21b) with $\theta = 30°$, is $\epsilon_{nt} = k$.
(c) The principal strains, given by Eqs. (3.8.3), are

$$\epsilon_1 = \epsilon_{xy} = 2k, \qquad \epsilon_2 = -\epsilon_{xy} = -2k.$$

The principal directions are given by Eq. (3.8.2), namely

$$\tan 2\theta = \frac{\epsilon_{xy}}{\frac{\epsilon_x - \epsilon_y}{2}} = \frac{2k}{0} \to \infty.$$

Since $\epsilon_{xy} > 0$, $2\theta_1 = 90°$ and therefore $\theta_1 = 45°$. The direction of n_2 is then given by $\theta_2 = 135°$.

The Mohr circle representing the state of strain is given in Fig. (3.8.3), which we recognize to be a strain state in pure shear.   □

Example 3.12: The state of plane strain at a point is given by $\epsilon_x = 10 \times 10^{-3}$, $\epsilon_y = -6 \times 10^{-3}$, $\epsilon_{xy} = 8 \times 10^{-3}$. Determine (a) the strain ϵ_n of a line segment at the point inclined at $\theta = 30°$ with respect to the x-axis, (b) the change in angles between two line segments originally oriented at 30° and 120° with respect to the x-axis and (c) the principal strains and directions.

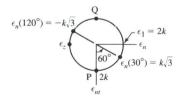

Figure 3.8.3

Solution:

(a) From the transformation law, Eq. (3.7.21a), with $\theta = 30°$,

$$\epsilon_n = \frac{\epsilon_x + \epsilon_y}{2} + \frac{\epsilon_x - \epsilon_y}{2} \cos 2\theta + \epsilon_{xy} \sin 2\theta$$

$$= \left[\frac{1}{2}(4) + \frac{1}{2}(16)(0.5) + 8(\sqrt{3}/2)\right] \times 10^{-3} = 12.93 \times 10^{-3}.$$

(b) The shear strain ϵ_{nt} between the two line segments originally at $30°$ and $120°$ with respect to the x-axis is $\epsilon_{nt}(\theta = 30°)$. Noting that $\sin 2\theta = \sqrt{3}/2$ and $\cos 2\theta = 0.5$, Eq. (3.7.21b) yields

$$\epsilon_{nt} = \epsilon_{xy} \cos 2\theta - \frac{\epsilon_x - \epsilon_y}{2} \sin 2\theta = \left[8(0.5) - \frac{16}{2}(\sqrt{3}/2)\right] \times 10^{-3}$$

$$= -2.93 \times 10^{-3}.$$

Therefore, the angle between these two line segments *increases* by 5.86×10^{-3} rad $= 0.34°$.

The line segments (n and t) before and after deformation (denoted by n^* and t^*) are shown (exaggerated) symbolically in Fig. (3.8.4a).

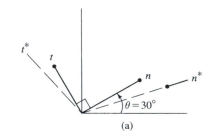

(a)

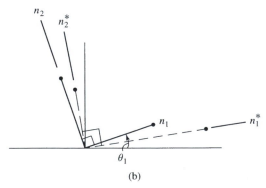

(b)

Figure 3.8.4

(c) The principal strains ϵ_1 and ϵ_2, given by Eqs. (3.8.3), are

$$\epsilon_{1,2} = \left[\frac{4}{2} \pm \sqrt{(16/2)^2 + 8^2}\right] \times 10^{-3} = [2 \pm \sqrt{128}] \times 10^{-3},$$

or

$$\epsilon_1 = 13.31 \times 10^{-3}, \qquad \epsilon_2 = -9.31 \times 10^{-3}.$$

The principal directions are given by the roots of Eq. (3.8.2), namely $\tan 2\theta = \frac{\epsilon_{xy}}{\frac{\epsilon_x - \epsilon_y}{2}} = \frac{8}{8} = 1$. Therefore $\theta_1 = 22.5°$.

The deformation of the line segments lying originally in the principal directions (n_1 and n_2) are shown symbolically in Fig. (3.8.4b). Note that these line segments remain orthogonal after deformation as opposed to the line segments of Fig. (3.8.4a).

The Mohr circle representing the state of strain at this point is given in Fig. (3.8.5).

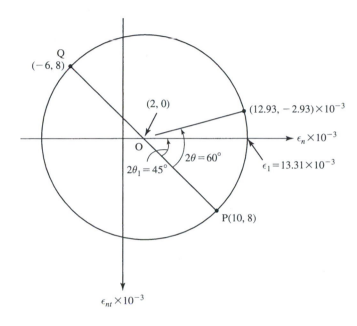

Figure 3.8.5

□

3.9 The strain rosette

One often wishes to obtain the two-dimensional state of strain at a point of a body experimentally in a laboratory. While it is impossible to measure the strain *at a point*, it is possible to measure the elongation or contraction of a short line segment in the vicinity of a point when a body undergoes deformation. In this case, we assume the strain state to be constant in the neighbourhood of the point.

We consider here the case of plane strain in the x–y plane. Now, to specify the state of strain for this case, one usually must know ϵ_x, ϵ_y and ϵ_{xy}. However, it is very difficult to measure experimentally the change in angle between two line segments. Nevertheless, as we shall see, it is possible to find the above three components by measuring the *extensional* strain components in any three arbitrary directions lying in the plane.

To show this, we recall the two-dimensional transformation law for strain, given in the form of Eq. (3.7.17a), namely

$$\epsilon_n(\theta) = \epsilon_x \cos^2 \theta + \epsilon_y \sin^2 \theta + 2\epsilon_{xy} \sin \theta \cos \theta, \qquad (3.9.1)$$

where $2\epsilon_{xy}$ has been substituted for γ_{xy}.

Assume now that we are able to measure the extensional strain of three arbitrary line segments a, b, c, as shown in Fig. (3.9.1). We denote the orientation of these

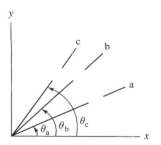

Figure 3.9.1

segments with respect to the x-axis by θ_a, θ_b, θ_c, respectively. Then it follows from Eq. (3.9.1) that

$$\epsilon_a \equiv \epsilon_n(\theta_a) = \epsilon_x \cos^2 \theta_a + \epsilon_y \sin^2 \theta_a + 2\epsilon_{xy} \sin \theta_a \cos \theta_a, \qquad (3.9.2a)$$

$$\epsilon_b \equiv \epsilon_n(\theta_b) = \epsilon_x \cos^2 \theta_b + \epsilon_y \sin^2 \theta_b + 2\epsilon_{xy} \sin \theta_b \cos \theta_b, \qquad (3.9.2b)$$

$$\epsilon_c \equiv \epsilon_n(\theta_c) = \epsilon_x \cos^2 \theta_c + \epsilon_y \sin^2 \theta_c + 2\epsilon_{xy} \sin \theta_c \cos \theta_c. \qquad (3.9.2c)$$

Now the values of ϵ_a, ϵ_b and ϵ_c as well as the orientation θ_a, θ_b and θ_c are known. Hence, we may consider Eqs. (3.9.2) to be three simultaneous equations in the three unknowns, ϵ_x, ϵ_y and ϵ_{xy}.

Various devices exist in the laboratory to determine experimentally the strains of line segments in three given directions. Such devices are called **strain rosettes**. Standard strain rosettes exist to measure strains along line segments that are oriented, for example, at $30°$, $45°$ or $60°$ with respect to each other [Fig. (3.9.2a,b,c)]. Based on such measurements, the strain components in x- and y-coordinate directions (ϵ_x, ϵ_y and ϵ_{xy}) can be found as is illustrated in the following example.

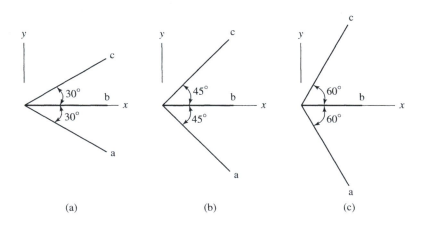

Figure 3.9.2

(a)   (b)   (c)

Example 3.13: Assume that strains are measured by a $45°$ rosette such that b is oriented along the x-axis and line segments a and c are at $-45°$ and $45°$ with respect to the x-axis [Fig. (3.9.2b)]. Let ϵ_a, ϵ_b and ϵ_c denote the measured strains. Determine the strain components ϵ_x, ϵ_y and ϵ_{xy}.

Solution: Since b is along the x-axis, then by definition, $\epsilon_x = \epsilon_b$.

Along a: $\sin(-45°) = -\sqrt{2}/2$, $\cos(-45°) = \sqrt{2}/2$

Along c: $\sin(45°) = \sqrt{2}/2 = \cos(45°) = \sqrt{2}/2$.

Substituting in Eqs. (3.9.2a and c):

$$0.5\epsilon_y - \epsilon_{xy} = \epsilon_a - 0.5\epsilon_b$$

$$0.5\epsilon_y + \epsilon_{xy} = \epsilon_c - 0.5\epsilon_b$$

Solving for the remaining components, ϵ_y and ϵ_{xy},

$$\epsilon_{xy} = \frac{1}{2}(\epsilon_c - \epsilon_a), \qquad \epsilon_y = \epsilon_a + \epsilon_c - \epsilon_b.$$

Thus the three components in the x, y coordinate directions have been obtained. □

3.10 Volumetric strain–dilatation

In our previous discussion, we defined two types of strain at a point: extensional strain and shear strain. Analogous to the definition of the extensional strain as the ratio of change in length to original length of a line segment, one may define a measure of the change of volume of an element existing at a point as the ratio of change of volume of an element to the original volume.

To this end, consider an elementary rectangular parallelepiped $dx\,dy\,dz$ as shown in Fig. (3.10.1). The volume of this element is then

$$dV = dx\,dy\,dz. \tag{3.10.1}$$

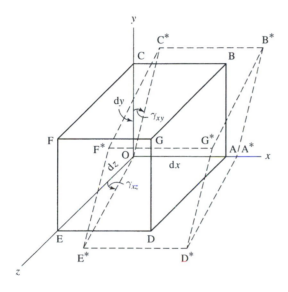

Figure 3.10.1

Let us assume that this element undergoes deformation where $A \to A^*$, $B \to B^*$, $C \to C^*$, $D \to D^*$, etc. such that $dx \to dx^*$, $dy \to dy^*$, $dz \to dz^*$. Then, by Eq. (3.3.2a),

$$dx^* = |OA^*| = (1 + \epsilon_x)\,dx, \tag{3.10.2a}$$

$$dy^* = |OC^*| = (1 + \epsilon_y)\,dy, \tag{3.10.2b}$$

$$dz^* = |OE^*| = (1 + \epsilon_z)\,dz. \tag{3.10.2c}$$

In addition to the changes of length of the sides, the element also distorts so that right angles no longer remain right angles, as shown in Fig. (3.10.1). For simplicity, let us assume that the plane $OA^*B^*C^*$ lies in the x–y plane. The volume dV^* of the deformed element is then given by

$$dV^* = [\text{Area}(OA^*B^*C^*)][dz^* \cos \gamma_{xz}] \tag{3.10.3a}$$

or

$$dV^* = [dx^*\,dy^* \cos \gamma_{xy}][dz^* \cos \gamma_{xz}] = dx^*\,dy^*\,dz^* \cos \gamma_{xy} \cos \gamma_{xz}. \tag{3.10.3b}$$

Substituting from Eqs. (3.10.2),

$$dV^* = (1 + \epsilon_x)(1 + \epsilon_y)(1 + \epsilon_z)\,dx\,dy\,dz \cos \gamma_{xy} \cos \gamma_{xz}. \tag{3.10.4}$$

We recall the Taylor series expansion for $\cos x$,

$$\cos x = 1 - \frac{x^2}{2!} + \frac{x^4}{4!} - \cdots .$$

Then, if $|\gamma_{xy}| \ll 1$, $|\gamma_{xz}| \ll 1$, upon dropping second-order infinitesimals, it follows that $\cos \gamma_{xy} \simeq 1$, $\cos \gamma_{xz} \simeq 1$. Hence after expanding Eq. (3.10.4), we obtain

$$dV^* = (1 + \epsilon_x + \epsilon_y + \epsilon_z + \epsilon_x \epsilon_y + \epsilon_y \epsilon_z + \epsilon_z \epsilon_x + \epsilon_x \epsilon_y \epsilon_z)\, dx\, dy\, dz. \quad (3.10.5a)$$

Dropping again all second-order infinitesimal terms, we obtain finally,

$$dV^* = (1 + \epsilon_x + \epsilon_y + \epsilon_z)\, dx\, dy\, dz. \quad (3.10.5b)$$

Now, analogously to the definition of extensional strain, we define Δ, the measure of volumetric strain, as

$$\Delta = \frac{dV^* - dV}{dV}. \quad (3.10.6)$$

Substituting Eqs. (3.10.1) and (3.10.5b),

$$\Delta = \frac{(1 + \epsilon_x + \epsilon_y + \epsilon_z)\, dx\, dy\, dz - dx\, dy\, dz}{dx\, dy\, dz} \quad (3.10.7a)$$

or

$$\Delta = \epsilon_x + \epsilon_y + \epsilon_z. \quad (3.10.7b)$$

This measure of volumetric strain, Δ, is called the **dilatation**. We emphasise that this expression for the dilatation is valid only for infinitesimal strains and rotations.

Using Eqs. (3.5.1a)–(3.5.1c), we note that for a given displacement field in a body, $\boldsymbol{u}(x, y, z) = u\boldsymbol{i} + v\boldsymbol{j} + w\boldsymbol{k}$, the dilatation Δ is given by

$$\Delta = \frac{\partial u}{\partial x} + \frac{\partial v}{\partial y} + \frac{\partial w}{\partial z}. \quad (3.10.8a)$$

This simple expression may be written in vector form as

$$\Delta = \nabla \cdot \boldsymbol{u}, \quad (3.10.8b)$$

where $\nabla = \frac{\partial}{\partial x}\boldsymbol{i} + \frac{\partial}{\partial y}\boldsymbol{j} + \frac{\partial}{\partial z}\boldsymbol{k}$. Thus, for a given displacement field, $\boldsymbol{u}(x, y, z)$, the dilatation is given by the divergence of $\boldsymbol{u}$.

Finally, upon comparing Eq. (3.10.7b) with Eq. (3.7.23), we observe that the invariant $I_{\epsilon 1}$ is equal precisely to the dilatation Δ. Hence this invariant has an immediate physical interpretation: namely, it represents the volumetric strain at a point. Since the invariant is independent of any particular coordinate system, we note that the volumetric strain at a point is a scalar quantity that does not depend on the coordinate system that has been chosen. Indeed, from a simple physical reasoning, we might have concluded that volumetric strain cannot depend on any particular directions.

PROBLEMS

Section 3

3.1: A 200-cm long copper wire is heated non-uniformly causing an extensional strain which is linearly proportional to the distance from one end of the wire. If the elongation of the wire is 1 cm. (a) What is the average extensional strain in the wire?

(b) What is the largest extensional strain in the wire? (c) What is the extensional strain at the centre of the wire? (d) If one end of the wire is fixed and x is the distance from this end, what is the displacement $u(x)$ at any point x?

3.2: A rigid rod AD is pinned at A and supported by a wire BC, as finite in Fig. (3P.2a), and subjected to a load P at D. Due to the load, the rod undergoes a finite rotation θ as shown in Fig. (3P.2b). (a) Determine the average strain $\bar{\epsilon}$ in the wire as a function of θ. (b) If $\theta \ll 1$, show that the average strain is given by $\bar{\epsilon} = \frac{L\theta}{a}$.

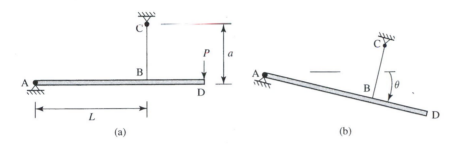

(a)   (b)   **Figure 3P.2**

3.3: A heavy vertical cable ABC of length $L = 200$ m is attached at the top, A, to a crane. Due to its own weight, the strain at any intermediate point B is proportional to the distance below B. When carrying a given load P at C, the strain is increased uniformly by 0.001. If due to its own weight and the load P, the cable undergoes a change in length $\Delta L = 60$ cm, what is the strain at the top?

3.4:* A segment of wire AB lies along the line $y = mx/b$, as shown in Fig. (3P.4). The wire is strained and displaced to lie along the line $y = nx/b\,(m < n)$ in such a way that any point originally at the coordinate x/b is displaced to $x^*/b = \frac{1}{2}(x/b)^2$. (a) Show that the extensional strain ϵ_n at any point of the wire is given by

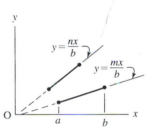

Figure 3P.4

$$\epsilon_n(x) = \sqrt{\frac{1+n^2}{1+m^2}}\left(\frac{x}{b}\right) - 1,$$

where x is the original coordinate of the point. (b) Determine the average strain in the wire.

3.5:* A segment of wire AB lies along a parabola $y = ax^2$. The wire is stretched to the shape of a parabola $y = bx^2$ $(b > a)$ such that any point originally at x lies at bx/a [see Fig. (3P.5)]. Determine the extensional strain ϵ_n as a function of x.

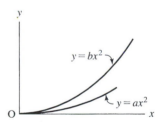

Figure 3P.5

3.6: The definition of extensional strain ϵ_n at a point P, given by Eq. (3.3.1), is known as a **Lagrangian** definition (since the reference is with respect to the original length Δs). On the other hand, one might define extensional strain with reference to the deformed length Δs^* (i.e. in a **Eulerian** sense) as

$$\tilde{\epsilon}_n(P) = \lim_{\substack{Q \to P \\ \Delta s^* \to 0}} \frac{\Delta s^* - \Delta s}{\Delta s^*},$$

where $Q^* \to P^*$ as $Q \to P$ [see Fig. (3.3.1b)]. Show that

$$\epsilon_n(P) - \tilde{\epsilon}_n(P) = \epsilon_n(P)\,\tilde{\epsilon}_n(P),$$

and hence if both $\epsilon_n(P) \ll 1$ and $\tilde{\epsilon}_n(P) \ll 1$, the difference of these two definitions is small, i.e. of second order.

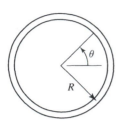

Figure 3P.7

3.7: A closed circular wire, as shown in **Fig.** (3P.7), is heated non-uniformly such that the extensional strain is given by $\epsilon_\theta = k\cos^2\theta$. Determine the increase in length of the wire.

3.8: The average circumference of the earth (considered as a perfect sphere) is assumed to be $L = 40,102$ km $= 40,102,000$ m. Imagine that a rope of length L is stretched to a length $L^* = 40,102,006$ m and then wound around the earth (at the equator) in the shape of a perfect circle. (a) Calculate the average strain in the rope; (b) Prior to making any further calculations, estimate intuitively which creature could crawl (or hop) under the rope: (i) an ant, (ii) a frog, (iii) a cat or (iv) a human being? (c) Based on a calculation, check the proper answer to (b).

3.9: A circular hoop 50 cm in diameter is heated uniformly so that the enclosed area is increased by 0.5π cm². To a first-order approximation, calculate the average extensional strain in the hoop.

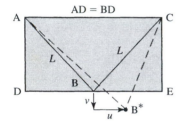

Figure 3P.10

3.10: Point B of a plate is displaced to B*, where the horizontal and vertical components of displacement are u and v, respectively, as shown in Fig. (3P.10). (a) Express the average extensional strains in AB and BC, in terms of u, v and L; (b) Determine an approximate expression for these average strains if $u \ll L$ and $v \ll L$.

3.11: A straight wire AB of length L is stretched and displaced to the position A*B*, as shown in Fig. (3P.11). Denoting the horizontal and vertical displacement components of point A and B as u_A, v_A and u_B and v_B, respectively, show that if $u_A \ll L$, $v_A \ll L$, $u_B \ll L$ and $v_B \ll L$, the average extensional strain in the wire is given by

$$\bar{\epsilon} = \frac{u_B - u_A}{L}\cos\alpha + \frac{v_B - v_A}{L}\sin\alpha.$$

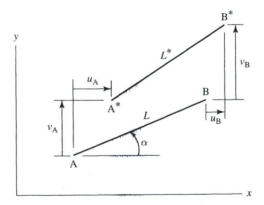

Figure 3P.11

Section 4

3.12: A thin triangular plate ABC, with edge AC fixed, is deformed to a shape AB*C, as shown in Fig. (3P.12). The uniform extensional strains along the n- and t-directions are given by $\epsilon_n = 0.006$ and $\epsilon_t = 0.005$, respectively. Noting that the strains are infinitesimal, determine, to first-order approximation, the change in angle, γ_{nt}, due to shear at point B.

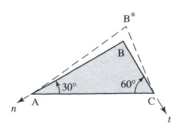

Figure 3P.12

3.13: The rectangular plate OABC, having dimensions $a \times b$, is deformed uniformly to OA*B*C (i.e., with constant strain $\epsilon_x = 0.04$ in the x-direction), as shown in Fig. (3P.13). Compute, to first-order approximation, the change in angle, γ_{nt}, at point P.

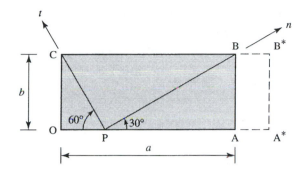

Figure 3P.13

3.14:* An L-shaped plate lying in the x–y plane, Fig. (3P.14a), undergoes uniform extensional strains ϵ_x and ϵ_y at all points such that all straight lines within the plate remain straight and all points along the x- and y-axes remain on the axes. Assuming that ϵ_x and ϵ_y are infinitesimals, determine the change in angle, γ_{nt}. Express the answer in terms of ϵ_x, ϵ_y and θ.

3.15: (a) Determine γ_{nt} of Problem 3.14, assuming that ϵ_x and ϵ_y are *not* necessarily infinitesimals. Express the answer in terms of ϵ_x, ϵ_y, a and b. (b) Simplify the expression obtained in (a) for the case where ϵ_x and ϵ_y are infinitesimals, and express the answer in terms of ϵ_x, ϵ_y and θ. *Note:* The following relations may prove to be useful:

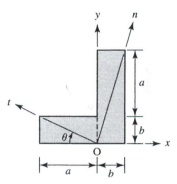

$$\tan^{-1}(x) \pm \tan^{-1}(y) = \tan^{-1}\left[\frac{x \pm y}{1 \mp xy}\right]; \qquad \tan^{-1}(x) \sim x \quad (x \ll 1).$$

Figure 3P.14

Section 5

3.16: In the following problems, points of a plate $(0 \leq x \leq 1, \ 0 \leq y \leq 1)$ lying in the x–y plane undergo displacements u and v in the x- and y-directions, respectively. For each case, (i) compute the strains $\epsilon_x \equiv \epsilon_{xx}$, $\epsilon_y \equiv \epsilon_{yy}$ and the change in angle, γ_{xy}; (ii) plot the deformed shape of the plate if $a = b = c = 0.1$ and indicate the coordinates of the corners of the deformed plate.
   (a) $u = ax + by, \ v = bx + cy, \ 0 < a, b, c \ll 1$
   (b) $u = ax, \ v = 0, \ 0 < a \ll 1$
   (c) $u = ax + bxy, \ v = -2ax, \ 0 < a, b \ll 1$.

3.17:* A square plate ($L \times L$), as shown in Fig. (3P.17a), is deformed as shown in Fig. (3P.17b) such that (i) points along the edges AB and AD do not move, (ii) CD is stretched uniformly in such a way that no point of the plate displaces in the y-direction and (iii) all vertical lines of the plate remain straight lines. Determine ϵ_x and γ_{xy} at any point of the plate in terms of δ, L and the original coordinates of the point.

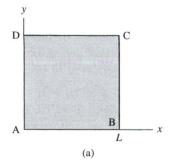

(a)

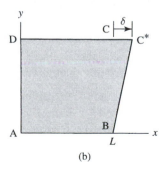

(b)

Figure 3P.17

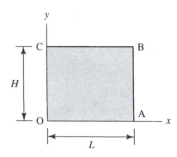

Figure 3P.18

3.18: The rectangular plate shown in Fig. (3P.18) undergoes, while deforming to OA*B*C*, changes in angle given by

$$\gamma_{xy} = ax/L + b/H + c(x/L)(y/H),$$

where $a, b, c \ll 1$. Compute, to first-order approximation, the difference in length between the lines O*A* and B*C* if all horizontal lines remain parallel to the x-axis.

3.19: Given a square plate ($L \times L$) lying in the x–y plane, which is deformed such that the extensional strains and changes in angle are given respectively by

$$\epsilon_x = \frac{ay}{L}, \qquad \epsilon_y = \frac{bx}{L}, \qquad \gamma_{xy} = \frac{ax}{L} + \frac{by}{L},$$

where $0 < a \ll 1$ and $0 < b \ll 1$ are known constants. The following boundary conditions on the displacements are specified : $u(0, 0) = v(0, 0) = 0$, $u(0, L) = e$. Determine the displacements $u(x, y)$ and $v(x, y)$ in the x- and y-directions, respectively. Express the answers in terms of a, b, e and L.

Section 7

Note: The symbol μ appearing in problems of this and subsequent sections denotes 10^{-6} – 'micron'; e.g., $\epsilon = 400\mu \equiv 400 \times 10^{-6} = 0.004$.

3.20: Verify the expression for the second invariant of (plane) strain given by Eq. (3.7.22b).

3.21: At a given point in a body in a state of plane strain with $\epsilon_{xy} = 2 \times 10^{-3}$, the ratio R of the invariants, $R \equiv \frac{I_{\epsilon_2}}{I_{\epsilon_1}} = 4 \times 10^{-3}$. If it is known that the stress $\epsilon_y = -4\epsilon_x$, determine possible values of ϵ_x.

3.22:* At a given point in a plate lying in the x–y plane with $\epsilon_z = \epsilon_{zx} = \epsilon_{zy} = 0$, the ratio R of the invariants, $R \equiv \frac{I_{\epsilon_2}}{I_{\epsilon_1}}$, is found to be equal to 4×10^{-3}. If it is known that the strain $\epsilon_y = -4\epsilon_x$, determine the possible range of values of ϵ_{xy} that can exist if this is to represent a state of plane strain.

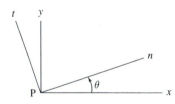

Figure 3P.23

3.23: On a plane passing through an arbitrary point P, two rectangular Cartesian systems, (x, y) and (n, t), are constructed as shown in Fig. (3P.23). For each of the plane strain cases listed below, determine the required quantities.

(a) $\epsilon_x = 200\mu, \epsilon_y = 400\mu, \epsilon_{xy} = 400\mu; \theta = 30°$. Find $\epsilon_n, \epsilon_t, \epsilon_{nt}$.
(b) $\epsilon_x = -400\mu, \epsilon_y = 0, \epsilon_{xy} = 300\mu; \theta = -30°$. Find $\epsilon_n, \epsilon_t, \epsilon_{nt}$.
(c) $\epsilon_x = 0, \epsilon_y = 0, \epsilon_{xy} = 300\mu; \theta = 45°$. Find $\epsilon_n, \epsilon_t, \epsilon_{nt}$.
(d) $\epsilon_x = 1.20 \times 10^{-3}, \epsilon_y = 0.80 \times 10^{-3}, \epsilon_{xy} = -0.80 \times 10^{-3}; \theta = 120°$. Find $\epsilon_n, \epsilon_t, \epsilon_{nt}$.
(e) $\epsilon_n = -100\mu, \epsilon_t = -50\mu, \epsilon_{nt} = 100\mu; \theta = 30°$. Find $\epsilon_x, \epsilon_y, \epsilon_{xy}$.
(f) $\epsilon_x = 0.20 \times 10^{-3}, \epsilon_y = 0.10 \times 10^{-3}, \epsilon_n = 0.05 \times 10^{-3}; \theta = 45°$. Find $\epsilon_{xy}, \epsilon_t, \epsilon_{nt}$.
(g) $\epsilon_n = 100\mu, \epsilon_y = 200\mu, \epsilon_{nt} = 0; \theta = 60°$. Find $\epsilon_x, \epsilon_{xy}, \epsilon_t$.
(h) $\epsilon_n = 100\mu, \epsilon_y = -200\mu, \epsilon_{nt} = 0; \theta = 60°$. Find $\epsilon_x, \epsilon_{xy}, \epsilon_t$.

3.24: Let n, s and t be three directions in a given x–y plane such that the n-direction lies along the x-axis, as shown in Fig. (3P.24). (a) Determine ϵ_{xy} in terms of ϵ_n, ϵ_s and ϵ_t if $\alpha = 45°$. (b) Determine ϵ_{xy} in terms of ϵ_n, ϵ_s and ϵ_t if $\alpha = 60°$.

3.25:* For each of the following plane strain states at a point lying in the x–y plane of Fig. (3P.23), determine the angle θ of the direction n with respect to the x-axis.

(a) $\epsilon_x = \epsilon_y = 100\mu, \epsilon_{xy} = 500\mu, \epsilon_n = 400\mu$.
(b) $\epsilon_x = 100\mu, \epsilon_y = -100\mu, \epsilon_{xy} = 150\mu, \epsilon_n = 0$.
(c) $\epsilon_x = -\epsilon_y = -\epsilon_{xy} = c$, ($c$ = constant), $\epsilon_n = 0$.

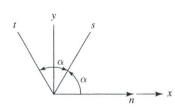

Figure 3P.24

Section 8

3.26:   Determine the principal strains and the corresponding principal directions with respect to the x–y coordinates of Fig. (3P.23) for the following given two-dimensional strain states. Draw the associated Mohr circle and show all critical quantities, i.e. ϵ_1, ϵ_2, $2\theta_1$, $2\theta_2$. Note the changes due to the $\pm$ values of the various strain components.

(a) $\epsilon_x = 20\mu$, $\epsilon_y = 40\mu$, $\epsilon_{xy} = 60\mu$.
(b) $\epsilon_x = 20\mu$, $\epsilon_y = 40\mu$, $\epsilon_{xy} = -60\mu$.
(c) $\epsilon_x = 20\mu$, $\epsilon_y = -40\mu$, $\epsilon_{xy} = 60\mu$.
(d) $\epsilon_x = 20\mu$, $\epsilon_y = -40\mu$, $\epsilon_{xy} = -60\mu$.
(e) $\epsilon_x = -20\mu$, $\epsilon_y = 40\mu$, $\epsilon_{xy} = 60\mu$.
(f) $\epsilon_x = -20\mu$, $\epsilon_y = 40\mu$, $\epsilon_{xy} = -60\mu$.
(g) $\epsilon_x = -20\mu$, $\epsilon_y = -40\mu$, $\epsilon_{xy} = 60\mu$.
(h) $\epsilon_x = -20\mu$, $\epsilon_y = -40\mu$, $\epsilon_{xy} = -60\mu$.

3.27:   For each of the plane strain cases listed below (with $\epsilon_z = \epsilon_{xz} = \epsilon_{yz} = 0$), (i) determine the principal strains ϵ_1 and ϵ_2, (ii) determine the principal directions as defined by θ_1 and θ_2 and (iii) sketch the appropriate Mohr circle showing ϵ_1, ϵ_2, $2\theta_1$ and $2\theta_2$ on the circle.

(a) $\epsilon_x = 60\mu$, $\epsilon_y = 0$, $\epsilon_{xy} = 40\mu$.
(b) $\epsilon_x = 0.20 \times 10^{-3}$, $\epsilon_y = -0.20 \times 10^{-3}$, $\epsilon_{xy} = -0.20 \times 10^{-3}$.
(c) $\epsilon_x = 900\mu$, $\epsilon_y = 100\mu$, $\epsilon_{xy} = 200\mu$.
(d) $\epsilon_x = 0.40 \times 10^{-3}$, $\epsilon_y = 0.80 \times 10^{-3}$, $\epsilon_{xy} = -0.60 \times 10^{-3}$.
(e) $\epsilon_x = -200\mu$, $\epsilon_y = -100\mu$, $\epsilon_{xy} = 200\mu$.
(f) $\epsilon_x = 2.0 \times 10^{-3}$, $\epsilon_y = 0.5 \times 10^{-3}$, $\epsilon_{xy} = -0.5 \times 10^{-3}$.
(g) $\epsilon_x = -120\mu$, $\epsilon_y = 40\mu$, $\epsilon_{xy} = -20\mu$.
(h) $\epsilon_x = 240\mu$, $\epsilon_y = 0$, $\epsilon_{xy} = 120\mu$.
(i) $\epsilon_x = -0.20 \times 10^{-3}$, $\epsilon_y = 0.10 \times 10^{-3}$, $\epsilon_{xy} = 0.32 \times 10^{-3}$.
(j) $\epsilon_x = 0$, $\epsilon_y = 240\mu$, $\epsilon_{xy} = 120\mu$.

3.28:   Let n, s and t be three unit vectors lying in a principal plane as shown in Fig. (3P.24).

(a) If $\epsilon_n = 100\mu$, $\epsilon_s = 50\mu$ and $\epsilon_t = 20\mu$, determine the principal strains and directions (with respect to the vector n) if $\alpha = 45°$.
(b) If $\epsilon_n = 100\mu$, $\epsilon_s = -20\mu$ and $\epsilon_t = 60\mu$, determine the principal strains and directions (with respect to the vector n) if $\alpha = 60°$.

3.29:* States of plane strain at a point, lying in the x–y plane of Fig. (3P.23), are given as follows:

(a) $\epsilon_x = 80\mu$, $\epsilon_y = -120\mu$. If $\epsilon_1 = 220\mu$, determine ϵ_2, θ_1 and θ_2 (i) if it is known that $\epsilon_{xy} > 0$ and (ii) if it is known that $\epsilon_{xy} < 0$.
(b) $\epsilon_x = 80\mu$, $\epsilon_y = 120\mu$. If $\epsilon_1 = 220\mu$, determine ϵ_2, θ_1 and θ_2 (i) if it is known that $\epsilon_{xy} > 0$ and (ii) if it is known that $\epsilon_{xy} < 0$.
(c) $\epsilon_y = 40\mu$, $\epsilon_{xy} = -30\mu$. If $\epsilon_1 = 80\mu$, determine ϵ_2, θ_1 and θ_2 (i).
(d) $\epsilon_y = 40\mu$, $\epsilon_{xy} = -30\mu$. If $\epsilon_2 = -80\mu$, determine ϵ_1, θ_1 and θ_2 (i).

Note: Verify answers via the appropriate Mohr circle.

Section 9

3.30:   Determine the state of (plane) strain, ϵ_x, ϵ_y, ϵ_{xy}, from the following θ-strain rosette measurements [see Fig. (3P.30)]:

(a) $\epsilon_a = 100\mu$, $\epsilon_b = 300\mu$, $\epsilon_c = -50\mu$; $\theta = 45°$, where a lies on the x-axis.
(b) $\epsilon_a = -600\mu$, $\epsilon_b = 200\mu$, $\epsilon_c = 0$; $\theta = 45°$, where b lies on the x-axis.

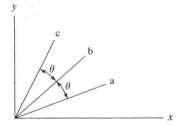

Figure 3P.30

(c) $\epsilon_a = 50\mu$, $\epsilon_b = -200\mu$, $\epsilon_c = -400\mu$; $\theta = 45°$, where c lies on the x-axis.

(d) $\epsilon_a = 100\mu$, $\epsilon_b = 200\mu\epsilon_c = 300\mu$; $\theta = 30°$, where a lies on the x-axis.

(e) $\epsilon_a = -300\mu$, $\epsilon_b = 0$, $\epsilon_c = 300\mu$; $\theta = 30°$, where b lies on the x-axis.

(f) $\epsilon_a = -500\mu$, $\epsilon_b = -1000\mu$, $\epsilon_c = -1500\mu$; $\theta = 30°$, where c lies on the x-axis.

3.31: Show that, for a 45° strain rosette, the principal strains can be given by

$$\epsilon_{1,2} = \frac{\epsilon_a + \epsilon_c}{2} \pm \frac{1}{2}\left[2\epsilon_a(\epsilon_a - 2\epsilon_b) + 2\epsilon_c(\epsilon_c - 2\epsilon_b) + 4\epsilon_b^2\right]^{1/2}.$$

3.32:* Show that the state of plane strain *cannot* be determined from measurements of three independent shear strains at a point.

3.33:* Original data from measurements of a strain rosette lying in the x–y plane have been lost. However, based on the original data, the following is known at a point: (i) the ratio of the invariants, $R \equiv \frac{I_{\epsilon_2}}{I_{\epsilon_1}} = 0.75 \times 10^{-3}$ and (ii) the ratio of the principal strains $\epsilon_1/\epsilon_2 = 3$. What are the possible values of the shear strain ϵ_{nt} in any arbitrary orthogonal directions, that can exist at this point?

Section 10

3.34: A plate whose area is A, and which lies in the x–y plane in the space ($0 \le x \le a$, $0 \le y \le b$) undergoes plane strain deformation. At any point P(x, y), the dilatation is given as

$$\Delta = \frac{k}{A}(x - a)^2(y - b)^2,$$

where $k \ll 1$ and has dimensions (1/m²). Determine, in terms of a and b, the change of area, δA, of the plate due to the deformation.

3.35: A square plate whose area is A and lies in the x–y plane in the space ($0 \le x \le L$, $0 \le y \le L$) undergoes plane strain deformation. The displacements in the x- and y-directions at any point P(x, y) are given as

$$u(x, y) = a\frac{xy}{L^2}, \qquad v(x, y) = b\frac{xy^2}{L^3},$$

where $a/L \ll 1$, $b/L \ll 1$ are known constants. Determine the change in area, δA, of the plate due to deformation in terms of a, b and L.

3.36: A plate lying in the x–y plane in the space ($-a/2 \le x \le a/2$, $-b/2 \le y \le b/2$) undergoes plane strain deformation due to non-uniform temperature changes $\delta T(x, y)$ given as

$$\delta T(x, y) = \delta T_0 + \delta T_1\left[\cos\left(\frac{\pi x}{a}\right)\cos^2\left(\frac{\pi y}{b}\right)\right].$$

Determine the change in area of the plate, δA, in terms of δT_0, δT_1, a, b, and the coefficient of thermal expansion, α.

Review and Comprehensive Problems

3.37:* A hollow cylinder, shown in Fig. (3P.37a), is deformed by rotating the outer surface through a small angle ϕ while holding the inner surface fixed as shown in Fig. (3P.37b). Assuming that the cylindrical surfaces remain circular and that all radial planes remain plane, determine to first-order the change in angle $\gamma_{r\theta}$ at any point

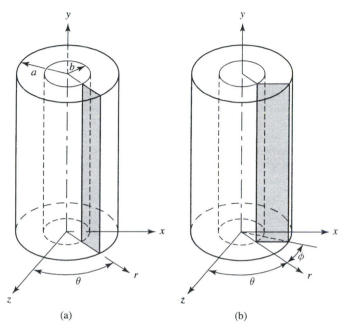

(a)   (b)

P, located at a radial distance r from the **axis**. Show specifically where the first-order approximation is used in the solution.

3.38:   A segment of a circular ring containing an arc $\overset{\frown}{nn'}$ having an original initial radius ρ, as shown in Fig. (3P.38a), is deformed into a segment such that the radius of the arc $\overset{\frown}{nn'}$ becomes R as shown in Fig. (3P.38b). Fibres along this arc $\overset{\frown}{nn'}$ are known to undergo no stretching (i.e. $\epsilon_\theta = 0$, where θ is in the circumferential direction). (The arc $\overset{\frown}{nn'}$ therefore is said to represent a 'neutral surface'.) Further, due to this deformation all cross-sections remain plane and perpendicular to the 'neutral surface', i.e. to the arc $\overset{\frown}{nn'}$. Show that the strain of any point P, measured a distance η from the neutral surface, is given by

$$\epsilon_\theta(P) = \left(\frac{\eta}{R}\right) \cdot \frac{1 - R/\rho}{1 + \eta/\rho}.$$

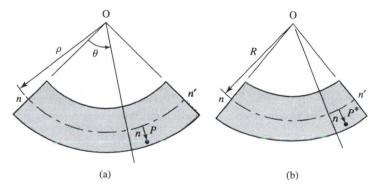

(a)   (b)   **Figure 3P.38**

Note: This expression reduces to $\epsilon_x(P) = \eta/R$ when $\rho \to \infty$, i.e., when the segment becomes an element of a straight beam lying in the x-direction.

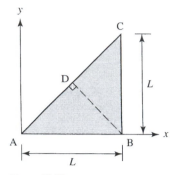

Figure 3P.39

3.39: Strains in the triangular plate ABC of Fig. (3P.39), lying in the x–y plane, are given as

$$\epsilon_x = ay, \qquad \epsilon_y = 2bx^2 y, \qquad \epsilon_{xy} = ax + 2bxy^2,$$

where $aL \ll 1$, $bL^3 \ll 1$. (a) Determine the increase in length of the edge BC; (b) determine ϵ_n along the edge AC; (c) determine the increase in length of the edge AC and (d) determine the angle between the lines DB and DC at point D, after deformation if $aL = 0.05$ and $bL^2 = a$. Express the answer in degrees.

3.40: The displacement of a given point of the plate ABCD [Fig. (3P.40)], lying in the x–y plane, is given as

$$\mathbf{u} = \frac{a}{L^2}(xy^2\,\mathbf{i} + x^2 y\,\mathbf{j})$$

where $a \ll 1$. (a) Determine the extensional strain at any point lying along the line AC; (b) determine the change in length of line AC; (c) determine the principal strains at point F; (d) Is AC a principal direction of strain at point F? Why?

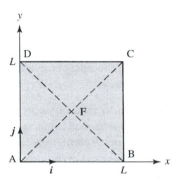

Figure 3P.40

3.41:* Points of a plate ($0 \le x \le 1$, $0 \le y \le 1$), lying in the x–y plane, undergo displacements $u = ax + 2by$, $v = cx + by$ in the x- and y-directions, respectively, where a, b and c are three undetermined constants, $0 < a$, b, $c \ll 1$. (a) Express the principal strains ϵ_1 and ϵ_2 at any point P(x, y) in terms of a, b and c; namely, express ϵ_1 as a function of the three variables, i.e. $\epsilon_1 = \epsilon_1(a, b, c)$; (b) express the direction of the principal direction θ_1 with respect to the x-axis in terms of a, b and c; (c) obtain the ratios a/c and b/c, which yield the extreme values of the extensional strains ϵ_1 and ϵ_2; (d) from the results of (c), determine this maximum value of ϵ_1 and the direction θ_1; (e) by means of a sketch (indicating the coordinates of the the corners of the deformed plate), show that the plate remains a square while undergoing rigid-body rotation and increasing in size. What is the increase in area, $\triangle A$, of the plate?

3.42: The displacements in the x- and y-directions of a given point P(x, y) of the plate ABCD, lying in the x–y plane [see Fig. (3P.42)], are given as

$$u = a\frac{x^2(y - L/2)^2}{L^4}, \qquad v = a\frac{x^2 y^3}{L^5}$$

where $a \ll L$. (a) Determine the strains ϵ_x, ϵ_y and ϵ_{xy} at any point P(x, y); (b) determine the change in length of line BC; (c) determine the change in length of line OC; (d) determine the angle at point C*, which exist between the edges B*C* and C*D after deformation, if $a/L = 0.02$. Express the answer in degrees; (e) denoting the original and final areas of the plate as A and A*, respectively, determine the ratio $\delta = \frac{A^* - A}{A}$.

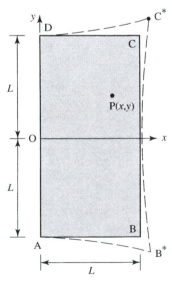

Figure 3P.42

3.43: A square plate ABCD undergoes plane strain to ABC*D, as shown in Fig. (3P.43), in such a way that points along the x- and y-axes remain fixed and such that the displacement of any point P(x, y) in the y-direction, $v = 0$. The displacement in the x-direction is given as $u(x, y) = kxy/L^2$, where $|k/L| \ll 1$. (a) Determine the extension Δ_{AC} of line AC; (b) determine the increase in length of the parabola $y = x^2/L$ passing through point E, due to the deformation; (c) determine the resulting average strain $\bar{\epsilon}$ of the diagonal AC and of the parabola AEC.

3.44: A rectangular plate ABCD [see Fig. (3P.44)], lying in the x–y plane, undergoes deformation such that (i) points lying on the y-axis undergo no displacement, (ii) the strain in the x-direction is constant, i.e. $\epsilon_x = c$, ($|c| \ll 1$) and (iii) the shear strain at any point P(x, y) is given by $\epsilon_{xy} = ax/L + by/L$ where $0 < b \leq a \ll 1$. (a) Determine the displacements $u(x, y)$ and $v(x, y)$ in the x- and y-directions, respectively; (b) determine the principal strains at any point P(x, y); (c) if $a = 2b = c$, along which line of the plate do points lie such that the x and y directions are the principal directions of strain? Show this line by means of a sketch. (d) Determine the principal strains for all points P along the line obtained in (c) above. At what point on this line does the maximum value $|\epsilon_2|$ occur? Show by means of a sketch and indicate the values of ϵ_1 and ϵ_2 at this point. (e) Sketch the deformed plate if $a = 0.2$, $b = 0.1$, $c = 0.05$ and indicate the coordinates at the corners of the deformed plate.

3.45: A square plate OABC, lying in the x–y plane, is deformed to OA*B*C* as shown in Fig. (3P.45). The following conditions (boundary conditions) are known: (i) The edges OA and OC remain on the x- and y-axes, respectively; (ii) the displacements of points A and C along the axes are $e \ll L$, as shown in the figure; (iii) the stretches along the axes are uniform; i.e., the strains ϵ_x and ϵ_y along the respective axes are constant and (iv) the displacement of point B in the y-direction is $2e$. Within the plate, the vertical lines are known to remain parallel to the y-axis and the shear strain at any point P(x, y) is given as $\epsilon_{xy} = axy/2L^3$, where a is a constant to be determined. (a) By integration of the strain–displacement relations, and making use of the boundary conditions given by (i)–(iv) above, show that the displacements $u(x, y)$ and $v(x, y)$ in the x- and y-directions, respectively, are

$$u = \frac{ex}{L}, \qquad v = e\left(\frac{x^2y}{L^3} + \frac{y}{L}\right)$$

(b) What is the average extensional strain $\bar{\epsilon}_n$ of the diagonal OB? Express the answer in terms of e and L, making use of the condition $e \ll L$. (c) Determine the strain ϵ_n of the line segment of OB at point D, the intersection of the two diagonals. (d) Determine the angle (in degrees) between the two diagonals after deformation if $e = 0.1L$.

3.46: A square plate OABC, lying in the x–y plane, as shown in Fig. (3P.46), is deformed to OA*B*C*, subject to the following conditions: (i) OA and OC remain on the x- and y-axes, respectively; (ii) the displacement of points A and C along the x- and y-axes, respectively, is e, as shown in the figure; (iii) the strains along the x- and y-axes are uniform, i.e. constant; (iv) the displacement of point B in the x-direction is $2e$; (v) within the plate, lines originally parallel to the x-axis remain parallel to the x-axis; (vi) the shear strain at any point P(x, y) is given by $\epsilon_{xy} = \frac{axy^2}{2L^4}$, where $a \ll L$ is an undetermined constant. (a) Express conditions (i) to (v) in mathematical terms. (b) By integrating the strain–displacement relations, show that the displacements u and v in

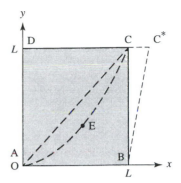

Figure 3P.43

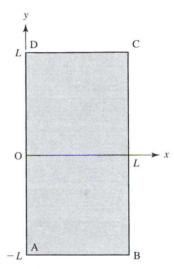

Figure 3P.44

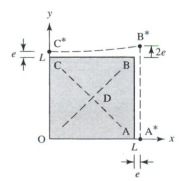

Figure 3P.45

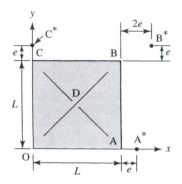

Figure 3P.46

the x- and y-directions are given, at any point $P(x, y)$, by

$$u = e\left(\frac{xy^3}{L^4} + \frac{x}{L}\right),$$ (I)

$$v = \frac{ey}{L}.$$ (II)

(c) Making use of Eqs. (I) and (II) above, determine the coordinates of B^* (x_{B^*}, y_{B^*}). Based on these coordinates, evaluate to first order the change in length Δ_{OB} of line OB subject to the condition $0 \le e/L \ll 1$. (d) Determine the extensional strain ϵ_n at any point P along OB and by integration, evaluate the change in length Δ_{OB} of line OB. (e) Determine (i) the displacement $|u|$ of point D and (ii) the component of displacement of point D in the direction of line OB. (f) Determine the change of area of the plate due to the deformation.

The following problems are to be solved using a computer.

3.47:   Using the transformation laws for plane strain [Eqs. (3.7.21)], write a computer program to determine ϵ_n, ϵ_t and ϵ_{nt} for any given state of strain, ϵ_x, ϵ_y, ϵ_{xy} and θ. Check the program by using some of the strain states given in Problem 3.23.

3.48:   Given a state of plane strain, ϵ_x, ϵ_y and ϵ_{xy}, write a program to determine the principal strains ϵ_1 and ϵ_2 and the principal directions θ_1 and θ_2. Check the program by using some of the strain states given in Problem 3.27.

3.49:   Write a computer program to determine the state of plane strain $(\epsilon_x, \epsilon_y, \epsilon_{xy})$ from data obtained from a θ-strain rosette [see Fig. (3.9.1)], i.e. from measurements of ϵ_a, ϵ_b and ϵ_c. Check the program using measurements given in Problem 3.30.

3.50:   Given the state of plane strain, $\epsilon_x = 2 \times 10^{-3}$, $\epsilon_y = 4 \times 10^{-3}$, $\epsilon_{xy} = 6 \times 10^{-3}$. In what direction (defined by the angle θ with respect to the x-axis) is the normal strain $\epsilon_n = 9 \times 10^{-3}$? (*Note*: The value θ can only be determined numerically.)

3.51:   Given the state of plane strain, $\epsilon_x = 400\mu$, $\epsilon_y = -300\mu$, $\epsilon_{xy} = 600\mu$. In what direction (defined by the angle θ with respect to the x-axis) is the normal strain $\epsilon_n = 700\mu$? (*Note*: The value θ can only be determined numerically.)

3.52:   Given the state of plane strain, $\epsilon_x = -200\mu$, $\epsilon_y = 150\mu$, $\epsilon_{xy} = 60\mu$. In what direction (defined by the angle θ with respect to the x-axis) is the shear strain $\epsilon_{nt} = 180\mu$? (*Note*: The value θ can only be determined numerically.)

4

Behaviour of materials: constitutive equations

4.1 Introduction

In our previous treatment, the laws of mechanics and kinematic relations (or more specifically geometric relations) that describe the deformation of a body have been developed. This led us to the definitions of two concepts: *stress* as a measure of intensity of internal forces and *strain* as a measure of the intensity of deformation; concepts that clearly are valid for any deformable body and thus independent of the material; the definitions and derived relations are as valid for a fluid as for a rod made of steel. Now a fluid and steel evidently behave quite differently. Therefore it is clear that to determine the behaviour of a given body, it is necessary to specify the general character of the mechanical behaviour of the material itself; this behaviour must be specified in mathematical terms. The mathematical equations describing the general behaviour of a material are known as **constitutive equations**. Thus, it is only when we introduce the constitutive equations in the problem that we specify the material under consideration.

Constitutive equations are equations that relate the various quantities (e.g. stress, strain, stress rate, etc.) governing the general behaviour of materials. However, these equations arc idealised equations since they take into account only certain effects. One could, for example, consider thermodynamic effects, electromagnetic effects, etc., on the behaviour of bodies. Since our goal here is to study the *mechanical* behaviour of bodies, we shall exclude these effects; we shall consider constitutive equations that relate only the *mechanical variables* in describing the behaviour.

It is important to observe that while the constitutive equations are simplified equations, they must nevertheless describe as accurately as possible the real behaviour of a material in nature. Thus we may say that constitutive equations describe the behaviour of a representative model that conforms with experimental data obtained for a given material.

4.2 Some general idealisations (definitions: 'micro' and 'macro' scales)

The idealisations that we make depend on the level at which a problem is to be studied and the purpose required. For example, all matter is known to be composed of atoms and corresponding molecules and/or crystals having characteristic dimensions measured in Angstroms or possibly microns. To understand certain phenomena, it is necessary to consider the behaviour at this level, be it at the atomic, such as in solid state physics or at the microscopic level. One studies materials at this level in order to understand *why* the material behaves in a particular way. Indeed this falls within the discipline of material sciences. However, in our treatment, we shall be interested in a more global approach, namely, in *how* a body behaves (e.g., under certain loading conditions). Our approach can therefore be said to be a phenomenological approach under a macroscopic scale. Hence, in using such an approach, the atomic or microscopic composition is 'blurred', and consequently we do not consider the atoms, molecules, etc., to lie at discrete points but instead consider the material to be distributed continuously at all points in a given space. The body is therefore said to constitute a 'continuum' since its particles are assumed to be located continuously at all points throughout a given (x, y, z) space.[†]

Before proceeding with a description of the mechanical behaviour, it is necessary to establish a precise terminology. We first observe that a body may be considered to be either *homogeneous* or *inhomogeneous*. A body is said to be **strictly homogeneous** if it possesses the *same material properties at all points* in the body. We shall refer to this definition as the definition on the 'micro-scale' since we refer here to the behaviour at various *points* in the body. That is, if we consider a body as shown in Fig (4.2.1), the material is micro-homogeneous if, given *any* two points in the body, P_1 and P_2, it has the same property at both points. If the material behaviour changes from point to point, then the material is said to be **inhomogeneous** on the micro-scale.

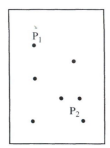

Figure 4.2.1

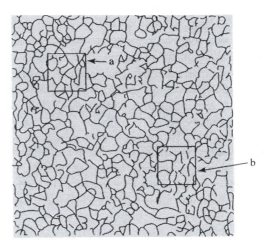

Figure 4.2.2

Let us now consider a real material such as steel, which is composed of iron crystals, of carbon (and possibly minute parts of other elements). If one examines steel under a microscope, it appears as is shown in Fig. (4.2.2); it is clear that the behaviour of the iron crystals will be different from that of the carbon. Thus the steel cannot, in fact, be said to be strictly a homogeneous material on the micro-scale. However, let us consider a small *representative element* of steel, for example, element 'a' as shown in Fig (4.2.2), which consists of a large number of *randomly oriented* crystals. In this case, we do not consider the material at a point but more globally, i.e. on a 'macro-scale'. It is clear that the several elements, e.g. elements 'a' and 'b', each containing crystals of iron and carbon, constitute the same

---

[†] It should be noted that in the previous treatment of stress and strain, the body was implicitly considered to be a continuum.

representative material. In this sense, although the steel is not micro-homogeneous, it can be said to be homogeneous on the macro-scale or 'macro-homogeneous'.

As another example, if we consider concrete (which is composed of cement, sand and gravel) [Fig. (4.2.3)], one does not require a microscope to observe that the material is not micro-homogeneous. Nevertheless, if we are interested in the *global behaviour* of the concrete, we may consider the material to be macro-homogeneous in the same sense as previously discussed; namely, the behaviour of a representative element is the same everywhere in the body.[†]

Now, let us consider the behaviour of a material *at a given point*. A material that behaves in such a way that its properties are the same in *all* directions is said to be **isotropic**. Thus, since an isotropic material exhibits the same behaviour at a given point in all directions, the material is said to have no 'preferred' directions. On the other hand, if the material exhibits a different behaviour, depending on the direction, it is said to be an **anisotropic** material. For example, wood clearly is an anisotropic material since it behaves differently if it is under tension in the direction of the grain or perpendicular to the grain [Figs. (4.2.4a and b)].

Now if we consider again, for example, steel or concrete, it is clear that neither material exhibits isotropic properties at a point; that is, they both are anisotropic on a micro-scale. However, observing Figs. (4.2.2) and (4.2.3), it is quite evident that the component parts in any finite representative element appear to be randomly oriented. Thus, for example, if we examine a representative element of concrete, it clearly does not have any preferred direction, and therefore, statistically, the properties are the same in all directions. Thus, we may consider the material to be isotropic on the macro-scale.

Since our subsequent treatment will be concerned with the global behaviour of bodies (e.g., subjected to external forces), our interest will be with the macro-behaviour of such bodies. Hence, in using the terms 'homogeneous material' or 'isotropic material' our reference to these properties will be on the macro-scale.

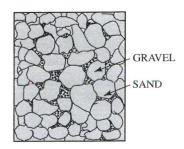

Figure 4.2.3

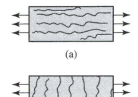

(a)

(b)

Figure 4.2.4

4.3 Classification of materials: viscous, elastic, visco-elastic and plastic materials

Based on tests conducted in the laboratory, there exist several broad classes of solid materials. These materials are best classified according to the different types of constitutive equations that characterise their behaviour. We may define these classes broadly as follows:

(a) Elastic material

An **elastic material** is one for which, at any given point, there exists a direct relation between the state of stress and state of strain. Denoting the stress and strain tensors symbolically by τ and ϵ, i.e.[‡]

$$\tau = \begin{bmatrix} \tau_{xx} & \tau_{xy} & \tau_{xz} \\ \tau_{yx} & \tau_{yy} & \tau_{yz} \\ \tau_{zx} & \tau_{zy} & \tau_{zz} \end{bmatrix} \qquad \epsilon = \begin{bmatrix} \epsilon_{xx} & \epsilon_{xy} & \epsilon_{xz} \\ \epsilon_{yx} & \epsilon_{yy} & \epsilon_{yz} \\ \epsilon_{zx} & \epsilon_{zy} & \epsilon_{zz} \end{bmatrix},$$

[†] Clearly, in more precise analyses of fracture in solids, or in the realm of solid state physics, where one must consider the material on a micro-scale, the concept of macro-homogeneity loses its validity.

[‡] Here we use the notation $\sigma_x \equiv \tau_{xx}, \sigma_y \equiv \tau_{yy}, \sigma_z \equiv \tau_{zz}$; $\epsilon_x \equiv \epsilon_{xx}$, etc.

respectively, we define an elastic material as one for which the constitutive equation is of the form

$$\tau = f(\epsilon), \qquad \tau(\epsilon = 0) = 0. \tag{4.3.1}$$

Note that implicit in its definition, the state of stress at points in the elastic body depends solely on the *final* deformed state of the body, that is on the final strain. Thus the stresses *do not* depend on the manner in which the deformation occurred.

In the following section, we shall elaborate at length on the properties of elastic materials.

(b) Viscous material

A **viscous material** is one for which the stress at a point is a function of the strain rate; i.e.,

$$\tau = f(\mathrm{d}\epsilon/\mathrm{d}t). \tag{4.3.2a}$$

For example, the relation might be a linear relation of the form

$$\tau = \alpha(\mathrm{d}\epsilon/\mathrm{d}t) \equiv \alpha\dot{\epsilon}, \tag{4.3.2b}$$

where α is a constant of viscosity. In this case, the material is said to exhibit linear viscosity. (Such materials are on the 'borderline' between a solid and fluid, depending on the viscous nature of the material.)

(c) Visco-elastic material

A **visco-elastic material** is one for which the constitutive equations express the stress and stress rates as a function of the strain and strain rates; thus they have the general form

$$f(\tau, \dot{\tau}, \ddot{\tau}, \ldots) = g(\epsilon, \dot{\epsilon}, \ddot{\epsilon}, \ldots). \tag{4.3.3a}$$

A material of this class having a linear relation, e.g., where the strain depends linearly not only on the stress but also on the stress rate, as in

$$\epsilon = \alpha\tau + \beta\dot{\tau}, \tag{4.3.3b}$$

where α and β are material constants, is said to be a **simple linear visco-elastic material**. Hence the deformation of a body of such a material will depend not only on the applied force but also on how fast or slowly the force is applied. The material is said to be **rate-sensitive**.

(d) Plastic material

A general definition of a **plastic material** is not quite as straightforward. It appears that the simplest definition would be that the stresses in a material undergoing plastic behaviour are such that they do not depend on the final state of strain but rather (as opposed to elastic materials) on the manner by which the state of strain was arrived at; that is, on the previous history of the material.

Because the majority of design and analysis problems encountered in engineering deal with elastic materials, in our subsequent treatment we shall limit our discussion mainly to elastic materials. However, where appropriate, we shall also consider plastic behaviour of materials.

4.4 Elastic materials

(a) Constitutive equations for elastic materials: general elastic and linear elastic behaviour, Hooke's law

(i) General elastic behaviour

As defined above, an elastic material is characterised by a relation between the state of stress and state of strain at a point.

For simplicity, let us first consider the case of uniaxial stress where all stress components with the exception of $\sigma_x \equiv \tau_{xx}$ vanish. Thus we consider a rod of length L and cross-section A under uniaxial tension. Then, as discussed in Chapter 1, by applying a tension force P of gradually increasing magnitude to the rod and measuring the change in length ΔL during a simple tension test, we calculate the stress $\sigma_x = P/A$ and $\epsilon_x = \Delta L/L$ and thus obtain the stress–strain curve of the material, as shown in Fig. (4.4.1).[†]

Now, let us say that we apply a load to the undeformed rod up to point B of Fig. (4.4.1) and then remove this load (i.e., we return to the point $\sigma_x = 0$). Clearly, if the rod is elastic, it will return to its undeformed shape, $\epsilon_x = 0$. If we now reapply the same load, we return to point B. Alternatively, if we reduce the load to, say, any point C and then reapply it, we arrive again at point B on the σ–ϵ curve. Thus, implicit in the definition of an elastic material as given above, an elastic material has the following properties:

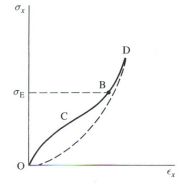

Figure 4.4.1

- The final stress state at a point depends solely on the final strain state (and vice versa); that is, it does not depend on the 'loading history'.
- The σ–ϵ curve defined by the constitutive equation must be a *unique* curve; e.g., the relation between the stress σ_x and strain ϵ_x must be a one-to-one relation. Thus, for example, a constitutive equation defined as $\sigma_x = k\epsilon_x^2$ [Fig. (4.4.2a)] cannot represent an elastic material since for any given σ_x, there exist two possible strains, namely $+\sqrt{\sigma_x/k}$ and $-\sqrt{\sigma_x/k}$. However, the constitutive equation, defined as $\sigma_x = k\epsilon_x^2$, only for positive strains, i.e. $\epsilon_x \geq 0$ [Fig. (4.4.2b)], does represent an elastic material since the strain is uniquely determined.

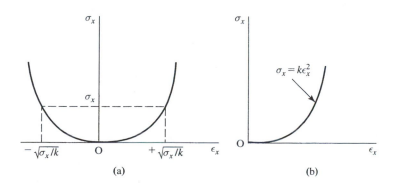

(a)   (b)

Figure 4.4.2

Let us imagine that, starting from the undeformed unstressed state, we now load the specimen represented in Fig. (4.4.1) successively by a series of gradual incremental loads. We would find that there is a maximum stress for which the material

---

[†] We assume here that, at all points in the rod, $\sigma_x = \bar{\sigma}_x$ and $\epsilon_x = \bar{\epsilon}_x$, where $\bar{\sigma}_x$ and $\bar{\epsilon}_x$ are the average stress and strain on a cross-section.

behaves elastically. We call this stress the **Elastic limit** and denote it by σ_E. Thus in Fig. (4.4.1), if the stress $\sigma_x \leq \sigma_E$, the material behaves as an elastic material; if the material is stressed to, say, point D, then it will not return to its initial shape; i.e., the rod will have a permanent deformation (or 'set') since it undergoes plastic deformation. Indeed, if the specimen is loaded to σ_D, 'unloading' will take place along a different curve as shown by the dashed line in the figure. (We discuss this unloading process in the next section.)

We now generalise the above ideas to materials where three-dimensional states of stress and strain, τ and ϵ, exist. We therefore define an elastic material as one whose constitutive equation is of the form

$$\tau = f(\epsilon), \qquad \tau(\epsilon = 0) = 0 \qquad (4.4.1a)$$

and which possesses a *unique inverse*, written symbolically as

$$\epsilon = f^{-1}(\tau), \qquad \epsilon(\tau = 0) = 0. \qquad (4.4.1b)$$

The general stress–strain curve of an elastic material is represented symbolically in Fig. (4.4.3). We note that if the state of strain at a point is given, the state of stress at the point is immediately known and vice versa. Furthermore, as a result of the unique one-to-one relationship given by Eq. (4.4.1), we again conclude that the behaviour of an elastic material is independent of the loading history; the material has but one 'memory', namely its initial undeformed state to which it returns when all stresses vanish.

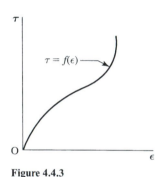

Figure 4.4.3

(ii) Linear elastic material: Hooke's law

A particular case of an elastic material, which is of great importance, is the case of a linear elastic material, namely one for which the stress and strain states are linearly related; i.e., f and f^{-1} are linear functions.

For convenience, let us again consider the simplest states of stress: uniaxial stress and the state of pure shear.

We consider first the uniaxial state of stress, $\sigma_x \equiv \tau_{xx}$, where all other stress components are zero. For this case, the linear relation is given by

$$\sigma_x = E\epsilon_x, \qquad (4.4.2)$$

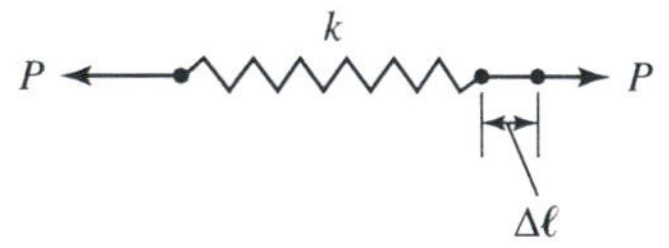

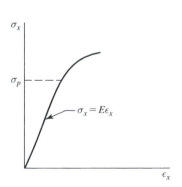

Figure 4.4.4

where $E(E > 0)$ is the **modulus of elasticity**. We note that the modulus of elasticity E represents the slope of the stress–strain curve. We note too that the relation is analogous to that of a linear spring having stiffness k (N/m). In this case, the relation between an applied force P and the change in length of the spring $\Delta\ell$ is $P = k\Delta\ell$. For this reason an elastic material is often represented by a model consisting of a spring, as in Fig. (4.4.4).

Indeed, many, but not all, materials behave initially as linear elastic Materials, provided the stress and strain are sufficiently small; that is, provided that σ_x does not exceed a certain value. Thus one may find that the $\sigma-\epsilon$ curve in uniaxial behaviour appears as shown in Fig. (4.4.5). The linear relation (4.4.2) then holds provided that $\sigma_x \leq \sigma_p$, where σ_p denotes the **proportional limit**.

Let us now consider the case of pure shear, for example, in the x–y plane. We note that if a shear stress τ_{xy} is applied to a linear elastic material, then any element will undergo shear deformation where we denote the angle measuring this deformation by γ_{xy}. The resulting $\tau-\gamma$ curve in shear is plotted as shown in

Figure 4.4.5

Fig. (4.4.6). The relation will be linear provided $\tau \leq \tau_p$ (where τ_p is the proportional limit in shear); i.e.,

$$\tau_{xy} = G\gamma_{xy} \quad \text{or} \quad \gamma_{xy} = \frac{\tau_{xy}}{G}. \tag{4.4.3}$$

Here the proportionality constant G ($G > 0$) is called the **shear modulus** or the **modulus of rigidity**. Note that both E and G are positive constants and have units of N/m^2 or Pa.

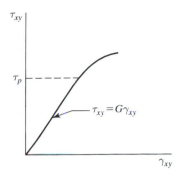

Figure 4.4.6

Having discussed the behaviour for the simplest states of stress, we turn again to consider the general three-dimensional state of stress and strain, where we recall that there exist six independent components of stress ($\tau_{xx}, \tau_{yy}, \tau_{zz}, \tau_{xy}, \tau_{yz}, \tau_{xz}$) and strain ($\epsilon_{xx}, \epsilon_{yy}, \epsilon_{zz}, \epsilon_{xy}, \epsilon_{yz}, \epsilon_{xz}$). Since for a linear elastic material, the state of stress is linearly related to the state of strain, it is reasonable to assume that, in principle, for an anisotropic material, any given stress component is linearly dependent on all six strain components. Thus we may write the general linear relation in the following form:

$$
\begin{aligned}
\tau_{xx} &= C_{11}\epsilon_{xx} + C_{12}\epsilon_{yy} + C_{13}\epsilon_{zz} + C_{14}\epsilon_{xy} + C_{15}\epsilon_{yz} + C_{16}\epsilon_{zx} \\
\tau_{yy} &= C_{21}\epsilon_{xx} + C_{22}\epsilon_{yy} + C_{23}\epsilon_{zz} + C_{24}\epsilon_{xy} + C_{25}\epsilon_{yz} + C_{26}\epsilon_{zx} \\
\tau_{zz} &= C_{31}\epsilon_{xx} + C_{32}\epsilon_{yy} + C_{33}\epsilon_{zz} + C_{34}\epsilon_{xy} + C_{35}\epsilon_{yz} + C_{36}\epsilon_{zx} \\
\tau_{xy} &= C_{41}\epsilon_{xx} + C_{42}\epsilon_{yy} + C_{43}\epsilon_{zz} + C_{44}\epsilon_{xy} + C_{45}\epsilon_{yz} + C_{46}\epsilon_{zx} \\
\tau_{yz} &= C_{51}\epsilon_{xx} + C_{52}\epsilon_{yy} + C_{53}\epsilon_{zz} + C_{54}\epsilon_{xy} + C_{55}\epsilon_{yz} + C_{56}\epsilon_{zx} \\
\tau_{zx} &= C_{61}\epsilon_{xx} + C_{62}\epsilon_{yy} + C_{63}\epsilon_{zz} + C_{64}\epsilon_{xy} + C_{65}\epsilon_{yz} + C_{66}\epsilon_{zx}
\end{aligned}
\tag{4.4.4}
$$

where the constants ($C_{11}, C_{12}, \ldots, C_{66}$) are material constants for any particular material. Alternatively, one might write the linear relations as strains in terms of stresses:

$$
\begin{aligned}
\epsilon_{xx} &= B_{11}\tau_{xx} + B_{12}\tau_{yy} + B_{13}\tau_{zz} + B_{14}\tau_{xy} + B_{15}\tau_{yz} + B_{16}\tau_{zx} \\
\epsilon_{yy} &= B_{21}\tau_{xx} + B_{22}\tau_{yy} + B_{23}\tau_{zz} + B_{24}\tau_{xy} + B_{25}\tau_{yz} + B_{26}\tau_{zx} \\
\epsilon_{zz} &= B_{31}\tau_{xx} + B_{32}\tau_{yy} + B_{33}\tau_{zz} + B_{34}\tau_{xy} + B_{35}\tau_{yz} + B_{36}\tau_{zx} \\
\epsilon_{xy} &= B_{41}\tau_{xx} + B_{42}\tau_{yy} + B_{43}\tau_{zz} + B_{44}\tau_{xy} + B_{45}\tau_{yz} + B_{46}\tau_{zx} \\
\epsilon_{yz} &= B_{51}\tau_{xx} + B_{52}\tau_{yy} + B_{53}\tau_{zz} + B_{54}\tau_{xy} + B_{55}\tau_{yz} + B_{56}\tau_{zx} \\
\epsilon_{zx} &= B_{61}\tau_{xx} + B_{62}\tau_{yy} + B_{63}\tau_{zz} + B_{64}\tau_{xy} + B_{65}\tau_{yz} + B_{66}\tau_{zx}
\end{aligned}
\tag{4.4.5}
$$

where the constants ($B_{11}, B_{12}, \ldots, B_{66}$) again are different material constants (but related to the constants, C), for the particular material.

The above represents the most general linear elastic stress–strain relation; this relation is referred to as the *generalised Hooke's Law for anisotropic materials*. It therefore appears that to describe a linear elastic material in its greatest generality, we would require 36 independent constants. Although it is beyond the scope of our present study, we mention here that, in fact, 15 of these constants are not independent and that therefore the most general anisotropic linear elastic material can be described by 21 *independent* constants.

The purpose of the above discussion of a general linear anisotropic material has been to consider the behaviour of materials in the framework of the general theory. Now, there exist various degrees of anisotropy in a body and evidently, as a material becomes less anisotropic, there will exist fewer number of independent constants. Thus the number of independent material constants required to describe a linear elastic material clearly diminishes as the material approaches isotropy; we shall limit our treatment below to these simplest materials, i.e. isotropic materials.[†]

Recalling that an isotropic material is one whose properties are the same in all directions, we therefore wish to develop the stress–strain relations for a linear isotropic elastic material when subjected to a general state of stress.

As in the previous discussion, we again start from the simple case of a uniaxial state of stress $\sigma_x \neq 0$, where we recall from Eq. (4.4.2) that the strain ϵ_x is given by $\epsilon_x = \sigma_x/E$. Let us assume that the element is subjected to a tensile stress $\sigma_x > 0$ in the x-direction. The resulting strain $\epsilon_x > 0$ therefore describes an extension of the given element in the x-direction. However, due to $\sigma_x > 0$, contractions take place in both the y- and z-lateral directions; i.e., the element undergoes lateral strains $\epsilon_y < 0$ and $\epsilon_z < 0$, as shown in Fig. (4.4.7). Moreover, since the material is isotropic, the strains ϵ_y and ϵ_z due to σ_x must necessarily be the same since neither the y-axis nor z-axis is a 'preferred direction'. In fact, upon measuring such strains in the laboratory, the strains ϵ_y and ϵ_z for a linear elastic material are found to be proportional to ϵ_x; that is,

$$\epsilon_y = -\nu\epsilon_x, \tag{4.4.6a}$$

$$\epsilon_z = -\nu\epsilon_x, \tag{4.4.6b}$$

where $\nu \geq 0$ is called the **Poisson ratio**. Thus, for a uniaxial state of stress, $\sigma_x \neq 0$, we have

$$\epsilon_x = \frac{\sigma_x}{E}, \tag{4.4.7a}$$

$$\epsilon_y = -\nu\frac{\sigma_x}{E}, \tag{4.4.7b}$$

$$\epsilon_z = -\nu\frac{\sigma_x}{E}. \tag{4.4.7c}$$

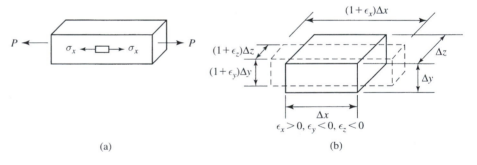

Figure 4.4.7   (a)   (b)

Note that the Poisson ratio ν is a non-dimensional constant. These equations express the fact that the strain in a direction perpendicular to an applied stress is always proportional (and of opposite sign) to the strain in the direction of the stress, the coefficient of proportionality being $-\nu/E$.

Consider now a general three-dimensional state of stress. Let us assume that the state of stress at the point is such that σ_x, σ_y and σ_z are all acting on an element. As we have seen, the strain ϵ_y in the direction that is perpendicular to a stress component σ_x is given by $\epsilon_y = -\nu\sigma_x/E$. Since there are no preferred directions

[†] We recall that by the term 'isotropic' we refer to an isotropic material on the macro-scale.

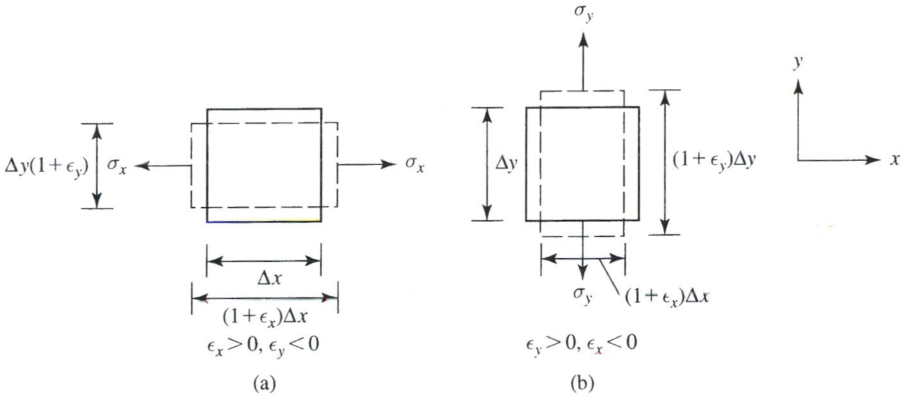

$\epsilon_x > 0, \epsilon_y < 0$

(a)

$\epsilon_y > 0, \epsilon_x < 0$

(b)

Figure 4.4.8

for an isotropic material, it follows that the effect on the strain ϵ_x due to the stress σ_y must be the same as the effect of σ_x upon ϵ_y; thus the strain ϵ_x due to σ_y is given by $-\nu\sigma_y/E$. The two-dimensional effect, neglecting the z-direction, is shown in Fig. (4.4.8). Similarly, ϵ_x due to σ_z will be given by $-\nu\sigma_z/E$. Since all the relations are linear, we may superimpose the effects of σ_x, σ_y and σ_z; thus we obtain

$$\epsilon_x = \frac{\sigma_x}{E} - \frac{\nu\sigma_y}{E} - \frac{\nu\sigma_z}{E}$$

or

$$\epsilon_x = \frac{1}{E}[\sigma_x - \nu(\sigma_y + \sigma_z)]. \qquad (4.4.8a)$$

Furthermore, since there are no preferred directions for an isotropic material, the effect of normal stresses acting in the y-, x- and z-directions on the strain ϵ_y in the y-direction must be the same as the effect of the normal stresses acting in the x-, y- and z-directions, respectively, on the strain ϵ_x in the x-direction. We may arrive at similar conclusions for the strain ϵ_z by using the same arguments based on the definition of isotropy. Therefore we have

$$\epsilon_y = \frac{1}{E}[\sigma_y - \nu(\sigma_z + \sigma_x)] \qquad (4.4.8b)$$

and

$$\epsilon_z = \frac{1}{E}[\sigma_z - \nu(\sigma_x + \sigma_y)]. \qquad (4.4.8c)$$

Note that, having established Eqs. (4.4.6) and (4.4.7), Eqs. (4.4.8a–4.4.8c) follow directly simply from the basic definition of isotropy.

If we examine the behaviour due to shear, we conclude that since the isotropic material has no preferred directions, the shear relations in the y–z and z–x planes must be of the same form as that in the x–y plane, as given by Eq. (4.4.3); thus

$$\gamma_{yz} = \frac{\tau_{yz}}{G} \qquad (4.4.9a)$$

and

$$\gamma_{zx} = \frac{\tau_{zx}}{G}. \qquad (4.4.9b)$$

Combining the above, and recalling that the shear strain is defined as half the change in angle (e.g., $\epsilon_{xy} = \gamma_{xy}/2$, etc.), the general stress–strain relations for a linear elastic material are

$$\epsilon_x = \tfrac{1}{E}[\sigma_x - \nu(\sigma_y + \sigma_z)] \qquad \epsilon_{xy} = \tfrac{\tau_{xy}}{2G}$$

$$\epsilon_y = \tfrac{1}{E}[\sigma_y - \nu(\sigma_z + \sigma_x)] \qquad \epsilon_{yz} = \tfrac{\tau_{yz}}{2G} \qquad (4.4.10\text{a--f})$$

$$\epsilon_z = \tfrac{1}{E}[\sigma_z - \nu(\sigma_x + \sigma_y)] \qquad \epsilon_{zx} = \tfrac{\tau_{zx}}{2G}$$

Note that the above represents six scalar relations between the six independent stress and strain components. We observe that any shear strain component is proportional only to the corresponding shear stress component and is independent of the normal stress components. These linear relations are known as *Hooke's law for a linear isotropic elastic material*. It therefore would appear that in order to represent the constitutive equation for a linear elastic material, we require three constants: E, G and ν. We now show that only two of these are *independent* constants.

We first note that Hooke's law is valid for *any* state of stress and strain. Let us therefore consider a two-dimensional state of stress with $\sigma_z = \tau_{xz} = \tau_{yz} = 0$, and furthermore, for this case of plane stress, let us consider, in particular, an element in a state of pure shear with $\tau_{xy} \neq 0$, $\sigma_x = \sigma_y = 0$ [Fig. (4.4.9)]. For this case of pure shear, the principal stresses $\sigma_1 = \tau_{xy}$ and $\sigma_2 = -\tau_{xy}$ are immediately determined. [The corresponding Mohr circle is shown in Fig. (4.4.10a). Note that this case of pure shear was treated in Example 2.11 of Chapter 2.] Using Hooke's law (with $\sigma_z = 0$), the strains in the corresponding directions, ϵ_1 and ϵ_2, are given by

$$\epsilon_1 = \frac{\sigma_1}{E} - \nu \frac{\sigma_2}{E}, \qquad (4.4.11\text{a})$$

$$\epsilon_2 = \frac{\sigma_2}{E} - \nu \frac{\sigma_1}{E}. \qquad (4.4.11\text{b})$$

Substituting for the values of σ_1 and σ_2, we obtain

$$\epsilon_1 = \frac{\tau_{xy}}{E}(1 + \nu), \qquad (4.4.12\text{a})$$

$$\epsilon_2 = -\frac{\tau_{xy}}{E}(1 + \nu). \qquad (4.4.12\text{b})$$

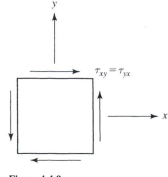

Figure 4.4.9

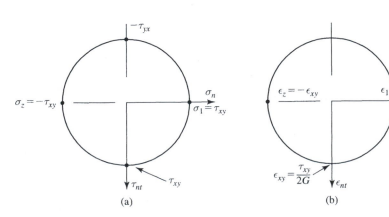

(a)   (b)

Figure 4.4.10

Recalling that the shear stress acting on any principal stress plane vanishes (here $\tau_{12} = 0$), it follows from Hooke's law that the shear strain with respect to the '1' and '2' directions also vanishes, i.e., $\epsilon_{12} = 0$. Hence, the shear strains ϵ_1 and ϵ_2 are actually principal shear strains.

Now, instead of first determining the principal stresses and then using Hooke's law to find the principal strains, let us reverse the process; i.e., we first use Hooke's law and then determine the principal strains.

Thus, for the given case of pure shear with $\sigma_x = \sigma_y = 0$, it follows that $\epsilon_x = \epsilon_y = 0$. Furthermore, using Hooke's law, the shear strain $\epsilon_{xy} = \frac{\tau_{xy}}{2G}$. Therefore, the strain state is also one of pure shear, and hence (e.g., as seen in the Mohr circle for strain [Fig. (4.4.10b)]), the principal strains $\epsilon_1 = \epsilon_{xy}$ and $\epsilon_2 = -\epsilon_{xy}$. It follows that

$$\epsilon_1 = \frac{\tau_{xy}}{2G}, \tag{4.4.13a}$$

$$\epsilon_2 = -\frac{\tau_{xy}}{2G}. \tag{4.4.13b}$$

Upon comparing Eqs. (4.4.12) with (4.4.13), we conclude that

$$G = \frac{E}{2(1 + \nu)}. \tag{4.4.14}$$

Thus, we have found that only two of the three elastic constants (E, G, ν) are independent: given any two constants, the third may be determined.

At this stage, it is useful to further define another constant. From Hooke's law, as given by Eqs. (4.4.10), it follows that

$$\epsilon_x + \epsilon_y + \epsilon_z = \frac{1}{E}(1 - 2\nu)(\sigma_x + \sigma_y + \sigma_z). \tag{4.4.15}$$

Now, from Chapter 3, we recall that the dilatation Δ, which represents the volumetric strain, is given by $\Delta = \epsilon_x + \epsilon_y + \epsilon_z$ [see Eq. (3.10.7b)] and that this is the first invariant of strain [Eq. (3.7.23)]. Furthermore, from Chapter 2, we note that the quantity appearing on the right-hand side of Eq. (4.4.15) is precisely the first invariant of stress [Eq. (2.6.10b)]. Thus, Eq. (4.4.15) is valid at any given point, irrespective of the chosen coordinate system. Let $\bar{\sigma}$ here define the **mean normal stress** at a point:

$$\bar{\sigma} = \frac{1}{3}(\sigma_x + \sigma_y + \sigma_z). \tag{4.4.16}$$

We may then write Eq. (4.4.15) as

$$\Delta = \frac{\bar{\sigma}}{\kappa}, \tag{4.4.17}$$

where

$$\kappa = \frac{E}{3(1 - 2\nu)}. \tag{4.4.18}$$

The constant κ is called the **bulk modulus**. Note that Eq. (4.4.17) expresses the volumetric strain at a point in terms of the mean stress existing at the point. For example, Eq. (4.4.17) permits one to determine the dilatation for a hydrostatic state of stress. Indeed, the bulk modulus is widely used in fluid mechanics. From physical reasoning, we conclude that $\kappa \geq 0$.

Finally, using the properties that $E > 0$, $G > 0$ and $\kappa > 0$, we can established certain limits on the Poisson ratio. Since $E > 0$, $G > 0$,

$$G = \frac{E}{2(1 + \nu)} \rightarrow \rightarrow 1 + \nu > 0 \rightarrow \rightarrow -1 < \nu.$$

Similarly, since $E > 0$, $\kappa \geq 0$,

$$\kappa = \frac{E}{3(1 - 2\nu)} \rightarrow \rightarrow 1 - 2\nu \geq 0 \rightarrow \rightarrow \nu \leq 0.5.$$

Therefore we have established bounds on the Poisson ratio; namely

$$-1 < \nu \leq 0.5. \tag{4.4.19}$$

While these are theoretical bounds, the Poisson ratio for real materials is found to fall in the range $0 \leq \nu \leq 0.5$. Typical values for ν and E are: steel, $\nu = 0.30$, $E = 200$ GPa; copper, $\nu = 0.35$, $E = 100$ GPa; aluminum, $\nu = 0.33$, $E = 70$ GPa. (Other typical values of mechanical properties for selected materials are given in Appendix D.)

It is interesting to note from Eq. (4.4.18), that as $\nu \rightarrow 0.5$, the bulk modulus $\kappa \rightarrow \infty$. Thus as ν approaches 0.5, $\Delta \rightarrow 0$; namely, the material becomes incompressible. Thus, whatever the state of stress, the volume of any given element tends to remain constant as $\nu \rightarrow 0.5$.

(b) Elastic strain energy

(i) Development of the concept

The concept of elastic strain energy, as energy 'stored' in a body due to deformation, can be introduced most simply by considering some very familiar examples.

- As a first example, let us consider the operation of a mechanical watch: in order for the hands of the watch to move, one winds a spring. In doing so, one deforms the spring from its relaxed state. The hands of the watch are seen to move as the spring unwinds: energy is transferred (in the form of kinetic energy) to the watch hands by the spring. In effect, energy was 'stored' in the spring due to its initial deformation.
- Another, more simple example, is that of a model airplane that flies under the action of a propeller. In such model airplanes, the propeller rotates due to the unwinding of a rubber band. Thus, by initially twisting the rubber band, one stores energy in it; this stored energy is then released to the propeller as the rubber band unwinds.

These two simple examples illustrate the idea of storage of energy in an elastic body by means of deformation; such energy is called **elastic strain energy**. Having considered this basic concept, we now define this form of energy more precisely in terms of known mechanical quantities.

To this end, let us consider a rod of length L and cross-sectional area $A(x)$ subjected to a uniaxial load P, as shown in Fig. (4.4.11), where the material properties

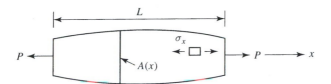

Figure 4.4.11

of the elastic rod are represented by the stress–strain curve shown in Fig. (4.4.12). We note that for the elastic material $\sigma_x = \sigma_x(\epsilon_x)$.

Let us assume that we apply the load P statically, i.e., we start from a zero force and gradually increase P until it reaches its final value $P = P^f$. For any intermediate value $0 \le P \le P^f$, the entire rod will lengthen, causing all elements in the bar to elongate. Consider a small element of cross-section ΔA and original length Δx. The force on this small element will then be $\Delta F = \sigma_x \Delta A$ and the resulting strain will be ϵ_x [Fig. (4.4.13)].

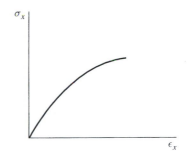

Figure 4.4.12

The length of the element under the load P is then given by $(1 + \epsilon_x)\Delta x$, and its elongation is $\epsilon_x \Delta x$ [see Eqs. (3.3.2)]. Consider now that P is increased by a small amount dP, causing an increase in the strain, $d\epsilon_x$; the element thus elongates by an additional amount $d\epsilon_x \Delta x$. The work done by the stress components when $P \to P + dP$ is then

$$d(\Delta W) = [\sigma_x(\epsilon_x)\,\Delta A](d\epsilon_x\,\Delta x)$$

or

$$(4.4.20)$$

$$d(\Delta W) = [\sigma_x(\epsilon_x)\,d\epsilon_x](\Delta A\,\Delta x).$$

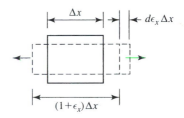

Figure 4.4.13

In the above, ΔW signifies that the work is done on a differential element and 'd' that the work is due to an increase in value of P by dP. Now, if we wish to determine the total work done by the stresses when P goes from zero to its final value P^f, we must sum up all the increments of $d(\Delta W)$. In the limit, this summation becomes an integral and the work done by the stress σ_x on this element is

$$\Delta W = \left[\int_0^{\epsilon_x^f} \sigma_x(\epsilon_x)\,d\epsilon_x\right]\Delta A\,\Delta x, \qquad (4.4.21)$$

where ϵ_x^f denotes the final strain occurring when $P = P^f$.

Letting $\Delta\Omega = \Delta A\,\Delta x$ denote a volume element, the total work W of the internal stresses is obtained by integrating over the volume of the body, V; thus

$$W = \iiint_V \left[\int_0^{\epsilon_x^f} \sigma_x(\epsilon_x)\,d\epsilon_x\right]\Delta\Omega. \qquad (4.4.22)$$

The term in bracket is the work done by the stresses per unit volume and we see that it is a function only of the final strain state, ϵ_x^f.

Now, for an elastic material, the work W is 'stored' within the body as energy. We call this stored energy the **elastic strain energy**, since it is stored in the body as a result of deformation. We shall call the strain energy per unit volume, **strain energy density** and denote it by U_0; thus

$$U_0 = \int_0^{\epsilon_x^f} \sigma_x(\epsilon_x)\,d\epsilon_x. \qquad (4.4.23)$$

The total strain energy stored in the body, U, is then given by

$$U = \iiint\limits_{V} U_0 \, d\Omega. \tag{4.4.24}$$

In passing, it is worth observing that it is possible to represent the strain energy density geometrically. From the known geometric representation of the integral of Eq. (4.4.23), it is clear that U_0 may be represented by the area under the stress–strain curve, as in Fig. (4.4.14a). Indeed, we note that the units of the area are $(N/m^2) \cdot (m/m)$; i.e. $(N\ m)/m^3$ or energy per unit volume.

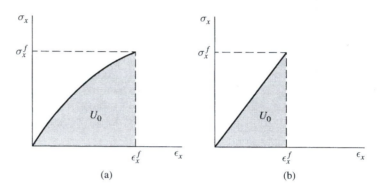

Figure 4.4.14
(a)   (b)

As an example, assume that the elastic material obeys the law $\sigma_x = k\epsilon_x^2$, $\epsilon_x \geq 0$, $k > 0$ [Fig. (4.4.2b)]. Then

$$U_0 = \int\limits_{0}^{\epsilon_x^f} k\epsilon_x^2 \, d\epsilon_x = \frac{1}{3}k\left(\epsilon_x^f\right)^3. \tag{4.4.25}$$

We now consider the special case of the linear isotropic elastic material. For the case of uniaxial stress, $\sigma_x = E\epsilon_x$, since all other stress components vanish for a slender bar. Then, substituting in Eq. (4.4.23),

$$U_0 = E\int\limits_{o}^{\epsilon_x^f} \epsilon_x \, d\epsilon_x = \frac{E}{2}\left(\epsilon_x^f\right)^2. \tag{4.4.26}$$

Using again the stress–strain relation, we may write

$$U_0 = \frac{E\left(\epsilon_x^f\right)^2}{2} = \frac{\sigma_x^f \epsilon_x^f}{2} = \frac{\left(\sigma_x^f\right)^2}{2E}. \tag{4.4.27}$$

These three alternate expressions are equally valid. We also observe that the strain energy density for a linear elastic material is represented by the triangular area in Fig. (4.4.14b).

At this point, we shall simplify our notation. For convenience we shall drop the superscript f and write Eq. (4.4.23) for the general uniaxial case as

$$U_0 = \int\limits_{0}^{\epsilon_x} \sigma_x(\epsilon_x) \, d\epsilon_x. \tag{4.4.28}$$

Similarly, Eq. (4.4.27) for the linear case becomes

$$U_0 = \frac{E\epsilon_x^2}{2} = \frac{\sigma_x \epsilon_x}{2} = \frac{\sigma_x^2}{2E}. \tag{4.4.29}$$

In the above, it must be clearly understood that the quantities represent the *final actual values* of the stress and strain components. Since the modulus $E > 0$, it follows from Eq. (4.4.29) that the strain energy density U_0 is always positive for any $\epsilon_x \neq 0$. Although the proof is beyond the scope of our treatment, we mention here that it can be shown that U_0 is always positive for *any* state of strain. The strain energy U_0 is therefore said to be 'positive definite'.

Our purpose here has been to introduce the concept of strain energy via a simple uniaxial state of stress. In Chapter 14, we shall consider strain energy under general states of stress and strain and shall find that strain energy proves not only to be an important concept in the study of solid mechanics, but also proves to be of great use in solving various types of problems.

There exists, in particular, a very fundamental principle for elastic materials, namely the principle of conservation of energy. We prove this principle here for the simple case of a uniaxial state of stress.

(ii) Conservation of energy

Consider a rod having a cross-sectional area A and length L, which is subjected to an axial load P, as shown in Fig. (4.4.15). The rod then undergoes displacements $u(x)$ under a state of uniaxial stress $\sigma_x = P/A$,[†] where all other stress components vanish. Using Eq. (4.4.29), we first express the strain energy density in the form

$$U_0 = \frac{\sigma_x \epsilon_x}{2}. \tag{4.4.30a}$$

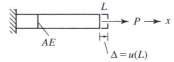

Figure 4.4.15

Substituting for σ_x, and using the strain–displacement relation, $\epsilon_x = \frac{\partial u(x)}{\partial x}$ [Eq. (3.5.18a)],

$$U_0 = \frac{P}{2A} \frac{\partial u(x)}{\partial x}. \tag{4.4.30b}$$

Noting that the elementary volume $d\Omega = A \cdot dx$, the total strain energy U in the body becomes

$$U = \frac{P}{2A} \iiint_V \frac{\partial u(x)}{\partial x}\, d\Omega = \frac{P}{2} \int_0^L \frac{\partial u(x)}{\partial x}\, dx. \tag{4.4.31}$$

Integrating, and observing from Fig. (4.4.15) that $u(0) = 0$, we obtain

$$U = \frac{P}{2} u(x)\big|_0^L = \frac{Pu(L)}{2}. \tag{4.4.32}$$

We recognise the displacement $u(L)$ as the elongation Δ of the rod, i.e. the displacement of the external force P [Fig. (4.4.15b)]. Hence, the right-hand side of Eq. (4.4.32) represents the work $W = \frac{P\Delta}{2}$ of the statically applied force P.[‡] Thus

---

[†] We assume, as before [see footnote p. 123] that $\sigma_x = \bar{\sigma}_x$.

[‡] For a linear elastic material, we may give the following heuristic explanation for the '1/2' term: If a load is applied *statically* to a linear body, then the 'average' force applied is equal to the sum of one-half the initial (zero) force and the final force P. The work done is then the product of the 'average' force and the displacement through which it acts.

we have

$$U = W. \tag{4.4.33}$$

This last relation leads to the statement of the *principle of conservation of energy*:

> If a linear elastic body is in equilibrium under an external force system, then the internal strain energy due to deformation is equal to the work of the externally applied force system.

This basic principle is, in fact, applicable to all elastic bodies. The above restricted proof was confined to a linear elastic body under a uniaxial state of stress. In Chapter 14, we shall provide a more general proof for a body under a general state of stress.

4.5 Mechanical properties of engineering materials

(a) Behaviour of ductile materials

We discuss here some of the mechanical properties of materials that are commonly encountered in engineering practice. While we recall that a description of the behaviour of materials was given in Chapter 1, we elaborate here on the behaviour and properties of materials and define certain terms in a systematic manner as they appear in our discussion.

A material commonly used in engineering practice is structural steel, which consists mainly of iron combined with a small percentage of carbon. We develop several ideas and definitions based on a description of a tension test on steel since the behaviour of this material is typical of a number of ductile metals.

> *Ductility*: The property of a material that enables it to undergo plastic deformation to a considerable extent and to sustain a load before fracture. A material that is not ductile is said to be 'brittle'.

As in our previous discussion, we describe a simple standard tension test on a specimen (having cross-sectional area A_0 and gauge length L_0) under a statically increasing applied load P. By measuring the elongation ΔL at incremental steps of P,[†] we obtain the typical $\sigma-\epsilon$ curve [Fig. (4.5.1)] where σ and ϵ are given by

$$\sigma = \frac{P}{A_0}, \qquad \epsilon = \frac{\Delta L}{L_0}. \tag{4.5.1}$$

For reasons that will become clear, the stress σ and strain ϵ, as calculated above, are called the **nominal stress** and **nominal strain**, respectively. The corresponding $\sigma-\epsilon$ curve is referred to as the **nominal stress–strain curve**.

Initially the $\sigma-\epsilon$ curve is linear and follows Hooke's law with $\sigma = E\epsilon$. The linear relation is valid for all stresses that do not exceed the 'proportional limit' σ_p. Note that E is represented by the slope of the $\sigma-\epsilon$ curve.

> *Proportional Limit* (σ_p): The largest stress which a material is capable of sustaining without deviating from Hooke's law.

Initially, the material behaves elastically but if stressed to some value that exceeds σ_E, the Elastic limit, the material ceases to behave elastically.

[†] We should note that in an actual tension test, one applies a deformation (i.e. strain) in incremental steps and, via a device called a **loading cell**, one then measures the applied load.

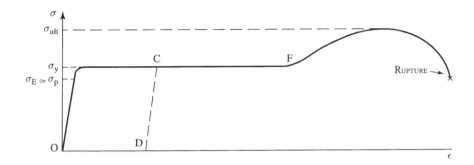

Figure 4.5.1

Elastic Limit (σ_E): The largest stress under which the material behaves elastically and for which no permanent deformation exists when a load is removed.

Note here that, by definition, σ_p and σ_E are *different* points. However, for steel and for many other ductile metals, the difference between them on the $\sigma - \epsilon$ curve is often indistinguishable so that $\sigma_E \simeq \sigma_p$. Hence, for such metals, they may be considered, in practice, to coincide.

As the material is stressed beyond σ_p, the curve deviates from a straight line and at some value of stress the curve becomes horizontal. The material is thus said to 'yield'; that is, with no apparent increase in stress, the material undergoes increasing deformation. The stress at which this yielding occurs is called the **yield point** or **yield stress**.

Yield stress or yield point (σ_y): The stress in a material at which there occurs a large increase in strain with no appreciable increase in stress.

Having reached the yield stress, the material undergoes increasing strain (several orders of magnitude greater than that at the onset of yielding, ϵ_y). This stage is referred to as *plastic deformation*. Now, let us imagine that the specimen yields until it reaches some strain, say point C of the $\sigma - \epsilon$ curve, and that upon reaching this strain, the load is slowly removed; that is, the stress is reduced to zero. The material is said to undergo 'unloading'. The unloading process is described by an 'unloading path', which is found to be parallel to the original linear elastic curve. [Thus, the unloading path in this case is from point C to point D of Fig. (4.5.1).] If the specimen is now reloaded, the reloading curve will follow the straight line DC, until it rejoins the original $\sigma - \epsilon$ curve. The specimen will then continue to yield along the horizontal segment of the $\sigma - \epsilon$ curve until some point, point F, after which an increase in stress is required for any further deformation. This latter phenomenon is known as **strain-hardening**. The stress is then observed to increase until it reaches some maximum (ultimate) value, the *nominal ultimate stress*, which we denote as σ_{ult}. Having reached this point, we observe that the material then yields rapidly under a *decreasing* stress until the specimen finally ruptures.

Nominal ultimate tensile stress (σ_{ult}): The maximum stress, in a tension test, calculated as $\sigma_{ult} = P_{max}/A_0$ (where A_0 is the undeformed original cross-sectional area), which a specimen is capable of sustaining.

An interesting and perplexing question can now be posed: why does the material apparently rupture under a stress that is less than the nominal ultimate tensile stress. The key to the answer lies in the term 'nominal'. To explain this apparent paradox, it is first necessary to describe more precisely the deformation of the specimen.

We recall that, according to Hooke's law for linear elastic behaviour, when an element is subjected to a tensile stress $\sigma_x > 0$, lateral contractions occurs in the cross-section of the element. Clearly, the same is true for the overall dimensions of the cross-section of the specimen. Although the lateral contractions are quite small while in the elastic range, with increasing deformation of a ductile material, these lateral contractions become quite large and important. As a result, the original cross-sectional area A_0 of the specimen decreases considerably in a region of large deformation. This effect, which starts as P approaches P_{max}, is known as 'necking' of the specimen [as shown in Fig. (4.5.2)] and becomes particularly significant as σ approaches σ_{ult}. As a result, a cross-section of initial area A_0 is reduced, in the region of necking, to an area A, which is considerably smaller than A_0. Consequently, the average stress, calculated according to $\sigma = P/A_0$, yields but a *nominal* value; we therefore now write $\sigma_{nom} \equiv \sigma_{ult} = P/A_0$.

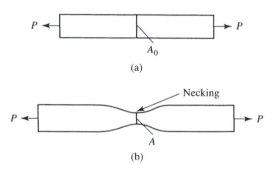

(a)

(b)

Figure 4.5.2

The more exact average stress, i.e. 'true' average stress in this region, given instead by $\sigma_{true} = P/A$, is a more accurate reflection of the state of stress.[†] Thus, specifically, while the nominal ultimate tensile stress is calculated as $\sigma_{ult} = P_{max}/A_0$, the true average stress under this load is $\sigma_{true} = P_{max}/A$, from which we note that $\sigma_{true} > \sigma_{ult}$. As the ratio A/A_0 during necking decreases (in fact just prior to rupture it may be of the order of 0.10), the ratio $\sigma_{true}/\sigma_{nom}$ can become quite large.

If we now examine the strains, the expression for the strain $\epsilon = \Delta L/L_0$, while quite valid for small strains (of the order of 10^{-3}), fails to provide a good or 'true' measure of deformation for larger deformations. We therefore refer to this strain as the 'nominal strain', i.e., $\epsilon_{nom} = \Delta L/L_0$. For large deformations, a more physically significant measure of the strain would be to consider the small change of strain occurring for each incremental increase of length $d\ell/\ell$ (where ℓ is the current length at any stage of the deformation). Then the 'true' strain ϵ_{true} is given by

$$\epsilon_{true} = \sum d\epsilon = \sum \frac{d\ell}{\ell} \tag{4.5.2a}$$

or in the limit

$$\epsilon_{true} = \int_{L_0}^{L} \left(\frac{d\ell}{\ell}\right) = \ln\left(\frac{L}{L_0}\right). \tag{4.5.2b}$$

Then, since the final length L, expressed in terms of the nominal strain ϵ_{nom}, is given

---

[†] We note that the stress σ_{true} is the *average* true stress. The actual state of stress in the region of necking and final rupture is found to be extremely complex and cannot be treated by the methods developed here.

by $L = (1 + \epsilon_{\text{nom}})L_0$, we find the relation between the nominal and true strain:

$$\epsilon_{\text{true}} = \ln(1 + \epsilon_{\text{nom}}). \tag{4.5.3}$$

Now, as mentioned at the beginning of the discussion, the $\sigma-\epsilon$ curve, as shown in Fig. (4.5.1), represents a *nominal* stress–strain curve. If we were to plot σ_{true} vs. ϵ_{true}, the curve would appear as the dashed curve in Fig. (4.5.3). Thus we have essentially answered the paradox: the specimen only *appears* to rupture under a stress smaller than the nominal ultimate stress; in fact, the true average stress at rupture is much larger than σ_{ult}. Note that the nominal and true $\sigma-\epsilon$ curves differ only in the region of relatively large strains.

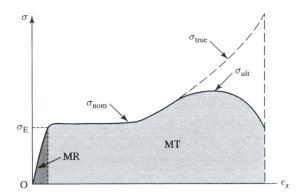

Figure 4.5.3

However, although it is clear that the nominal stress does not accurately reflect the true stress state during plastic flow (yielding), the nominal ultimate tensile stress σ_{ult} is nevertheless a useful quantity since it serves as a nominal measure of the stress that the material is capable of sustaining. Thus, for example, after calculating the maximum stress due to a given load on a structure or body, it is possible to determine whether the calculated stress is acceptable by comparing with the σ_{ult} according to some established criteria (e.g., a factor of safety). Moreover, although σ_{ult} is but a nominal quantity, it provides us with a measure of the *relative* strengths of various materials; thus, for example, one may state that steel (with $\sigma_{\text{ult}} = 400$ MPa) is far stronger in tension than cast iron (with $\sigma_{\text{ult}} = 170$ MPa).

We recall now that the area under the $\sigma-\epsilon$ curve represents the energy of deformation per unit volume. As a result, the area under the $\sigma-\epsilon$ curve up to σ_{E}, the elastic limit, represents the maximum *elastic* strain energy of deformation (per unit volume). Accordingly, the following property is defined:

Modulus of resilience (MR): The modulus of resilience is the greatest strain energy (per unit volume) that a body can absorb without undergoing any permanent deformation. It is calculated as the area under the $\sigma-\epsilon$ curve for $\sigma \leq \sigma_{\text{E}}$ in a tension test [Fig. (4.5.3)]. (For the common case $\sigma_{\text{E}} \simeq \sigma_{\text{p}}$, MR $= \frac{\sigma_{\text{E}}^2}{2E}$.)

Similarly, we define another quantity related to energy absorption of a material:

Modulus of toughness (MT): The modulus of toughness is calculated as the total area under the *nominal* stress–strain curve of a material from its undeformed state until rupture. It represents the nominal maximum strain energy (per unit volume) that a body can absorb before fracture, or conversely, the strain energy (per unit volume) required to cause a material to fracture.

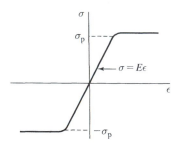

Figure 4.5.4

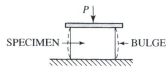

Figure 4.5.5

It is evident that MT, as defined above, is a rather fictitious quantity and does not represent the real energy of deformation of any given material. However, it provides a useful measure of the *relative* ability of various materials to absorb energy. Thus, for two different materials of the same strength, the MT of a very ductile material will be much greater than that of a moderately ductile material.

From the above discussion, it is clear that the behaviour of materials in the range beyond the elastic limit is considerably more complicated than the initial elastic behaviour. To treat problems in this range, certain simplifying models are used. These will be discussed in the next section.

Having described the behaviour of steel as a ductile material under tension, we now consider the behaviour in compression. From laboratory tests, we observe that initially, in the elastic range, the σ–ϵ relation is the mirror image of tension; that is, the behaviour follows Hooke's law as shown in Fig. (4.5.4). Note that the modulus of elasticity E is the same in compression as in tension. Furthermore, for large compressive strains, no necking occurs; instead local bulging, which causes a slight increase in the cross-sectional area, may occur [Fig. (4.5.5)]. However, this bulging effect is not as significant as the necking effect and therefore the nominal and true σ–ϵ curves in compression are approximately the same.

In the above discussion of the behaviour of steel as a ductile material in tension, we note that a sharp yield stress σ_y exists. However, we mention here that there also exist materials, such as aluminum alloys which, although they exhibit ductile behaviour, do not possess a definite yield stress; their σ–ϵ curve appears as in Fig. (4.5.6a).

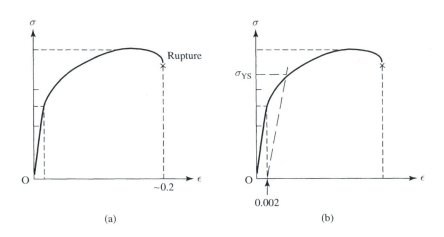

Figure 4.5.6

(a)   (b)

If such a material is to be used in engineering design, it is necessary to define some measure of an allowable design stress. Since, as we have observed no yield point exists, one must therefore choose a value arbitrarily; this value is called the **yield strength** or alternatively the **proof stress** and will be denoted by σ_{YS} (to distinguish it from the measured yield stress σ_y). A standard method of defining the yield strength is to first choose a strain arbitrarily and, from this point, draw a straight line parallel to the initial slope of the σ–ϵ curve. The stress at which this line intersects the σ–ϵ curve is then defined as the yield strength [Fig. (4.5.6b)]. This common method is referred to as the 'offset method' and one therefore refers to the 'yield strength for a given percent offset' or the 'percent proof stress'. For example, the yield strength shown in the figure is for a typical offset of 0.2%.

Finally, we mention here that some ductile materials may indeed initially be-have elastically but not linearly. The results of a standard tension test can then be expressed by nonlinear empiric expressions, which represent closely the resulting $\sigma-\epsilon$ curve. One such typical expression is the Ramberg–Osgood equation, valid for loading behaviour of ductile materials; namely

$$\epsilon = \frac{\sigma}{C_1} + C_2\left(\frac{\sigma}{C_3}\right)^n, \qquad (4.5.4)$$

where C_1, C_2, C_3 are constants and n is an integer.

Another approximation for ductile materials, for example copper, is the so-called **sinh law**, given as

$$\epsilon = \epsilon_0 \sinh(\sigma/\sigma_0), \qquad (4.5.5)$$

where ϵ_0 and σ_0 are prescribed constants.

(b) Behaviour of brittle materials

Brittle materials are characterised by their inability to undergo large deformation; hence a material which is not ductile, is said to be brittle. Cast iron, concrete, stone and ceramics are typical examples of brittle materials. The $\sigma-\epsilon$ curve of a brittle material in a standard tension test has the usual form as shown in Fig. (4.5.7a). We note that this curve is characterised by the absence of a yield point. The material ruptures at the maximum value of the attained stress; we observe that the largest strain ϵ which a brittle material attains is quite small and that no necking occurs. Thus, in a tension test for brittle materials, no distinction is made between the nominal and true $\sigma-\epsilon$ curves.

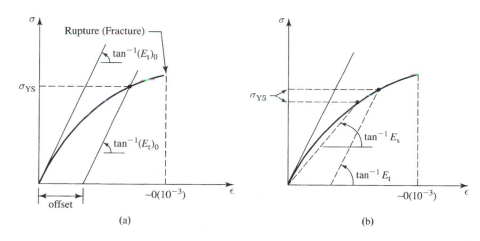

(a)　　　　(b)　　　　**Figure 4.5.7**

The maximum strain is often given as a percentage. Thus, the brittleness (or conversely the ductility) of the material is often measured by the maximum strain that a material can undergo before fracturing; this quantity is often de-fined as the **percentage of elongation (or strain) in a specified (original) gauge length**:

$$\frac{L - L_0}{L_0} \cdot 100,$$

where L and L_0 are the final and original gauge lengths of the specimen.

Moreover, very often, the initial behaviour of brittle materials, while approximating a linear behaviour, is not always truly linear. In this case, it is convenient to describe the initial behaviour as following Hooke's law, using an approximate value for the modulus of elasticity. Recalling that E for a linear elastic material is represented by the slope of the $\sigma-\epsilon$ curve, we define the 'tangent modulus' E_t as the tangent to the $\sigma-\epsilon$ curve, i.e., $E_t = d\sigma/d\epsilon$ at a given point on the curve. In particular, at the point ($\sigma = 0$, $\epsilon = 0$), we obtain the *initial* tangent modulus $(E_t)_0$ [see Fig. (4.5.7a)].

As discussed above for the case of ductile materials having no definite yield stress (yield point), if a brittle material is to be used in engineering design, it is also necessary to define arbitrarily a yield strength or proof stress, σ_{YS}, as a measure of an allowable design stress. Since, as we have observed, no yield stress exists for brittle materials, one must choose a value arbitrarily. As previously described, the yield strength σ_{YS} can be defined as the intersection of the $\sigma-\epsilon$ curve with a straight line parallel to the initial tangent of the $\sigma-\epsilon$ curve that passes through the arbitrarily chosen strain offset [Fig. (4.5.7b)]. Note that here one must first determine the initial tangent modulus $(E_t)_0$.

Another method for establishing the yield strength is to arbitrarily decide on its value as a fraction of the ultimate strength of the material. For cast iron with an ultimate strength in tension of $\sigma = 170$ MPa, one might choose to arbitrarily define the yield strength, say, as $\sigma_{YS} = 100$ MPa. Corresponding to this stress value, we may also arbitrarily define the **secant modulus of elasticity**, E_s, as the slope of a straight line between the origin ($\sigma = \epsilon = 0$) and the intersection of the $\sigma-\epsilon$ curve at σ_{YS} [Fig. (4.5.7b)].[†] Using this method, one first chooses arbitrarily the strain offset and then determines the secant modulus E_s.

It is clear that the modulus of toughness (as represented by the area of the $\sigma-\epsilon$ curve up to the point of rupture) of a brittle material is far less than that for ductile materials. Indeed, a main characteristic of brittle materials is their inability to absorb energy of deformation. This explains, for example, why if a piece of chalk – a typical brittle material – is dropped from a relatively low height, it will fracture immediately upon hitting a rigid surface, while the same piece, if made of rubber, will deform without breaking.

The modulus of resilience and modulus of toughness, being measures of the ability of a material to absorb energy, are important factors in designing structural parts to resist impact or dynamic loads.

The general shape of the stress–strain curve for brittle materials in compression resembles closely that of a tension test with a significant exception: the compressive stress at which fracture occurs is far greater than the maximum tensile stress. Indeed, it may often be greater by an order of magnitude. It is a characteristic of brittle materials that they are relatively strong in compression and particularly weak in tension.

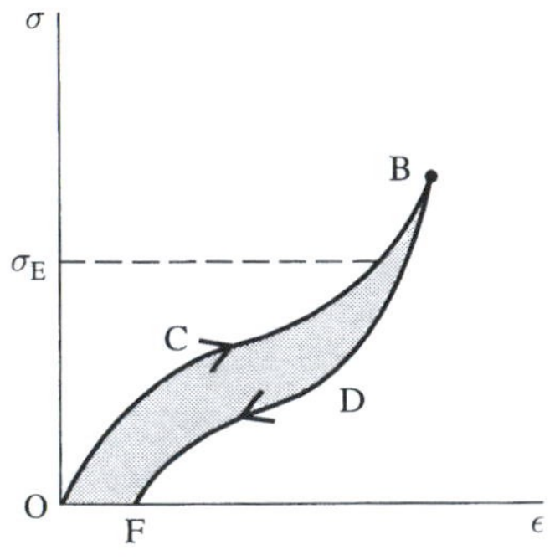

(c) Behaviour of rubber-like materials

The initial behaviour of plastics or rubber-like materials is generally elastic but nonlinear and can be described by a stress–strain curve as shown in Fig. (4.5.8). We note the absence of a yield point. Thus, if a material having a given elastic

Figure 4.5.8

[†] It is evident that the values of σ_{YS} obtained by these two methods are not necessarily the same since they are determined arbitrarily. The two different values of σ_{YS} are shown in Fig. (4.5.7b).

limit σ_E, as in the figure, is subjected to a stress $\sigma \leq \sigma_E$, the behaviour will be elastic.

At this point, it is worthwhile to recall again that the area under the $\sigma-\epsilon$ curve within the elastic range represents the elastic energy absorbed by a unit element of the material. According to our definition of elastic behaviour, if the stress $\sigma < \sigma_E$ is now removed, the material will 'unload' along the unique $\sigma-\epsilon$ curve. In the process of unloading, energy is retrieved; the material is said to 'give off' or 'return' all the stored energy.

However, let us now consider the case where the specimen is loaded along the $\sigma-\epsilon$ curve to a stress $\sigma > \sigma_E$, say along the curve OCB. Since $\sigma > \sigma_E$, it will then unload along the curve BDF (since, by definition it no longer behaves elastically). As we have observed, energy is retrieved from the material during the unloading process. However, we note that the area under the curve BDF is less than that under the original elastic curve OCB. Thus we conclude that some of the energy is not retrieved, namely that represented by the area OCBDF (shown shaded in the figure). We therefore conclude that this area represents *dissipation* of energy.

From the above discussion, we reach an important conclusion: following its basic definition, an elastic material is one for which no dissipation of energy can take place. Note that this conclusion follows directly from the unique one-to-one relation [Eqs. (4.4.1)] between stress and strain of the material.

4.6   Plastic behaviour: idealised models

As we have seen, the behaviour in the plastic range is much more complex than the simple relations governing elastic materials: for example, Hooke's law. Thus if we consider the stress–strain curve of Fig. (4.5.1), it is clear that some simplifying assumptions must be made if one is to treat a problem of such a material outside the elastic range. One therefore must model the material in such a way that it adequately approximates the behaviour of the material. Noting that the stress–strain curve is approximately horizontal for a large range of strains, for example, $10^{-3} \leq \epsilon \leq 0.1$ (i.e. for about two orders of magnitude), and observing that the yield point σ_y does not deviate greatly from the σ_p, a reasonable model is to assume that the $\sigma-\epsilon$ curve is as shown in Fig. (4.6.1) with σ_0 representing the yield stress. Note that in this case, we assume implicitly that $\sigma_0 = \sigma_p = \sigma_y$. This simplification permits one to obtain reasonable solutions to many problems that would otherwise prove to be intractable. The material represented by this $\sigma-\epsilon$ curve is called an **ideal elastic–plastic** material or a **perfect elastic–plastic** material. Note too that in this model the unloading path is parallel to the initial elastic load path.

A further simplification can also be made. Since the elastic behaviour of the material occurs within a small range of the strains (i.e., the major behaviour takes place in the plastic range) one chooses, at times, to neglect the elastic range completely. The resulting $\sigma-\epsilon$ curve representing this model then appears as in Fig. (4.6.2). Thus, according to this model, the material undergoes no deformation (i.e. it remains rigid) provided $\sigma < \sigma_0$. Hence such a material is referred to as a **rigid plastic** material.

The phenomenon of strain-hardening, discussed in Section 5, can also be treated by means of a simplifying model. Depending on the given material, one can, for example, approximate the $\sigma-\epsilon$ curve by means of two straight lines as shown in Fig. (4.6.3). In this case, the material is known as an *elastic strain-hardening* material or a *linear-hardening material*. Such a material is typically described by

Figure 4.6.1

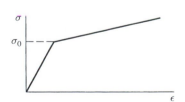

Figure 4.6.2

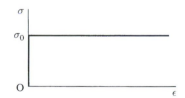

Figure 4.6.3

the relations

$$\epsilon = \begin{cases} \dfrac{\sigma}{E} & \sigma \le \sigma_0 \\ \dfrac{1+\alpha}{\alpha E}\sigma - \dfrac{\sigma}{\alpha E}, & \sigma \ge \sigma_0, \end{cases} \tag{4.5.6}$$

where α is a constant, which depends on the material.

We thus observe that one may choose various models to describe the material. The type and choice of the model will clearly depend on the type and range of behaviour that is to be studied.

PROBLEMS

4.1: The state of stress in a steel plate lying in the x–y plane ($E = 200$ GPa, $\nu = 0.3$) is given as $\sigma_x = 20$ MPa, $\sigma_y = -30$ MPa, $\tau_{xy} = 40$ MPa, $\sigma_z = \tau_{yz} = \tau_{xz} = 0$. Determine the principal strains and the principal directions with respect to the x-axis.

4.2: Show that the principal directions of strain are normal to the principal stress planes at any point of a linear isotropic elastic material.

4.3: Show that for a state of plane stress in the x–y plane ($\sigma_z = \tau_{xz} = \tau_{yz} = 0$), Hooke's law relating the extensional strain components to the extensional stresses may be written as

$$\sigma_x = E\,\frac{\epsilon_x + \nu\epsilon_y}{1 - \nu^2}, \qquad \sigma_y = E\,\frac{\epsilon_y + \nu\epsilon_x}{1 - \nu^2},$$

and that

$$\epsilon_z = -\frac{\nu}{1-\nu}(\epsilon_x + \epsilon_y).$$

4.4: Recalling that for a state of plane strain in the x–y plane, $\epsilon_z = 0$, show that the extensional strains, ϵ_x and ϵ_y, from Hooke's law are given by

$$\epsilon_x = \frac{1}{E^*}[\sigma_x - \nu^*\sigma_y], \qquad \epsilon_y = \frac{1}{E^*}[\sigma_y - \nu^*\sigma_x],$$

where

$$E^* = \frac{E}{(1-\nu^2)} \quad \text{and} \quad \nu^* = \frac{\nu}{(1-\nu)}.$$

and that $\sigma_z = \nu(\sigma_x + \sigma_y)$

4.5: From strain rosette measurements at a point on the surface of a thin aluminum plate ($E = 70$ GPa, $\nu = 0.30$) lying in the x–y plane, the following strain components are known:

$$\epsilon_x = 60\mu, \qquad \epsilon_y = 30\mu, \qquad \epsilon_{xy} = 15\mu.$$

Using the results of Problem 4.3, determine the principal stresses σ_1 and σ_2.

4.6: A linear elastic plate with modulus of elasticity E and Poisson ratio ν is subject to a uniform compressive stress σ_0, as shown in Fig. (4P.6), such that at all points the only non-zero stress is $\sigma_x = \sigma_0$. (a) Show that the change in slope of line AC, $\Delta \equiv \tan(\alpha + \delta\alpha) - \tan\alpha$, is given by

$$\Delta = \frac{b}{a}\left[\frac{1 + (\nu\sigma_0/E)}{1 - (\sigma_0/E)} - 1\right]$$

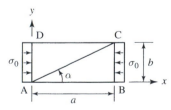

Figure 4P.6

(b) If $\sigma_0/E \ll 1$ show that $\Delta \approx (b/a)(1 + \nu)(\sigma_0/E)$. (c) What is Δ if the material is incompressible? (d) Calculate (i) the change in slope and (ii) the change of *angle*, $\delta\alpha$ (in degrees), if $a = 2b$, $\nu = 0.25$ and $\epsilon_x = 10^{-3}$. (e) Derive an expression for the change of *angle*, $\delta\alpha$ in terms of ν, and the ratios b/a and σ_0/E. (f) Re-evaluate $\delta\alpha$ using the numerical values given in (d) above.

4.7:   Hooke's law can be written in the form

$$\sigma_x = \lambda\Delta + 2\mu\epsilon_x, \qquad \sigma_y = \lambda\Delta + 2\mu\epsilon_y, \qquad \sigma_z = \lambda\Delta + 2\mu\epsilon_z,$$
$$\tau_{xy} = 2\mu\epsilon_{xy}, \qquad \tau_{yz} = 2\mu\epsilon_{yz}, \qquad \tau_{zx} = 2\mu\epsilon_{zx},$$

where $\Delta = \epsilon_x + \epsilon_y + \epsilon_z$ is the dilatation and λ and μ are called the **Lamé constants**. Show that the following relations exist (i) between E, ν and λ, μ and (ii) between E, G and λ, μ:

$$\text{(i)} \quad \lambda = \frac{E\nu}{(1 - 2\nu)(1 + \nu)}, \qquad \mu = \frac{E}{2(1 + \nu)}$$

$$\text{(ii)} \quad \mu \equiv G, \qquad \lambda = \frac{G(E - 2G)}{3G - E}.$$

4.8:   Using the results of Problem 4.7, show that an alternative expression for the bulk modulus κ as defined in Eq. (4.4.18), written in terms of λ and μ, is $\kappa = \frac{3\lambda + 2\mu}{3}$.

4.9:   A hard rubber cylinder ($E = 1.5$ MPa, $\nu = 0.40$), inserted in a pressurised tank, is subjected to a hydrostatic pressure p of 10 MPa; i.e., $\sigma_x = \sigma_r = \sigma_\theta = -p$ at all points within the body. If the cylinder is 20 cm in diameter and has a height h of 50 cm, determine the change in (a) diameter, (b) height of the cylinder and (c) volume. *Note:* Assume that the cylinder behaves as a linear isotropic elastic material.

4.10:   The data given in the table below was obtained from a tensile test of a 1.50-cm diameter specimen of a magnesium alloy. A 5-cm gauge length extensometer was used. (a) Plot the stress–strain curve; (b) determine the proportional limit and the elastic modulus; (c) determine the yield strength for a 0.2% offset; (d) determine both the tangent modulus E_t and the secant modulus E_s, for a yield strength $\sigma_{YS} = 260$ MPa.

Load (N)	Extensometer reading (cm)	Load (N)	Extensometer reading (cm)
0	0	43,649	0.0300
3,890	0.0025	36,370	0.0325
7,770	0.0050	46,370	0.0350
11,650	0.0075	50,165	0.0375
15,540	0.0100	51,345	0.0400
19,430	0.0125	52,030	0.0425
23,310	0.0150	52,215	0.0475
27,200	0.0175	52,340	0.0525
31,080	0.0200	52,430	0.0575
34,440	0.0225	52,460	0.0625
37,790	0.0250	52,490	0.0700
40,780	0.0275	Data stopped	No rupture

4.11:   Show that $\epsilon_{\text{true}} > \epsilon_{\text{nom}}$, where ϵ_{true} is defined in Eq. (4.5.2b) and where $\epsilon_{\text{nom}} = \Delta L/L_0 \ll 1$.

4.12:   The Ramberg–Osgood equation describing the stress–strain curve of a material during loading is given as

$$\epsilon = \frac{\sigma}{C_1} + C_2 \left(\frac{\sigma}{C_3} \right)^3,$$

where $C_1 = 1.5 \times 10^{11}, C_2 = 200, C_3 = 2.5 \times 10^{10}$. (a) Determine (i) the initial tangent modulus $(E_t)_0$, (ii) the tangent modulus E_t and the secant modulus E_s for a given yield strength $\sigma_{YS} = 175$ MPa. (b) Determine the elastic energy U_0 per unit volume stored in the material (N m/m³) if the material is loaded to a stress $\sigma = 175$ MPa.

4.13:   Given a material whose loading curve is represented by the Ramberg–Osgood equation [Eq. (4.5.4)], with C_1, C_2, C_3 positive. Show, for any stress σ, that the secant modulus is always greater than the tangent modulus, i.e., $E_s/E_t > 1$ for any integer value $n > 1$.

The following problems are to be solved using a computer.

4.14:   Write a program to plot the ratio κ/E as a function of Poisson's ratio ν and plot this ratio for $0 \leq \nu < 0.5$.

4.15:   Using the results of Problem 4.7, write a program to plot the ratio λ/E as a function of Poisson's ratio ν and plot this ratio for $0 \leq \nu < 0.5$.

5

Summary of basic results and further idealisations: solutions using the 'mechanics-of-materials' approach

5.1 Introduction

In this brief chapter, we review and summarise the previously developed results. In particular, we recall that three fundamental relations have been derived and developed, namely (a) the equations of mechanics, (b) the kinematic equations and (c) the constitutive equations. These relations must, in general, be satisfied at all points within a body.

In the following chapters, we shall use the derived relations to analyse a number of problems of practical interest. However, since a major portion of our future study will be devoted to the analysis of linear elastic members – rods, beams, shafts, etc. – it is worthwhile and instructive to first discuss these problems from an overall, or general, point of view.

We recall that, in principle, our goal in solid mechanics is to determine internal forces and describe the deformation of a body when subjected, say, to external forces; specifically, we wish to determine the following quantities at all points P in a body [Fig. (5.1.1)]:

- three displacement components: u, v, w
- six strain components: ϵ_x, ϵ_y, ϵ_z, ϵ_{xy}, ϵ_{yz}, ϵ_{zx}
- six stress components: σ_x, σ_y, σ_z, τ_{xy}, τ_{yz}, τ_{zx}

We thus observe that, essentially, there exist 15 unknown quantities at each point of a body. However, as shown in Table 5.1, there also exist 15 equations for these unknowns. It is therefore reasonable to assume, in principle, that one can theoretically solve for the unknowns. If unknowns are found which satisfy all the given equations as well as the boundary conditions on the body (e.g., the applied forces), the solution is then said to be an *exact* solution to a given problem. This approach is that of the *theory of elasticity*, and in particular, the equations in Table 5.1 are referred to as the *equations of linear elasticity*. However, this approach is usually quite mathematical and, at this stage, is beyond the scope of our present study.

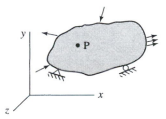

Figure 5.1.1

145

Table 5.1   Summary of unknowns and relations – linear isotropic elastic bodies

Unknowns	Equations
Displacements	**Equations of equilibrium**
u, v, w (3 unknowns)	$\frac{\partial \sigma_x}{\partial x} + \frac{\partial \tau_{yx}}{\partial y} + \frac{\partial \tau_{zx}}{\partial z} + B_x = 0$
	$\frac{\partial \tau_{xy}}{\partial x} + \frac{\partial \sigma_y}{\partial y} + \frac{\partial \tau_{zy}}{\partial z} + B_y = 0$
	$\frac{\partial \tau_{xz}}{\partial x} + \frac{\partial \tau_{yz}}{\partial y} + \frac{\partial \sigma_z}{\partial z} + B_z = 0$
Strain components	**Kinematic equations: strain–displacement relations**
$\epsilon_x, \epsilon_y, \epsilon_z, \epsilon_{xy}, \epsilon_{yz}, \epsilon_{zx}$	$\epsilon_x = \frac{\partial u}{\partial x}, \quad \epsilon_y = \frac{\partial v}{\partial y}, \quad \epsilon_z = \frac{\partial w}{\partial z}$
(6 unknowns)	$\epsilon_{xy} = \frac{1}{2}\left(\frac{\partial v}{\partial x} + \frac{\partial u}{\partial y}\right), \quad \epsilon_{yz} = \frac{1}{2}\left(\frac{\partial w}{\partial y} + \frac{\partial v}{\partial z}\right),$
	$\epsilon_{zx} = \frac{1}{2}\left(\frac{\partial u}{\partial z} + \frac{\partial w}{\partial x}\right).$
Stress components	**Constitutive equations: Hooke's law**
$\sigma_x, \sigma_y, \sigma_z, \tau_{xy}, \tau_{yz}, \tau_{zx}$	$\epsilon_x = \frac{1}{E}[\sigma_x - \nu(\sigma_y + \sigma_z)]$
(6 unknowns)	$\epsilon_y = \frac{1}{E}[\sigma_y - \nu(\sigma_z + \sigma_x)]$
	$\epsilon_z = \frac{1}{E}[\sigma_z - \nu(\sigma_x + \sigma_y)]$
	$\epsilon_{xy} = \frac{\tau_{xy}}{2G}, \quad \epsilon_{zx} = \frac{\tau_{zx}}{2G}, \quad \epsilon_{yz} = \frac{\tau_{yz}}{2G}$
Number of unknowns: 15	**Number of equations: 15**

Nevertheless, there exists another fruitful, but simpler approach; namely that of *mechanics of materials*. Using this approach, instead of attempting to satisfy all the relevant equations at *all* points of the body, we seek solutions that satisfy the relevant equations globally; for example, at the cross-sections of a rod or a beam. This approach can then lead to either exact or approximate solutions, depending on the problem at hand. Indeed, one is often satisfied to obtain approximate but reasonably accurate solutions, which are of practical importance for a wide range of engineering structures; such solutions are often called **engineering solutions**. In the case of approximate solutions, one can then determine the degree of accuracy by substituting the solution back in the exact equations of elasticity.

In the following chapters, we shall apply the mechanics-of-materials approach to various types of problems and, in particular, we shall study the behaviour of simple bodies that are subjected to various loading conditions. As we have seen previously, in practice, most solids undergo rather small deformations while in the elastic range. Since this is particularly true of most engineering structures encountered in practice, in the subsequent chapters, we therefore shall generally limit our treatment to bodies undergoing small strains (and rotations). Consequently, in addition to the equations of equilibrium, and Hooke's law, we note that the strain–displacement relations are also linear.[†] Thus, as a result of this limitation, the equations governing the behaviour of the mechanical system (e.g., those shown in Table 5.1) are all linear and the mechanical system itself is therefore said to be linear. Linear systems possess an important property, namely the property of superposition. We demonstrate below that, subject to the above conditions, this property, known as the **principle of superposition**, can be applied for both strains and stresses.

[†] We recall that infinitesimal strains also imply linearity since the strains are expressed in terms of linear spatial derivatives of the displacements of the body where all quadratic spatial derivative of the displacements are neglected. (See footnote p. 87).

5.2 Superposition principles

(a) Superposition of infinitesimal strains

Consider a body that, due to various causes, undergoes deformations where all strains are infinitesimal. If such a body is subjected only to infinitesimal strains, it follows that the strains are simply additive and that we may therefore superimpose the strains. To show this, let us, for simplicity, again consider the case of a rod that undergoes axial elongation. For example, assume that the rod is first heated and that due to the temperature increase, the strain of any element Δx is given by $\epsilon_x^{(1)}$ [Fig. (5.2.1a)]. The length of the element therefore becomes [see Eq. (3.3.2a)]

$$\Delta x^* \simeq \left[1 + \epsilon_x^{(1)}\right]\Delta x. \tag{5.2.1a}$$

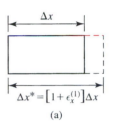

(a)

Now, let us assume that in addition the rod is also subjected to an axial force P such that each element undergoes an *additional* strain, $\epsilon_x^{(2)}$ [Fig. (5.2.1b)]. Clearly, due to $\epsilon_x^{(2)}$, the element will change length from Δx^* to Δx^{**}; thus we have

$$\Delta x^{**} \simeq \left[1 + \epsilon_x^{(2)}\right]\Delta x^*. \tag{5.2.1b}$$

Substituting Eq. (5.2.1a),

$$\Delta x^{**} \simeq \left[1 + \epsilon_x^{(2)}\right]\left[1 + \epsilon_x^{(1)}\right]\Delta x$$

or

$$\Delta x^{**} \simeq \left[1 + \epsilon_x^{(1)} + \epsilon_x^{(2)} + \epsilon_x^{(1)}\epsilon_x^{(2)}\right]\Delta x. \tag{5.2.1c}$$

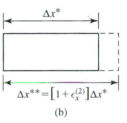

(b)

Figure 5.2.1

If the strains are infinitesimal, the last terms above are infinitesimals of higher order and therefore, keeping only the linear terms, we have

$$\Delta x^{**} \simeq \left[1 + \epsilon_x^{(1)} + \epsilon_x^{(2)}\right]\Delta x. \tag{5.2.2}$$

Now, using the definition of extensional strain given in Chapter 3 [see Eq. (3.3.1)], the total strain with respect to the initial undeformed element, Δx, is given by

$$\epsilon_x = \lim_{\Delta x \to 0} \frac{\Delta x^{**} - \Delta x}{\Delta x}. \tag{5.2.3}$$

Comparing Eqs. (5.2.2) and (5.2.3), we observe that

$$\epsilon_x = \epsilon_x^{(1)} + \epsilon_x^{(2)}. \tag{5.2.4a}$$

Thus, in this case we conclude that if a body undergoes infinitesimal strains, we may then determine the total strains by simple addition; we say that the strains can be superimposed'.

It is clear that if strains ϵ_y, ϵ_{xy}, etc. exist in the body, then similarly, they can also be superimposed by simple addition; i.e.,

$$\epsilon_y = \epsilon_y^{(1)} + \epsilon_y^{(2)}, \tag{5.2.4b}$$

$$\epsilon_{xy} = \epsilon_{xy}^{(1)} + \epsilon_{xy}^{(2)}. \tag{5.2.4c}$$

It is important to emphasise that in referring to the superposition of the strains, we can only superimpose the *same* components of the strain tensor; thus, for example, we may *not* superimpose $\epsilon_x^{(1)} + \epsilon_y^{(2)}$.

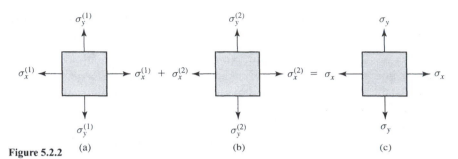

Figure 5.2.2   (a)   (b)   (c)

(b) Basic principle of superposition for linear elastic bodies

We consider here a two-dimensional body under plane stress ($\sigma_z = \tau_{xz} = \tau_{yz} = 0$) under a set of forces such that at any point the state of stress is $\sigma_x^{(1)}, \sigma_y^{(1)}$ [Fig. (5.2.2a)]. Using Hooke's law, Eq. (4.4.10), the strains $\epsilon_x^{(1)}, \epsilon_y^{(1)}$ at the point are given by

$$\epsilon_x^{(1)} = \frac{1}{E}\left[\sigma_x^{(1)} - \nu\sigma_y^{(1)}\right], \qquad \epsilon_y^{(1)} = \frac{1}{E}\left[\sigma_y^{(1)} - \nu\sigma_x^{(1)}\right]. \tag{5.2.5a}$$

Let us say that due to some other forces, a second state of stress, $\sigma_x^{(2)}, \sigma_y^{(2)}$, exists at the point [Fig. (5.2.2b)]; then

$$\epsilon_x^{(2)} = \frac{1}{E}\left[\sigma_x^{(2)} - \nu\sigma_y^{(2)}\right], \qquad \epsilon_y^{(2)} = \frac{1}{E}\left[\sigma_y^{(2)} - \nu\sigma_x^{(2)}\right]. \tag{5.2.5b}$$

If all the strains are infinitesimal, then according to Eqs. (5.2.4a), the strains are simply additive; thus for the total strain, we may write

$$\epsilon_x = \epsilon_x^{(1)} + \epsilon_x^{(2)}$$

or

$$\epsilon_x = \frac{1}{E}\left[\sigma_x^{(1)} + \sigma_x^{(2)}\right] - \nu\left[\sigma_y^{(1)} + \sigma_y^{(2)}\right]. \tag{5.2.6a}$$

Similarly,

$$\epsilon_y = \frac{1}{E}\left[\sigma_y^{(1)} + \sigma_y^{(2)}\right] - \nu\left[\sigma_x^{(1)} + \sigma_x^{(2)}\right]. \tag{5.2.6b}$$

Now, the total stress state, σ_x and σ_y, due to the combined applied stress state, is given by [Fig. (5.2.2c)]

$$\sigma_x = \sigma_x^{(1)} + \sigma_x^{(2)}, \qquad \sigma_y = \sigma_y^{(1)} + \sigma_y^{(2)}. \tag{5.2.7}$$

It follows that

$$\epsilon_x = \frac{1}{E}[\sigma_x - \nu\sigma_y], \tag{5.2.8a}$$

$$\epsilon_y = \frac{1}{E}[\sigma_y - \nu\sigma_x]. \tag{5.2.8b}$$

Thus we observe that the total strain at the point due to two separate 'causes', '1' and '2', can be found by simple addition of the two effects. Note that to do so, we require that the strains be infinitesimal and that the elastic stress–strain relation be linear. The principle is therefore referred to as the **principle of linear superposition**. We shall find that the use of this principle leads to considerable simplifications in the analysis of problems in mechanics since it permits us to analyse separately the behaviour due to any particular cause. Thus, for example, when a body is subjected to a complex loading system, the principle also permits one to identify the effect due to any specific load.

In the following example, we demonstrate that the principle of superposition becomes invalid for an elastic material whose constitutive law is nonlinear.

Example 5.1: Consider a rod of uniform cross-section A subjected to an axial load as shown in Fig. (5.2.3a). The stress–strain relation of the material is given as [Fig. (5.2.3b)]

$$\sigma_x = k\sqrt{\epsilon_x}, \quad \epsilon_x \geq 0. \tag{5.2.9a}$$

Determine the total elongation ΔL of the rod, assuming that the stress σ_x at any point of the rod is given by the average stress $\overline{\sigma}_x = P/A$; that is, $\sigma_x = P/A$.

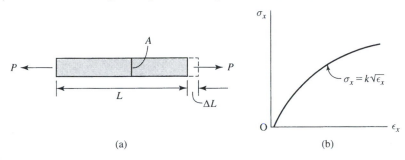

(a)   (b)   **Figure 5.2.3**

Solution: From Eq. (5.2.9a), $\epsilon_x = (\sigma_x/k)^2$ and hence

$$\epsilon_x = \left(\frac{P}{kA}\right)^2 \tag{5.2.9b}$$

Then, using Eq. (3.3.6),

$$\Delta L = \int_o^L \epsilon_x(x)\,dx = \left(\frac{P}{kA}\right)^2 L = cP^2, \tag{5.2.10}$$

where $c = L/(kA)^2$. Thus we note that for any given force P the elongation varies with the square of P.

Let us consider the application of three separate forces: P_1, P_2 and $P_3 = P_1 + P_2$.

$$\text{Due to } P_1: \quad \Delta L_1 = cP_1^2$$
$$\text{Due to } P_2: \quad \Delta L_2 = cP_2^2$$
$$\text{Due to } P_3: \quad \Delta L_3 = cP_3^2$$

Therefore $\Delta L_3 = c(P_1 + P_2)^2 = (P_1^2 + P_2^2 + 2P_1P_2) \neq \Delta L_1 + \Delta L_2$.

Thus, the elongation of the bar due to the force P_3 cannot be obtained by superimposing the elongations due to P_1 and P_2, respectively; the principle of superposition clearly is not valid here since we cannot simply add the effects separately.

Finally we re-emphasise that the principle of superposition is valid only under conditions of (a) infinitesimal strains and (b) linear stress–strain relations.[†] We note too that these are *necessary*, but not sufficient, conditions.   □

[†] See previous footnote.

5.3 The principle of de Saint Venant

The principle of de Saint Venant, first based mainly on physical intuition and stated in 1855, is of great practical importance and is used repeatedly in solid mechanics. We find it appropriate to first introduce the principle by means of a simple example. To this end, let us consider a bar of uniform cross-section subjected to a tensile force P [Fig. (5.3.1)]. If the load is a point load as shown in the figure, a very complex stress state will exist at all points in the vicinity of the point of application. For example, it is clear that near and on the right end, the stresses will be very great at points lying on the x-axis, while for all points ($x = L$, $y \neq 0$, $z \neq 0$) the surface traction on the end surface, $T_x = 0$ and hence at these end points, $\sigma_x = \tau_{xy} = \tau_{xz} = 0$.

Figure 5.3.1

If we imagine the undeformed rod to be composed of elements as shown in Fig. (5.3.2a), the deformation will appear as shown in Fig. (5.3.2b). We observe, however, that the complex deformation pattern at the right end is highly localised. Indeed, at points away from the vicinity of load application, the deformation, and therefore the distribution of stresses appear to be quite uniform. Thus, since the complex stress state is highly localised, we conclude that at points sufficiently far away from the applied loads, the strain and stress states do not depend on the precise manner in which the force is applied. Having developed these ideas, we now state the *principle of de Saint Venant:*

> Two different distributions of force acting on the same portion of a body have essentially the same effect on those parts of the body that are sufficiently distant from the region of load application provided that the applied force distributions represent *equivalent* force systems (namely, they possess the same resultants that pass through the same line of action).

By 'sufficiently distant from the region of load application' we shall mean at distances roughly greater than the largest dimension of the surface acted upon. Moreover, by 'essentially the same effect', we mean that any *difference* in the

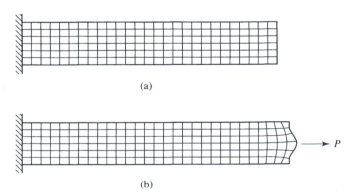

(a)

(b)

Figure 5.3.2

stresses (or strains) due to the two separately applied distributions is less than these calculated stresses (or strains) by several orders of magnitude. Or, in other words, the stress and strain fields at points sufficiently distant from the region of application are essentially the same due to the two distributions

To illustrate the principle, let us consider specifically a rod of rectangular cross-section $b \times h$, $(b < h)$, loaded as shown in Figs. (5.3.3a and b). Near the end, the strain and stress states will be quite different in each case. However, at points located roughly at a distance greater than h, the three different loading systems will produce the same effect.

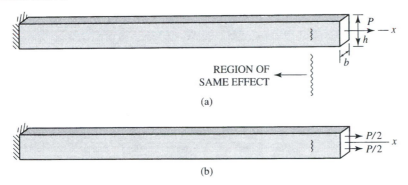

Figure 5.3.3

It is of interest to observe that, as a corollary to the principle, a self-equilibrating system (i.e., one whose resultant $R = 0$) produces no stress or deformation in a region away from the points of application. This is illustrated in Figs. (5.3.4a and b).

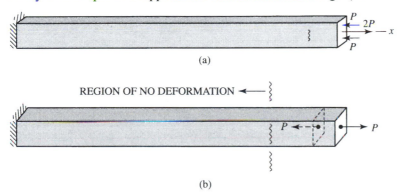

Figure 5.3.4

We mention here that although first enunciated in 1855, no exact or complete proof exists to the principle of de Saint Venant. However, it has been verified repeatedly, for various bodies and loading conditions, by sophisticated analyses, numerical solutions and laboratory experiments. Its justification should therefore be accepted mainly based on sound empirical evidence although it is also clear that its acceptance can be based largely on physical intuition.[†]

In the following chapters, we shall tacitly apply the principle of de Saint Venant in the analysis of all problems. However, we note that in applying the principle

---

[†] Note that the principle was given only for linear elastic solids. However, based purely on intuition and using the same type of reasoning, one should expect the principle to be valid for a body undergoing nonlinear elastic or even plastic behaviour. In practice, the principle is therefore usually also applied to such bodies.

to bodies subjected to concentrated forces, we are able, using the mechanics-of-materials approach, to determine the behaviour only at points far away from the applied loads; implicitly, *we neglect all localised effects near the points of load application*. Therefore, the solutions that we shall obtain will be valid only if the regions of localised complex stress–strain states represent but a small portion of the entire body. As a result, the mechanics-of-materials solutions can be valid only for relatively long slender bodies (e.g. rods, shafts or beams) where the major portion of the body is distant from any applied loads.

Applications to simple elements

6

Axial loadings

6.1 Introduction

In this chapter, we study the behaviour of an element where one dimension, in the 'longitudinal' direction, is considerably greater than the other two, namely the dimensions defining the cross-section. One refers to such an element as a 'rod' (or at times a 'bar'). In particular, we study here the behaviour of a rod that is subjected to an axial force acting in the longitudinal direction. Although this represents the simplest possible case and loading condition, the resulting relations permit us to treat several interesting types of problems that are encountered in practice.

We shall discuss mainly elastic behaviour but at a later stage will consider the behaviour when the material enters the plastic range.

6.2 Elastic behaviour of prismatic rods: basic results

Consider a long prismatic elastic rod, i.e., a rod of constant cross-sectional area A and of length L whose longitudinal axis lies along the x-axis. The rod is assumed to be linearly elastic with modulus of elasticity E and Poisson ratio v. Let the rod be subjected to an axial load P, which acts along this x-axis [Fig. (6.2.1a)]. Note that we have not specified the exact location of this axis; we know only that this axis intersects the cross-sections (lying in the y–z plane) at some point O as shown in Fig. (6.2.1a).

From equilibrium, the resultant internal force system at any cross-section consists of an axial force $F = P$ (as in Chapter 2), which acts along the x-axis [Fig. (6.2.1b)]. Furthermore, the moments about the y- and z-axes of the cross-section are necessarily zero; thus the internal force system acting on any cross-section is given by

$$F = P, \qquad M_y = 0, \qquad M_z = 0. \tag{6.2.1}$$

Now, it is reasonable to assume that due to this applied load, the rod will undergo extension in the axial x-direction. We therefore make the following assumptions on the deformation based on physical reasoning:

(a) the axis remains straight after deformation, and
(b) all plane cross-sections remain plane and perpendicular to the x-axis.

It is these kinds of assumptions, namely plausible assumptions on the nature of the deformation, which are typical of the approach generally referred to as **mechanics of materials**.

As a result of the above assumptions, all points in a given y–z plane have the same displacements in the x-direction. Thus, if we consider a small segment as shown in Fig. (6.2.1c), any line segment (or 'fibre') AB undergoes the same

155

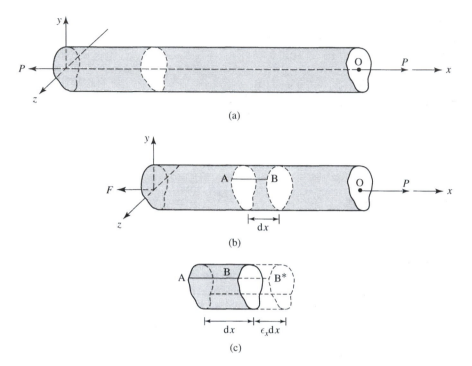

(a)

(b)

(c)

Figure 6.2.1

strain ϵ_x; therefore ϵ_x cannot be a function of y or z but, at most, is a function only of x; that is, $\epsilon_x = \epsilon_x(x)$. Moreover, as a result of assumptions (a) and (b) above, the shear strains $\epsilon_{xy} = 0$ and $\epsilon_{xz} = 0$ throughout the rod. Since $\tau_{xy} = 2G\epsilon_{xy}$, $\tau_{xz} = 2G\epsilon_{xz}$, this assumption leads us to conclude that the shear stress components

$$\tau_{xy} = 0, \qquad \tau_{xz} = 0 \qquad (6.2.2)$$

at all points of the rod.

Since we are studying a linear elastic isotropic bar, the stress–strain relations for normal stress and strain components are governed by Hooke's law [Eqs. (4.4.10)]:

$$\epsilon_x = \frac{1}{E}[\sigma_x - \nu(\sigma_y + \sigma_z)],$$

$$\epsilon_y = \frac{1}{E}[\sigma_y - \nu(\sigma_z + \sigma_x)], \qquad (6.2.3)$$

$$\epsilon_z = \frac{1}{E}[\sigma_z - \nu(\sigma_x + \sigma_y)].$$

For simplicity, let us for the moment, consider a rod having a rectangular cross-section, as shown in Fig. (6.2.2). Clearly, since no external forces are acting on the top and bottom faces of the rod, $\sigma_y = 0$ and $\tau_{yz} = 0$ on these surfaces. Similarly, $\sigma_z = 0$ and $\tau_{zy} = 0$ on the two lateral surfaces of the rod; thus, in particular,

$$\sigma_y(y = \pm d/2) = 0, \qquad \sigma_z(z = \pm b/2) = 0, \qquad \tau_{yz}(y = \pm d/2) = 0,$$

$$\tau_{zy}(z = \pm b/2) = 0. \qquad (6.2.4)$$

We now limit our analysis to that of a *long* rod, namely one for which $b \ll L$ and $d \ll L$, that is, a rod for which the lateral dimensions are small relative to the length L. Since these stress components vanish at the boundary of the cross-section, and

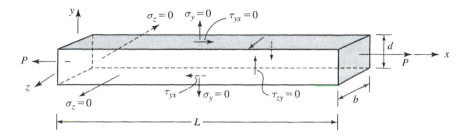

Figure 6.2.2

since the distance between the lateral surfaces is relatively small, it is reasonable to assume that the stresses σ_y, σ_z and τ_{yz} cannot vary very much from top to bottom or from one side to another within the cross-section. Therefore, we make the following reasonable *assumption*: the stress components $\sigma_y = 0$, $\sigma_z = 0$, $\tau_{zy} = 0$ at all points in the rod (in addition to $\tau_{xy} = \tau_{xz} = 0$ as previously established). Note that we are able to make this assumption only for relatively long thin rods. Clearly, if the rod is short and stubby, the above reasoning does not hold and therefore the results obtained below will not be valid for relatively short rods.

Based on the above assumption, the stress–strain relations, Eqs. (6.2.3), reduce to

$$\epsilon_x = \frac{\sigma_x}{E}, \tag{6.2.5a}$$

$$\epsilon_y = -\nu \frac{\sigma_x}{E}, \tag{6.2.5b}$$

$$\epsilon_z = -\nu \frac{\sigma_x}{E}. \tag{6.2.5c}$$

In particular, $\sigma_x = E\epsilon_x$ and since ϵ_x can only be a function of x, we note that, at most, $\sigma_x = \sigma_x(x)$ also.

We now consider the cross-section to be composed of a large number of incremental areas $\mathrm{d}A$. Then, as in Section 5 of Chapter 2, on each area $\mathrm{d}A$ an incremental force $\mathrm{d}F = \sigma_x\,\mathrm{d}A$ acts [Fig. (6.2.3)] and consequently the total normal force F is

$$F = \iint_A \sigma_x(x)\,\mathrm{d}A. \tag{6.2.6}$$

Since σ_x is independent of y and z, $F = \sigma_x \iint_A \mathrm{d}A = \sigma_x A$ and therefore we have the simple relation

$$\sigma_x = \frac{F}{A}. \tag{6.2.7}$$

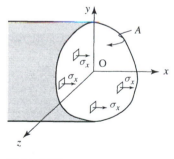

Figure 6.2.3

Furthermore, we recall from Chapter 2 [Eq. (2.5.8a)], that $\mathrm{d}M_y = \sigma_x \cdot z\,\mathrm{d}A$; hence

$$M_y = \iint_A \sigma_x \cdot z\,\mathrm{d}A = \sigma_x \iint_A z\,\mathrm{d}A. \tag{6.2.8}$$

However, from Eq. (6.2.1), $M_y = 0$. Therefore

$$\iint_A z\,\mathrm{d}A = 0. \tag{6.2.9a}$$

Similarly, from Eq. (2.5.8b), $M_z = -\sigma_x \iint_A y \, dA = 0$, and therefore

$$\iint\limits_A y \, dA = 0. \tag{6.2.9b}$$

Since the integrals appearing in Eqs. (6.2.9) vanish, it follows by definition, that point O is the centroid of the cross-section. Thus, consistent with our assumptions (namely that the x-axis does not bend, and that $\sigma_y = \sigma_z = 0$), we have established that the longitudinal x-axis must be a *centroidal axis*. Hence we conclude that a uniform stress distribution of σ_x will exist only if P passes through the centroid; only in this case, is it true that $\sigma_x = \frac{F}{A}$.

Now, from Eq. (6.2.5a), the strain at any cross-section x of the rod is

$$\epsilon_x = \frac{\sigma_x}{E} = \frac{F}{EA}. \tag{6.2.10}$$

The elongation $d\Delta$ of any element dx [Fig. (6.2.4a)] is then, according to Chapter 3 [Eq. (3.3.2b)], $d\Delta = \epsilon_x \, dx = \frac{F}{EA} \, dx$. Hence the total elongation Δ of the rod is

$$\Delta = \int_0^L d\Delta = \int_0^L \frac{F}{EA} \, dx. \tag{6.2.11}$$

If A and $F = P$ are constants, we have finally [Fig. (6.2.4b)]

$$\Delta = \frac{P}{EA} \int_0^L dx = \frac{PL}{EA}. \tag{6.2.12}$$

The following example serves to provide an idea of the order of magnitude of the elongation of an elastic rod as encountered in engineering practice.

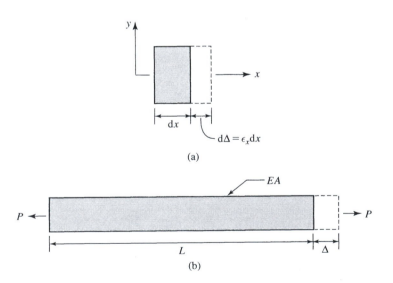

Figure 6.2.4

Example 6.1: A steel rod ($E = 200$ GPa, $\nu = 0.25$) with cross-sectional area $A = 4 \, cm^2$ and $L = 240$ cm is subjected to an axial force $P = 50,000$ N. (a) Determine the elongation of the rod. (b) If the cross-section is a circle, what is the change in the original diameter D?

Solution:

(a) From Eq. (6.2.12),

$$\Delta = \frac{(50,000)(2.40)}{(200 \times 10^9)(4 \times 10^{-4})} = 1.5 \times 10^{-3} \text{ m} = 0.15 \text{ cm}.$$

We observe that for such a rod, the total elongation is very small indeed. Note that the behaviour remains elastic since $\sigma_x = \frac{50,000}{4 \times 10^{-4}} = 125 \times 10^6 \text{ N/m}^2 = 125$ MPa does not exceed the yield point of steel, $\sigma_y = 200$ MPa.

(b) The original diameter $D = 2\sqrt{A/\pi}$. From Eq. (6.2.5b), $\epsilon_y = -0.25 P/EA = -0.156 \times 10^{-3}$. Therefore the change in diameter, $dD = \epsilon_y D = 2\epsilon_y \cdot \sqrt{A/\pi} = -0.353 \times 10^{-3}$ cm. Note that the negative sign of dD indicates a shortening of the diameter. We observe that the ratio $|dD|/\Delta = 0.0024$; that is, the change in the dimensions of the cross-section is much smaller than the elongation of the rod. $\qquad\square$

Example 6.2: As in Example 2.3 of Chapter 2, a magnet is attached at the free end of an iron rod of length L and cross-sectional area A, as shown in Fig. (6.2.5). The magnetic force of attraction can be represented by the function $f(x) = c\,e^{-x/L}$ (where c is a constant with units N/m³). Assuming that the rod behaves as a linear elastic material with modulus of elasticity E determine the extension due to the attractive magnetic force.

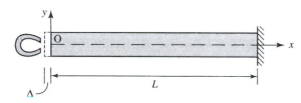

Figure 6.2.5

Solution: The normal force $F(x)$ acting on any cross-section was found, in Example 2.3, to be [Eq. (2.2.4)] $F(x) = cAL[1 - e^{-x/L}]$. Using Eq. (6.2.11), simple integration yields

$$\Delta = \frac{cL}{E} \int_0^L \left[1 - e^{-x/L}\right] dx = \frac{cL^2}{Ee}.$$

$\qquad\square$

6.3 Some general comments

(a) In the development of expressions for axial loading of a rod, we observe that there exist two key points in the derivation: (i) the basic assumptions on the deformation pattern and (ii) the assumption that the stresses σ_y and σ_z vanish at all points in the interior of the rod. Once these assumptions had been made, we arrived at the simple expressions for σ_x and the axial elongation, and we *concluded* that the stress σ_x is uniformly distributed over the cross-section only if the axial force acts through the centroid. It is important to observe that we did not initially assume the x-axis to be a centroidal axis.

(b) We also recall that in Chapter 1 we defined the average stress σ_x on a cross-section as $\overline{\sigma}_x = \frac{P}{A}$. From the derived expression of the preceding section, $\sigma_x = \frac{F}{A}$, we now see this gives the true stress at all points in the cross-section when the axial force acts through the centroid. Indeed, this simple expression is known to be an 'exact' expression according to the linear theory of elasticity. [We may further check the validity by verifying that all points of the rod are in equilibrium according to Eqs. (2.4.4).] Note that for the *prismatic* rod considered, the cross-sectional areas are constant and hence σ_x is not a function of x.

(c) From Eq. (6.2.12), we observe that the elongation Δ is linearly proportional to the applied force P and inversely proportional to the quantity 'EA'. We therefore refer to EA as the **axial rigidity** of the rod since, for a rod of given length L subjected to a given force P, the elongation will decrease as EA increases. We observe that the axial rigidity is a function of the material property E and of the geometric property A, the cross-sectional area.

(d) We note, according to Eqs. (6.2.5b) and (6.2.5c), that non-zero strains ϵ_y and ϵ_z exist in the rod and since $\nu \leq 0.5$, $|\epsilon_y/\epsilon_x| < 1$ and $|\epsilon_z/\epsilon_x| < 1$. From these equations, we also observe that ϵ_y and ϵ_z are of opposite sign to ϵ_x; therefore, as expected, for a rod in tension the lateral dimensions contract while, if the rod is in compression, the lateral dimensions increase [Fig. (6.3.1)]. However,

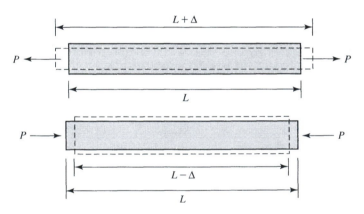

Figure 6.3.1

we recall that for a thin rod the lateral dimensions of the cross-section are, by definition, much smaller than the longitudinal dimension. Therefore, due to the application of an axial load, the changes in dimensions of the cross-section will be much smaller than the elongation (or shortening) of the rod. Letting A^* denote the cross-sectional area of the *deformed* rod due to such changes, we may write

$$A^* = A(1 + \mathrm{d}A/A), \tag{6.3.1}$$

where $\mathrm{d}A$, representing the change in area, is given by[†]

$$|\mathrm{d}A| = \iint\limits_{A} |\epsilon_y + \epsilon_z|\, \mathrm{d}A = |\epsilon_y + \epsilon_z| \iint\limits_{A} \mathrm{d}A = |\epsilon_y + \epsilon_z| \cdot A \tag{6.3.2}$$

since ϵ_y and ϵ_z are assumed to be only functions of x. (Note that for a tensile force with $\sigma_x > 0$, $\mathrm{d}A < 0$.)

[†] Note that $|\mathrm{d}A| \simeq |\iint_A [(1 + \epsilon_x)(1 + \epsilon_y) - 1]\, \mathrm{d}A| = |\iint_A [\epsilon_x + \epsilon_y + \epsilon_x \epsilon_y]\, \mathrm{d}A| \simeq \iint_A |(\epsilon_x + \epsilon_y)|\, \mathrm{d}A$ since $|\epsilon_x| \ll 1$, $|\epsilon_y| \ll 1$.

Thus, for infinitesimal strains [say of order $O(10^{-3})$], $dA/A = (\epsilon_y + \epsilon_z) \ll 1$ is an infinitesimal. Now, since the deformed cross-sectional area is A^*, it follows, in principle, that a uniform distribution of the stress σ_x would require that σ_x be given by the expression $\sigma_x = F/A^*$. However, using the binomial expansion

$$\frac{1}{1+x} = 1 - x + \frac{x^2}{2} - \frac{x^3}{3} + \cdots, \quad |x| \ll 1,$$

we may write, using Eq. (6.3.1),

$$\sigma_x = \frac{F}{A^*} = \frac{F}{A}\left[1 - \frac{dA}{A} + \frac{1}{2}\left(\frac{dA}{A}\right)^2 + \cdots\right]. \tag{6.3.3}$$

Since dA/A is an infinitesimal, we may drop such terms and thus recover Eq. (6.2.7), namely $\sigma_x = F/A$. This expression for σ_x is thus seen to be consistent within the accuracy of our (first-order) linear theory.

(e) We point out here that, throughout this book, our treatment will be confined to bodies that undergo small strains and changes in geometry. Therefore, *although in principle, we examine all bodies in their deformed state, we neglect infinitesimal changes in geometry (with respect to the original geometry) and therefore we write all expressions in terms of the given original geometry* (lengths, areas, etc.). This procedure, consistent with the 'linear theory' as discussed in Chapter 5, will be followed in all subsequent developments in this book.

(f) Finally, it should be remembered that in the above analysis we have implicitly invoked the principle of de Saint Venant. Clearly, as discussed in Chapter 5, this principle is valid in the case of axial loading of a rod only for long thin rods and fails to have any validity for short rods.

The use of the principle of de Saint Venant is particularly useful in the analysis of a rod where more than a single force is applied or for rods consisting of more than one component where an abrupt change in cross-section occurs. For example, for the rod with applied forces as shown in Fig. (6.3.2), we obtain, from the free-body diagram, the resultant axial force $F = P_1$ in BC while in the region CD, $F = P_2$; i.e., our simple free-body analysis leads to a discontinuous axial force at the cross-section C. Furthermore, at C, we note that there is a discontinuity in the cross-sectional areas. As a result of these discontinuities, there can no longer be a uniform distribution of stresses in the region of C. However, for a long thin rod, we may consider the behaviour in this region to be a *localised* effect. For such rods, these localised effects are usually neglected by implicitly invoking the Principle of de Saint Venant.

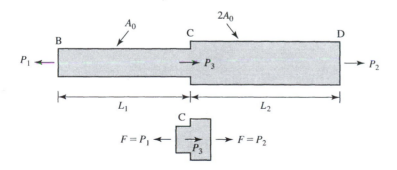

Figure 6.3.2

Example 6.3: If the rod shown in Fig. (6.3.2) is composed of two elements having cross-sectional areas A_0 and $2A_0$, determine the total extension of the rod due to applied axial forces acting through the centroid at points B, C and D.

Solution: From a free-body diagram, the axial force in element BC is $F_{BC} = P_1$ and in CD, $F_{CD} = P_2$. Note that both segments are under tension. The total lengthening is $\Delta = \Delta_{BC} + \Delta_{CD}$ or

$$\Delta = \frac{P_1 L_1}{A_0 E} + \frac{P_2 L_2}{2 A_0 E} = \frac{1}{E A_0}[P_1 L_1 + 0.5 P_2 L_2] \qquad \square$$

We observe that in this problem, we have disregarded the local effects in the region of C when calculating Δ. This is permissible only if each segment of the rod is sufficiently long.

6.4 Extension of results

According to the discussion of the previous section, the expressions given by Eqs. (6.2.7) and (6.2.12) are 'exact' for the case of a prismatic rod. Let us now consider the case of a non-prismatic rod, that is, a rod for which $A = A(x)$ [Fig. (6.4.1)]. Clearly σ_x will then be a function of x. Now, recalling that the analysis for the prismatic beam was based entirely on assumptions (a) and (b) of Section 2, it is evident that if we accept the same assumptions for the present case of the non-prismatic rod, and follow the development of Section 2 step by step, we conclude that the distribution of σ_x over the cross-section is uniform, with

$$\sigma_x = \frac{F}{A(x)}. \tag{6.4.1}$$

However, we show now that, in particular, assumption (b), namely that plane sections remain plane and perpendicular to the x-axis is no longer valid for the case of non-prismatic rods. We recall that from this assumption it follows that the stresses $\tau_{zx} = \tau_{yx} = 0$ [see Eqs. (6.2.2)]. Hence, if we can demonstrate that for a non-prismatic rod there must exist non-zero shear stresses, we will have shown that our basic assumption is no longer valid for non-prismatic rods.

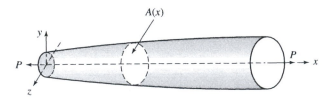

Figure 6.4.1

To do so, let us consider the simple case of a rod with varying depth but whose width b is constant with respect to x, as shown in Fig. (6.4.2a). We first isolate a small wedge-shaped element as in Fig. (6.4.2b). Now, on the right face (having area $b \cdot \Delta y$), there exists a force $[\sigma_x(b \cdot \Delta y)]$. Clearly, if no shear stresses exist, the wedge cannot be in equilibrium in the x-direction. Thus we see that for a non-prismatic rod, shear stresses must necessarily exist and therefore plane sections will

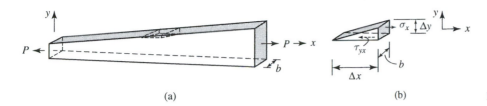

(a)   (b)   **Figure 6.4.2**

no longer remain plane as in our basic assumption. To examine the magnitude of such shear stresses, we consider equilibrium in the x-direction:

$$\sum F_x = \sigma_x(b \cdot \Delta y) - \tau_{yx}(b \cdot \Delta x) = 0 \qquad (6.4.2a)$$

so that

$$\tau_{yx} = \sigma_x(x)\frac{\Delta y}{\Delta x}, \qquad (6.4.2b)$$

where $y(x)$ represents the variation of the depth with x.

Taking the limit as $\Delta x \to 0$, we note that $\tau_{yx} \to 0$ as $\Delta y/\Delta x \to 0$, i.e., as the slope of the upper surface of the rod tends to zero. However, if $0 < \Delta y/\Delta x \ll 1$, then $0 < |\tau_{yx}/\sigma_x| \ll 1$. Therefore we conclude that for a rod with a slowly varying cross-section, our basic assumption will have a small error. Consequently, Eq. (6.4.1) is a good approximation if $A(x)$ is a slowly varying function of x, e.g., for rods having a relatively small taper as in Fig. (6.4.2a). For rods with a strong variation of $A(x)$, Fig. (6.4.3), Eq. (6.4.1) may lead to highly inaccurate results.[†]

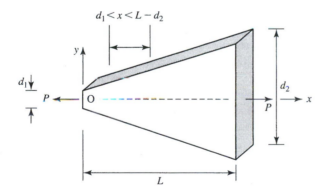

Figure 6.4.3

The above analysis can also be applied to rods containing a notch or cut-out [Fig. (6.4.4a)]. If we examine, for example, a small wedge in the region of the notch where $\Delta y/\Delta x$ is not small [Fig. (6.4.4b)], we arrive at the conclusion that shear stresses of the same order of magnitude as σ_x will exist; consequently Eq. (6.4.1) will not yield a good approximation for σ_x in this localised region. The equation will simply give some average σ_x acting on the cross-section in the sense of our

---

[†] From the viewpoint of the principle of de Saint Venant, we are led to the same conclusion. From Fig. (6.4.3), we observe that for a rod with a large taper, d_2 is necessarily of the same order of magnitude as L. We recall, following our discussion in Chapter 5 (Section 3) that according to de Saint Venant's principle, a solution for stresses due to applied concentrated forces, as applied here, is valid only at distances $d_1 < x < L - d_2$. In this case, the range of validity of x is insignificant compared to the entire rod. It therefore is evident that the principle cannot be invoked for rods with a strong variation in $A(x)$.

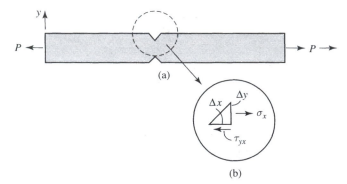

Figure 6.4.4

Figure 6.4.4

discussion in Chapter 1.[†] However, invoking again de Saint Venant's principle, the stresses at distances far removed from these localised regions are given by Eq. (6.2.7).

For rods where Eq. (6.4.1) yields a good approximation for σ_x, it follows that the elongation of the rod Δ is given by Eq. (6.2.11):

$$\Delta = \int_0^L \mathrm{d}\Delta = \int_0^L \frac{F}{E\,A(x)}\,\mathrm{d}x. \tag{6.4.3}$$

6.5 Statically indeterminate axially loaded members

Up to this point in the previous developments, it has always been possible to determine the internal forces in a body by means of the equations of statics. However, as we shall see, it is not always possible to do so, in general, for all systems; that is, there exist systems for which the equations of statics are not sufficient to permit one to obtain all forces: such mechanical systems are said to be **statically indeterminate**. The simplest statically indeterminate systems, encountered in the case of axial loading, are examined here. To illustrate these ideas, we consider the following specific problem.

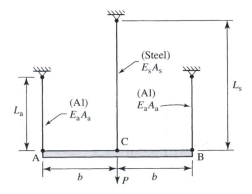

Figure 6.5.1

Consider a *rigid* (but weightless) plate, which is suspended by three symmetrically placed wires at A, B and C, as shown in Fig. (6.5.1). The centre wire is steel

---

[†] More sophisticated analyses, based on the theory of elasticity, are possible, but are beyond the scope of our study. We simply mention here that in these localised regions the stress field usually consists of high stresses defined by *stress concentration factors*.

(with modulus of elasticity, $E_s = 200$ GPa and cross-sectional area A_s) and the two outer aluminium wires each have cross-sectional areas A_a and modulus of elasticity $E_a = 70$ GPa. Note that the lengths of the steel and aluminium wires are L_s and L_a, respectively. A load P is assumed to be applied at the centre point C. We wish to determine (a) the resisting force in each wire and (b) the downward displacement of the plate, assuming elastic behaviour.

We denote the resisting force in the steel wire by F_s and let F_{a1} and F_{a2} be the forces in the aluminium wires, as shown in the free-body diagram of the plate [Fig. (6.5.2)].

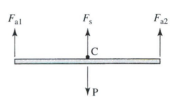

Figure 6.5.2

From the equations of equilibrium, we have

$$\sum F_y = F_{a1} + F_{a2} + F_s - P = 0. \qquad (6.5.1\text{a})$$

It is important to observe that, according to the free-body diagram, all the resisting forces in the wires have been assumed to be under *tension*.

Taking moments about point C,

$$\sum M_c = bF_{a1} - bF_{a2} = 0 \qquad (6.5.1\text{b})$$

from which $F_{a1} = F_{a2}$. We therefore denote the force in each of the aluminium wires by F_a; i.e., $F_a \equiv F_{a1} = F_{a2}$.[†]

Equation (6.5.1a) then becomes

$$2F_a + F_s = P. \qquad (6.5.2)$$

Note that we have used all the equations of equilibrium ($\sum F_x = 0$ is satisfied identically) but are unable to determine the two unknowns, F_a and F_s, from the single equation, Eq. (6.5.2). It is for this reason that the problem is said to be **statically indeterminate**, since the equations of statics are not sufficient to yield the solution. Clearly, we require another equation to solve for the two unknowns.

Now, if we consider the deformation for this system, we observe that due to the symmetry of the problem, the rigid plate must necessarily remain horizontal. It follows that the downward displacement Δ at all points is the same [Fig. (6.5.3)]; specifically, the elongation of the steel and aluminium wires must be identical. Thus we write

$$\Delta_s = \Delta_a. \qquad (6.5.3)$$

[Note that in Fig. (6.5.3), the downward displacement Δ corresponds to elongations of the wires. This elongation is consistent with the assumed tension of the wires. We also remark that Eq. (6.5.3) is an equation that represents the *geometric compatibility* of the system. Although this comment may appear here to be superfluous, it is, as we shall see, an essential feature in the solution of statically indeterminate problems.]

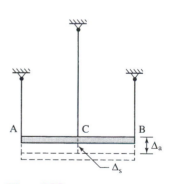

Figure 6.5.3

Now, from Eq. (6.2.12), we write, for the aluminium and steel wires,

$$\Delta_a = \frac{F_a L_a}{E_a A_a}, \qquad (6.5.4\text{a})$$

$$\Delta_s = \frac{F_s L_s}{A_s E_s}, \qquad (6.5.4\text{b})$$

[†] Note that we might have initially concluded from symmetry that the two aluminium wires carry the same load; however, we should observe that this is not an independent conclusion since, in fact, it is derivable from equations of mechanics.

respectively. Then using the equation of geometric compatibility,

$$\frac{F_a L_a}{E_a A_a} = \frac{F_s L_s}{E_s A_s} \qquad (6.5.5a)$$

from which

$$F_a = \left(\frac{L_s}{L_a}\frac{E_a A_a}{E_s A_s}\right) F_s. \qquad (6.5.5b)$$

Substituting in the equation of equilibrium, Eq. (6.5.2),

$$\left[1 + 2\left(\frac{L_s}{L_a}\frac{E_a A_a}{E_s A_s}\right)\right] F_s = P \qquad (6.5.6a)$$

or

$$F_s = \frac{\frac{L_a}{L_s}\left(\frac{E_s A_s}{E_a A_a}\right)}{\left(2 + \frac{L_a}{L_s}\frac{E_s A_s}{E_a A_a}\right)} P. \qquad (6.5.6b)$$

We now consider a numerical case; let us assume that $L \equiv L_s = L_a = 30$ cm and that $A_a = 0.5 A_s$ with $A_s = 0.05$ cm^2. Then,

$$F_s = \frac{20}{27} P = 0.741 P. \qquad (6.5.7a)$$

From Eq. (6.5.2), we obtain

$$F_a = 0.5(P - F_s) = 0.130 P. \qquad (6.5.7b)$$

The stresses σ_s and σ_a in the steel and aluminium, respectively, are, from Eq. (6.2.7),

$$\sigma_s = \frac{F_s}{A_s} = 14.82 P \ (\text{N/cm}^2), \qquad \sigma_a = \frac{F_a}{A_a} = 5.20 P \ (\text{N/cm}^2) \qquad (6.5.8)$$

To find the displacement of the plate, we use either of Eqs. (6.5.4); e.g.,

$$\Delta = \frac{F_s L_s}{E_s A_s} = \frac{0.741 P \cdot (30 \times 10^{-2})}{(200 \times 10^9)(0.05 \times 10^{-4})} = 2.22 \times 10^{-7} P \text{ m},$$

$$= 2.22 \times 10^{-5} P \text{ cm} \ (P \text{ in Newtons}).$$

We observe, for the numerical example considered above, with $L_a = L_s$, that 74% of the load P is carried by the steel wire and only 13% is carried by each of the aluminium wires.

A further examination of the above results leads us to greater physical insight. For the present case ($L_a = L_s$) we have, from Eq. (6.5.6b),

$$\frac{F_s}{P} = \frac{\frac{E_s A_s}{E_a A_a}}{\left(2 + \frac{E_s A_s}{E_a A_a}\right)}. \qquad (6.5.9a)$$

Then, since $F_a = 0.5(P - F_s)$,

$$\frac{F_a}{P} = \frac{\frac{E_a A_a}{E_s A_s}}{\left(1 + 2\frac{E_a A_a}{E_s A_s}\right)}. \qquad (6.5.9b)$$

Note that for $L_a = L_s$, Eq. (6.5.5a) yields

$$\frac{F_s}{F_a} = \frac{E_s A_s}{E_a A_a}. \qquad (6.5.10)$$

Thus we see that the *resisting forces in the wires are proportional to their axial rigidities* as defined in Section 3. A plot of Eqs. (6.5.9) is shown in Fig. (6.5.4). We observe that for the case of a relatively thin steel wire, e.g., $E_s A_s / E_a A_a = 0.3$, we have $F_s/P = 0.1304$ and $F_a/P = 0.4348$. For a large ratio, e.g., $E_s A_s / E_a A_a = 9.0$, we find $F_s/P = 0.8182$ and $F_a/P = 0.0909$. When $E_s A_s / E_a A_a = 1$, we have $F_s/P = F_a/P = 0.333$, that is, each wire carries an equal portion of the load. In the limiting case, as $E_s A_s / E_a A_a = 0$, it is clear that $F_s = 0$; that is, the entire load is carried by the aluminium wires alone.

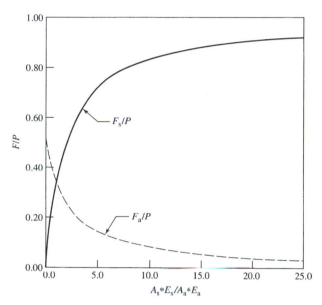

Figure 6.5.4

From an analysis of this simple problem, we arrive at an important and generally valid conclusion: *in a statically indeterminate structure consisting of several components, each component tends to resist applied loads in proportion to its relative stiffness*. This very general principle provides us with a physical insight that proves to be very useful in understanding the behaviour of more elaborate and complex indeterminate systems encountered in structural mechanics.

To illustrate further some of the ideas in the solution of statically indeterminate problems, we consider the following example.

Example 6.4: A rod consisting of two rigidly connect elements, '1' and '2', is rigidly held at the top and bottom at points B and D, as shown in Fig. (6.5.5). The cross-sectional areas and moduli of elasticity are A_1, E_1 and A_2, E_2, respectively. A force P is applied along the 'collar' at C such that its resultant passes through the centroid. Determine the resisting reaction at B and D.

Solution: Let us assume that the reactions at B and D are both upward; we denote these reactions by R_B and R_D, respectively, as shown in the free body of Fig. (6.5.6a).

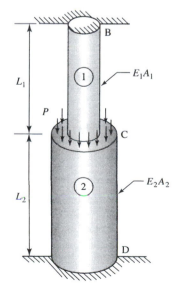

Figure 6.5.5

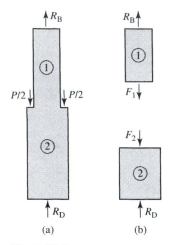

(a)   (b)

Figure 6.5.6

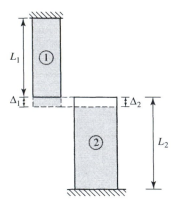

Figure 6.5.7

[While the assumed upward directions of both of the reactions follow logically from physical intuition, it should be noted that the assumed positive direction is completely arbitrary.] From Fig. (6.5.6b), with the above assumed reactions, rod 1 is clearly under tension since $F_1 = R_B$ while rod 2 is in compression since $F_2 = R_D$.

From equilibrium, we write

$$\sum F_y = R_B + R_D - P = 0 \qquad (6.5.11a)$$

or

$$F_1 + F_2 = P. \qquad (6.5.11b)$$

Again, we evidently require a second equation to solve for the two unknowns. As in the previous example, *this additional equation is an equation of geometric compatibility*.

Now, in order to be consistent with our assumed forces, in considering the deformation of the rods, we must assume that rod 1 will elongate; similarly, we must assume that rod 2 shortens (since, according to the free-body diagram, it has already been assumed that it is in compression). We denote the assumed elongation of rod 1 by Δ_1 and the shortening of rod 2 by Δ_2.

However, from the physics of the problem, rods 1 and 2 do not detach from one another, nor do they overlap. Thus, the assumed elongation of rod 1 must be equal to the assumed shortening of rod 2. The geometry of deformation is shown in Fig. (6.5.7), where for pictorial clarity, we have offset the two separate bars.

The condition of geometric compatibility is therefore

$$\Delta_1 = \Delta_2. \qquad (6.5.12)$$

Now, from Eq. (6.2.12),

$$\Delta_1 = \frac{F_1 L_1}{E_1 A_1}, \qquad \Delta_2 = \frac{F_2 L_2}{E_2 A_2}. \qquad (6.5.13)$$

Substituting in Eq. (6.5.12),

$$\frac{F_1 L_1}{E_1 A_1} = \frac{F_2 L_2}{E_2 A_2} \qquad (6.5.14)$$

and since $R_B = F_1(+F_1 \rightarrow$ tension) and $R_D = F_2(+F_2 \rightarrow$ compression), we have

$$F_1 = \left(\frac{E_1 A_1}{E_2 A_2} \frac{L_2}{L_1} \right) F_2. \qquad (6.5.15)$$

From the equation of equilibrium, Eq. (6.5.11b), we find

$$\left[1 + \frac{E_1 A_1}{E_2 A_2} \frac{L_2}{L_1} \right] F_2 = P \qquad (6.5.16)$$

from which

$$R_D \equiv F_2 = \left[\frac{E_2 A_2 L_1}{E_1 A_1 L_2 + E_2 A_2 L_1} \right] P. \qquad (6.5.17a)$$

Then, substituting back in Eq. (6.5.11b),

$$R_B \equiv F_1 = \left[\frac{E_1 A_1 L_2}{E_2 A_2 L_1 + E_1 A_1 L_2} \right] P. \qquad (6.5.17b)$$

For the case where $L_1 = L_2 \equiv L$,

$$R_B = \left[\frac{E_1 A_1}{E_1 A_1 + E_2 A_2} \right] P, \qquad R_D = \left[\frac{E_2 A_2}{E_1 A_1 + E_2 A_2} \right] P. \qquad (6.5.17c)$$

Having found $F_1 = R_B$, $F_2 = R_D$, the stress σ in each rod is given by

$$\sigma_1 = \frac{R_B}{A_1}, \qquad \sigma_2 = \frac{R_D}{A_2}. \qquad (6.5.18)$$

Note again that since F_1 represents a tension force, positive σ_1 is a tensile stress; similarly, since R_D represents a compressive force, positive σ_2 represents a compressive stress.

As mentioned above, the assumed positive sense for each of the unknown forces R_B and R_D may be chosen arbitrarily. In order to emphasise this point and to clarify some aspects of the solution, we shall solve this same problem under different assumptions.

Alternative solution: We consider the identically same problem as above [Fig. (6.5.5)]. Let us now choose the unknown reactive force R_B to be in the downward direction and R_D to be in the upward direction, as shown in Fig. (6.5.8a). Clearly, this now implies that rods 1 and 2 are both in compression with $F_1 = R_B$ ($+F_1 \rightarrow$ compression) and $F_2 = R_D$ ($+F_2 \rightarrow$ compression) [Fig. (6.5.8b)]. From equilibrium,

$$\sum F_y = R_D - R_B - P = 0 \qquad (6.5.19a)$$

and hence

$$F_2 - F_1 = P. \qquad (6.5.19b)$$

We now consider the deformation. Since both bars have implicitly been assumed to be in compression to maintain consistency, they must both be assumed to shorten as shown in Fig. (6.5.9) (where again for pictorial simplicity the bars have been drawn offset). Now, from the physics of the problem, there can be no separation of the two rod elements. Therefore, from Fig. (6.5.9), we have

$$\Delta_1 + \Delta_2 = 0. \qquad (6.5.20)$$

Here, again

$$\Delta_1 = \frac{F_1 L_1}{E_1 A_1}, \qquad \Delta_2 = \frac{F_2 L_2}{E_2 A_2}. \qquad (6.5.21)$$

Note that, just as the assumed positive values of the forces F_1 and F_2 signify compression, so do $\Delta_1 > 0$ and $\Delta_2 > 0$ signify contraction. (This is in contrast to the Δ_1 and Δ_2 of Eq. (6.5.13) where $\Delta_1 > 0$ represented elongation, while $\Delta_2 > 0$ represented a contraction.) Substituting Eq. (6.5.21) in Eq. (6.5.20), we have

$$F_1 = -\frac{E_1 A_1}{E_2 A_2} \frac{L_2}{L_1} F_2 \qquad (6.5.22)$$

and hence using Eq. (6.5.19b),

$$\left[1 + \frac{E_1 A_1}{E_2 A_2} \frac{L_2}{L_1} \right] F_2 = P \qquad (6.5.23)$$

from which

$$R_D \equiv F_2 = \left[\frac{E_2 A_2 L_1}{E_1 A_1 L_2 + E_2 A_2 L_1} \right] P. \qquad (6.5.24a)$$

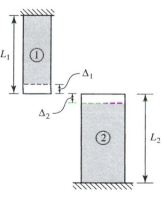

Figure 6.5.8

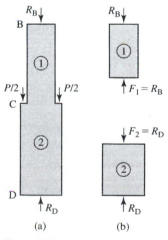

Figure 6.5.9

Then, substituting back in Eq. (6.5.22),

$$R_B \equiv F_1 = -\left[\frac{E_1 A_1 L_2}{E_2 A_2 L_1 + E_1 A_1 L_2}\right] P. \qquad (6.5.24b)$$

The stresses in rods 1 and 2 are then given, as before, by

$$\sigma_1 = \frac{F_1}{A_1}, \qquad \sigma_2 = \frac{F_2}{A_2}. \qquad (6.5.25)$$

Note, however, that $F_2 > 0$ while $F_1 < 0$. Since positive $F_1 \equiv R_B$ was taken to be compression, the negative value obtained in the solution indicates that physically the normal force F_1 is under tension (which agrees with the physics of the problem and with the previous solution). □

For emphasis, we summarise here the general features that characterise the procedure for solving statically indeterminate problems subjected to axial loads:

- Equations of equilibrium are written in terms of unknown external forces which are chosen as positive in an arbitrary direction.
- The positive internal normal forces are then determined according to the free body diagrams of each element. (These will then be either tension or compression).
- The equation of geometric compatibility must be written in terms of elongations (or shortening) of an element which are *consistent with the assumed tension (or compression)* of the element.[†]

6.6 Temperature problems: thermal stresses

An interesting class of problems whose solutions can be obtained quite simply using the relations developed in Section 2 of this chapter occurs in problems due to temperature changes of a rod.

Consider first a bar of length L, which is subjected to a change of temperature ΔT. Due to such a temperature change, the bar, in general, will undergo a change of length Δ^T, given by [Fig. (6.6.1)]

$$\Delta^T = \alpha L \cdot \Delta T, \qquad (6.6.1)$$

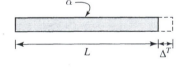

Figure 6.6.1

where α is the coefficient of thermal expansion. For example, for steel $\alpha = 11.7 \times 10^{-6}$ cm/cm/°C.

Now, if there is no restraining force, for example, if the bar is resting on a frictionless table, it will expand or contract freely (depending if $\Delta T > 0$ or $\Delta T < 0$, respectively) and hence no internal stresses will be induced. However, if there is a restraining force, that is, a force which prevents a free expansion or shortening, internal stress will occur. The internal stresses induced by these restraining forces are called **thermal stresses**. We now illustrate this idea by means of a simple example.

Example 6.5: A steel rod of cross-sectional area A, length L, modulus of elasticity E ($E = 200$ GPa) and coefficient of thermal expansion α undergoes a

---

[†] We mention that the procedure as outlined here is, in principle, a general procedure used in the analysis of any statically indeterminate system. Since the unknowns appearing in the resulting equations are forces, this general procedure is known, in structural analysis, as the *force method*. The ideas of the force method will be used in Chapters 7 and 9 to solve indeterminate problems due to torsion of rods and bending of beams.

change of temperature ΔT. Determine the resulting stress if the bar is held between two rigid walls as shown in Fig. (6.6.2a).[†]

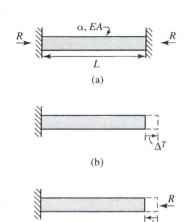

(a)

(b)

(c)

Figure 6.6.2

Solution: Solutions to this class of problems are easily obtained using the superposition principle; i.e., we first imagine that the rod expands freely with no restraint by an amount Δ^T [Fig. (6.6.2b)]. For this simple case, it is clear that the walls actually exert a compressive force so as to prevent the rod from expanding in the longitudinal direction [Fig. (6.6.2c)]. Thus, if we imagine that the process takes place in two stages, the wall will then exert a compressive force R to 'push back' the rod to its original length. Denoting the shortening effect of the reaction R by Δ^R, $\Delta^R = \frac{RL}{AE}$,[‡] we have the simple relation

$$\Delta^R = \Delta^T. \qquad (6.6.2a)$$

This relation is again, in fact, a trivial example of an equation of geometric compatibility, which, written explicitly, is

$$\frac{RL}{AE} = \alpha L \cdot \Delta T \qquad (6.6.2b)$$

or

$$R = AE\alpha \cdot \Delta T. \qquad (6.6.2c)$$

The axial stress is therefore $\sigma = E\alpha \cdot \Delta T$. We observe that for this problem, the solution is independent of the length L. It is important to note here that we have implicitly assumed positive R to be compression and hence the positive stress σ here corresponds to a compressive stress. Assuming a temperature increase of 50°C, the stress in the rod is $\sigma = (200 \times 10^9) \cdot (11.7 \times 10^{-7}) \cdot (50) = 11.7 \times 10^7 \text{ N/m}^2 = 117 \text{ MPa}$. □

Example 6.6: Consider two bars of different materials (brass and steel), having the same cross-sectional area A, which are held rigidly at B and C and are initially separated by a gap 'δ' [Fig. (6.6.3)]. The temperature of the entire system is increased by an amount ΔT, which is greater than that required to close the gap. Determine the resulting axial stress.

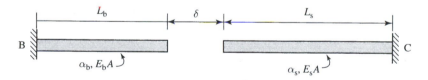

Figure 6.6.3

Solution: As the temperature is increased, the rods will first elongate freely, according to Eq. (6.6.1), until the gap is closed. Clearly, since the existing supports at B and C are rigidly fixed in space, the supports will then exert forces that tend to restrain any subsequent elongation due to a further increase in temperature.

In solving this problem, we again make use of the principle of superposition; i.e., we first determine the free elongation due to the temperature increase disregarding

---

[†] We assume implicitly that the wall is frictionless and therefore that the rod is free to deform *laterally* with no constraints.

[‡] Note that although the length of the bar after the temperature change is imagined to be $L^* = L + \Delta^T$ ($\Delta^T \ll L$) [see Fig. (6.6.2b)], in calculating Δ^R, we use the length of the original geometry, L [see comment (e) of Section 6.3].

any restraints. To these elongations, we then superimpose the effect of the restraints. We denote the free-temperature effect by Δ^T and the restraint effect by Δ^R.

Thus, assuming that the bars are allowed to expand freely, the free expansions of the brass and steel bars are Δ_b^T and Δ_s^T, respectively.

However, since the two supports are rigid, after the gap has been closed, these supports clearly exert compressive resisting forces that tend to prevent any further elongation of the system. Moreover, since no other external forces are acting on this system, it follows from equilibrium that the reacting forces must evidently be the same [Fig. (6.6.4a)]; we denote these (unknown) reactive forces by R. Thus both the steel bar and the brass bar are subjected to the same compressive axial force $F = R$, as shown in the free-body diagrams of Fig. (6.6.4b).

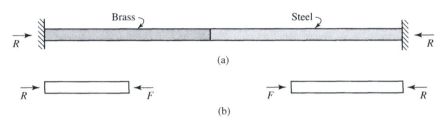

(a)

(b)

Figure 6.6.4

Due to the resisting compressive axial force R, the *shortening* of the two bars is given by

$$\Delta_b^R = \frac{RL_b}{E_b A}, \qquad \Delta_s^R = \frac{RL_s}{E_s A}. \qquad (6.6.3)$$

Then, using superposition, the final changes in length are

$$\Delta_b = \Delta_b^T - \Delta_b^R, \qquad \Delta_s = \Delta_s^T - \Delta_s^R. \qquad (6.6.4)$$

The deformation of the bars, showing the temperature and reactive effects of the bars, is shown in Fig. (6.6.5). From the simple geometric relations of this figure, we write

$$\Delta_b + \Delta_s = \delta. \qquad (6.6.5)$$

Note that this relation is the basic *geometric compatibility equation* governing the deformation of this system. Substituting Eqs. (6.6.4),

$$\left(\Delta_b^T - \Delta_b^R\right) + \left(\Delta_s^T - \Delta_s^R\right) = \delta, \qquad (6.6.6a)$$

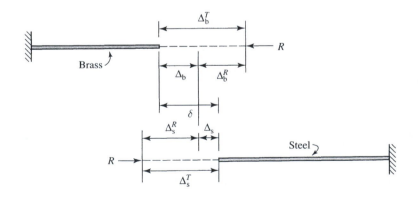

Figure 6.6.5

which we rewrite as

$$\Delta_b^R + \Delta_s^R = \Delta_b^T + \Delta_s^T - \delta. \tag{6.6.6b}$$

Now, from Eq. (6.6.1), we have

$$\Delta_b^T = \alpha_b L_b \cdot \Delta T, \qquad \Delta_s^T = \alpha_s L_s \cdot \Delta T \tag{6.6.7}$$

Substituting Eqs. (6.6.3) and (6.6.7) in Eq. (6.6.6b), we have

$$\frac{R}{A}\left[\frac{L_b}{E_b} + \frac{L_s}{E_s}\right] = (\alpha_b L_b + \alpha_s L_s) \cdot \Delta T - \delta \tag{6.6.8}$$

from which we finally obtain

$$R = \frac{A E_b E_s}{L_s E_b + L_b E_s}[(\alpha_b L_b + \alpha_s L_s) \cdot \Delta T - \delta]. \tag{6.6.9}$$

We consider a numerical example that proves to be instructive. Let

$$A = 3\,\text{cm}^2, \qquad L \equiv L_b = L_s = 100\,\text{cm}, \qquad \Delta T = 80°\text{C}.$$

Note that we have not yet assigned here a value to δ. Furthermore, we recapitulate the properties of the two materials

$$E_b = 120\,\text{GPa}, \qquad\qquad E_s = 200\,\text{GPa},$$
$$\alpha_b = 18.7 \times 10^{-6}\,°\text{C}^{-1}, \qquad \alpha_s = 11.7 \times 10^{-6}\,°\text{C}^{-1}.$$

From Eq. (6.6.9),

$$R = \frac{A E_b E_s}{E_b + E_s}[(\alpha_b + \alpha_s) \cdot \Delta T - \delta/L]. \tag{6.6.10}$$

Substituting the above values,

$$R = \frac{(3 \times 10^{-4}) \cdot (2.4 \times 10^{22})}{3.2 \times 10^{11}}[(30.4 \times 10^{-6}) \cdot 80 - \delta/L]$$

or

$$R = 22.5 \times 10^6(2.432 \times 10^{-3} - \delta/L)$$

If, for example, $\delta = 0.2$ cm, then with $L = 100$ cm, $R = 9720$ N from which we find $\sigma = 32.4$ MPa (compression) in both bars.

It is instructive to compare this result with the case $\delta = 0$, that is, when no gap exists. The reactive force is then $R = 54,720$ N and $\sigma = 182.4$ MPa. □

From the numerical results of the above problem, we observe that the thermal stresses induced by the given temperature changes are reduced from $\sigma = 182.4$ MPa to $\sigma = 32.4$ MPa (i.e. by 82%) by the mere introduction of a very small gap of $\delta = 0.2$ cm in the system.

The important influence of the gap in a system may be seen more clearly if we examine the above problem where both rods are of the same material. Thus letting $E_b = E_s = E$ and $\alpha_b = \alpha_s = \alpha$, Eq. (6.6.10) becomes

$$R = AE(\alpha\Delta T - \delta/2L) \tag{6.6.11a}$$

and therefore the stress σ in the system is

$$\sigma = E\alpha\Delta T\left(1 - \frac{\delta/L}{2\alpha\Delta T}\right). \tag{6.6.11b}$$

We note that the quantity '$\alpha \Delta T$' is usually very small (generally of the order of 10^{-4} for $\Delta T \leq 100°C$). Hence we observe that for a small gap δ with δ/L of this same order, the stress σ is reduced considerably.

Indeed, the magnitude of stresses in any statically indeterminate mechanical or structural system is very sensitive to 'gaps' existing in the system. With this knowledge, one often purposely introduces, if possible, such gaps in the design of a structure, in order to minimise induced thermal stresses. This result has many practical applications in the construction of structures. For example, 'construction joints' in bridges and roadways are usually introduced to prevent high induced stresses due to temperature changes.

The above problem, representing a typical example of the evaluation of thermal stresses due to axial loading, is in fact a statically indeterminate problem. We emphasise again that, as is true for this class of problems, we require an appropriate equation of geometric compatibility in addition to the appropriate equation(s) of equilibrium.

6.7 Elastic–plastic behaviour: residual stresses

At this stage, we have considered only elastic behaviour under axial loadings. However, as discussed in Chapter 4, it is clear that if the loads acting on a body are sufficiently large, the body may cease to behave elastically and may enter the plastic range. We introduce here a simple problem to illustrate the analysis of a system in which elastic–plastic behaviour occurs.

Consider a system consisting of a *rigid* (weightless) plate, supported symmetrically by steel and (hard drawn) copper wires, as shown in Fig. (6.7.1a). The copper and steel wires each have the same cross-sectional area A. A force P is applied at the centre. The steel and copper wires are each assumed to behave as an ideal elastic–plastic material with $\sigma–\epsilon$ diagrams as given in Figs. (6.7.2a and b), respectively.

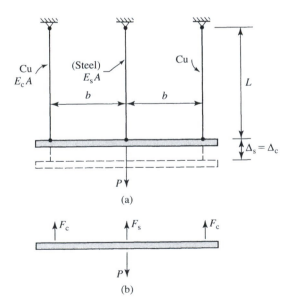

(a)

(b)

Figure 6.7.1

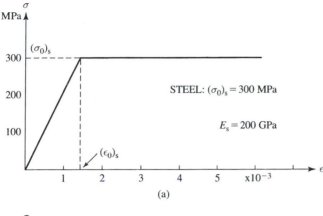

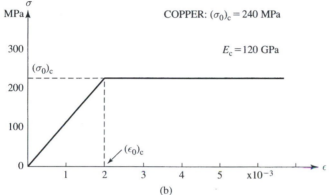

Figure 6.7.2

We consider the case as a force P, of gradually increasing magnitude, is applied (*loading*) and which is removed (*unloading*) after the system undoes plastic deformation. Specifically, we wish to determine

- the load $P = P_y$ at which yielding first occurs and the corresponding vertical displacement Δ_y;
- the ultimate load P_{ult} that the system can carry.
- The vertical displacement Δ_{ult} as P approaches P_{ult}.
- The permanent deformation after unloading.

Using the condition of symmetry, the same force F_c exists in each of the copper wires. (Note that by taking $\sum M = 0$ about the midpoint of the plate, we arrive at the same result.) From equilibrium, $\sum F_y = 0$, we have [Fig. (6.7.1b)]

$$2F_c + F_s = P, \tag{6.7.1}$$

where F_s and F_c are the forces in the steel and copper wires, respectively. Note that this equation is independent of the material properties and therefore remains always valid.

Due to the rigidity of the plate and the symmetry of the system, we immediately write the geometric compatibility equation

$$\Delta_s = \Delta_c. \tag{6.7.2a}$$

We find it here more appropriate to express the compatibility equation in terms of the axial strains in the wire; thus (since the wires are of the same length L), we have

$$\epsilon_s = \epsilon_c. \tag{6.7.2b}$$

Evidently, the initial behaviour of the system is elastic provided $P < P_y$. Yielding first takes place when the stress in one of the wires, steel or copper, reaches the value $(\sigma_0)_s$ or $(\sigma_0)_c$, respectively. From Fig. (6.7.2), this is equivalent to stating that yielding occurs when the strain first reaches $(\epsilon_0)_s$ or $(\epsilon_0)_c$. Using the given numerical values as shown in this figure and the $\sigma-\epsilon$ relations, $\sigma = E\epsilon$, these strains are $(\epsilon_0)_s = 1.5 \times 10^{-3}$ and $(\epsilon_0)_c = 2.0 \times 10^{-3}$. Since $(\epsilon_0)_s < (\epsilon_0)_c$, it follows that yielding will first take place in the steel wire when the strain ϵ_s reaches $(\epsilon_0)_s$.

From Eq. (6.7.2b), the strain in the copper $\epsilon_c = (\epsilon_0)_s$ at this first yielding and thus the corresponding stress in the copper wire is

$$\sigma_c = E_c\epsilon_c = (120 \times 10^9) \cdot (1.5 \times 10^{-3}) = 180 \times 10^6 \text{ N/m}^2. \tag{6.7.3}$$

Hence the forces in the wires are

$$F_s = (300 \times 10^6) \cdot A \quad \text{and} \quad F_c = (180 \times 10^6) \cdot A, \tag{6.7.4}$$

respectively. (We note that this value of F_s is the maximum force $F_s|_{max}$ that the steel wire can carry.)

Then by Eq. (6.7.1), the force P at the first yielding is

$$P_y = (300 \times 10^6) \cdot A + 2(180 \times 10^6) \cdot A = (660 \times 10^6) \cdot A \tag{6.7.5a}$$

and the corresponding displacement is

$$\Delta_y = (\epsilon_0)_s \cdot L = (1.5 \times 10^{-3}) \cdot L. \tag{6.7.5b}$$

As P increases beyond P_y, the stress in the steel remains constant, $(\sigma_0)_s$. However, the stress in the copper increases gradually with increasing P until it reaches the value $(\sigma_0)_c$ with a corresponding strain $(\epsilon_0)_c$. Thus, the maximum force that the copper wire can carry is $F_c|_{max} = (240 \times 10^6) \cdot A$. At this point, the force P has reached its ultimate value, P_{ult}, given by

$$P_{ult} = F_s|_{max} + 2F_c|_{max} = (300 \times 10^6)A + 2(240 \times 10^6)A = (780 \times 10^6)A. \tag{6.7.6a}$$

As P reaches $P = P_{ult}$ (which occurs just as the copper yields), the displacement is

$$\Delta_{ult} = (\epsilon_0)_c L = 2.0 \times 10^{-3}L. \tag{6.7.6b}$$

Thereafter the system continues to yield under P_{ult}.

A plot of the load–displacement relation is shown in Fig (6.7.3). The original line OB represents purely elastic behaviour; the line BC represents partly elastic behaviour (of the copper) and plastic behaviour (of the steel); the horizontal line CG represents purely plastic behaviour (yielding) of the entire system. We note that the slope of the line OB is considerably greater than BC. Recalling that the slope is a measure of the stiffness, a physical explanation is clear: initially, both the copper and steel wires offer resistance to deformation resulting in a relatively stiff system. Once the steel wire has yielded, the only resistance to increased deformation is due to the copper wire and hence the system is less stiff, as reflected by the lower slope of BC. At $P = P_{ult}$ the copper yields and assuming that the force $P = P_{ult}$ is

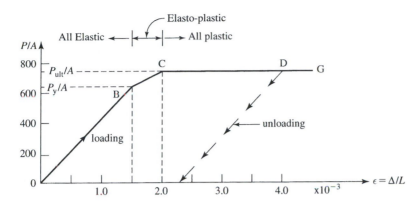

Figure 6.7.3

thereafter maintained, all wires having yielded, the entire system undergoes increasing deformation as represented by the horizontal line $P = P_{ult}$.

Let us assume that the load is slowly removed when the displacement reaches $\Delta = \Delta_D$ (with the corresponding strain, $\epsilon_D = \Delta_D/L$ in the wires). The system is said to undergo 'unloading'. We recall from Chapter 4 [Fig. (4.6.1)], that for a given material having ideal elastic–plastic behaviour, the unloading is elastic and is represented by a line on the $\sigma-\epsilon$ diagram, which is parallel to the original loading curve.

We now wish to study the behaviour of the system during unloading. The respective $\sigma-\epsilon$ curves for the steel and copper (representing both loading and unloading) are shown in Fig. (6.7.4). It is clear that since the system has undergone plastic deformation, we cannot expect that it will return to its original position: that is, the system will undergo a permanent deformation after the load is completely removed. We denote the resulting permanent displacement by δ and denote $\epsilon^F = \delta/L$ as the corresponding (final) permanent strain. Note that the compatibility condition, Eq. (6.7.2b), remains valid; i.e., the strain in all the wires must be identical for this symmetric system. We wish to determine δ.

We observe that after removal of the load, the equilibrium equation, Eq. (6.7.1), with $P = 0$, is

$$2F_c^F + F_s^F = 0 \tag{6.7.7a}$$

and hence

$$\sigma_s^F = -2\sigma_c^F, \tag{6.7.7b}$$

where the superscript F indicates the final value after removal of the load. Thus if both $\sigma_s^F \neq 0$, $\sigma_c^F \neq 0$, the two stresses must necessarily be of opposite sign; either the copper is under tension and the steel in compression or vice versa.[†]

Now, the general $\sigma-\epsilon$ equations of the straight (unloading) lines for the copper and steel are, respectively [see Fig. (6.7.4)],

$$\epsilon_c = \epsilon_D - \left[\frac{(\sigma_0)_c - \sigma_c}{E_c}\right], \tag{6.7.8a}$$

$$\epsilon_s = \epsilon_D - \left[\frac{(\sigma_0)_s - \sigma_s}{E_s}\right]. \tag{6.7.8b}$$

[†] Since the slope of the steel is greater than the corresponding slope of the copper, it is clear that unloading will take place at a faster rate in the steel than in the copper. Furthermore, for the given numerical values of the material properties as shown in Figs. (6.7.2a and b), $\frac{E_s}{E_c} > \frac{(\sigma_0)_s}{(\sigma_0)_c}$. We therefore can anticipate that $\sigma_c^F > 0$ and $\sigma_s^F < 0$ when the load P is removed.

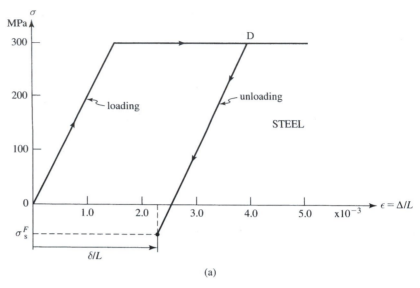

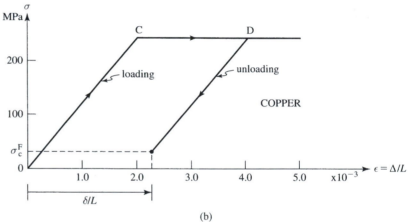

Figure 6.7.4

In particular, the final strains existing in the copper and steel, ϵ_c^F and ϵ_s^F, are given by Eqs. (6.7.8) with $\sigma_c = \sigma_c^F$ and $\sigma_s = \sigma_s^F$, respectively. But, from compatibility, $\epsilon_c^F = \epsilon_s^F$; we therefore have

$$\frac{(\sigma_0)_c - \sigma_c^F}{E_c} = \frac{(\sigma_0)_s - \sigma_s^F}{E_s} \tag{6.7.9a}$$

from which it follows that

$$\sigma_s^F - \sigma_c^F \cdot (E_s/E_c) = (\sigma_0)_s - (\sigma_0)_c \cdot (E_s/E_c). \tag{6.7.9b}$$

Using Eq. (6.7.7b), we find, upon solving for σ_c^F,

$$\sigma_c^F = \frac{(\sigma_0)_c \cdot (E_s/E_c) - (\sigma_0)_s}{2 + (E_s/E_c)}. \tag{6.7.10}$$

Substituting the appropriate numerical values, $\sigma_c^F = 27.3 \times 10^6 \ \text{N/m}^2 = 27.3 \ \text{MPa}$ (tension). Then from Eq. (6.7.7b), $\sigma_s^F = -54.6 \ \text{MPa}$ (compression).

The permanent deformation δ is then found by substituting in either of Eqs. (6.7.8): e.g.,

$$\delta = \epsilon_s^F \cdot L = \Delta_D - \left[\frac{(\sigma_0)_s - \sigma_s^F}{E_s} \right] \cdot L \qquad (6.7.11)$$

Let us assume that $\epsilon_D = 4.0 \times 10^{-3}$. Then $\delta = 4.0 \times 10^{-3}L - (1.77 \times 10^{-3}) \cdot L = 2.23 \times 10^{-3}L$.

Thus we note that if a system undergoes plastic deformation, not only will there be a permanent state of deformation (after removing all external loads) but also a non-zero state of stress may thereafter exist in the unloaded system. The stresses σ_s^F and σ_c^F are therefore called **residual stresses**. We emphasise, however, that such residual stresses as calculated above can only exist in a statically indeterminate system.

PROBLEMS

Note: In all problems below, the material behaviour of the members is assumed to be linear elastic unless specified otherwise. Neglect all localised effects in the solution of problems.

The following constants are to be used in solving the problems.

Steel	$E_s = 200$ GPa,	$\alpha_s = 11.7 \times 10^{-6}$ (°C^{-1})
Aluminium	$E_a = 70$ GPa,	$\alpha_a = 23.6 \times 10^{-6}$ (°C^{-1})
Brass	$E_b = 120$ GPa,	$\alpha_b = 18.7 \times 10^{-6}$ (°C^{-1})
Bronze	$E_{br} = 105$ GPa,	$\alpha_{br} = 18 \times 10^{-6}$ (°C^{-1})
Copper	$E_{cu} = 120$ GPa,	$\alpha_{cu} = 16.9 \times 10^{-6}$ (°C^{-1})

Sections 2–4

6.1: A cylindrical steel rod of length $L = 50$ cm and cross-sectional area A is subjected to an axial tensile force $P = 12$ kN. If the allowable tensile stress is $\sigma_{allow} = 120$ MPa, and the maximum permitted elongation is $\Delta L = 200\mu = 200 \times 10^{-6}$ m, determine the minimum required diameter d.

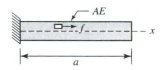

Figure 6P.2

6.2: A rod, consisting of two segments AB and BC with moduli of elasticity E_{AB} and E_{BC}, respectively, is subjected to axial loads, as shown in Fig. (6P.2). If C is not permitted to displace, determine the required ratio A_{AB}/A_{BC}.

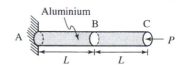

Figure 6P.3

6.3: A cylindrical rod, part of a sensitive instrument, consists of two segments AB and BC bonded to each other, each having the same cross-sectional area $A = 140$ mm^2, as shown in Fig. (6P.3). Segment AB is aluminium. As a design specification, it is required that under a compressive axial load $P = 12$ kN, the displacement of C is not to exceed 1 mm nor be less than 0.9 mm. Which material(s) can be used for segment BC – steel, aluminium, brass, bronze or copper?

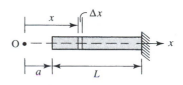

Figure 6P.4

6.4: A linear isotropic elastic cylindrical rod with modulus of elasticity E is fixed at one end and subjected, as shown in Fig. (6P.4), to a force located at point O, which attracts any given element of thickness Δx with a (body) force given by $\Delta f = \frac{k}{x^2} A \Delta x$, where A is the cross-sectional area of the rod, x is the distance from O to the element Δx and k is a constant. Determine the elongation of the rod.

6.5: Body forces, varying as $f(x) = C e^{x/a}$ (where C is constant having units N/m^3), act in the x-direction on a prismatic rod having axial rigidity AE and length a, as shown in Fig. (6P.5). (a) Show that the resulting stress σ_x at any cross-section, x, is $\sigma_x = Ca(e - e^{x/a})$ and (b) determine the change in length ΔL of the rod.

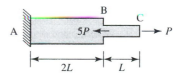

Figure 6P.5

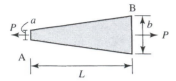

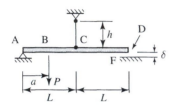

Figure 6P.7

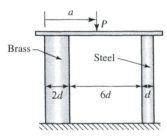

Figure 6P.8

Figure 6P.9

6.6: A copper–nickel alloy tube 40 cm in length, whose cross-sectional area is $A = 60\,mm^2$, is subjected to an axial tensile force of 8000 N. The material behaviour of the copper is governed by the 'sinh law' [given by Eq. (4.5.5) of Chapter 4] with $\sigma_0 = 100$ MPa and $\epsilon_0 = 0.006$. Determine the change in length ΔL.

6.7: A tapered rod AB with modulus of elasticity E and thickness t, having a width varying linearly from a to $b\,(a \le b)$ and length L, is subjected to axial forces P, as shown in Fig. (6P.7). Assuming that the taper is sufficiently small (i.e., $\frac{b-a}{L} \ll 1$), the normal stress can then be assumed to be uniformly distributed over the cross-section, i.e., $\sigma = \frac{P}{A(x)}$ is a reasonable approximation to the true state of stress. Using this approximation, show that the change in length of the rod is $\Delta L = \frac{PL}{Eta}\frac{\ln(b/a)}{b/a-1}$ or in non-dimensional terms, $\frac{\Delta L}{PL/A_0 E} = \frac{\ln(b/a)}{b/a-1}$, where A_0 is the cross-sectional area at A.

(*Note:* See computer-related Problem 6.42.)

6.8: A rigid rod ABCD is simply supported at A and by a steel wire at C, whose cross-sectional area, $A_s = 15\,mm^2$. Under zero load, a gap $\delta = 2$ mm exists at the right end between D and F, as shown in Fig. (6P.8). (a) Determine, in terms of E_s, A, δ, h, L and P, the distance from A (i.e., the distance a) at which a load P should be applied such that the right end makes contact with point F. (b) Sketch the position a/L as a function of P in terms of the non-dimensional quantity $\frac{E_s A \delta}{Ph}$. If $h = 2$ m and the maximum load P that can be applied is $P = 900$ N, what is the shortest distance a/L?

6.9: A rigid plate is supported by cylindrical steel and brass rods whose diameters are as shown in Fig. (6P.9). Determine a, the position of the load P, if the rigid plate remains horizontal.

6.10: By means of a rigid end plate, an axial force, $P = 120$ kN, passing through the x-axis is applied to the composite member consisting of a steel core and an outer aluminium cylindrical shell, as shown in Fig. (6P.10). Determine (a) the axial stress in the core and in the shell and (b) the change of length ΔL.

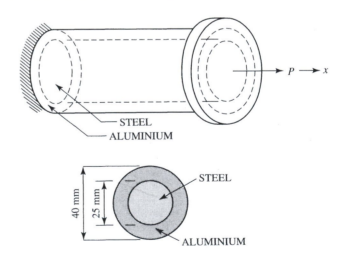

Figure 6P.10

6.11: A structure consists of a rod BC, which is simply supported at C and supported by a steel wire BD, whose cross-sectional area is $A_s = 1.2\,cm^2$. A load $P = 18$ kN is applied at point B, as shown in Fig. (6P.11). Assuming that the rod BC is rigid, determine v_B, the vertical component of displacement of point B.

6.12:* Determine the horizontal and vertical components of displacement of point B, u_B and v_B respectively, of the structure of Fig. (6P.11) if the rod BC is made of aluminium and has a cross-sectional area, $A_a = 3\,cm^2$.

6.13: A compressive axial force $P = 500\,kN$ is applied, by means of a rigid plate, to a concrete column 3 m in height and having cross-sectional dimensions 20 cm × 30 cm [see Fig. (6P.13a)]. The column contains 8 steel reinforcing bars, placed uniformly within the column, each of whose cross-sectional area is $A_s = 2\,cm^2$ [see Fig. (6P.13b)]. The modulus of elasticity of the concrete is $E_c = 20\,GPa$. If the plate remains horizontal, determine (a) the compressive stress in both the concrete and the steel bars and (b) the shortening, ΔL, of the column.

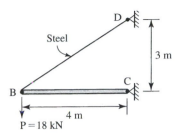

Figure 6P.11

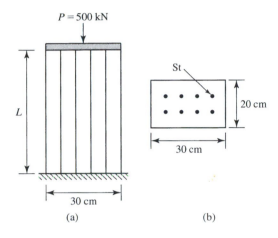

(a)   (b)

Figure 6P.13

6.14: Rigid plates are connected to the ends of a composite rod, made of two materials, as shown in Fig. (6P.14). The moduli of elasticity of the two components of the rod are E_1 and E_2, where $E_1 > E_2$. Determine the value of e (measured from the interface) at which an axial load must be applied to produce a uniform extensional strain throughout any cross-section.

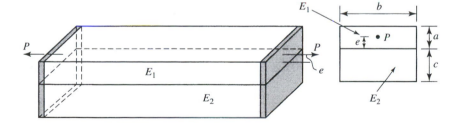

Figure 6P.14

6.15:* A structure, consisting of an elastic rod AB having axial rigidity $E_0 A_0$, is supported by means of a wire at A (whose axial rigidity is $A_s E_s$) and simply supported at B. A force P, making an angle α with the rod, acts at point A, as shown in Fig. (6P.15). Determine the ratios u_A/L and v_A/L of the horizontal and vertical components of displacement of point A, as a function of α.

(*Note:* See computer-related Problem 6.43.)

6.16: A composite rod of length L, having a cross-section as shown in Fig. (6P.16), is made of two materials with moduli of elasticity E_1 and E_2. A load P is applied at

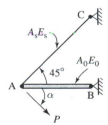

Figure 6P.15

the end plates such that the strain is uniform throughout any cross-section. Determine (a) the value of e, as shown in the figure and (b) the elongation of the composite rod.

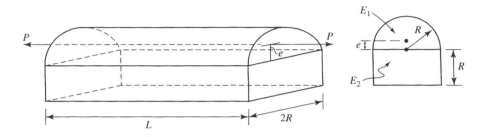

Figure 6P.16

Section 5

6.17: A rigid rod ABCD is suspended by means of three identical wires, as shown in Fig. (6P.17), and carries a load P at B. (a) Determine the tensile force in each wire as a function of P and the ratio b/a. (b) If $b=3a$, $P=19$ kN and the allowable tensile stress in the wires is $\sigma = 100$ MPa, what are the required diameters of the wires?

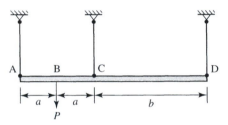

Figure 6P.17

6.18: A pin-connected truss is composed of three rods of the same material. The cross-sectional area of rod BD is A_0 and that of AD and CD is A_1. The truss is subjected to a force P at D, as shown in Fig. (6P.18). (a) Show that if $A_0 = A_1$, the axial forces are given by

$$F_{AD} = F_{CD} = \frac{P \cos^2 \beta}{1 + 2 \cos^3 \beta}, \qquad F_{BD} = \frac{P}{1 + 2 \cos^3 \beta} \quad (0 < \beta < 90°).$$

(b) For this case, namely if $A_0 = A_1$, determine the required cross-sectional area if the allowable tensile stress in the rods is $\sigma = 120$ MPa, if $P = 50$ kN and if $a = 3$ m and $h = 4$ m.

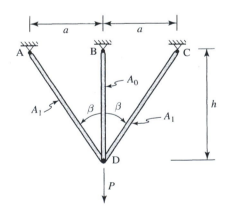

Figure 6P.18

6.19:* The rods of the truss shown in Fig. (6P.18) have different cross-sectional areas and are made of different materials: rods AD and CD are made of steel with $A_1 = A_s$ and rod BD is made of aluminium with $A_0 = A_a$. (a) If the allowable stresses in the steel and aluminium are given as $\sigma_s|_{all} = 120\,\text{MPa}$ and $\sigma_a|_{all} = 90\,\text{MPa}$, determine the angle β such that each rod is stressed to its maximum allowable value when subjected to the load P, as shown in the figure. [Note that this value provides the optimal configuration for the given material properties and cross-section of the rods.] (b) If $A_0 = 2\,\text{cm}^2$ and $A_1 = 4\,\text{cm}^2$, what is the maximum allowable load P.

6.20: A tapered member having constant thickness t is welded at A and B to rigid supports, as shown in Fig. (6P.20). Assuming that the taper is small (i.e., $a - b \ll L$) such that the axial stress can be considered to be uniform at any cross-section, (a) determine (in terms of a, b and P) the reactions at A and B due to an axial load P acting at the centre and (b) show that if $b \to a$, $R_A = R_B = P/2$.

6.21: A tapered member, as shown in Fig. (6P.21), having constant thickness t is rigidly attached to supports at A and B. Assuming that the taper is small (i.e., $a - b \ll L$) such that the axial stress can be considered to be uniform at any cross-section, (a) determine (in terms of a, b and P) the reactions at A and B due to an axial load P acting at the centre, C and (b) show that if $a \to b$, $R_A = R_B = P/2$.

6.22: A tapered member, as shown in Fig. (6P.22), having constant thickness t is rigidly attached to supports at A and B. An axial force P is applied at section D located a distance αL from A ($0 < \alpha \le 0.5$) as shown in the figure. Assume that the taper is small (i.e., $a - b \ll L$) such that the axial stress is uniform at any cross-section. (a) Show that the reactions at A and B are given, respectively, by

$$R_A = P\,\frac{\ln[(1 - 2\alpha)a + 2\alpha b] - \ln a}{2[\ln b - \ln a]}$$

$$R_B = P\,\frac{2\ln b - \ln a - \ln[(1 - 2\alpha)a + 2\alpha b]}{2[\ln b - \ln a]}$$

(b) Show that if $\alpha = 0.5$, $R_A = R_B = P/2$.

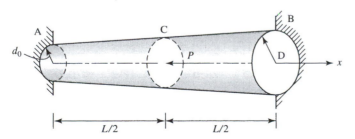

6.23: A conical rod ACB whose diameter varies linearly from $d_0 > 0$ to D is rigidly attached to supports at A and B, as shown in Fig. (6P.23). A force P is applied at the centre C. Determine the reactions at A and B in terms of d_0, D and P.

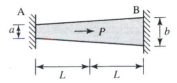

Figure 6P.20

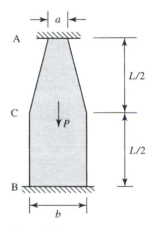

Figure 6P.21

Figure 6P.22

Figure 6P.23

Section 6

6.24: A bronze and an aluminium rod, having cross-sectional areas $A_a = 1800\,\text{mm}^2$ and $A_{br} = 1500\,\text{mm}^2$ respectively, are bonded together as shown in Fig. (6P.24) and are placed between two rigid supports such that a gap $\delta = 0.6\,\text{mm}$ exists at the left end. Determine (a) the compressive reaction on the rods due to a uniform temperature increase of 150°C, (b) the stress in each rod and (c) the resulting change in length of each rod.

6.25: An assembly consists of a steel bolt whose cross-sectional area is $A_s = 0.6\,\text{cm}^2$ surrounded by a hollow aluminium cylinder of cross-sectional area $A_a = 1.5\,\text{cm}^2$, as shown in Fig. (6P.25). Determine (a) the stress in each member and (b) the change in length if the entire assembly undergoes an increase of temperature $\Delta T = 50°\text{C}$.

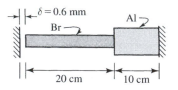

Figure 6P.24

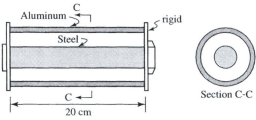

Figure 6P.25

6.26: An assembly is designed to be used at very high temperatures. The assembly consists of a titanium rod whose cross-sectional area is $A_T = 0.6\,\text{cm}^2$ surrounded by a hollow monel alloy cylinder of cross-sectional area $A_M = 1.5\,\text{cm}^2$. The assembly is bounded by rigid end plates, as shown in Fig. (6P.26a). The titanium as well as the monel are assumed to be elastic–perfectly plastic with stress–strain curves as shown in Fig. (6P.26b). Determine the change in temperature ΔT at which first yielding occurs. (<u>Given</u>: $\alpha_T = 9.5 \times 10^{-6}\,°\text{C}^{-1}$, $\alpha_M = 13.9 \times 10^{-6}\,°\text{C}^{-1}$.)

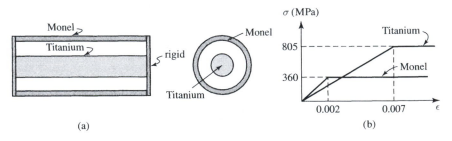

Figure 6P.26

(a)   (b)

6.27: An assembly, consisting of a rigid plate, is supported by means of two steel rods each having a cross-sectional area $A_s = 5\,\text{cm}^2$ and length $h = 2\,\text{m}$. A copper rod with $A_{cu} = 10\,\text{cm}^2$ is inserted at the centre where a gap $\delta = 0.50\,\text{mm}$ exists between the rod and plate, as shown in Fig. (6P.27). If the temperature of the entire system is increased by 60°C, determine (a) the axial stress in the copper and steel rods, (b) the deflection of the rigid plate.

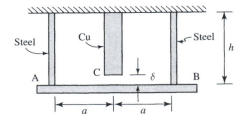

Figure 6P.27

6.28:* Given that all rods of the truss of Fig. (6P.18) have identical axial rigidities (i.e., $A_0 E_0 = A_1 E_1 = AE$) and undergo the same increase in temperature ΔT. (a) Determine the axial force in each member due to this increase in temperature in terms of the geometry, and the coefficient of thermal expansion α. Show that the force in rod BD is a maximum when $\beta = 64.4°$.

(*Note:* See computer-related Problem 6.44.)

6.29:* The square frame shown in Fig. (6P.29) consists of four aluminium rods that are pinned at the corners and braced by two diagonal steel wires. The ratio of the cross-sectional area of the aluminium rods to that of the steel wires is given as 20:1. Determine the axial stress in both the rods and wires if the entire frame is subjected to an increase in temperature of 40°C.

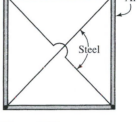

Figure 6P.29

Section 7

6.30: A rigid plate BCDF is simply supported at B and by two wires whose cross-sectional areas are $A = 0.5\,\text{cm}^2$ at C and D, as shown in Fig. (6P.30). The wires are assumed to behave as elastic–perfectly plastic materials with a yield stress of σ_0 and modulus of elasticity E. A load P is applied at point F. Determine, in terms of σ_0, A and h, (a) the force P_y at which yielding first takes place, (b) the displacement of point F when $P = P_y$, (c) the ultimate load P_{ult}, (d) the displacement of point F as P reaches P_{ult} and (e) plot P vs. Δ_F and show the values at all critical points.

6.31: A rod BCD of cross-sectional area A and made of an elastic–perfectly plastic material (with modulus E and yield point σ_0) is rigidly attached to supports at B and D. Initially, the bar is free of all stresses. An axial force P is applied at C, as shown in Fig. (6P.31). (a) Determine the displacement δ_C of the cross-section at C as P increases from zero to its ultimate value, P_{ult}. (b) Plot the results P vs. δ_C, and show all critical points on the graph.

6.32:* Given three rods of equal length L and cross-sectional area A. The rods are fixed at A and D, are connected at B and C and are subjected to a slowly applied axial load P, as shown in Fig. (6P.32a). The rods are assumed to behave as elastic–perfectly plastic materials; the modulus of elasticity of rods AB and CD is given as E_1 and that of BC is $E_2 = E_1/4$, where the stress-strain curves of the two materials are shown in Fig. (6P.32b). (a) Determine (in terms of σ_0, A, L and E) the relation between the displacement δ_C of point C and P for values $0 \le P < P_{ult}$. (b) Plot P vs. δ_C showing all critical values on the graph. (c) If the road is loaded until $\delta_C = 6\sigma_0 L/E$, what is the permanent displacement $\delta_C|_{\text{perm}}$ after the loads are removed.

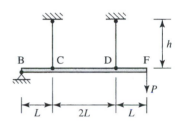

Figure 6P.30

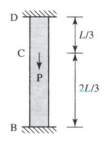

Figure 6P.31

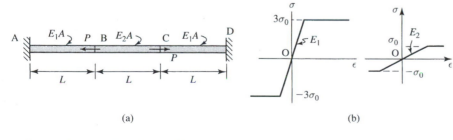

(a)   (b)   **Figure 6P.32**

Review and comprehensive problems

6.33:* Given a plate ($a \times b$) lying in the x–y plane, as shown in Fig. (6P.33), subjected to a distribution of body forces $f = Ce^{x/a}$ (where C is a constant having units N/m³), which act in the x-direction. The material properties of the plate are given as E and ν, respectively. (a) Show, by means of a simple sketch, that due to the symmetry of the applied loads, $\tau_{xy}(x, 0) = 0$. (b) If $b \ll a$, one may then assume, upon making use of this given geometry, that $\tau_{xy}(x, y) = 0$ at all points of the plate. Explain the reasoning. (c) Making use of the assumption of (b), show that the solution to the stress equations of equilibrium for plane stress yields a stress field $\sigma_y = 0$ and σ_x, as given in Problem 6.5. (d) Determine the angle $\angle ABD$ *after* deformation where the distances between A, B and D are assumed to be infinitesimal. (e) Determine the displacements in the x- and y-directions of point B.

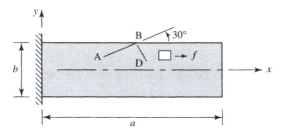

Figure 6P.33

6.34: Making use of the principle of conservation of energy for a linear elastic body (see under Hooke's law in Chapter 4), solve part (a) of Problem 6.8.

6.35: A wire AB of length L is pre-stressed to a given tension T_0 and attached to rigid supports at the two ends. A weight W is attached at some intermediate point, located at a distance $a = \alpha L$ ($0 \leq \alpha \leq 1$), from one end, as shown in Fig. (6P.35). Show that the tensile reactions R_A and R_B at A and B, respectively, are

$$R_A = W(\alpha - 1 + \gamma), \qquad R_B = W(\alpha + \gamma),$$

where $\gamma = T_0/W$.

(*Note*: See computer-related Problem 6.45.)

6.36:* An elastic cylindrical rod of diameter D, modulus of elasticity E and length L is inserted in a bore having the same diameter. To lower the rod, an axial force P must be applied at the top, as shown in Fig. (6P.36), to counteract the frictional force $f(y)$ along the lateral surface, which is found to vary as $f = ky^2$, where k has units N/m⁴. (a) Assuming that plane cross-sections in the rod remain plane, determine the the axial stress $\sigma(y)$ at any cross-section. (b) Determine the change in length ΔL of the rod when the force P is applied.

6.37:* A circular cylindrical rod of length L whose material density is ρ (N/m³), hangs from a rigid support, as shown in Fig. (6P.37). The radius of the rod varies parabolically as $r(y) = r_a + (r_b - r_a)(y/L)^2$, where $r_b - r_a \ll L$ such that the variation of the cross-sectional area is 'slow'. (a) Assuming that all cross-sections remain plane, determine the (average) axial stress σ_y at any cross-section y. (b) Show that the elongation ΔL of the rod is given by

$$\frac{\Delta L}{\rho L^2/E} = \int_0^1 \xi \left(\frac{1 + 2\beta \xi^2/3 + \beta^2 \xi^4/5}{1 + 2\beta \xi^2 + \beta^2 \xi^4} \right) d\xi,$$

where $\beta = r_b/r_a - 1$ and $\xi = y/L$. (c) What is an appropriate criterion for β, that permits use of the above approximation?

(*Note*: See computer-related Problem 6.46.)

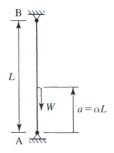

Figure 6P.35

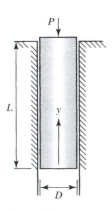

Figure 6P.36

6.38:* A shock absorber consists of a steel rod (with modulus of elasticity E_s) of diameter $d = 4$ cm and length $L = 30$ cm, surrounded by a rubbery material which is encased in a rigid cylindrical shell whose inner diameter is $D = 50$ cm, as shown in Fig. (6P.38a). When subjected to an axial force P, the system deforms as shown in Fig. (6P.38b). The rubbery material is assumed to behave elastically with a shear modulus $G = 6$ MPa. (a) Determine the displacement δ of the bottom of the rod, in terms of d, D, G and P. (b) Evaluate (a) for the given numerical values of the parameters of the problem if $P = 100$ kN. (c) Assuming that all cross-sections of the rod remain plane, determine the shortening ΔL, of the rod, in terms of d, D, E_s, L and P. (d) Evaluate ΔL numerically.

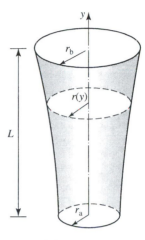

Figure 6P.37

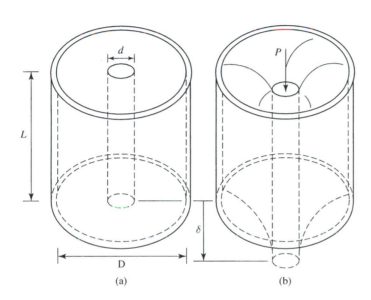

(a)   (b)

Figure 6P.38

6.39: An elastic rod of length L and uniform cross-sectional area A, whose material behaviour is elastic–perfectly plastic with modulus of elasticity E, yield point σ_0 and coefficient of thermal expansion α, is welded to two rigid supports while at a temperature T_0 after which it undergoes an increase in temperature ΔT. (a) What is the largest increase in temperature $\Delta T = \Delta T_E$ for which the rod remains elastic? (b) Determine the residual stress if the rod is first subjected to a uniform temperature increase $\Delta T > \Delta T_E$ and then cooled down to its original temperature T_0. Indicate if in tension or compression. (c) Evaluate the results numerically if the rod is made of steel, with $\sigma_0 = 250$ MPa and $\Delta T = 60°$C.

6.40: Composite materials are often made of thin carbon graphite fibres or high-strength glass fibres (each having a cross-sectional area A_f and modulus of elasticity E_f), which are embedded in a 'soft matrix' (usually consisting of an epoxy) whose modulus of elasticity is E_m. E_f is usually orders of magnitude greater than E_m, i.e. $E_f \gg E_m$. A representative element, having an area A, of the cross-section of this material is shown in Fig. (6P.40).

(a) Assuming that a perfect bond exists at the fibre/matrix interfaces, and that all cross-sections remain plane under axial loading and undergo a strain ϵ, determine the ratio of the axial stress in the fibre, σ_f, to that in the matrix, σ_m.

(b) If n fibres exist within the representative cross-sectional element, determine the axial resultant force P, which exists on the area A in terms of σ_f, σ_m, A_f, A_m and n, where A_m is the cross-sectional area of the matrix within the element.

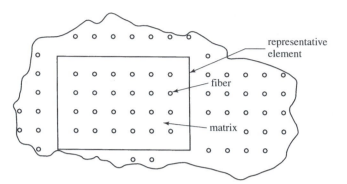

Figure 6P.40

(c) Defining the *fibre volume fraction*, v_f, of the fibres as $v_f = nA_f/A$ and that of the matrix as $v_m = A_m/A$ (where A_m is the net area of the matrix) (note that $v_f + v_m = 1$), show that the "average" axial stress $\bar{\sigma}$ on the cross-section, defined as $\bar{\sigma} = \frac{P}{A}$, in terms of σ_f, σ_m, v_f and v_m, is given as $\bar{\sigma} = v_f \sigma_f + v_m \sigma_m$.

(d) Upon defining the effective modulus of elasticity for this material as $E_{eff} = \frac{\bar{\sigma}}{\epsilon}$, show that E_{eff} can be expressed as $E_{eff} = v_f E_f + v_m E_m$.

(e) Typical material constants for carbon fibres embedded in an epoxy matrix are $E_f = 300$ GPa and $E_m = 2.4$ GPa, respectively. (i) Using these values, evaluate E_{eff} for a composite material with a typical value $v_f = 0.2$, (ii) determine the average stress $\bar{\sigma}$ and the change of length of a rod if an axial force $P = 6000$ kN acts in the direction of the fibres on a rod whose cross-sectional area is $A = 4$ cm^2 and whose length is $L = 4$ m and (iii) determine the stresses σ_f and σ_m in the fibres and matrix, respectively.

6.41: Using the results of Problem 6.40, repeat part (e) of Problem 6.40 for a typical glass/epoxy composite having the following properties: $E_f = 72$ GPa, $E_m = 2.4$ GPa with $v_f = 0.45$.

The following problems are to be solved using a computer

6.42: Write a computer program to evaluate the elongation of the slightly tapered rod of Problem 6.7, having length L, in terms of the non-dimensional quantity, viz. in terms of $\frac{\Delta L}{PL/A_0 E}$ in the range $1 \leq b/a \leq 10$ and plot the results. For what range of b/a are the results meaningful?

6.43: Write a computer program (a) to evaluate the forces F_{AB}/P and F_{AC}/P of the structure of Problem 6.15 as a function of α and (b) the displacements u_A and v_A in non-dimensional form (i.e., $u_A A_0 E_0/PL$, $v_A A_0 E_0/PL$) and plot the results for values $0 \leq \alpha \leq 180°$C.

6.44: Plot the forces F_{AD} and F_{BD} in rod AD and BD, respectively, of the truss of Problem 6.28 in non-dimensional terms (i.e., $F_{AD}/AE\alpha\Delta T$, etc.) as a function of β, and determine numerically the maximum/minimum values of the defined ratios.

6.45: (a) Using the results given for the pre-stressed wire of length L of Problem 6.35, express the ratio $\beta = R_B/R_A$ in terms of the position of the weight, $\alpha = a/L$ and $\gamma = T_0/W$. (b) For several discrete values of β ($1 < \beta < \infty$), plot a family of curves for γ as a function of α [this does not require a computer]. For what position of α does $R_B = 2R_A$ and $R_B = 3R_A$ if $\gamma = 1.5$ and $\gamma = 1$, respectively. (c) Alternatively (for several discrete values of γ, $1 < \gamma < 10$) plot a family of curves for β as a function of α.

(d) What are, respectively, the required pre-tensions (as measured by γ) in order that $\beta > 2$ and $\beta < 3$ irrespective of the position of the weight, α. (e) What conclusions may be drawn from the above curves.

6.46:   Using the results given for Problem 6.37, evaluate numerically the integral that yields the change in length (given in non-dimensinal form, $\frac{\Delta L}{\rho L^2/E}$) of the parabolically tapered rod and plot as a function of $\beta = r_b/r_a - 1$, where r_a and r_b are shown in Fig. (6P.37).

7

Torsion of circular cylindrical rods: Coulomb torsion

7.1 Introduction

In this chapter, we study the behaviour of slender elastic rods, which are subjected to moments about their longitudinal axis. We limit our study to rods that have the shape of a circular cylinder with cross-sections as shown in Fig. (7.1.1). Due to these moments, it is evident that the rod will twist: the rod is then said to be in torsion and the applied moment is referred to as a **torsional moment** or **torque**. We shall use these two terms interchangeably.

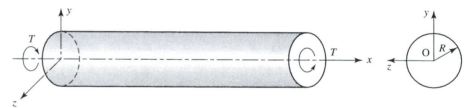

Figure 7.1.1

Circular rods under this force system are, in practice, referred to as *shafts*, as they are often used to transfer energy from engines, for example, in automobiles, aircraft or other machinery.

We shall be interested in determining the internal stresses and the rotation due to applied torques. We first study elastic behaviour and, at a later stage, will consider elastic–plastic behaviour of such rods.

7.2 Basic relations for elastic members under pure torsion

(a) Deformation analysis: conclusions based on axi-symmetry of the rod

We consider a prismatic circular rod of radius R whose longitudinal axis lies along the x-axis, which passes through the centroids of the cross-sections, point O. Applied torsional moments $M_x \equiv T$ are assumed to act at the two ends, as shown in Fig. (7.1.1). Note that the same moment then exists at all cross-sections of the rod; that is, the resisting torque is not a function of x. The rod is therefore said to be in a state of 'pure torsion'.

In our study, we adopt the following sign convention for the torque T: $T > 0$ if it acts on a positive x-face of the rod in a counterclockwise direction. Note that

190

this is equivalent to stating that $T > 0$ if its vector representation acts on a positive (negative) x-face in the positive (negative) x-direction according to the right-hand rule [Fig. (7.2.1)]. Correspondingly, a positive rotation of a section occurs in the counterclockwise direction when viewed from the positive x-axis.

According to the discussion of Chapter 5, and as seen in the previous analysis of axial loadings, we recall that to obtain the solution, we are required to satisfy three types of equations: (a) equations of equilibrium, (b) strain–displacement relations and (c) stress–strain relations (Hooke's law) governing elastic behaviour.

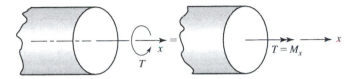

Figure 7.2.1

Following the methodology discussed in Chapter 5 and used in the previous chapter, we start our analysis by considering the possible deformation pattern of the rod, based on plausible physical reasoning.

For any given circular cylindrical rod, we observe that the rod is symmetric about the x-axis, which passes through the centres O of the cross-section; there are clearly no 'preferred directions' in the plane of the cross-sections. The system is therefore said to be **axi-symmetric** about the x-axis. It follows that the cross-sections will rotate about point O (i.e., about the x-axis of symmetry). We refer to point O as the *centre of twist* of the cross-section.

We first investigate whether plane cross-sections remain plane or whether they warp under the applied torque. To do so, let us assume that the sections actually warp so that the rod appears as in Fig. (7.2.2). Since the rod is prismatic, i.e., all cross-sections are identical, and since the same moment $M_x = T$ acts throughout the rod, the warping must be identical at all cross-sections. In particular, we note from the assumed deformation shown in the figure that the right end 'bulges out'. Furthermore, if the rod is observed from a point along the x-axis to the right of the rod, the applied moment appears to be acting in the counterclockwise direction. Let us imagine that we now observe the rod from a point on the x-axis to the left of the rod. From this vantage point, the torque T again appears to be acting in the counterclockwise direction. However, the left-hand cross-section appears to be 'bulging inward'. Now, since we have established that due to the symmetry of the member, all cross-sections deform *identically*, it is not possible that if viewed from the right the cross-section bulges out and if viewed from the left it bulges inward under the same torque.[†] We therefore conclude that no bulging can occur; that is, all cross-sections must remain plane and do not warp. Moreover, due to the axi-symmetry of the cross-section, all planes cross-sections remain perpendicular to the longitudinal x-axis.

Figure 7.2.2

Let us now investigate the deformation *within* the cross-section. Due to axi-symmetry, it is evident that in twisting, all points at the outer edge of a given

---

[†] Note that, as viewed from either end, the torques are seen to act in the same sense; in this case, as counterclockwise. This is essential to the arguments of symmetry.

cross-section will rotate through the same angle. We now examine the deformation of radial lines emanating from point O. Let us assume that if viewed from the right, straight radial lines as in Fig. (7.2.3) deform as shown in Fig. (7.2.4a). However if viewed from the left, the deformed lines appear as in Fig. (7.2.4b). Now, we observe that the torque T is acting in the same sense (here, counterclockwise) in both Figs. (7.2.4a and b). Since all sections deform identically, the two patterns must show an identical appearance if viewed from either the right or left ends. Therefore, noting that an identical appearance can exist only if the radial lines remain straight, we conclude that all (straight) radial lines must remain straight lines after deformation (twisting).

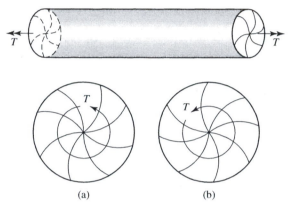

Figure 7.2.3

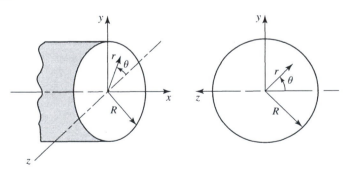

(a)   (b)

Figure 7.2.4

Thus, based on simple arguments of symmetry, it is possible to conclude that (a) cross-sections rotate with respect to the centre of twist lying along the x-axis, (b) cross-sections remain plane, (c) all plane cross-sections remain perpendicular to the longitudinal x-axis and (d) radial lines remain straight. We emphasise that these are *not assumptions*. Finally, we should mention that these conclusions are valid only for a circular cross-section, since perfect axi-symmetry can exists only for such sections.

Having established the basic deformation pattern, let us now consider the resulting strains in the rod. Before determining these strains, we first remark that since we are examining a circular member, it is clearly more natural to use a polar coordinate system (r, θ, x), as shown in Fig. (7.2.5), rather than a Cartesian system. We refer to the r-coordinate as the *radial* coordinate and the θ-coordinate as the *circumferential* coordinate.

Figure 7.2.5

As a result of conclusions (b) and (c) above, namely that all cross-sections remain plane and perpendicular to the longitudinal x-axis, we conclude that line segments in the x- and r-directions remain orthogonal and therefore, since there is no change in angle, $\epsilon_{xr} = 0$. From conclusion (d) above, all radial lines remain orthogonal to the circumferential direction and hence $\epsilon_{r\theta} = 0$. Therefore, by Hooke's law, $\tau_{xr} = 2G\epsilon_{xr} = 0$ and $\tau_{r\theta} = 2G\epsilon_{r\theta} = 0$. Thus, the only non-zero shear stress in the rod is $\tau_{x\theta} = \tau_{\theta x}$.

(b) Basic relations

Due to the applied torque, the cylinder will twist and any line originally parallel to the x-axis (a 'generator' of the cylinder) will assume the shape of a helix [Fig. (7.2.6)]. Since we are interested in determining the strain at any point within the circle, let us consider an imaginary circle within the rod of arbitrary radius r and let P be some point on the circle through which a generator of the imaginary cylinder is drawn, passing through all the cross-sections, as shown in Fig. (7.2.7a).

Figure 7.2.6

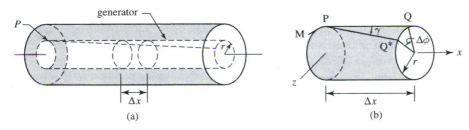

(a)   (b)   Figure 7.2.7

For convenience we consider the left end, $x = 0$, to be fixed against rotation, and let $\phi(x)$ denote the rotation at any section x. Since ϕ is not constant, we examine an element Δx within the rod, as shown in Fig. (7.2.7b), and let $\Delta\phi$ represent the *relative* rotation of the two cross-sections, Δx apart. We denote two points of a typical generator, as P and Q, at the two ends of the element. Since we are considering the relative rotation, we examine the geometry assuming the section at P is fixed. Due to the rotation $\Delta\phi$, point Q will then rotate to Q*, as in Fig. (7.2.7b), and the line segment PQ $\rightarrow$ PQ*. Thus we note that the line PQ* is no longer parallel to the x-axis and, in particular, after deformation PQ is no longer perpendicular to the segment PM lying in the circumferential direction θ. The change of angle differs by γ, where $\gamma \equiv \angle QPQ^*$. We recall from its definition, that the change in angle between two line segments, which were originally orthogonal, is represented by the shear strain. Since PQ and PM were originally in the x- and θ-directions, we therefore denote this angle by $\gamma_{x\theta}$.[†] From the geometry of Fig. (7.2.7b), $\widehat{QQ^*} = r \cdot \Delta\phi$. Furthermore, from this figure, may write $\widehat{QQ^*} \simeq \gamma \cdot \Delta x$. Hence we have

$$\gamma_{x\theta} \cdot \Delta x \simeq r \cdot \Delta\phi. \qquad (7.2.1)$$

Dividing through by Δx and taking the limit,

$$\gamma_{x\theta} = \lim_{\Delta x \to 0} r \frac{\Delta\phi}{\Delta x} = r \frac{d\phi}{dx}. \qquad (7.2.2a)$$

[†] Here we have referred loosely to $\gamma \equiv \gamma_{x\theta}$ as the *shear strain*.

Thus, based on the displacement pattern, we find that the shear strain $\gamma_{x\theta}$ is proportional to the radial distance from the centre of twist. Note that Eq. (7.2.2a) expresses a strain–displacement relation. Recalling that we defined the shear strain as $\gamma/2$, we have

$$\epsilon_{x\theta} = \frac{1}{2} r \frac{d\phi}{dx}, \qquad (7.2.2b)$$

which, we observe, is a purely geometric relation and is independent of the material behaviour of the rod.

Since we are studying the behaviour of an elastic member, we now apply Hooke's law, namely Eq. (4.4.10),

$$\tau_{x\theta} = 2G\epsilon_{x\theta}, \qquad (7.2.3)$$

and therefore

$$\tau_{x\theta} = Gr \frac{d\phi}{dx}. \qquad (7.2.4)$$

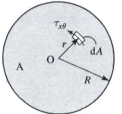

Figure 7.2.8

Thus, the shear stress is proportional to the radial distance from the centre of twist O and acts in a circumferential direction, as shown in Fig. (7.2.8). The shear deformation of typical elements is shown for a portion of the imaginary cylinder in Fig. (7.2.9).

From equilibrium, the internal resisting moment M_x at any cross-section must be equal to the applied torque T, i.e., $M_x = T$. Moreover, the resisting torque at the cross-sections is, in effect, the stress resultant of the shear stresses (as defined in Chapter 2). Now, from Fig. (7.2.8), the incremental moment about the x-axis due to the shear stresses $\tau_{x\theta}$ acting on an element dA at a distance r from the x-axis is [see also Eq. (2.5.12c)]

Figure 7.2.9

$$\Delta T = r \cdot \tau_{x\theta}\, dA. \qquad (7.2.5a)$$

Substituting Eq. (7.2.4),

$$\Delta T = Gr^2 \frac{d\phi}{dx}\, dA. \qquad (7.2.5b)$$

Therefore, the total moment T acting on the cross-section of area A is given by

$$T = G \iint_A r^2 \frac{d\phi}{dx}\, dA, \qquad (7.2.6a)$$

and since $d\phi/dx$ is independent of dA,

$$T = G\frac{d\phi}{dx} \iint_A r^2\, dA. \qquad (7.2.6b)$$

We denote the integral by J, i.e., $J = \iint_A r^2\, dA$, which we recognise as the **polar moment** of the area about the x-axis.[†] Thus,

$$\frac{d\phi}{dx} = \frac{T}{GJ}. \qquad (7.2.7)$$

[†] This is also referred to loosely as the *polar moment of inertia* (see Appendix A.1).

Now, $\frac{d\phi}{dx}$ represents the rate of change of the angle of twist. Since we are considering the case of pure torsion, i.e. $T \neq T(x)$, it follows that for this case, $\frac{d\phi}{dx}$ is also constant through the rod. We denote this constant by Θ and therefore write

$$\frac{d\phi}{dx} = \Theta. \tag{7.2.8}$$

From its definition, we observe that Θ represents the 'unit angle of twist'; that is, Θ represents the relative rotation of two cross-sections a unit distance apart. Note that Θ has units of 'radians per length' (e.g., rad/m).

Thus, from Eq. (7.2.7), the unit angle of twist is given by the simple expression

$$\Theta = \frac{T}{GJ}. \tag{7.2.9}$$

Substituting Eqs. (7.2.8) in Eq. (7.2.4),

$$\tau_{x\theta} = Gr\frac{d\phi}{dx} = Gr\Theta, \tag{7.2.10}$$

and therefore, by Eq. (7.2.9), the shear stress is finally given by

$$\tau_{x\theta} = \frac{Tr}{J}. \tag{7.2.11}$$

This last expression relates the shear stress to the torque existing at the section. It is of interest to note that if T is known, the shear stress $\tau_{x\theta}$ is *independent of the material properties*; that is, the stress is the same for a rod of any given material. For convenience, in our subsequent development, we shall denote the shear stress due to torsion by τ; i.e., we let $\tau \equiv \tau_{x\theta}$.

From Eq. (7.2.11), we observe that the maximum stress occurs at the outer edge $r = R$, i.e.

$$\tau_{max} = \frac{TR}{J}. \tag{7.2.12}$$

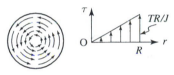

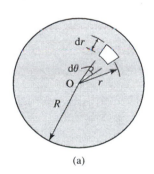

Figure 7.2.10

The shear stress distribution of $\tau \equiv \tau_{x\theta}(r)$ is shown in Fig. (7.2.10). We note that the shear components act in the circumferential direction at all points within the cross-section. From Eq. (7.2.12), we note too that a rod of material having a yield point τ_0 in shear, will behave elastically provided the torque T does not exceed the *elastic torque* T_E given by[†]

$$T_E = \frac{\tau_0 J}{R}. \tag{7.2.13}$$

At this stage we calculate the polar moment of the area, J, for a solid rod. Noting that an incremental area dA is given by $dA = r\, d\theta \cdot dr$ [Fig. (7.2.11a)], we have

$$J = \int_0^{2\pi}\left(\int_0^R r^3 dr\right) d\theta = \frac{R^4}{4}\int_0^{2\pi} d\theta = \frac{\pi R^4}{2}. \tag{7.2.14}$$

(a)

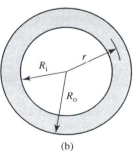

(b)

Figure 7.2.11

[†] We assume here effectively that $\tau_0 = \tau_E$, i.e., that the proportional limit, the elastic limit and the yield point in shear τ_0 coincide; namely, the material is an ideal elastic–plastic material (see Chapter 4, Section 6).

Substituting this value in Eq. (7.2.12), we find that the maximum shear stress in a rod is given by

$$\tau_{\text{max}} = \frac{2T}{\pi R^3};$$ (7.2.15)

that is, the maximum shear stress occurring in a solid rod is inversely proportional to R^3.

Using Eq. (7.2.7), we may now determine the relative rotation of any two cross-sections, say $x = x_0$ and $x = x_1$ ($x_0 < x_1$), by simple integration:

$$d\phi = \frac{T(x)}{GJ} \, dx.$$ (7.2.16)

Therefore

$$\int_{\phi_0}^{\phi_1} d\phi = \frac{1}{GJ} \int_{x_0}^{x_1} T(x) \, dx.$$

or

$$\phi_1 - \phi_0 = \frac{1}{GJ} \int_{x_0}^{x_1} T(x) \, dx.$$ (7.2.17)

For pure torsion, $T = \text{constant}$,

$$\phi_{1|0} \equiv \phi_1 - \phi_0 = \frac{T \cdot (x_1 - x_0)}{GJ}.$$ (7.2.18)

Note here that the notation $\phi_{1|0}$ represents the rotation of the section x_1 with respect to section x_0. Using this notation, it is clear that

$$\phi_{0|1} = -\phi_{1|0} \equiv -(\phi_1 - \phi_0).$$ (7.2.19)

If $\phi_0 = 0$ and $x_1 = L$,

$$\phi(L) = \frac{TL}{GJ},$$ (7.2.20)

which is the rotation of a cross-section at $x = L$ when the end $x = 0$ is held fixed under a state of pure torsion.

We note, by comparing Eqs. (7.2.9) and (7.2.20), that the angle of twist ϕ for the case of pure torsion is

$$\phi = \Theta \cdot L.$$ (7.2.21)

We observe that the unit angle of twist Θ is proportional to the torque T and inversely proportional to GJ. We therefore refer to the quantity GJ as the *torsional stiffness* or *torsional rigidity* of the circular rod. Note that the torsional rigidity depends on a material property (here, the shear modulus G) and the geometry of the cross-section.

Now, the above development was not necessarily limited to solid rods. Indeed, the only restriction placed on the development is that the rods be circular. Thus, all the relations remain valid for a hollow rod having inner and outer radii R_i and R_o, respectively, as shown in Fig. (7.2.11b). However, instead of the polar moment of the cross-sectional area for the solid rod, given by Eq. (7.2.14), we now have

$$J = \int_0^{2\pi} \left(\int_{R_i}^{R_o} r^3 \, dr \right) d\theta = \frac{1}{4}(R_o^4 - R_i^4) \int_0^{2\pi} d\theta = \frac{\pi}{2}(R_o^4 - R_i^4). \quad (7.2.22)$$

The above development, first derived by Coulomb in 1784, is known as the **Coulomb torsion solution** for circular cylinders.

Example 7.1: (a) What is the maximum torque T_E that can be applied to a solid steel cylindrical shaft 8 cm in diameter [Fig. (7.2.12)] if the shaft is to remain elastic? The elastic limit in shear and the shear modulus are $\tau_0 = 145$ MPa and $G = 76$ GPa, respectively. (b) Determine the relative rotation of the two ends due to this torque if $L = 2.0$ m.

Solution:

(a) From Eq. (7.2.13), the maximum elastic torque T_E is $T_E = \frac{\tau_0 J}{R}$.

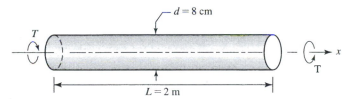

$$L = 2 \text{ m}$$

Figure 7.2.12

For $R = 4$ cm, $J = \pi R^4/2 = 128\pi$ cm^4, and therefore $T_E = (145 \times 10^2)(128\pi)/4 = 14.58 \times 10^5$ N-cm $= 14{,}580$ N-m.

(b) From Eq. (7.2.9),

$$\Theta = \frac{T_E}{GJ} = \frac{14{,}580}{(76 \times 10^9)(128\pi \times 10^{-8})} = 0.048 \text{ rad/m} = 2.7°/\text{m}.$$

Therefore, the relative rotation is $\phi = \Theta L = 5.4°$.   □

We observe that for usual materials encountered in engineering practice, the unit angle of twist Θ is indeed very small.

7.3 Some comments on the derived expressions: extension of the results and approximations

(a) Comments on the solution

(i) The expressions derived in the preceding section, namely[†]

$$\tau = \frac{Tr}{J}, \quad (7.3.1a)$$

$$\Theta = \frac{T}{GJ}, \quad (7.3.1b)$$

[†] In the rest of this chapter, the shear stress τ, when written without subscripts, denotes $\tau_{x\theta}$; i.e., $\tau \equiv \tau_{x\theta}$.

are valid only for prismatic circular cylinders under pure torsion. Since the shear stress $\tau \equiv \tau_{x\theta}$ acts in the circumferential direction at all points within the cross-section, in particular, at the outer edge, $r = R$, the shear stresses act tangentially to the circle defining the cylinder.

We recall from the previous section that as a result of the deformation pattern resulting from the axi-symmetry of the problem, we determined that the stress component in the radial direction is $\tau_{xr} = 0$ at all points, $r \leq R$, within the rod. We might also have reached this conclusion by consideration of the stress state existing at the boundary $r = R$ of the rod. Let us therefore consider the case assuming that *non-zero* components τ_{xr} are acting on the cross-section at the outer edge, $r = R$. Since shear stress components always exist in conjugate pairs, stress components $\tau_{rx} = \tau_{xr}$ would then exist on the outer lateral cylindrical surface of the rod, as shown in Fig. (7.3.1a). However, the $r = R$ lateral surface – i.e., the 'r-face' – is a *free surface*, and consequently no stress component can exist on it; i.e., $\tau_{rx} = 0$ [Fig. (7.3.1b)]. Therefore, it necessarily follows that $\tau_{xr} = 0$ in the cross-section at $r = R$. We therefore conclude that *at all points of an edge corresponding to a free lateral surface, the resultant shear stress in the cross-section must always act tangentially to the edge* [Fig. (7.3.1c)].[†] Moreover, because of axial symmetry of the rod, $\tau_{xr} = 0$ at the point O. Recalling that the above treatment pertains to relatively slender rods (i.e., rods whose diameters are small compared to their length), and since $\tau_{xr} = 0$ at the ends of the diameter as well as at the centre, we expect that any variation of τ_{xr} over a relatively small diameter would necessarily be small, that is infinitesimal. Hence, we assert that $\tau_{xr} = 0$ everywhere throughout the rod. Moreover, since the lateral surface $r = R$ is a free surface, $\tau_{rr} = 0$ must also be true. Following the same above reasoning for τ_{rx} given for slender rods, we conclude that $\tau_{rr} = 0$ everywhere throughout the rod.[‡] While it appears, from the above argument, that this latter assertion is merely an assumption, we mention here that this assertion is indeed correct and conforms with an exact solution found according to the *Theory of Elasticity*. Moreover, according to this exact solution, all normal stresses vanish on the coordinate surfaces, that is $\tau_{rr} = \tau_{\theta\theta} = \tau_{xx} = 0$ throughout the rod.[§]

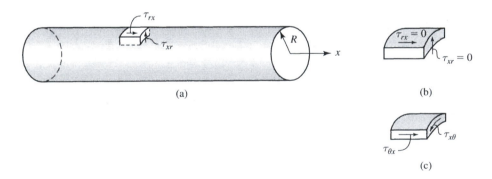

(a)

(b)

(c)

Figure 7.3.1

[†] This statement is not limited to rods subjected to torsion but is a general statement valid for all bodies at points existing at a free surface.

[‡] Note that a similar argument (leading to the conclusion that $\sigma_y = \sigma_z = 0$ throughout the rod) was used in Section 2 of Chapter 6 for slender rods under axial loadings.

[§] These conclusions will be shown to be true in Chapter 12.

(ii) We observe that when stating that 'the rod is subjected to a torque T at a particular section', we have not specified in what manner this torque is applied. For example, it might be applied by means of a gear acting on the outer lateral surface. On the other hand, it might be applied at the longitudinal axis by means of a wrench, as in Fig. (7.3.2). Irrespective of the manner by which the torque is applied, it is clear that a complex state of stress (in equilibrium with the given applied torque) will exist near the section of application. Such a complex stress state cannot be described by the expressions (7.3.1) above. However, as mentioned in Chapter 5, provided this stress distribution has a resultant equal to T, these expressions will be valid, according to the principle of de Saint Venant, at points sufficiently distant from the point of the applied torque. For example, for the shaft shown in Fig. (7.3.3), the calculated stresses $\tau_{x\theta}$ are only valid at a distance $s > d$, from section B. Thus, by invoking the principle of de Saint Venant, we assert that, *for a sufficiently long shaft, $L \gg d$, the Coulomb solution describes the behaviour of the shaft* except in these localised regions. In practice, we therefore essentially neglect these localised effects and assume that the behaviour of the entire rod is described by the Coulomb solution.

Figure 7.3.2

The same reasoning applies if a torque is applied at a particular interior section, for example, at $x = x_C$ (i.e., at section C) of Fig. (7.3.3); we then disregard effects in the localised region $x_C - d < x < x_C + d$.

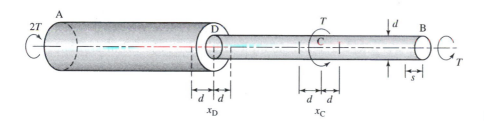

Figure 7.3.3

The above comments are equally valid for a rod whose cross-section changes abruptly, such as section D of Fig. (7.3.3), where again, a more complex stress distribution exists. Therefore, in treating such problems, we again implicitly neglect such localised effects.

Having found that the shear stress varies linearly from the centre of twist for circular cross-sections, one might expect the same to be true for non-circular cross-sections – for example, for a rectangular cross-section, as shown in Fig. (7.3.4), whose centre of twist coincides with its centroid. However, we have found that the *shear stress must always act tangentially at all points located at the edge of a cross-section.* It follows that for this cross-section, the shear stresses τ_{xy} and τ_{xz} at a corner (e.g., point B) must necessarily be zero, for otherwise non-zero stress τ_{yx} and τ_{zx} would be acting on the lateral surfaces. Now, if the shear stresses are proportional to the distance from the centre of twist of this section, they clearly will not be zero at the corners since the corners are the farthest points from the centre, point O. Thus we conclude that for this rectangular section, shear stresses do not vary linearly with the distance from the centre of twist. Indeed, this will be shown to be true for *any non-circular cross-section*. Solutions for the torsion of non-circular cross-sections are far more complex than the Coulomb solution; these will be treated later in Chapter 12.

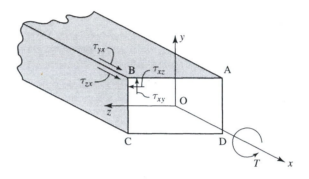

Figure 7.3.4

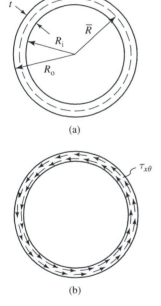

Figure 7.3.5

(b) An approximation for thin-wall circular tubular cross-sections

We derive here an approximate relation for the average shear stress existing in a thin-wall tubular circular cross-section. Consider a closed thin-wall section having inner and outer radii R_i and R_o, respectively. By a 'thin wall', we mean a section whose thickness t is small with respect to the mean radius $\overline{R}$ of the tube, i.e., $t/\overline{R} \ll 1$, where $\overline{R} = (R_o + R_i)/2$ [Fig. (7.3.5a)].

We first observe that for a thin-wall section, the shear stress, as given by Eq. (7.3.1a), cannot vary greatly across the thickness of the wall in the small interval $R_i \leq r \leq R_o$. According to Eq. (7.2.22), the polar moment of the cross-sectional area for this cross-section is

$$J = \frac{\pi}{2} \left(R_o^4 - R_i^4 \right);$$ (7.3.2)

the shear stress is therefore given by

$$\tau = \frac{2Tr}{\pi \left(R_o^4 - R_i^4 \right)} = \frac{2Tr}{\pi \left(R_o^2 - R_i^2 \right) \cdot \left(R_o^2 + R_i^2 \right)}.$$ (7.3.3)

We first note that

$$\left(R_o^2 - R_i^2 \right) = (R_o + R_i)(R_o - R_i) = 2\overline{R} \cdot t$$

and

$$\left(R_o^2 + R_i^2 \right) = [\overline{R} + t/2]^2 + [\overline{R} - t/2]^2 = 2\overline{R}^2 + t^2/2$$
$$= 2\overline{R}^2[1 + (t/2\overline{R})^2] \simeq 2\overline{R}^2$$

since $t/\overline{R} \ll 1$. It follows that $\tau = Tr/2\pi\overline{R}^3 t$. The average stress τ at $r = \overline{R}$ is then given by

$$\tau = \frac{T}{2\pi\overline{R}^2 t} \tag{7.3.4a}$$

or

$$\tau = \frac{T}{2At}, \tag{7.3.4b}$$

where $A = \pi\overline{R}^2$ represents the area within the circle of radius $\overline{R}$.

This expression is thus an approximation for the *average* shear stress acting in the circumferential direction within the thin wall, as shown in Fig. (7.3.5b). Since we have observed that the shear stress has a small variation throughout the wall thickness, this simple expression proves to be an excellent approximation, provided $t/\overline{R} \ll 1$.

In Chapter 12, it will be shown that Eq. (7.3.4b) yields the average shear stress in a closed thin-wall tubular section having *any arbitrary geometry*.

(c) Extension of the results: engineering approximations
(i) Torsion of non-prismatic rods

We recall that the expressions developed in the preceding section are based on the conclusions that all plane sections remain plane and all radial lines remain straight lines. These fundamental conclusions were established using arguments of symmetry, which are valid only for rods whose cross-sections do not vary with the longitudinal coordinate x, that is for *prismatic circular rods*.

Let us now consider a circular rod with a varying radius $R = R(x)$, as shown in Fig. (7.3.6). Clearly, since the cross-sections are not identical for such a non-prismatic rod, the arguments of symmetry of the previous section cannot be used and therefore the conclusions are no longer valid; that is, for non-prismatic rods, we can no longer deduce that plane cross-sections remain plane, nor can we assert that radial lines remain straight. However, the deviation from the behaviour of the prismatic rod evidently depends on how sharply the radius $R(x)$ varies with x. If this variation is relatively small, then the deviation from our conclusions will be relatively small. Hence, for rods whose cross-sections vary slowly with x, it is reasonable to expect that the above expressions will yield a good approximation to the true solution. Thus instead of Eqs. (7.2.9) and (7.2.11), we write

$$\tau = \frac{Tr}{J(x)}, \tag{7.3.5a}$$

$$\Theta(x) = \frac{T}{GJ(x)}, \tag{7.3.5b}$$

again with the clear understanding that these are but approximations.

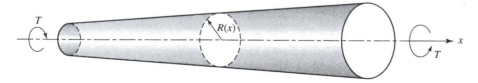

Figure 7.3.6

(ii) Rods subjected to a general torsion, $T = T(x)$

The above analysis has been concerned with a circular rod in *pure torsion*. Now let us assume that the rod is subjected to torsional moments, not only at the ends but continuously along its axis; i.e., there exists a distribution of torques, say $t(x)$ (having units N-m/m), which is a function of x, as shown in Fig. (7.3.7), such that the torque at any section is $T = T(x)$. Then, again, the arguments of symmetry, which were used to establish the basic displacement pattern, no longer hold true: since one can no longer state that all cross-sections deform identically, one cannot assert that plane sections remain plane and that all radial lines remain straight lines. Consequently, the results given by Eqs. (7.2.9)–(7.2.21) are no longer exact. However, if we now *assume* that any warping of the sections is small and that straight lines do not deviate much from straight lines in the deformed state, then, starting from this basic premise, we arrive at the same expressions, except that T now is a function of x. It then follows that, in lieu of Eqs. (7.2.9) and (7.2.11), we have the approximate expressions

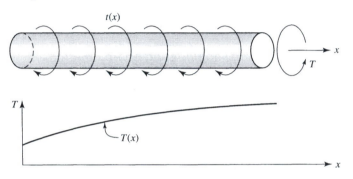

Figure 7.3.7

$$\tau(x) = \frac{T(x)r}{J}, \tag{7.3.6a}$$

$$\Theta(x) = \frac{T(x)}{GJ}. \tag{7.3.6b}$$

Example 7.2:   An aluminium shaft (with yield strength in shear, $\tau_0 = 55$ MPa, and shear modulus $G = 26$ GPa) is composed of two segments AB and BC, as shown in Fig. (7.3.8a). Externally applied torques, $T_1 = 8,000$ N-m and $T_2 = 5,000$ N-m are applied at B and C, respectively. Determine the maximum shear stress in each section and the total angle of twist at C if the shaft is held fixed at A.

Solution: From the free-body diagrams [Fig. (7.3.8b)], we find that $T_{AB} = 13,000$ N-m in sector AB and $T_{BC} = 5000$ N-m in sector BC. The torque $T(x)$ is plotted as a function of x in Fig. (7.3.8c).

From Eq. (7.2.14), with $R = 6$ cm, $J_{AB} = \pi R^4/2 = 648\pi$ cm^4. Similarly, in sector BC, $J_{BC} = 128\pi$ cm^4.

In sector AB: $\tau_{x\theta} = \frac{T_{AB}R}{J_{AB}} = (13 \times 10^5) \cdot \frac{6}{648\pi} = 3830$ N/cm$^2 = 38.3$ MPa. Using Eq. (7.2.18),

$$\phi_B - \phi_A = \frac{T_{AB}L_{AB}}{GJ_{AB}} = (13 \times 10^3) \cdot \frac{2.0}{(26 \times 10^9)(648\pi \cdot 10^{-8})} = 4.91 \times 10^{-2} \text{ rad.}$$

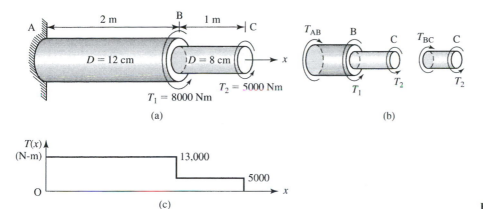

(a)

(b)

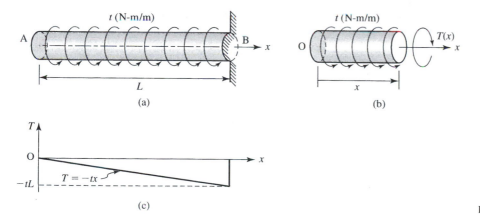

Figure 7.3.8

In sector BC: $\tau_{x\theta} = \frac{T_{BC}R}{J_{BC}} = (50 \times 10^4) \cdot \frac{4}{128\pi} = 4970$ N/cm$^2 = 49.7$ MPa.

$$\phi_C - \phi_B = \frac{T_{BC}L_{BC}}{GJ_{BC}} = (5 \times 10^3) \cdot \frac{1.0}{(26 \times 10^9)(128\pi \cdot 10^{-8})} = 4.78 \times 10^{-2} \text{ rad.}$$

Noting that $\phi_A = 0$, the total angle of twist ϕ_C is therefore

$$\phi_C = (\phi_B - \phi_A) + (\phi_C - \phi_B) = (4.91 + 4.78) \times 10^{-2} = 0.0969 \text{ rad} = 5.55°.$$

Note that we have implicitly neglected the localised effects at sections A, B, and C.

$\square$

Example 7.3: A series of rotating gears acts along the surface of a circular shaft of radius R and length L, producing a torque t (N-m/m) per unit length along the shaft, as shown in Fig. (7.3.9a). Determine the maximum shear stress in the shaft and the angle of twist ϕ at $x = 0$. Express the answer in terms of t, R, J, L and G.

Figure 7.3.9

Solution: Using a free-body diagram [Fig. (7.3.9b)], we note that the torque $T(x)$ at any section x is given by $T(x) = -t \cdot x$. The linear variation of $T(x)$ is shown in Fig. (7.3.9c). The maximum torque therefore occurs at the right end and is given by $T = -tL$.

From Eqs. (7.3.6),

$$|\tau_{max}| = \frac{TR}{J} = \frac{tLR}{J},$$
(7.3.7a)

$$\Theta(x) = \frac{-tx}{GJ}.$$
(7.3.7b)

Since, here, the torque $T = T(x)$, we can no longer use Eq. (7.2.18) or (7.2.20) to determine the rotation, but must instead use Eq. (7.2.17); thus

$$\phi_B - \phi_A = \frac{1}{GJ} \int_0^L -tx \, dx = -\frac{tL^2}{2GJ}.$$
(7.3.8a)

Since $\phi_B = 0$, we find that the section at $x = 0$ rotates through an angle

$$\phi_A = \frac{tL^2}{2GJ}.$$
(7.3.8b)

Note that $\phi_A > 0$ indicates a counterclockwise rotation.   □

7.4 Some practical engineering design applications of the theory

The results obtained in the previous section find particular use in the application to several engineering design problems. We consider two such applications in the examples below.

Example 7.4: It is required to design a shaft such that when subjected to a torque, the maximum shear stress and unit angle of twist Θ should be kept to a minimum. Because of space limitations, the maximum permitted diameter is 5 cm. Two members are available: a solid shaft having a radius $R = 1.5$ cm and a hollow tubular section with inner radius $R_i = 2.0$ cm and outer radius $R_o = 2.5$ cm [Fig. (7.4.1)]. Determine the ratio of maximum shear stress of the solid to hollow section, $\frac{\tau_{max_s}}{\tau_{max_h}}$, and the ratio Θ_s/Θ_h, where Θ_s and Θ_h denote the unit angle of twist for the solid and hollow shafts, respectively.

Solution: We note first that the cross-sectional areas of the two shafts are the same; namely $A_s = \pi R^2 = 2.25\pi$ and $A_h = \pi(R_o^2 - R_i^2) = \pi(6.25 - 4) = 2.25\pi$.

For the solid shaft,

$$\tau_{max_s} = \frac{TR}{J_s},$$
(7.4.1a)

$$\Theta_s = \frac{T}{GJ_s},$$
(7.4.1b)

while for the hollow shaft,

$$\tau_{max_h} = \frac{TR_0}{J_h},$$
(7.4.1c)

$$\Theta_h = \frac{T}{GJ_h}.$$
(7.4.1d)

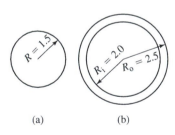

(a)   (b)

Figure 7.4.1

If the two shafts are subjected to the same torque, then

$$\frac{\tau_{max_s}}{\tau_{max_h}} = \frac{R}{R_0} \frac{J_h}{J_s},$$ (7.4.2a)

$$\frac{\Theta_s}{\Theta_h} = \frac{J_h}{J_s}.$$ (7.4.2b)

For the given values,

$$\frac{J_h}{J_s} = \frac{\pi \left(R_o^4 - R_i^4\right)/2}{\pi R^4/2} = \frac{R_o^4 - R_i^4}{R^4} = \frac{(2.5)^4 - 2^4}{1.5^4} = 4.51.$$

Therefore

$$\frac{\tau_{max_s}}{\tau_{max_h}} = \frac{1.5}{2.5} \times 4.51 = 2.72, \qquad \frac{\Theta_s}{\Theta_h} = 4.51.$$

We therefore conclude that for two members having the *same* cross-sectional area, the hollow one is much more efficient than the solid one. The solid member is less efficient because stresses in the section close to the centre of rotation contribute little to the resisting torque as the 'lever arm' (with respect to the x-axis) at these points is relatively small. In the case of the hollow shaft, all points in the cross-section have relatively large lever arms. Thus, in the hollow cylinder, the stresses provide a greater contribution to the resisting torque. (Although, as we have observed, a hollow shaft is indeed more efficient than a solid one, we mention here that other design criteria, such as stability of the shaft, may nevertheless often require that a solid shaft be used in the design.)

Example 7.5: Two steel shafts are to be connected by means of eight bolts acting through the flanges of the shafts, as shown in Figs. (7.4.2a and b). The bolts are evenly spaced and are located along a circle whose diameter is $D_b = 20$ cm. If the average permissible shear stress in the bolts is $\tau = 40$ MPa, determine the minimum required diameter d of each bolt if a torque $T = 5600$ Nm must be transferred between the shafts.

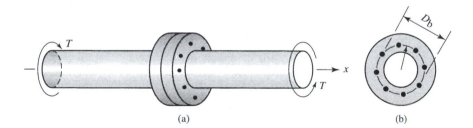

(a)

(b)

Figure 7.4.2

Solution: Since the function of the bolts is to transfer the torque, they must each exert a force, in the circumferential direction, $F_b = 2T/nD_b$ [Fig. (7.4.3)]. Clearly this force is transmitted via shear stresses, τ_b, in the bolts. Assuming that the average shear stress in the bolt is evenly distributed throughout its cross-section, we write $\tau_b = F_b/A_b$, where A_b is the cross-sectional area of each bolt. Therefore we have

$$A_b = \frac{2T}{nD_b \cdot \tau_b}$$ (7.4.3)

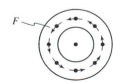

Figure 7.4.3

and hence

$$A_b = \frac{2 \cdot (5600 \times 10^2)}{8 \cdot (20) \cdot (40 \times 10^2)} = 1.75 \, \text{cm}^2.$$

The required diameter is therefore $d = 2\sqrt{A_b/\pi} = 1.49$ cm. In practice one would use bolts with a diameter $d = 1.5$ cm.

It is worthwhile to remark on the use of the term 'average permissible shear stress in the bolt'. From previous observations, we recall that shear stresses at the edge of a cross-section, which represents a free lateral surface, must *necessarily* be zero. Therefore, in using the above relation $\tau_b = F_b/A_b$, we obtained but a rough average value; it clearly does not represent the true shear stress distribution found in the cross-section. Such an expression is a typical example of crude approximations, which are often used for engineering design purposes. □

Circular shafts evidently are widely used in transmitting power in machinery. For example, in an automobile, power of the engine is transferred to the wheel axes by means of a circular shaft. Similarly, shafts are used to transfer power in electric motors. We consider a practical application of the theory to such a problem.

Example 7.6: The solid steel shaft of Example 7.1 ($D = 8$ cm, $\tau_o = 145$ MPa) is to be used as a transmission shaft to transmit power from an electric motor. Determine the maximum power that the shaft can deliver and remain elastic if it rotates with $N = 1200$ rpm (revolutions per minute). Express the answer in Watts. [Note: Watt = 1 N-m/s.]

Solution: In Example 7.1, the maximum elastic torque T_E was found to be $T_E = 14,580$ Nm.

Recalling that power represents work per unit time, we first calculate the work done by the torque T_E. Now, since the work of a torque (torsional moment) is given by $W = T\alpha$, where α is the angle through which the torque rotates, the work done in one revolution is $W = 2\pi T$. If the shaft rotates, for example, at N revolutions per minute, the power P (work per minute) of the torque is then

$$P = 2\pi N T \quad (N = \text{rpm}). \tag{7.4.4a}$$

Alternatively, power can also be expressed as work per second in terms of Hertz (Hz = frequency per second) according to the relation

$$P = 2\pi f T \quad (f = \text{Hz}). \tag{7.4.4b}$$

Substituting $T = T_E$ in this last expression (with $f = 1200/60 = 20$ Hz),

$$P = 2\pi f T_E = 2\pi(20)(14,580) = 1832 \times 10^3 \, \text{N-m/s} = 1832 \, \text{kW}.$$

Thus we find that the given shaft can transmit up to 1832 kW and still remain elastic. □

7.5 Circular members under combined loads

Consider a solid shaft of radius R (with polar moment of area, J) subjected to a torque T as well as an axial load P acting along the centroidal x-axis, as shown in Fig. (7.5.1). From Eq. (6.2.7), the axial stress is $\sigma_x = \frac{P}{A}$, while from Eq. (7.2.12), the maximum shear stress is $\tau_{x\theta} = \frac{TR}{J}$.

From the previous results, we have determined that for elastic behaviour, the strains and rotations are indeed quite small. Consequently, since the relations are also

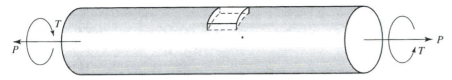

Figure 7.5.1

linear, according to the discussion of Chapter 5, use of the principle of superposition is permissible.

We now examine an infinitesimal element near the outer edge of the rod, as shown in Fig. (7.5.2a), with the stresses σ_x and $\tau_{x\theta}$ acting upon this element. If the element is sufficiently small, its curvature can be neglected and therefore the state of stress may be considered to be two-dimensional, as in Fig. (7.5.2b). Consequently, the stress components acting in any given direction can be calculated using the stress transformations laws, Eqs. (2.6.6) and (2.6.8) of Chapter 2 [where the coordinate θ corresponds to $-y$ of Fig. (2.6.1)].

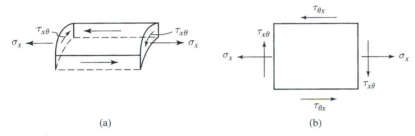

(a)   (b)   **Figure 7.5.2**

In particular, if $P = 0$, the element is in a state of 'pure shear', as shown in Fig. (7.5.3a). Therefore, according to the results of Chapter 2 (see Example 2.11), the principal planes will, for this case, be oriented at $45°$ to the x-axis with principal stresses $\sigma_1 = \tau$ and $\sigma_2 = -\tau$, as shown in Fig. (7.5.3b). Recalling that brittle materials are weakest in resisting tensile stresses, we expect a rod made of brittle material to fracture along such $45°$ lines. Indeed, from simple experiments, we find that brittle rods subjected to torsion fracture along a 'helicoidal' surface [Fig. (7.5.4)].

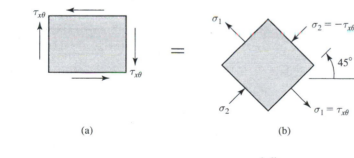

(a)   (b)   **Figure 7.5.3**

Figure 7.5.4

7.6 Statically indeterminate systems under torsion

Consider a rod fixed at two ends A and C against rotation and subjected to a torque T at B, as shown in Fig. (7.6.1a). We wish to determine (a) the stress in sector AB and BC and (b) the angle of rotation ϕ_B.

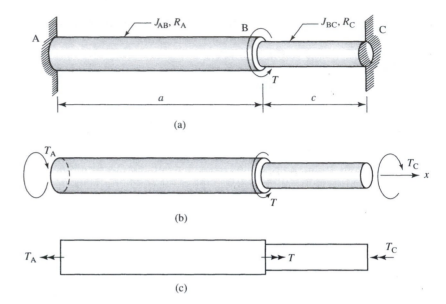

Figure 7.6.1

Let T_A and T_C denote the resisting moments at A and C, respectively [Fig. (7.6.1b)]. Note that we have assumed the two resisting torques T_A and T_C to be acting in the clockwise directions, as seen from the right. [An equivalent vector representation of the torques is shown in Fig. (7.6.1c)]. Then, for equilibrium $\sum M_x = 0$, we write, according to the free-body diagram of Fig. (7.6.1b or c),

$$T_A + T_C = T. \tag{7.6.1}$$

Considering portions of sectors AB and BC as free bodies [Fig. (7.6.2a)], we note that within AB, $T = T_A$ and within BC, $T = T_C$. The variation of T with x is shown in Fig. (7.6.2b). Note that we have used the adopted sign convention for torsional moments.

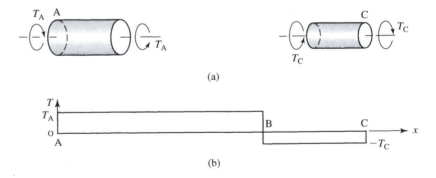

Figure 7.6.2

Since there exist two unknowns, T_A and T_C, and only a single equation of equilibrium, the system is statically indeterminate. Consequently, we require a second equation to solve for the two unknowns.

Let us therefore consider the geometry of the deformation. Clearly, the rotation of section B with respect to A must be the same as that with respect to C; that is,

$$\phi_{B|A} = \phi_{B|C}, \tag{7.6.2}$$

where $\phi_{B|A}$ and $\phi_{B|C}$ denote respectively the rotation of section B relative to A and C.[†]

We observe that this is essentially an *equation of geometric compatibility* as discussed in the previous chapter.

Using Eq. (7.2.18) with these values, we may write

$$\phi_{B|A} = \frac{T_A \cdot a}{G J_{AB}}, \qquad (7.6.3a)$$

$$\phi_{B|C} = \frac{T_C \cdot c}{G J_{BC}}. \qquad (7.6.3b)$$

Substituting in Eq. (7.6.2),

$$\frac{T_A \cdot a}{J_{AB}} = \frac{T_C \cdot c}{J_{BC}}, \qquad (7.6.4)$$

which is the equation of geometric compatibility written explicitly in terms of the torques.

Thus we now have two simultaneous equations [Eqs. (7.6.1) and (7.6.4)], in the two unknowns. From Eq. (7.6.4), we find

$$T_C = T_A \frac{J_{BC}}{J_{AB}} \frac{a}{c}, \qquad (7.6.5a)$$

and substituting in Eq. (7.6.1),

$$T_A = \frac{T}{1 + \frac{J_{BC}}{J_{AB}} \frac{a}{c}}. \qquad (7.6.5b)$$

Then, since $T_C = T - T_A$,

$$T_C = \frac{T}{1 + \frac{J_{AB}}{J_{BC}} \frac{c}{a}}. \qquad (7.6.5c)$$

Finally, the maximum shear stresses in each sector are, according to Eq. (7.2.12),

$$\tau_{AB} = \frac{T_{AB} R_A}{J_{AB}}, \qquad (7.6.6a)$$

$$\tau_{BC} = \frac{T_{BC} R_C}{J_{BC}}, \qquad (7.6.6b)$$

where $T_{AB} = T_A$ and $T_{BC} = -T_C$.

Substituting in either of Eqs. (7.6.3), the rotation of section B is given by

$$\phi_{B|A} = \phi_{B|C} = \frac{Tac}{G(cJ_{AB} + aJ_{BC})}. \qquad (7.6.7)$$

It is instructive to consider the particular case where the two sectors AB and BC have the same length; i.e., $a = c$. Using Eq. (7.6.5a) for this case,

$$\frac{T_A}{T_C} = \frac{J_{AB}}{J_{BC}}. \qquad (7.6.8)$$

[†] From Eq. (7.6.2), $\phi_{B|A} - \phi_{B|C} = 0$. However from Eq. (7.2.19), $\phi_{B|C} = -\phi_{C|B}$ and therefore $\phi_{C|B} + \phi_{B|A} = 0$; that is, $\phi_{C|A} = 0$. Thus, Eq. (7.6.2) is equivalent to stating that the two support sections A and C do not rotate with respect to each other.

Now, recalling that the quantities GJ_{AB} and GJ_{BC} represent the torsional stiffness of the two components AB and BC, we observe that the torque resisted by each component is proportional to its stiffness. Thus we find confirmation of the general remark given in Section 5 of Chapter 6, namely 'in a statically indeterminate system consisting of several components, each component tends to resist applied loads in proportion to its relative stiffness'.

7.7   Elastic–plastic torsion

In the preceding sections we have considered only torsion of cylindrical rods, which behave elastically where the shear stress and the unit angle of twist Θ are, respectively,

$$\tau = \frac{Tr}{J}, \qquad \Theta = \frac{T}{GJ}. \tag{7.7.1}$$

We now wish to study the problem as the rod enters the plastic range. It is clear that the T–Θ relation given above is valid only for elastic behaviour; we determine here the corresponding relation when the rod behaves plastically.

To this end, we consider a rod of radius R subjected to pure torsion, whose material behaviour is elastic–perfectly plastic as described by the τ–γ curve of Fig. (7.7.1), where τ_0 is the yield point in shear.[†]

If the torque T is applied statically, the initial behaviour will be elastic with shear stresses given by Eq. (7.7.1). We note that the rod will behave elastically provided the applied torque does not exceed the elastic torque T_E given by

$$T_E = \frac{\tau_0 J}{R} \tag{7.7.2a}$$

or, expressed explicitly in terms of the radius,

$$T_E = \frac{\pi R^3}{2} \tau_0. \tag{7.7.2b}$$

Thus, Eq. (7.7.1) is valid in the range $0 \leq T \leq T_E$. Specifically, when the torque T reaches the value $T = T_E$, the stress in the outer fibres of the cross-section, $r = R$, is $\tau = \tau_0$ with γ_0, the corresponding strain at the outer edge, as shown in Fig. (7.7.1).

We first recall from Eq. (7.2.2a) that the strain $\gamma \equiv \gamma_{x\theta}$, at any point within the rod, is given by

$$\gamma = r \frac{d\phi}{dx} = r \cdot \Theta. \tag{7.7.3}$$

Since the unit angle of twist Θ is constant with respect to r, γ varies linearly with r. We emphasise that this relation for the strain is a geometric relation and therefore *independent of the material behaviour*. Consequently, Eq. (7.7.3) remains valid for plastic as well as elastic behaviour; that is, for *all* values of $|T| \geq 0$.

Let us now assume that the torque is increased such that $T > T_E$. Under such torques the rod will rotate further; that is, Θ will increase. As T increases beyond T_E, the strains γ, while remaining linear with r, will increase according to Eq. (7.7.3). Hence, when $T > T_E$, the outermost fibres ($r = R$) enter the plastic range; we thus note that plastic behaviour occurs initially in the outer region of the rod while the inner zone behaves elastically [Fig. (7.7.2)]. For successively

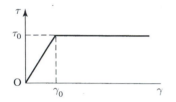

Figure 7.7.1

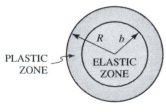

PLASTIC ZONE

ELASTIC ZONE

R b

Figure 7.7.2

[†] In this section, we shall refer to $\gamma \equiv \gamma_{x\theta}$ as the *shear strain*.

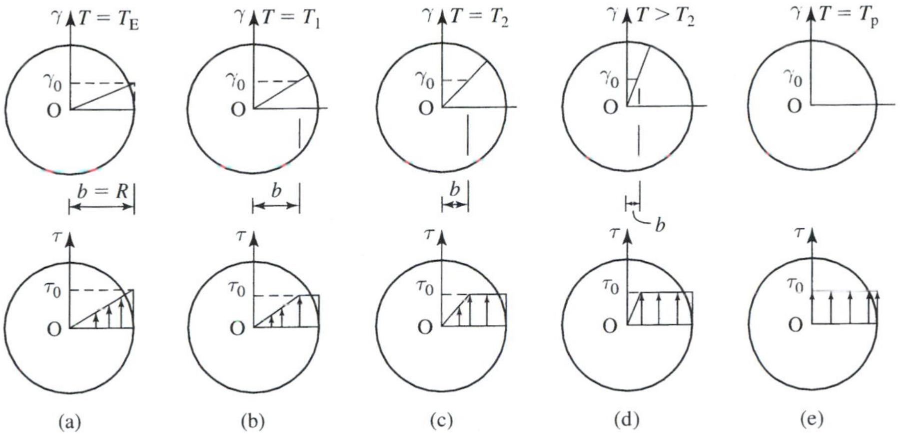

Figure 7.7.3

increasing torques, $T_E < T_1 < T_2 < T_3$, etc., the strain distribution will appear as in Figs. (7.7.3b–e). Noting that for $\gamma \geq \gamma_0$, $\tau = \tau_0$, the corresponding stress distribution will appear as shown in this figure. We observe that with increasing T, the plastic zone increases and the elastic zone decreases. Let b denote the location of the elastic–plastic interface. From Fig. (7.7.3), we observe that $\gamma(r = b) = \gamma_0$ and therefore $\tau(r = b) = \tau_0$. Since the stress in the elastic zone is proportional to the strain γ [according to the relation $\tau(r) = G\gamma(r) = G\Theta r$], it therefore also varies linearly with r; hence the stress distribution in the rod for any $T > T_E$ is given by

$$\tau(r) = \begin{cases} \frac{r}{b}\tau_0, & r \leq b \\ \tau_0, & b \leq r \leq R. \end{cases} \tag{7.7.4}$$

Furthermore, we note that at the interface $r = b$, $\gamma_0 = b\Theta$ and hence

$$\Theta = \frac{\tau_0}{Gb}. \tag{7.7.5}$$

The stress resultant representing the torque is given by

$$T = \iint\limits_A \tau(r) \cdot r \, dA = \int_0^{2\pi} \int_0^R r^2 \tau(r) \, dr \, d\theta, \tag{7.7.6}$$

as in Eq. (2.5.12c).

Substituting Eq. (7.7.4) in the appropriate zones,

$$T = 2\pi \left[\int_0^b r^2 \left(\frac{r\tau_0}{b} \right) dr + \int_b^R \tau_0 r^2 \, dr \right],$$

which, upon integration, yields

$$T = 2\pi \tau_0 \left[\frac{R^3}{3} - \frac{b^3}{12} \right]. \tag{7.7.7}$$

As T continues to increase, b decreases, and in the limit $b \to 0$; that is, the entire cross-section behaves plastically. The corresponding torque is then called the **ultimate plastic torque**, denoted by T_p; thus

$$T_p = \frac{2\pi}{3} R^3 \tau_0. \qquad (7.7.8)$$

We observe from Eq. (7.7.7) that this is the largest possible torque which the rod can sustain; as $T \to T_p$, the rod continues to yield freely.

In passing, we note from Eqs. (7.7.8) and (7.7.2b), the ratio

$$\frac{T_p}{T_E} = \frac{4}{3}. \qquad (7.7.9)$$

Thus we observe that the rod is capable of sustaining a torque T_p, which is 33% greater than the elastic torque T_E.

Clearly, $b = R$ when $T \le T_E$ and, as shown above, $b \to 0$ when $T \to T_p$. The location of the elastic–plastic interface when $T_E \le T \le T_p$, namely the relation $b = b(T)$, is readily obtained from Eq. (7.7.7):

$$b = \left[4R^3 - \frac{6T}{\pi \tau_0} \right]^{1/3} = R \left[4 \left(1 - \frac{3T}{2\pi R^3 \tau_0} \right) \right]^{1/3}. \qquad (7.7.10)$$

Substituting Eq. (7.7.8), we may write this in a more convenient form as

$$b = R \left[4 \left(1 - \frac{T}{T_p} \right) \right]^{1/3}, \quad T_E \le T < T_p. \qquad (7.7.11)$$

The variation of $b(T)$ is shown in Fig. (7.7.4); we observe that the elastic–plastic interface approaches the centre O of the rod very rapidly as T approaches T_p.

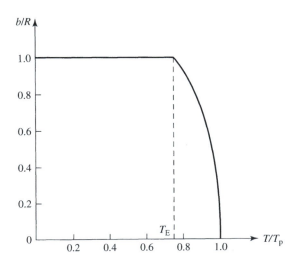

Figure 7.7.4

Substituting Eq. (7.7.11) in Eq. (7.7.5),

$$\Theta = \frac{\tau_0}{Gb} = \frac{\tau_0}{GR[4(1 - T/T_p)]^{1/3}}, \quad T_E \le T < T_p, \qquad (7.7.12a)$$

Since $\tau_0 = T_{\mathrm{E}} R / J$, we obtain finally

$$\Theta = \frac{T_{\mathrm{E}}}{G J [4(1 - T / T_p)]^{1/3}}, \quad T_{\mathrm{E}} \le T < T_{\mathrm{p}}, \qquad (7.7.12b)$$

which, using Eq. (7.7.9), we can write in the alternative form,

$$\Theta = \frac{3 T_{\mathrm{p}}}{4 G J [4(1 - T / T_{\mathrm{p}})]^{1/3}}, \quad T_{\mathrm{E}} \le T < T_{\mathrm{p}}. \qquad (7.7.12c)$$

Thus, the T–Θ relations for plastic behaviour is seen to differ considerably from the simple elastic relation of Eq. (7.7.1). In particular, as $T \to T_{\mathrm{p}}$, $\Theta \to \infty$, indicating that the rod is yielding (rotating) freely. From the non-dimensional T–Θ relation, plotted in Fig. (7.7.5), we observe that the slope of the curve, which represents the torsional stiffness of the rod, decreases sharply as the rod starts to exhibit plastic behaviour.

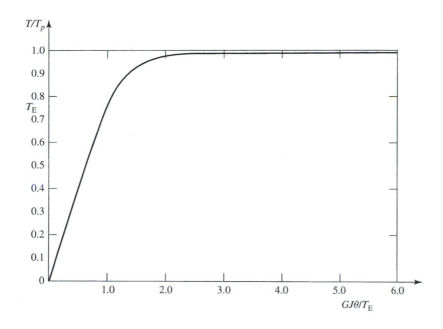

Figure 7.7.5

PROBLEMS

In all the following problems, neglect localised effects near junctures and at points of load application. Assume linear elastic behaviour for all problems unless otherwise specified.

The following constants are to be used in solving the problems.

Steel	$E_s = 200$ GPa,	$G_s = 76.0$ GPa
Aluminium	$E_a = 70$ GPa,	$G_a = 26.0$ GPa
Brass	$E_b = 120$ GPa,	$G_b = 42$ GPa
Bronze	$E_{br} = 105$ GPa,	$G_{br} = 45$ GPa

Section 2

7.1: A rod of length $L = 120$ cm, and whose diameter is 1.6 cm, is subjected to a torsional moment $T = 3000$ N-cm. It is required that the relative rotation of the two ends be greater than 3.5° and not to exceed 7.2°. What possible material(s), of those listed above, can be used for the rod?

7.2: A hollow steel shaft whose outer diameter is 60 mm is subjected to a torque $T = 3000$ N-m. What is the required thickness of the shaft if the maximum shear stress is not to exceed 100 MPa.

7.3: It is required to replace a solid shaft of diameter d by a hollow shaft (i.e., a cylindrical tube) of the same material such that in both cases, the maximum shear stress does not exceed a given allowable stress when subjected to the same torsional moment T. Determine the outer diameter D of the tube if the wall thickness is $D/20$.

7.4: A solid shaft of diameter D is made of a material whose stress–strain relation in shear is given as $\tau = c\sqrt{\gamma/2}$ with $0 \leq \tau$, $0 \leq \gamma$. Determine the relation between the unit angle of twist Θ and an applied torsional moment T.

7.5: A circular cylindrical rod is composed of two or more homogeneous materials, say 'a' and 'b', as shown in Fig. (7P.5), each having different material properties. Is it necessary to make an *assumption* that, when subjected to a torque T, all cross-sections remain plane and radial lines remain radial or is this a valid *conclusion*, as in the case (considered in Section 2 of this chapter) of a rod consisting of a homogeneous material?

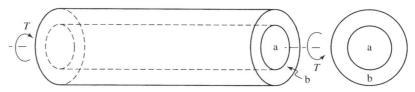

Figure 7P.5

7.6: A solid shaft of diameter d and made of a homogeneous material is subjected to a torsional moment T. What percentage of this torque is resisted by the material of the inner core, i.e., within a radius $0 \leq r \leq d/4$ and within the outer sector $d/4 \leq r \leq d/2$.

7.7: A solid aluminium shaft, 2-m long, is to be subjected to a torque $T = 6000$ N-m. If the allowable shear stress is $\tau_{all} = 40$ MPa and the relative rotations of the ends of shaft is not to exceed 1.5°, determine the minimum required diameter.

Section 3

7.8: A cylindrical shaft ABC consists of a steel segment AB with an allowable shear stress $\tau_{allow} = 100$ MPa, rigidly connected, as shown in Fig. (7P.8), to segment BC, which is made of brass and has a diameter of 50 mm. The brass has an allowable shear stress $\tau_{allow} = 40$ MPa. Determine (a) the maximum permissible torsional moment T that can be applied without exceeding the allowable shear stress in BC and (b) the minimum required diameter d of AB under this torque T.

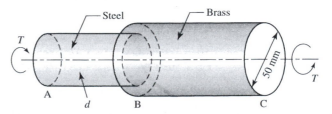

Figure 7P.8

7.9: A solid steel shaft having a diameter of 80 mm is subjected to torques by means of gears, as shown in Fig. (7P.9), which are equally spaced 1.25 m apart. Determine

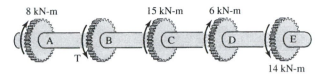

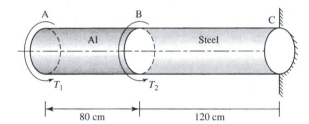

Figure 7P.9

(a) the maximum shear stress in the shaft and indicate in which segment this occurs, (b) the rotation of section D with respect to B and (c) the rotation of section E with respect to A.

7.10: A solid shaft is composed of two segments: AB (made of aluminium) and BC (made of steel), as shown in Fig. (7P.10). The shaft is subjected to a torques $T_1 = 600$ N-m and $T_2 = 1000$ N-m. Determine the required diameter if the allowable shear stresses in the steel and aluminium are given as $\tau_s = 80$ MPa and $\tau_a = 50$ MPa, respectively, and if the free end is not to rotate more than $\phi = 2.5°$.

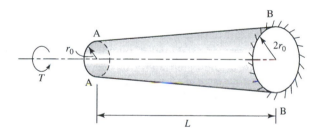

Figure 7P.10

7.11: A solid tapered shaft AB of length L is subjected to a torsional moment T at A and held fixed against rotation at B, as shown in Fig. (7P.11). Determine the angle of rotation ϕ_A if $r_0 \ll L$. (*Note*: Assume that all cross-sections remain plane and that all radial lines remain radial.)

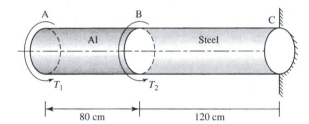

Figure 7P.11

Section 4

7.12: A gear–shaft system consists of identical steel shafts having diameter d, as shown in Fig. (7P.12), where A is fixed against rotation. The diameters of gears B and C are 250 and 100 mm, respectively. A torsional moment $T = 1000$ N-m is applied at D. Assuming no slippage of the gears, determine the required diameter of the shafts if the allowable shear stress is $\tau = 60$ MPa and if the rotation of the cross-section D is not to exceed 1.5°.

7.13: Determine the maximum shear stress in an elastic rotating shaft, 40 mm in diameter, which transmits 100 kW of power at a speed of 800 rpm.

7.14: Determine the required diameter of a solid steel shaft if the shearing stress is not to exceed 90 MPa and if it is to transmit 400 kW of power at a speed of 50 Hz.

7.15: A shaft is designed to transmit 600 kW of power from an electric generator. The diameter of the shaft is 4 cm and the maximum allowable shear stress is 80 MPa. What is the required speed (rpm) of the generator?

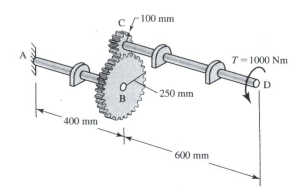

Figure 7P.12

Section 5

7.16: A solid circular steel shaft AD of diameter $d = 3\,\text{cm}$ and length L is placed between two frictionless rigid walls such that they provide no constraint against rotation. Torsional moments $T_0 = 500\,\text{N-m}$ are applied at B and C, as shown in Fig. (7P.16). If the temperature of the entire shaft is increased by $\Delta T = 50°\text{C}$, determine the principal stresses and the principal directions θ_1 and θ_2 (with respect to the longitudinal x-axis), which exist in segment BC at the outer lateral surface, $r = d/2$. (*Given*: $E = 200\,\text{GPa}$, $\alpha = 11.7 \times 10^{-6}\,°\text{C}^{-1}$.)

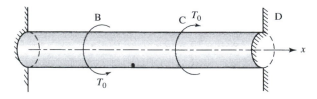

Figure 7P.16

7.17: A hollow circular shaft having an outer radius R_o and inner radius $R_i = R_o/2$ is subjected to a torque T and an axial load P, where $T = PR_o/4$. Determine (a) the principal stresses that exist at the inner lateral surface $r = R_i$ and the outer surface $r = R_o$. (Express answers in terms of P and R_o.) and (b) the principal directions θ_1 and θ_2 (with respect to the longitudinal x-axis) and evaluate in degrees.

Section 6

7.18: A steel and a bronze rod are rigidly bonded to form a shaft of length $L = 4\,\text{m}$, which is fixed against rotation at the two ends A and C, as shown in Fig. (7P.18). A torsional moment T is applied at B. If the allowable shear stress in the steel and bronze is 125 and 40 MPa, respectively, determine (a) the maximum torque T that can be applied and (b) the angle of rotation (in degrees) of section B.

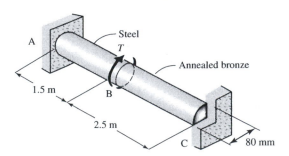

Figure 7P.18

7.19:* A gear–shaft system, consisting of two shafts connected by a gear B of radius $b = 100$ mm and a gear C of radius $c = 40$ mm, is fixed against rotation at A and D, as shown in Fig. (7P.19). A torsional moment $T = 5000$ N-m is applied to gear B. (a) Show that the ratio of the torque $(T_B)_C$ (i.e., the torque exerted by gear C on gear B to that of the torque $(T_C)_B$ (representing the torque exerted by gear B on gear C) is given as $\frac{(T_B)_C}{(T_C)_B} = \frac{b}{c}$. (b) Determine the maximum shear stress in shaft AB and CD.

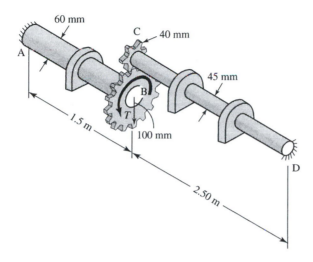

60 mm

C 40 mm

A

45 mm

B

1.5 m

T

100 mm

2.50 m

D

Figure 7P.19

7.20: The shaft shown in Fig. (7P.20) consists of a steel segment AC rigidly connected to a bronze segment CD. The shaft is restrained at A and D against rotation. If torques $T = 20$ kN-m are applied at B and C, determine (a) the maximum shear stress in both the steel and bronze segments and (b) the angle of rotation (in degrees) of section B.

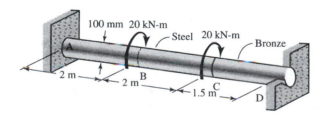

100 mm 20 kN-m

Steel 20 kN-m

Bronze

2 m

B

2 m

1.5 m

C

D

Figure 7P.20

7.21:* A cylindrical shaft AC of a material with shear modulus G and having diameter d and length L is fixed at both ends against rotation. A distributed torsional moment $t(x)$ (N-m/m), which varies along the axis as $t(x) = t_0[1 + (x/L)^2]$, is applied as shown in Fig. (7P.21). Assuming that plane cross-sections remain plane and that all radial lines remain radial, determine (a) the reactive torques at A and C and (b) the rotation at the cross-section at the midpoint B.

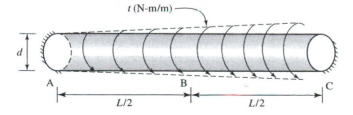

t (N-m/m)

d

A

B

C

$L/2$

$L/2$

Figure 7P.21

7.22:* A torque $T = 10$ kN-m is applied to a steel shaft 80 mm in diameter. While the shaft is under the torque, a brass sleeve of thickness $t = 10$ mm is slipped into place and rigidly attached to the steel shaft at its ends, as shown in Fig. (7P.22), after which the torque is released. Determine (a) the resulting shear stress in the brass and the steel and (b) the relative rotation of the two ends of the steel shaft.

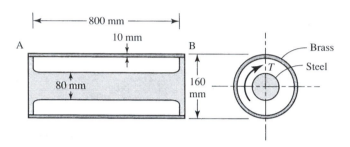

Figure 7P.22

Section 7

7.23: The solid steel shaft of radius $R = 40$ mm is assumed to behave as an elastic–ideal plastic material with a yield point $\tau_0 = 145$ MPa. Determine (a) the maximum moment T_E, which can be applied prior to initial plastic behaviour; (b) the radius of the elastic zone, $r = b$, if the rod is subjected to a torque $T = 18$ kN; (c) the unit angle of twist Θ (in deg/m).

7.24: Given a hollow rod with inner and outer radii R_i and R_o, respectively, made of a material that is assumed to behave as an elastic–ideal plastic material with a yield point in shear τ_o and shear modulus G during elastic behaviour. Determine (a) the torque T_E under which yielding first occurs; (b) the unit angle of twist Θ when $T = T_E$, (c) the limiting torque T_p and (d) Θ when $T \rightarrow T_p$.

7.25:* A hollow shaft of length 120 cm, whose cross-section is shown in Fig. (7P.25), is subjected to a torsional moment T. The shaft is made of steel, which is assumed to behave as an elastic–perfectly plastic material with $\tau_0 = 180$ MPa. Determine (a) the torque $T = T_y$ at first yielding, (b) the relative rotation of the two ends, ϕ, when yielding first takes place, (c) the torque T when the radius of the interface r_b separating the elastic and plastic zones is 30 mm, (d) the ultimate plastic moment T_p and (e) the unit angle of twist Θ as the rod becomes fully plastic, i.e. as $T \rightarrow T_p$.

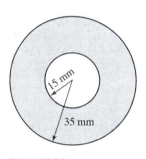

Figure 7P.25

7.26: A torsional moment is applied to a solid cylindrical rod of radius R whose material behaviour is elastic–pefectly plastic with a yield stress τ_0. The torque is applied until the rod becomes fully plastic (i.e., $T = T_p$) and is then removed. Show (a) that residual shear stresses within the rod vary linearly with the radial coordinate r and (b) that the residual stress at $r = 3R/4$ is zero.

7.27:* A steel shaft of diameter $d = 75$ mm is tapered over a length $L = 180$ mm to a diameter of 60 mm, as shown in Fig. (7P.27), and is subjected to a torsional moment $T = 8000$ N-m. Assume the steel behaves as an ideal elastic–plastic material with $\tau_o = 145$ MPa. Determine (a) the radius of the elastic zone within the segment AB, (b) the length of the segment (i.e., as measured from C) that behaves fully elastically, (c) the ultimate torsional moment T_p that can be applied to the shaft and (d) the length of the segment (as measured from C) that behaves fully elastically when T_p is applied.

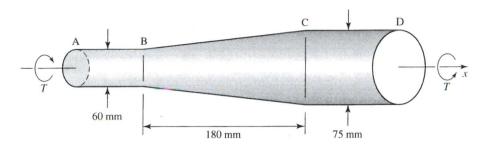

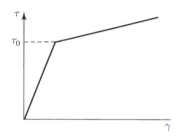

Figure 7P.27

7.28:* A solid cylindrical rod of radius R is made of a linear strain-hardening material having the following stress–strain relation in shear [see Fig. (7P.28)]:

$$\gamma = \begin{cases} \tau/G, & \tau \leq \tau_0 \\ \frac{1+c}{cG}\tau - \frac{\tau_0}{cG}, & \tau \geq \tau_0, \end{cases}$$

where $c > 0$ is a constant and where τ_0 is the yield stress.

Figure 7P.28

Show that the relation between the unit angle of twist, Θ, of the rod and an applied torque $T \geq T_E$ (where T_E is the maximum elastic torsional moment) is given by

$$\frac{T}{T_E} = \frac{1}{3(1+c)}[4 + 3c\beta - 1/\beta^3],$$

where $\beta = GR\Theta/\tau_0 \geq 1$ is a non-dimensional parameter defining the unit angle of twist. (*Note:* See computer-related Problem 7.48.)

Review and comprehensive problems

7.29: A composite cylindrical rod, whose cross-section is shown in Fig. (7P.29), is composed of material 'a' that serves as an inner core, and an outer surrounding material 'b'. The shear moduli of the two materials are G_a and G_b, respectively. Assuming perfect bonding between the two materials, determine the unit angle of twist Θ in terms of R_1, R_2, G_a and G_b if the rod is subjected to a torque T.

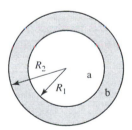

Figure 7P.29

7.30: A hollow tapered shaft of constant thickness t, having a mean radius $0 < \bar{R}_a \leq \bar{R}(x) \leq \bar{R}_b = 2\bar{R}_a$, is subjected to torsional moments, as shown in Fig. (7P.30). Assuming

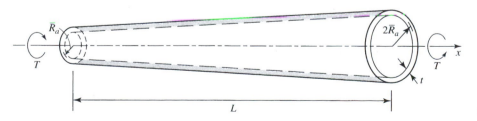

Figure 7P.30

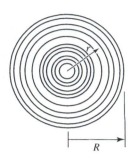

Figure 7P.31

the taper is small (i.e., $\overline{R}_a \ll L$) and that $t \ll \overline{R}_a$, determine the relative rotation of the two ends, $\phi_{B|A}$, in terms of T, $\overline{R}_a$, t, L and G.

7.31: Two composite shafts A and B, each of outer radius R, are constructed by gluing together as laminates, successive thin cylindrical cylinders, each consisting of an elastic material having a different stiffness [see Fig. (7P.31)]. In shaft A, laminates of increasing stiffness are glued together; the resulting radial variation of the shear modulus of the cylinder is expressed as $G(r) = G_0 r/R$, where G_0 is the shear modulus of the outermost laminate. In shaft B, laminates of decreasing stiffness are glued together resulting in a shear modulus expressed as $G(r) = G_0[1 - r/R]$. Both shafts are subjected to the same torsional moment, T. (a) Determine the unit angle of twist Θ for both shafts A and B; (b) determine the ratio Θ_A/Θ_B; (c) provide a physical explanation for this ratio.

7.32:* A hollow shaft is made by rolling a 5-mm thick plate into a cylindrical shape with an outer diameter $d = 120$ mm. The edges are then welded together along the resulting helical seams, which are oriented by an angle $\theta = 60°$ with respect to the x-axis, as shown in Fig. (7P.32). What is the maximum torsional moment that can be applied to the shaft if the allowable shear and tensile stresses in the weld are 50 and 100 MPa, respectively.

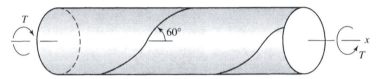

Figure 7P.32

7.33: A solid cylindrical rod of radius R is subjected to a torsional moment T. The material behaviour of the rod is governed in shear by the relation $\tau = k\gamma^n$, where k is a given constant, n is an integer and γ is the change in angle between two fibres originally oriented in the longitudinal x and circumferential directions. (a) Show that the nonlinear relation between T and Θ, the unit angle of twist, is given by

$$T = \frac{2\pi k R^{n+3}}{n+3} \Theta^n.$$

(b) Determine the maximum shear stress in the rod in terms of T and R.

7.34: A circular cylindrical shaft AB fixed at B, is subjected to a uniformly distributed torsional moment t_0 (N-m/m), as shown in Fig. (7P.34). The material behaviour of the shaft in shear is given as $\tau = k\gamma^2$, $0 \le \gamma$. Assuming plane sections remain plane and radial lines remain radial, determine (a) the shear stress $\tau_{x\theta} = \tau_{x\theta}(r, x)$ and (b) the rotation of the free end A.

t_0 (N-m/m)

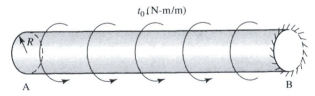

Figure 7P.34

7.35:* A solid composite cylindrical rod of length L, made of two materials, a core 'a' and a sleeve 'b', having shear moduli G_a and G_b respectively, is subjected to a torsional moment T, as shown in Fig. (7P.29).

    (a) Determine the ratio $\frac{(\tau_a)_{max}}{(\tau_b)_{max}}$, where τ_a and τ_b are the shear stress in a and b, respectively.

    (b) Determine the unit angle of rotation if $G_a = 6G_b$ and $R_1 = \sqrt{2}R_2/2$.

(c) For the values given in part (b), what part of the total applied torque T does the core a resist?

Note: Give answers to the above in terms of G_a and R_1.

(d) If, prior to the application of the torque, a thin wire (with modulus of elasticity E_s) is, as shown in Fig. (7P.35), wound around the cylinder in the form of a helix (making $45°$ with the x-axis), determine the stress in the wire after a torque T is applied.

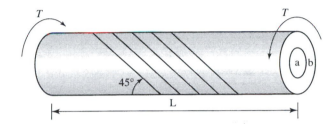

Figure 7P.35

7.36: A 60-mm diameter monel rod ($G = 65\,$GPa) is used as a solid shaft. To increase the stiffness of the shaft a stainless steel tube ($G = 86\,$GPa) with inner diameter 60 mm is placed over the monel shaft so as to form a composite member. Assuming perfect bonding between the steel and monel core, determine the required thickness of the tube if the sleeve is to decrease the unit angle of twist of the shaft by 60%.

7.37: A shaft ABC, made up of two segments AB and BC, is fixed at A and subjected to a torque T, as shown in Fig. (7P.37). The modulus of rigidity of the material and the yield stress in shear of the shaft material are G and τ_0, respectively. Determine (a) the largest torque T_E under which the shaft behaves elastically, (b) the rotation ϕ_C of section C when $T = T_E$, (c) the magnitude of the torque $T = T_b$ ($T_E \leq T_b \leq T_p$) that initiates plastic behaviour in segment AB and the ratio T_E/T_b and (d) the radius $r = b$ of the elastic zone in segment BC when $T = 0.15T_b$.

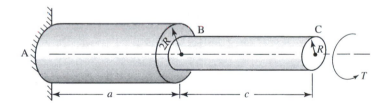

Figure 7P.37

7.38:* An assembly, consisting of a solid steel alloy cylindrical core, encased within a hollow aluminium shaft is fixed against rotation at A. A rigid plate at the end B is bonded to the steel and aluminium, as shown in Fig. (7P.38a). The two materials are

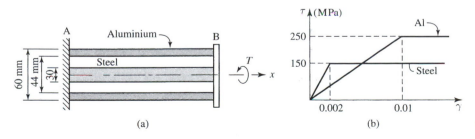

(a)   (b)

Figure 7P.38

assumed to behave as linearly elastic–perfectly plastic materials whose stress–strain relations in shear are shown in Fig. (7P.38b). A torque T is applied to the assembly. Determine (a) the maximum torque $T = T_E$ that may be applied such that the behaviour of the entire assembly remains elastic and (b) the maximum ultimate torque T_p that the system can sustain.

7.39: A steel rod, 20 mm in diameter, is subjected to a torsional moment T, as shown in Fig. (7P.39). Assuming the steel behaves as a perfectly elastic–plastic material with $\tau_0 = 150$ MPa, determine (a) the torque required to cause the free end to twist by 25° and (b) the radius b of the elastic core under this torque.

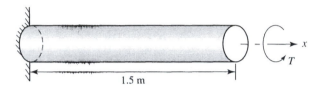

Figure 7P.39

1.5 m

7.40:* Two solid cylindrical steel shafts, each having a diameter of 40 mm, are connected by means of gears B and C, as shown in Fig. (7P.40). The shaft AB is fixed at A and plates D and E provide no rotational constraints. (a) If the allowable shear stress is 90 MPa, what is the largest torque T that can be applied at F. (b) Determine the rotation ϕ_F due to the torque T.

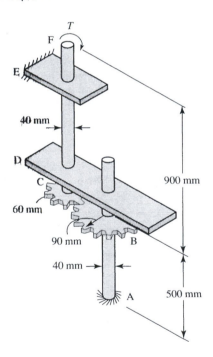

Figure 7P.40

7.41: Given two shafts made of the same material – a solid and a hollow shaft – each having the same cross-sectional area, as shown in Fig. (7P.41). The two shafts are subjected to torsional moments T_s and T_h, respectively. Assuming $\beta = R_i/R_o$ ($\beta \le 1$), show that (a) the maximum stress in the two shafts is the same if $\frac{T_s}{T_h} = \frac{\sqrt{1-\beta^2}}{1+\beta^2}$ and (b) the unit angle of twist Θ is the same if $\frac{T_s}{T_h} = \frac{(1-\beta^2)}{(1+\beta^2)}$.

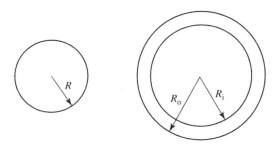

Figure 7P.41

7.42:* A composite cylinder whose cross-section, as shown in Fig. (7P.42a), composed of material A in the core surrounded by material B, is subjected to a torque T. Both the materials are perfectly bonded and are assumed to behave as ideal elastic–plastic materials; the stress–strain curves in shear are shown in Fig. (7P.42b) and the shear moduli are given as $G_A = 25\,\text{GPa}$ and $G_B = 80\,\text{GPa}$, respectively. Determine (a) the torque $T = T_y$ at which first yielding takes place, (b) the unit angle of twist when $T = T_y$ and (c) the ultimate torsional moment $T = T_p$.

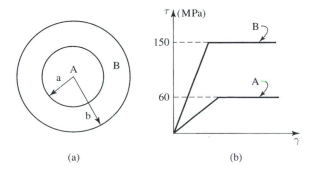

(a) (b)

Figure 7P.42

7.43:* A shaft consists of segment AB of an aluminium alloy, which is connected to a steel alloy segment BD by means of a flange coupling, attached by four bolts, as shown in Fig. (7P.43). The diameters d of both segments are 75 mm. The bolts, each having a cross-sectional area of 150 mm², are placed on a circle of diameter $D = 200$ mm. The allowable shear stress in the bolts is given as 60 MPa. The assembly is fixed at A and D against rotation and a torque is applied at section C as shown. Determine (a) the maximum torque T that can be applied, (b) the maximum shearing stress in segments AB, BC and CD and (c) the rotation of the flange coupling.

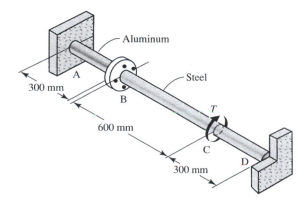

Figure 7P.43

7.44:   An oil drill of length $L = 2600$ m is composed of a bit, which is attached to a solid steel shaft whose diameter is 30 cm. While drilling the well, the bit, which is attached to the bottom of the shaft, starts to rotate only after the shaft has made three complete revolutions at the surface. Determine the maximum shearing stress in the shaft.

The following problems are to be solved using a computer

7.45:   It is required to replace a solid shaft of diameter d by a hollow shaft (i.e., a cylindrical tube) of the same material such that in both cases, the maximum shear stress does not exceed a given allowable stress when subjected to the same torsional moment T.

(a) Determine the outer diameter D of the tube if the wall thickness is D/k, where $2 \le k$, and plot D/d as a function of k.

(b) Plot the cross-sectional area of the tube A in terms of the non-dimensional ratio $\frac{A}{\pi d^2/4}$ as a function of k.

(c) What conclusions can be drawn from the curves?

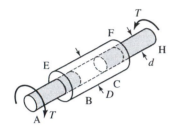

Figure 7P.46

7.46:   Two solid steel rods AB and CH, each having a diameter d, are rigidly connected to a hollow shaft EF whose outer diameter is D, as shown in Fig. (7P.46). If torques T are applied to the ends A and H, determine (a) the ratio D/d if the unit angle of twist in the segment EF is the same as in the rods and (b) the ratio D/d (to six or more significant figures) if the maximum shear stress in both the shaft EF and the rods is the same.

7.47:*   A hollow circular shaft having an outer radius R_o and inner radius $R_i = kR_o$ $(0 \le k < 1)$ is subjected to a torque T and an axial load P, where $T = \beta P R_o$ $(0 \le \beta)$.

(a) Show that the principal stresses σ_1 and σ_2 and the principal direction θ_1 (with respect to the longitudinal x-axis), which exist at the outer surface $r = R_o$, are given respectively by

$$\sigma_{1,2} = \frac{P}{2\pi R_o^2(1-k^2)}\left\{1 + \frac{1}{1+k^2}[(1+k^2)^2 + 16\beta^2]^{1/2}\right\}, \quad \theta_1 = \frac{1}{2}\tan^{-1}\left(\frac{4\beta(1-k^2)}{1-k^4}\right).$$

(b) By means of a computer, plot a family of curves for the non-dimensional ratios $\sigma_1 R_o^2/P$ and $\sigma_2 R_o^2/P$ as a function of k for several discrete values of β.

(c) By means of a computer, evaluate the principal direction θ_1 (in degrees), and plot θ_1 as a function of k for several discrete values of β.

7.48:   For the rod made of a strain-hardening material with a stress–strain relation as given in Problem 7.28, the relation between the applied torque $T \ge T_E$ and the unit angle of twist is given as

$$T/T_E = \frac{1}{3(1+c)}[4 + 3c\beta - 1/\beta^3],$$

where $\beta = GR\Theta/\tau_0 \ge 1$ is a non-dimensional parameter defining the unit angle of twist. Plot the values T/T_E as a function of β for several values of c: $c = 1, 2, 3, \ldots$.

8

Symmetric bending of beams – basic relations and stresses

8.1 Introduction

An important element very often encountered in engineering structures is a 'beam element'. Geometrically, a beam is characterised in the same manner as a rod, which was previously studied: namely, it is an element where one dimension, called the *longitudinal dimension*, is considerably greater than the remaining two dimensions that define the cross-section. However, as opposed to a rod, we refer to an element as a 'beam' if there exist components of the applied forces, which are perpendicular to the longitudinal axis. We then refer to these applied forces as *transverse forces* or *lateral forces*. Due to such forces, the beam no longer remains straight but will deform and undergo bending. The beam is then said to be in a state of 'flexure'.

Our goal in this chapter is to establish the basic relations governing the flexure of beams and to determine the resulting stresses due to the flexural deformation.

These relations are dependent on the internal force resultants existing at the cross-sections of the beam. Although force resultants were considered with some generality in Chapter 2, it is necessary to reconsider these quantities more specifically, and to some greater depth, as they apply to beams.

8.2 Resultant shear and bending moments – sign convention

(a) Some simple examples

We consider a beam whose longitudinal axis coincides with the x-axis. Furthermore, we shall restrict our study, at this stage, to beams for which all applied transverse loads lie in the x–y plane and for which all couples act about the z-axis [Figs. (8.2.1a and b)]. It follows that all internal forces in the z-direction must vanish as must all

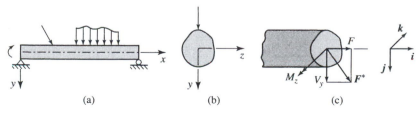

| (a) | (b) | (c) | **Figure 8.2.1** |

225

moments about the x- and y-axes. Consequently, the internal resultant force system acting on any cross-section will then consist only of a force $\boldsymbol{F}^*$ and a moment $\boldsymbol{M}^*$ whose sole component acts in the z-direction [Fig. (8.2.1c)], i.e.[†]

$$\boldsymbol{F}^* = F\boldsymbol{i} + V_y\boldsymbol{j}, \tag{8.2.1a}$$

$$\boldsymbol{M}^* = -M_z\boldsymbol{k}. \tag{8.2.1b}$$

We note that this is a particular case of the general expression for internal forces given in Chapter 2 [Eqs. (2.2.2)], where $n \to i$, $s \to j$ and $t \to k$.

Since we are interested in studying the effect of flexure, we shall generally consider only those cases where the transverse applied forces not only lie in the $x-y$ plane but act only in the y-direction so that F, the axial force, vanishes at all cross-sections. Further, noting that there is but one shear force component V_y and a single moment component M_z for ease of notation, we shall set $V \equiv V_y$ and $M \equiv M_z$. We now consider two simple cases, given in the following examples.

Example 8.1: A simply supported beam AB is subjected to a constant uniformly applied load $q(x) = w$ acting downward, as shown in Fig. (8.2.2a). Determine the shear force $V(x)$ and the moment $M(x)$ at all cross-sections.

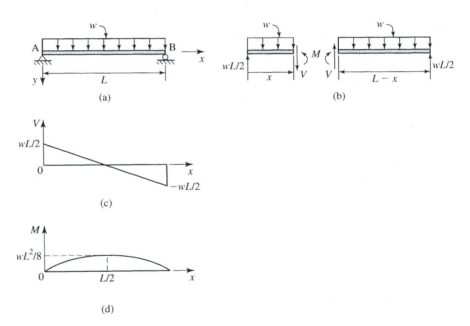

Figure 8.2.2

Solution: From the equilibrium equations $\sum F_y = 0$ and $\sum M_z = 0$, we find that the reactions at A and B are $R = wL/2$. Proceeding as in Chapter 2, we make a cut at some general cross-section located at the coordinate x, and isolate the two segments of the beam. We may then analyse either of these segments as a free body. Choosing arbitrarily the left-hand segment [Fig. (8.2.2b)], the equilibrium equations

---

[†] Note that in Eq. (8.2.1b), $\boldsymbol{M}^*$ is defined with a minus sign such that $+M_z$ points in the negative z-direction. This is done in order to be consistent with the sign convention to be adopted in subsection (b) of this section.

are then

$$+ \uparrow \sum F_y = 0 \rightarrow\rightarrow wL/2 - V(x) - wx = 0 \rightarrow\rightarrow V(x) = w(L/2 - x),$$

$$\curvearrowleft + \sum (M_z)_x = 0 \rightarrow\rightarrow -(WL/2)x + (wx)(x/2) + M(x)$$

$$= 0 \rightarrow\rightarrow M(x) = \frac{wx}{2}(L - x),$$

where $\curvearrowleft + \sum (M_z)_x$ indicates that the sum of the moments is taken about the z-axis passing through the section at x. Note too that the above expressions are valid for all $0 \le x \le L$.

We may now plot the expressions for V and M as a function of x. These are shown in Figs. (8.2.2c and d), respectively. We observe that the shear V varies linearly with x while the moment M is a quadratic function of x. The maximum value, which clearly occurs at $x = L/2$, is $M(L/2) = wL^2/8$.

Such graphs of $V(x)$ and $M(x)$ plotted as a function of the position x are commonly called **shear and moment diagrams**.   □

Example 8.2: Determine the resultant shear $V(x)$ and moment $M(x)$ for the simply supported beam loaded by a concentrated force P acting at the centre, $x = L/2$, as shown in Fig. (8.2.3a), and plot the shear and moment diagrams.

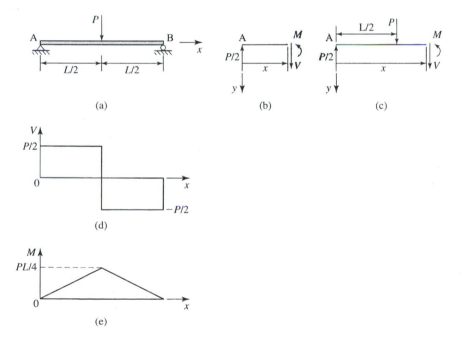

(a)   (b)   (c)

(d)

(e)

Figure 8.2.3

Solution: From equilibrium, the reactions at A and B are $R_A = R_B = P/2$. As in the previous example, we make a cut at a section x and write the equilibrium equations for the isolated segment as a free body. However, here, we note that we obtain a different result if we make a cut at a section to the left or to the right of P [Fig. (8.2.3b) or (8.2.3c), respectively]. Thus it is necessary to analyse two separate cases.

For $0 < x < L/2$ [Fig. (8.2.3b)], we obtain

$$+ \uparrow \sum F_y = 0 \rightarrow\rightarrow -P/2 + V(x) = 0 \rightarrow\rightarrow V(x) = P/2,$$

$$\curvearrowleft + \sum (M_z)_x = 0 \rightarrow\rightarrow -(P/2)x + M(x) = 0 \rightarrow\rightarrow M(x) = Px/2.$$

Similarly for $L/2 < x < L$ [Fig. (8.2.3c)], we obtain

$$+ \uparrow \sum F_y = 0 \rightarrow\rightarrow -P/2 + P + V(x) = 0 \rightarrow\rightarrow V(x) = -P/2,$$

$$\curvearrowleft + \sum (M_z)_x = 0 \rightarrow\rightarrow -(P/2)x + P(x - L/2) + M(x)$$

$$= 0 \rightarrow\rightarrow M(x) = \frac{P}{2}(L - x).$$

The corresponding shear and moment diagrams are shown in Figs. (8.2.3d and e), respectively.　　　　　　　　　　　　　　　　　　　　　　　　　　　□

We note that the moment diagram is linear in the two regions of x and that M is a maximum at $x = L/2$, where $M(x = L/2) = PL/4$. On the other hand, the shear is constant in the two regions. However, our solution leads to a discontinuity in v at $x = L/2$! Clearly this requires an explanation, for, in general, we would not expect such a discontinuity in nature for a static problem. (However, we mention here that in dynamic cases, one may encounter discontinuities such as at wave fronts of propagating waves in solids.)

To explain this discontinuity, it is necessary to recall the definition of a concentrated force as developed in Chapter 1. From that discussion, it was seen that a concentrated force is an idealisation which represents a force of high intensity, acting over a very small area; indeed, a concentrated force does not exist in nature. Thus, in reality, we are essentially examining a beam subject to a force system, as shown in Fig. (8.2.4a), where $\Delta \to 0$. For this system, with $\Delta \neq 0$, the shear diagram appears as in Fig. (8.2.4b). Hence, the discontinuity, as shown in Fig. (8.2.3d), is but a result of our idealisation. With this interpretation, we shall continue to make use of this idealisation, but with the clear understanding that the true results, as found in nature, do not lead to such discontinuous solutions.

We note that in the above examples, the shear force V and moment M are linearly dependent on the applied forces w and P. Following our discussion in Chapter 5, if we assume that the beam undergoes small deformations, the combined effect of w and P on the beam may be obtained by linear superposition of the two solutions. These are shown in Fig. (8.2.5).

In obtaining the above solutions, the direction of positive shear forces and moments were chosen arbitrarily as in Figs. (8.2.2b) and (8.2.3b and c). Now, in order to be consistent, it is necessary to adopt a sign convention for these forces and moments as well as for applied forces.

Finally, before establishing the sign convention, it is appropriate at this point to consider the effect of moments and shears on the deformation of the beam. From the basic principles of mechanics, we are aware that, in general, moments tend to cause rotation of a body. In particular, in the case of beams, the moment will tend to cause each segment of the beam to rotate and deform; the global effect on the entire beam will then be to cause the beam to undergo bending, as shown in Fig (8.2.6a). On the other hand, as was discussed in Chapters 3 and 4, shear forces tend to cause rectangular segments to undergo distortion and assume the shape of

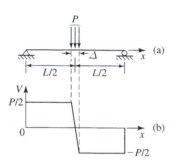

Figure 8.2.4

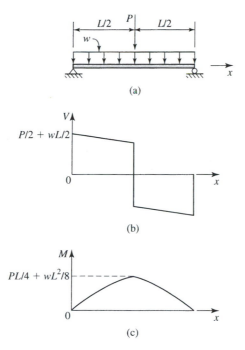

(a)

(b)

(c)

Figure 8.2.5

parallelograms [Fig. (8.2.6b)]. In our study below, we shall consider these two effects separately.

(b) Sign convention

To establish properly a sign convention for beams, it is first necessary to establish a coordinate system since the sign convention depends directly on the coordinate system used. To this end, we define a right-hand coordinate system xyz with x pointing to the right and positive y pointing *downward* [Fig. (8.2.7)]. The beam is oriented with respect to this coordinate system such that its longitudinal axis coincides with the x-axis.[†] The cross-sections of the beam therefore lie in the y–z plane.

We now recall the standard sign convention (as defined in Chapter 2) that defines the faces of the cross-section: a positive (negative) face is one for which the outward normal is acting in the positive (negative) coordinate direction.

The sign convention for the shear forces $V \equiv V_y$ is then defined as follows:

- A shear force $V \equiv V_y$ is said to be positive if it acts on a positive (negative) x-face of the cross-section in the positive (negative) y-direction [Fig. (8.2.8a)]. Conversely, a negative shear force V acts on a positive (negative) face in the negative (positive) y-direction.

Recalling also that $M \equiv M_z$, i.e. M represents the moment about the z-axis, the following sign convention is adopted for the moment M:

- A moment M is said to be positive if it tends to cause the beam to bend such that extension occurs in those fibres defined by positive y-coordinates [Fig. (8.2.8b)].

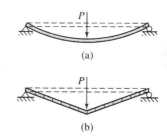

(a)

(b)

Figure 8.2.6

Figure 8.2.7

[†] Note that here and in all subsequent treatment of flexure of beams, the x-coordinate is measured from the extreme left end of the beam.

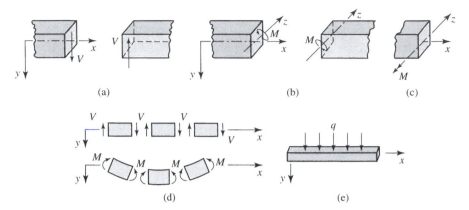

Figure 8.2.8

It is emphasised that in using this sign convention, the concept of positive or negative moments as being clockwise or counter-clockwise is totally irrelevant. For example, the vector representation for positive moments acting on a positive face is, according to the right-hand rule, shown in Fig. (8.2.8c). Moreover, we note that the vector representing positive M points in the negative z-direction. Thus we observe that the adopted *sign convention is also independent of a vector sign convention*; it is often referred to as being a *mechanics-of-materials* sign convention. For further clarity, positive V and M are also shown in plane view in Fig. (8.2.8d).

The following sign convention is adopted for applied lateral loads $q(x)$:

■ Positive (negative) forces $q(x)$ act in the positive (negative) y-direction [Fig. (8.2.8e)].

In anticipation of this sign convention, the direction of positive forces and moments appearing in Examples 8.1 and 8.2 were assumed consistent with this adopted convention.

8.3 Differential relations for beams

Consider the beam shown in Fig. (8.3.1a), which is subjected to transverse forces $q(x)$. Clearly, due to the applied loads, shear forces V and moments M will exist at any cross-section. In general these will not be constant but will vary with x; that is, $V = V(x)$ and $M = M(x)$. Let us consider an arbitrary element of width Δx such that the left cross-section is located at the coordinate x and the right face is at coordinate $x + \Delta x$. On the right side, the shear and moments are then $V(x) + \Delta V$ and $M(x) + \Delta M$, respectively, as shown in Fig. (8.3.1b). (Note that we have taken all forces and moments to be positive according to the adopted sign convention.)

We now consider the equilibrium of the element in the deflected state.

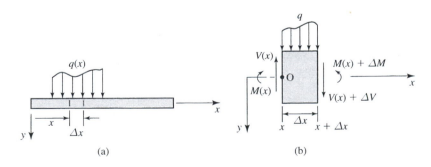

Figure 8.3.1

From equilibrium in the y-direction, $+\downarrow \sum F_y = 0$,

$$-V(x) + [V(x) + \Delta V] + q(\bar{x})\Delta x = 0,$$

where $x \leq \bar{x} < x + \Delta x$. (Note that in writing $q(\bar{x})\Delta x$ we have implicitly used the mean-value theorem to represent the total lateral applied force acting on the element.) Therefore, upon dividing through by Δx and taking the limit as $\Delta x \rightarrow 0$,

$$\frac{dV(x)}{dx} = -q(x) \tag{8.3.1}$$

since $\Delta V \rightarrow 0$ and $\bar{x} \rightarrow x$ as $\Delta x \rightarrow 0$.

Moment equilibrium about point O yields

$$\curvearrowleft + \left(\sum M_z\right)_0 = -M(x) + [M(x) + \Delta M] - [V(x) + \Delta V]\Delta x \\ - [q(\bar{x})\Delta x]\alpha \Delta x = 0,$$

where $0 < \alpha < 1$. (Note here that the location of the resultant of the lateral force is unknown; the parameter α is introduced to indicate that the resultant passes somewhere between x and $x + \Delta x$.)

Dividing through by Δx and taking the limit as $\Delta x \rightarrow 0$,

$$\lim_{\Delta x \rightarrow 0} \frac{\Delta M}{\Delta x} = \lim_{\Delta x \rightarrow 0} [V(x) + \Delta V + \alpha q(\bar{x})\Delta x]$$

and therefore

$$\frac{dM(x)}{dx} = V(x). \tag{8.3.2}$$

Combining Eqs. (8.3.1) and (8.3.2), we find

$$\frac{d^2 M(x)}{dx^2} = -q(x). \tag{8.3.3}$$

At this point, the reader is urged to return to Examples 8.1 and 8.2 and observe that Eqs. (8.3.2) and (8.3.3) are satisfied. Note that in Example 8.1, the lateral load $q(x) = w$, a constant, for all $0 < x < L$, while in Example 8.2, $q(x) = 0$ for all $x \neq L/2$.

Finally, since from Eq. (8.3.2), $\frac{dM(x)}{dx} = V(x)$, we arrive at an important conclusion: *the moment is stationary (i.e. has a maximum or minimum value) at those cross-sections where the shear force $V = 0$.*

8.4 Some further examples for resultant forces in beams

We present here some relatively elementary examples. While the solutions are rather simple, the reader is urged to read carefully the comments and conclusions to better understand some of the important features of the general results.

Note that in all subsequent problems, positive directions of unknown V and M are taken in accordance with the adopted sign convention.

Example 8.3: A simply supported beam AB of length L is subjected to an applied couple M_0 at A, as shown in Fig. (8.4.1a). Determine the shear $V(x)$ and moment $M(x)$ and draw the shear and moment diagrams.

Solution: Since no component of the applied loading system exists in the x-direction, it is clear that there exist no reactive components in the x-direction; i.e., $\sum F_x = 0$

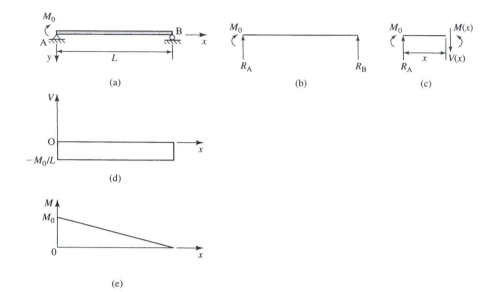

Figure 8.4.1

is satisfied identically. The reactions R_A and R_B in the y-direction [Fig. (8.4.1b)] are first obtained from the remaining equilibrium conditions for the entire beam:

$$\curvearrowright + \sum (M_z)_B = 0 \rightarrow\rightarrow M_0 + R_A L = 0 \rightarrow\rightarrow R_A = -\frac{M_0}{L}, \quad (8.4.1a)$$

$$+ \uparrow \sum F_y = 0 \rightarrow\rightarrow R_A + R_B = 0 \rightarrow\rightarrow R_B = -R_A = \frac{M_0}{L}. \quad (8.4.1b)$$

Note that the negative sign appearing for R_A indicates that the reactive force is downward.

To obtain the internal shear and moment, we make a cut at an arbitrary section x, $0 < x < L$ [Fig. (8.4.1c)]. Then, using equilibrium of either segment (here we again arbitrarily choose the left segment) as a free body, we have

$$+ \uparrow \sum F_y = 0 \rightarrow\rightarrow R_A - V(x) = 0 \rightarrow\rightarrow V(x) = R_A = -\frac{M_0}{L}, \quad (8.4.2a)$$

$$\curvearrowright + \sum (M_z)_x = 0 \rightarrow\rightarrow M_0 + R_A x - M(x) = 0 \rightarrow\rightarrow M(x)$$
$$= M_0 + R_A x = M_0 \left(1 - \frac{x}{L}\right). \quad (8.4.2b)$$

From this trivially simple example, we first make several observations:

- We note that the shear force $V(x)$ is negative; i.e., it is acting physically in the negative y-direction on the positive x-face.
- Furthermore, the shear force V is constant throughout the beam. Noting that no lateral applied forces are applied to the beam, i.e. $q(x) = 0$, we observe from Eq. (8.3.1) that the shear force must indeed be constant.
- From the above expressions, we also observe that the derivative of the $M(x)$ is precisely $V(x)$, in accordance with Eq. (8.3.2).

The shear and moment diagrams are shown in Figs. (8.4.1d and e). Note too that the moment $M(x = L) = 0$, i.e. the moment at B vanishes. This is necessarily so since, by definition, a simple support provides no moment reaction against any rotational constraint. □

Example 8.4:   A cantilever beam AB of length L is loaded by a transverse load, varying linearly from w_0 (N/m) at A to zero at B, as shown in Fig. (8.4.2a). Determine $V(x)$ and $M(x)$ and draw the shear and moment diagrams.

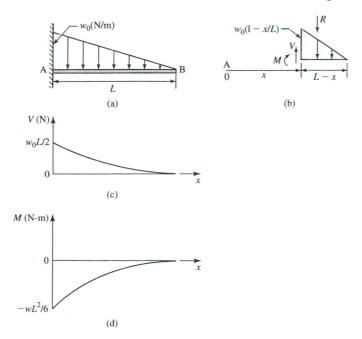

(a)   (b)

(c)

(d)   **Figure 8.4.2**

Solution: Since the lateral load q varies linearly, we obtain immediately an expression for $q(x)$; namely

$$q(x) = w_0(1 - x/L) \quad \text{(N/m)}. \tag{8.4.3a}$$

To obtain $V(x)$ and $M(x)$, we make a cut at any arbitrary section x, $0 < x < L$, and arbitrarily isolate the right-hand side [Fig. (8.4.2b)]. Note that by using this segment as a free body, no unknown reactive components appear. Consequently, it is not necessary to first find the reactions (at A) in order to solve this problem. (This is in contrast to the previous example, Example 8.3, where it was first necessary to determine the reaction.)

Isolating the free body, we first observe that the total downward resultant force of the applied load, represented by the small triangle of Fig. (8.4.2b), is $(w_0/2) \cdot (1 - x/L)(L - x) = (w_0L/2)(1 - x/L)^2$. The location of this resultant is immediately known, namely it occurs at the third point from the cross-section located at x, namely at $x + (L - x)/3 = (L + 2x)/3L$ [Fig. (8.4.2b)].[†] We then have for equilibrium of the free body

$$+\uparrow \sum F_y = 0 \rightarrow\rightarrow V(x) - (w_0L/2)(1 - x/L)^2 = 0$$

$$\rightarrow\rightarrow V(x) = \frac{w_0}{2L}(L - x)^2, \tag{8.4.3b}$$

$$\curvearrowright + \sum (M_z)_x = 0 \rightarrow\rightarrow M(x) + \frac{w_0(L - x)^2}{2L}\frac{L - x}{3} = 0$$

$$\rightarrow\rightarrow M(x) = \frac{w_0(x - L)^3}{6L}. \tag{8.4.3c}$$

Note here again, that Eqs. (8.3.1) and (8.3.2) are satisfied.

[†] Note that x is measured from point A, the extreme left end of the beam. See previous footnote, p. 229.

The shear and moment diagrams are given in Figs. (8.4.2c and d). The shear is seen to vary quadratically and the moment to be a function of x^3. We observe that at A, $x = 0$, the shear and moments reach their maximum absolute values; namely $V_A = w_0 L/2$ (N) and $M_A = -w_0 L^2/6$ (N-m). We note that these values represent the force and moment acting on the negative x-face at the section of A; that is, they represent the reactions which the *wall exerts on the beam* at A. Physically, the positive value of V_A indicates that the wall exerts an *upward* force on the beam; note too that the reactive moment M_A acting on the negative x-face of the beam is, in fact, counter-clockwise.   □

At this stage, it is worth making a further observation: since $\frac{dM(x)}{dx} = V(x)$, it is evident that if at any point x, $V > 0$, then the slope of the moment diagram will be positive; if $V < 0$, the slope of $M(x)$ is then negative. Furthermore, if the slope of $V(x)$ is positive (negative) at any point x, then the second derivative of $M(x)$ will be positive (negative). Since the sign of the second derivative of a function defines the sign of the curvature, we conclude that if the *slope* of V is positive (negative) at any point x, then the curvature of $M(x)$ at this point will be positive (negative). These conclusions are obviously of great assistance in checking the shear and moment diagrams and in preventing possible errors.

Example 8.5: A beam ABCD is loaded at point A and D, as shown in Fig. (8.4.3a). Determine $V(x)$ and $M(x)$ for all $0 < x < L + 2a$ and draw shear and moment diagrams.

Solution: From equilibrium of the entire beam [Fig. (8.4.3b)], we obtain $R_B = R_C = P$. Isolating the segment $0 < x < a$ [Fig. (8.4.3c)] we have, from equilibrium,

$$\uparrow \sum F_y = 0 \rightarrow \rightarrow -P - V(x) = 0 \rightarrow \rightarrow V(x) = -P \quad \text{(a constant)},$$

$$\curvearrowright + \sum (M_z)_x = 0 \rightarrow \rightarrow M(x) + Px = 0 \rightarrow \rightarrow M(x) = -Px.$$

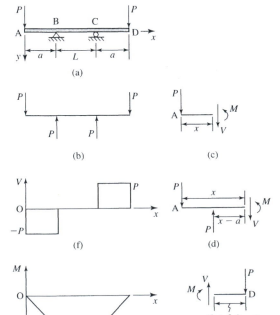

Figure 8.4.3

For an arbitrary cross-section between B and C ($a < x < L + a$) [Fig. (8.4.3d)],

$$+\uparrow \sum F_y = 0 \rightarrow\rightarrow -P + R_B - V(x) = 0.$$

Since

$$R_B = P \rightarrow\rightarrow V(x) = 0$$

$$\curvearrowright + \sum (M_z)_x = 0 \rightarrow\rightarrow Px - R_B(x-a) + M(x)$$

$$= 0 \rightarrow\rightarrow M(x) = -Pa \quad \text{(a constant)}.$$

Similarly, upon cutting the beam at any arbitrary point between C and D and isolating the segments [noting that here it is preferable to analyse the right-hand segment for equilibrium; see Fig. (8.4.3e)] we obtain $V(x) = P$, $M(x) = P[(L + 2a) - x]$, $L + a < x < L + 2a$.

The shear and moment diagrams are shown in Figs. (8.4.3f and g). We observe that between B and C, $V = 0$ and $M = -Pa$. Thus the moment does not vary in the beam between B and C. This is consistent with our general results, Eq. (8.3.2); namely, if the shear is zero, the moment must be constant. Within this span (BC) the beam is said to be in *pure bending*. The state of *pure bending* is an important one in the development of beam theory as we shall see below. □

Example 8.6: A beam ABCD, as shown in Fig. (8.4.4a), is fixed at A and simply supported at D. The two segments ABC and CD are connected by a pin at C. The beam is loaded as shown, where the segment BE is assumed to be rigidly attached to the beam at B. Determine expressions for $F(x)$, $V(x)$, $M(x)$ and plot their variation with x.

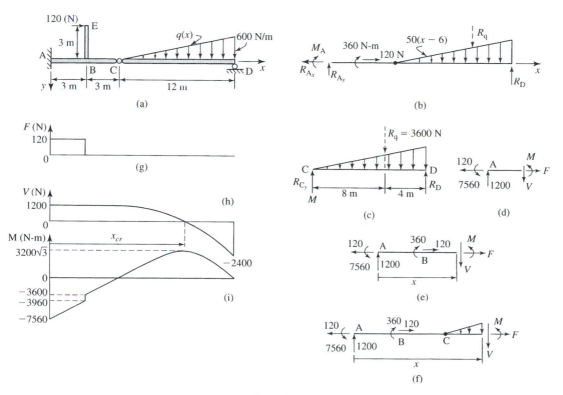

Figure 8.4.4

Solution: First, we replace the horizontal force acting at E by an equivalent force system consisting of a horizontal force at B and a couple of 360 N-m [Fig. (8.4.4b)]. We also observe from this figure, that there exist four reactive components. However, we recall, from Chapter 2, that the pin provides no constraint against relative rotation of the two beam segments AC and CD. Therefore, as the pin cannot transmit any moment from one segment to the other at point C, it follows that $M_C = 0$.

Recalling that x is measured from point A, an expression for the vertically applied lateral load is immediately obtained; thus, using the property of similar triangles, we have

$$\frac{q(x)}{x-6} = \frac{600}{12} \rightarrow\rightarrow q(x) = 50(x-6) \text{ N/m}, \quad 6 < x \leq 18$$

Moreover, the total resultant force of this vertically applied load is $R_q = \frac{600\,(12)}{2} = 3600$ N; this resultant force acts 4 m to the left of point D.

We are now in a position to determine the reactions. Since a horizontal force 120 N is acting, it is clear that the horizontal reaction at A must be $R_{Ax} = 120$ N (to the left).

To obtain the vertical reactions, we proceed as follows: we consider the entire structure ABCD. From equilibrium:

$$\curvearrowright + \sum (M_z)_D = 0 \rightarrow\rightarrow M_A + 18R_{Ay} + 360 - 4(3600) = 0,$$

$$+\uparrow \sum F_y = 0 \rightarrow\rightarrow R_{Ay} + R_D - 3600 = 0.$$

Note that at this stage it is not possible to solve for the three unknowns, M_A, R_{Ay} and R_D using these two equations. However, making use of the fact that $M_C = 0$, we now isolate segment CD. From equilibrium [Fig. (8.4.4c)]

$$\curvearrowright + \sum (M_z)_C = 0 \rightarrow\rightarrow 12R_D - 8(3600) = 0 \rightarrow\rightarrow R_D = 2400 \text{ N}.$$

Substituting into the preceding equation, we find $R_{Ay} = 1200$ N; it then follows from the first equation that $M_A = -7560$ N-m.

Having found all the reactions, one may, as in the previous examples, determine the internal forces, (F, V and M) in the various segments of the beam, AB, BC and CD, by making suitable cuts and using the equations of equilibrium on the appropriate free bodies, as shown in Figs. (8.4.3d–f) respectively. One thus obtains

Segment AB: $0 < x < 3$

$$F = 120, \quad V = 1200, \quad M = 1200x - 7560$$

Segment BC: $3 < x < 6$

$$F = 0, \quad V = 1200, \quad M = 1200x - 7200$$

Segment CD: $6 < x < 18$

$$F = 0, \quad V = -25(x-6)^2 + 1200, \quad M = -\frac{25}{3}(x-6)^3 + 1200x - 7200$$

The resulting variations of F, $V(x)$ and $M(x)$ are shown in Figs. (8.4.4g–i), respectively. It is worthwhile to point out several features as shown in the diagrams:

■ A constant axial force F exists in the segment AB and is zero from B to D.
■ A discontinuity of 360 N m exists in the moment diagram at point B($x = 3$). This discontinuity is due to the applied couple of the same magnitude. (Thus we note that just as a concentrated force acting within a beam causes a discontinuity in the shear V, a concentrated moment – a couple – causes a discontinuity in the moment M).

- The moment at C is zero. This provides a check on our calculations since a pin is known to exist at C.
- At point A, $V = 1200$ N and $M = -7560$ N m.

These quantities correspond to the reactive vertical force and moment of the wall on the beam. Thus the wall exerts an upward reactive force on the beam and a counter-clockwise moment (which is required to prevent rotation of the beam at A). Note that these directions are immediately known since they were chosen consistent with the adopted sign convention for shear and moment.

- From the expression for $V(x)$ in segment CD, as given above, we may immediately determine the cross-section $x = x_{cr}$ at which $V = 0$; thus

$$-25(x_{cr} - 6)^2 + 1200 = 0 \rightarrow\rightarrow (x_{cr} - 6)^2 = 48 \rightarrow\rightarrow x_{cr} - 6$$

$$= 4\sqrt{3} \rightarrow\rightarrow x_{cr} = 4\sqrt{3} + 6$$

As was concluded in Section 3, the moment M will have a stationary value at this section. Substituting $x_{cr} = 4\sqrt{3} + 6$ in the expression for $M(x)$ given above, we find

$$M(x_{cr}) = -\frac{25}{3}(x_{cr} - 6)^3 + 1200x_{cr} - 7200 = 3200\sqrt{3} \text{ N-m.}$$

From the moment diagram it is clear that this is actually a maximum value. However, we may establish analytically that this value must be a maximum, using the general expression $\frac{d^2 M(x)}{dx^2} = -q(x)$ as given by Eq. (8.3.3). Noting that in this example, $q(x)$ is positive for $x > 6$, we find that $\frac{d^2 M(x)}{dx^2} < 0$, thus indicating that $M(x)$ is maximum at $x = x_{cr}$.   □

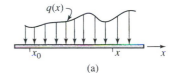

(a)

(b)

Figure 8.4.5

We observe that in all the previous examples, Examples 8.1–8.6, simple expressions for the shear forces and moments were immediately obtainable since the resultants of the applied forces as well as their location were readily known. Although this is usually true for simple variations of force (e.g., linear, quadratic, etc.), in general one does *not* know, a priori, the resultant and its location for arbitrarily varying given loads $q(x)$, as shown in Fig. (8.4.5). It is therefore useful to establish explicit relations for the shear forces and moments for such cases; these relations are expressed by integral expressions as derived below.

8.5   Integral relations for beams

Consider a straight beam subjected to an arbitrary lateral load $q(x)$, as shown in Fig. (8.4.5a). At this stage, we consider only continuous applied loads and exclude concentrated forces and moments. Let us assume that the shear force and moment are *known* at some point $x = x_0$, i.e. $V(x_0)$ and $M(x_0)$ are given. We wish to obtain the shear force $V(x)$ and moment $M(x)$ at some other arbitrary section x, $x > x_0$, a finite distance apart [Fig. (8.4.5b)]. From the differential expression, Eq. (8.3.1), we have

$$dV(x) = -q(x)\,dx. \tag{8.5.1a}$$

Taking the integral on both sides,

$$\int_{V(x_0)}^{V(x)} dV = -\int_{x_0}^{x} q(x)\,dx, \tag{8.5.1b}$$

we obtain

$$V(x) = V(x_0) - \int_{x_0}^{x} q(x)\,dx. \tag{8.5.1c}$$

Recalling that the variable appearing within the integral is but a dummy variable, we rewrite Eq. (8.5.1c) as

$$V(x) = V(x_0) - \int_{x_0}^{x} q(\eta)\,d\eta. \tag{8.5.2}$$

It is important to note that $V(x)$ is a function of the upper limit x appearing on the right-hand side of Eq. (8.5.2).

Now, similarly, from Eq. (8.3.2),

$$dM(x) = V(x)\,dx. \tag{8.5.3a}$$

Integrating both sides,

$$\int_{M(x_0)}^{M(x)} dM = \int_{x_0}^{x} V(x)\,dx \equiv \int_{x_0}^{x} V(\xi)\,d\xi. \tag{8.5.3b}$$

Here, to avoid confusion, we have called the dummy variable, appearing on the right-hand side, ξ.

Substituting Eq. (8.5.2), we obtain

$$M(x) - M(x_0) = \int_{x_0}^{x} \left[V(x_0) - \int_{x_0}^{\xi} q(\eta)\,d\eta \right] d\xi \tag{8.5.4a}$$

or, upon noting that $V(x_0)$ is a constant,

$$M(x) = M(x_0) + V(x_0) \cdot (x - x_0) - \int_{x_0}^{x} \left(\int_{x_0}^{\xi} q(\eta)\,d\eta \right) d\xi. \tag{8.5.4b}$$

Now, the double integral appearing on the right-hand side, may be treated as follows. Recalling the expression from the differential calculus,

$$\int u\,dv = uv - \int v\,du, \tag{8.5.5a}$$

we let

$$u = \int_{x_0}^{\xi} q(\eta)\,d\eta, \tag{8.5.5b}$$

$$dv = d\xi, \tag{8.5.5c}$$

from which

$$du = q(\xi)\,d\xi, \qquad (8.5.5d,)$$

$$v = \xi. \qquad (8.5.5e)$$

Hence, making use of Eq. (8.5.5a), we have

$$\int_{x_0}^{x}\left(\int_{x_0}^{\xi}q(\eta)\,d\eta\right)d\xi = \left[\xi\int_{x_0}^{\xi}q(\eta)\,d\eta\right]_{x_0}^{x} - \int_{x_0}^{x}\xi q(\xi)\,d\xi$$

$$= x\int_{x_0}^{x}q(\eta)\,d\eta - \int_{x_0}^{x}\xi q(\xi)\,d\xi$$

$$= \int_{x_0}^{x}(x-\xi)q(\xi)\,d\xi. \qquad (8.5.6)$$

Therefore, finally, substituting in Eq. (8.5.4b),

$$M(x) = M(x_0) + V(x_0)\cdot(x-x_0) - \int_{x_0}^{x}(x-\xi)q(\xi)\,d\xi. \qquad (8.5.7)$$

Equations (8.5.2) and (8.5.7) are thus explicit expressions for the shear $V(x)$ and moment $M(x)$ at any cross-section $x_0 < x$, provided the shear and moment at x_0 are known.

These equations lend themselves to a physical interpretation if we recall, from the calculus, that the definition of the Riemann integral is given by

$$\int_{a}^{b}f(\xi)\,d\xi = \lim_{\substack{\Delta\xi\to 0 \\ n\to\infty}}\sum_{i=1}^{n}f(\xi_i)\,\Delta\xi. \qquad (8.5.8)$$

Thus from Eq. (8.5.2), we have

$$V(x) = V(x_0) - \lim_{\substack{\Delta\xi\to 0 \\ n\to\infty}}\sum_{i=1}^{n}q(\xi_i)\,\Delta\xi, \qquad (8.5.9)$$

while from Eq. (8.5.7),

$$M(x) = M(x_0) + V(x_0)\cdot(x-x_0) - \lim_{\substack{\Delta\xi\to o \\ n\to\infty}}\sum_{i=1}^{n}(x-\xi_i)q(\xi_i)\Delta\xi. \qquad (8.5.10)$$

These sums are readily interpreted as being the sum of the individual effect of small forces, represented by infinitesimal rectangles (each located at a different ξ_i on the shear and moments [see Fig. (8.5.1a)]). Note, for example, that the quantity $(x-\xi_i)q(\xi_i)\Delta\xi$ [essentially the product of the incremental force $q(\xi_i)\Delta\xi$ located at ξ_i, times $(x-\xi_i)$, its lever arm to x] represents the contribution of this force to the moment M at x.

With this interpretation in mind, we may now immediately write down the expression for the shear and moments if, in addition to $q(x)$, a number n of concentrated forces P (positive downward) acting at x_j $(x_0 < x_j < x)$ and a number

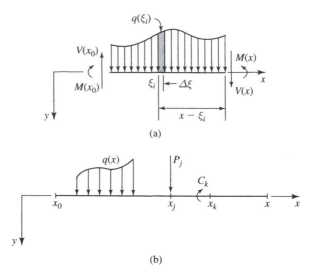

Figure 8.5.1

(a)

(b)

m of concentrated moments (couples) C_k acting at x_k ($x_0 < x_k < x$) are applied [Fig. (8.5.1b)]:

$$V(x) = V(x_0) - \int_{x_0}^{x} q(\xi)\,d\xi - \sum_{j=1}^{n} P_j, \qquad (8.5.11a)$$

$$M(x) = M(x_0) + V(x_0) \cdot (x - x_0) - \int_{x_0}^{x} (x - \xi)\, q(\xi)\,d\xi$$

$$- \sum_{j=1}^{n} P_j \cdot (x - x_j) + \sum_{k=1}^{m} C_k. \qquad (8.5.11b)$$

Finally, it is worth observing that the integral expressions are, in effect, equations that represent the equilibrium conditions of any segment of a beam, whose two end cross-sections are a finite distance apart.

We illustrate the use of Eqs. (8.5.2) and (8.5.7) by means of the following example.

Example 8.7: A simply supported beam AB of length L is subjected, as shown in Fig. (8.5.2a), to an exponentially decaying lateral load given by $q(x) = We^{-x/L}$, where W is a constant. Determine $V(x)$ and $M(x)$, draw the shear and moment diagrams and find the maximum value of $M(x)$.

Solution: Since the supports at A and B are simple supports which provide no constraint against rotation, the moment $M = 0$ at $x = 0$ and $x = L$. Furthermore, as explained in the previous examples, the shear at A represents the upward reaction R_A [Fig. (8.5.2b)].

Now, the integral relations developed above relate the shear forces and moments at any two cross-sections separated by a finite distance. Thus, since the load expression $q(x)$ is valid throughout the beam, we set $x_0 = 0$ and $x = L$ in Eq. (8.5.7); hence, this equation becomes explicitly

$$0 = 0 + R_A L - W \int_{0}^{L} (L - \xi)\, e^{-\xi/L}\,d\xi$$

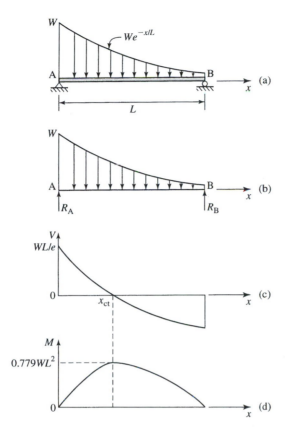

Figure 8.5.2

or

$$R_A = \frac{W}{L} \left(L \int_0^L e^{-\xi/L} \, d\xi - \int_0^L \xi \, e^{-\xi/L} \, d\xi \right)$$

$$= \frac{W}{L} \left\{ -L^2 e^{-\xi/L} \big|_0^L - L^2 \left[e^{-\xi/L} (-\xi/L - 1) \right] \big|_0^L \right\}.$$

Upon evaluating at the upper and lower limits, we obtain

$$R_A = \frac{WL}{e}.$$

The shear $V(x)$ and moment $M(x)$ can now be evaluated directly using Eqs. (8.5.2) and (8.5.7), respectively, with $x_0 = 0$ [with $V(x_0) = R_A$ as calculated above] and with x representing any cross-section; thus

$$V(x) = \frac{WL}{e} - W \int_0^x e^{-\xi/L} \, d\xi = WL \left(\frac{1}{e} + e^{-x/L} - 1 \right),$$

$$M(x) = \frac{WL}{e} x - W \int_0^x (x - \xi) e^{-\xi/L} \, d\xi = WL \left[L \left(1 - e^{-x/L} \right) + x \left(\frac{1}{e} - 1 \right) \right].$$

The shear and moment diagrams are shown in Figs. (8.5.2c and d), respectively. To find the maximum value of M, we determine the value $x = x_{cr}$ at which $V = 0$; thus from the expression for $V(x)$, $\frac{(x_{cr})}{L} = 1 - \ln(e - 1) = 0.459$

Substituting in the expression for $M(x)$, we find $M_{max} = M(x_{cr}) = 0.7794WL^2$.

$\square$

As we have seen, the use of the integral expression for beams permits us to establish expressions for $V(x)$ and $M(x)$ for any arbitrary applied force. Thus we need not know, a priori, the resultants nor the location of the resultants of the applied lateral forces.

However, Eqs. (8.5.2) and (8.5.7) have other important uses. Assume, for example, that instead of being given an analytic expression for the lateral applied loads, these loads are prescribed numerically in tabular form. The integral expressions can then be applied directly, where the integrals may be evaluated numerically by means of a computer. It is therefore evident that, given their generality, the integral expressions may prove quite useful in practice.

From our study in the above sections, we are in a position to determine the total resultants (shear forces and moments) for a beam subjected to arbitrary lateral loads, which are applied in the x–y plane. We now consider the deformation and stresses resulting from the bending of the beam.

8.6 Symmetrical bending of beams in a state of pure bending

(a) Some preliminary definitions and limitations – deformation analysis

Consider a prismatic beam, that is, one having a constant cross-section of area A, which is initially straight with a longitudinal axis lying along the x-axis [Fig. (8.6.1a)]. We note that at this stage of the analysis, the x-axis is *not necessarily* the centroidal axis.

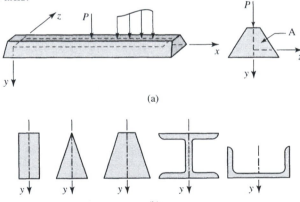

(a)

(b)

Figure 8.6.1

Since we wish here to investigate solely the effect of bending (i.e., we wish to exclude the possibility of twisting effects) we impose the following limitations in our study; that is, we restrict our analysis to cases for which

(i) The y-axis, passing through the cross-sections, is an axis of symmetry. (The x–y plane is then a plane of symmetry). For example, cross-sections of this class may be rectangular, triangular, or may be such sections as are found in engineering practice: e.g., I-sections, channel sections, etc., as shown in Fig. (8.6.1b).

(ii) All applied loads lie in the $x-y$ plane of symmetry and all applied couples act about the z-axis only. Thus, as in the previous sections, we consider beams for which the moments are $M = M_z$.

Bending that conforms to conditions of (i) and (ii) is called **symmetrical bending**. It follows, from (i) and (ii), that because of this symmetry, the beam will bend *without twisting*; thus, all points lying in the $x-y$ plane remain in this plane. However, due to the bending of the beam, the longitudinal axis of the initially straight beam assumes a curved configuration with a curvature denoted by $\kappa(x)$. As a result, all points lying in the $x-y$ plane displace in the y-direction; we refer to the y-displacements of these points as the lateral displacements (or deflections) of the beam.

In addition, at this stage, we wish to eliminate the effect of shear in the beam; we therefore impose a further limitation to our study:

(iii) We investigate the case where the beam is in a state of *pure bending*, namely $M =$ const., i.e. $M \neq M(x)$; hence, by Eq. (8.3.2), the shear $V = 0$ throughout the beam.

As a result of this last limitation, and the prismatic property of the beam, it follows that the deformation caused by the constant moment is the same at all cross-sections of the beam; thus the curvature κ of the beam does not depend on the x-coordinate. The reciprocal of the curvature κ is the radius of curvature, R, of the deformed longitudinal axis; i.e., $R = 1/\kappa$, where for pure bending, $R \neq R(x)$; i.e. $R =$ const. [Fig. (8.6.2)].

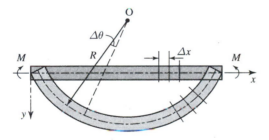

Figure 8.6.2

We now examine the deformation of the cross-sections resulting from the limitation of (iii) above. Let us assume that due to the bending, the beam deforms as shown in Fig. (8.6.3a), where the beam is viewed from the negative z-axis.

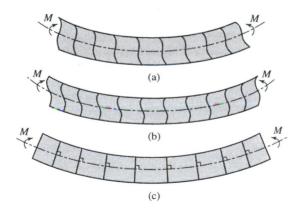

(a)

(b)

(c)

Figure 8.6.3

Note that *in a state of pure bending, all cross-sections must be assumed to deform, as shown in the figure, in exactly the same manner*. If we now consider the same beam as viewed from the positive z-axis, the deformation appears as in Fig. (8.6.3b). Now, clearly the deformed beam cannot appear both as in Fig. (8.6.3a) and Fig. (8.6.3b) simultaneously; the appearance of the deformed beam must be the same since it is subjected to the same load. Figures (8.6.3a and b) will be identical only if all cross-sections, originally plane in the undeformed state, remain plane after deformation [Fig. (8.6.3c)] (i.e., if *no warping of the cross-section* takes place) and if the plane cross-sections remain perpendicular to the deformed longitudinal axis.

Thus we reach an important conclusion: *a beam in a state of pure bending deforms in such a way that all cross-sections remain plane and remain perpendicular to the deformed longitudinal axis*. It is to be emphasised that this pattern of deformation, here, is *not* based on an assumption; *it is a conclusion* based on physical reasoning. However, it should be remembered that the arguments used in reaching this conclusion are valid only for prismatic beams in a state of pure bending.

This conclusion forms the basis for the development of expressions that relate the deformation and stresses to the forces existing in a beam.

(b) Moment–curvature relations and flexural stresses in an elastic beam under pure bending: Euler–Bernoulli relations

We consider a long slender straight beam with a (constant) cross-section symmetric with respect to the y-axis and subject to pure bending with moment $M = \text{const.}$ [Figure (8.6.2) represents the beam in the undeformed and deformed state.] The beam is made of a linear elastic material with modulus of elasticity E and Poisson ratio ν .

We now examine a typical element, originally of length Δx in its undeformed state [Fig. (8.6.4a)]. Due to the bending, the element assumes the shape shown in Fig. (8.6.4b). Since the two planes defined by the end cross-sections of the deformed element are no longer parallel, their intersection in the x–y plane must be along a line passing through some point O, the **centre of curvature** [see Fig. (8.6.4b)]. We denote the (constant) radius of curvature of the deformed longitudinal axis by R and let $\Delta\theta$ be the subtended angle between the two end cross-sections of the element after deformation. Clearly, in the deformed state, some fibres ($\widehat{mm'}$) elongate and others ($\widehat{tt'}$) shorten. There must exist, therefore, some fibres ($\widehat{nn'}$), originally lying in the plane P of the undeformed element, which neither shorten nor elongate. The x-axis is taken to lie along this plane and thus, because of symmetry, this plane is the original x–z plane [Fig. (8.6.4a)]. (Note, however, that although it is known to intersect the cross-section, its location has not yet been established.) Let N denote the intersection of the plane P, with the cross-section. After deformation, the plane P becomes a curved surface P′, with typical fibres lying along the arc $\widehat{nn'}$ of Fig. (8.6.4b); for these fibres, the extensional strain is, by definition, $\epsilon_x = 0$. The surface P', which contains fibres for which $\epsilon_x = 0$, is called the **neutral surface**. The x–y plane of the deformed element is shown in Fig. (8.6.4c).

Clearly, as previously defined, let R be the radius of curvature to the deformed $\widehat{nn'}$ fibre. Because of the symmetric nature of the cross-section and loading, the line N in the undeformed element, falls on the z-axis as shown. This line, which represents points for which $\epsilon_x = 0$, is called the **neutral axis**. The fibres m′, n′, t′ and the neutral axis, as they appear in a cross-sectional view, are shown in Fig. (8.6.4d).

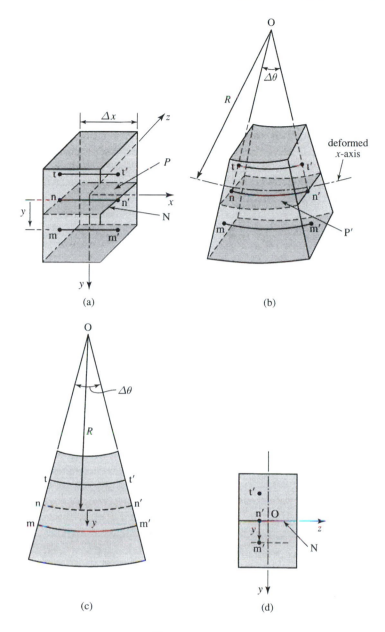

(a)

(b)

(c)

(d)

Figure 8.6.4

From geometry, the arc length $\widehat{nn'} = R\Delta\theta$. But since $\epsilon_x = 0$ for the nn' fibres,

$$R\Delta\theta = \Delta x. \tag{8.6.1a}$$

Consider now the fibres mm' located at some distance y of the undeformed element, measured perpendicularly from the neutral axis, i.e. from the line N [Figs. (8.6.4a and c)]. After deformation, the arc length $\widehat{mm'} = (R + y)\Delta\theta$. But since all fibres in the element were initially of length Δx, the strain ϵ_x in the mm' fibres is given by

$$\epsilon_x = \frac{\widehat{mm'} - \Delta x}{\Delta x} = \frac{(R + y)\Delta\theta - R\Delta\theta}{R\Delta\theta}, \tag{8.6.1b}$$

i.e.

$$\epsilon_x = \frac{y}{R}. \tag{8.6.1c}$$

From Eq. (8.6.1c), we observe that the strain ϵ_x varies with the perpendicular distance from the neutral axis. Note that Eq. (8.6.1c) describes the variation of ϵ_x in the cross-section; it is a geometric relation and in this sense, it is a kinematic equation. It is important to observe too that the above relation is valid for beams made of any material since we have not considered any material properties in the analysis. However, we now introduce the constitutive equations for the linear elastic (and isotropic) beam material, which, as we have seen in Chapter 4, are in fact Hooke's law [Eqs. (4.4.10)]:

$$\epsilon_x = \tfrac{1}{E}[\sigma_x - \nu(\sigma_y + \sigma_z)],$$

$$\epsilon_y = \tfrac{1}{E}[\sigma_y - \nu(\sigma_z + \sigma_x)], \tag{8.6.2}$$

$$\epsilon_z = \tfrac{1}{E}[\sigma_z - \nu(\sigma_x + \sigma_y)].$$

Furthermore, since shear effects have been eliminated for this case of pure bending, it follows that all shear stresses ($\tau_{xy}, \tau_{xz}, \tau_{yz}$) and strains ($\epsilon_{xy}, \epsilon_{xz}, \epsilon_{yz}$) vanish throughout the beam.

Now, the above relations are considerably simplified if we recall that we are considering long slender beams. As in our analysis of thin rods under axial loads, we first observe that $\sigma_y = 0$ at the top and bottom lateral surfaces. (For example, if the beam is of a rectangular cross-section [Fig. (8.6.5)], this is immediately evident since these top and bottom surfaces are free surfaces.) Similarly on the two lateral side surfaces, $\sigma_z = 0$. Now, we recall that a long beam is one for which the lateral dimensions are small with respect to the longitudinal dimension. Consequently, the distance between the top and bottom, and between the lateral side surfaces are relatively small. It is therefore reasonable to assume that there cannot be any great variation of σ_y and σ_z between the corresponding surfaces; i.e., σ_y and σ_z must remain relatively small. As a result of our limitation to long beams, we therefore neglect σ_y and σ_z with respect to σ_x and thus, in effect, *assume* that $\sigma_y = \sigma_z = 0$ at all points throughout the beam. (Notice that this is an assumption based on physical reasoning and not a conclusion.) Hence the stress–strain relations, Eqs. (8.6.2), become

$$\epsilon_x = \frac{1}{E}\sigma_x, \tag{8.6.3a}$$

$$\epsilon_y = -\frac{\nu}{E}\sigma_x, \tag{8.6.3b}$$

$$\epsilon_z = -\frac{\nu}{E}\sigma_x. \tag{8.6.3c}$$

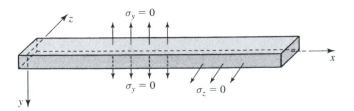

Figure 8.6.5

Combining Eq. (8.6.3a) and Eq. (8.6.1c), we obtain

$$\sigma_x = \frac{Ey}{R}. \tag{8.6.4}$$

Thus along the line N representing the nn' fibres ($y = 0$), i.e. the neutral axis, the stress $\sigma_x = 0$. From Eq. (8.6.4), we note that the stress also varies linearly with the perpendicular distance from the neutral axis. The stress σ_x is referred to as the *flexural stres* or *bending stress* since it results from bending of the beam.

The moment about the neutral axis, which we have established to coincide with the z-axis, is given by [see Eq. (2.5.7b)][†]

$$M = \iint_A y\sigma_x \, dA. \tag{8.6.5a}$$

Substituting Eq. (8.6.4),

$$M = \iint_A y\left(\frac{Ey}{R}\right) dA = \frac{E}{R} \iint_A y^2 \, dA. \tag{8.6.5b}$$

However, $I = \int_A\!\!\int y^2 \, dA$ is the second moment of the cross-sectional area *about the neutral axis*. Hence we obtain the moment–curvature relation

$$M = \frac{EI}{R} \tag{8.6.6a}$$

or since the curvature $\kappa = 1/R$, we may write

$$M = EI\kappa. \tag{8.6.6b}$$

This moment-curvature relation is known as the Euler–Bernoulli relation for elastic beams and is *always valid when M and I are taken about the neutral axis.* (Note that at this stage, we still have not yet established the location of the neutral axis since the location of the z-axis is still unknown.)

This last equation may be interpreted as the moment required to cause the beam to bend to a curvature κ; we observe that this moment is linearly proportional to κ and to the quantity EI. This latter quantity is called the **flexural rigidity**; it depends on the stiffness of the material E and on the given geometric property of the cross-section, I.

Consider now the resultant force F acting normal to the cross-section in the x-direction, and given by

$$F = \iint_A \sigma_x \, dA. \tag{8.6.7a}$$

Substituting Eq. (8.6.4) for σ_x,

$$F = \iint_A \frac{Ey}{R} \, dA = \frac{E}{R} \iint_A y \, dA. \tag{8.6.7b}$$

Since for pure bending, the resultant normal force on the cross-section must vanish, i.e. $F = 0$, it follows that $\int_A\!\!\int y \, dA = 0$, which, by definition, defines the z-axis as a centroidal axis. Thus, since the neutral axis lies on the z-axis, we have now

---

[†]   In Chapter 2 positive y-direction was taken upward, while here positive y-direction is downward. This accounts for the difference in sign.

established that *the neutral axis must always pass through the centroid of the cross-section*. Hence fibers lying initially along the x-centroidal longitudinal axis undergo no extension (or contraction). The longitudinal axis of a beam passing through the centroids, when subjected to flexure, is therefore said to undergo 'inextensional deformation'.

Combining Eq. (8.6.4) and the moment-curvature relation, Eq. (8.6.6a), the expression for the flexural stress becomes

$$\sigma_x(y) = \frac{My}{I}. \tag{8.6.8}$$

Because the neutral axis coincides with the z-axis in the case of pure symmetric bending, we shall rewrite Eqs. (8.6.6a) and (8.6.8) explicitly as

$$M_z = \frac{E I_{zz}}{R}, \tag{8.6.9a}$$

$$\sigma_x(y) = \frac{M_z y}{I_{zz}} \tag{8.6.9b}$$

with the clear understanding that *although we have prescribed I_{zz} and M_z as being about the z-axis, we are in fact taking these quantities about the neutral axis*.

The above expressions for the flexural stress, Eq. (8.6.9b), and the Euler-Bernoulli relation, Eq. (8.6.6), are basic relations that govern the flexure of beams under pure bending.

Substituting Eq. (8.6.9b) in Eqs. (8.6.3), we obtain the simple expressions for the strains, namely

$$\epsilon_x = \frac{M_z y}{E I_{zz}}, \tag{8.6.9c}$$

$$\epsilon_y = -\nu \frac{M_z y}{E I_{zz}}, \tag{8.6.9d}$$

$$\epsilon_z = -\nu \frac{M_z y}{E I_{zz}}. \tag{8.6.9e}$$

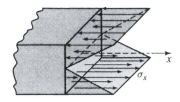

Figure 8.6.6

The linear variation in the cross-section of the flexural stress σ_x with y is shown in Fig. (8.6.6). We observe that the maximum flexural stress occurs at the points farthest from the neutral axis. Note, however, that since we have made use of Eqs. (8.6.2), the derived expressions are valid only for linear elastic behaviour of the beam, provided that the flexural stress is less than the proportional limit, i.e. $|\sigma_x| \le (\sigma_x)_p$.

Returning now to the expression for the flexural stress, it should be noted that for $M > 0$, $\sigma_x > 0$ (tension) for $y > 0$, and $\sigma_x < 0$ (compression) for $y < 0$. (The reverse is clearly true for $M < 0$.) While we observed that the expression is valid for a flexure stress below the elastic and proportional limit, it is of interest to note that it is *independent of any material constants*. Thus, for a given M, σ_x is the same for all linearly elastic beams having the same cross-section, irrespective of the material.

Finally, since, as previously noted, the magnitude of σ_x is greatest at the point farthest from the neutral axis, we may write

$$|\sigma_x|_{\max} = \frac{|M_z y_{\max}|}{I_{zz}} = \frac{|M_z c|}{I_{zz}}, \tag{8.6.10a}$$

where $c = |y|_{\max}$.

Alternatively,

$$|M_z| = \frac{|\sigma_x|_{max}\, I_{zz}}{c}.$$ (8.6.10b)

This leads us to the definition of the *section modulus S*, given as

$$S = \frac{I_{zz}}{c}$$ (8.6.10c)

from which we have

$$|\sigma_x|_{max} = \frac{|M_z|}{S}$$ (8.6.10d)

or

$$|M_z| = S\,|\sigma_x|_{max}.$$ (8.6.10e)

Note that the section modulus S is a geometric property of the cross-section (whose dimensions are m^3); this section property is found to be very useful in the engineering design of beams.

The subscripts z were used here to specify clearly that the quantities are to be taken about the z-axis. However, since we are considering only moments about this axis, we shall, in general, hereafter adopt (with the exception of Section 12 below) the simplified notation, M and I, with the clear understanding that $M \equiv M_z$ and $I \equiv I_{zz}$.

We illustrate two interesting applications of the above relations by means of the following examples.

Example 8.8:   A beam of rectangular cross-section, as shown in Fig. (8.6.7a), is made of steel, with $E = 200$ GPa. The maximum permissible stress σ_x is given as $\sigma_x = 120$ MPa. (a) Determine the maximum permissible moment that the cross-section can withstand. (b) What is the radius of curvature R of the beam if all cross-sections are subjected to the same moment.

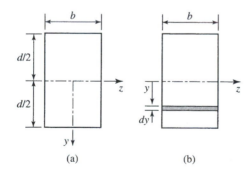

(a)   (b)   **Figure 8.6.7**

Solution:

(a) For this cross-section, the second moment of the area $I \equiv I_{zz}$ is calculated as follows [Fig. (8.6.7b)]:

$$I = \iint_A y^2\, \mathrm{d}A = b \int_{-d/2}^{d/2} y^2\, \mathrm{d}y = b\left[\frac{y^3}{3}\right]_{-d/2}^{d/2} = \frac{bd^3}{12}.$$ (8.6.11)

With $b = 2$ cm and $d = 6$ cm, $I = \frac{2(6)^3}{12} = 36$ cm^4 $= 36 \times 10^{-8}$ m^4. Therefore,

$$M_{max} = \frac{(\sigma_x)_{max} I}{y_{max}} = \frac{(120 \times 10^6)(36 \times 10^{-8})}{3 \times 10^{-2}} = 1440 \text{ N-m}.$$

(b) From the Euler-Bernoulli relation,

$$R = \frac{EI}{M} = \frac{(200 \times 10^9)(36 \times 10^{-8})}{14.4 \times 10^2} = 50.0 \text{ m}.$$

It is worthwhile noting that the radius of curvature of this relatively stiff (steel) beam is very large.   □

Example 8.9:   A thick cable is composed of individual strands of copper wire each of diameter $d = 1.0$ mm and behaves as an ideal elastic–plastic material. The originally straight cable is to be wound about a spool of diameter D [Fig. (8.6.8)]. Determine the smallest diameter D of the spool such that the stress in the strands should not exceed σ_0. (Note that this requirement is necessary if the wire is to be straight after it unwinds.)

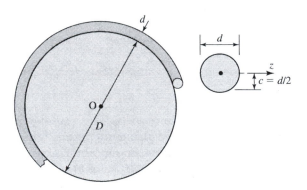

Figure 8.6.8

Solution: From Eq. (8.6.10b) or (8.6.10e),

$$M = \frac{\sigma_0 I}{d/2}$$

is the maximum moment that the wires can sustain and still remain elastic. (Note that the section modulus is $S = \frac{I}{d/2}$.)

The moment in the strands, as a function of curvature is given, according to the moment-curvature relation, Eq. (8.6.6), by

$$M = \frac{EI}{R} = \frac{EI}{D/2}.$$

Equating the above two expressions,

$$D = \frac{E}{\sigma_0} d.$$

Using typical values for copper ($\sigma_0 = 80$ MPa, $E = 100$ GPa),

$$D = \frac{(100 \times 10^9)(1.0 \times 10^{-3})}{80 \times 10^6} = 1.25 \text{ m}.$$

It should be observed that for a relatively stiff material (i.e., large E), one requires a relatively large diameter spool; increasing d has the same effect.   □

Example 8.10:   A simply supported elastic beam AB, having an arbitrary symmetric cross-section and whose flexural rigidity is EI, is subjected to end couples M at each end, as shown in Fig. (8.6.9a). The depth of the beam is given as d. Determine (a) the lateral displacement Δ at the mid-point [Fig. (8.6.9b)] for any given applied end couples M, assuming elastic behaviour, (b) the largest moment M_E for which the beam remains elastic. Assume that the maximum stress for elastic behaviour is $\sigma_x = \sigma_0$ [Fig. (4.6.1)] and (c) the lateral displacement Δ at the mid-point of the beam when subjected to the end couples $M = M_E$.

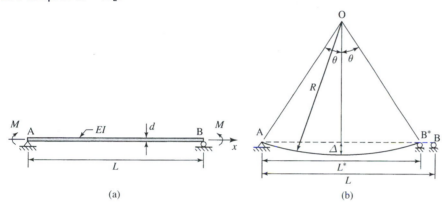

Figure 8.6.9

Solution:

(a) For the given applied end couples M, the moment at any cross-section will be M and thus the beam is in a state of pure bending with constant radius of curvature R. Therefore, the shape of the beam in the deformed state will be a portion of an arc of a circle whose subtended angle we denote as 2θ. Moreover, as we have seen, the strain of fibres originally along the x-axis is $\epsilon_x = 0$; i.e., the fibres AB are said to be **inextensional**. Hence, the length AB* along the arc is $L = 2R\theta$ or $\theta = \frac{L}{2R}$ [Fig. (8.6.9b)].

Now, from geometry,

$$\Delta = R(1 - \cos\theta) \tag{8.6.12a}$$

or

$$\Delta = R\left[1 - \cos\left(\frac{L}{2R}\right)\right]. \tag{8.6.12b}$$

Substituting from the general Euler-Bernoulli relation $R = \frac{EI}{M}$,

$$\Delta = \frac{EI}{M}\left[1 - \cos\left(\frac{ML}{2EI}\right)\right]. \tag{8.6.13}$$

We note that this is an 'exact' result in accordance with our theory. We shall use this result later in evaluating the accuracy of a linearised theory for deflections of beams.

(It is worthwhile to evaluate the displacement Δ numerically in order to establish an order of magnitude of Δ under typical conditions. For example, let us assume, for simplicity, that the beam is of length $L = 2$ m and is made of aluminium, with $E_A = 70$ GPa and $\sigma_0 = 50$ MPa. Furthermore, let us assume that the beam is of rectangular cross-section with $b = 2$ cm and $d = 2c = 6$ cm and that end moments $M = 300$ N-m are applied. Then the displacement Δ, given by Eq. (8.6.13), is $\Delta = 0.595$ cm. Note that the ratio $\Delta/L = 2.976 \times 10^{-3}$, which is indeed very small. Such small displacements are typical of many beams encountered in engineering practice).

(b) The beam will behave elastically provided $|\sigma_x| \leq \sigma_0$. Therefore, the maximum elastic moment $M = M_E$ that can be applied, according to Eq. (8.6.10b), is

$$M_E = \frac{\sigma_0 I}{c} = \frac{\sigma_0 I}{\alpha d}, \quad 1/2 \leq \alpha < 1. \tag{8.6.14}$$

since, for any arbitrary symmetric cross-section of depth d, c can always be expressed as $c = \alpha d$, where $1/2 \leq \alpha < 1$. [For the case of a rectangular cross-section, $\alpha = 1/2$. Hence for the numerical values given previously in part (a), $M_E = 600$ N-m. Therefore, the behaviour of the beam, when subjected to end couples $M = 300$ N-m, as in the numerical case of part (a), is seen to be within the elastic range.]

(c) Substituting Eq. (8.6.14) in Eq. (8.6.13), the maximum deflection Δ due to M_E becomes

$$\Delta = \frac{E\alpha d}{\sigma_0}\left[1 - \cos\left(\frac{\sigma_0}{E}\frac{L}{2\alpha d}\right)\right] \tag{8.6.15}$$

It should be noted that Eq. (8.6.15) represents the largest possible mid-span deflection of the beam in the elastic range. It is of interest to observe that this maximum deflection depends on the stiffness of the material (through the ratio E/σ_0) and on the depth of the beam (through the ratio L/d) but *not* on the specific shape of the cross-section.

Again, we observe that this is an exact result consistent with Euler-Bernoulli beam theory.

Substituting the same numerical values as above in Eq. (8.6.15), we obtain for the rectangular cross-section (using $\alpha = 1/2$), the maximum elastic deflection at the mid-span, $\Delta_{max} = 1.190$ cm, and hence $\Delta/L = 5.95 \times 10^{-3}$. □

◊(c) Axial displacements of beams under pure bending

As noted previously, in a beam under flexure, fibres lying along the neutral surface undergo no extension. It is of interest to determine the axial displacement of such fibres in a beam under pure bending and, in particular, the axial displacement at the end points. Let us therefore consider the beam of Example 8.10 where we observed, from Fig. (8.6.9b), that point B has moved to B*. We wish to determine the displacement $\Delta_x \equiv \overline{BB^*}$ and, specifically, the ratios Δ_x/L and Δ_x/Δ. From simple geometry, $\Delta_x = L - 2R\sin\theta$. Since $R = EI/M$ and $\theta = L/2R$, we obtain

$$\Delta_x = L - \frac{2EI}{M}\sin\left(\frac{ML}{2EI}\right) \tag{8.6.16a}$$

or

$$\Delta_x = L\left[1 - \frac{2EI}{ML}\sin\left(\frac{ML}{2EI}\right)\right]. \tag{8.6.16b}$$

Clearly, Δ_x increases with increasing moment M and therefore reaches its maximum value under elastic behaviour when $M = M_E$ as given by Eq. (8.6.14). Therefore, under this moment,

$$\Delta_x = L\left[1 - \frac{2E\alpha d}{\sigma_0 L} \sin\left(\frac{\sigma_0 L}{2E\alpha d}\right)\right]. \tag{8.6.17a}$$

Letting $\gamma = \frac{\sigma_0 L}{E\alpha d}$, we may write

$$\Delta_x = L\left[1 - \frac{2}{\gamma} \sin\frac{\gamma}{2}\right]. \tag{8.6.17b}$$

Similarly from Eq. (8.6.15),

$$\Delta = \frac{L}{\gamma}\left[1 - \cos\frac{\gamma}{2}\right]. \tag{8.6.18}$$

Now, for most engineering materials within the elastic range, $\sigma_0/E = O(10^{-3})$, and consequently γ is an infinitesimal, i.e. $\gamma \ll 1$. Using the series representations for sin and cos, we therefore have

$$\Delta_x = L\left[1 - \frac{2}{\gamma}\left(\frac{\gamma}{2} - \frac{\gamma^3}{48} + \cdots\right)\right] \approx \frac{\gamma^2 L}{24} \tag{8.6.19a}$$

and

$$\Delta = \frac{L}{\gamma}\left[1 - \left(1 - \frac{\gamma^2}{8} + \cdots\right)\right] \approx \frac{\gamma L}{8} \tag{8.6.19b}$$

from which

$$\frac{\Delta_x}{L} = \frac{\gamma^2}{24} \tag{8.6.20a}$$

and

$$\frac{\Delta_x}{\Delta} = \frac{\gamma}{3}. \tag{8.6.20b}$$

Letting $\alpha = 1/2$, and using $\sigma_0/E = 10^{-3}$, we obtain, say for $L/d = 20$, the upper bounds

$$\frac{\Delta_x}{L} = 1.7 \times 10^{-6} \quad \text{and} \quad \frac{\Delta_x}{\Delta} = 6.7 \times 10^{-3}.$$

Similarly for $L/d = 100$,

$$\frac{\Delta_x}{L} = 4.2 \times 10^{-4} \quad \text{and} \quad \frac{\Delta_x}{\Delta} = 3.3 \times 10^{-2}.$$

Noting that the ratio Δ_x/L is an infinitesimal of second order, we thus observe that the projected length L^* on the x-axis is $L^* = L(1 - \Delta_x/L) \approx L$. Hence, consistent with a linearised theory, we make no distinction between L^* and L and therefore use the original length L, for example, in writing the equations of equilibrium. [See comment (e) of Section 3 in Chapter 6.] While the above has been shown to be true for the specific example of pure bending considered here, a more general expression, derived later in Chapter 9, yields the same result for cases other than pure bending.

(d) Comments on the solution – exactness of the solution

We first collect together the results obtained above for the elastic beam in a state of pure symmetric bending:

$$M = \frac{EI}{R}, \qquad \sigma_x = \frac{M_z y}{I_{zz}},$$

$$\sigma_y = \sigma_z = 0, \qquad \tau_{xy} = \tau_{yz} = \tau_{zx} = 0, \qquad (8.6.21)$$

$$\epsilon_x = \frac{M_z y}{EI}, \qquad \epsilon_y = \epsilon_z = -\nu \frac{M_z y}{EI}, \qquad \epsilon_{xy} = \epsilon_{yz} = \epsilon_{zx} = 0.$$

Now, clearly, the above quantities satisfy the stress–strain relations. Furthermore, if we substitute the stresses in the equations of equilibrium, Eqs. (2.4.4) (with all body forces $\boldsymbol{B} = \boldsymbol{0}$), we find that these equations are satisfied identically. In fact, all the relevant equations (called **the equations of linear elasticity**) are satisfied. Hence we may conclude that the solution obtained is 'exact' within the context of linear elastic theory.

(e) Methodology of solution – the methodology of mechanics of materials

It is appropriate, at this point, to reflect on the methodology that has been followed to derive the above relations. Indeed, as may already be evident, the methodology used in deriving the relations for pure bending of a beam is precisely the same as that used previously in the derivations of the relations for axial deformation and for torsion.

We recall that, in general, it is necessary to satisfy three sets of equations: (a) the geometric (kinematic) relations of deformation, (b) the stress–strain relations and (c) the equations of equilibrium.

In considering axial behaviour, torsion and flexure, respectively, the investigations proceeded basically along the same five steps:

(a) The fundamental first step was to establish a deformation pattern, based either on a physically plausible assumption (as in the case of axial deformation) or a physical conclusion (as in the case of torsion and bending).

(b) From the deformation pattern, the variation of strain in the cross-section was obtained from simple geometric considerations; that is, from strain–displacement relations. These are essentially geometric relations, which define the strain in terms of a global deformation quantity of the cross-section, namely $\epsilon_x = $ const., θ or $1/R$ (for axial, torsional and flexural behaviour respectively).

(c) Then, making some reasonable assumptions – again based on plausible physical reasoning – on the stresses, we introduced the stress–strain relations; this then yielded the variation of stress in the cross-section in terms of the global deformation quantity.

(d) Upon using equations of mechanics, namely the relation between stresses in the cross-section and the 'stress resultant' (F, T, M, respectively), the global deformation of the cross-section was expressed in terms of the stress resultant.

(e) Upon substituting back in (c), we obtained explicit expressions for the variation of stresses in terms of the stress resultants and a geometric property of the cross-section.

This methodology is shown clearly in the following block diagram for the three types of phenomena previously considered.

Strain–displacement relations		Stress–strain Hooke's law	Eqs. of mechanics	Back substitution
(a)	(b)	(c)	(d)	(e)
Deformation pattern	Resulting strain distribution	Resulting stress distribution	Deformation in terms of stress result	Stresses in terms of stress result
Cross-sections remain plane	$\epsilon_x = \epsilon_x(x),$ $\gamma_{x\theta} = r\dfrac{d\phi}{dx} \equiv r\Theta$ $\epsilon_x = \dfrac{y}{R}$	$\sigma_x = E\epsilon_x$ $\tau_{x\theta} = G\Theta r$ $\sigma_x = \dfrac{Ey}{R}$	$\epsilon_x = \dfrac{F}{EA}$ $\Theta = \dfrac{T}{GJ}$ $\dfrac{1}{R} = \dfrac{M}{EI}$	$\sigma_x = F/A$ $\tau_{x\theta} = \dfrac{Tr}{J}$ $\sigma_x = \dfrac{My}{I}$

We observe that the quantities EA, GJ, EI are analogous; namely, they represent the axial, torsional and flexural rigidity of a member, respectively. In each case the rigidity is a product of a material property (representing stiffness of the material) and a geometric property of the cross-section.

8.7 Flexure of beams due to applied lateral loads – Navier's hypothesis

In our previous discussion, we developed the expressions for flexure of beams that are in a state of pure bending, $M \neq M(x)$; in particular, we obtained the Euler-Bernoulli relations and expressions for the flexural stress,

$$M = \frac{EI}{R}, \qquad \sigma_x(y) = \frac{My}{I},$$

which we have seen are 'exact' solutions. It is important to re-emphasise that these expressions were derived based on the fundamental conclusion: plane sections remain plane and perpendicular to the deformed longitudinal axis.

Consider now a beam subjected to typical arbitrary transverse loads acting in the x–y plane of symmetry [Fig. (8.7.1a)]. In this case, the moment $M = M(x)$ and $V(x) \neq 0$ [Figs. (8.7.1b and c)] and thus we no longer have a state of pure bending. Consequently, we can no longer conclude rigorously from the fundamental arguments of symmetry that plane cross-sections remain plane in the deformed state. However, careful experiments performed in a laboratory show that, although some warping of the cross-section does take place, such warping is extremely small for long slender beams and indeed is quite negligible.

We therefore make the following important assumptions: we assume (a) that even under such loading conditions, plane sections still remain plane after deformation and that they remain perpendicular to the deformed longitudinal axis. This assumption is called **Navier's hypothesis**. Furthermore, we shall assume (b) that the lateral stresses σ_y and σ_z are small compared to σ_x.

If we accept these assumptions, we again arrive at both Eqs. (8.6.1c) and (8.6.4). Then, proceeding from this point in the derivation, we consequently obtain the same Euler-Bernoulli relation and the same expression for the flexural stresses σ_x.

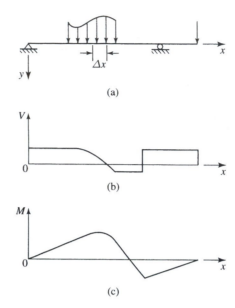

(a)

(b)

(c)

Figure 8.7.1

However, since $M = M(x)$, we then have

$$M_z(x) = \frac{EI}{R(x)} \tag{8.7.1a}$$

and

$$\sigma_x(x, y) = \frac{M(x)y}{I}. \tag{8.7.1b}$$

Nevertheless, we recognise that the above relations, Eqs. (8.7.1a) and (8.7.1b), are not exact; they are merely very good approximations, which yield errors of a very small percentage. Consequently, they are widely used in engineering analyses of beams since they provide very accurate solutions to a wide variety of engineering problems. [Equations (8.7.1) are often referred to as *engineering beam formulas*.]

We shall, however, show, at a later stage, that the expression for the flexural stress is exact not only for $M = $ const. but also when M is a linear function of x.

Example 8.11: A wooden member of length $L = 3$ m, having a rectangular cross-section 3 cm $\times$ 6 cm, is to be used as a cantilever beam with a load $P = 240$ N acting at the free end [Figs. (8.7.2a and b)]. Can the member carry this load if the allowable flexural stress, both in tension and in compression, is $(\sigma_x)_{\text{all}} = 50$ MPa?

Solution: The moment $M \equiv M_z$ at any cross-section is given as $M(x) = -Px$ [Fig. (8.7.2c)].

From Eq. (8.6.11), the second moment of the area about the z-axis is $I_{zz} = \frac{bd^3}{12}$, where b and d are the dimensions in the z- and y-directions, respectively. Substituting in Eq. (8.7.1b), the stress in any cross-section of the bottom fibres A [located at $y = d/2$ of Fig. (8.7.2b)] is

$$\sigma_x = \frac{My}{I} = \frac{M(x)(d/2)}{bd^3/12} = \frac{6M(x)}{bd^2}.$$

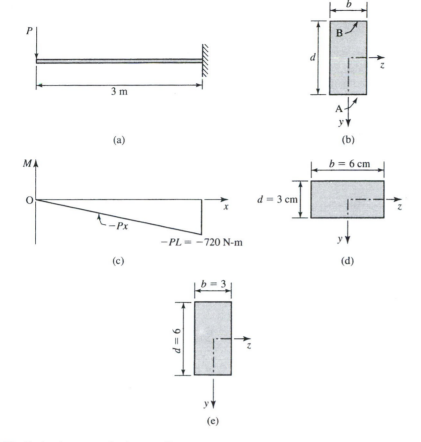

Figure 8.7.2

Similarly, the stress in the top fibres B (with $y = -d/2$) is [Fig. (8.7.2b)]

$$\sigma_x = \frac{My}{I} = \frac{M(x)(-d/2)}{bd^3/12} = -\frac{6M(x)}{bd^2}.$$

The largest (absolute) value of the stresses will occur at the fixed end, $x = L$, where $M(L) = -PL$. Therefore, at $x = L$,

$$\sigma_x|_A = -\frac{6PL}{bd^2} \quad \text{and} \quad \sigma_x|_B = \frac{6PL}{bd^2}.$$

We observe that at A the fibres are in compression while at B they are in tension. It is also worthwhile to note that the stresses are inversely proportional to the square of the depth of the section. This implies immediately that in using such a member, it is preferable to use the beam oriented as shown in Fig. (8.7.2e), rather than as in Fig. (8.7.2d).

If the beam is oriented as in Fig. (8.7.2d), with $b = 6$ cm and $d = 3$ cm, the allowable load is

$$P_{\text{all}} = \frac{(\sigma_x)_{\text{all}} bd^2}{6L} = \frac{(50 \times 10^6)(6 \times 10^{-2})(3 \times 10^{-2})^2}{6 \times 3} = 150 \text{ N}.$$

For a beam oriented as in Fig. (8.7.2e), with $b = 3$ cm and $d = 6$ cm,

$$P_{\text{all}} = \frac{(\sigma_x)_{\text{all}} bd^2}{6L} = \frac{(50 \times 10^6)(3 \times 10^{-2})(6 \times 10^{-2})^2}{6 \times 3} = 300 \text{ N}.$$

Thus the beam can carry the given load $P = 240$ N only when oriented as in Fig. (8.7.2e).　□

Example 8.12:　A simply supported beam of length $L = 4$ m is subjected to a uniform load $w = 400$ N/m over its entire span [Fig. (8.7.3a)]. The cross-section of the beam is made of two pieces of wood, which are glued together by a strong adhesive so as to form a monolithic T-section, as shown in Fig. (8.7.3b). (The allowable tension and compressive stress of the wood is given as $\sigma_{all} = 25$ MPa.) Determine the largest compressive and tensile flexural stress in the beam, assuming the beam behaves elastically.

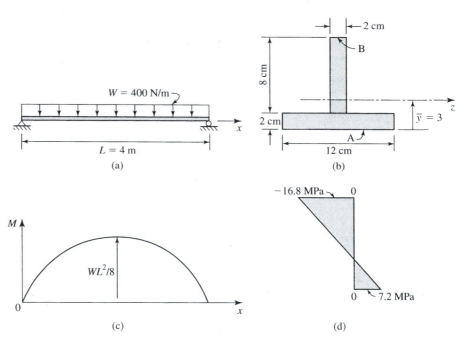

Figure 8.7.3

Solution:　In Example 8.1, the moment $M(x)$ in the beam was determined. From the moment diagram, shown again in Fig. (8.7.3c), the maximum moment $M = \frac{wL^2}{8} = \frac{(400)(4)^2}{8} = 800$ N-m.

Since the terms in the expression for the flexural stress, $\sigma_x = M_z y / I_{zz}$, are taken about the centroidal z-axis of the cross-section, it is first necessary to locate the centroidal axis. Letting $\bar{y}$ be the distance of the centroidal z-axis measured from the bottom of the section, and considering the T-section as a combination of the two rectangular component elements, we have

$$\bar{y} = \frac{[(2 \times 12)(1) + (2 \times 8)(6)]}{(2 \times 12) + (2 \times 8)} = \frac{120}{40} = 3 \text{ cm}.$$

Using the parallel axis theorem (see Appendix A.1) (and recalling that for each of the rectangular components, the second moment of the cross-sectional area about its own centroidal z-axis, $I_{zz} = bd^3/12$ [see Eq. (8.6.11)]), we calculate $I \equiv I_{zz}$ of the entire section:

$$I = \left[\frac{12 \times 2^3}{12} + 24 \times 2^2\right] + \left[\frac{2 \times 8^3}{12} + 16 \times 3^2\right] = \frac{1000}{3} \text{ cm}^4 = \frac{10^{-5}}{3} \text{ m}^4.$$

The stress at the bottom fibres (A) of the section is then

$$\sigma_x|_A = \frac{My_A}{I} = \frac{(800)(3 \times 10^{-2})}{1/3 \times 10^{-5}} = 72.0 \times 10^5 \, \text{N/m}^2 = 7.2 \, \text{MPa}.$$

Similarly, the stress at the top fibres (B) of the section is

$$\sigma_x|_B = \frac{My_B}{I} = \frac{(800)(-7 \times 10^{-2})}{1/3x10^{-5}} = -168.0 \times 10^5 \, \text{N/m}^2 = -16.8 \, \text{MPa}.$$

Note here that the stress σ_A is a tensile stress and σ_B is compressive. The variation (with y) of σ_x in the cross-section is shown in Fig. (8.7.3d). Since the magnitudes of both $\sigma_x|_A$ and $\sigma_x|_B$ are less than the prescribed $(\sigma_0)_{\text{all}} = 25$ MPa, the beam is in the elastic range. □

As we have seen, if $M = M(x)$, then from Eq. (8.3.2) is clear that $V(x) \neq 0$; that is, a shear force exists, in general, on any given cross-section. It necessarily follows that shear stresses must exist on these cross-sections. We derive expressions for these shear stresses in the following section.

8.8 Shear stresses in beams due to symmetric bending

(a) Derivation

We consider a prismatic elastic beam having flexural rigidity EI and subjected to lateral loads such that $M = M(x)$ [Fig. (8.7.1)], where, at any cross-section, the flexural stresses are given by Eq. (8.7.1b). Now since $M = M(x)$, the stresses vary from section to section, i.e. $\sigma_x = \sigma_x(x, y)$. Let us consider an element of the beam of width Δx, which we isolate as a free body, as shown in Fig. (8.8.1). If, for example, $M(x)$ varies positively with x, then for any given y, $\sigma_x(x) < \sigma_x(x + \Delta x)$. We note moreover that since the resultant thrust F on each of the two end cross-sections of the element vanishes, the element is in equilibrium in the x-direction.

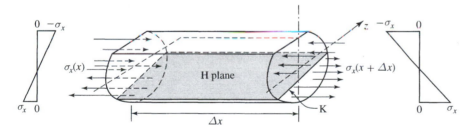

Figure 8.8.1

Let us make a cut of the element by means of an *arbitrary* plane H (whose normal n is perpendicular to the x-axis), as shown in Fig. (8.8.2a), and let b denote the length of the cut (the line K) in the cross-section [Fig. (8.8.2b)]. Note that the area of this plane is then $b \times \Delta x$.

We now isolate the two portions of the element and consider them as free bodies. We choose arbitrarily to examine the equilibrium of the bottom part, and denote the area of the cross-section of this bottom portion by $\overline{A}$ [Fig. (8.8.2a)]. Now, clearly, this isolated portion of the element cannot be in equilibrium in the x-direction under the two flexural stresses $\sigma_x(x)$ and $\sigma_x(x + \Delta x)$ alone. However, we recall that whenever we make a cut in a body, we must consider the stresses that act upon this cut. In particular, shear stresses τ_{nx} exist that act tangentially along the plane H. Let $\tau \equiv \tau_{nx}$ denote the *average shear stress*, which acts in the plane H, where we

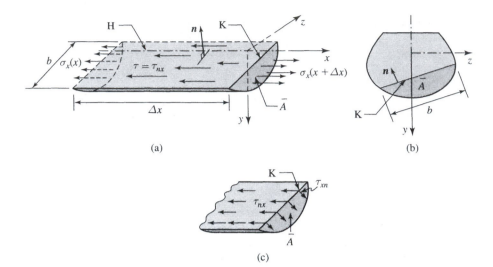

Figure 8.8.2

observe, we have taken *positive τ to be acting toward the left* [Fig. (8.8.2a)]. (Note that at this point we have abandoned the standard sign convention for stresses since we have taken positive τ_{nx} to act in the negative x-direction.)

We now wish to satisfy equilibrium in the x-direction. Taking $\overset{\rightarrow}{+} \sum F_x = 0$, we have

$$\iint\limits_{\overline{A}} \sigma_x(x + \Delta x, y)\, \mathrm{d}A - \iint\limits_{\overline{A}} \sigma_x(x, y)\, \mathrm{d}A - \tau(b\Delta x) = 0 \qquad (8.8.1\text{a})$$

or

$$b\tau = \frac{1}{\Delta x} \left[\iint\limits_{\overline{A}} \sigma_x(x + \Delta x, y)\, \mathrm{d}A - \iint\limits_{\overline{A}} \sigma_x(x, y)\, \mathrm{d}A \right]. \qquad (8.8.1\text{b})$$

But, by Eq. (8.7.1b), $\sigma_x(x, y) = \frac{M(x)y}{I}$. Substituting in the above and noting that $M(x)$ and I are not functions of $\mathrm{d}A$, we have

$$\tau = \frac{1}{I}\frac{1}{\Delta x} \left[M(x + \Delta x) \iint\limits_{\overline{A}} y\, \mathrm{d}A - M(x) \iint\limits_{\overline{A}} y\, \mathrm{d}A \right] \qquad (8.8.1\text{c})$$

or recombining

$$\tau = \frac{1}{Ib}\frac{[M(x + \Delta x) - M(x)]}{\Delta x} \iint\limits_{\overline{A}} y\, \mathrm{d}A. \qquad (8.8.1\text{d})$$

Taking the limit as $\Delta x \to 0$

$$\tau = \frac{1}{Ib} \lim_{\Delta x \to 0} \left(\frac{[M(x + \Delta x) - M(x)]}{\Delta x} \right) \iint\limits_{\overline{A}} y\, \mathrm{d}A = \frac{1}{Ib}\frac{\mathrm{d}M(x)}{\mathrm{d}x} \iint\limits_{\overline{A}} y\, \mathrm{d}A$$

$$(8.8.1\text{e})$$

since, by definition, the limiting process yields the derivative.

Finally, using Eq. (8.3.2), $dM(x)/dx = V$, we have

$$\tau = \frac{V}{Ib} \iint\limits_{\overline{A}} y \, dA. \qquad (8.8.2)$$

We recall now that the above integral is, by definition, the (first) moment of the area $\overline{A}$ about the z-axis; we denote this by the symbol $Q \equiv Q_z$, that is, we let

$$Q = \iint\limits_{\overline{A}} y \, dA. \qquad (8.8.3)$$

Hence we finally obtain

$$\tau = \frac{VQ}{Ib}. \qquad (8.8.4a)$$

For emphasis, recalling that $V \equiv V_y$, $I \equiv I_{zz}$ and $Q \equiv Q_z$, we rewrite Eq. (8.8.4a) explicitly as

$$\tau = \frac{V_y Q_z}{I_{zz} b}. \qquad (8.8.4b)$$

Now we recall that in the above derivation, $\tau \equiv \tau_{nx}$ was defined as the average shear stress acting on the plane H of area $b \times \Delta x$. However, in the process of the derivation, we have taken the limit as $\Delta x \rightarrow 0$ [see Eq. (8.8.1e)]. Hence τ no longer represents the average stress on the plane H but rather yields the *average shear stress* existing at points along K, the line of the cut. However, from the equality of the conjugate shear stresses, it is also true that $\tau \equiv \tau_{nx} = \tau_{xn}$ [Fig.(8.8.2c)]. This latter term defines the shear stress acting on the plane of the cross-section. Consequently, we conclude that the expression for τ, given by Eqs. (8.8.4), represents *the average shear stress in the cross-section that acts at points along K, the line of the cut of length b, and that acts perpendicular to this line.*

Since the usual sign convention for stresses was abandoned, it is necessary to define what we mean by a positive (and negative) sign for τ. We recall that τ, as it appears in the derivation, was originally taken to be acting to the left in Fig. (8.8.2a). Hence *a positive value of τ, acting on the cross-section (having a positive x-face), as calculated by Eq. (8.8.4), represents a stress that is directed inward toward the area of the isolated portion represented by the area $\overline{A}$. Similarly, a negative value represents a stress component directed out of this area $\overline{A}$.*

We illustrate the use of Eq. (8.8.4) by means of the following examples.

Example 8.13: A beam of rectangular cross-section of width b and depth $d = 2c$, as shown in Fig. (8.8.3a), is subjected to a positive shear force $V \equiv V_y$. Determine the shear stress distribution for τ along any line D–D, which is parallel to the z-axis; i.e., determine $\tau = \tau(y)$.

Solution: Since we are interested in τ along D–D, we imagine that we make a cut along this line and isolate the two portions [Fig. (8.8.3b)]; we choose to apply Eq. (8.8.4) to the lower portion. (Note that we have arbitrarily chosen the lower portion; we may equally choose the upper portion – the final physical result must clearly be

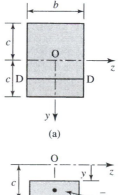

(a)

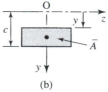

(b)

Figure 8.8.3

the same.) Thus we have

$$\tau = \frac{V Q(y)}{I b};$$

(8.8.5)

that is, the y-dependency is a function of $Q(y)$ alone.

Now, because of the simple geometry of the area $\overline{A}$, we need not use here the integral expression, Eq. (8.8.3), to calculate Q, which represents the moment of this area; instead we obtain the moment of $\overline{A}$ as follows:

$$Q = [b(c - y)] \cdot \left[y + \frac{1}{2}(c - y) \right] = \frac{b}{2}(c^2 - y^2).$$

Noting that $I = \frac{bd^3}{12} = \frac{2bc^3}{3}$, we obtain

$$\tau(y) = \frac{3}{4}\frac{V}{bc^3}(c^2 - y^2).$$

(8.8.6a)

Finally, since the cross-sectional area $A = 2bc$,

$$\tau(y) = \frac{3}{2}\frac{V}{A}\left(1 - \frac{y^2}{c^2}\right).$$

(8.8.6b)

From Eqs. (8.8.6), we make several observations:

(i) At $y = \pm c$, $\tau = 0$; that is, the average shear stress vanishes at points along the top and bottom of the cross-section. Now, since the shear stress τ acts on the plane of the cross-section, the condition $\tau(y = \pm c) = 0$ is a physical necessity; if this were not so, we would violate the physical boundary conditions for the beam. It is worthwhile to explain this point in some detail. Assume, for a moment, that $\tau(y = \pm c)$ were not equal to zero. Then recalling that $\tau \equiv \tau_{xn} = \tau_{nx}$, we would then have $\tau_{nx}(y = \pm c) \neq 0$. Now τ_{nx} at $y = \pm c$ represents the shear stress on the bottom and top lateral surfaces, respectively [Fig. (8.8.4)]. Clearly, the shear stresses on these two 'free' surfaces must vanish. Thus the physical condition leads to the required boundary condition $\tau(y = \pm c) = 0$, which is satisfied by our solution.

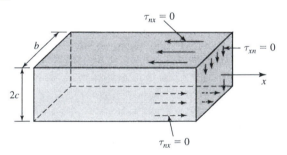

Figure 8.8.4

(ii) For $V > 0$, $\tau(y) \geq 0$ since $y^2 \leq c^2$. Thus the directions of the shear stress are everywhere directed into the area $\overline{A}$ (downward in this case), as shown in Fig. (8.8.5a).

(iii) The distribution of the average shear stress in the cross-section is parabolic with y [Fig. (8.8.5b)]. Furthermore, τ is a maximum at $y = 0$, i.e., along the z-axis; this maximum value is $\tau = \frac{3}{2}\frac{V}{A}$. We note that it is always possible to express the maximum shear stress as

$$\tau = k\frac{V}{A},$$

(8.8.7)

where k is a constant, which depends on the geometry of the cross-section, and A is the cross-sectional area. Thus for the rectangular cross-section considered here, $k = 3/2$; for a circular cross-section, $k = 4/3$. For these two geometries, the maximum shear stress acts at points along the neutral axis. It should be pointed out, however, that while this occurs very often, it is not necessarily true that the maximum shear stress always lies along the neutral axis for all cross-sections. Thus, although the value for k is the same, namely $k = 3/2$, for a cross-section having a triangular shape, in that case, the maximum shear stress does not occur at the neutral axis. □

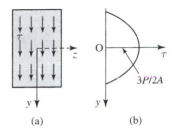

(a)   (b)

Figure 8.8.5

Example 8.14:   Consider a beam whose cross-section is symmetric about both the y- and z-axes, as shown in Fig. (8.8.6a). Assume that a shear force $V \equiv V_y > 0$ is acting on the cross-section. (The cross-section component (ii), ABC (or DEF), is called the *flange* of the section; the component (iii), BE, is referred to as the *web*.) Determine the average shear stress τ acting along the lines a–a, c–c, and d–d in terms of V and I. (Note that the line a–a is just below the top flange and line d–d is immediately to the left of the web.)

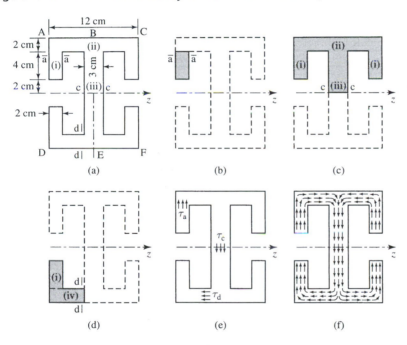

(a)   (b)   (c)

(d)   (e)   (f)

Figure 8.8.6

Solution: Since V and I are constants, the shear stress along the three given lines will depend solely on Q and b.

Along a–a: We isolate the rectangular portion denoted by (i) in Fig. (8.8.6b). Then, $b = 2$ cm; $Q = (2 \times 4) \times (-4) = -32$ cm$^3 \rightarrow\rightarrow \tau = -\frac{32}{2}\frac{V}{I} = -16\frac{V}{I}$.

Note that for $V > 0$, $\tau < 0$, which indicates that the stress is physically pointing out of the area (i), i.e. upward. This result is shown in Fig. (8.8.6e).

Along c–c: We isolate the entire upper part of the cross-section above the z-axis. Letting (i), (ii) and (iii) represent the rectangular components of the areas [Fig. (8.8.6c)], we have $b = 3$ cm and $Q = 2[(2 \times 4) \times (-4)] + (2 \times 12) \times (-7) + (6 \times 3) \times (-3) = -286$ cm$^3 \rightarrow\rightarrow \tau = -\frac{286}{3}\frac{V}{I}$.

Again, for $V > 0$, a negative τ indicates that the stress is acting out of the area [in this case downward, as shown in Fig. (8.8.6e)].

Along d–d: Here, upon making a cut along d–d, we find it convenient to choose the L-shaped portion defined by the rectangular components (i) and (iv) [Fig. (8.8.6d)]. We then have $b = 2$ cm; $Q = (2 \times 4) \times 4 + (4.5 \times 2) \times (7) = 95$ cm$^3 \rightarrow\rightarrow \tau = \frac{95}{2}\frac{V}{I} = 47.5\frac{V}{I}$.

Note that since $\tau > 0$, the shear stress is pointing into the area $\overline{A}$, i.e. to the left in Fig. (8.8.6e).

The shear stresses (due to a positive shear $V_y > 0$) acting at points throughout the section are shown in Fig. (8.8.6f). Such a distribution of shear stresses is, for obvious reasons, often referred to as the *shear flow*.

We observe that the idea of 'left' and 'right' or 'up' and 'down' are totally irrelevant in determining the proper direction in which the shear stress acts. The direction is determined solely from a consideration of the sign with respect to the area $\overline{A}$ which was used as the isolated portion. (Since the use of Eq. (8.8.4) implies isolating a portion $\overline{A}$ of the cross-section, one may always choose either portion as the isolated free body; one usually chooses the more convenient area, as was done in calculating τ at d–d above. However, as was pointed out previously, the choice is irrelevant; both choices lead to the same physical solution). $\qquad\square$

(b) Limitations on the derived expression

■ We first recall that τ, as given by Eq. (8.8.4), yields the *average* value of the shear stress along points lying on a line of length b oriented in some arbitrary direction. We note too that the end points of this line always lie on lateral surfaces of the beam. Because Eq. (8.8.4) leads only to average values of τ, one may obtain, in certain cases, results that violate the actual physical boundary conditions of a problem. Consider, for example, a circular cross-section, as shown in Fig. (8.8.7a), subjected to a shear force V. Use of Eq. (8.8.4) then leads to an average value of τ along line a–a, which acts perpendicularly to this line. Now, it is clear that the shear stress on any lateral surface must vanish. Hence, because of the equality of the conjugate shear stresses, the shear stress τ in the cross-section at points along the outer boundary (e.g., point B) must act *tangentially* to the circumference in order to satisfy this condition. This reasoning follows precisely the same as discussed in comment (i) of Example 8.13. (Recall also that this was found to be true in our discussion of torsion in Chapter 7.) Thus the average τ, as calculated by Eq. (8.8.4), does not satisfy the boundary conditions in this case. In fact, the boundary condition will *only be satisfied if the lateral surfaces are perpendicular to the line K at the two end points*. (Note that in

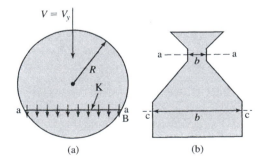

Figure 8.8.7

(a)   (b)

Examples 8.13 and 8.14, the boundary conditions were satisfied along all lateral surfaces.)

■ Since Eq. (8.8.4) leads to average values of the shear stress along points lying on a line of length b, we should expect the accuracy of the results to depend on the length b. Now, if b is relatively small (with respect to other dimensions of the cross-section), one may expect only a small variation of τ along this line. Hence, if b is relatively small, the calculated value of τ will be reasonably close to the true value for any given point along the line. However, if b is relatively large, Eq. (8.8.4) may lead to relatively inaccurate results. This is illustrated for the cross-section shown in Fig. (8.8.7b), where it is clear that the use of Eq. (8.8.4) will lead to far more accurate results for τ along line a–a than along line c–c.

The use of Eq. (8.8.4) therefore must be used with great care. Nevertheless, although it provides but average values, this expression proves to be very useful in yielding reasonably accurate values for the shear stress in a large variety of problems.

We shall use the derived expression for τ (as well as for σ_x) in a subsequent section in applications to the design of beams as encountered in typical engineering practice. However, we first continue to analyse some implications of the derived expressions in order to better understand the limitations and applicability of these expressions.

(c) Shear effect on beams – warping of the cross-sections due to shear

As we have seen, for all states, except for the case of pure bending, there exist shear forces $V(x) \neq 0$ and consequently shear stresses τ exist throughout the cross-section. Now, as discussed in Chapter 4, such shear stresses lead to shear deformation; that is, any rectangular element subjected to shear stresses will undergo shear strain and deform into a parallelogram. If we consider an element Δx in the $x–y$ plane, between two adjacent cross-sections, each small sub-elemental rectangle will deform.[†]

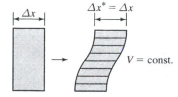

Figure 8.8.8

For the case where $V = V(x)$, each cross-section will deform, i.e. warp, differently. As a result the length of any fibre between the cross-sections will be affected by this difference in warping. However, if $V = $ const., it is clear that all cross-sections deform identically and consequently, for this case, the length of fibres between any two cross-sections does not change, as shown in Fig. (8.8.8). Thus, we may conclude that when $V = $ const., *fibres originally parallel to the x-axis will undergo no extension due to warping of the cross-sections.*

8.9 Re-examination of the expression for flexural stress $\sigma_x = My/I$: further engineering approximations

(a) Examination of equilibrium state

We recall that the expression for the flexural stress was derived for a state of pure bending in the absence of warping of the cross-sections. By invoking Navier's hypothesis, we have seen that the expression may then also be applied to beams

---

[†] Warping at any given cross-section thus depends on the shear strain ε_{xy} which is a function of y as well as of x. [For example, for a beam having a rectangular cross-section (see Example 8.12), $\varepsilon_{xy} = 3V/4AG$ at $y = 0$ and $\varepsilon_{xy} = 0$ at $y = \pm c$ with sub-elements near the neutral axis undergoing the largest deformation while at $y = \pm c$, the sub-elements undergo no deformation.] The overall warping deformation of the cross-section may be represented by a weighted average in terms of a shape factor α which depends on the shape of the cross-section; one then refers to the *average shear strain* in the cross-section as $\varepsilon_{xy}(x) = \alpha V(x)/2AG$ where A is the cross-sectional area.

subjected to lateral forces. We examine here the implications of this generalised engineering use of the expression.

Now, we recall from Chapter 2 that (in the absence of body forces $\boldsymbol{B}$) the stresses in a body must satisfy the equations of equilibrium, Eqs. (2.4.4), namely

$$\frac{\partial \sigma_x}{\partial x} + \frac{\partial \tau_{yx}}{\partial y} + \frac{\partial \tau_{zx}}{\partial z} = 0, \tag{8.9.1a}$$

$$\frac{\partial \tau_{xy}}{\partial x} + \frac{\partial \sigma_y}{\partial y} + \frac{\partial \tau_{zy}}{\partial z} = 0, \tag{8.9.1b}$$

$$\frac{\partial \tau_{xz}}{\partial x} + \frac{\partial \tau_{yz}}{\partial y} + \frac{\partial \sigma_z}{\partial z} = 0. \tag{8.9.1c}$$

Here we use the notation $\sigma_x \equiv \tau_{xx}$, $\sigma_y \equiv \tau_{yy}$, $\sigma_z \equiv \tau_{zz}$.

We recall that in the derivation of the expression for the flexural stress it was also assumed, for long slender beams, that $\sigma_y = \sigma_z = 0$. We now show that even for the case $V \neq 0$ we may assume τ_{yz} to be zero for such beams. (To visualise better the physical situation, let us concentrate on a beam of rectangular cross-section [Fig. (8.9.1)].) We first observe that along the lateral surfaces of the beam, $\tau_{yz}(y = \pm c) = 0$ and $\tau_{zy}(z = \pm b/2) = 0$. Thus, on *all* lateral surfaces, $\tau_{zy} = 0$. Moreover, since we are concerned here only with symmetrical bending of beams of symmetrical cross-sections, it follows that at all points along the y-axis, $\epsilon_{yz} = 0$ and hence $\tau_{yz} = 0$. Furthermore, since the cross-sectional dimensions of long slender beams are relatively small, it is therefore reasonable to *assume* that $\tau_{zy} = 0$ at all points in the beam. [Note that τ_{xy} and τ_{xz}, which act in the cross-section, cannot be assumed to vanish everywhere – indeed these are the stresses arising due to the resultant shear forces V.]

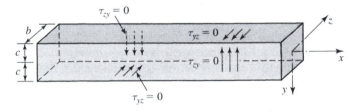

Figure 8.9.1

We therefore study the implications of the assumptions

$$\sigma_y = \sigma_z = \tau_{zy} = 0 \tag{8.9.2}$$

for the case where shear stresses are acting on the cross-sections.

Substituting Eqs. (8.9.2) in Eqs. (8.9.1b) and (8.9.1c), we find

$$\frac{\partial \tau_{xy}}{\partial x} = 0, \tag{8.9.3a}$$

$$\frac{\partial \tau_{xz}}{\partial x} = 0; \tag{8.9.3b}$$

hence

$$\tau_{xy} = f_1(y, z), \tag{8.9.4a}$$

$$\tau_{xz} = f_2(y, z), \tag{8.9.4b}$$

where $f_1(y, z)$ and $f_2(y, z)$ are arbitrary functions. Substituting back in Eq. (8.9.1a),

$$\frac{\partial \sigma_x}{\partial x} = -\left(\frac{\partial \tau_{yx}}{\partial y} + \frac{\partial \tau_{zx}}{\partial z}\right) = g(y, z), \tag{8.9.5}$$

where $g(y, z)$ is again any arbitrary function of y and z.

Therefore, upon integrating,

$$\sigma_x = x g(y, z) + C, \tag{8.9.6}$$

where C is either a function of y and z, or an arbitrary constant (with respect to x). Thus we observe that σ_x is, at most, a linear function of x.

Therefore, we immediately conclude that to be consistent with the assumptions of Eqs. (8.9.2), the stress σ_x, as given by $\sigma_x = \frac{M(x)y}{I}$, will satisfy Eqs. (8.9.1) only if $M(x)$ has the form

$$M(x) = ax + b; \tag{8.9.7}$$

that is, the moment M can either be a constant or a linear function of x. For any other variation of $M(x)$, the equations of equilibrium are not satisfied!

We now give a physical interpretation to this apparently contradictory result. Let us recall that the expression for the flexural stress σ_x was rigorously derived for a state of pure bending ($M = $ const., $V = 0$) for which all cross-sections remained plane. Why then does this expression appear to be equally valid for a linear variation with x? We first note, however, that if $M = ax$, then $V = a$ (a constant); i.e., the same shear force acts on all cross-sections. Now, in the previous section, we observed that if $V = $ const., the longitudinal fibres undergo no extension due to shear since, in this case, all cross-sections warp identically; thus, the strain ϵ_x in the fibres is not affected by shear deformation. Hence, in this case, the extension of the fibres depends only on bending of the beam, and consequently the expression $\epsilon_x = y/R$ remains valid. As this is the fundamental starting point in the derivation, the expression for flexural stress remains valid when M is a linear function of x.

For beams with $M(x)$ not of the form given by Eq. (8.9.7), *we shall nevertheless continue to invoke Navier's hypothesis* with the knowledge (as discussed in Section 7) that any warping effects are negligible. In the following sections, we therefore continue to use the expressions for flexural stress and shear stresses, Eqs. (8.7.1b) and (8.8.4) respectively, and apply them to the design of beams as encountered in engineering practice.

(b) Flexural stress in a non-prismatic beam – an engineering approximation

As we have observed, the expressions for the flexural stress, Eq. (8.7.1b), are exact only for an elastic beam with constant cross-section where the moment varies as $M(x) = ax + b$.

Consider now a non-prismatic member, such that $A = A(x)$ and $I = I(x)$ [Fig. (8.9.2)], subjected to a state of pure bending. In this case, since we cannot

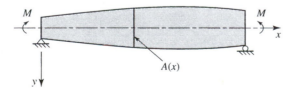

$A(x)$

Figure 8.9.2

invoke arguments of symmetry, it is clear, from the derivation, that we cannot expect Eqs. (8.7.1) to provide an exact solution even for pure bending. Nevertheless, if the cross-sectional properties do not vary sharply, that is, if the variation of $I(x)$ is not great, we may assume that an approximate value for the stress σ_x is given by

$$\sigma_x = \frac{My}{I(x)}. \qquad (8.9.8)$$

Equation (8.9.8) provides a good engineering approximation for such a beam with small variations of the cross-section. To justify this assertion, consider, e.g., a linearly tapered beam having, for simplicity, rectangular cross-sections with constant width b, as shown in Fig. (8.9.3a). Let us assume that the flexural stress σ_x has been calculated according to Eq. (8.9.8). We now isolate a small triangular portion of width Δx and height Δy [Fig. (8.9.3b)]. Acting on the cross-sectional area of this element, as represented by $\Delta y \cdot b$, there exists a stress σ_x as calculated above. However, we note that the triangular element cannot be in equilibrium under this stress alone, since $\sum F_x = 0$ is not satisfied. Therefore, there must exist shear stresses τ acting on the horizontal surface $b\Delta x$. For equilibrium, we require that $\tau \Delta x = \sigma_x \Delta y$ so that $\tau = \sigma_x \frac{\Delta y}{\Delta x}$. In the limiting case, as $\Delta y / \Delta x \to 0$, $\tau \to 0$, which is a correct result for pure bending. However, if $\frac{\Delta y}{\Delta x}$ is relatively small, then $|\tau/\sigma_x| \ll 1$. Hence, from this simple analysis, we may conclude that although there is a shear effect, it will be relatively small, and consequently, for a beam with a small taper [i.e., one for which the depth $d(x)$ varies slowly], Eq. (8.9.8) will yield a good approximation to the flexural stress. It is clear that for a beam with a relatively large

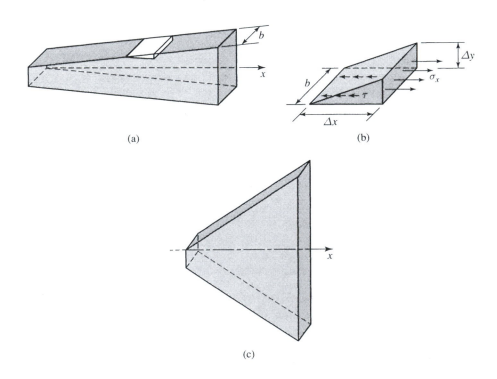

(a)

(b)

(c)

Figure 8.9.3

taper [Fig. (8.9.3c)], Eq. (8.9.8) will yield results of substantial error; therefore the expression cannot be used for such beams.[†]

If in addition, we again invoke Navier's hypothesis, Eq. (8.9.8) may then be used as an engineering approximation for the flexural stress in beams subject to arbitrary lateral loads, i.e. when $M = M(x)$ is a general function of x.

8.10   Engineering design applications for beams

We present here, by means of the following examples, typical design problems as encountered in engineering practice.

Example 8.15:   A structural steel member whose cross-section is given as a S102 × 14 section, as shown in Fig. (8.10.1a), spans a length $L = 2$ m. (This represents a typical manufactured cross-section called a **rolled section**. The geometric dimensions and properties of such standard sections are given in tables, such as in Appendix E.) The dimensions of the specified section, as found in the table (see p. 705), are as follows: $d = 101.6$ mm, $t_f = 7.4$ mm, $b_f = 71$ mm, $t_w = 8.3$ mm.[‡] In addition, the second moment of the area of the cross-section and the section modulus, as found in the same table, are $I = 2.83 \times 10^6$ mm^4 and $S = 55.6 \times 10^3$ mm^3, respectively. The member is to be used as a simple supported beam with a span of length $L = 2$ m in order to carry a concentrated load P at the centre. The allowable flexural and shear stresses are given as $\sigma_{all} = 150$ MPa and $\tau_{all} = 100$ MPa, respectively. What is the maximum force P that the beam can support?

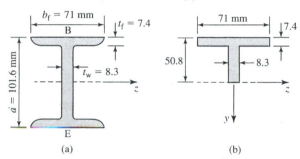

(a)   (b)   **Figure 8.10.1**

Solution: To determine the maximum value of P, we must consider both flexure and shear effects due to $M(x)$ and $V(x)$ as obtained in Example 8.2.

Considering flexure, we have, from Eq. (8.6.10e),

$$|M_z|_{all} = S|\sigma_x|_{max} = (55.6 \times 10^3)(150) = 83.4 \times 10^5 \text{ N-mm} = 8340 \text{ N-m}.$$

Since $M_{max} = \frac{PL}{4}$, $P_{all} = \frac{4M_{max}}{L} = 16,680$ N is the maximum allowable force as governed by the flexure criterion.

To consider the shear criterion, we note that the maximum shear stress here occurs at points along the z-axis. Therefore, following Eq. (8.8.4b), the allowable shear force

---

[†] We note too that for a beam having a large taper, de Saint Venant's principle can no longer be applied since, according to its geometry, the largest depth is of the same order of magnitude as the span length.
[‡] In designating a steel section as Sxx×yy, or Wxx×yy, etc., xx designates the nominal depth of the beam and yy designates the mass per unit length, namely kg/m. It is noted that the properties appearing in the standard tables are given in mm. These units prove to be convenient since 1 N/mm^2 = 1 N/(10^{-6} m^2) = 10^6 N/m^2 = 1 MPa.

on the section is $V_{all} = \tau_{all} \frac{Ib}{Q}$, where Q, the moment of the isolated area as shown in Fig. (8.10.1b), is

$$Q = (71 \times 7.4)[-(50.8 - 3.7)] + (8.3 \times 43.3)(-21.7) = -32.55 \times 10^3 \text{ mm}^3.$$

Therefore, $|V_{all}| = (100)[\frac{(2.83 \times 10^6) \times 8.3}{32.55 \times 10^3}] = 0.722 \times 10^5 \text{ N} = 72{,}200 \text{ N}.$

From the results of Example 8.2, $V_{max} = \frac{P}{2}$ and therefore $P_{all} = 2V = 144{,}400 \text{ N}.$

Thus, in this case, the allowable force P is clearly governed by the moment criterion; i.e., the allowable load on the beam is $P = 16{,}680 \text{ N}$.   □

It is worthwhile to mention that in general, as the length of a span increases, the flexure criterion tends to predominate while for relatively shorter beams, the shear criterion tends to predominate. (It is important to recall that all expressions derived in this chapter are valid only for long slender beams and therefore are inapplicable to short beams. However, when referring to 'relatively short' beams, one still means beams such that the ratio of length to depth is large).

We observe that the area A_w of the web BE is given approximately by $A_w = t_w d_w$, where d_w is the depth of the web. (For the given section here, $d_w = 101.6 - 14.8 = 86.8$ mm.) It is instructive to consider the expression $\tau = \frac{V}{A_w}$, which represents the average shear stress in the web. If we use this simple formula, we note that we obtain, in this case, $V = \tau A_w = (100)(86.8 \times 8.3) = 72{,}000 \text{ N}$. Comparing with the value $V = 72{,}200 \text{ N}$ obtained above, we note that this simple formula yields, for such sections, a very good approximation relating the shear force to maximum shear stress in the web; consequently, it is often used to obtain shear stresses in the web for this type of cross-section. Justification of this empirical formula follows from a comparison with the actual shear stress distribution in the web, as shown in Fig. (8.10.2). Thus we observe that the shear force V_y in this type of cross-section is carried essentially by the web.

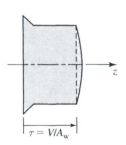

$\tau = V/A_w$

Figure 8.10.2

Example 8.16: For the beam of the cross-section of Example 8.15, previously considered [Fig. (8.10.1a)], the maximum allowable centre-span load, as governed by the flexure criterion, was calculated to be $P = 16{,}660 \text{ N}$.

Two steel plates, each of width 71 mm and thickness $t = 6$ mm are now attached, as shown in Fig. (8.10.3a), to the top and bottom flanges by means of rivets (each having a cross-sectional area $A_r = 95 \text{ mm}^2$) to form a monolithic 'built-up' section, as shown in Fig. (8.10.3b). Determine (a) the maximum allowable midspan load if the maximum allowable stresses are as given in Example 8.15; namely $\sigma_{all} = 150$ MPa, $\tau_{all} = 100$ MPa and (b) the maximum permissible spacing s of the pairs of rivets [Fig. (8.10.3a)] if the allowable shear stress in the rivets is $\tau_r = 53$ MPa.

Solution:

(a) By attaching the plates to the two flanges, it is clear that the plates contribute to increase the second moment of the area I_{zz} of the section. To calculate $I \equiv I_{zz}$ we make use of the parallel axis theorem; thus

$$I = 2.83 \times 10^6 + 2\left\{ \frac{71 \times 6^3}{12} + (71 \times 6)\left[\frac{101.6}{2} + 3\right]^2 \right\} = 5.30 \times 10^6 \text{ mm}^4.$$

From Eq. (8.6.10b),

$$M_{max} = \frac{\sigma_0 I}{d/2} = \frac{(150)(5.30 \times 10^6)}{56.8} = 14.0 \times 10^6 \text{ N-mm} = 14{,}000 \text{ N-m},$$

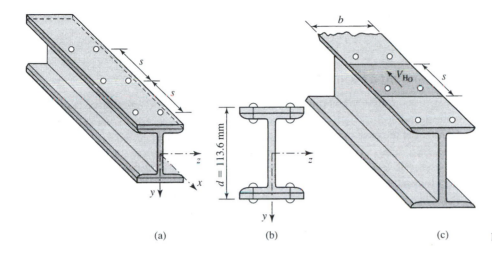

(a)   (b)   (c)   **Figure 8.10.3**

where, in the above, the depth of the built-up section is $d = 113.6$ mm.
Hence, $P_{\text{all}} = \frac{4M}{L} = 28{,}000$ N.

We note that by attaching the two plates, the maximum permissible moments that can be carried by the section has been increased from 8340 to 14,000 N m.

(b) The above calculations are based on the assumption that the section behaves as a monolithic unit; i.e., there is no slippage between the plates and the flanges. (If there were such slippage, then the above calculations would have no validity.) We recognise too that any slippage is prevented by the rivets, which hold the two parts of the section together, thus creating the monolithic section. The rivets must therefore carry the shear between the flange and the attached plate. To prevent such slippage it is therefore necessary that the rivets be sufficiently strong in shear. Now, treating the section as an equivalent monolithic unit, the shear stress at the plate–flange interface, according to Eq. (8.8.4) [with $Q = -(71 \times 6) \cdot (50.8 + 3) = -71 \times 6 \times 53.8$ mm^3], is

$$|\tau| = \left| \frac{VQ}{Ib} \right| = \frac{71 \times 6 \times 53.8V}{(5.3 \times 10^6)(71)} = 60.9 \times 10^{-6} V \text{ N/mm}^2 = 60.9 \times 10^{-6} V \text{ MPa}.$$

Noting, from Example 8.2, that $V = P/2$, it follows that for $P_{\text{all}} = 28{,}000$ N [as found in part (a)] the equivalent average shear stress $\tau = (30.45 \times 10^{-6}) \cdot P = 0.853$ MPa.

While τ is the average shear τ_{xn} acting in the cross-section, since $\tau_{nx} = \tau_{xn}$, it also represents the average 'equivalent shear stress' acting in the plane of the interface. Hence, the equivalent horizontal shear force V_H acting over a segment $(b \times s)$ of the interface is $V_H = 0.853 \times (71 \times s) = 60.6s$ N [Fig. (8.10.3c)], where s has units of mm.

Now, let us assume that each rivet can carry a force in shear given by $F_r = \tau_r A_r = (53)(95) = 5035$ N. (Note that this too is an engineering approximation, used rather empirically, since $\tau_r = F_r/A_r$ yields only an approximate average value of the shear stress in the rivet.)

Since any pair of rivets must carry this shear force V_H, we equate the two; therefore from $2F_r = V_H$, we obtain

$$s = \frac{2 \times 5035}{60.6} = 166 \text{ mm} = 16.6 \text{ cm}.$$

In engineering design, one would specify conservatively a maximum distance of 16 cm.   □

8.11   Bending of composite beams

The expressions developed previously for bending of beams were derived under the assumption that the beam is composed of a single homogeneous material. Let us now consider the case of beams that are made of two or more materials: such beams are referred to as *composite beams*. In analysing the behaviour of these beams, it is necessary to modify our derivation slightly although, as we shall see below, the basic ideas remain the same.

For convenience, we consider a beam subjected to pure bending, which is made up of two materials, say materials '1' and '2', having moduli of elasticity E_1 and E_2 respectively, as shown in the cross-section of Fig. (8.11.1). We denote the cross-sectional area of each material by A_1 and A_2, respectively, such that the total cross-sectional area $A = A_1 + A_2$.

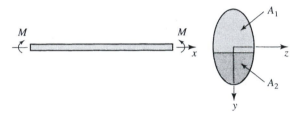

Figure 8.11.1

From Section 6, we recall our conclusion that for a beam undergoing pure bending ($M = $ const.), all cross-sections remain plane and perpendicular to the deformed longitudinal axis. From an analysis of the geometry, the strain ϵ_x was found to be given by Eq. (8.6.1c), namely

$$\epsilon_x = \frac{y}{R},\tag{8.11.1}$$

where $R \equiv 1/\kappa$ is the radius of curvature. We emphasise here again that this relation is independent of the material.

Using Hooke's law and assuming, as previously, that for long beams $\sigma_y = \sigma_z = 0$ at all points within the beam, the flexural stresses in materials '1' and '2' are then given, respectively, by

$$\sigma_{x_1} = E_1 \epsilon_x = \frac{E_1 y}{R},\tag{8.11.2a}$$

$$\sigma_{x_2} = E_2 \epsilon_x = \frac{E_2 y}{R}.\tag{8.11.2b}$$

The moment about the neutral axis, is given by Eq. (8.6.5a), namely

$$M = \iint_A y\sigma_x \, dA.\tag{8.11.3a}$$

Substituting Eqs. (8.11.2),

$$M = \frac{1}{R}\left[E_1 \iint_{A_1} y^2 \, dA + E_2 \iint_{A_2} y^2 \, dA\right]\tag{8.11.3b}$$

or

$$M = \frac{1}{R}\left(E_1 I_1 + E_2 I_2\right), \tag{8.11.3c}$$

where

$$I_1 = \iint\limits_{A_1} y^2 \, dA, \tag{8.11.4a}$$

$$I_2 = \iint\limits_{A_2} y^2 \, dA \tag{8.11.4b}$$

are the second moments of the areas A_1 and A_2, respectively, about the neutral axis such that I, the second moment of the area of the entire cross-section, is given by $I = I_1 + I_2$.

Equation (8.11.3c) thus represents the Euler-Bernoulli relation for the composite beam. We note, however, that the location of the neutral axis has not yet been determined. To find its location, we recall that the resultant normal force F, acting normal to the cross-section in the x-direction, is given by

$$F = \iint\limits_{A} \sigma_x \, dA. \tag{8.11.5a}$$

Substituting Eqs. (8.11.2), we find

$$F = \frac{1}{R}\left(E_1 \iint\limits_{A_1} y \, dA + E_2 \iint\limits_{A_2} y \, dA\right). \tag{8.11.5b}$$

Now, since for the case of pure bending, the normal force $F = 0$, we then have

$$E_1 \iint\limits_{A_1} y \, dA + E_2 \iint\limits_{A_2} y \, dA = 0. \tag{8.11.6}$$

This last relation then defines the location of the neutral axis.

We observe that Eqs. (8.11.2) and (8.11.3c) correspond to Eqs. (8.6.4) and (8.6.6a) respectively (which were derived for a homogeneous beam), and degenerate to these equations when $E_1 = E_2 = E$.

Let us now consider the case when $E_2 = nE_1$, where n ($n \geq 1$) is some constant. Equations (8.11.2), (8.11.3c) and (8.11.6) then become

$$\sigma_{x_1} = E_1 \epsilon_x = \frac{E_1 y}{R}, \tag{8.11.7a}$$

$$\sigma_{x_2} = nE_1 \epsilon_x = \frac{nE_1 y}{R}, \tag{8.11.7b}$$

$$M = \frac{E_1}{R}(I_1 + nI_2), \tag{8.11.7c}$$

$$\iint\limits_{A_1} y \, dA + n \iint\limits_{A_2} y \, dA = 0. \tag{8.11.7d}$$

The above expressions lead to a very simple and physical interpretation of the results. We first note, from Eqs. (8.11.2a) and (8.11.2b), that the incremental forces over an incremental element area ΔA in the two materials is given by

$$\Delta F_1 = \frac{E_1}{R} y \cdot \Delta A, \tag{8.11.8a}$$

$$\Delta F_2 = \frac{n E_1}{R} y \cdot \Delta A, \tag{8.11.8b}$$

which we may rewrite as

$$\Delta F_1 = \frac{E_1}{R} y \cdot \Delta A, \tag{8.11.8c}$$

$$\Delta F_2 = \frac{E_1}{R} y \cdot (n \Delta A). \tag{8.11.8d}$$

We may interpret this last expression for ΔF_2 as an incremental force *located at a point y from the neutral axis*, which acts on an equivalent fiber of material 1 over an element having an equivalent area, $n \Delta A$ [Fig. (8.11.2)]. With this interpretation in mind, we recognise that we may consider the actual cross-section [Fig. (8.11.3a)] as being equivalent to a section composed of a single homogeneous material (in this case with $E = E_1$) whose geometry is as shown in Fig. (8.11.3b). Such an equivalent cross-section that consists solely, e.g., of material 1, is often referred to as a *transformed cross-section. Thus the behaviour of the composite beam can be obtained by analysing the equivalent transformed section.* Denoting the second moment of area of the transformed cross-section section by $\overline{I}$, we note that

$$\overline{I} = I_1 + n I_2. \tag{8.11.9}$$

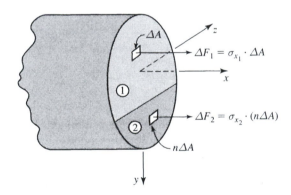

Figure 8.11.2

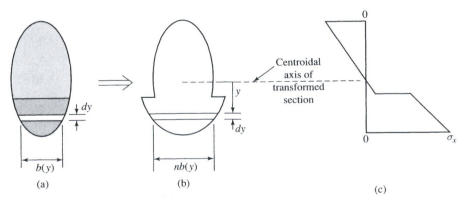

Figure 8.11.3

It then follows from Eq. (8.11.7c) that the curvature of the beam κ is given by

$$\kappa \equiv \frac{1}{R} = \frac{M}{E_1 \overline{I}}. \tag{8.11.10}$$

Substitution in Eqs. (8.11.7a) and (8.11.7b) then leads to

$$\sigma_{x_1} = \frac{My}{\overline{I}}, \tag{8.11.11a}$$

$$\sigma_{x_2} = \frac{nMy}{\overline{I}}. \tag{8.11.11b}$$

The resulting stress distribution in the actual cross-section is shown in Fig. (8.11.3c). Noting that

$$\overline{Q} = \iint_{A_1} y \, dA + n \iint_{A_2} y \, dA \tag{8.11.12}$$

represents the first moment of the transformed cross-section about the neutral axis, we conclude, from Eq. (8.11.7d), that the neutral axis of the actual cross-section coincides with the centroidal axis of the transformed cross-section. The location is shown symbolically in Fig. (8.11.3).

Example 8.17: A beam consisting of steel and brass (with $E_s = 200$ GPa and $E_b = 100$ GPa respectively), bonded together to form a rectangular cross-section as shown in Fig. (8.11.4a), is subjected to a given moment $M = 25$ N-m. Determine (a) the maximum stress in the brass and steel, (b) the flexural stress in the brass and steel at the interface and (c) the radius of curvature R of the beam at the cross-section.

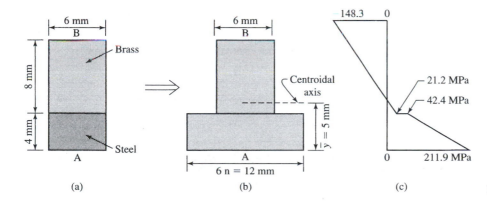

Figure 8.11.4

Solution: For the given materials, the ratio $n = 200/100 = 2$. Therefore the equivalent brass transformed cross-section is as shown in Fig. (8.11.4b).

The location of the neutral axis, $\overline{y}$ (i.e., the centroid of the transformed cross-section), is

$$\overline{y} = \frac{(6 \times 8)(8) + (12 \times 4)(2)}{48 + 48} = 5.0 \text{ mm}.$$

Using the parallel axis theorem,

$$\overline{I} = \left[\frac{6 \times 8^3}{12} + (48)(3.0)^2\right] + \left[\frac{12 \times 4^3}{12} + (48)(3.0)^2\right]$$

$$= 1184\,\text{mm}^4 = 1.18 \times 10^{-9}\,\text{m}^4$$

(a) The maximum stress in the brass at B (with $y = -7.0$ mm) is

$$\sigma_{xb} = \frac{My}{\overline{I}} = \frac{(25)(-7.0 \times 10^{-3})}{1.18 \times 10^{-9}} = -148.3 \times 10^6\,\text{N/m}^2$$

$$= -148.3\,\text{MPa (compression)}.$$

The maximum stress in the steel at A (with $y = 5.0$ mm) is, for $n = 2$,

$$\sigma_{xs} = \frac{2My}{\overline{I}} = \frac{2(25)(5.0 \times 10^{-3})}{1.18 \times 10^{-9}} = 211.9 \times 10^6\,\text{N/m}^2 = 211.9\,\text{MPa (tension)}.$$

(b) At the interface, $y = 1$ mm, the stresses in the brass and steel are $\sigma_{xb} = 21.2$ MPa and $\sigma_{xs} = 42.4$ MPa, respectively. Note that here $\sigma_{xs} = n\sigma_{xb}$ and that both are tensile stresses. The stress distribution in the actual cross-section is shown in Fig. (8.11.4c).

(c) Using Eq. (8.11.10), the radius of curvature of the beam $R = E\overline{I}/M$ is then

$$R = \frac{(100 \times 10^9)(1.18 \times 10^{-9})}{25} = 4.72\,\text{m}. \qquad \square$$

8.12   Combined loads

We have, up to now, considered only forces acting in an x–y plane of symmetry such that the moments at a cross-section are $M = M_z$. We now consider the case of a prismatic beam of cross-sectional area A such that both the y- and z-axes are centroidal axes of symmetry (with second moments I_{zz} and I_{yy} respectively). Furthermore, we assume that the beam is subjected to both moments M_y as well as M_z [Figs. (8.12.1a and b)]. (Note that since we are considering in this section moments about both the y- and z-axes, we must abandon the simplified notation, $M \equiv M_z$ and $I \equiv I_{zz}$). Clearly, in this case, we obtain the flexural stress due to M_y by interchanging the y- and z-subscripts in Eq. (8.6.9b) and hence due to M_y, we obtain $\sigma_x = \frac{M_y z}{I_{yy}}$. Analogous to the sign convention for M_z (as established in Section 8.2b) we define a positive moment M_y as a moment that tends to cause tension in fibres with positive z-coordinates.

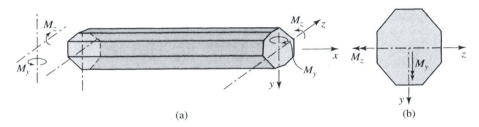

(a)   (b)

Figure 8.12.1

Now, if the beam is linearly elastic and undergoes small strains, then, as discussed in Chapter 5, we may use the principle of linear superposition. Thus we may write

$$\sigma_x = \frac{M_z y}{I_{zz}} + \frac{M_y z}{I_{yy}}. \tag{8.12.1}$$

We observe that the neutral axis always passes through the centroid of the section ($y = z = 0$). Note too that, following the discussion of Section 8.6d, the above expression is 'exact' for the case of pure bending as well as when M_y and M_z are linear functions of x (see Section 8.9).

[While Eq. (8.12.1) is valid, as mentioned above, for the case where both the y- and z-axes are axes of symmetry, we shall show, in Chapter 13, that this requirement is overly restrictive. Indeed, it will be shown that it is only necessary that the y- and z-axes be principal axes of the cross-section for Eq. (8.12.1) to remain valid].

Invoking again Navier's hypothesis, Eq. (8.12.1) may also be used if the beam is subjected to moments M_y and M_z due to applied loads that act in a plane inclined by an angle θ with respect to the y-axis, as shown in Fig. (8.12.2). Since clearly any load P can be resolved into components in the y- and z-directions ($P_y = P \cos \theta$ and $P_z = P \sin \theta$), M_y is due to lateral forces P_z and M_z is due, as before, to lateral P_y forces.

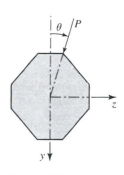

Figure 8.12.2

We may proceed further with the use of the principle of linear superposition. Assume now that in addition to these lateral forces, the beam is subjected to an axial load F acting through the centroid [Fig. (8.12.3)]. Then, using the results of Chapter 6, we have

$$\sigma_x = \frac{F}{A} + \frac{M_z y}{I_{zz}} + \frac{M_y z}{I_{yy}}. \tag{8.12.2}$$

Note that, in addition to lateral forces, axial forces that do not pass through the centroid of the cross-section cause moments M_y and M_z in the beam. This feature is illustrated in the following example.

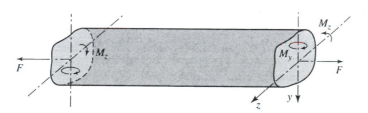

Figure 8.12.3

Example 8.18: A member having a circular cross-section of radius R is subjected to a compressive force P applied with an eccentricity e with respect to the longitudinal x-axis [Fig. (8.12.4a)]. Determine the largest value of e such that no fibres are in tension, i.e. such that for all points in the section, $\sigma_x \leq 0$.

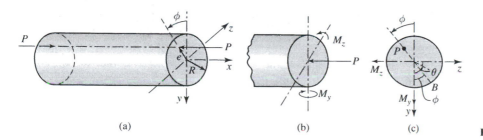

(a)   (b)   (c)

Figure 8.12.4

Solution: We first replace the given force system, as shown in Fig. (8.12.4a), by an equivalent force system [Fig. (8.12.4b)] consisting of a compressive axial force P

acting through the origin, and moments,

$$M_y = Pe\sin\phi, \tag{8.12.3a}$$

$$M_z = Pe\cos\phi, \tag{8.12.3b}$$

where we note that for $0 < \phi < \pi/2$, $M_y > 0$ and $M_z > 0$.

Therefore, from Eq. (8.12.2),

$$\sigma_x = -\frac{P}{A} + \frac{(Pe\cos\phi)y}{I_{zz}} + \frac{(Pe\sin\phi)z}{I_{yy}} \tag{8.12.3c}$$

$$= -\frac{P}{A} + \frac{Pe}{I}(y\cos\phi + z\sin\phi), \tag{8.12.3d}$$

where, due to the axial symmetry of the cross-section, $I \equiv I_{yy} = I_{zz} = \frac{\pi R^4}{4}$.

We observe that the moments M_y and M_z will tend to cause fibres with positive y- and z-coordinates to be in tension. Letting $y = r\cos\theta$, $z = r\sin\theta$, where θ is measured from the positive y-axis [Fig. (8.12.4c)],

$$\sigma_x = -\frac{P}{A} + \frac{Per}{I}(\cos\theta\,\cos\phi + \sin\theta\,\sin\phi). \tag{8.12.4b}$$

Clearly, from Eq. (8.12.4b), points on the edge $r = R$ have the greatest tendency to be in tension. To determine the specific critical point, we take the derivative

$$\frac{\mathrm{d}\sigma_x}{\mathrm{d}\theta} = \frac{PeR}{I}(-\sin\theta\cos\phi + \cos\theta\sin\phi) = \frac{PeR}{I}\sin(\phi - \theta). \tag{8.12.5}$$

Setting $\frac{\mathrm{d}\sigma_x}{\mathrm{d}\theta} = 0$, as anticipated, the critical point is given by $\theta = \phi$, namely the point B of Fig. (8.12.4c). Substituting in Eq. (8.12.4b) with $r = R$, we find

$$\sigma_x = P\left(-\frac{1}{A} + \frac{eR}{I}\right). \tag{8.12.6}$$

The required condition $\sigma_x \le 0$ then leads to

$$e \le \frac{I}{AR} = \frac{R}{4} \tag{8.12.7}$$

since, for a circle, $I/A = R^2/4$.

Thus we have established that if an eccentric compressive force acts at points within the quarter point from the centroidal x-axis, no tensile stresses will exist within the circular member. The shaded area representing these points, as shown in Fig. (8.12.5), is often referred to as the 'core' of the section. □

We remark that the 'core' of a section is a geometric property and depends on the particular geometry of the cross-section. In this example, the core is axially symmetric; that is, $e \neq e(\phi)$. For other geometries, the core will also possess a symmetry but will depend on the y- and z-axes. For example, the core of a rectangular section is as shown in Fig. (8.12.6).

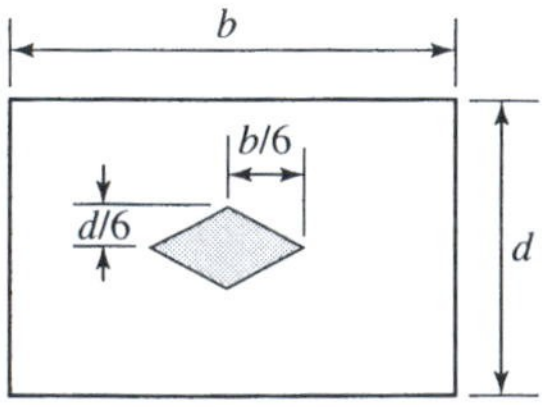

8.13 Elastic–plastic behaviour

(a) Fully plastic moments – location of the neutral axis

In the preceding analyses of this chapter, we considered only elastic beams, and found for elastic behaviour, that (a) the neutral axis coincides with the centroidal axis

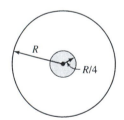

Figure 8.12.5

Figure 8.12.6

and (b) the flexural strain and stress vary linearly with the perpendicular direction from the neutral axis.

As will become evident, when the material enters the plastic range, the above conclusions cease to be valid. We investigate here a beam subjected to (positive) end moments $M \equiv M_z$ [Fig. (8.13.1a)] (such that a state of pure bending exists throughout the beam) and assume that the moments, applied statically from $M = 0$, are progressively increased.

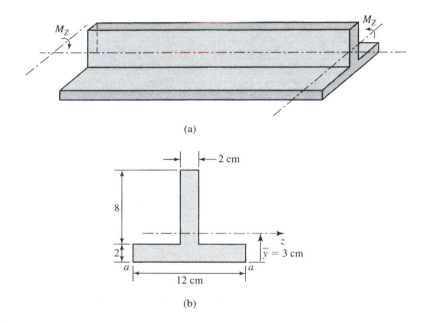

(a)

(b)

Figure 8.13.1

We consider a beam having the T-shaped cross-section of Example 8.12, where, for this section, we recall that $A = 40$ cm^2, $I_{zz} = 1000/3$ cm^4 and the location of the centroidal z-axis is given by $\bar{y} = 3$ cm [Fig. (8.13.1b)].

We assume here that the member is composed of an elastic–perfectly plastic material having a stress–strain ($\sigma \equiv \sigma_x$, $\epsilon \equiv \epsilon_x$) curve, as shown in Fig. (8.13.2), where the yield point σ_0 (assumed here, for convenience, to have units of N/cm^2) is the same in tension and compression.

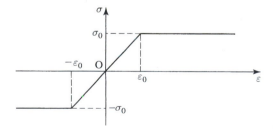

Figure 8.13.2

As M is initially applied, the beam behaves elastically and the stress and strain distributions vary linearly with x [according to Eqs. (8.6.9b) and (8.6.9c)], as shown in Fig. (8.13.3a).

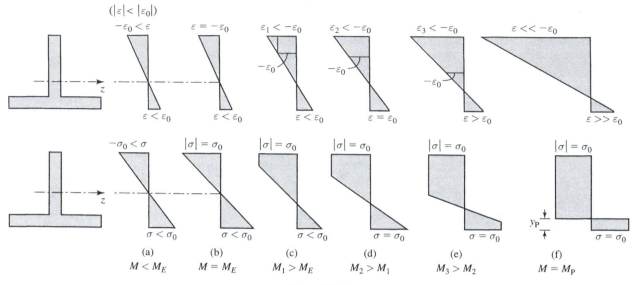

Figure 8.13.3

From Eq. (8.6.10b), the maximum moment M_E (called the **elastic moment**) that the section can sustain in the elastic range is

$$M_E = \frac{\sigma_0 I}{c} = \frac{\sigma_0 \left(\frac{1000}{3}\right)}{7} = 47.62\sigma_0 \text{ N-cm}. \tag{8.13.1}$$

Thus for $M \leq M_E$ the beam behaves elastically. When the moment reaches the limiting value $M = M_E$, the strain and stress distribution still varies linearly with y. We note that the top fibres ($y = -7$), being farthest from the neutral axis, are the first to reach the plastic range, $|\epsilon| = \epsilon_0$ and $|\sigma| = \sigma_0$ [Fig. (8.13.3b)].

We now assume that the loading is increased such that the applied moment $M > M_E$. Such moments tend to increase the bending of the beam, that is, they cause a greater curvature κ. Now, we recall that the strain ϵ in the beam, given by $\epsilon = \kappa y \equiv y/R$, is *independent of the material*. (This geometric relation follows from the conclusion that plane sections remain plane after deformation – a conclusion valid for a state of pure bending and for any material). Consequently this expression remains valid irrespective of the elastic or plastic material behaviour. Clearly, as the curvature κ increases the strain ϵ increases according to the relation $\epsilon = \kappa y$. Hence the cross-sectional zone for which $|\epsilon| > \epsilon_0$ becomes larger as values of $M/M_E > 1$ increase.

Consequently, for successively increasing moments, $M_E < M_1 < M_2 < M_3 \ldots$, the strain and stress distribution will appear as in Figs. (8.13.3c–e). Note that while the upper fibres first enter the plastic range, the bottom fibres initially continue to behave elastically. However, for some given value of M, the strain in the bottom fibres reaches ϵ_0 after which the lower fibres also enter the plastic range.

Thus, as the moment M increases, the plastic zones increase and the elastic zone decreases. In the limiting case [Fig. (8.13.3f)], the entire cross-section exhibits plastic behaviour; for positive moments, the entire lower portion of the beam is clearly in tension with $\sigma = \sigma_0$ while the upper portion is in compression with $\sigma = -\sigma_0$.

The moment M at the cross-section for this limiting case is called the **fully plastic moment** and is denoted by M_P.

We now wish to consider the location of the neutral axis for this limiting case, $M = M_P$. We recall that the conclusion that the neutral axis coincides with the centroidal axes was based purely on an elastic analysis. We therefore have no reason to assume that this is so during plastic behaviour.

Now, since the beam is solely in flexure (no axial forces are applied), we may locate the neutral axis in the limiting case from the equations of equilibrium, $\sum F_x = 0$. Noting that

$$\sum F_x = \iint_A \sigma \, dA = 0, \tag{8.13.2a}$$

we have, from the given stress distribution for this case,

$$\sigma_0 \iint_{A_{tens}} dA - \sigma_0 \iint_{A_{comp}} dA = 0 \tag{8.13.2b}$$

by which it follows that

$$A_{tens} = A_{comp}, \tag{8.13.2c}$$

where A_{tens} and A_{comp} represent the areas of the tension and compression zones, respectively.

Thus we note that when the fully plastic moment M_P acts, the neutral axis no longer necessarily lies along the centroidal axis; instead it lies on *an axis that bisects the total area of the cross-section into equal tension and compressive zones*.

For the section considered, when subjected to a fully plastic moment, the neutral axis is therefore located at $y_P = \frac{20}{12} = 1.66$ cm from the bottom [Fig. (8.13.3f)]. The uniform stresses $\pm\sigma_0$ acting on the section are shown in Fig. (8.13.4).

It should be mentioned, however, that the shift of the neutral axis, from its centroidal axis position during elastic behaviour, to its final location when $M = M_P$ does not occur suddenly. It is clear that the position changes continuously with $M > M_E$; for this section, it moves progressively downward [see Fig. (8.13.3)]. Although, its location, as a function of M can be established in principle for any given cross-section, we shall not pursue this topic here.

Having located the neutral axis in the limiting case, the fully plastic moment M_P can be readily found by calculating this moment about the neutral axis. However, recalling that a force system consisting only of a moment (couple) produces the *same moment about any set of parallel axes*, we find it more convenient to calculate M_P about the axis a–a [Fig. (8.13.1b)], which passes through the lower face of the section. Hence we obtain M_P as follows:

$$M_P = -\sigma_0 \left(\frac{10}{6} \times 12 \right) \times \frac{5}{6} + \sigma_0 \left[(16 \times 6) + \left(12 \times \frac{1}{3} \right) \times \frac{11}{6} \right]$$

$$= \frac{260}{3}\sigma_0 = 86.66\,\sigma_0.$$

It is of interest to note that $\frac{M_P}{M_E} = \frac{86.66}{47.62} = 1.82$; that is, the fully plastic moment is

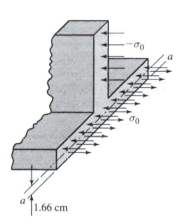

Figure 8.13.4

82% greater than the elastic moment. Indeed, M_P is the largest possible moment that a cross-section can carry; for this reason, M_P is often referred to as the *ultimate moment*.

(b) Moment–curvature relation for beams of rectangular cross-section in the plastic range

We consider here an elastic–perfectly plastic beam of rectangular cross-section $(b \times d)$ whose material behaviour is, as before, shown in Fig. (8.13.2). In this case, the centroidal z-axis also bisects the area and hence the neutral axis remains on the centroidal axis during plastic behaviour. (It is for this reason that we investigate this simpler case of the rectangular beam since we then need not consider any change in the location of the neutral axis.)

For this rectangular section, the elastic moment M_E is readily found from Eq. (8.6.10b):

$$M_E = \frac{\sigma_0 I}{d/2} = \frac{bd^2}{6}\sigma_0 \tag{8.13.3}$$

since the stress at the top and bottom fibres is $\mp\sigma_0$, respectively.

Thus initially, for $M \leq M_E$, the strain and stress distributions are as shown in Figs. (8.13.5a and b). Let us now consider that the applied moments are increased such that $M_E < M < M_P$ (where M_P represents the fully plastic moment, as defined above). Clearly, as discussed previously, the strain and stress distributions for such an intermediate value of M are as shown in Figs. (8.13.6a and b), respectively. When the moment reaches the fully plastic moment, i.e. $M = M_P$, the elastic zone disappears and the flexural stress is $\sigma_x = \pm\sigma_0$, respectively, in the remaining tensile and compressive plastic zones of the cross-section.

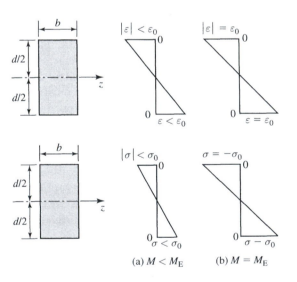

Figure 8.13.5

(a) $M < M_E$ (b) $M = M_E$

Now for values $M_E < M < M_P$, it is clear that the Euler–Bernoulli relation for elastic beams, $\frac{1}{R} = \frac{M}{EI}$, is no longer valid. As in the preceding example, with $M > M_E$, the central zone of the cross-section exhibits elastic behaviour while in the top and bottom zones the material behaves plastically. Let $y = \pm\ell$ represent

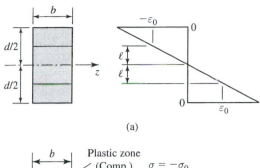

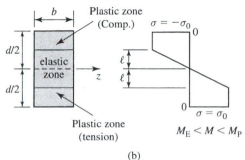

(b)

Figure 8.13.6

the elastic–plastic zone interface. Thus $|y| \leq \ell$ defines the elastic zone while $\ell < |y| < d/2$ defines the plastic zone. By definition, at the interface, $|\epsilon| = \epsilon_0$, i.e. $|\epsilon|_{y=\pm\ell} = \epsilon_0$.

We now investigate the moment-curvature relation for $M > M_E$ and determine $\ell = \ell(M)$.

Since the relation $\epsilon = y/R$ is still valid for $-d/2 \leq y \leq d/2$ (i.e. *throughout the cross-section*) and since by definition $|\epsilon| = \epsilon_0$ at the interface, we write

$$\epsilon_0 = \ell/R. \tag{8.13.4}$$

The strain distribution in the cross-section, for $M_E \leq M < M_P$, is then given by [Fig. (8.13.6a)]

$$\epsilon < -\epsilon_0, \quad -d/2 < y < -\ell, \quad \text{plastic zone} \tag{8.13.5a}$$

$$\epsilon(y) = \frac{y}{\ell}\epsilon_0, \quad |y| \leq \ell, \quad \text{elastic zone} \tag{8.13.5b}$$

$$\epsilon > \epsilon_0, \quad \ell < y < d/2, \quad \text{plastic zone.} \tag{8.13.5c}$$

The stress distribution, for $M_E \leq M < M_P$, is then [Fig. (8.13.6b)]

$$\sigma = -\sigma_0, \quad -d/2 < y < -\ell, \quad \text{plastic zone} \tag{8.13.6a}$$

$$\sigma = E\epsilon = \frac{y}{\ell}E\epsilon_0 = \frac{y}{\ell}\sigma_0, \quad -\ell < y < \ell, \quad \text{elastic zone} \tag{8.13.6b}$$

$$\sigma = \sigma_0, \quad \ell < y < d/2, \quad \text{plastic zone.} \tag{8.13.6c}$$

The stress resultant representing the moment $M \equiv |M_z|$ is given by [see Eq. (8.6.5a)]

$$M = \iint\limits_{A} y\,\sigma_x\,\mathrm{d}A = b\int\limits_{-c}^{c} y\sigma_x(y)\,\mathrm{d}y,$$ (8.13.7a)

where $c \equiv d/2$.

Substituting Eqs. (8.13.6),

$$M = b\left[\int\limits_{-c}^{-\ell} -\sigma_0 y\,\mathrm{d}y + \frac{1}{\ell}\int\limits_{-\ell}^{\ell} \sigma_0 y^2\,\mathrm{d}y + \int\limits_{\ell}^{c} \sigma_0 y\,\mathrm{d}y\right]$$

$$= b\sigma_0\left\{-\frac{y^2}{2}\Big]_{-c}^{-\ell} + \frac{1}{3\ell}y^3\Big]_{-\ell}^{\ell} + \frac{y^2}{2}\Big]_{\ell}^{c}\right\}$$

or

$$M = \frac{b\sigma_0}{12}(3d^2 - 4\ell^2).$$ (8.13.7b)

Solving for ℓ in terms of M, we obtain

$$\ell = \sqrt{3}\sqrt{\left(\frac{d}{2}\right)^2 - \frac{M}{b\sigma_0}}.$$ (8.13.8a)

Now it is more meaningful to express ℓ in terms of the ratio M/M_E. From Eq. (8.13.3), $b\sigma_0 = \frac{6M_\mathrm{E}}{d^2}$ and therefore we write

$$\ell = \sqrt{3}\sqrt{\left(\frac{d}{2}\right)^2 - \frac{Md^2}{6M_\mathrm{E}}} = d\sqrt{3}\sqrt{\frac{1}{4} - \frac{M}{6M_\mathrm{E}}}.$$ (8.13.8b)

Note that for $M = M_\mathrm{E}$, Eq. (8.13.8b) yields $\ell = d/2$; for this value of ℓ we recover, from Eq. (8.13.6b), the stress distribution of Fig. (8.13.5b).

Now, the fully plastic moment is reached when $\ell = 0$ [Fig. (8.13.7)]. Substituting $\ell = 0$ in Eq. (8.13.7b), we find

$$M_\mathrm{P} = \sigma_0\frac{bd^2}{4}.$$ (8.13.9)

Recalling Eq. (8.13.3), we therefore have $\frac{M_\mathrm{E}}{M_\mathrm{P}} = \frac{2}{3}$. Substituting back in Eq. (8.13.8b),

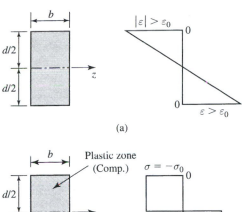

(a)

(b)

Figure 8.13.7

we may now express ℓ in terms of the ratio $\frac{M}{M_P}$:

$$\ell = \frac{d}{2}\sqrt{3}\sqrt{\left[1 - \frac{M}{M_P}\right]}, \quad M_E = 2/3\,M_P \leq M \leq M_P. \qquad (8.13.10)$$

The variation of $\frac{\ell}{d/2}$ vs. $\frac{M}{M_P}$ is shown in Fig. (8.13.8), where we note that for $M \leq M_E$, $\frac{\ell}{d/2} = 1$.

The moment-curvature relation is then readily established by noting from Eq. (8.13.4) that $\epsilon_0 = \frac{\ell}{R}$ so that $\sigma_0 = E\epsilon_0 = \frac{E\ell}{R}$. Hence

$$\frac{1}{R} = \frac{\sigma_0}{E\ell}. \qquad (8.13.11)$$

Substituting for ℓ as given by Eq. (8.13.10), leads to

$$\frac{1}{R} = \frac{\sigma_0}{E\sqrt{3}(d/2)}\frac{1}{\sqrt{1 - M/M_P}}, \quad M_E < M < M_P. \qquad (8.13.12a)$$

Finally, using Eq. (8.13.9), we may obtain an expression for EI/R in terms of the moment ratio, namely

$$\frac{EI}{R} = \frac{2}{3\sqrt{3}}\frac{M_P}{\sqrt{1 - M/M_P}}, \quad 2/3\,M_E < M < M_P. \qquad (8.13.12b)$$

The variation of $\kappa \equiv 1/R$ with the moment ratio M/M_P is shown in Fig. (8.13.9). We observe that for $M \leq M_E$, EI/R varies linearly with M according to the Euler–Bernoulli relation.

It is of interest to note from the analytical expression, Eqs. (8.13.12), that as $M \to M_P$, $1/R \to \infty$; that is, the radius of curvature $R \to 0$ when the entire section behaves plastically. This limiting case represents a 'plastic hinge' in the beam: at a

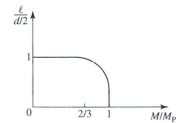

Figure 8.13.8

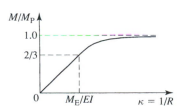

Figure 8.13.9

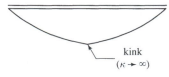

Figure 8.13.10

section for which $M = M_P$, there is a sudden change in the slope (a kink) of the beam [Fig. (8.13.10)]. Physically this corresponds to a sudden yielding of all the fibres in the cross-section in accordance with ideal elastic–plastic material behaviour.

PROBLEMS

For all problems below, the x-axis is to be taken at the extreme left end of the structure. Assume elastic behaviour unless specified otherwise.

Sections 2–4

8.1:   Sketch the shear and moment diagrams for the beams shown in Figs. (8P.1a–d) and give the values of all critical ordinates (in terms of a, b and L). Indicate in which segment (if any) of the beam a state of pure bending exists.

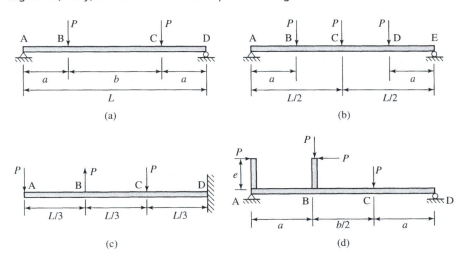

Figure 8P.1

8.2:   The beam shown in Fig. (8P.2) is subjected to two vertical forces and an eccentric horizontal force, each having the same magnitude P. (a) Sketch the shear and moment diagrams and show all critical ordinates (in terms of a, b and/or c). (b) What is the required value of e if segment BC is to be in a state of pure bending? Sketch the resulting moment diagram.

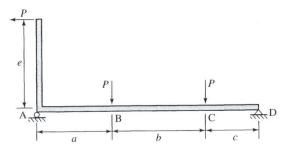

Figure 8P.2

8.3:   Express the shear force $V(x)$ and moment $M(x)$ as a function of x for the beams shown in Figs. (8P.3a–j). Sketch the variation with x showing all critical values (maxima and minima) and verify that the expressions satisfy the relations $dM(x)/dx = V(x)$, $dV(x)/dx = -q(x)$ and $d^2 M(x)/dx^2 = -q(x)$.

8.4:   Express the shear force $V(x)$ and moment $M(x)$ as a function of x in terms of P_1, P_2, a_1 and a_2 in the two regions $0 \le x < a_1$ and $a_1 < x \le a_2$, for the beams shown in Fig. (8P.4) and verify that the expressions satisfy the differential relations for beams.

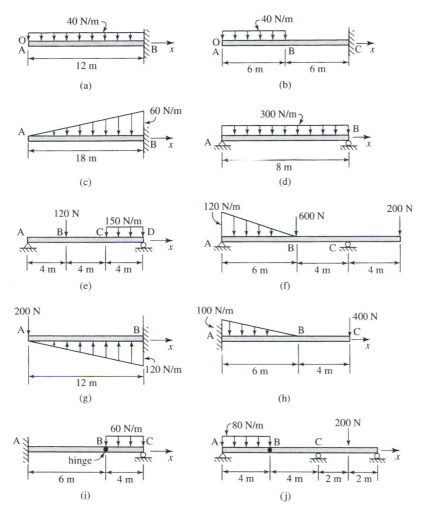

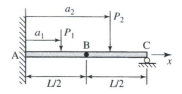

Figure 8P.3

Figure 8P.4

Figure 8P.5

8.5: Express the shear force $V(x)$ and moment $M(x)$ as a function of x in terms of w and P for the beam shown in Fig. (8P.5). Sketch the variation with x if $P = wL$ and verify that the expressions satisfy the differential relations for beams.

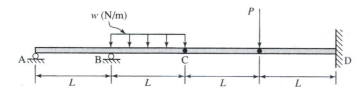

8.6: Determine the axial force F and express the shear force $V(x)$ and moment $M(x)$ within the span AC as a function of x for the beams shown in Figs. (8P.6a–e). Sketch the variation with x and verify that the expressions satisfy the differential relations for beams.

8.7:* Beam ABC is subjected to an eccentric horizontal force P and a uniformly distributed load w (N/m), as shown in Fig. (8P.7), where $0 \le P \le wL^2/8e$. (a) Determine the expressions for $V(x)$ and $M(x)$ in segments AB and BC. (b) Sketch the shear and moment diagrams and show all critical values. (c) What value of P (in terms of w, L and e) will yield the (algebraically) smallest maximum moment in segment AB, i.e. in $0 \le x \le L/2$.

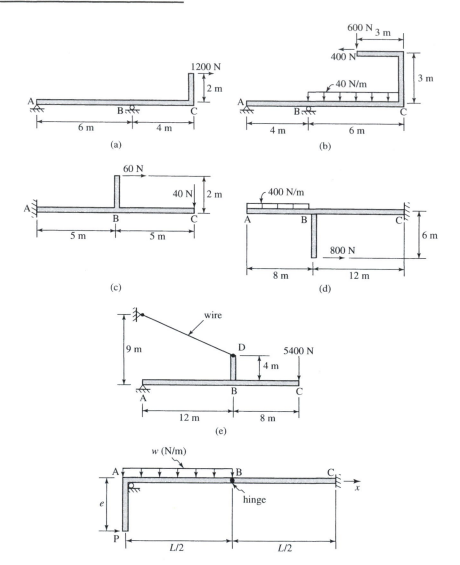

Figure 8P.6

Figure 8P.7

Evaluate this maximum moment and sketch the shear and moment diagrams for this value of P.

8.8:   For the structure ABCD containing a hinge at B, as shown in Fig. (8P.8), (a) determine $V(x)$ and $M(x)$ and (b) draw the shear and moment diagrams.

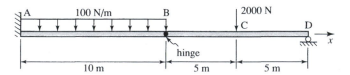

Figure 8P.8

8.9:   The structure shown in Fig. (8P.9) consists of two beams ABC and DEF, respectively, which are connected by means of a roller. Sketch the appropriate shear and moment diagrams for each beam and show all critical ordinates.

8.10:   The structure shown in Fig. (8P.10) consists of two beams ABC and DEF, containing a hinge at B and E, respectively, and connected by means of a roller. Sketch the appropriate shear and moment diagrams for each beam and show all critical ordinates.

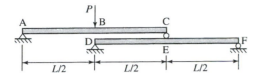

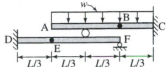

Figure 8P.9

8.11: Shear and moment diagrams are given for each of the beam structures shown in Figs. (8P.11a–c). Determine the loadings on the beam for each case.

Figure 8P.10

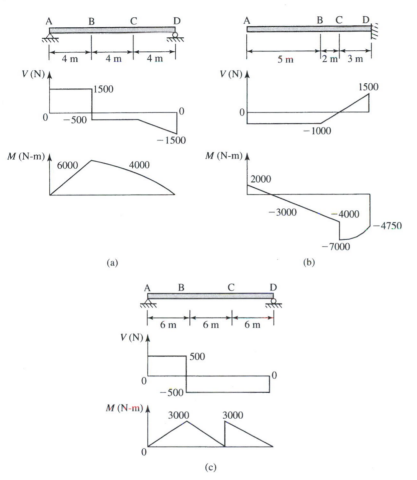

(a)

(b)

(c)

Figure 8P.11

Section 5

8.12: A cantilever beam AB, as shown in Fig. (8P.12), free at A ($x = 0$) and fixed at B ($x = L$), is subjected to a loading $q(x) = q_0(x/L)^n$, where $n \geq 0$ is an integer. (a) Prior to solving this problem, estimate whether one should expect the reactions at B to increase or decrease with increasing values of n, (b) determine the shear $V(x)$ and moment $M(x)$, (c) find the reactions R_B and M_B and (d) sketch the shear and moment diagrams for $n = 0$, 0.5, 1 and 2.

8.13: For the loading $q(x) = q_0 \sin(\pi x/2L)$ of the simply supported beam shown in Fig. (8P.13), (a) determine the reactions at A and B, (b) determine $V(x)$ and $M(x)$ and (c) sketch the shear and moment diagrams.

8.14:* A simply supported beam AB is subjected to a variable loading, as shown in Fig. (8P.14), which is given by the approximating expression $q_a(x) = q_0(1 - e^{-\alpha x/L})$.

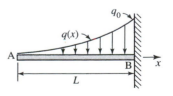

Figure 8P.12

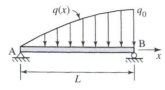

Figure 8P.13

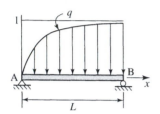

Figure 8P.14

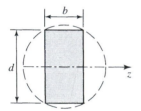

Figure 8P.17

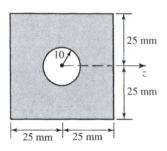

Figure 8P.19

Figure 8P.20

(a) Evaluate the reaction $(R_A)_a$ due to q_a. (b) Determine expressions for $V(x, \alpha)$ and $M(x, \alpha)$. (c) Determine expressions for $V(x)$ and $M(x)$ as $\alpha \to \infty$. To what loading does this correspond? (d) If the loading, instead, is expressed by the approximation $q_b(x) = q_0 \tanh(\alpha x/L)$, based on an analysis of the relative magnitudes of q_a and q_b [i.e. without evaluating the resulting reaction $(R_A)_b$], estimate if $(R_A)_b$ is greater or less than $(R_A)_a$.

Section 6

8.15: A beam of depth d is subjected to end couples. Upon bending, the resulting extensional strain ϵ_x at the bottom of the beam is $\epsilon_x = \epsilon_t > 0$ and $\epsilon_x = \epsilon_c < 0$ at the top. (a) Determine the radius of curvature $\overline{R}$ of the middle surface (i.e., the surface midway between the bottom and top of the beam) in terms of ϵ_c, ϵ_t and d. (b) What is the strain at the middle surface? Express the answer in terms of $\overline{R}$ and R, the radius of curvature to the neutral axis. (c) If the neutral axis lies on the middle surface, show that $\epsilon_t = -\epsilon_c$.

8.16: A long high-strength copper wire ($E = 120$ GPa), 3 mm in diameter, is wound about a 1.5-m diameter drum. Determine (a) the bending moment in the wire and (b) the maximum flexural stress in the wire.

8.17: * Material along the lateral surfaces of a long circular cylindrical log is to be sawed off to form a rectangular cross-section $b \times d$, as shown in Fig. (8P.17). If the resulting beam is to be subjected to a moment about the z-axis, determine (a) the optimal ratio d/b that minimises the curvature of the beam and (b) the ratio d/b that minimises the maximum flexural stress.

8.18: A beam having a square cross-sectional area of 64 mm^2 is subjected to a pure moment. The strain ϵ_x at the top of the cross-section is found to be 1600μ. Determine (a) the radius of curvature of the deformed beam and (b) the moment acting on any cross-section if the beam is made of a high-strength steel ($E = 200$ GPa) and (c) the maximum flexural stress.

8.19: A beam, whose cross-section is as shown in Fig. (8P.19), is subjected to a moment M about the z-axis. If the allowable stress is 120 MPa, determine the maximum permissible moment that can be applied.

8.20: A beam whose cross-section consists of a semi-circle of radius R, as shown in Fig. (8P.20), is subjected to a moment about the horizontal axis. (a) Determine the maximum flexural stress (in absolute value), i.e. $|\sigma_x|_{max}$ in terms of M and R. (b) What are the maximum tensile and compressive stresses if $R = 2$ cm and $M = 10,000$ N-cm.

8.21: A beam having a rectangular cross-section (i.e, width b and depth d) is subjected to a moment M_z. The beam is made of a material whose stress–strain relation is given by $\sigma = \alpha \epsilon^n$, where $\alpha > 0$ is a constant and $0 < n$ is an odd integer, i.e., $n = 1, 3, 5, \ldots$. (a) Show that the flexural stress σ is given by $\sigma = \left[\frac{2^{n+1}(n+2)}{bd^{n+2}}\right] M y^n$. (b) Show that the relation between the moment and the curvature $\kappa = 1/R$ of the beam is given by $\kappa = \frac{2[2(n+2)]^{1/n}}{d(\alpha bd^2)^{1/n}} M^{1/n}$. (c) Show that (a) and (b) above yield the Euler–Bernoulli relations if $\alpha \equiv E$ and $n = 1$.

8.22: A beam having a rectangular cross-section (with width b and depth d) is subjected to a moment M. The beam is made of a material whose behaviour can be described by

$$\sigma = \begin{cases} \alpha \epsilon^n, & \epsilon \geq 0 \\ -\alpha |\epsilon|^n, & \epsilon \leq 0 \end{cases}$$

where $\alpha \geq 0$ is a constant and $0 < n \leq 1$. Show that the expressions for the flexural stress σ and the radius of curvature κ for this case, $0 < n \leq 1$, are identical to those of Problem 8.21 for the case $1 \leq n$, n odd.

8.23: A prismatic beam is composed of two or more homogeneous materials, say 'a' and 'b', as shown in Fig. (8P.23), each having different material properties. Is it necessary to make an *assumption* that, when subjected to a bending moment M, all cross-sections remain plane and perpendicular to the deformed longitudinal axis or is this a valid *conclusion*, as in the case (considered in Section 6 of this chapter) of a beam consisting of a homogeneous material.

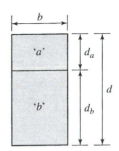

Figure 8P.23

8.24: A rectangular beam ($b \times d$), as shown in Fig. (8P.23), is composed of two materials, 'a' and 'b', having moduli of elasticity E_a and E_b respectively. Determine the location of the neutral axis as measured from the interface of the two materials.

8.25:* A rectangular beam ($b \times d$), as shown in Fig. (8P.25), is composed of thin laminates each having a different modulus of elasticity. The resulting inhomogeneous beam can be considered as having a varying modulus of elasticity, approximated by the expression $E(\eta) = E_0[1 + \beta(\eta/d)]$, where η is measured from the top of the beam and β is a constant. Determine (a) the location $\bar{y}$ of the neutral axis (measured from the top of the beam) and (b) the radius of curvature of the neutral surface if the beam is subjected to a moment M about the horizontal axis.

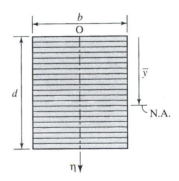

Figure 8P.25

8.26: The cross-section of a steel beam ($E = 200$ GPa) is constructed my means of plates, which are welded together to form an I-section, as shown in Fig. (8P.26).
   (a) Determine the section modulus $S = I/c$ of the cross-section. Compare the results with a similar standard wide flange section, e.g. a W203×36 section, as given in the tables of Appendix E (see footnote p. 269).
   (b) If the maximum allowable flexural stress in the beam is 150 MPa, what is the maximum permissible moment M that can be applied about the z-axis?
   (c) What is the resultant axial force in the flange abc under this positive moment M?
   (d) What is the resultant axial force in the upper part, bd, of the web under this moment?
   (e) What part of the total moment M is resisted by the two flanges and what part by the web?

What conclusions can be drawn from these answers?

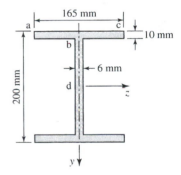

Figure 8P.26

8.27:* A beam having a trapezoidal cross-section, as shown in Fig. (8P.27), is subjected to a given moment M about the horizontal axis. Determine the maximum elastic moment $M = M_E$ under which the beam behaves elastically if σ_E is the elastic limit.

Sections 7 and 8

8.28: The overhanging beam shown in Fig. (8P.28a) is simply supported at B, fixed at E and contains a hinge at C. The cross-section and location of the neutral axis are shown in Fig. (8P.28b). The moment of inertia about the neutral axis is given as $I = 720$ cm⁴. (a) Draw the shear and moment diagrams and show all critical ordinates. (b) Determine the maximum flexural tensile and compressive stresses σ_x and indicate, by means of a sketch, at which cross-sections and at which points within the cross-section they occur. (c) Determine the maximum average shear stress $|\tau_{xy}|$, which occurs along the line c–c in the cross-section and indicate the direction of the shear stress acting on a positive x-plane.

8.29: The wooden beam, shown in Fig. (8P.29a), is subjected to several concentrated loads. The cross-section of the beam consists of four wooden components, which are

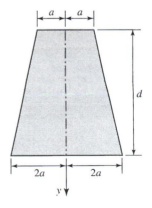

Figure 8P.27

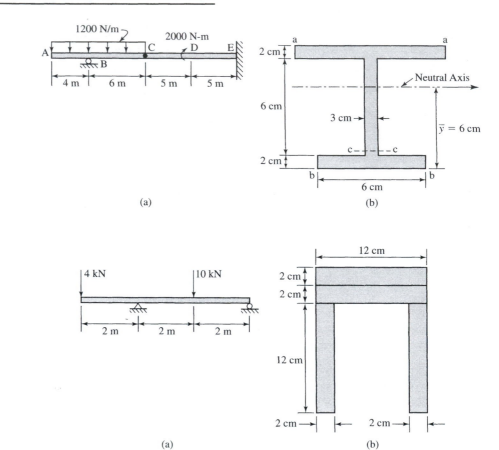

Figure 8P.28

(a) (b)

Figure 8P.29

(a) (b)

glued together, as shown in Fig. (8P.29b). Determine (a) the maximum flexural tensile stress existing in the beam, (b) the maximum compressive stress, (c) the maximum shear stress in the beam and (d) the maximum shear stress in the glue.

8.30: The wooden beam, subjected to two vertical forces applied at points B and C [Fig. (8P.30a)], has a cross-section as shown in Fig. (8P.30b). Determine (a) the maximum tensile and compressive flexural stresses σ_x in segment BC and indicate, by means of a figure, where they occur in the cross-section and (b) the average shear stress along line c–c, which exists in segment CD. Indicate, by means of a figure, the direction of τ acting on a positve x-plane.

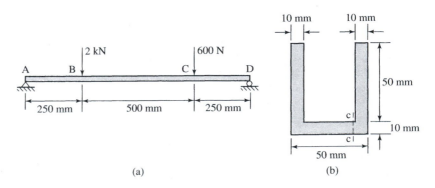

(a) (b)

Figure 8P.30

8.31: The S305 × 47 steel beam ($E = 200\,$GPa) shown in Fig. (8P.31) is subjected to two vertical forces, $P = 50\,$kN, which are applied at points B and C. Determine (a) the maximum tensile and compressive flexural stresses σ_x in segment BC, (b) the average shear stress along line c–c, which exists in segment AB, (c) the average shear stress in the web at the neutral axis (i) using the expression of Eq. (8.8.4) and (ii) assuming that the shear stress is distributed uniformly over the web of the section. What is the percentage difference in the two answers? and (d) the radius of curvature of the beam within segment BC.

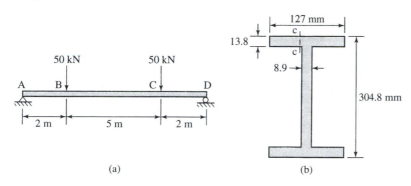

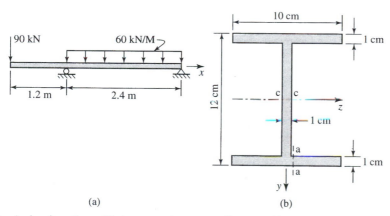

(a)   (b)   **Figure 8P.31**

8.32: Determine the average maximum shear stress components along lines a–a and c–c for the beam shown in Fig. (8P.32). Indicate the directions of the shear stress acting on a positive x-plane by means of a sketch.

(a)   (b)   **Figure 8P.32**

8.33:* A circular pipe with inner and outer radius a and b, respectively, is used as a cantilever beam. A vertical force P, acting along the axis of symmetry, as shown in Fig. (8P.33), is applied at the free end of the beam. (a) Based on symmetry considerations, what conclusions can be drawn about the shear flow in the cross-section? (b) Determine the average shear stress in the pipe, $\tau_{x\theta}$, along the line c–c, as a function of θ. (c) If the pipe is a thin wall section with $t = b - a \ll \overline{R}$, where $\overline{R} = (b + a)/2$ is the mean radius, show that the shear stress $\tau_{x\theta}$ is given by the approximate relation $\tau_{x\theta} = V \sin\theta / \pi \overline{R} t$. (d) Show that the resultant of the shear stresses existing in the entire cross-section is in equilibrium with the vertical load P.

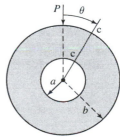

Figure 8P.33

8.34:* A beam having a triangular cross-section, as shown in Fig. (8P.34), is subjected to a positive shear force V_y. (a) Determine the maximum average shear stress τ_{xy} in the cross-section and the location of the line c–c along which this stress occurs. Express

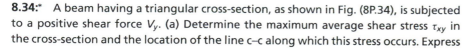

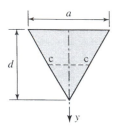

Figure 8P.34

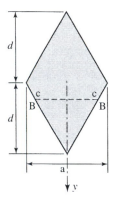

Figure 8P.35

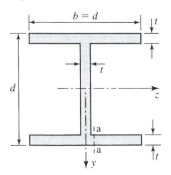

Figure 8P.36

Figure 8P.37

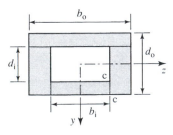

Figure 8P.38

Figure 8P.39

the answer as $\tau_{xy} = k\frac{V_y}{A}$, where A is the cross-sectional area of the beam. (b) Are the boundary conditions satisfied at the end points of the line c–c? Explain.

8.35: Repeat Problem 8.34 for a diamond-shaped cross-section shown in Fig. (8P.35).

8.36: The cross-section of a beam shown in Fig. (8P.36) consists of a flange whose width is equal to the depth of the beam ($b = d$), both having the same thickness t. The beam is subjected to a shear force acting in the y-direction. Determine the possible range of the ratio $K = \frac{(\tau)_{NA}}{(\tau)_{a-a}}$, namely the ratio of the average shear stress in the web at the neutral axis to that along line a–a.

8.37: A beam is made up of four wooden components, which are glued together to form a cross-section, as shown in Fig. (8P.37). (a) Determine the average shear stress in the glue, $(\tau)_y$ due to a vertical shear force V_y acting in the y-direction and $(\tau)_z$ due to a shear force V_z acting in the z-direction. (b) If the cross-section is square (i.e., $b_o = d_o$ and $b_i = d_i$) and if $V_y = V_z$, show that $\frac{(\tau)_y}{(\tau)_z} = \frac{b_o}{b_i}\left(\frac{b_o^2 + b_i^2}{b_o^2 - b_i^2}\right)$.

8.38: A beam made up of four wooden components, which are glued together to form a cross-section, as shown in Fig. (8P.38), is subjected to a shear force V_y acting in the y-direction. Determine the average shear stress in the glue along the line c–c.

8.39:* A beam ABC, having a triangular cross-section, is simply supported at points A and B (where $a \leq L$), as shown in Fig. (8P.39), and is subjected to a uniformly distributed load w. (a) Determine the ratio a/L such that the largest tensile stress existing at points c of the cross-section is equal to the largest tensile stress existing at points d. (b) Based on the value a/L obtained above, sketch the resulting shear and moment diagrams and show all critical ordinates. (c) Based on the same value of a/L, determine the maximum shear stress $|\tau_{xy}|$ existing in the beam if the cross-section is an equilateral triangle with sides b. Express the answer in terms of w, L and b. [See comment (iii) following Example 8.13.]

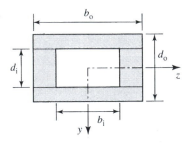

8.40: Two C305 × 45 channels are connected to two plates, each 260 mm × 15 mm, by means of bolts, as shown in Fig. (8P.40), to form the cross-section of a simply supported beam. If the bolts are spaced 80 cm apart, and each bolt can carry an allowable force in shear of 450 N, determine the maximum load P that can be applied at the centre of the beam (assuming that this shear criterion is the governing criterion).

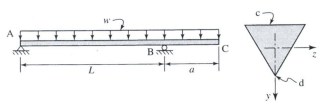

8.41: Four angles L102 × 102 × 9.5 are connected to a plate 12-mm thick by means of two 20-mm diameter bolts (spaced 100 mm apart along the longitudinal axis) to form a cross-section, as shown in Fig. (8P.41). The section is used as a cantilever beam loaded at the free end by a force $P = 240$ kN. Assuming the shear stress is distributed uniformly over the cross-section of each bolt, determine the average shear stress in the bolts.

8.42: The cantilevered beam shown in Fig. (8P.42) has a rectangular cross-section of constant width but of varying depth $h(x)$. The beam is loaded at the free end. (a) Assuming that cross-sections remain plane and perpendicular to the deformed longitudinal axis, determine the required variation of the depth $h(x)$ in order that the maximum value of the flexure stress at all cross-sections be constant over the length of the beam, $0 \leq x < L$. (Such a beam is referred to as a *beam of constant strength*.) (b) Recalling that the expression $\sigma_x = My/I(x)$ is only a good approximation if the lateral surfaces of the beam have a small slope θ with respect to the x-axis, say $|\theta| = 5°$, use this criterion and the answer from part (a) to determine the range of x/L for which the given expression for σ_x is a good approximation for a given beam with $h_0/L = 1/15$.

8.43:* A simply supported wooden beam whose constant width is b and whose depth $h(x)$ varies with x is to be used as a beam of constant strength such that the maximum flexural stress is constant over its entire length L, $0 < x < L$. The beam is subjected to a distributed load $q(x) = q_0 \sin(\pi x/L)$, as shown in Fig. (8P.43). The maximum permissible flexural stress is given as σ_0. (a) Determine the maximum required depth h_0 at $x = L/2$ in terms of the given parameters and loading. (b) Determine the required variation of $h(x)$ in terms of h_0. (c) Referring to the limitation given in part (b) of Problem 8.42, and using the same criterion, determine the range of the loading q_0 that can be applied to the beam while satisfying the given criterion over its entire length. (d) Using the result of (c) above, determine the maximum loading q_0 and the required depth h_0 if $\sigma_0 = 10$ MPa, $b = 10$ cm and $L = 3$ m.

Section 10

8.44: A cantilevered steel beam whose cross-section is specified as a W254×45 section, is loaded as shown in Fig. (8P.44). Determine the total allowable load on the beam if the allowable flexural stress and shear stress are specified as $\sigma = 150$ MPa and $\tau = 80$ MPa, respectively. (Assume that the shear stress in the web is uniformly distributed over the area of the web.)

8.45: A simply supported beam of length L [whose cross-section is, say, a standard (S), or channel (C) section] is to carry a load W (N), uniformly distributed along its length. Assume that all properties of the cross-section are known, namely the section modulus S, the depth of the beam d and the dimensions of the flange and web. The allowable flexural and shear stress are given as σ_{allow} and τ_{allow}. (a) Determine, the allowable load $(W)_f$ according to the flexure criterion and $(W)_s$ according to the shear criteria in terms of the given parameters. (Assume that the shear force is resisted by the web and that the shear stress in the web is uniformly distributed over the area of the web.) (b) Determine the ranges of L for which the flexure criterion governs and for which the shear criterion governs the design. What conclusion can be drawn from this result? (c) For a steel beam of length 5 m with a cross-section S203×34, determine the allowable load W if $\sigma_{allow} = 200$ MPa and $\tau_{allow} = 100$ MPa.

8.46: A wooden beam ABC, containing a hinge at B, is loaded, as shown in Fig (8P.46a), by means of a uniformly distributed load between A and B. Several wooden components are nailed together to form a cross-section, as shown in Fig. (8P.46b), with

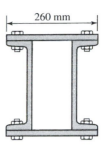

260 mm

Figure 8P.40

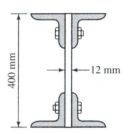

400 mm — 12 mm

Figure 8P.41

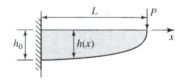

L | P
h_0 | $h(x)$ | x

Figure 8P.42

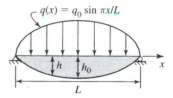

$q(x) = q_0 \sin \pi x/L$

h | h_0 | x

L

Figure 8P.43

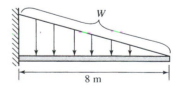

W

8 m

Figure 8P.44

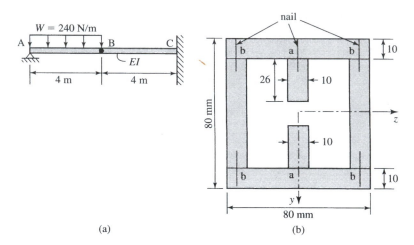

Figure 8P.46

(a)   (b)

$I_{zz} = 2 \times 10^6$ mm⁴. The maximum shear force that can be carried by each nail is given as $f = 300$ N. What is the maximum permitted spacing of (a) nails 'a' in the segment BC and (b) nails 'b' in the segment BC?

8.47: A beam is made up of four component parts, which are glued together to form a hollow box section, as shown in Fig. (8P.47). The allowable flexural stress of the wood is given as 25 MPa and the allowable stress in the glue is given as 2 MPa. The beam is to be used as a simply supported beam 2 m in length, and subjected to a load W (N), which is uniformly distributed over its entire length. Determine the maximum load W that the beam can carry.

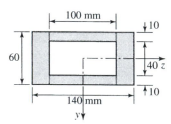

Figure 8P.47

8.48: A W254×67 steel beam CD, 6 m in length, is subjected to a force of 200 kN, which is to be applied via a steel beam AB of length L placed symmetrically as shown in Fig. (8P.48). The allowable flexural and shear stress are given as 140 MPa and 85 MPa, respectively. (a) Determine the shortest permissible length L of the beam AB. (b) Choose the most suitable structural W-section (namely a beam having the least weight per unit length) from the tables (see Appendix E) that may be used for the beam AB.

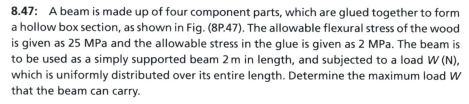

Figure 8P.48

8.49:* Consider a steel beam having either a W- or S-section with given I_0, section modulus S_0 and depth d. Assume, as in Example 8.16, that it is required to strengthen the beam in flexure by adding plates having dimensions $b \times t$ at the top and bottom flanges [see Fig. (8.10.3b)] thus increasing the section modulus to $S > S_0$. Show that if the thickness $t \ll d$, the increased modulus is given by the approximate expression $S = S_0 + btd$.

8.50: A steel beam ABCD, containing a hinge at C, is simply supported at B and fixed at D, as shown in Fig. (8P.50). Using an allowable flexure stress $\sigma = 180$ MPa, an S-section of minimum weight was originally chosen to support a load $P = 60$ kN at A. (a) What

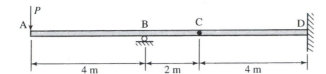

section was chosen? (b) Due to changes of the loading, it is now required to strengthen the beam to resist flexure in the segment CD in order to carry a load $P = 120$ kN at A by attaching plates (having the same width as the flanges) to the top and bottom flanges by means of two bolts at each flange, as shown in Fig. (8.10.3). Using the result given in Problem 8.49, determine the required thickness of the plates. Check that the resulting flexural stress is within the allowable limits. (c) Determine the maximum permissible spacing of the bolts, s, along the longitudinal axis if the thickness of the plates is 25 mm, if the bolts are 20 mm in diameter and if the allowable shear stress in the bolts is $\tau = 50$ MPa. (Assume the shear stress in the bolts is the average stress distributed uniformly over the cross-section.)

8.51:* A beam is made up of two identical angles, which are connected by means of two bolts, as shown in Fig. (8P.51). When the beam is subjected to vertical loads acting along the y-axis, should one expect the bolts to undergo shear? Justify the answer (i) by physical reasoning and (ii) by analytical reasoning based on Eq. (8.8.4).

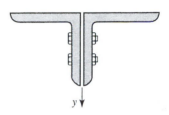

8.52: A cantilever beam is made up of two L127 × 127×9.5 angles and a plate (254 mm × 10 mm). The angles and plate are connected by means of 20-mm diameter bolts 'a' and 'b' as shown in Fig. (8P.52). A load $P = 2000$ N acts at the free end of the beam in the y-direction. Determine (a) the maximum flexural tensile and compressive stress in the beam and (b) the maximum shear stress in bolts 'a' and 'b' (assuming that the shear stress is uniformly distributed over their cross-sections) if the bolts are spaced at 40-cm intervals along the longitudinal axis.

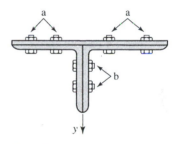

Section 11

8.53: By bonding two aluminium bars ($E = 70$ GPa) to two brass bars ($E = 105$ GPa), it is possible to form two different composite cross-sections of a beam, as shown in Figs. (8P.53a and b). If a moment $M = 2$ kN-m acts about the z-axis, determine the maximum flexural stress in the aluminium and the brass for each case.

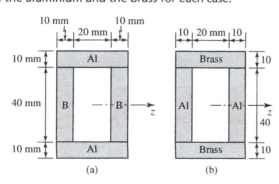

(a)   (b)

8.54: An aluminium bar ($E = 70$ GPa) and a steel bar ($E = 200$ GPa) are bonded together to form a composite cross-section, as shown in Fig. (8P.54). Determine (a) the maximum flexural stress in the steel and aluminium if the beam is subjected to a pure positive moment $M = 1200$ N m about the horizontal axis and (b) the radius of curvature of the deformed beam.

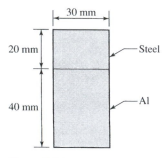

Figure 8P.54

8.55: A rectangular wooden beam (150 mm × 250 mm) with $E = 12$ GPa is reinforced by means of two steel plates ($E = 200$ GPa), as shown in Fig. (8P.55). (a) If the allowable stress in the wood and steel are given as 12 and 200 MPa, respectively, determine the maximum permissible moment that can act about the z-axis. (b) Determine the radius of curvature of the beam under this moment.

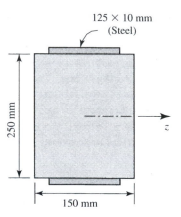

Figure 8P.55

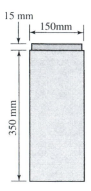

Figure 8P.56

8.56: A wooden beam ($E = 10$ GPa) having a rectangular shape (150 mm × 350 mm) is reinforced by means of a steel plate ($E = 200$ GPa), which is fastened securely to its top face, as shown in Fig. (8P.56). Determine the maximum flexural stress in the wood when the maximum stress in the plate is 75 MPa.

8.57:* The cross-section of a rectangular reinforced concrete beam, reinforced by steel rods as shown in Fig. (8P.57a), can be considered as a composite of two materials: concrete and steel. As is characteristic of all brittle materials, concrete being such a material, is very weak in tension and cracks form in the tension zone when the beam undergoes flexure. It is therefore usual, when designing reinforced concrete beams according to elastic theory, to assume that the concrete can withstand compression but no tension, i.e. the concrete is assumed to withstand only flexural stresses $\sigma \leq 0$. Tension in the beam is therefore assumed to be carried only by the steel rods whose (total) cross-sectional area is denoted as A_s. Based on this assumption, the transformed section is as shown in Fig. (8P.57b).

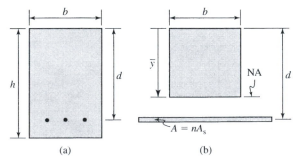

Figure 8P.57

(a) Denoting the modulus of elasticity of the concrete and steel as E_c and E_s, respectively, and letting $E_s = nE_c$, show that the location of the neutral axis $\bar{y}$ of the rectangular beam is given by the expression

$$\bar{y} = \frac{nA_s}{b}\left[\left(1 + \frac{2bd}{nA_s}\right)^{1/2} - 1\right].$$

(b) Show that if the beam is subjected to a positive moment M about the neutral axis, the maximum stress in the concrete and steel is given by

$$\sigma_c = -\frac{6M}{b\bar{y}(3d - \bar{y})}, \qquad \sigma_s = \frac{3M}{A_s(3d - \bar{y})}.$$

8.58: A reinforced concrete beam, 8 m in length, has a rectangular cross-section as shown in Fig. (8P.57a), with $b = 200$ mm and $h = 350$ mm. Four steel rods, each 15 mm in diameter, are placed 75 mm from the bottom of the beam. A concentrated force $P = 10$ kN, located at the centre of the beam, acts in the y-direction. If $E_s = 200$ GPa and $E_c = 15$ GPa, determine the maximum flexural stress in the concrete and the steel.

8.59:* In designing a rectangular reinforced concrete beam, it is often desired to achieve a 'balanced design', namely one for which the maximum stress in the concrete and steel are equal to the maximum allowable stress in the two materials, σ_c and σ_s, respectively. Show that for such a balanced design, the location $\bar{y}$ of the neutral axis is given by

$$\bar{y} = \frac{d}{1 + \sigma_s E_c / \sigma_c E_s},$$

where $\bar{y}$ and d are as shown in Fig. (8P.57b) and E_c and E_s are the respective moduli of elasticity.

Section 12

8.60: A cantilevered beam is subjected to axial and transverse loads, as shown in Fig. (8P.60). Determine the flexural and shear stresses at points a, b and c due to the applied loading.

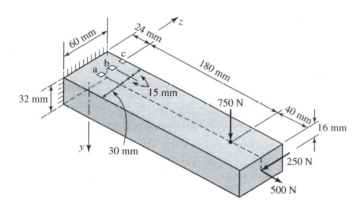

Figure 8P.60

8.61: An eccentric vertical load P is applied (with an eccentricity e) to the free end of a cantilever beam having a circular cross-section as shown in Fig. (8P.61). Determine the maximum shear stress that exists in the beam and indicate, by means of a sketch, where it occurs in the cross-section.

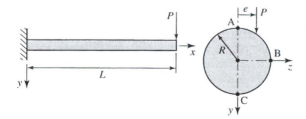

Figure 8P.61

8.62:* An elastic circular bar of radius R and length L, fixed at $x=0$, is subjected to an eccentric force P at its free end. The force, lying in the x–y plane and inclined at an angle α with respect to the x-axis, acts at point B ($x=L$, $y=e$) as shown in Fig. (8P.62a). (a) Locate the position of the neutral axis at the section $x=0$ in terms of the given parameters of the problem, i.e. determine $y_0 = y_0(L, R, e, \alpha)$. (b) If P acts in the x-direction ($\alpha=0$), where must the force be applied (i.e., what must be the value of e) such that $\sigma_x=0$ at point C of Fig. (8P.62a)? (c) If P, acting in the x-directions is applied at point B [as found in (b) above], and an additional torque $T = Pe$ is applied, as shown in Fig. (8P.62b), what is the maximum shear stress τ_{max} at point D? On which plane (defined by its normal n with respect to the x-axis) does τ_{max} act?

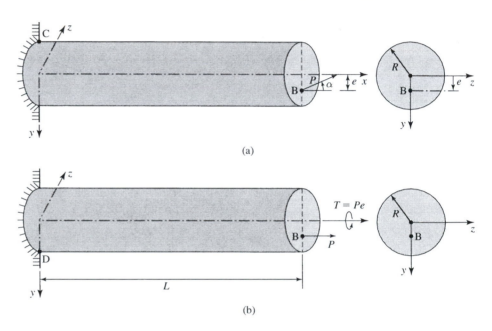

(a)

(b)

Figure 8P.62

8.63: A simply supported steel beam ABC is loaded as shown in Fig. (8P.63), where $P = 200\,kN$ is inclined by an angle of $30°$ with respect to the x-axis. In addition, an axial compressive force acts at the end C. Determine the maximum tensile and compressive stresses σ_x that exists (i) immediately to the left of point B and (ii) immediately to the right of B if the cross-section is a W305×97 section.

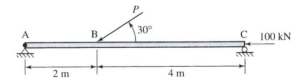

Figure 8P.63

8.64:* An eccentric compressive force P is applied at point B in the x-direction with an eccentricity with respect to the centroidal longitudinal axis of a beam having a rectangular cross-section $b \times d$, as shown in Fig. (8P.64). Show that the resulting stress σ_x is compressive (i.e., $\sigma_x \leq 0$) throughout the cross-section, provided that the force is applied within the 'core' of the cross-section, as shown in Fig. (8.12.6).

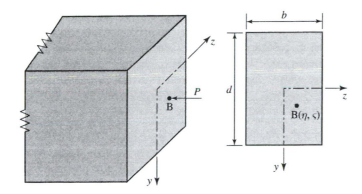

Figure 8P.64

8.65:   A beam ABCD having a square cross-section with sides b is supported at points E and F and is loaded as shown in Fig. (8P.65). Determine the maximum flexural stresses and the maximum shear stress that exist in segments BC and CD. *Note: $b \ll L$.*

8.66:   A force $P = 500$ N is applied at point A to a bent rod, as shown in Fig. (8P.66). Determine the normal and shear stresses at points B and C.

Figure 8P.65

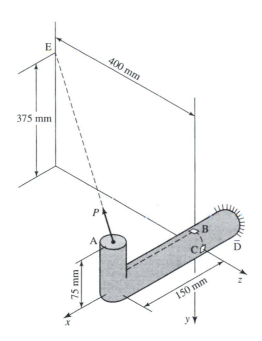

Figure 8P.66

8.67:*   Three strain gauges are attached to an aluminium rod ($E = 70$ GPa), 1 cm × 1 cm in cross-section at points A, B and C, as shown in Fig. (8P.67). When P and F are applied, the following readings are obtained: $\epsilon_A = 550\mu$, $\epsilon_B = 400\mu$ and $\epsilon_C = -300\mu$. Determine the magnitude of P, F and the position b of P as measured from B.

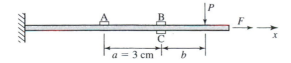

Figure 8P.67

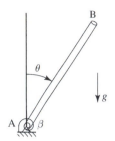

Figure 8P.68

Figure 8P.69

Figure 8P.70

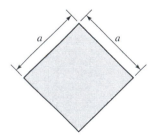

Figure 8P.73

8.68:   Given a cylindrical rod AB of length L, diameter d and whose material density is ρ (N/m^3). As shown in Fig. (8P.68), the rod is supported at A by means of a linear torsional spring having constant β (N-m/rad) such that when the rod is inclined by an angle θ with respect to the vertical axis, the spring exerts a moment $M = \beta\theta$. Determine the minimum angle θ with respect to the vertical axis at which a tensile stress occurs in the rod.

8.69:   A circular wire having diameter d is wound as a helix (with radius R) to form a coiled spring, as shown in Fig. (8P.69). If a tensile force P is exerted at the two ends, determine the maximum shear stress within the wire. (*Note*: The average maximum shear stress in a rod of circular cross-section of area A, when subjected to a shear force V, is $\tau = \frac{4}{3}\frac{V}{A}$.)

Section 13

8.70:   The cross-section of a beam, whose material behaves as an ideal elastic–plastic material with yield stress in tension and compression $\pm\sigma_0$ [see Fig. (8P.70b)], has a triangular shape as shown in Fig. (8P.70a). Determine (a) the maximum moment M_E that can be applied about a horizontal axis of the cross-section for its behaviour to remain elastic, (b) the location y_p of the neutral axis (measured from the apex) as the moment reaches the fully plastic moment M_P and (c) the value of M_P and the ratio M_P/M_E.

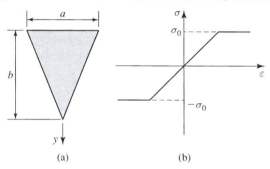

(a)   (b)

8.71:   A steel beam, assumed to behave as an elastic–perfectly plastic material (with $E = 200$ GPa and yield stress $\sigma_0 = 200$ MPa), has a square cross-section with area $A = 100$ cm^2. The beam is bent by end couples, which cause strains $\epsilon = 0.004$ at the top of the beam. Determine (a) the depth of yielding, d_y, within the cross-section and (b) the magnitude of the applied bending couples.

8.72:   Determine the maximum elastic and plastic moments for a W762 × 196 steel beam, which is assumed to behave as an elastic–perfectly plastic material with $\sigma_0 = 250$ MPa.

8.73:   Determine the maximum elastic and plastic moments for a beam, which is assumed to behave as an elastic–perfectly plastic material with yield point $\pm\sigma_0$ if the cross-section is a square with sides a and the moment is applied about its diagonal, as shown in Fig. (8P.73).

8.74:*   A beam with a cross-section, shown in Fig. (8P.74), is subjected to bending about the horizontal axis. Determine the ultimate plastic moment M_P if the beam material is elastic–perfectly plastic with yield stress $\pm\sigma_0$.

8.75:*   A beam of rectangular cross-section consisting of a linearly elastic core (with modulus of elasticity E) bonded to top and bottom plates, as shown in Fig. (8P.75), is subjected to a moment about the z-axis. The plates consist of a rigid-plastic material with a yield stress $\pm\sigma_0$ [see Fig. (4.6.2)]. (a) What is the smallest value of the bending moment M_{cr} that will cause the beam to deform? (b) Assuming that plane cross-sections remain plane and perpendicular to the deformed longitudinal axis, obtain the bending-curvature relation.

8.76: * A beam, in a state of pure bending, having a rectangular cross-section $b \times d$ is subjected to a moment about the z-axis, as shown in Fig. (8P.76a). The linear strain-hardening behaviour of the beam is described by the stress–strain curve of Fig. (8P.76b), where E_1 and E_2 are the moduli of elasticity in the two regions.

   (a) Show that if the flexural strains at the bottom and top of the cross-section are $\pm\epsilon_m$ respectively, the moment M acting on the beam is given as

$$M = \frac{bd^2}{12}\sigma_0\left\{\left[3 - \left(\frac{\epsilon_0}{\epsilon_m}\right)^2\right]\left(1 - \frac{E_2}{E_1}\right) + 2\frac{E_2}{E_1}\left(\frac{\epsilon_m}{\epsilon_0}\right)\right\}.$$

   (b) (i) Determine M if the strain $\epsilon_m = \pm 3.6 \times 10^{-3}$ and if $b = 20\,\text{mm}$, $d = 30\,\text{mm}$, $E_1 = 100\,\text{GPa}$, $E_2 = 50\,\text{GPa}$, $\sigma_0 = 120\,\text{MPa}$ and (ii) sketch the stress distribution in the cross-section due to the moment.

   (c) The beam specified in (b) is now unloaded, i.e. the moment is reduced to zero. Sketch the stress distribution, σ_{unload} acting on the cross-section due to this unloading. Determine the residual stresses σ_{res} and sketch their distribution in the cross-section.

Figure 8P.74

Review and comprehensive problems

8.77: The differential relations for a beam subjected to a distributed load $q(x)$ were found to be [see Eqs. (8.3.1)–(8.3.3)]

$$\frac{dM(x)}{dx} = V(x), \quad \frac{dV(x)}{dx} = -q(x) \quad \text{and} \quad \frac{d^2M(x)}{dx^2} = -q(x).$$

Show, by a similar derivation that if, in addition, a distributed moment $m(x)$ (N-m/m) also acts on the beam about the z-axis [see Fig. (8P.77)], the differential relations are

$$\frac{dM(x)}{dx} = V(x) + m(x), \quad \frac{dV(x)}{dx} - -q(x) \quad \text{and} \quad \frac{d^2M(x)}{dx^2} = -q(x) + \frac{dm(x)}{dx}.$$

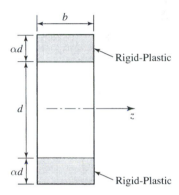

Figure 8P.75

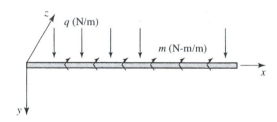

Figure 8P.77

8.78: A composite material of infinite length is made of periodically spaced laminates having different moduli of elasticity, as shown in Fig. (8P.78a). When subjected to loading and/or temperature changes in the x-direction, it is found that shear stresses $\tau = \mp A\sin(2\pi x/L)$ (where $A \geq 0$ is a constant), acting in the x-direction, exist along the ($\pm$) interface of a typical laminate over a given length L within the composite, as shown in Fig. (8P.78b). (a) If the thickness of the typical laminate is h, determine the resulting moment $M(x)$ at any cross-section of the fibres. (*Note:* Assume that $M = 0$ at $x = 0$.) (b) Determine the curvature $\kappa(x)$ of the laminate if its modulus of elasticity is E and, based on the curvature, sketch the approximate deformation of the laminate assuming it undergoes no deflection in the vertical direction at $x = 0$ and $x = L$.

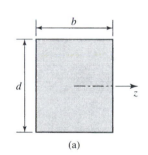

Figure 8P.76

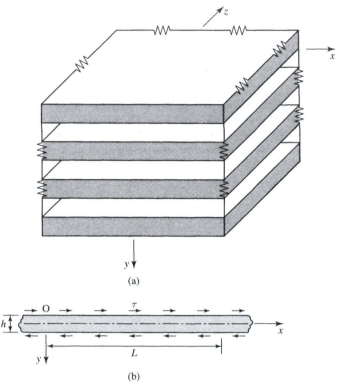

Figure 8P.78

8.79: A beam having a cross-section, as shown in Fig. (8P.79), is made of an ideal elastic–plastic material such that the yield stress is $\pm\sigma_0$ in tension and compression, respectively, and τ_0 in shear. Determine (a) the maximum elastic moment M_E about the z-axis in terms of σ_0 and R, (b) the maximum shear V that can be exerted for the beam to remain elastic and (c) the ultimate plastic moment M_P in terms of σ_0 and R.

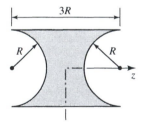

Figure 8P.79

8.80:* A beam of length L, consisting of a material for which the allowable stress is σ_{allow}, rests on a frictionless surface and is subjected to two symmetrically applied loads P at positions from the ends represented by $a > 0$, as shown in Fig. (8P.80). The beam is to be designed using various available square cross-sections, $b \times b$. (a) Assuming that the reactive pressure exerted by the surface on the beam is constant over the length L, sketch the shear and moment diagrams and label all maxima and minima in terms of P, L and a. (b) Segment BD is coated at the top of the beam with a thin brittle adhesive (namely one which it is assumed cannot support any tension). Based on a flexural criterion, determine the location of the loads that leads to an optimal design of the beam, i.e. which minimises the dimension b. What is the required dimension b?

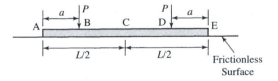

Figure 8P.80

8.81: A simply supported beam, having a linearly varying depth $d(x)$, as shown in Fig. (8P.81), carries a load P at the centre. Determine the cross-section at which the maximum flexural stress σ_x occurs and find $(\sigma_x)_{max}$. (Assume that the taper is sufficiently small so that warping of any cross-section is negligible.)

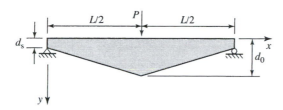

Figure 8P.81

8.82: A wooden cantilever beam of length L, having a rectangular cross-section with constant width b but varying depth $h(x)$, carries a total load W (N), which varies linearly as shown in Fig. (8P.82). The maximum allowable flexural stress is 10 MPa. Assume that plane cross-sections remain plane. (a) Determine (i) the required depth h_0 at the support B if $L/b = 40$ and $W = 5$ kN and (ii) the required variation $h(x)$ if the maximum flexural stress in the beam is to be the same at all cross-sections. (b) It is known that the expression $\sigma_x = My/I(x)$ is only a good approximation if the lateral surfaces of the beam have a small slope with respect to the x-axis, θ, say $|\theta| = 5°$. Using this criterion and the answer of part (a), determine the range of x/L for which the given expression for the flexural σ_x is a good approximation when the beam is subjected to the given load $W = 5$ kN if $L = 2$ m. (c) For the same ratio b/L, allowable stress σ and total load W as given above, determine the required length L of the beam if the above criterion is satisfied at all cross-sections.

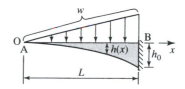

Figure 8P.82

8.83: Repeat part (a) of Problem 8.82 for the same loading W and ratio b/L if the average shear stress reaches the same maximum value $\tau = 2$ MPa at all cross-sections.

8.84: A composite beam whose cross-section [as shown in Fig. (8P.84)] consists of two materials 'A' and 'B' having different moduli of elasticity, $E_B \geq E_A$, is subjected to a moment about the z-axis. Determine the maximum flexural stress in each of the materials.

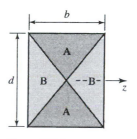

Figure 8P.84

8.85: A pipe consists of an external steel ($E = 200$ GPa) pipe bonded to an internal aluminium ($E = 70$ GPa) pipe to form a composite cross-section, as shown in Fig. (8P.85). Determine (a) the maximum stress in the aluminium and steel if the cross-section is subjected to a moment of 2500 N-m and (b) the radius of curvature of the composite beam at the cross-section.

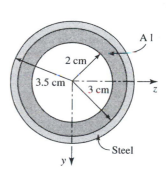

Figure 8P.85

8.86:* A beam, having a rectangular cross-section $b \times d$ [Fig. (8P.86b)] (whose second moment about its horizontal centroidal axis is I), is made of a material having different moduli of elasticity in tension and compression, namely E_t and E_c respectively, as shown in Fig. (8P.86a). The beam is subjected to a positive bending moment M about the z-axis. (a) Show that the curvature–moment relation of the deformed beam can be written as $\kappa = \frac{M}{E^* I}$,

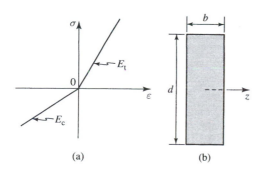

Figure 8P.86

where E^*, the *equivalent modulus of elasticity*, is

$$E^* = \frac{4E_t E_c}{[\sqrt{E_t} + \sqrt{E_c}]^2}.$$

(b) Determine the maximum flexural tensile stress.

8.87:* A beam, whose cross-section consists of a semi-circle, is subjected to a vertical shear force V acting in the y-direction. Show that the average shear stress τ_{xy} along the line a–a of the cross-section shown in Fig. (8P.87), is given by

$$\tau_{xy} = \frac{24\pi V}{(9\pi^2 - 64)R^2 \sin\theta}\left[\sin^3\theta - \frac{1}{\pi}(2\theta - \sin 2\theta)\right].$$

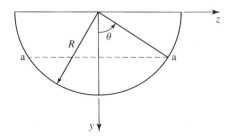

Figure 8P.87

(*Note*: See computer-related Problem 8.97.)

8.88:* Given an elastic beam with modulus of elasticity E, whose cross-section is a rectangle $(b \times d)$ and which has an initial radius of curvature ρ, as shown in Fig. (8P.88a). A moment is applied in order to straighten out the beam [see Fig. (8P.88b)].

  (a) Noting that while the beam undergoes deformation, there exists a surface for which the strain $\epsilon = 0$ (namely, the neutral surface represented by the line N–N'), and letting ρ be the radius of curvature to the neutral surface, show that when the beam is straightened the strain at any arbitrary fibre is $\epsilon = -\frac{\eta}{\rho+\eta}$, where η denotes the perpendicular distance from this surface to an arbitrary fibre.
  (b) Show that the neutral surface does *not* pass through the centroid of the cross-section; i.e. $c_1 \neq c_2$, where c_1 and c_2 are shown in the figure. (Note that $c_1 + c_2 = d$.)
  (c) (i) Show that the location of the neutral axis (which depends on the ratio $\alpha \equiv d/\rho$) is given by

$$\frac{c_2}{d} = \frac{(1-\alpha)e^\alpha - 1}{\alpha(1 - e^\alpha)}.$$

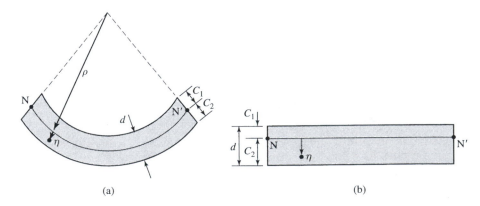

(a)   (b)

Figure 8P.88

(ii) Show that the moment required to straighten out the beam is given by the expression

$$M = E b \rho^2 \left[\frac{d}{\rho} - \frac{d(c_2 - c_1)}{2\rho^2} - \ln\left(1 + \frac{c_2}{\rho}\right) + \ln\left(1 - \frac{c_1}{\rho}\right) \right].$$

(d) Obtain a simplified expression,

$$\frac{c_2}{d} = \frac{1}{2}\left(1 + \frac{d}{6\rho}\right)$$

for the location of the neutral axis, c_2/d, if the curvature is relatively small, i.e. if $d/\rho \ll 1$.

(e) Show that if the curvature is relatively small, the moment required to straighten the beam is given by $|M| = EI/\rho$, namely the same moment as is required to cause a straight beam to be bent into a curve with radius ρ. (We conclude that the Euler–Bernoulli relations may be applied to a beam with initial curvature $\kappa = 1/\rho$ for the case $d/\rho \ll 1$. Clearly, this is not so if the condition $d/\rho \ll 1$ is not satisfied.)

[*Hint*: For parts (d) and (e), use appropriate series expansions for the logarithmic and exponential terms appearing in (c) above.

(*Note*: See computer-related Problem 8.101.)

8.89:* Two separate thin elastic strips 'A' and 'B', having the same rectangular cross-section $b \times 2c$, (with moment of inertia I_0) are bent by end couples M_A and M_B, as shown in Fig. (8P.89), such that the radius of curvature of the common interface I–I is R_0 and such that no separation exists at the interface. (a) Show that the required relation between the moments is $M_A = M_B[1 + 2c/R_0 + 2(c/R_0)^2 + \cdots]$ and therefore if $c \ll R_0$, $M_A \simeq M_B$. (b) The two strips are subsequently glued together and the moments are then removed. Assuming that $c \ll R_0$, (i) determine the resulting flexural stress in each strip and (ii) sketch the stress distribution in the combined cross-section. (*Note* : For the case $c \ll R_0$, the Euler–Bernoulli relation, $M = EI/R$, remains valid [see part (e) of Problem 8.88].) (c) Determine the resulting final radius of curvature of the interface, R, of the combined strips in terms of R_0.

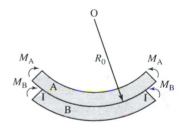

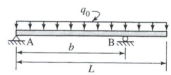

Figure 8P.89

8.90:* A beam of length L, simply supported at A and at a variable point B as shown in Fig. (8P.90), is subjected to a uniform load q_0 over its entire length. It is desired to optimise the design of the beam by minimising the (absolute) value of the maximum moment in the beam. (a) Determine the position of B (i.e., find b), that yields this optimal solution. (b) Determine the maximum resulting (absolute) value of the moment in the beam. (c) Sketch the variation of $|M|_{max}$ between A and B and that of $|M_B|$ as a function of b/L for $0 < b/L \leq 1$.

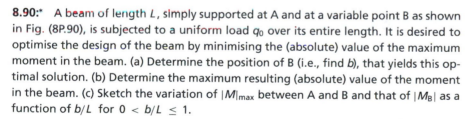

Figure 8P.90

8.91: An elastic beam having a square cross-section ($a \times a$) is subjected to a moment $M \equiv M_z$, about the z-axis, as shown in Fig. (8P.91a). (a) Determine the maximum flexural stress $|(\sigma_x)_a|$. (b) By removing material at the top and bottom corners, a cross-section as shown by solid lines in Fig. (8P.91b) is obtained. Show that the resulting maximum stress $(\sigma_x)_b$ is given as

$$(\sigma_x)_b = \frac{3M}{c^3}\left[\frac{1 - \alpha}{1 - \alpha^2(3\alpha^2 - 8\alpha + 6)}\right]$$

[where α represents the cut over a fraction of the depth, see Fig. (8P.91b)] and thus obtain the ratio $R_\sigma \equiv \frac{(\sigma_x)_a}{(\sigma_x)_b} = (\alpha - 1)^2(3\alpha + 1)$. (c) For which values of α is $R_\sigma > 1$ and $R_\sigma < 1$? What does this imply physically?

(*Note*: See computer-related Problem 8.100.)

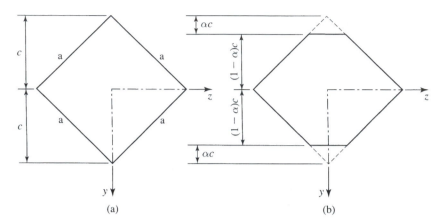

Figure 8P.91

(a)   (b)

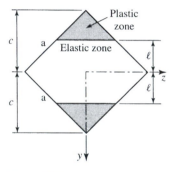

Figure 8P.92

8.92: A moment $M_E \le M \le M_P$ causing plastic behaviour is applied about the diagonal of the cross-section of a square beam whose diagonal length is $d = 2c$, as shown in Fig. (8P.92). The beam material is elastic–perfectly plastic with yield point $\pm\sigma_0$. (a) Determine the relation between M and the location of ℓ, the interface of the elastic and plastic zones. (b) Verify that $M_E = \frac{\sigma_0 d^3}{24}$ and $M_P = \frac{\sigma_0 d^3}{12}$ when $\ell = d/2$ and $\ell = 0$, respectively. (c) Expressing M/M_P as a function of ℓ/d, plot the non-dimensional ratio M/M_P vs. ℓ/d.

8.93: In deriving the expression for the flexural stress under pure bending, $\sigma_x = \frac{My}{I}$, the stress σ_y is assumed to be negligible; i.e. $\sigma_y \ll \sigma_x$. Consider now, for example, a simply supported beam of length L, having a rectangular cross-section $b \times d$, subjected to a uniformly distributed load q_0, as shown in Fig. (8P.93). It is clear that near the ends $x = 0$ and $x = L$, the moment M is very small and therefore, as x approaches the end points, $\sigma_x \to 0$. Since the maximum compressive value of σ_y directly under the load is $|\sigma_y| = q_0/b$, the assumption $|\sigma_y/\sigma_x| \ll 1$ is clearly contradicted in this region. (a) Show that if $L/d \gg 1$, the region where this contradiction occurs is negligible with respect to the length L and may therefore be disregarded in an analysis. For example, estimate $|\sigma_y/\sigma_x|$ directly under the load q_0 for the cases $\Delta/L = 0.05$ and 0.10 with $L/d = 20$ and 100, where Δ is a small distance away from the ends. (b) Determine $|\sigma_y/\sigma_x|$ directly under the load q_0 at $x/L = 0.5$ if $L/d = 20$.

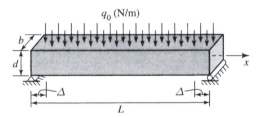

Figure 8P.93

8.94:* In deriving the expression for the flexural stress under pure bending, $\sigma_x = \frac{My}{I}$, the stress σ_y is assumed to be negligible; i.e., $|\sigma_y| \ll |\sigma_x|$. Consider now, for example, a beam of rectangular cross-section, $b \times 2c$, subjected to lateral loads $q(x)$, as shown in Fig. (8P.94a). Clearly since the stress $\sigma_y = q/b$ at the top of the beam and $\sigma_y = 0$ at the bottom, $\sigma_y = f(y)$, i.e σ_y varies with y and is not zero throughout the beam. Note that here positive σ_y is *compressive*.

We first isolate an element $(b \times 2c \times dx)$ [see Fig. (8P.94b)]; shear forces V and $V + \Delta V$ act on the two faces as shown. Let us now isolate a portion of the element, $b \times (1 - \alpha)c \times dx$, where $y = -\alpha c$ ($0 \le \alpha \le 1$), as shown in Fig. (8P.94c). Since $q(x) \ne 0$,

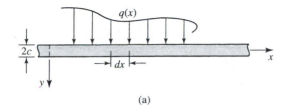

(a)

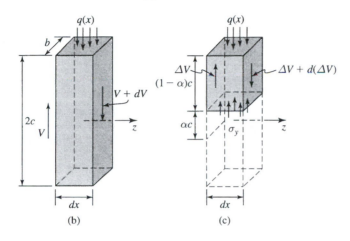

(b)   (c)   **Figure 8P.94**

$V = V(x)$ on any cross-section; therefore, acting on this portion of the element there exist, in addition to $q(x)$, (i) shear forces (which are due to the shear stresses τ_{xy}) on the left and right faces, which we denote as ΔV and $d(\Delta V)$, respectively, and (ii) a stress σ_y. (Note that here 'Δ' refers to the portion of the cross-section and 'd' refers to the difference of the left and right faces.)

(a) From equilibrium considerations, show that

$$\sigma_y = \frac{1}{b}\left[q(x) + \frac{d(\Delta V)}{dx}\right]$$

and hence, using the relation $dV(x)/dx = -q(x)$,

$$\sigma_y = \frac{q(x)}{b}\left[1 - \frac{1}{I}\int_{-c}^{-\alpha c} Q(y)\,dy\right],$$

where $Q(y)$ is the first moment about the centroidal z-axis of the portion of the cross-section $b \times (1 - \alpha)c$.

(b) By evaluating the integral, obtain an expression for σ_y, namely

$$\sigma_y(x, \alpha) = \frac{q(x)}{4b}[2 + 3\alpha - \alpha^3]$$

or

$$\sigma_y(x, y) = \frac{q(x)}{4b}\left[2 - 3\frac{y}{c} + \frac{y^3}{c^3}\right].$$

(c) Plot σ_y as a function of y in the beam. What is σ_y at the neutral axis?

(d) Show that directly under the load q, the ratio $R_\sigma \equiv \frac{|\sigma_y|_{max}}{|\sigma_x|_{max}} = \frac{4}{3}(\frac{d}{L})^2$ at the centre of a simply supported beam subjected to a uniformly distributed load q_0 and therefore $R_\sigma \ll 1$ if $d/L \ll 1$.

The following problems are designed to require the use of a computer.

8.95: The following loads, acting on a cantilever beam of length L, free at $x = 0$ and fixed at $x = L$, were measured along its length during an experiment.

x/L	Load (N)	x/L	Load (N)
0	0	0.50	0.4621
0.05	0.0500	0.55	0.5005
0.10	0.0997	0.60	0.5370
0.15	0.1489	0.65	0.5717
0.20	0.1974	0.70	0.6044
0.25	0.2449	0.75	0.6351
0.30	0.2913	0.80	0.6640
0.35	0.3364	0.85	0.6911
0.40	0.3799	0.90	0.7163
0.45	0.4219	0.95	0.7398
		1.00	0.7616

(a) By means of a computer, plot the load as a function of x/L. (b) Integrating Eqs. (8.5.2) and (8.5.7) numerically, determine the shear V and moment M at the points x/L.

8.96: The following loads, acting on a simply supported beam of length L, were measured along its length during an experiment.

x/L	Load (N)	x/L	Load (N)
0	0	0.50	0.3935
0.05	0.0488	0.55	0.4231
0.10	0.0952	0.60	0.4512
0.15	0.1393	0.65	0.4780
0.20	0.1813	0.70	0.5034
0.25	0.2212	0.75	0.5276
0.30	0.2592	0.80	0.5507
0.35	0.2953	0.85	0.5753
0.40	0.3297	0.90	0.5934
0.45	0.3624	0.95	0.6133
		1.00	0.6321

(a) By means of a computer, plot the load as a function of x/L. (b) Integrating Eqs. (8.5.2) and (8.5.7) numerically, determine the shear V and moment M at the points x/L. (c) Based on a numerical analysis of the curve obtained in (a) for the given load, it appears that the loading could be represented analytically by the expression $q(x) = 1 - e^{-x/L}$. Assuming this is correct, obtain the shear and moment by integrating this expression analytically according to Eqs. (8.5.2) and (8.5.7). (d) Using a computer, plot the results of (b) and (c) on the same graph and compare the results.

8.97: The average shear stress τ_{xy} due to a shear force V acting in the y-direction on a semi-circular cross-section is given as (see Problem 8.87)

$$\tau_{xy} = \frac{24\pi V}{(9\pi^2 - 64)R^2 \sin\theta} \left[\sin^3\theta - \frac{1}{\pi}(2\theta - \sin 2\theta)\right],$$

where θ is as shown in Fig. (8P.87). (a) Derive the transcendental equation whose roots determine the location of the line along which the maximum shear stress occurs and solve the equation numerically. (b) (i) Evaluate the location of this line as measured by $\tilde{y}$, the distance from the top of the beam and (ii) determine numerically the value of k according to the relation $\tau_{max} = k\frac{V}{A}$, where A is the area of the cross-section.

8.98:* A solid rod having a cross-section of radius a with modulus of elasticity E and yield point $\pm\sigma_0$ is assumed to behave as an elastic–perfectly plastic material. The rod is subjected to a bending moment M, which increases incrementally from zero to M_E, the maximum elastic moment, and subsequently to $M = M_P$, the fully plastic moment [Fig. 8P.98].

Denoting the elastic–plastic interface by ℓ,

(a) determine analytically the expression for the moment M as a function of the ratio ℓ/a;
(b) verify that M, as obtained in (a) above, yields M_E and M_P as $\ell/a = 1$ and $\ell/a \to 0$ respectively;
(c) rewrite the expression obtained in (a) above in terms of M/M_P and ℓ/a;
(d) by means of a computer, solve numerically for ℓ/a in terms of the ratio M/M_P for values $M_E/M_P \le M/M_P \le 1$;
(e) using a computer, plot a curve, ℓ/a vs. M/M_P for values $0 \le M/M_P \le 1$;
(f) (i) write an expression for the curvature of the rod $\kappa \equiv 1/R$ in terms of EI, M_P and $f(M/M_P)$, where $\ell/a = f(M/M_P)$ is the function that was determined numerically in (d) above and (ii) using a computer, plot M/M_P vs. EI/R for $0 \le M/M_P \le 1$.

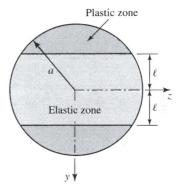

Figure 8P.98

8.99: In Problems 8.21 and 8.22, two beams having a rectangular cross-section (with width b and depth d) and subjected to a moment M were considered. In the first beam, the stress–strain relation was given as

$$\sigma = \alpha\epsilon^n, \quad n = 1, 3, 5...,$$

where $\alpha \ge 0$ is a constant. In the second beam, the stress–strain relation was given as

$$\sigma = \begin{cases} \alpha\epsilon^n, & \epsilon \ge 0 \\ -\alpha|\epsilon|^n, & \epsilon \le 0, \end{cases}$$

where $0 < n \le 1$.

For both beams the expression for the curvature κ was found to be identical, namely

$$\kappa = \frac{2[2(n+2)]^{1/n}}{d(\alpha bd^2)^{1/n}} M^{1/n}.$$

(a) Using a computer, (i) plot the stress–strain curve, i.e. σ/α vs. ϵ, for the cases $n = 0.5, 1, 2$ and (ii) plot the relation κ vs. M for the cases $n = 0.5, 1, 2$. (b) For which beam, governed by $n = 0.5$ or $n = 2$, would one expect the beam to be stiffer when (i) M is very small and (ii) M is relatively large? (c) For what value of M would the two beams bend with the same curvature?

8.100: Referring to Problem 8.91, the ratio $R_\sigma \equiv \frac{(\sigma_x)_a}{(\sigma_x)_b}$ of the maximum flexural stress in the two cross-sections of Figs. (8.91a and b), respectively, when subjected to the same moment M is given as $R_\sigma = (\alpha - 1)^2(3\alpha + 1)$. It is desired to optimise the design by removing material at the top and bottom corners in order to minimise the maximum stress in the beam. (a) Determine the optimal values of α and the resulting value of R_σ. (b) Plot R_σ vs. α.

8.101:   Referring to Problem 8.88, the location of the neutral axis in the rectangular cross-section of a beam having depth d and initial radius of curvature ρ, is

$$\frac{c_2}{d} = \frac{(1-\alpha)e^\alpha - 1}{\alpha(1 - e^\alpha)},$$

where $\alpha = d/\rho$ and where c_2 is as shown in Fig. (8P.88a). A simplified expression for the case $d/\rho \ll 1$ is

$$\frac{c_2}{d} = \frac{1}{2}\left(1 + \frac{d}{6\rho}\right).$$

Using a computer, plot the two expressions as functions of $\alpha = d/\rho$ on the same graph and determine the range of d/ρ for which the difference is within 10%.

9

Symmetric bending of beams: deflections, fundamental solutions and superposition

9.1 Introduction

We consider a linearly elastic prismatic beam having a symmetric cross-section such that its longitudinal axis lies along the x-axis and define the y-axis (whose positive direction is, as in Chapter 8, taken downward) to be the axis of symmetry of the cross-sections. Furthermore, we consider only symmetric bending, i.e., bending due to moments that act only about the z-axis [Fig. (9.1.1)].

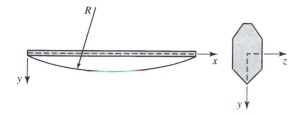

Figure 9.1.1

In Chapter 8, it was established that for such a beam, the neutral axis coincides with the centroidal z-axis; we shall therefore make no further distinction between these two axes although conceptually, when referring to the z-axis, we shall mean the neutral axis.

As previously derived, the basic relation governing the bending of beams is the Euler–Bernoulli relation,

$$EI\kappa = M, \qquad (9.1.1a)$$

where EI ($I \equiv I_{zz}$) represents the flexural rigidity of the beam and where the curvature $\kappa(|\kappa| \equiv 1/R)$ lies in a plane perpendicular to the neutral axis, namely in the x–y plane.

This relation, derived for a state of pure symmetric bending, was seen to be exact. By the use of Navier's hypothesis, the expression, generalised to apply to beams subjected to bending due to arbitrary lateral symmetric loads, becomes

$$EI\kappa(x) = M(x). \qquad (9.1.1b)$$

313

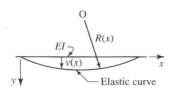

Figure 9.1.2

Based on this equation, we investigate in this chapter, the lateral deflection of the longitudinal axis for beams subjected to transverse loads. For a beam subjected to loads that lie in the x–y plane, the deflection, which is perpendicular to the neutral axis, will be in the y-direction.

Since we are only concerned with the deflection of the longitudinal x-axis, we represent the beam by means of a single line; thus, the beam is represented here by a one-dimensional element. We denote the lateral deflection of the longitudinal axis of the beam by $v(x)$ [Fig. (9.1.2)]. The curve defined by $v(x)$ is called the **elastic curve**.

9.2   Linearised beam theory

From the previous discussion, a beam will bend, under any given loading condition, in the shape of the elastic curve $v(x)$. Now, from the calculus, the curvature $|\kappa| = \frac{1}{R}$, for any curve lying in the x–y plane, is given by (see Appendix B.1)

$$|\kappa(x)| \equiv \frac{1}{R(x)} = \left| \frac{d^2v(x)/dx^2}{\{1 + [dv(x)/dx]^2\}^{3/2}} \right|. \qquad (9.2.1a)$$

In order to be compatible with the adopted sign convention for beams, as given in Chapter 8, we define positive curvature $\kappa > 0$ as[†]

$$\kappa(x) = -\frac{d^2v(x)/dx^2}{\{1 + [dv(x)/dx]^2\}^{3/2}}. \qquad (9.2.1b)$$

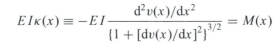

Figure 9.2.1

With this definition, positive (negative) curvature κ then corresponds to positive (negative) moments as shown in Fig. (9.2.1), and is therefore in agreement with Eqs. (9.1.1). Substituting the above equation in Eq. (9.1.1b), we have

$$EI\kappa(x) \equiv -EI \frac{d^2v(x)/dx^2}{\{1 + [dv(x)/dx]^2\}^{3/2}} = M(x)$$

or

$$EI \frac{d^2v(x)}{dx^2} = -\left[\{1 + [dv(x)/dx]^2\}^{3/2} \right] M(x). \qquad (9.2.2)$$

We observe that this is a highly nonlinear equation whose integration, in general, is rather complicated.

Now, we limit our investigation to beams having a relatively large flexural rigidity EI; that is, to relatively stiff beams such that their maximum deflection is small compared to the span length. It follows that the slope of the deformed beam, $|\frac{dv(x)}{dx}| \ll 1$. Therefore, this term can be considered an infinitesimal with respect to unity and the expression for the curvature reduces to

$$\kappa = -\frac{d^2v(x)}{dx^2}. \qquad (9.2.3)$$

[†] Note that in Appendix B.1 positive y is taken in the upward direction. Since, in deriving the beam relations, positive v is taken in the downward direction, a positive curvature κ must be defined with a minus sign. With this definition, a curve $v(x)$ with a positive curvature κ will then have, as its centre of curvature, points O which tend to lie in the negative y half-plane.

Hence, for beams undergoing small rotations (and displacements), the Euler–Bernoulli relation of Eqs. (9.1.1) becomes[†]

$$EI(x)\frac{d^2v(x)}{dx^2} = -M(x). \qquad (9.2.4)$$

We observe that Eq. (9.2.4) is thus the linearised form of the fundamental equation, Eq. (9.2.2), and consequently, we refer to this equation as the governing equation for linear beam theory.

Equation (9.2.4) is the basic equation by which we shall calculate the deflections of beams. We note that although the restriction to small displacements may seem severe, it is applicable to a host of engineering problems for which the flexural stiffness is sufficiently great such that all displacements are very small compared to the span length.

It is obvious that the integration of Eqs. (9.2.4) yields deflections due solely to bending; deflections due to shear deformation in the beam are not considered here. We shall show, in a later treatment, that the effect of such shear deformation on the deflections is negligible for relatively long beams.

Now, upon recognising that Eq. (9.2.4) is essentially a second-order differential equation, it is evident that two constants of integration must necessarily appear in its solution. Therefore, to complete its solution for any particular beam, we require appropriate boundary conditions that describe the support conditions of the beam; these boundary conditions (associated with a second order differential equation) can involve only the function itself or the first derivative, dv/dx, or a combination of these.[‡]

We consider here several cases:

For a simply supported beam as shown in Fig. (9.2.2a), the displacement at the two ends A and B vanish. Thus the appropriate boundary conditions are $v(0) = 0$ and $v(L) = 0$.

For the cantilever beam of Fig. (9.2.2b), the support at $A(x = 0)$ provides a constraint against both displacement and rotation. Hence the appropriate boundary conditions are $v(0) = 0$ and $\frac{dv(0)}{dx} = 0$.

We recognise that the boundary conditions are conditions on the geometry of the beam; hence the appropriate boundary conditions associated with the second-order differential equation of a beam are referred to as *geometric boundary conditions*.[§] Finally, it is worthwhile to note that Eq. (9.2.4) is readily integrable, provided the expression for $M(x)$ is known.[‖] For the case of statically determinate beams, one can always write an explicit expression for $M(x)$ in terms of *known* quantities from the equations of statics. For indeterminate beams one must resort to other means, as we shall see below.

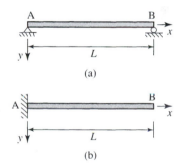

Figure 9.2.2

[†] Note that, according to the theory of differential equations, Eq. (9.2.4) is defined in an open domain, not at the boundary of the domain.

[‡] We recall, in general, that associated boundary conditions for an nth-order differential equation can involve, at most, derivatives of order $(n - 1)$.

[§] In Section 12 of this chapter, we will encounter other types of boundary conditions, namely 'mechanical boundary conditions'.

[‖] It is emphasised here that the second-order differential equation is integrable in terms of analytic functions only if $\frac{dv(x)}{dx}$ and $\frac{dv^2(x)}{dx^2}$ are *continuous* within the domain of validity of the equation. Thus, for example, a 'hinge' may not exist *within* the domain since this implies a discontinuity in the slope of the beam. We emphasise that the continuity conditions are *not* required at the end points of the domain.

◇9.3 Accuracy of the linearised beam theory

Although it was stated that it is permissible to use a linearised beam theory to obtain the deflections $v(x)$ of relatively stiff beams, we wish to establish an order of the error incurred if we are to use the linearised theory with some confidence.

It is clear that the error depends on the difference between the linearised expression for the curvature [Eq. (9.2.3)], $\kappa = -d^2v(x)/dx^2$, and the exact expression for κ as given by Eq. (9.2.1b). This difference clearly increases with increased bending and therefore with increased moment M. Hence the greatest error occurs for a beam subjected to the largest possible moment such that elastic behaviour is maintained; thus the greatest error in using linearised theory occurs for a beam subjected to $M = M_E$, the elastic moment, as defined in Chapter 8.

An estimate of the error can be obtained by comparing the deflection from the linearised theory with the exact deflection obtained using the relation of Eq. (9.1.1). We now recall that for a simply supported beam of flexural rigidity EI, length L and depth d subjected to a constant moment $M = M_E$ [see Fig. (8.6.9) or (9.3.1a)], the *exact* mid-span displacement Δ was found in Example 8.10 to be [Eq. (8.6.15)]

$$\Delta = \frac{E\alpha d}{\sigma_0}\left[1 - \cos\left(\frac{\sigma_0}{2E}\frac{L}{\alpha d}\right)\right], \quad 1/2 \leq \alpha < 1, \tag{9.3.1}$$

where σ_0 represents the elastic limit (or proportional limit, in the model), and $\alpha, (1/2 \leq \alpha < 1)$ is a parameter defining the distance of the furthest point in the cross-section from the neutral axis [Fig. (9.3.1b)].

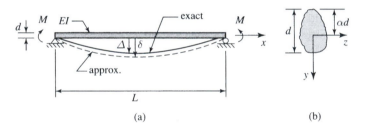

Figure 9.3.1

(a)　　　　　(b)

Let us therefore calculate the displacements of this same beam according to the linearised theory. Since, for this beam, $M = $ constant over the entire span, we integrate Eq. (9.2.4) [i.e. here, $EId^2v(x)/dx^2 = -M$, constant] subject to the boundary conditions $v(0) = v(L) = 0$. Upon integrating, we obtain the general solution

$$EIv(x) = -\frac{Mx^2}{2} + Ax + B, \tag{9.3.2}$$

where A and B are constants of integration. From the condition $v(0) = 0$, it follows that $B = 0$. From $v(L) = 0$, we obtain $A = \frac{ML}{2}$. Therefore the elastic curve, which we note is a parabola, is

$$v(x) = \frac{Mx}{2EI}(L - x). \tag{9.3.3}$$

Clearly, this is an approximation resulting from the linearisation, for the true shape is known to be a portion of a circular arc with a (constant) radius of curvature R [see Fig. (9.3.1a)]. To examine the error, we calculate the mid-span deflection at

$x = L/2$. Letting $\delta \equiv v(L/2)$, Eq. (9.3.3) yields

$$\delta = \frac{ML^2}{8EI}. \tag{9.3.4}$$

Now, the maximum error occurs, as we have seen, for the beam subjected to the maximum elastic moment M_E which, by Eq. (8.6.14), is $M_E = \frac{\sigma_0 I}{\alpha d}$. Hence under this moment,

$$\delta = \frac{1}{8\alpha} \frac{\sigma_0}{E} \frac{L^2}{d}. \tag{9.3.5}$$

We now compare the exact deflection Δ [Eq. (9.3.1)] with the deflection δ, obtained from the linearised theory. To do so, we expand the cosine term appearing in Eq. (9.3.1) in its Taylor series (upon recognising that $\frac{\sigma_0}{E} \ll 1$) and retain the first three terms:

$$\cos\left(\frac{\sigma_0}{2E} \frac{L}{\alpha d}\right) = 1 - \frac{1}{2}\left(\frac{\sigma_0}{2E} \frac{L}{\alpha d}\right)^2 + \frac{1}{24}\left(\frac{\sigma_0}{2E} \frac{L}{\alpha d}\right)^4 + \cdots. \tag{9.3.6}$$

Therefore Eq. (9.3.1) becomes:

$$\Delta = \frac{E\alpha d}{\sigma_0}\left[\frac{1}{2}\left(\frac{\sigma_0}{2E} \frac{L}{\alpha d}\right)^2 - \frac{1}{24}\left(\frac{\sigma_0}{2E} \frac{L}{\alpha d}\right)^4\right] + \cdots,$$

that is,

$$\Delta = \frac{1}{8} \frac{\sigma_0}{E} \frac{L^2}{\alpha d}\left[1 - \frac{1}{48}\left(\frac{\sigma_0}{E}\right)^2\left(\frac{L}{\alpha d}\right)^2\right] + \cdots. \tag{9.3.7}$$

We then have, from Eq. (9.3.5),

$$\Delta = \delta\left[1 - \frac{1}{48}\left(\frac{\sigma_0}{E}\right)^2\left(\frac{L}{\alpha d}\right)^2\right] + \cdots \quad 1/2 \leq \alpha < 1. \tag{9.3.8}$$

We observe immediately that the exact deflection Δ is less than δ, calculated according to the linearised theory.

To obtain the accuracy of the linearised deflection δ, we calculate the percentage error (using $\alpha = 1/2$) noting again that $\sigma_0/E \ll 1$; thus

$$\left|\frac{\delta - \Delta}{\Delta}\right| = \frac{1}{12}\left(\frac{\sigma_0}{E}\right)^2\left(\frac{L}{d}\right)^2. \tag{9.3.9}$$

According to our previous discussion, Eq. (9.3.9) provides us with an *upper bound* on the error. We observe that this bound error depends on two ratios: the ratio $\frac{\sigma_0}{E}$ and the length-to-depth ratio. (Note that it *does not depend on the specific shape of the cross-section*.)

Now typical values for the ratio $\frac{\sigma_0}{E}$ for materials encountered in engineering practice are $O(10^{-3})$. For long slender beams, say with $\frac{L}{d} \simeq 100$, the error, according to Eq. (9.3.9), is then of the order $O(10^{-3})$, i.e. about 0.1%. For less slender beams, e.g. $L/d = 20$, the error is of the order of 0.01%.

Thus from this simple analysis, we may conclude that the linearised theory will yield displacements of great accuracy for beams as encountered in usual engineering practice.

Having established an estimate for the accuracy of the linearised theory, we illustrate the method by means of some simple examples.

9.4   Elastic curve equations for some 'classical' cases

Among statically determinate beams, we find, in particular, two common 'classical' cases: simply supported beams and cantilever beams. In both cases, the equation of the elastic curve is readily found by integration of the linear second-order differential equation given above. We illustrate here the method of solution for several loading cases.

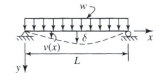

Figure 9.4.1

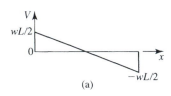

(a)

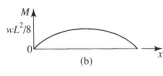

(b)

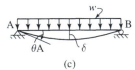

(c)

Figure 9.4.2

Example 9.1: A simply supported beam of flexural rigidity EI and length L is subjected to a uniformly distributed load w (N/m), as shown in Fig. (9.4.1). (a) Determine the equation of the elastic curve $v(x)$. (b) Evaluate the displacement $\delta = v(L/2)$ if the beam is made of wood ($E = 10$ GPa), has a rectangular cross-section (with $b = 6$ cm and $d = 12$ cm such that $I = 864$ cm⁴) and if $w = 1200$ N/m and $L = 4$ m.

Solution:

(a) The moment $M(x)$, from Example 8.1, is given by $M(x) = \frac{w}{2}x(L - x), 0 < x < L$. [Shear and moment diagrams are shown in Figs. (9.4.2a and b).] Substituting $M(x)$ in Eq. (9.2.4), we have[†]

$$EIv''(x) = -\frac{w}{2}(Lx - x^2), \quad 0 < x < L. \tag{9.4.1}$$

The associated boundary conditions (B.C.) are $v(0) = v(L) = 0$.
    Integrating the above equation, we obtain

$$v(x) = \frac{w}{2EI}\left(-\frac{Lx^3}{6} + \frac{x^4}{12}\right) + Ax + B, \tag{9.4.2a}$$

where A and B are constants of integration. From B.C. $v(0) = 0$, $B = 0$. From the second boundary condition $v(L) = \frac{w}{2EI}(-\frac{L^4}{6} + \frac{L^4}{12}) + AL = 0$, we find $A = \frac{wL^3}{24EI}$. Therefore

$$v(x) = \frac{w}{24EI}(x^4 - 2Lx^3 + L^3x) \tag{9.4.2b}$$

represents the equation of the elastic curve. Note that this is a quartic equation. The maximum displacement, which occurs at the centre $x = L/2$, is

$$\delta \equiv v(x = L/2) = \frac{5wL^4}{384EI}. \tag{9.4.3}$$

Furthermore the slope $\theta = \tan^{-1} v' \simeq v'(x)$ of the deflected beam at any point x is given by

$$v'(x) = \frac{w}{24EI}(4x^3 - 6Lx^2 + L^3); \tag{9.4.4a}$$

the largest slope occurs at $x = 0$ and $x = L$; at point A,

$$\theta_A = \frac{wL^3}{24EI}. \tag{9.4.4b}$$

[†] The simplified prime notation, $v'(x) \equiv \frac{dv(x)}{dx}$, $v''(x) \equiv \frac{d^2v(x)}{dx^2}$, etc., is used here and throughout this chapter. Note, too, that Eq. (9.4.1) is defined only in the open region, $0 < x < L$. (See footnote p. 315.)

The shape of the elastic curve is shown in Fig. (9.4.2c).

(b) For the given numerical values,

$$v(L/2) = \frac{5(1200)(4)^4}{(384)(10 \times 10^9)(864 \times 10^{-8})} = 4.63 \times 10^{-2} = 0.0463 \text{ m}.$$

We note in passing that the ratio $\delta/L = \frac{4.63}{400} = 1.12 \times 10^{-2}$ is very small. Moreover, using the same numerical values, the largest slope of the beam, from Eq. (9.4.4b), is found to be $\theta_A = 0.03704 \text{ rad} = 2.12°$. These numerical results thus conform with the assumptions leading to linearised theory.   □

Example 9.2:   A force P acts at the free end of a cantilever beam AB of length L and flexural rigidity EI, as shown in Fig. (9.4.3). Determine the equation of the elastic curve and the displacement of B under the load.

Solution: From the equations of statics, the reactions at A are $R_A = P$ and $M_A = -PL$. Therefore, the moment $M(x)$ is

$$M(x) = -PL + Px = P(x - L), \quad 0 \le x \le L \tag{9.4.5a}$$

and hence the governing equation, $EIv''(x) = -M(x)$, becomes

$$EIv''(x) = -P(x - L), \quad 0 < x < L, \tag{9.4.5b}$$

with the associated boundary conditions $v(0) = v'(0) = 0$. Integrating the above equation, we obtain

$$EIv'(x) = -P(x^2/2 - Lx) + A, \tag{9.4.6a}$$

and using the B.C. $v'(0) = 0$, $A = 0$. Integrating once more,

$$EIv(x) = -P(x^3/6 - Lx^2/2) + B; \tag{9.4.6b}$$

from B.C. $v(0) = 0$, $B = 0$. Therefore

$$v(x) = -\frac{Px^2}{6EI}(x - 3L). \tag{9.4.7a}$$

We note that here the elastic curve, shown in Fig. (9.4.3), is defined by a cubic equation with the maximum deflection, occurring at point B($x = L$),

$$v(L) = \frac{PL^3}{3EI}. \tag{9.4.7b}$$

□

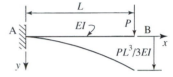

Figure 9.4.3

Having found the results for a cantilever beam of constant flexural rigidity EI, we wish to examine some cases where the flexural rigidity varies with x, i.e., $I = I(x)$, in order to examine the effect on the maximum displacement. We do so in the following example for two particular cases.

Example 9.3:   A force P acts at the free end of a cantilever beam AB of length L but having variable flexural rigidity $EI(x)$ [Fig. (9.4.4)]. Let us assume that the beam has a rectangular cross-section, with a variable width $b(x)$. Consequently, since $I(x) = b(x)d^3/12$, $I(x)$ and $b(x)$ have the same dependency on x. We examine two cases: Beam (i), where $I(x)$ varies linearly along the length as $I(x) = I_0(1 - x/L)$; Beam (ii), where $I(x)$ is assumed to vary parabolically as $I(x) = I_0(1 - x^2/L^2)$. The variations of $b(x)$ are shown in Figs. (9.4.5a and b), respectively. We wish to determine $v(x)$ and the deflection at the free end.

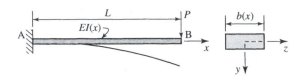

Figure 9.4.4

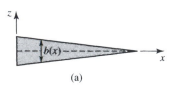

(a)

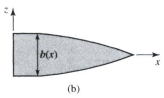

(b)

Figure 9.4.5

Solution: Since the beam is statically determinate, the moment $M(x)$ is independent of the cross-section and is therefore the same as in Example 9.2 [Eq. (9.4.5a)]. Consequently, the governing equation is

$$EI(x)v''(x) = PL(1 - x/L), \quad 0 < x < L. \tag{9.4.8}$$

The boundary conditions are, as in the previous example, $v(0) = v'(0) = 0$.

Solution for Beam (i): $I(x) = I_0(1 - x/L)$
We recognise that $I(x)$ and $M(x)$ have the same dependency on x. Therefore, the linearised Euler–Bernoulli equation becomes simply

$$EI_0 v''(x) = PL. \tag{9.4.9a}$$

Integrating twice and using the above boundary conditions, yields

$$EI_0 v(x) = PLx^2/2. \tag{9.4.9b}$$

The maximum deflection at the free end is therefore

$$v(L) = \frac{PL^3}{2EI_0}. \tag{9.4.9c}$$

Solution for Beam (ii): $I(x) = I_0(1 - x^2/L^2)$
Substituting in Eq. (9.2.4), we find

$$EI_0 v''(x) = PL \cdot \frac{1 - x/L}{1 - x^2/L^2}$$

or

$$EI_0 v''(x) = \frac{PL}{1 + x/L}. \tag{9.4.10a}$$

Integrating once, we obtain

$$EI_0 v'(x) = PL^2 \ln(1 + x/L) + A. \tag{9.4.10b}$$

From B.C. $v'(0) = 0$, $A = 0$. Recalling the indefinite integral,

$$\int \ln(z)\,dz = z[\ln(z) - 1],$$

a second integration yields

$$EI_0 v(x) = PL^3(1 + x/L)[\ln(1 + x/L) - 1] + B. \tag{9.4.10c}$$

From B.C. $v(0) = 0$, $B = PL^3$. Therefore the elastic curve is given by

$$v(x) = \frac{PL^3}{EI_0}\{(1 + x/L)[\ln(1 + x/L) - 1] + 1\} \tag{9.4.11a}$$

and hence the deflection at the free end is

$$v(L) = \frac{PL^3}{EI_0}(2\ln 2 - 1) = 0.38629\frac{PL^3}{EI}. \tag{9.4.11b}$$

$\square$

It is interesting to compare the results of this example with that of Example 9.2. For the beam of constant EI_0, we found $\Delta_B = \frac{PL^3}{3EI_0} = 0.333\frac{PL^3}{EI_0}$, while for Beam (ii) $\Delta_B = 0.38629PL^3/EI_0$ and for Beam (i), $\Delta_B = 0.5PL^3/EI_0$. Clearly, the increased displacements are due to the lower overall flexural rigidity of Beams (i) and (ii). While we observe that the maximum deflection depends on the flexural rigidity of a beam, we point out that for statically determinate beams subjected to applied loads, the maximum moments are *independent of the flexural rigidity*.

Finally, it is worthwhile here to make a general comment. Having examined several beam systems in this chapter, we observe specifically that *the magnitude of the deflections of a beam is inversely proportional to the flexural stiffness, EI*. Indeed, this relation is always found to be true for any beam governed by the linear Euler–Bernoulli relations.

Example 9.4:   A simply supported beam AB having flexural rigidity *EI* and length *L* is subjected to a couple M_A at point A, as shown in Fig. (9.4.6a). Determine (a) the equation of the elastic curve, (b) the displacement at the centre point ($x = L/2$), (c) the location and magnitude of the maximum deflection and (d) the slope of the beam at A.

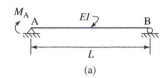

(a)

Solution:

(a) From equilibrium, the reactions are $R_A = M_A/L$ (downward) and $R_B = M_A/L$ (upward). The resulting moment is therefore

$$M(x) = M_A(1 - x/L), \quad 0 \leq x \leq L \qquad (9.4.12a)$$

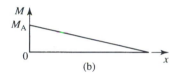

(b)

and hence the governing equation is

$$EIv''(x) = -M_A(1 - x/L), \quad 0 < x < L \qquad (9.4.12b)$$

subject to the boundary conditions $v(0) = v(L) = 0$. [The moment diagram is shown in Fig. (9.4.6b).]

Integrating the above equation twice,

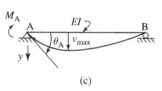

(c)

Figure 9.4.6

$$EIv'(x) = -M_A(x - x^2/2L) + A \qquad (9.4.13a)$$

$$EIv(x) = -M_A(x^2/2 - x^3/6L) + Ax + B. \qquad (9.4.13b)$$

From BC $v(0) = 0$, it follows that $B = 0$. Then, from $v(L) = 0$, we find $A = M_A L/3$. Hence

$$v(x) = \frac{M_A}{6EIL}(x^3 - 3Lx^2 + 2L^2x). \qquad (9.4.13c)$$

(b) The mid-point deflection at $x = L/2$ is therefore

$$v(L/2) = \frac{M_A L^2}{16EI}. \qquad (9.4.13d)$$

(c) Setting $v'(x) = 0$, the maximum displacement occurs at x given by the solution of the equation, $3x^2 - 6Lx + 2L^2 = 0$, whose (relevant) root is $x = (3 - \sqrt{3})L/3$. Substituting this in Eq. (9.4.13c), we find the maximum displacement to be

$$v_{max} = \frac{\sqrt{3}}{27}\frac{M_A L^2}{EI}. \qquad (9.4.13e)$$

(d) The slope $\theta_A = v'(0)$ at point A is given by

$$\theta_A = \frac{M_A L}{3EI}. \tag{9.4.14}$$

The elastic curve and calculated quantities are shown in Fig. (9.4.6c). $\quad\square$

Example 9.5: The simply supported beam of length L and having flexural rigidity EI is subjected to a concentrated force P acting a distance a from the left end, point A, as shown in Fig. (9.4.7a). Determine (a) the equation of the elastic curve and (b) the maximum deflection if the load is applied at the centre ($a = L/2$).

Solution: From statics, the reactions are $R_A = P(L-a)/L$ and $R_B = Pa/L$. The resulting shear and moment diagrams are shown in Figs. (9.4.7b and c). We observe, from the moment diagram, that the moment $M(x)$ in the beam is represented by two different functions of x, which are as follows:

$$M(x) = \frac{P(L-a)}{L}x, \quad 0 \le x \le a \tag{9.4.15a}$$

$$M(x) = \frac{Pa}{L}(L-x), \quad a \le x \le L. \tag{9.4.15b}$$

Thus we note that there exist two separate domains AC and CB which we shall call domains D_1 and D_2, respectively. Mathematically, for the open domains, we write $D_1: \{x \mid 0 < x < a\}$; $D_2: \{x \mid a < x < L\}$. We therefore must write the Euler–Bernoulli equation separately for each domain. Denoting the deflections in the two domains by $v_1(x)$ and $v_2(x)$, respectively, we have

Domain 1 $(0 < x < a)$ $\qquad\qquad$ Domain 2 $(a < x < L)$

$$EIv_1''(x) = -\frac{P(L-a)}{L}x \qquad\qquad EIv_2''(x) = -\frac{Pa}{L}(L-x)$$
$$\tag{9.4.16a,b}$$

Integrating successively,

$$EIv_1'(x) = -\frac{P(L-a)}{2L}x^2 + A_1 \qquad EIv_2'(x) = \frac{Pa}{2L}(L-x)^2 + A_2$$
$$\tag{9.4.16c,d}$$

$$EIv_1(x) = -\frac{P(L-a)x^3}{6L} + A_1x + B_1 \qquad EIv_2(x) = -\frac{Pa}{6L}(L-x)^3 + A_2x + B_2$$
$$\tag{9.4.16e,f}$$

We note that there then exist four constants of integration, A_1, A_2, B_1 and B_2. We therefore require four associated boundary conditions. Clearly, for the two end points, we have

$$v_1(0) = 0, \tag{9.4.17a}$$

$$v_2(L) = 0 \tag{9.4.17b}$$

The remaining two boundary conditions are determined from the physics of the beam: the deflections v_1 and v_2 of the beam must clearly be the same at $x = a$. Furthermore, since no hinge exists at $x = a$, there can be no finite relative rotation of the two segments of the beam at $x = a$; that is, no 'kink' exists in the beam at this point. Therefore the slope of the beam at $x = a$ must be the same, and hence, we write the

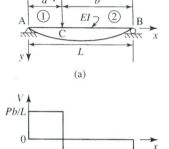

(a)

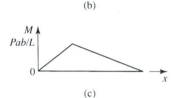

(b)

(c)

Figure 9.4.7

boundary conditions as

$$v_1(a) = v_2(a) \tag{9.4.17c}$$

$$v_1'(a) = v_2'(a). \tag{9.4.17d}$$

These two boundary conditions are thus essentially 'continuity conditions'.
From Eqs. (9.4.17a and b), we find

$$B_1 = 0 \quad \text{and} \quad B_2 = -A_2 L.$$

Equations (19.4.17c and d) become

$$-\frac{P(L-a)a^3}{6L} + aA_1 = -\frac{Pa(L-a)^3}{6L} - (L-a)A_2$$

$$-\frac{P(L-a)a^2}{2L} + A_1 = \frac{Pa(L-a)^2}{2L} + A_2.$$

These are two simultaneous equations in the two unknowns A_1 and A_2, whose solutions are

$$A_1 = -\frac{Pa(L-a)(2L-a)}{6L}, \qquad A_2 = -\frac{Pa(L^2-a^2)}{6L}.$$

Substituting in Eqs. (9.4.16e) and (9.4.16), we arrive at the expressions for the deflections

$$v_1(x) = \frac{P(L-a)x}{6EIL}[a(2L-a) - x^2], \quad 0 \le x \le a \tag{9.4.18a}$$

$$v_2(x) = \frac{Pa(L-x)}{6EIL}[x(2L-x) - a^2], \quad a \le x \le L. \tag{9.4.18b}$$

For the particular case of the beam loaded at the centre ($a = L/2$) [Fig. (9.4.8)], the deflection becomes

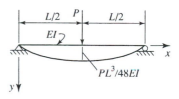

$$v_1(x) = \frac{Px}{48EI}(3L^2 - 4x^2), \quad 0 \le x \le L/2 \tag{9.4.19a}$$

$$v_2(x) = \frac{P(L-x)}{48EI}[3L^2 - 4(L-x)^2], \quad L/2 \le x \le L. \tag{9.4.19b}$$

Figure 9.4.8

The deflection at $x = L/2$ under the load for this case is then

$$v(x = L/2) = \frac{PL^3}{48EI}. \tag{9.4.20}$$

The elastic curve for this case is shown in Fig. (9.4.8). The slope of the beam in the domain $0 \le x \le L/2$ is readily found

$$v'(x) = \frac{P}{16EI}(L^2 - 4x^2). \tag{9.4.21}$$

Note that $v'(L/2) = 0$ as it must be for this symmetric case.   □

Although the solution to this problem was quite straightforward, we observe that the calculations become somewhat complicated since there exist two domains. If, for example, we have a beam subjected to loads as in Fig. (9.4.9), we observe that

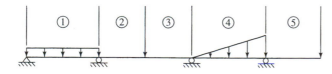

Figure 9.4.9

there exist five domains; in each we must integrate the differential equation and it becomes necessary to evaluate 10 constants of integration from 10 boundary conditions.

Now, there exists a class of functions, called **singularity functions**, which permits us to extend the domain of validity of an equation and thus leads to substantial simplifications in the calculations. We shall consider these functions and their application in Section 7.

◇9.5   Axial displacements due to flexure of a beam under lateral loads

In Chapter 8, the upper bound for the axial displacement of a simply supported beam of length L $(0 \leq x \leq L)$ under *pure bending* was shown to be an infinitesimal of second order. Specifically, the axial displacement, Δ_x, of a roller support at $x = L$ was found to be [see Eq. (8.6.20a)]

$$\Delta_x = \frac{\gamma^2}{24}. \tag{9.5.1}$$

Here

$$\gamma = \frac{\sigma_0 L}{E\alpha d}, \tag{9.5.2}$$

where σ_0 is the maximum elastic stress, d is the depth of the beam and α, $(1/2 < \alpha < 1)$ is a coefficient as previously defined.

We wish to establish a general expression for the axial displacement $u(x)$ at any point of the elastic curve of the beam due to lateral loading, namely for cases where $M = M(x)$.

Let us therefore consider a beam for which the displacement in the x-direction is zero at $x = 0$; for example, either a pinned support or a clamped support [see Fig. (9.5.1a or b)]. Let $v(x)$ represents the deflection due to lateral loads.

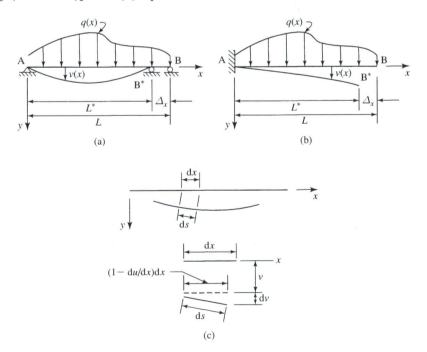

Figure 9.5.1

Recalling that the elastic curve is inextensible, it is possible to establish an expression for Δ_x in terms of $v(x)$ and more specifically in terms of $v'(x)$. We proceed as follows.

We first denote the displacement *to the left* of any point of the elastic curve by $u(x)$; i.e., we take here positive $u(x)$ in the *negative x-direction*.

Consider now a typical element of length dx of the elastic curve, as shown in Fig. (9.5.1c). In the deformed state, the length is given by

$$ds^2 = \left[\left(1 - \frac{du}{dx}\right) dx\right]^2 + dv^2, \tag{9.5.3a}$$

or

$$ds^2 = \left[\left(1 - \frac{du}{dx}\right)^2 + \left(\frac{dv}{dx}\right)^2\right] dx^2. \tag{9.5.3b}$$

Now since flexural deformation is *inextensional* – that is, there is no change in the length of the fibre lying on the elastic curve namely originally on the longitudinal centroidal axis [see Section 6b of Chapter 8], upon setting $ds^2 = dx^2$,

$$\left(1 - \frac{du}{dx}\right)^2 + \left(\frac{dv}{dx}\right)^2 = 1. \tag{9.5.4a}$$

Solving for du/dx,

$$\frac{du}{dx} = 1 - \sqrt{1 - \left(\frac{dv}{dx}\right)^2}. \tag{9.5.4b}$$

Now, by the binomial theorem,

$$[1 - (v')^2]^{1/2} = 1 - \frac{1}{2}(v')^2 - \frac{1}{8}(v')^4 + \cdots. \tag{9.5.5a}$$

Consistent with linear theory, recalling that $|v'| \ll 1$, we neglect all terms higher than the quadratic

$$[1 - (v')^2]^{1/2} \simeq 1 - \frac{1}{2}(v')^2, \tag{9.5.5b}$$

and therefore, from Eq. (9.5.4b),

$$\frac{du}{dx} = 1 - \left[1 - \frac{1}{2}(v')^2\right] = \frac{1}{2}(v')^2. \tag{9.5.6}$$

Integrating,

$$u(x) - u(0) = \frac{1}{2} \int_0^x (v')^2 \, dx. \tag{9.5.7}$$

Since $u(0) = 0$, we finally obtain for $\Delta_x \equiv u(L)$,

$$\Delta_x = \frac{1}{2} \int_0^L (v')^2 \, dx. \tag{9.5.8}$$

We now consider a specific case, namely a simply supported beam of length L subjected to a uniformly distributed load w (N/m), as shown in Fig. (9.4.1). Due to this loading, the slope $v'(x)$ was found to be [see Eq. (9.4.4a)]

$$v'(x) = \frac{w}{24EI}(4x^3 - 6Lx^2 + L^3). \tag{9.5.9}$$

Substituting this in Eq. (9.5.8), we get

$$\Delta_x = \frac{w^2}{2(24EI)^2} \int_0^L (4x^3 - 6Lx^2 + L^3)^2 \, dx \tag{9.5.10a}$$

and integrating the polynomial, we obtain

$$\Delta_x = \frac{17w^2L^7}{70(24EI)^2}. \tag{9.5.10b}$$

Recalling that the maximum moment in this case, occurring at $x = L/2$, is $M_{max} = wL^2/8$, and noting from Eq. (8.6.14) that $M_E = \frac{\sigma_0 I}{\alpha d}$, we have $wL^2 = 8\sigma_0 I/\alpha d$. Substituting in Eq. (9.5.10b), we find

$$\frac{\Delta_x}{L} = \frac{17}{630}\left(\frac{\sigma_0}{E}\right)^2\left(\frac{L}{\alpha d}\right)^2. \tag{9.5.11a}$$

For an upper bound, we let $\alpha = 1/2$ and therefore obtain

$$\frac{\Delta_x}{L} = \frac{17}{630}\left(\frac{\sigma_0}{E}\right)^2\left(\frac{L}{d}\right)^2. \tag{9.5.11b}$$

As was mentioned previously, for relatively stiff beams as encountered in engineering practice, $\sigma_0/E = O(10^{-3})$. Therefore, taking $\sigma_0/E = 10^{-3}$, we obtain for the cases $L/d = 20$ and 100,

$$\frac{\Delta_x}{L} = 1.08 \times 10^{-5} \quad \text{and} \quad \frac{\Delta_x}{L} = 2.69 \times 10^{-4},$$

respectively. Thus we observe that these are infinitesimals of the second order and hence

$$L^* = L(1 - \Delta_x/L) \approx L. \tag{9.5.12}$$

This justifies using the original length L when analysing problems according to the linear theory of beams. As the reader will notice, this has been the case in the solution of all examples previously studied.

It is worthwhile to point out here that since axial displacements of beams undergoing flexure due to transverse loads are of second order, in analysing a beam according to linear theory, we neglect stretching of an elastic curve. As a result, for beams supported, e.g. as in Fig. (9.5.2), the reactive axial forces are neglected in a linear analysis since they are only second-order effects.

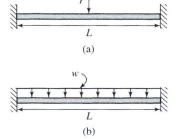

Figure 9.5.2

9.6   Deflections due to shear deformation

In determining the deflections of beams, displacements were obtained by integrating the Euler–Bernoulli equations; thus we considered deformation only due to bending. For cases other than pure bending, i.e. when $V \neq 0$, it is clear that a beam undergoes deformation due to shear, and consequently, it would seem appropriate to also calculate the deflections due to such deformation. In practice, however, displacements due to shear are very small compared to those due to flexure and are therefore generally neglected. We now seek to justify such practice.

To do so, we consider two cases: a cantilever beam of length L subjected to a concentrated force P at the end [Fig. (9.6.1a)] and a simply supported beam of length L subjected to a uniformly distributed load w (N/m), as shown in Fig. (9.6.2a). The displacements due to bending, which we denote here by v_b for the two cases, are given by Eqs. (9.4.7b) and (9.4.3), respectively: namely

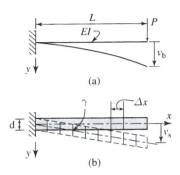

$$v_b = \frac{PL^3}{3EI} \qquad (9.6.1a)$$

and

$$v_b = \frac{5wL^4}{384EI}, \qquad (9.6.1b)$$

Figure 9.6.1

where EI is the flexural rigidity of the beam. These are shown in the respective figures.

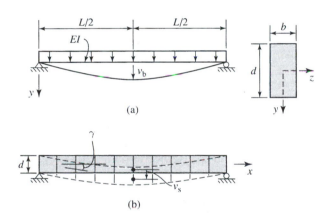

(a)

(b)

Figure 9.6.2

Let us, for simplicity, consider the beams to have rectangular cross-sections $b \times d$ with $I = bd^3/12$. In this case, the deflections are given explicitly by

$$v_b(L) = \frac{4PL^3}{Ebd^3} \qquad (9.6.2a)$$

and

$$v_b(L/2) = \frac{5wL^4}{32Ebd^3}. \qquad (9.6.2b)$$

We now derive approximate required expressions for the corresponding deflections due to shear deformation.[†] If we consider the beam to be composed of a series of elements, dx, then due to the shear stress, any rectangular element deforms into a parallelogram and, according to Hooke's law, the angle changes, γ, are given by

---

[†] We point out here that since we do not consider the variation of the shear stress in the cross-section (see Section 8c, Chapter 8), the present analysis, is but approximate. A correction to this analysis, which takes the variation into account, is obtained by use of a 'shape factor'. More accurate displacements due to shear deformation are determined using methods given in Chapter 14.

$\gamma = \tau/G$, where G is the shear modulus. Recalling from Eq. (8.8.6b) that the shear stress along the z-axis of a rectangular section is given by $\tau = \frac{3V}{2A}$, it follows that

$$\gamma(x) = \frac{3V(x)}{2AG},\qquad (9.6.3)$$

and hence the relative displacement, dv_s, of any two cross-sections, dx apart, is [Fig. (9.6.3)]

$$dv_s = \gamma(x) \cdot dx = \frac{3V(x)}{2AG} \cdot dx.\qquad (9.6.4a)$$

Therefore, assuming $v_s(x = 0) = 0$, integration yields

$$v_s(x) = \frac{3}{2AG} \int_0^x V(\xi)\,d\xi.\qquad (9.6.4b)$$

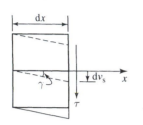

Figure 9.6.3

Let us first consider the case of the cantilever beam where $V = P$ (constant). For this case, the shear deformation of the beam is independent of x and appears as in Fig. (9.6.1b). The resulting deflection $v_s(L)$ of the free end is then given by

$$v_s(L) = \frac{3P}{2AG} \int_0^L dx = \frac{3PL}{2AG}\qquad (9.6.5a)$$

and since $A = b \cdot d$, we have

$$v_s(L) = \frac{3PL}{2bdG}.\qquad (9.6.5b)$$

For the case of the simply supported beam, the shear force $V = w(L/2 - x)$. Substituting in Eq. (9.6.4b), the mid-span deflection [Fig. (9.6.2b)] is given by

$$v_s(L/2) = \frac{3w}{2AG} \int_0^{L/2} (L/2 - x)\,dx = \frac{3wL^2}{16AG}\qquad (9.6.6a)$$

or

$$v_s(L/2) = \frac{3wL^2}{16bdG}.\qquad (9.6.6b)$$

We now calculate the ratio $\frac{v_s}{v_b}$; that is, the ratio of the displacement due to shear to that due to flexure. From Eqs. (9.6.1), (9.6.5b) and (9.6.6b), the ratios for the cantilever beam and simply supported beam are, respectively,

$$\frac{v_s}{v_b} = \frac{3}{8}\frac{E}{G}\left(\frac{d}{L}\right)^2\qquad (9.6.7a)$$

and

$$\frac{v_s}{v_b} = \frac{6}{5}\frac{E}{G}\left(\frac{d}{L}\right)^2.\qquad (9.6.7b)$$

Recalling Eq. (4.4.14), $G = \frac{E}{2(1+v)}$, we finally have

$$\frac{v_s}{v_b} = \frac{3(1+v)}{4}\left(\frac{d}{L}\right)^2\qquad (9.6.8a)$$

and

$$\frac{v_{\mathrm{s}}}{v_{\mathrm{b}}} = \frac{3(1+\nu)}{5}\left(\frac{d}{L}\right)^2. \tag{9.6.8b}$$

We observe that this ratio is quite small for $L \gg d$, namely for long beams where the length of the beam is much greater than its depth. For example, if $L = 20d$, for the case of the cantilever beam, $v_{\mathrm{s}}/v_{\mathrm{b}} = 1.88(1+\nu) \times 10^{-3}$ while for the case of the simply supported beam, $v_{\mathrm{s}}/v_{\mathrm{b}} = 1.5(1+\nu) \times 10^{-3}$. Recalling that $\nu \le 0.5$, we observe that for these cases, $v_{\mathrm{s}}/v_{\mathrm{b}} \simeq 1/500$. From these results, it is clear that the longer the beam, the smaller the influence of shear deformation.

We thus conclude that in calculating the deflection of long beams, one is generally quite justified in neglecting displacements due to shear deformation.

9.7 Singularity functions and their application

(a) Definition of singularity functions

Singularity functions are defined in such a way that they enable us to write a *single* expression for a polynomial function of any degree n $(0 \le n)$ over various domains.

The singularity function is denoted by means of brackets '$\langle \cdots \rangle$', called **Macaulay brackets**, and is defined as follows:

$$\langle x - k \rangle^n = \begin{cases} (x-k)^n, & k \le x, \quad n \ge 0 \\ 0, & x < k, \quad n \ge 0. \end{cases} \tag{9.7.1a}$$

In particular,[†]

$$\langle x - k \rangle^0 = \begin{cases} 1, & k \le x \\ 0, & x < k. \end{cases} \tag{9.7.1b}$$

The singularity functions $\langle x - k \rangle^n$, $n = 1$ and 2, are shown in Fig. (9.7.1a); $\langle x - k \rangle^0$ is shown in Fig. (9.7.1b). Thus whenever the argument within the brackets $\langle \cdots \rangle$ is negative or zero, the singularity function is equal to zero. From its basic definition, we observe the interesting property

$$\langle x - k \rangle^n = (x-k)^n - (-1)^n \langle k - x \rangle^n, \quad n \ge 0. \tag{9.7.2}$$

Note that $\langle x - k \rangle^n \ne -\langle k - x \rangle^n$.

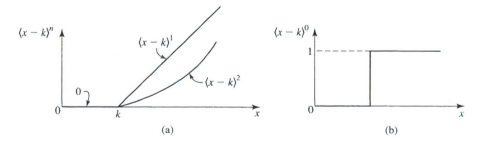

(a)   (b)   **Figure 9.7.1**

[†] From its definition, we observe that when $n = 0$ we obtain, as a particular case of the singularity function, the 'Heaviside function', conventionally denoted by $H(x - k)$ [i.e., $\langle \cdots \rangle^0 \equiv H(\cdots)$]. We shall see later that the *Dirac-delta function* (known also as the **impulse function**) is another particular case of the singularity function.

The singularity function $\langle x - k \rangle^n$ is *defined* to behave according to the ordinary rules of scalar algebra. In particular, among its operational properties the singularity function satisfies the elementary rules of integration of polynomials; that is,[†]

$$\int \langle x - k \rangle^n \, dx = \frac{\langle x - k \rangle^{n+1}}{n + 1} + C, \quad n \geq 0. \tag{9.7.3}$$

Before proceeding with some applications, it is important to emphasise that the use of singularity functions, as developed here, will be limited to beams where the moment $M(x)$ is expressed in terms of polynomial functions.

(b) Applications

We first illustrate the application of singularity functions to the problem previously considered in Example 9.5. In that problem, we observed that there exist two expressions for $M(x)$; namely Eqs. (9.4.15a) and (9.4.15b). Using the definition of the singularity function as given by Eq. (9.7.1a), we may now combine these two equations into the single equation,

$$M(x) = \frac{P(L - a)}{L} x - P\langle x - a \rangle, \quad 0 \leq x \leq L, \tag{9.7.4}$$

i.e. an equation whose domain extends over the entire length of the beam.

Substituting Eq. (9.7.4) in Eq. (9.2.4), we have (with $b = L - a$)

$$EIv''(x) = -\frac{Pbx}{L} + P\langle x - a \rangle, \quad 0 < x < L \tag{9.7.5}$$

subjected to the boundary conditions $v(0) = v(L) = 0$.

We note that since there now exists a *single differential equation that is valid over the entire domain of the beam*, its integral leads to only two constants of integration. Indeed, this is the main advantage in using singularity functions. Thus, since Eq. (9.7.5) is valid in $0 < x < L$, the only boundary conditions that exist are at $x = 0$ and $x = L$ and these are sufficient to solve for the two constants of integration.[‡] (Note that this is in contrast to the calculations of Example 9.5, where two additional boundary conditions existed at $x = a$ [and which were necessary for evaluating the constants of integration].)

Integrating Eq. (9.7.5), and using the property of Eq. (9.7.3), we obtain

$$EIv'(x) = -\frac{Pbx^2}{2L} + \frac{P}{2}\langle x - a \rangle^2 + C_1 \tag{9.7.6a}$$

and

$$EIv(x) = -\frac{Pbx^3}{6L} + \frac{P}{6}\langle x - a \rangle^3 + C_1 x + C_2. \tag{9.7.6b}$$

From B.C. $v(0) = 0$, the constant $C_2 = 0$ since, according to the definition of Eq. (9.7.1a), the term in the brackets $\langle \cdots \rangle$ vanishes when $x < a$. Then using again

---

[†] We should note that we have not attempted to give a rigorous mathematical treatment of the singularity function. It is sufficient for our purposes to define the function and give its operational properties. A rigorous development of singularity functions and the related Heaviside and Delta-dirac functions is treated in the branch of mathematics known as **distribution theory**.

[‡] Note that since v and its derivatives, v' and v'', are implicitly assumed to be continuous *within* the domain of validity (see footnote, p. 315), as opposed to the treatment of this problem in Example 9.5, boundary conditions at $x = a$ are superfluous and, in fact, inappropriate.

the definition when $x > a$, from $v(L) = 0$, we have

$$-\frac{PbL^3}{6L} + \frac{P}{6}(L - a)^3 + C_1 L = 0$$

or

$$C_1 = -\frac{Pb^3}{6L} + \frac{PbL}{6}. \qquad (9.7.7)$$

Substituting C_1 back in Eq. (9.7.6b), we get

$$v(x) = \frac{P}{6EI}\left[-\frac{bx^3}{L} + \langle x - a \rangle^3 - \frac{b^3 x}{L} + bLx\right]. \qquad (9.7.8)$$

The reader is urged to check that this expression is identical to the deflection as given by Eqs. (9.4.18).

The following examples illustrate other aspects and techniques in the use of singularity functions.

Example 9.6: Determine the deflection of an 'overhanging beam' of flexural rigidity EI, that is subjected to a uniformly distributed load as shown in Fig. (9.7.2a).

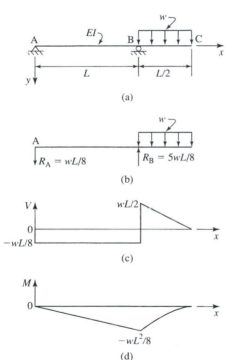

(a)

(b)

(c)

(d)

Figure 9.7.2

Solution: From the equations of statics, reactions are immediately found [see Fig. (9.7.2b)] to be a downward reaction $R_A = wL/8$ and an upward reaction $R_B = 5wL/8$. The moment expressions are then

$$M(x) = -\frac{wL}{8}x, \quad 0 \le x \le L, \qquad (9.7.9a)$$

$$M(x) = -\frac{wL}{8}x + \frac{5wL}{8}(x - L) - \frac{w}{2}(x - L)^2, \quad L \le x \le 3L/2. \qquad (9.7.9b)$$

The shear and moment diagrams are shown in Figs. (9.7.2c and d).

Using singularity functions, the moment expressions can be combined into a single expression, namely

$$M(x) = -\frac{wL}{8}x + \frac{5wL}{8}\langle x - L \rangle - \frac{w}{2}\langle x - L \rangle^2, \quad 0 \leq x \leq 3L/2, \quad (9.7.9c)$$

which we note is valid throughout the length of the beam. Substituting Eq. (9.7.9c) in $EIv''(x) = -M(x)$,

$$EIv''(x) = \frac{wL}{8}x - \frac{5wL}{8}\langle x - L \rangle + \frac{w}{2}\langle x - L \rangle^2, \quad 0 < x < 3L/2. \quad (9.7.10)$$

Integrating this, we obtain

$$EIv'(x) = \frac{wL}{16}x^2 - \frac{5wL}{16}\langle x - L \rangle^2 + \frac{w}{6}\langle x - L \rangle^3 + A, \quad 0 < x < 3L/2,$$
(9.7.11a)

$$EIv(x) = \frac{wL}{48}x^3 - \frac{5wL}{48}\langle x - L \rangle^3 + \frac{w}{24}\langle x - L \rangle^4 + Ax + B, \quad 0 < x < 3L/2.$$
(9.7.11b)

From the boundary condition $v(0) = 0$, we find $B = 0$. Similarly, the boundary condition $v(L) = 0$ leads to $\frac{wL^4}{48} + AL = 0$, from which $A = -wL^3/48$.

Substituting B and A back in Eq. (9.7.11b) and simplifying, we obtain

$$v(x) = \frac{w}{48EI}[-Lx(L^2 - x^2) - 5L\langle x - L \rangle^3 + 2\langle x - L \rangle^4]. \quad (9.7.12)$$

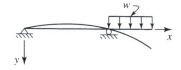

The shape of the elastic curve is shown in Fig. (9.7.3). As expected, we observe that the beam deflects upward in the region $0 \leq x \leq L$, and deflects downward in the over-hanging part. From the figure, we note also that the curvature of the beam ($\kappa \simeq -v'' < 0$) is negative throughout the beam [see Fig. (9.2.1)]. This is in agreement with the relation $v''(x) = -\frac{M(x)}{EI}$, since from the moment diagram, we note that $M(x) < 0$ for all x. $\square$

Figure 9.7.3

Example 9.7: Determine the deflection $v(x)$ of a cantilever beam of flexural rigidity EI and length L due to a load P acting at any arbitrary point $x = \zeta$, as shown in Fig. (9.7.4a).

Solution: Denoting the unknown reactions as in Fig. (9.7.4b), from equilibrium, $\sum F_y = 0$, $R_A = P$, $M_A = -P\zeta$, and hence the shear and moment expressions are[†]

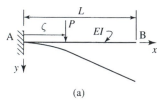

(a)

$$V(x) = P\langle \zeta - x \rangle^0, \quad (9.7.13a)$$

$$M(x) = P[-\zeta + x - \langle x - \zeta \rangle] = -P\langle \zeta - x \rangle, \quad (9.7.13b)$$

upon using the property of Eq. (9.7.2). The beam is therefore governed by the equation

$$EIv''(x) = P(\zeta - x) + P\langle x - \zeta \rangle, \quad 0 < x < L \quad (9.7.14a)$$

or alternatively by

$$EIv''(x) = P\langle \zeta - x \rangle, \quad (9.7.14b)$$

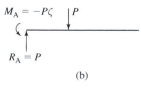

(b)

Figure 9.7.4

and subjected to boundary conditions $v(0) = v'(0) = 0$.

[†] Differentiating Eq. (9.7.13b), $\frac{dM(x)}{dx} = P[1 - \langle x - \zeta \rangle^0]$. Noting from Eq. (9.7.2) that $\langle x - \zeta \rangle^0 = 1 - \langle \zeta - x \rangle^0$, we observe, in comparing Eqs. (9.7.13a) and (9.7.13b), that the known relation $\frac{dM(x)}{dx} = V(x)$ of Eq. (8.3.2) holds true.

Following the procedures as developed previously, upon integrating Eq. (9.7.14a) and using the above boundary conditions, we obtain

$$v(x, \zeta) = \frac{P}{6EI}[3\zeta x^2 - x^3 + \langle x - \zeta \rangle^3]. \tag{9.7.15a}$$

Alternatively, integrating Eq. (9.7.14b) yields

$$v(x, \zeta) = \frac{P}{6EI}[\langle \zeta - x \rangle^3 + 3\zeta^2 x - \zeta^3]. \tag{9.7.15b}$$

We observe that the above expressions are equivalent.

We note here that we have denoted the deflection v as being a function of two variables: x, the coordinate of the cross-section, and ζ, the position of the load P. Thus, from eqs. (9.7.15),

$$v(x, \zeta) = \frac{P}{6EI}(3\zeta x^2 - x^3), \quad x \le \zeta$$

$$v(x, \zeta) = \frac{P}{6EI}(3\zeta^2 x - \zeta^3), \quad \zeta \le x. \tag{9.7.15c}$$

Hence, while $v(x)$ is represented by a cubic expression in x when $x < \zeta$, the elastic curve is described by a linear equation in x for $\zeta < x$; i.e., the beam has zero curvature in this latter region. Indeed, this must be so since the moment $M = 0$ in this region.

The deflection at point B ($\zeta \le x = L$), $\Delta_B \equiv v(L, \zeta)$, is then

$$\Delta_B = \frac{P}{6EI}(3L\zeta^2 - \zeta^3). \tag{9.7.16a}$$

If the load acts at the free end ($\zeta = L$), the displacement is

$$\Delta_B = \frac{PL^3}{3EI}. \tag{9.7.16b}$$

We observe that this is the same result as given by Eq. (9.4.7b).   □

In the following example, we illustrate an additional technique in the use of singularity functions.

Example 9.8: Determine the deflection of the elastic curve for the cantilever beam ABC of flexural rigidity EI and length L, as shown in Fig. (9.7.5a), when subjected to a uniformly distributed load w (N/m) over a length $0 \le x \le a$.

Solution: From simple statics, the upward reaction at A is $R_A = wa$ and the moment $M_A = -wa^2/2$. The resulting shear and moment diagrams are shown in Figs. (9.7.5b and c). Furthermore, the expression for the moment is

$$M(x) = -\frac{wa^2}{2} + wax - \frac{wx^2}{2}, \quad 0 \le x \le a, \tag{9.7.17}$$

while for $a \le x \le L, M = 0$.

We again wish to represent the moment $M(x)$ by means of a single expression for all x, $0 \le x \le L$. To do so, we imagine that upward and downward distributed loads of magnitude w exist between B and C, as shown in Fig. (9.7.5d). Clearly, the addition of these imaginary loads cancel out and therefore the reaction R_A and moment M_A

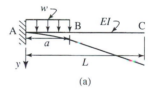

(a)

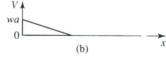

(b)

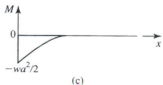

(c)

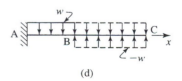

(d)

Figure 9.7.5

are as given above. From Fig. (9.7.5d), the moment expression for the entire beam is

$$M(x) = -\frac{wa^2}{2} + wax - \frac{wx^2}{2} + \frac{w}{2}\langle x - a\rangle^2, \quad 0 \le x \le L, \quad (9.7.18a)$$

and hence we have the equation

$$EIv''(x) = \frac{wa^2}{2} - wax + \frac{wx^2}{2} - \frac{w}{2}\langle x - a\rangle^2, \quad 0 < x < L \quad (9.7.18b)$$

subject to the boundary conditions $v(0) = 0$, $v'(0) = 0$. Integrating the equation, we get

$$EIv'(x) = \frac{wa^2x}{2} - \frac{wax^2}{2} + \frac{wx^3}{6} - \frac{w}{6}\langle x - a\rangle^3 + C_1, \quad 0 < x < L. \quad (9.7.18c)$$

From $v'(0) = 0$, it follows that $C_1 = 0$. Integrating once more, we have

$$EIv(x) = \frac{wa^2x^2}{4} - \frac{wax^3}{6} + \frac{wx^4}{24} - \frac{w}{24}\langle x - a\rangle^4 + C_2, \quad 0 < x < L. \quad (9.7.18d)$$

Now, using the boundary condition $v(0) = 0$, $C_2 = 0$. Upon simplifying, we obtain

$$v(x) = \frac{w}{24EI}[6a^2x^2 - 4ax^3 + x^4 - \langle x - a\rangle^4]. \quad (9.7.19)$$

It is worthwhile to consider the deflection in each region. In the region AB,

$$v(x) = \frac{w}{24EI}[6a^2x^2 - 4ax^3 + x^4], \quad (9.7.20a)$$

and thus we note that the elastic curve assumes the shape of a quartic equation.
   In the region BC,

$$v(x) = \frac{w}{24EI}[6a^2x^2 - 4ax^3 + x^4 - (x - a)^4],$$

which after simplification, becomes

$$v(x) = \frac{wa^3}{24EI}(4x - a), \quad a \le x \le L. \quad (9.7.20b)$$

Observe that in this region, the deflection is a linear function of x, that is, the elastic curve is represented by a straight line and hence the curvature $\kappa = 0$. [The elastic curve of the beam is shown in Fig. (9.7.5a).] This is in accord with the Euler–Bernoulli relation $EI\kappa = M$ since, as was seen previously, $M = 0$ in the region BC.
   Finally, at $x = L$,

$$v(L) = \frac{wa^3}{24EI}(4L - a). \quad (9.7.21)$$

From the general solution above, upon setting $a = L$, we recover the solution for the classical case of a cantilever beam uniformly loaded over its entire length L [Fig. (9.7.6)], namely

$$v(x) = \frac{wx^2}{24EI}(6L^2 - 4Lx + x^2) \quad (9.7.22a)$$

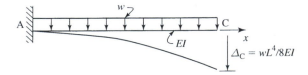

Figure 9.7.6

and the deflection Δ at point $C(x = L)$ is

$$\Delta = \frac{wL^4}{8EI}.$$

(9.7.22b)

$\square$

Example 9.9: A beam ABC, having flexural rigidity EI and length $3L$, is subjected to a load, as shown in Fig. (9.7.7a). Determine the equation of the elastic curve $v(x)$.

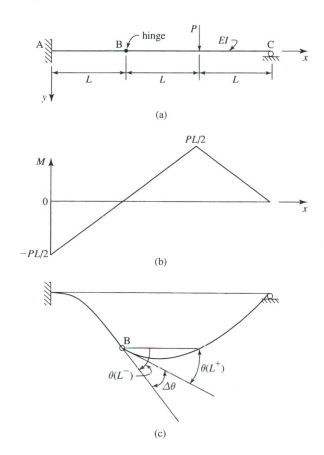

Figure 9.7.7

Solution: Taking moments about point B of the element BC, we find $R_C = P/2$. From equilibrium of the entire beam ABC, $M_A = -PL/2$ and $R_A = P/2$. The resulting moment diagram for $M(x)$, given by

$$M(x) = -\frac{PL}{2} + \frac{P}{2}x - P\langle x - 2L \rangle, \quad 0 \le x \le 3L,$$

(9.7.23)

is shown in Fig. (9.7.7b).

We note that since a hinge exists at point B, the slope of the beam, θ, and hence the first derivative v', is *discontinuous* at this point. Now, we recall that Eq. (9.2.4) is integrable in terms of analytic functions only if the first and second derivatives of $v(x)$ are continuous *within* the domain of the equation. Therefore, to integrate the equation in terms of such functions, point B may only *lie at the boundary of a domain* (see footnote, p. 315). Thus, although the domain of validity for $M(x)$ given above is $0 \le x \le 3L$, we may not use Eq. (9.2.4) over this entire domain; instead, we must write separate equations in the two open domains, $D_1: \{x \mid 0 < x < L\}$; $D_2: \{x \mid L < x < 3L\}$, namely

$$EIv_1''(x) = \frac{PL}{2} - \frac{P}{2}x, \tag{9.7.24a}$$

$$EIv_2''(x) = \frac{PL}{2} - \frac{P}{2}x + P\langle x - 2L \rangle, \tag{9.7.24b}$$

where v_1 and v_2 represent the deflections in the two respective domains.

The boundary conditions are clearly

$$v_1(0) = 0, \tag{9.7.25a}$$

$$v_1'(0) = 0, \tag{9.7.25b}$$

$$v_2(3L) = 0, \tag{9.7.25c}$$

$$v_1(L) = v_2(L). \tag{9.7.25d}$$

(Note that, as opposed to Example 9.5, we do not impose a continuity condition on v' at the boundary between the two domains, here $x = L$.)

Integrating Eq. (9.7.24a) and using boundary conditions, Eqs. (9.7.25a) and (9.7.25b), we find

$$EIv_1(x) = \frac{PLx^2}{4} - \frac{Px^3}{12}. \tag{9.7.26a}$$

Integration of Eq. (9.7.24b) yields

$$EIv_2(x) = \frac{PLx^2}{4} - \frac{Px^3}{12} + \frac{P}{6}\langle x - 2L \rangle^3 + Ax + B. \tag{9.7.26b}$$

From the boundary conditions $v_2(3L) = 0$ and $v_1(L) = v_2(L)$, we find, respectively,

$$3AL + B = -\frac{PL^3}{6}, \tag{9.7.27a}$$

$$AL + B = 0. \tag{9.7.27b}$$

Solving Eqs. (9.7.27) for the two constants of integration

$$A = -\frac{PL^2}{12}, \qquad B = \frac{PL^3}{12}. \tag{9.7.27c}$$

Hence, after simplifying, the elastic curve is given by the expressions,

$$v_1(x) = \frac{P}{12EI}(-x^3 + 3Lx^2) \tag{9.7.28a}$$

and

$$v_2(x) = \frac{P}{12EI}[-x^3 + 3Lx^2 + 2\langle x - 2L \rangle^3 - L^2x + L^3]. \tag{9.7.28b}$$

We now verify the assumed discontinuous character of the slope, $\theta \simeq v'$, at the hinge $B(x = L)$. Differentiating Eqs. (9.7.28) and setting $x = L$, we find

$$v_1'(L^-) = \frac{PL^2}{4EI} \tag{9.7.29a}$$

and

$$v_2'(L^+) = \frac{PL^2}{6EI},\qquad(9.7.29b)$$

where $x = L^-$ and $x = L^+$ signify the values in the two contiguous domains at B. Thus, at the hinge, there exists a discontinuity in the slope, $\Delta\theta \equiv v_2'(L^+) - v_1'(L^-) = -\frac{PL^2}{12EI}$. The elastic curve and the discontinuity $|\Delta\theta|$ are shown in Fig. (9.7.7c). $\square$

9.8 Solutions for statically indeterminate beams by integration of the differential equation

In the previous sections of this chapter, as well as in Chapter 8, we have considered only statically determinate beams. As in the case of statically indeterminate rods under axial or torsional loading (discussed in Chapters 6 and 7), there also can exist beam structures that are statically indeterminate, that is, beams for which it is not possible to obtain unique solutions for the reactive forces (reactions and moments) from the equations of statics. In such cases, we require additional equations: such additional equations are, as we previously have seen, equations of geometric compatibility. There exist several methods for analysing statically indeterminate beams. We first consider a method of solution via integration of the differential equation of beams. With this in mind, we study the following problem.

Consider the beam AB of length L, fixed at point A, simply supported at B and subjected to a uniformly distributed load w, as shown in Fig. (9.8.1a). There then exist three unknown reactions, R_A, M_A and R_B, as shown in Fig. (9.8.1b). However, we have but two equations of equilibrium, namely

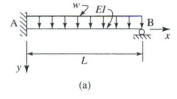

(a)

$$\sum F_y = R_A + R_B - wL = 0\qquad(9.8.1a)$$

$$\sum M|_B = M_A + R_A L - \frac{wL^2}{2} = 0.\qquad(9.8.1b)$$

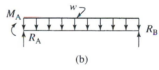

(b)

Figure 9.8.1

Clearly, it is not possible to obtain a unique solution for the three unknowns since we have here an 'extra' unknown.[†] Nevertheless, it is possible to write the following expression for the moment $M(x)$ in terms of the unknowns, namely

$$M(x) = M_A + R_A x - \frac{wx^2}{2}.\qquad(9.8.2a)$$

Substituting this in Eq. (9.2.4), we have

$$EIv''(x) = -M_A - R_A x + \frac{wx^2}{2}.\qquad(9.8.2b)$$

Now the boundary conditions for this problem are

$$v(0) = 0,\qquad(9.8.3a)$$

$$v'(0) = 0,\qquad(9.8.3b)$$

$$v(L) = 0.\qquad(9.8.3c)$$

Before proceeding with the solution, we note that upon integrating the differential

---

[†] Since we have here three unknowns and two equations, there exist an infinity of solutions to these equations. We require, however, a unique solution; namely one which governs the actual behaviour of the beam.

equation (9.8.2b), two unknown constants of integration will appear, say C_1 and C_2. However, from Eqs. (9.8.3), we observe that three boundary conditions exist; i.e., we now have an 'extra' equation. Thus the number of equations now balances the number of unknowns. [Observe that at this stage there exist five unknowns (R_A, R_B, M_A, C_1 and C_2) and five equations, Eqs. (9.8.1a) and (9.8.1b) and Eqs. (9.8.3a)–(9.8.3c). We shall return to this point after completing the solution.]

To proceed with the solution, upon integrating Eq. (9.8.2b) once, we have

$$EIv'(x) = -M_A x - R_A \frac{x^2}{2} + \frac{wx^3}{6} + C_1. \tag{9.8.4a}$$

From $v'(0) = 0$, $C_1 = 0$. Integrating once more,

$$EIv(x) = -M_A \frac{x^2}{2} - R_A \frac{x^3}{6} + \frac{wx^4}{24} + C_2. \tag{9.8.4b}$$

From $v(0) = 0$, $C_2 = 0$. Finally, from the boundary condition $v(L) = 0$, we obtain

$$-M_A \frac{L^2}{2} - R_A \frac{L^3}{6} + \frac{wL^4}{24} = 0. \tag{9.8.5}$$

Equations (9.8.1) and (9.8.5) are thus three equations sufficient to solve for the remaining three unknowns R_A, R_B and M_A. These equations yield

$$R_A = \frac{5wL}{8}, \tag{9.8.6a}$$

$$R_B = \frac{3wL}{8}, \tag{9.8.6b}$$

$$M_A = -\frac{wL^2}{8}. \tag{9.8.6c}$$

Observing the forces acting on the beam, shown as a free body in Fig. (9.8.2a), we note that the (negative) moment M_A is required to prevent any rotation of the beam

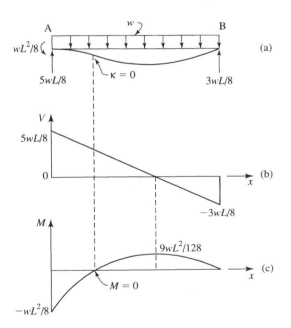

Figure 9.8.2

at point A. Substituting R_A, M_A and C_2 back in Eq. (9.8.4b) and simplifying, we obtain the expression for the elastic curve:

$$v(x) = \frac{wx^2}{48EI}(2x^2 - 5Lx + 3L^2). \qquad (9.8.7a)$$

The deflection at the centre $x = L/2$ is then

$$v(L/2) = \frac{wL^4}{192EI}. \qquad (9.8.7b)$$

The elastic curve is shown in Fig. (9.8.2a). Once the reactions have been obtained, we may readily find explicit expressions for the shear $V(x)$ and moment $M(x)$, namely

$$V(x) = \frac{w}{8}(5L - 8x) \qquad (9.8.8a)$$

$$M(x) = \frac{w}{8}(-L^2 + 5Lx - 4x^2). \qquad (9.8.8b)$$

The corresponding shear and moment diagrams are shown in Figs. (9.8.2b and c).

We note that $M = 0$ at $x = 0.25L$. This corresponds exactly to the point of zero curvature of the elastic curve, as may be seen by comparing Figs. (9.8.2a and c). Such a point is often referred to as a point of *contraflexure*. □

We return to reconsider the number of unknowns and the number of equations; in doing so, we can interpret the meaning of the 'extra' equation mentioned above. As we observed previously, integration of the differential equation leads to two unknown constants C_1 and C_2. Now, there clearly exist three boundary conditions in this problem, namely Eqs. (9.8.3). We observe also that had the beam been a simple cantilever beam, we would have had but two boundary conditions, $v(0) = v'(0) = 0$; clearly, these would have been sufficient to solve for C_1 and C_2 *for a cantilever beam*. However, since we have a fixed-end beam, which is supported at point B (i.e., providing a constraint against vertical displacement at B), there exists an additional boundary condition $v(L) = 0$. Thus the 'extra' boundary condition, $v(L) = 0$, is an additional 'statement' that expresses the geometric condition of the beam. Hence, as in our previous discussion of indeterminate systems, we again observe that *the extra equation is, in fact, an equation of 'geometric compatibility'*.

9.9 Application of linear superposition in beam theory

In the previous sections, we derived expressions for the deflection of the elastic curve, using linear beam theory, namely the relation $EIv''(x) = -M(x)$. In all cases, the resulting displacements were found to be functions of the geometry and inversely proportional to the flexural rigidity EI. Moreover, the displacements were found to be *linearly dependent on the applied loads*.

Let us consider, for example, the case of the cantilever beam AB under two separate loading conditions such as a uniformly distributed load w and an end load P as shown in Figs. (9.9.1b and c), respectively. Let us denote the maximum deflection at the free end B due to these two loading conditions by Δ_B^w and Δ_B^P, respectively. From Eqs. (9.7.22b) and (9.4.7b), these are

$$\Delta_B^w = \frac{wL^4}{8EI} \qquad (9.9.1a)$$

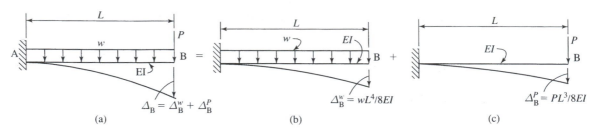

Figure 9.9.1

and

$$\Delta_B^P = \frac{PL^3}{3EI}. \tag{9.9.1b}$$

Since the displacements are linear functions of the applied load, it follows that the principle of linear superposition is applicable;[†] thus the deflection Δ under the combined loading w and P [Fig. (9.9.1a)] is given by $\Delta_B = \Delta_B^w + \Delta_B^P$ or

$$\Delta_B = \frac{wL^4}{8EI} + \frac{PL^3}{3EI}. \tag{9.9.1c}$$

Similarly, for the case of a simply supported beam ACB, the deflection due to a concentrated load applied at C and due to a uniformly distributed load w [Figs. (9.9.2a and b)] are, from Eqs. (9.4.3) and (9.4.20),

$$\Delta_C^w = \frac{5wL^4}{384EI} \tag{9.9.2a}$$

and

$$\Delta_C^P = \frac{PL^3}{48EI}. \tag{9.9.2b}$$

Figure 9.9.2

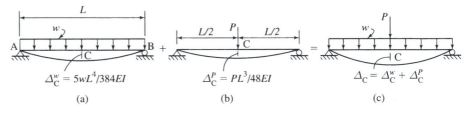

Again, using the principle of linear superposition, the deflection under the combined loading system [Fig. (9.9.2c)] is

$$\Delta_C = \frac{5wL^4}{384EI} + \frac{PL^3}{48EI}. \tag{9.9.2c}$$

As was previously mentioned, the use of the principle of superposition proves quite useful for elastic bodies satisfying a linear theory, since one may analyse each case separately and superimpose the results.

[†] We note also that since the radii of curvature are large, the strains in the beam, $\epsilon_x = y/R$, are necessarily small. This necessary requirement for application of linear superposition, as developed in Chapter 5 (Section 2), is therefore also satisfied.

It is worthwhile here to reconsider the indeterminate beam [Fig. (9.8.1a)] analysed in Section 8, in the context of the principle of superposition. From the free body of Fig. (9.8.1b), we may consider the beam to be subjected to the following forces: the applied load w, the reactions R_A and R_B and the moment $M_A = -wL^2/8$. Now, the basic difference between the given indeterminate beam and an equivalent simply supported beam is the restraint against rotation at the fixed end that is provided by the moment M_A. Thus we may consider the given indeterminate beam to be an equivalent simply supported beam subjected to the applied load w and a (negative) moment $M_A = -wL^2/8$. Since all the relations are linear, we apply the principle of superposition, as shown symbolically in Figs. (9.9.3a–c). From Eqs. (9.4.3) and (9.4.13d), we have, respectively,

$$\Delta_C^w = \frac{5wL^4}{384EI} \quad \text{and} \quad \Delta_C^M = \frac{M_A L^2}{16EI}. \tag{9.9.3a}$$

Therefore, from our known result, $M_A = -wL^2/8$, we may readily calculate the deflection Δ_C as

$$\Delta_C = \frac{5wL^4}{384EI} + \frac{(-wL^2/8)L^2}{16EI} = \frac{wL^4}{192EI}. \tag{9.9.3b}$$

Note that this is precisely the same result obtained in Section 8 [Eq. (9.8.7b)].

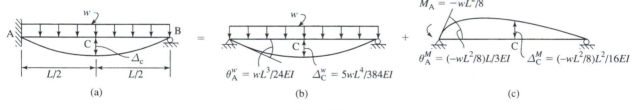

Figure 9.9.3

We thus observe that one may obtain solutions to more complex problems by superimposing the separate solutions of more simple problems. However, even if the latter solutions are not known, one can nevertheless often gain great physical insight on the *qualitative* behaviour of a more complex structure by superposing the effects of several individual forces. For example, for the present beam structure considered, it required but little experience to reason that a *negative* moment M_A is necessary to maintain zero slope at A. Consequently, from Figs. (9.9.3), one might immediately have concluded *qualitatively* that the deflection at the mid-span of the given indeterminate beam is less than that of an equivalent simply supported beam.[†]

In this section, we confined the discussion of the principle of superposition to concentrated forces and moments (couples). We shall return to a more general discussion of superposition in Section 11 of this chapter and will develop its application for more general loadings.

---

† The *quantitative* result for the deflection Δ_C, given in Eq. (9.8.7b), however, required previous knowledge of the magnitude of M_A (as was obtained in Section 8).

As we shall presently see, the principle of superposition, as developed in this section, proves to be particularly useful in analysing indeterminate beams.

9.10   Analysis of statically indeterminate beams: the force method

In Section 8, a method of solution for statically indeterminate structures was developed via integration of the governing differential equation together with the associated boundary conditions. We consider here another method, known as the **force method**, to solve for indeterminate structures.

(a) Development of the force method

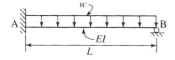

Figure 9.10.1

Let us again consider the indeterminate beam, fixed at one end and simply supported at the other, subjected to a uniformly distributed load w [Fig. (9.10.1)]. This problem was previously solved in Section 8 [see Fig. (9.8.1)] by integration of the governing differential equation. We recall that there exist three unknown reactions, R_A, M_A and R_B, but only two equations of equilibrium, $\sum F_y = 0$, $\sum M = 0$. However, it is clear that if, for example, the reaction R_B did not exist (i.e., if $R_B = 0$), the beam would still be capable of carrying the applied load w. Indeed the beam would then be a statically determinate cantilever beam. Thus, the additional support existing at B of the indeterminate beam may be considered to be an 'extra' reaction. Such an unknown reaction is referred to as a *redundant* reaction ('redundant' in the sense that if this reaction were absent, it would still be possible to maintain the beam in equilibrium). However, although the equations of equilibrium can be satisfied with $R_B = 0$, this value of the redundant *will not provide the necessary constraint against a vertical displacement at B.*

Thus, choosing R_B as the redundant reaction, we consider the beam AB to be subjected to the uniformly applied distributed load w and to an *unknown* force R_B, as shown in Fig. (9.10.2a). Note that, at this stage, we do *not* state that the beam cannot displace at point B; that is, we essentially *release the constraint.*

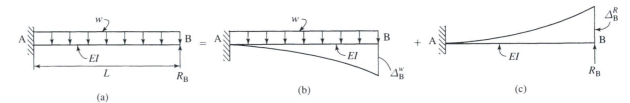

Figure 9.10.2

The resulting cantilever beam is therefore called the **released beam** (or in general terms, the **released structure**) corresponding to the given indeterminate beam. To complete the problem, it is necessary to reimpose the constraint at a later stage. Now, as in the previous section, we make use of the principle of superposition; that is, we consider the beam to be subjected to two separate systems such as (a) the uniform load w and (b) an arbitrary upward force R_B, as shown in Figs. (9.10.2b and c). Clearly, the displacement of point B, Δ_B, is then given by $\Delta_B = \Delta_B^w + \Delta_B^R$. From the previous section, Eq. (9.9.1c), we note that for the given w and (upward) force R_B, the displacement *of the released beam* is given by

$$\Delta_B = \frac{wL^4}{8EI} - \frac{R_BL^3}{3EI}. \tag{9.10.1}$$

However, since the beam is, in fact, simply supported at B, we now impose the necessary constraint; that is, we stipulate the geometric compatibility of the system, namely

$$\Delta_B = 0. \tag{9.10.2}$$

It is important to observe that the equation of geometric compatibility is written on a displacement that corresponds to the chosen redundant.

Therefore, from Eq. (9.10.2), we obtain[†]

$$R_B = \frac{3wL}{8}, \tag{9.10.3}$$

which is the same result as found in Section 8 [Eq. (9.8.6b)].

Having determined R_B, the remaining two reactions are found from the equations of statics, $\sum F_y = 0$ and $\sum M = 0$, to be $R_A = \frac{5wL}{8}$ and $M_A = -\frac{wL^2}{8}$.

In solving this problem, we chose R_B to be the redundant reaction. However, the choice of the redundant reaction *is not unique*. Indeed, one may choose any reactive force to be redundant, provided the remaining reactions on the released structure are sufficient to maintain the system in equilibrium under applied loads. We illustrate this by means of the following example.

Example 9.10: Determine the reaction for the indeterminate beam of Fig. (9.10.1) by choosing the reacting moment M_A as the redundant.

Solution: Upon choosing M_A as the redundant, we have essentially released the constraint against rotation at A and therefore the 'released structure' now is a simply supported beam, as shown in Fig. (9.10.3a), which is subjected to the uniform load w and (at this stage) an arbitrary moment M_A. The rotation θ_A at A due to w and M_A [given, respectively, by Eqs. (9.4.4b) and (9.4.14)] denoted by θ_A^w and θ_A^M, respectively, is shown in Figs. (9.10.3b and c). Using superposition, we obtained

$$\theta_A = \theta_A^w + \theta_A^M = \frac{wL^3}{24EI} + \frac{M_AL}{3EI}. \tag{9.10.4}$$

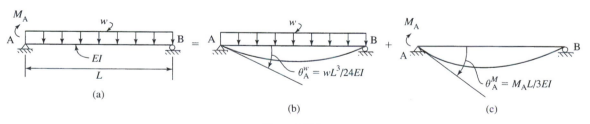

Figure 9.10.3

[†] Note that, since here R_B is assumed to be *upward*, as opposed to Eq. (9.9.2c), a minus sign appears in Eq. (9.10.1) for the deflection due to R_B. We emphasise that had we instead chosen R_B to be downward, Eq. (9.10.1) would have contained a positive sign and hence we would have obtained $R_B = -3wL/8$, thus indicating that R_B is, in fact, upward.

To satisfy geometric compatibility of the given beam, we stipulate as before the geometric compatibility on the 'displacement' that corresponds to the chosen redundant 'force'. Thus, here, we stipulate that

$$\theta_A = 0. \tag{9.10.5}$$

Therefore, from Eq. (9.10.4),[†]

$$M_A = -\frac{wL^2}{8}, \tag{9.10.6}$$

which we note is the same result as obtained in Section 8 [Eq. (9.8.6c)]. As before, the remaining unknown reactions R_A and R_B are obtained from the equations of equilibrium.   □

Example 9.11:   Using the force method, determine the reactions at A, B and C for the beam having flexural rigidity *EI*, as shown in Fig. (9.10.4a). Obtain the expression for the shear *V(x)* and moments *M(x)* and draw the shear and moment diagrams showing all critical values.

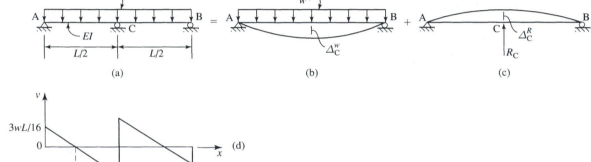

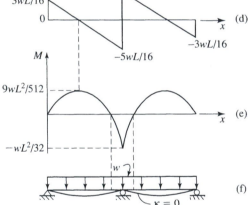

Figure 9.10.4

Solution: Let us choose the centre reaction R_C as the redundant (upward) force. Hence the released structure consists of a simply supported beam subjected to the uniformly distributed load *w* and an unknown force R_C [Figs. (9.10.4b and c)]. The equation of geometric compatibility on the displacement that corresponds to the chosen redundant is therefore

$$\Delta_C = 0. \tag{9.10.7}$$

[†] Note that the negative sign here indicates that a counter-clockwise moment M_A is acting on the beam.

From Eq. (9.9.2c), the combined deflection at C due to the two force systems is

$$\Delta_C = \frac{5wL^4}{384EI} - \frac{R_C L^3}{48EI}, \tag{9.10.8}$$

where $0 < \Delta_C$ denotes a downward deflection. Using Eq. (9.10.7), we find

$$R_C = \frac{5wL}{8}. \tag{9.10.9}$$

It follows from the equations of equilibrium that $R_A = R_B = \frac{3wL}{16}$. The resulting shear and moment expressions are then readily written as

$$V(x) = 3wL/16 - wx, \qquad M(x) = 3wLx/16 - wx^2/2, \quad 0 \le x \le L/2,$$

$$V(x) = -3wL/16 + w(L-x), \qquad M(x) = 3wL(L-x)/16 - w(L-x)^2/2,$$

$$L/2 \le x \le L.$$

Shear and moment diagrams are shown in Figs. (9.10.4d and e). The elastic curve is shown in Fig. (9.10.4f). We note that there is a change in the curvature at the cross-section where $M = 0$. □

In the following example, we illustrate the use of the force method for the case where there is but a partial restraint to the displacement.

Example 9.12: Using the force method, determine the reaction of the linear spring (having constant k) acting on the beam ACB having flexural rigidity EI when subjected to the downward uniformly load w, as shown in Fig. (9.10.5a).

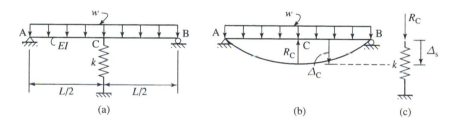

(a)   (b)   (c)   **Figure 9.10.5**

Solution: We denote the (upward) force that the spring exerts on the beam by R_C. We observe from Fig. (9.10.5b) that since there exist three unknown reactions, the system remains statically indeterminate.

Due to the applied load w and the upward force R_C acting on the beam, the deflection of point C of the released beam is, as before [Eq. (9.9.2c)],

$$\Delta_C = \frac{5wL^4}{384EI} - \frac{R_C L^3}{48EI}, \tag{9.10.10a}$$

where we recall that $0 < \Delta_C$ denotes a downward deflection of point C of the beam. Now, since it is assumed that the spring exerts an upward force on the beam, it follows that the beam exerts a downward force R_C on the spring. Denoting the compression of the spring by Δ_s, the spring will compress by an amount

$$\Delta_s = \frac{R_C}{k}. \tag{9.10.10b}$$

Since the spring and the beam do not separate, the required geometric compatibility equation is

$$\Delta_C = \Delta_s. \tag{9.10.11}$$

Substituting Eqs. (9.10.10a) and (9.10.10b) in Eq. (9.10.11), we find

$$\frac{5wL^4}{384EI} - \frac{R_C L^3}{48EI} = \frac{R_C}{k}$$

and hence

$$R_C = \frac{5wL}{8} \frac{1}{1 + 48EI/kL^3}. \tag{9.10.12a}$$

We observe that R_C depends on the relative stiffness of the spring to the flexural stiffness of the beam, i.e. on the non-dimensional parameter

$$\alpha = \frac{kL^3}{48EI}. \tag{9.10.12b}$$

Thus we may write[†]

$$R_C = \frac{5wL}{8} \frac{\alpha}{1 + \alpha}. \tag{9.10.12c}$$

The variation of R_C with α is shown in Fig. (9.10.6). We observe that R_C approaches the value $R_C = 5wL/8$ asymptotically with increasing values of α. Moreover, we

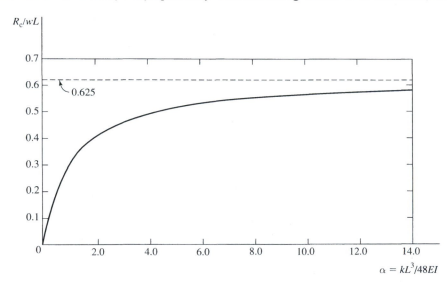

Figure 9.10.6

[†] It is significant to observe that in all previous problems solved in this chapter, solutions have yielded reactions that are independent of the stiffness of the system and, in fact, independent of the material properties of the system. However, we note here that R_C is a function of the *relative stiffness* of the component parts of the system. To explain this difference, we first observe that in all previous problems, the displacement of the chosen redundant was zero. For the present problem, however, the redundant (here, R_C) undergoes a displacement. It is this difference in behaviour, which accounts for the dependency of the redundant on the relative stiffness of the structure components. Thus, when a chosen redundant undergoes (an unknown) non-zero displacement, the redundant will always be found to depend on the relative stiffness of the structure components. [This explanation is valid in general for all structures having a linear force–displacement behaviour. It is given here *ex cathedra* since it is beyond the scope of our treatment to provide a general proof; the proof falls within the realm of structural mechanics.]

observe that as $k \rightarrow \infty$, we recover the value $R_C = 5wL/8$, which is the reaction exerted by a simple support at C, as found in Example 9.11. Note also that for $k = 0$, which represents a spring having no rigidity, $R_C = 0$.

Example 9.13:   Determine the reaction at B for the indeterminate beam shown in Fig. (9.10.7a), which is subjected to a force P acting at some point ζ to the right of point A.

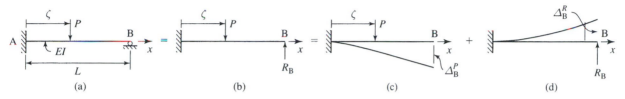

Figure 9.10.7

Solution: We again release the constraint at B and choose the reaction R_B as the redundant reaction. We denote the *downward* displacement of point B of the released structure due to the force P by Δ_B^P. Similarly, we arbitrarily denote Δ_B^R as the *upward* displacement of B of the released structure due to the unknown force R_B [Figs. (9.10.7b–d).[†] The equation of geometric compatibility then becomes

$$\Delta_B^P - \Delta_B^R = 0. \qquad (9.10.13)$$

From Eqs. (9.7.16),

$$\Delta_B^P = \frac{P}{6EI}(3L\zeta^2 - \zeta^3), \qquad \Delta_B^R = \frac{R_B L^3}{3EI}. \qquad (9.10.14)$$

Substituting Eqs. (9.10.14) in the compatibility equation (9.10.13), we find

$$\frac{P}{6EI}(3L\zeta^2 - \zeta^3) - \frac{R_B L^3}{3EI} = 0 \qquad (9.10.15a)$$

and therefore

$$R_B = \frac{P}{2L^3}(3L\zeta^2 - \zeta^3). \qquad (9.10.15b)$$

We observe that $R_B = R_B(\zeta)$ is clearly a function of the position of the applied load P but varies linearly with the magnitude of P. A plot of R_B as a function of ζ/L is shown in Fig. (9.10.8).   □

[†] Note that here we have chosen a positive direction for Δ_B^R to be upward. It is important to note that the choice of positive directions is totally arbitrary. Clearly, the sign of Δ_B^R appearing in the compatibility equation [here, Eq. (9.10.13)] will then depend on the chosen positive direction.

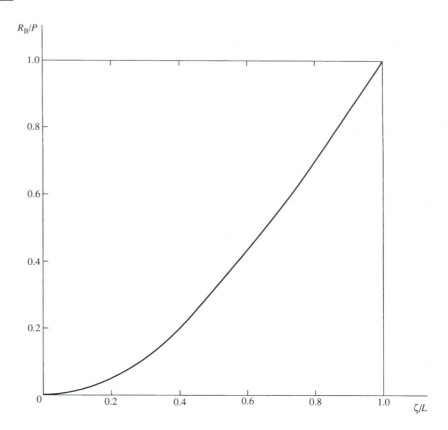

Figure 9.10.8

(b) Comments on the force method

Having developed the force method for indeterminate beams and investigated several problems, it is worthwhile, at this stage, to observe several features and comment on the method.

- In order to implement the method, the separate displacements at a given point *of the released structure* corresponding to the chosen redundant must be known (i) due to the applied loading and (ii) due to the redundant force itself. (If these quantities are unknown, then they must necessarily be calculated. Thus the method requires that these quantities be known or given.) While, at this stage of our study, these quantities can be obtained only by integration of the basic governing differential equations, we mention here that there exist other methods for finding these quantities. (These methods are considered in Chapter 14.)
- A statically indeterminate structure is, by definition, one for which there exist more unknowns than *independent* equations of equilibrium. Therefore, it is possible to find an infinite number of solutions to these equations. However, there exists but a unique solution that satisfies both equilibrium *and* the geometric conditions of the structure. The equation of geometric compatibility thus is the necessary additional equation required to yield the unique solution for the unknowns.
- The method developed in this section is referred to as the *Force method* since the basic equations of compatibility are written in terms of unknown forces. (This method is often also referred to as the *flexibility method of analysis*.)[†]

[†] It should be mentioned that there exist other methods for analysing statically indeterminate structures in which the required equations are written in terms of unknown displacements. Such methods are referred to as *displacement methods* or *stiffness methods*.

■ Finally, we should mention that in our study, we have discussed only cases where there exists but one redundant force; that is, the number of unknowns is greater than the number of equilibrium equations by one. The resulting structures analysed here are therefore said to have 'one degree of indeterminacy'. In practice, there exist structures for which there may be many degrees of indeterminacy. While the concepts of the force method, as developed above, are valid for all such structures, a treatment of the force method for systems having an n-degree of indeterminacy is beyond the scope of our study and lies within the realm of structural mechanics.

◇9.11 Superposition – integral formulation: the fundamental solution and Green's functions

In Section 10, we discussed the principle of superposition for linear elastic beams in the context of concentrated forces and moments (couples). We develop here a general integral formulation of the principle of superposition as applicable to beams. We shall find that the principle has wide applicability and, for a given (statically determinate or indeterminate) beam, we can, using superposition, obtain solutions *due to any arbitrary loading condition* if a specific solution (of the given beam), referred to as the *fundamental solution*, is known.

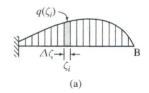

Figure 9.11.1

(a) Development and applications

Let us consider, for example, the cantilever beam, shown in Fig. (9.11.1), subjected to any arbitrary varying load $q(x)$. We wish to find the deflection of point B. Now, it is clear that the given applied load may be represented as being made up of a set of n infinitesimal loads, each situated at a point $x = \zeta_i$ and having magnitude $q(\zeta_i)\,\Delta\zeta$, where $\Delta\zeta = L/n$ [Fig. (9.11.2a)] and each of the (small) loads produces a displacement Δv [Fig. (9.11.2b)] of point B.

We recall from Eq. (9.7.16a) or (9.10.14) that for a force P located at $x = \zeta_i$, the deflection at B is given by

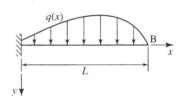

(a)

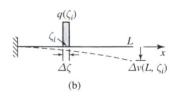

(b)

$$v(L, \zeta_i) = \frac{P}{6EI}\left(3L\zeta_i^2 - \zeta_i^3\right). \qquad (9.11.1)$$

Note that here we have used two variables L and ζ, the first to indicate that the deflection is at L and the second to indicate that the deflection is due to a force P located at ζ_i.

Since the displacement is linearly dependent on the applied load, it is clear that due to a unit load $P = 1$, the deflection at $x = L$ is [see Fig. (9.11.2c)]

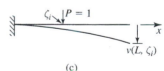

(c)

Figure 9.11.2

$$v(L, \zeta_i)|_{P=1} = \frac{1}{6EI}\left(3L\zeta_i^2 - \zeta_i^3\right). \qquad (9.11.2)$$

Therefore, using linear superposition, the displacement Δv due to an incremental load $q(\zeta_i)\,\Delta\zeta$ acting at $x = \zeta_i$ is [Fig. (9.11.2b)]

$$\Delta v(L, \zeta_i) = [q(\zeta_i)\,\Delta\zeta] \cdot \left[\frac{1}{6EI}\left(3L\zeta_i^2 - \zeta_i^3\right)\right]. \qquad (9.11.3)$$

Hence, the displacement $v(L)$ due to the set of loads $q(\zeta_i)\Delta\zeta$ acting along the beam is

$$v(L) = \sum_{i=1}^{n} \Delta v\,(L, \zeta_i) \qquad (9.11.4a)$$

or

$$v(L) = \sum_{i=1}^{n} \left[\frac{q(\zeta_i)}{6EI}(3L\zeta_i^2 - \zeta_i^3) \right] \Delta\zeta. \qquad (9.11.4b)$$

Upon letting $n \to \infty$ such that $\Delta\zeta \to 0$, we recognise that the limiting sum is, by definition, the integral from 0 to L. Thus we obtain

$$v(L) = \int_0^L q(\zeta) \left[\frac{1}{6EI}(3L\zeta^2 - \zeta^3) \right] d\zeta. \qquad (9.11.5)$$

It is convenient to define the displacement $v(L)$ due to a unit load acting at $x = \zeta$ by $G_v(L, \zeta) = v(L, \zeta)|_{P=1}$. Hence, from Eq. (9.11.2), we have explicitly

$$G_v(L, \zeta) = \frac{1}{6EI}(3L\zeta^2 - \zeta^3). \qquad (9.11.6)$$

We may now write Eq. (9.11.5) as

$$v(L) = \int_0^L q(\zeta)G_v(L, \zeta)\,d\zeta. \qquad (9.11.7)$$

We thus observe that the displacement at point B due to any arbitrary varying load distribution may be obtained by simple integration as in Eq. (9.11.7) *if* $G_v(L, \zeta)$, *the displacement of point B($x = L$) due to a unit load acting at* ζ, *is known.* The above integral is therefore often referred to as the *superposition integral.*

We mention here that in mathematics, $G_v(L, \zeta)$ is called a **Green's function** for the deflection at B since it represents the effect due to a unit loading. When appearing in the integral it is referred to as the 'kernel' of the integral.[†]

In the above, we have found the Green's function for the deflection at B. The more general Green's function that represents the deflection of any point x in the beam due to a unit load $P = 1$ acting at ζ may now be readily written by setting $P = 1$ in Eq. (9.7.15a); thus the Green's function for the deflection at any point x is

$$G_v(x, \zeta) = \frac{1}{6EI}[3\zeta x^2 - x^3 + \langle x - \zeta \rangle^3], \quad 0 \le x \le L, \;\; 0 \le \zeta \le L. \qquad (9.11.8)$$

The deflection of any point x in the beam due to an arbitrary varying load $q(\zeta)$ is

---

[†] In structural mechanics, the function $G_v(L, \zeta)$ is also called the **influence function** for the displacement at B.

then given by[†]

$$v(x) = \int_0^L q(\zeta)G_v(x, \zeta)\,d\zeta. \tag{9.11.9}$$

Use of the above formulation is illustrated in the following example.

Example 9.14: Using the Green's function for the deflection at B [Eq. (9.11.6)], determine the deflection of the free end of the cantilever beam due to (i) a uniformly distributed load w_0, (ii) a linearly varying distributed load and (iii) a sinusoidal load, as shown in Figs. (9.11.3a–c), respectively.

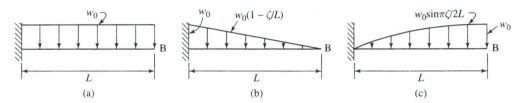

(a) (b) (c)

Figure 9.11.3

Solution:

(i) Substituting the uniformly distributed load, $q(\zeta) = w_0$, in Eq. (9.11.7),

$$v(L) = \frac{w_0}{6EI} \int_0^L (3L\zeta^2 - \zeta^3)\,d\zeta, \tag{9.11.10}$$

which upon simple integration yields

$$v(L) = \frac{w_0 L^4}{8EI}. \tag{9.11.11}$$

We observe that this is the same result as in Eq. (9.7.22b).

(ii) Substituting the linearly varying load of Fig. (9.11.3b), $q(\zeta) = w_0(1 - \zeta/L)$, in Eq. (9.11.7),

$$v(L) = \frac{w_0}{6EI} \int_0^L (1 - \zeta/L)(3L\zeta^2 - \zeta^3)\,d\zeta, \tag{9.11.12}$$

which, by a simple integration, yields

$$v(L) = \frac{w_0 L^4}{30EI}. \tag{9.11.13}$$

[†] Using the conventional notation for Green's functions, the first variable defines the point at which the behaviour occurs while the second variable describes the location of the unit force causing the response, i.e. the 'source' of the response.

(iii) Substituting the sinusoidal load of Fig. (9.11.3c), $q(\zeta) = w_0 \sin \pi \zeta / 2L$, in Eq. (9.11.7),

$$v(L) = \frac{w_0}{6EI} \int_0^L \sin \frac{\pi \zeta}{2L} \cdot (3L\zeta^2 - \zeta^3) \, d\zeta. \qquad (9.11.14)$$

Integrating and substituting the upper and lower limits, we obtain

$$v(L) = \frac{2w_0 L^4}{\pi^4 EI} (\pi^2 - 4\pi + 8). \qquad (9.11.15)$$

$\square$

(b) Generalisation: Green's functions for shears, moments, etc. in beams

We observe that the application of the integral formulation of superposition provides a very simple means to obtain the response of a *linear* elastic system to any arbitrary loading system, once the response to a unit load, that is, the Green's function, is known. While we have developed here the Green's function for the deflection, clearly one may also determine the Green's function for, say the moment or shear at any point x of a beam.[†] For example, the moment $M(x)$ in the cantilever beam of Fig. (9.7.4a) due to a load P acting at a point $x = \zeta$ [Eq. (9.7.13b)] is

$$M(x) = P[-\zeta + x - \langle x - \zeta \rangle] = -P\langle \zeta - x \rangle. \qquad (9.11.16)$$

Hence by its definition, the Green's function for the moment, which we denote by $G_M(x, \zeta)$, is given by

$$G_M(x, \zeta) = -\langle \zeta - x \rangle. \qquad (9.11.17)$$

A plot of $G_M(x, \zeta)$, known as the **influence line** for $M(x)$, is shown in Fig. (9.11.4) as a function of ζ for any given x. Note that when $x > \zeta$, $G_M(x, \zeta) = 0$.

Using this as the kernel of the superposition integral, the moment at any point x due to an arbitrarily varying load $q(\zeta)$ is then given by

$$M(x) = \int_0^L q(\zeta) G_M(x, \zeta) \, d\zeta. \qquad (9.11.18)$$

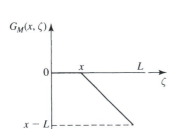

Figure 9.11.4

Thus, to calculate the moment due to a varying load $q(\zeta) = w_0(1 - \zeta/L)$, we integrate as follows:

$$M(x) = -w_0 \int_0^L (1 - \zeta/L)\langle \zeta - x \rangle \, d\zeta \qquad (9.11.19a)$$

[†] In Chapter 14, we develop a simple means to obtain such Green's functions in statically determinate beams.

Using the property of Eq. (9.7.1a), the bracketed term $\langle \zeta - x \rangle = 0$ for $\zeta < x$. Therefore we have

$$M(x) = -w_0 \int_x^L (1 - \zeta/L)(\zeta - x)\, d\zeta. \qquad (9.11.19b)$$

Noting the indefinite integral,

$$\int (1 - \zeta/L)(\zeta - x)\, d\zeta = -x\zeta + \frac{1}{2}(1 + x/L)\zeta^2 - \frac{\zeta^3}{3L} + C,$$

where C is a constant of integration, and substituting the appropriate upper and lower limits, we find

$$M(x) = w_0\{(xL/2 - L^2/6) - (x^2/2 - x^3/6L)\}$$

or

$$M(x) = \frac{w_0}{6}(x^3/L - 3x^2 + 3Lx - L^2) = \frac{w_0}{6L}(x - L)^3. \qquad (9.11.19c)$$

Note that this same result [Eq. (8.4.3c)] was found from the equations of statics in Example 8.4 of Chapter 8.

As a further example, let us assume that the linearly varying load, $q(\zeta) = w_0(1 - \zeta/L)$, of Example 9.14 is applied over a length $0 \le x \le c$, $c \le L$ [Fig. (9.11.5)] and that we wish to find the shear force $V(x)$ for any point under this load; i.e., $0 \le x \le c$. We may easily find $V(x)$ using the Green's function $G_V(x, \zeta)$ according to the relation

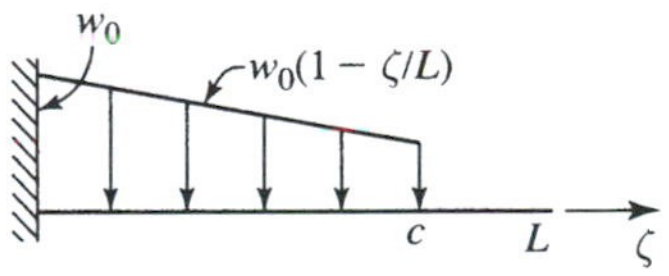

$$V(x) = \int_A^B q(\zeta)G_V(x, \zeta)\, d\zeta, \qquad (9.11.20a)$$

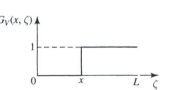

Figure 9.11.5

where the load is applied between arbitrary points A and B. For our problem, we have

$$V(x) = \int_0^c q(\zeta)G_V(x, \zeta)\, d\zeta \qquad (9.11.20b)$$

We note from Eq. (9.7.13a) that $G_V(x, \zeta)$ is given by

$$G_V(x, \zeta) = \langle \zeta - x \rangle^0. \qquad (9.11.20c)$$

The Green's function $G_V(x, \zeta)$, i.e. the influence line for $V(x)$, is shown in Fig. (9.11.6) as a function of ζ for any given x. Note that $G_V(x, \zeta) = 0$ for $x > \zeta$. From Eq. (9.11.20b), we write

$$V(x) = w_0 \int_0^c (1 - \zeta/L)\langle \zeta - x \rangle^0\, d\zeta. \qquad (9.11.21a)$$

Figure 9.11.6

Using the property of Eq. (9.7.1b),

$$V(x) = w_0 \int_x^c (1 - \zeta/L)\,d\zeta = w_0[\zeta - \zeta^2/2L]\Big|_x^c, \qquad (9.11.21b)$$

and evaluating, we obtain

$$V(x) = \frac{w_0}{2}[x^2/L - 2x + c(2 - c/L)], \quad 0 \le x \le c. \quad (9.11.22a)$$

If the load is applied over the entire length AB ($c = L$),

$$V(x) = \frac{w_0}{2}(x^2/L - 2x + L) = \frac{w_0 L}{2}(1 - x/L)^2, \qquad (9.11.22b)$$

which is identical with the result given by Eq. (8.4.3b).

Example 9.15: *Part A*: (i) Determine the Green's function, $G_{M_C} \equiv G_M(2L, \zeta)$, of the moment at C, for the statically determinate beam shown in Fig. (9.11.7a). (ii) Plot the influence line for M_C, G_{M_C}, as a function of ζ. (iii) Using the Green's function, determine the moment M_C due to a uniformly distributed load w (N/m) acting over the entire length of the structure. *Part B*: Repeat Part (A) for the upward reaction R_C at point C.

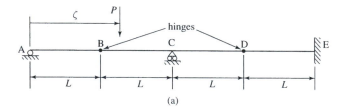

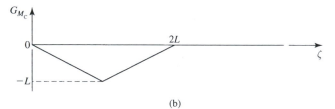

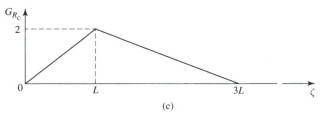

Figure 9.11.7

Solution: *Part A*: (i) In order to determine G_{M_C}, we must satisfy equilibrium conditions for any ζ; the function therefore will depend on which region of the beam the load is situated. We proceed as follows:

For $0 \le \zeta \le L$:
Isolating AB as a free body, [Fig. (9.11.8a)]

$$\sum M_B = 0 \rightarrow\rightarrow R_A L - P(L - \zeta) = 0 \rightarrow\rightarrow R_A = \frac{P}{L}(L - \zeta). \quad (9.11.23a)$$

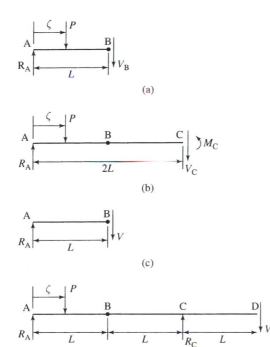

Figure 9.11.8

Now, treating AC as a free body [Fig. (9.11.8b)],

$$\sum M_C = 0 \rightarrow\rightarrow M_C = 2R_A L - P(2L - \zeta) \qquad (9.11.23b)$$

and therefore, from Eq. (9.11.23a),

$$M_C = 2P(L - \zeta) - P(2L - \zeta) = -P\zeta. \qquad (9.11.23c)$$

For $L \leq \zeta \leq 2L$:
Isolating AB as a free body [Fig. (9.11.8c)], we find from $\sum M_B = 0 \rightarrow\rightarrow R_A = 0$. [Note that $R_A = 0$ for all $L \leq \zeta \leq 4L$]. Therefore, again using AC as a free body [Fig. (9.11.8b)], we obtain, from Eq. (9.11.23b), $M_C = -P(2L - \zeta)$.

For $2L \leq \zeta \leq 4L$:
Here it is clear that $R_A = 0$. Hence $M_C = 0$.
 Upon setting $P = 1$ in Eqs. (9.11.23), we write the Green's function as

$$G_M(2L, \zeta) = \begin{cases} -\zeta, & 0 \leq \zeta \leq L, \\ \zeta - 2L, & L \leq \zeta \leq 2L, \\ 0, & 2L \leq \zeta \leq 4L. \end{cases} \qquad (9.11.24)$$

(ii) A plot of $G_M(2L, \zeta)$ is shown in Fig. (9.11.7b).
(iii) The moment due to a uniformly distributed load w acting over the entire region, $0 \leq \zeta \leq 4L$, is

$$M_C \equiv M(2L) = w \int_0^{4L} G_M(2L, \zeta)\, d\zeta. \qquad (9.11.25)$$

We recognise here that the integral represents the area under the G_{M_C} function. Therefore, $M_C = -wL^2$.

Part B: (i) To determine R_C, we again consider each sector of ζ separately.

For $0 \leq \zeta \leq 3L$:

Treating AD as a free body [Fig. (9.11.8d)],

$$\sum M_D = 0 \to\to 3L\,R_A + L\,R_C - P(3L - \zeta) = 0$$

from which,

$$R_C = -3R_A + \frac{P}{L}(3L - \zeta). \tag{9.11.26a}$$

For $0 \leq \zeta \leq L$:

We have from Eq. (9.11.23a), $R_A = \frac{P}{L}(L - \zeta)$ and hence

$$R_C = -\frac{3P}{L}(L - \zeta) + \frac{P}{L}(3L - \zeta) = \frac{2P\zeta}{L}. \tag{9.11.26b}$$

For $L \leq \zeta \leq 3L$:

$R_A = 0$, and therefore

$$R_C = \frac{P}{L}(3L - \zeta). \tag{9.11.26c}$$

For $3L \leq \zeta \leq 4L$:

It is clear, from $\sum M_D = 0$ for the free body AD that $R_C = 0$.

The Green's function G_{R_C} is therefore given by

$$G_R(2L, \zeta) = \begin{cases} 2\zeta/L, & 0 \leq \zeta \leq L, \\ 3 - \zeta/L, & L \leq \zeta \leq 3L, \\ 0, & 3L \leq \zeta \leq 4L. \end{cases} \tag{9.11.27}$$

(ii) A plot of $G_R(2L, \zeta)$ is shown in Fig. (9.11.7c).

(iii) The reaction R_C due to a uniformly distributed load w acting over the entire region, $0 \leq \zeta \leq 4L$, is

$$R_C \equiv R(2L) = w \int_0^{4L} G_R(3L, \zeta)\,d\zeta. \tag{9.11.28}$$

We recognise here that the integral represents the area under the G_{R_C} function. Therefore, $R_C = 3wL$. $\qquad\square$

(c) Some general comments

■ As we have just seen, once the Green's function has been obtained (for a deflection, moment, shear, etc.) it is then possible to determine the corresponding response (deflection, moment, shear, etc.) due to any arbitrary loading by using the appropriate Green's function as the kernel in the superposition integral. Thus the Green's function may be said to provide a 'key' to solutions under general loading conditions. Hence the Green's function $G(x, \zeta)$, which is the solution at x due to a unit force acting at ζ, is said to represent the *fundamental solution* for a physical problem.

■ While, in the above, we have developed Green's functions for one-dimensional cases, we mention here that Green's functions exist also for two or three-dimensional bodies. Nevertheless, the basic ideas remain the same. However, it is important to emphasise that *Green's functions can only be used to obtain solutions for a system whose behaviour is governed by linear equations such that the resulting solutions depend linearly on the applied loads.*

9.12 The fourth-order differential equation for beams

(a) Development and applications

Consider a beam under any arbitrary loading $q(x)$, as shown in Fig. (9.12.1). Assuming small rotations, $|v'| \ll 1$, the beam was found to be governed by the Euler–Bernoulli equation [Eq. (9.2.4)],

$$EI(x)\frac{\mathrm{d}^2 v(x)}{\mathrm{d}x^2} = -M(x). \qquad (9.12.1\text{a})$$

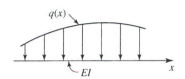

Figure 9.12.1

We also recall that equations of equilibrium for an element, $\sum F_y = 0$, $\sum M = 0$, yielded Eq. (8.3.3), namely

$$\frac{\mathrm{d}^2 M(x)}{\mathrm{d}x^2} = -q(x). \qquad (9.12.1\text{b})$$

Therefore, upon differentiating Eq. (9.12.1a) twice with respect to x and using Eq. (9.12.1b), we find

$$\frac{\mathrm{d}^2}{\mathrm{d}x^2}\left[EI(x)\frac{\mathrm{d}^2 v(x)}{\mathrm{d}x^2}\right] = q(x). \qquad (9.12.2\text{a})$$

For the case where $EI = $ constant, we obtain[†]

$$EI\frac{\mathrm{d}^4 v(x)}{\mathrm{d}x^4} = q(x). \qquad (9.12.2\text{b})$$

For the present we shall assume $q(x)$ to be a *smooth function*; hence we exclude here concentrated forces and couples. (We postpone treatment of these loading cases.)

We thus observe that a beam may be considered to be governed either by a second-order differential equation [Eq. (9.12.1a)] or a fourth-order differential equation [Eqs. (9.12.2)]. Let us first consider the advantages and disadvantages of these two equations.

In our previous solutions, we have used the former equation. We note, however, that in using this equation, it is first necessary to obtain, using equations of statics, an expression for $M(x)$ at all points in the beam. For beams under relatively 'conventional' loads, finding $M(x)$ is a relatively simple matter. However, for more complex loadings, finding the correct expression for $M(x)$ may prove to be a tedious task. For example, for the (indeterminate) beam shown in Fig. (9.12.2), one must first use the equations of statics to find relations existing among the reactive forces at A and B to write an expression for $M(x)$.

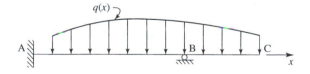

Figure 9.12.2

[†] We mention here that for the fourth-order differential equation to be integrable in terms of analytic functions, it is necessary that the first four derivatives be *continuous within* the domain of validity of the equation. Such continuity conditions need not necessarily be satisfied at the end points of the domain.

However, as a distinct advantage, solutions to the second-order differential equation require, as we have seen, only two integrations and yield two constants of integration. Appropriate boundary conditions for any given specified beam are then used to solve for the unknowns. As we noted previously in Section 9.2, these boundary conditions can only involve either the deflection v itself or its derivative dv/dx. In either case, these derivatives describe *geometric* boundary conditions.

In using the fourth-order equation, Eq. (9.12.2), we first note that the right-hand side, representing the applied loading, is a *known quantity*. Thus, as a distinct advantage, one may proceed to integrate the equation directly, since it is not necessary to first solve equations of equilibrium. However, as a disadvantageous feature, we require four integrations in which four constants of integration appear.

The four constants of integration are then determined using four boundary conditions. These may depend on v, dv/dx, d^2v/dx^2 and/or d^3v/dx^3.

While v and dv/dx at a point represent geometric boundary conditions, the latter two represent 'mechanical' boundary conditions.

This is clear since, by Eq. (9.12.1a), $EI \frac{d^2v}{dx^2} = -M$. Thus, if the moment M is known at a particular point, this provides a condition on the second derivative.

Recalling Eq. (8.3.2), $dM/dx = V$, we observe, upon taking the derivative of Eq. (9.12.1a), that

$$\frac{d}{dx}\left[EI(x)\frac{d^2v}{dx^2}\right] = -V \tag{9.12.3a}$$

and if $EI = $ constant,

$$EI\frac{d^3v}{dx^3} = -V. \tag{9.12.3b}$$

Thus, if the shear V is known at a point, this provides a condition on the third derivative.

As an example, for the beam of Fig. (9.12.2), the following boundary conditions are required when using the second-order equation:

$$\text{At A:} \quad v = 0, \qquad v' = 0$$
$$\text{At B:} \quad v = 0$$

However, when using the fourth-order equation, we require the following boundary conditions:

$$\text{At A:} \quad v = 0, \qquad v' = 0$$
$$\text{At B:} \quad v = 0$$
$$\text{At C:} \quad v'' = 0, \qquad v''' = 0$$

The latter boundary conditions express the mechanical conditions that at the free end C (with no load applied), the moment $M = 0$ and the shear $V = 0$.

We now illustrate the use of the fourth-order equation in the following example.

Example 9.16: Determine the deflection $v(x)$ of the elastic curve for the simply supported beam of Fig. (9.12.3) subjected to a load

$$q(x) = q_0 \sin\frac{\pi x}{L}. \tag{9.12.4}$$

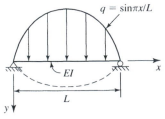

Figure 9.12.3

Solution: Using the fourth-order equation, we have

$$EI\frac{d^4 v(x)}{dx^4} = q_0 \sin\frac{\pi x}{L}, \quad 0 < x < L. \tag{9.12.5}$$

Noting that the moments $M = 0$ at simple supports, the associated boundary conditions for this problem are

$$v(0) = 0, \quad v''(0) = 0, \quad v(L) = 0, \quad v''(L) = 0. \tag{9.12.6}$$

Integrating Eq. (9.12.5) successively, we obtain

$$EIv'''(x) = -\frac{q_0 L}{\pi}\cos\frac{\pi x}{L} + A, \tag{9.12.7a}$$

$$EIv''(x) = -q_0(L/\pi)^2 \sin\frac{\pi x}{L} + Ax + B, \tag{9.12.7b}$$

$$EIv'(x) = q_0(L/\pi)^3 \cos\frac{\pi x}{L} + \frac{Ax^2}{2} + Bx + C, \tag{9.12.7c}$$

$$EIv(x) = q_0(L/\pi)^4 \sin\frac{\pi x}{L} + \frac{Ax^3}{6} + \frac{Bx^2}{2} + Cx + D, \tag{9.12.7d}$$

where A, B, C and D are constants of integration. From $v''(0) = 0$, $B = 0$ and therefore from $v''(L) = 0$, $A = 0$. From $v(0) = 0$, $D = 0$ and therefore from $v(L) = 0$, $C = 0$. Therefore, the deflection is given by

$$v(x) = \frac{q_0 L^4}{\pi^4 EI}\sin\frac{\pi x}{L}. \tag{9.12.8a}$$

The elastic curve is also shown in Fig. (9.12.3). We observe that the maximum deflection at $x = L/2$ is

$$v(L/2) = \frac{q_0 L^4}{\pi^4 EI}. \tag{9.12.8b}$$

Having obtained the deflection, viz. Eq. (9.12.8a), we may readily obtain the moments and shears at any point, using Eqs. (9.12.1a) and (9.12.3b), respectively. For example, the shear forces at the end points A and B($x = 0, L$) are, from Eqs. (9.12.7a) and (9.12.3b) with $A = 0$, respectively,

$$V = \pm\frac{q_0 L}{\pi}. \tag{9.12.9}$$

As we have noted previously, these shear forces clearly represent the reactions at A and B.   □

We observe that had we used the second-order differential equation, it would have been necessary first to obtain the reactions at A and B and then to write explicit expressions for the moment $M(x)$ before proceeding with the integration. Thus, in solving this problem, it is clearly preferable to use a fourth-order equation.

(b) The fourth-order differential equation for concentrated force and couple loadings

In differentiating twice to obtain the fourth-order equation, Eq. (9.12.2),

$$\frac{d^2}{dx^2}\left[EI(x)\frac{d^2 v(x)}{dx^2}\right] = q(x),$$

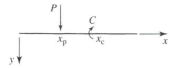

Figure 9.12.4

we assumed implicitly that $q(x)$ was a smooth function. We now wish to treat the case of applied concentrated forces and couples. To this end, let us assume that a concentrated force P and a couple C are applied at points x_p and x_c, as shown in Fig. (9.12.4).

However, how does one represent such applied loads in the fourth-order equations, Eq. (9.12.2a) or (9.12.2b)? Let us first treat the case of the concentrated load P. We first note that for an applied load P, $q(x) = 0$ at all points $x \neq x_p$. We therefore wish to represent P by means of a function that is zero for all $x \neq x_p$ but which is not zero at $x = x_p$. Furthermore, we recall from Chapter 1 [see Eq. (1.2.3)] that a concentrated force acting at point x_p may be considered as the limiting case of a load $q(x_p) \to \infty$ over a small interval Δx, as $\Delta x \to 0$.

Now there exists a function, known as a Dirac-delta function, denoted by δ, which has the following properties:

$$\int_0^x \delta(\zeta - x_p)\,d\zeta = \begin{cases} 0, & x \leq x_p \\ 1, & x > x_p. \end{cases} \qquad (9.12.10)$$

It is worthwhile to consider the character of the function δ since, being defined by its integral property, it is not a function in the ordinary sense. Assume, momentarily, that $x = x_p + \epsilon$ for small ϵ, i.e. for $0 < \epsilon \ll 1$. Then, from its definition, the integral equals unity. Furthermore, since the integral has the same value, namely unity, for *all* values $x > x_p$, we conclude that the integrand $\delta(\zeta - x_p) = 0$ whenever $x < x_p$ or $x > x_p$. Thus, as the integrand contributes nothing to the integral for these values of x, we observe that the entire contribution to the integral occurs in the infinitesimal region about x_p as $\epsilon \to 0$. It follows that $\delta \to \infty$ in the range $x_p < x < x_p + \epsilon$ as $\epsilon \to 0$ [Fig. (9.12.5)].

We note that this agrees precisely with the idealisation, mentioned above, which led to a definition of a concentrated load. Thus we may represent a concentrated force having magnitude P and acting downward at x_p by $P\delta(x - x_p)$. Recalling the definition of the singularity function with $n = 0$, Eq. (9.7.1b), we may rewrite Eq. (9.12.10) more simply as[†]

Figure 9.12.5

$$\int_0^x \delta(\zeta - x_p)\,d\zeta = \langle x - x_p \rangle^0. \qquad (9.12.11)$$

We now treat the representation of a couple acting at a point x_c. First we note that a couple can always be considered as a system of two forces $C/\Delta x$ acting in opposite directions a distance Δx apart, for example, at x_c and at $x_c + \Delta x$ [Fig. (9.12.6)].

Now, as we have just noted, each of these forces can be represented by means of the δ-function; i.e., $-\frac{C}{\Delta x} \cdot \delta(x - x_c)$ and $\frac{C}{\Delta x} \cdot \delta[x - (x_c + \Delta x)]$, respectively. Hence the system, consisting of these two forces, is given by

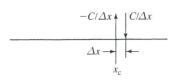

Figure 9.12.6

$$C \left\{ \frac{\delta[x - (x_c + \Delta x)] - \delta(x - x_c)}{\Delta x} \right\}$$

or

$$C \left\{ \frac{\delta[(x - \Delta x) - x_c] - \delta(x - x_c)}{\Delta x} \right\}.$$

[†] Thus the Dirac-delta function, $\delta(x - x_p)$, may be considered as the derivative of the singularity function $\langle x - x_p \rangle^0$.

Upon taking the limit as $\Delta x \to 0$, we note that this last expression becomes, by definition,[†] the derivative $-C\frac{d}{dx}\delta(x - x_c)$. Hence we may represent a *positive* couple $C > 0$, having magnitude C [i.e., positive as shown in Fig. (9.12.4)] by

$$-C\frac{d}{dx}\delta(x - x_c) \equiv -C\delta'.$$

Using the property of integration, we note that[‡]

$$\int \frac{d}{dx}[\delta(x - x_c)]\,dx = \delta(x - x_c). \qquad (9.12.12)$$

Finally, we mention here that the Dirac-delta function, $\delta(x - x_p)$, has another interesting but very useful property. Having observed that $\delta(x - x_p) \to \infty$ as $x \to x_p$ and that otherwise $\delta = 0$ [see Fig. (9.12.5)], the product of any finite function $f(x)$ with $\delta(x - x_p)$ appearing in an integrand cannot contribute to the integral at points other than $x_p - \epsilon < x < x_p + \epsilon$. Thus, making use of the mean-value theorem in the generalised sense,[§]

$$\int_{-\infty}^{\infty} f(x)\delta(x - x_p)\,dx = f(x_p)\int_{-\infty}^{\infty} \delta(x - x_p)\,dx = f(x_p). \qquad (9.12.13)$$

Note that the above is equally true if the limits $]-\infty, \infty[$ are replaced by any finite values, say $[a, b]$, provided $a < x_p < b$.

We now apply our results in the following example.

Example 9.17:   A concentrated load P and a couple C are applied to a simply supported beam of length L, as shown in Fig. (9.12.7). Determine the equation of the resulting elastic curve.

Solution: Since the given loading is represented by

$$q(x) - P\delta(x - a) - C\delta'(x - b), \qquad (9.12.14a)$$

the governing equation in $0 < x < L$ is

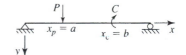

Figure 9.12.7

$$EI\frac{dv^4(x)}{dx^4} = P\delta(x - a) - C\delta'(x - b), \qquad (9.12.14b)$$

subject to the boundary conditions $v(0) = (L) = v''(0) = v''(L) = 0$.

Integrating, and using the properties of Eqs. (9.12.11) and (9.12.12), we obtain

$$EIv'''(x) = P\langle x - a\rangle^0 - C\delta(x - b) + A_1 \qquad (9.12.15a)$$

$$EIv''(x) = P\langle x - a\rangle^1 - C\langle x - b\rangle^0 + A_1x + A_2. \qquad (9.12.15b)$$

[†]   Note that the appearance of the minus sign before Δx in the numerator of the above expression leads, in the limiting case, to $-\frac{d\delta}{dx}$.

[‡]   The derivative of the Dirac-delta function is sometimes called a **doublet**. It is worthwhile mentioning that the Dirac-delta function δ, as well as its derivative δ', are not functions (and certainly not analytical functions) in the ordinary sense. They, as well as singularity functions, are 'functions' only when considered within the theory of distributions, as was mentioned in the footnote on p. 330. Note also that we have limited ourselves here to simply giving their operational properties.

[§]   Since the Dirac-delta function is not a function in the usual sense, the 'derivation' here is clearly only heuristic and is not meant to be rigorous.

From $v''(0) = 0$, $A_2 = 0$ while $v''(L) = 0$ yields

$$A_1 = \frac{1}{L}[C - P(L - a)]. \qquad (9.12.16)$$

Integrating again,

$$EIv'(x) = \frac{P}{2}\langle x - a\rangle^2 - C\langle x - b\rangle^1 + \frac{A_1 x^2}{2} + A_3 \qquad (9.12.17a)$$

and

$$EIv(x) = \frac{P}{6}\langle x - a\rangle^3 - \frac{C}{2}\langle x - a\rangle^2 + \frac{A_1 x^3}{6} + A_3 x + A_4. \qquad (9.12.17b)$$

From $v(0) = 0$, $A_4 = 0$ while from $v(L) = 0$, we obtain

$$\frac{P}{6}(L - a)^3 - \frac{C}{2}(L - b)^2 + \frac{A_1 L^3}{6} + A_3 L = 0. \qquad (9.12.18)$$

Substituting the value of A_1 and solving for A_3, yields

$$A_3 = \frac{P(L - a)}{6L}[L^2 - (L - a)^2] + \frac{C}{6L}[3(L - b)^2 - L^2]. \qquad (9.12.19)$$

Thus the deflection is given by

$$v(x) = \frac{1}{EI}\left[\frac{P}{6}\langle x - a\rangle^3 - \frac{C}{2}\langle x - b\rangle^2 + \frac{A_1 x^3}{6} + A_3 x\right] \qquad (9.12.20)$$

with A_1 and A_3 given by Eqs. (9.12.16) and (9.12.19), respectively.

Let us consider the effect of load P acting alone at the mid-point $a = L/2$. Then, with $C = 0$,

$$v(x) = \frac{P}{2EI}\left[\frac{1}{3}\langle x - L/2\rangle^3 - \frac{x^3}{6} + \frac{L^2 x}{8}\right]. \qquad (9.12.21)$$

The shape of the elastic curve is shown in Fig. (9.12.8a). We note that the deflection at $x = L/2$ is $v(L/2) = \frac{PL^3}{48EI}$, which agrees with the result given in Eq. (9.4.21).

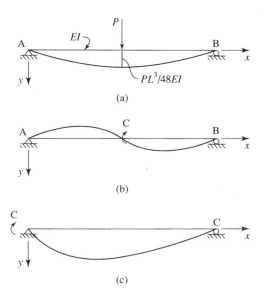

Figure 9.12.8

If the couple C is acting alone at the point $b = L/2$, then with $P = 0$,

$$A_1 = \frac{C}{L} \tag{9.12.22a}$$

and

$$A_3 = \frac{-CL}{24}. \tag{9.12.22b}$$

Hence for this case,

$$v(x) = \frac{C}{EI}\left[-\frac{1}{2}\langle x - L/2\rangle^2 + \frac{x^3}{6L} - \frac{Lx}{24} \right]. \tag{9.12.23a}$$

The shape of the elastic curve for this case, which we observe is anti-symmetric, is shown in Fig. (9.12.8b). The largest deflection in the left-hand region occurs at $x = \sqrt{3}L/6$ and has magnitude

$$|v| = \frac{\sqrt{3}}{216}\frac{CL^2}{EI}. \tag{9.12.23b}$$

If C is acting alone at A (i.e., $b = 0$), then with $P = 0$,

$$A_1 = \frac{C}{L} \quad \text{and} \quad A_3 = \frac{CL}{3}. \tag{9.12.24a}$$

Substituting these in Eq. (9.12.20), we find

$$v(x) = \frac{C}{6EIL}(x^3 - 3Lx^2 + 2L^2x). \tag{9.12.24b}$$

The shape of the elastic curve is shown in Fig. (9.12.8c). Note that this is the same result [Eq. (9.4.13c)] as obtain in Example 9.4 for a simply supported beam subjected to an end couple. □

Example 9.18:   A concentrated force P is applied at a variable point η to a cantilever beam having flexural rigidity EI, as shown in Fig. (9.12.9). Determine the deflection $v(x)$ of the elastic curve.

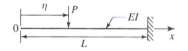

Figure 9.12.9

Solution: The load $q(x)$ is therefore represented as $P\delta(x - \eta)$. Hence the differential equation, $EIv^{(\mathrm{iv})} = q(x)$, becomes

$$EI\frac{\mathrm{d}^4v(x)}{\mathrm{d}x^4} = P\delta(x - \eta), \quad 0 < x < L \tag{9.12.25}$$

subject to the boundary conditions $v(L) = v'(L) = 0$ and $v''(0) = v'''(0) = 0$.
   Integrating, and using the property of Eq. (9.12.11),

$$EIv'''(x) = P\langle x - \eta\rangle^0 + A. \tag{9.12.26a}$$

Since $v'''(0) = 0$, $A = 0$. Integrating again and using the boundary condition $v''(0) = 0$, we have

$$EIv''(x) = P\langle x - \eta\rangle. \tag{9.12.26b}$$

Upon subsequent integrations, we obtain

$$EIv'(x) = \frac{P}{2}\langle x - \eta\rangle^2 + C \tag{9.12.26c}$$

and

$$EIv(x) = \frac{P}{6}\langle x - \eta \rangle^3 + Cx + D. \tag{9.12.26d}$$

Boundary condition $v'(L) = 0$ yields $C = -\frac{P}{2}(L - \eta)^3$ while $v(L) = 0$ yields $D = \frac{P}{6}(l - \eta)^2(2L + \eta)$.

Hence, we obtain

$$v(x) = \frac{P}{6EI}[\langle x - \eta \rangle^3 + (L - \eta)^2(2L + \eta - 3x)]. \tag{9.12.26d}$$

Note that upon letting $\eta = L - \zeta$, we recover the expressions for $v(x)$, as given by Eq. (9.7.15b), keeping in mind that x here is measured from the free end of the beam.

□

Throughout this chapter, it has been emphasised that the domain of validity of the governing differential equation is an *open* domain; i.e., it excludes the boundary points. Although it may have appeared that this feature may is not relevant in the solutions obtained till now, the following example illustrates specifically the relevance and importance of such considerations.

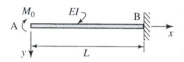

Figure 9.12.10

Example 9.19: A cantilever beam AB is subjected to a couple M_0 at the free end, as shown in Fig. (9.12.10). Determine the displacement $v(x)$ of the beam.

Solution: Since no applied load $q(x)$ exists in the region $0 < x < L$, the governing equation $EIv^{iv}(x) = q(x)$ is, in effect, simply

$$EI\frac{d^4v(x)}{dx^4} = 0, \quad 0 < x < L. \tag{9.12.27}$$

We note here that one might be tempted to represent the applied couple loading by means of a doublet, namely as $-M_0\,\delta'(x)$ [see, e.g., Eq. (9.12.14a)], as was done in Example 9.17. We observe, however, that in that example, the moment and forces were applied *within* the open domain. In the present example, this is not possible since the applied couple M_0 is applied at $x = 0$, which is *not within the given domain* of the differential equation, $0 < x < L$.

The appropriate boundary conditions, compatible with Eq. (9.12.27), are

$$EIv'''(0) = 0, \tag{9.12.28a}$$

$$EIv''(0) = -M_0, \tag{9.12.28b}$$

$$v'(L) = 0, \tag{9.12.28c}$$

$$v(L) = 0. \tag{9.12.28d}$$

Integrating Eq. (9.12.27), we obtain $EIv'''(x) = A$, and therefore $A = 0$ according to Eq. (9.12.28a). Integrating once more, we obtain $EIv''(x) = B$. From Eq. (9.12.28b), we have $B = -M_0$ and hence $EIv''(x) = -M_0$. Upon integrating again twice and using the remaining boundary conditions, we find

$$EI(v) = M_0(-x^2/2 + Lx - L^2/2) \tag{9.12.29a}$$

or

$$v(x) = -\frac{M_0(L - x)^2}{2EI}. \tag{9.12.29b}$$

Thus the displacement is a parabola with $v_A = -M_0L^2/2EI$.

□

◇9.13 Moment–area theorems

In the above sections, integration of the Euler–Bernoulli beam equation, Eq. (9.2.4) (which we repeat here as),

$$EI(x)\frac{d^2v(x)}{dx^2} = -M(x), \qquad (9.13.1a)$$

yielded the equation of the elastic curve, that is, the deflection of the beam. We now develop two relations, known as the **moment–area theorems**, which express the displacement and slope of a beam at any cross-section, located at x, in terms of those quantities that exist at another cross-section, say at x_0. In the following, we assume that $x > x_0$ [Fig. (9.13.1)]. While the solutions to beam problems obtained by application of these theorems lead to the same solutions obtainable by integration of Eq. (9.13.1a), the moment–area theorems often simplify considerably the calculations. As we shall find, the moment–area theorems are particularly useful, for example, in determining deflections for beams where $I = I(x)$.

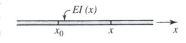

Figure 9.13.1

To develop these theorems, we first note that Eq. (9.13.1a) can be written as

$$EI(x)\frac{d}{dx}\left[\frac{dv(x)}{dx}\right] = -M(x). \qquad (9.13.1b)$$

Recalling that Eqs. (9.13.1a) and (9.13.1b) are only valid provided $|v'(x)| \ll 1$, it follows that the slope $|\theta| \ll 1$ at any x is given by

$$\theta(x) \simeq \tan\theta = \frac{dv}{dx}. \qquad (9.13.2)$$

Thus we may write Eq. (9.13.1b) as

$$EI(x)\frac{d\theta(x)}{dx} = -M(x), \qquad (9.13.3a)$$

which we recognise as a first-order differential equation on $\theta(x)$.[†]

Dividing through by $EI(x)$, we then have

$$d\theta = -\frac{M(x)}{EI(x)}\,dx. \qquad (9.13.3b)$$

Taking the integral on both sides,

$$\int_{\theta(x_0)}^{\theta(x)} d\theta = -\int_{x_0}^{x}\frac{M(x)}{EI(x)}\,dx, \qquad (9.13.4a)$$

we obtain

$$\theta(x) = \theta(x_0) - \int_{x_0}^{x}\frac{M(x)}{EI(x)}\,dx. \qquad (9.13.4b)$$

Recalling that the variable appearing within the integral is but a dummy variable,

---

[†] Note that Eq. (9.13.3a) is therefore integrable in terms of analytic functions only if $d\theta/dx$ is continuous within the domain of interest, i.e. here between x_0 and x as shown in Fig. (9.13.1). [See footnotes following Eqs. (9.2.4) and (9.12.2).] We therefore exclude the presence of a 'hinge' *within* this domain.

we rewrite Eq. (9.13.4b) as

$$\theta(x) = \theta(x_0) - \int_{x_0}^{x} \frac{M(\eta)}{EI(\eta)}\,d\eta. \tag{9.13.5}$$

It is important to note that, since x_0 is fixed (i.e., a number), and $\theta(x_0)$ is assumed to be known, $\theta(x)$ is a function only of the upper limit x appearing on the right-hand side of Eq. (9.13.5).

Now, from Eq. (9.13.2),

$$dv = \theta(x)\,dx. \tag{9.13.6a}$$

Integrating on both sides,

$$\int_{v(x_0)}^{v(x)} dv = \int_{x_0}^{x} \theta(x)\,dx = \int_{x_0}^{x} \theta(\xi)\,d\xi, \tag{9.13.6b}$$

where here (to avoid confusion in its use below) we have called the dummy variable, appearing on the right-hand side, ξ.

Substituting Eq. (9.13.5), we obtain

$$v(x) - v(x_0) = \int_{x_0}^{x} \theta(x_0) \cdot d\xi - \int_{x_0}^{x} \left[\int_{x_0}^{\xi} \frac{M(\eta)}{EI(\eta)}\,d\eta \right] d\xi \tag{9.13.7}$$

and, upon noting again that $\theta(x_0)$ is a constant,

$$v(x) = v(x_0) + \theta(x_0) \cdot (x - x_0) - \int_{x_0}^{x} \left[\int_{x_0}^{\xi} \frac{M(\eta)}{EI(\eta)}\,d\eta \right] d\xi. \tag{9.13.8}$$

Now, the double integral appearing on the right-hand side may be simplified as follows. Recalling, from the differential calculus, the expression for 'integration by parts',[†]

$$\int u\,dv^* = uv^* - \int v^*\,du, \tag{9.13.9a}$$

we let

$$u = \int_{x_0}^{\xi} \frac{M(\eta)}{EI(\eta)}\,d\eta, \tag{9.13.9b}$$

$$dv^* = d\xi \tag{9.13.9c}$$

from which

$$du = \frac{M(\xi)}{EI(\xi)}\,d\xi, \tag{9.13.9d}$$

$$v^* = \xi. \tag{9.13.9e}$$

---

† In keeping the conventional notation used for integration by parts, we denote here the variable by v^* to avoid confusion with the displacement $v(x)$.

Hence, making use of Eq. (9.13.9a), we have

$$\int_{x_0}^{x}\left[\int_{x_0}^{\xi}\frac{M(\eta)}{EI(\eta)}\mathrm{d}\eta\right]\mathrm{d}\xi = \left[\xi\int_{x_0}^{\xi}\frac{M(\eta)}{EI(\eta)}\mathrm{d}\eta\right]_{x_0}^{x} - \int_{x_0}^{x}\xi\frac{M(\xi)}{EI(\xi)}\,\mathrm{d}\xi$$

$$= x\int_{x_0}^{x}\frac{M(\eta)}{EI(\eta)}\mathrm{d}\eta - \int_{x_0}^{x}\xi\frac{M(\xi)}{EI(\xi)}\mathrm{d}\xi = \int_{x_0}^{x}(x-\xi)\frac{M(\xi)}{EI(\xi)}\mathrm{d}\xi$$

$$(9.13.10)$$

Finally, substituting in Eq. (9.13.8),

$$v(x) = v(x_0) + \theta(x_0)\cdot(x-x_0) - \int_{x_0}^{x}(x-\xi)\frac{M(\xi)}{EI(\xi)}\,\mathrm{d}\xi. \qquad (9.13.11)$$

Equations (9.13.5) and (9.13.11), known as the moment–area theorems, are explicit expressions for the slope $\theta(x)$ and displacement $v(x)$ at any cross-section $x_0 < x$, provided that $\theta(x_0)$ and $v(x_0)$ are known at x_0.

The above relations lend themselves to a simple and immediate geometric interpretation. Instead of drawing the moment diagram for a given beam, we first plot instead $\frac{M(x)}{EI(x)}$ as a function of x thus obtaining an '$\frac{M(x)}{EI(x)}$ diagram'. Since the definite integral represents the area under a function between the lower and upper limits, we may therefore interpret Eq. (9.13.5) as follows: the difference in the slopes of the beam, $\theta(x_0) - \theta(x)$, is represented by the area under the $\frac{M}{EI}$ diagram between the two points, x_0 and x ($x > x_0$) [Fig. (9.13.2a)].

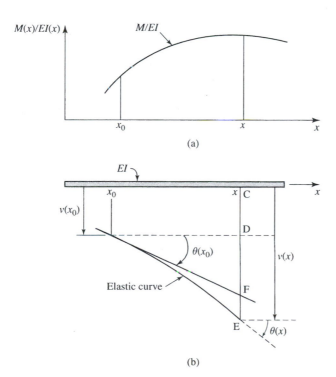

(a)

(b)

Figure 9.13.2

If we denote the area of the $\frac{M}{EI}$ diagram between x_0 and x by $A|_{x_0}^x$, we may therefore rewrite Eq. (9.13.5) symbolically as

$$\theta(x) = \theta(x_0) - A|_{x_0}^x. \tag{9.13.12}$$

Equation (9.13.11) may also be interpreted geometrically. We first note that the difference of displacements, $v(x) - v(x_0)$, is expressed by two terms: (i) $\theta(x_0) \cdot (x - x_0)$ and (ii) the integral $-\int_{x_0}^x (x - \xi) \frac{M(\xi)}{EI(\xi)} \, d\xi$. We consider each term separately.

- Since $|\theta| \ll 1$, we may replace the term $\theta(x_0) \cdot (x - x_0)$ by $(x - x_0) \cdot \tan \theta(x_0)$. We therefore observe that this term represents the distance DF at the cross-section x, as shown in Fig. (9.13.2).
- The integral $\int_{x_0}^x [(x - \xi) \cdot M(\xi)/EI(\xi)] \, d\xi$ represents the 'first moment' of the area under the 'M/EI diagram' between x_0 and x, *when taken about the point* x; we denote this term by $Q_x|_{x_0}^x$. We then interpret $-Q_x|_{x_0}^x$ as a measure of the (downward) vertical distance FE, i.e. between the beam at x and the tangent to the beam at x_0, as shown in Fig. (9.13.2b).

[Note that since $A|_{x_0}^x$ and $Q_x|_{x_0}^x$ depend on the sign of $M(x)$, they may be positive or negative.] Hence the geometric representation of Eq. (9.13.11),

$$v(x) = v(x_0) + \theta(x_0) \cdot (x - x_0) - \int_{x_0}^x (x - \xi) \frac{M(\xi)}{EI(\xi)} \, d\xi$$

is

$$CE = CD + DF + FE.$$

Using the symbolic notation $Q_x|_{x_0}^x$ as defined above, we rewrite Eq. (9.13.11) as

$$v(x) = v(x_0) + \theta(x_0) \cdot (x - x_0) - Q_x|_{x_0}^x. \tag{9.13.13a}$$

We note also that with $x_0 < x$, we have the relation

$$v(x_0) = v(x) - \theta(x) \cdot (x - x_0) - Q_{x_0}|_{x_0}^x, \tag{9.13.13b}$$

where, here, $Q_{x_0}|_{x_0}^x$ is the first moment of the area under the 'M/EI diagram' between x_0 and x, *when taken about the point* x_0. We observe that this relation is consistent with Fig. (9.13.2b).

It is clear that in addition to the moment–area theorems, the solution to any given beam problem requires the use of the appropriate boundary conditions. We illustrate the use of the moment–area theorems in the following examples where we shall develop several simplifying techniques.

Example 9.20: Two cantilever beams, I and II, each of length L and fixed at $x = 0$, are subjected to a load P at the free end. Beam I has a constant flexural rigidity EI and Beam II has flexural rigidity as shown in Figs. (9.13.3a and b). Determine the deflection and the slopes of the two beams at the free end.

Solution: Since the beams are statically determinate, the moment in both beams is identical; namely $M(x) = -PL + Px$.

Solution for Beam I:
The resulting M/EI diagram is shown in Fig. (9.13.3c).

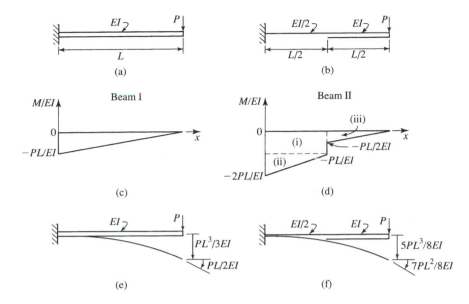

Figure 9.13.3

We let $x_0 = 0$ and $x = L$ in the first moment-area theorem. The boundary conditions are then $v(x_0) = \theta(x_0) = 0$. Recognising that $\cdot A|_0^L = -PL^2/2EI$, Eq. (9.13.12) becomes

$$\theta(L) = \theta(x_0) - A|_{x_0}^L = \frac{PL^2}{2EI}. \qquad (9.13.14a)$$

Similarly, noting that the moment about $x = L$ of the triangular area appearing in Fig. (9.13.3c) is $\cdot Q_L|_0^L = (-PL^2/2EI)(2L/3) = -PL^3/3EI$, Eq. (9.13.13a) becomes

$$v_B \equiv v(L) = v(x_0) + \theta(x_0) \cdot L - Q_L|_0^L = \frac{PL^3}{3EI}. \qquad (9.13.14b)$$

We note that this is the same result obtained in Example 9.7 [Eq. (9.7.16b)].

Solution for Beam II:
For this beam, the boundary conditions are the same; i.e., $v(x_0 = 0) = \theta(x_0 = 0) = 0$. However, the resulting M/EI diagram is as shown in Fig. (9.13.3d). Decomposing the M/EI diagram into component regions [(i), (ii), (iii), according to the dashed lines shown], we calculate the required quantities as follows:

$$A|_0^L = \left(-\frac{PL}{EI}\right) \cdot \frac{L}{2} + \frac{1}{2} \cdot \left(-\frac{PL}{EI}\right) \cdot \frac{L}{2} + \frac{1}{2} \cdot \left(-\frac{PL}{2EI}\right) \cdot \left(\frac{L}{2}\right)$$

$$= \frac{1}{EI} \left[-\frac{PL^2}{2} - \frac{PL^2}{4} - \frac{PL^2}{8} \right]$$

or

$$A|_0^L = -\frac{7PL^2}{8EI}. \qquad (9.13.15a)$$

Then

$$Q_L|_0^L = \left(-\frac{PL^2}{2EI}\right) \cdot \left(\frac{3L}{4}\right) + \left(-\frac{PL^2}{4EI}\right) \cdot \left(\frac{5L}{6}\right) + \left(-\frac{PL^2}{8EI}\right) \cdot \left(\frac{L}{3}\right)$$

or

$$Q_L|_0^L = -\frac{5PL^3}{8EI}. \tag{9.13.15b}$$

Substituting in Eqs. (9.13.12) and (9.13.13), with $x = L$, we find

$$\theta_B = \frac{7PL^2}{8EI} \tag{9.13.16a}$$

and

$$v_B = \frac{5PL^3}{8EI}. \tag{9.13.16b}$$

The elastic curves for the two beams are sketched in Figs. (9.13.3e and f), respectively. Comparing the results, we note as expected, that the deflection of Beam (II) is considerably larger than that of Beam I.   □

Since application of the moment–area theorems requires calculations of the areas $A|_{x_0}^x$ and $Q_x|_{x_0}^x$, it is evident that these theorems are of practical value mainly in cases where the areas of M/EI (or their component parts) are simple geometric shapes since, for such cases (e.g., rectangles, triangles, parabolas, etc.), both the surface areas and the location of their centroids are known quantities.

For beams subjected to either concentrated loads, or distributed loads $q(x)$, which are expressed in terms of polynomials of any degree [e.g., $q(x) = w$, wx or wx^2], one can decompose the M/EI diagram into these simple component shapes by drawing the moment diagram 'by parts'. This technique, which is also particularly useful for indeterminate beams, is illustrated in the following two examples.

Example 9.21: A simply supported beam AB of constant flexural rigidity EI and length L is subjected to a uniformly distributed load w [Fig. (9.13.4a)]. Determine (a) the deflection at $x = L/2$, (b) the slope of the beam at $x = L/4$ and (c) the deflection at $x = L/4$.

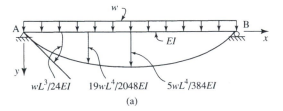

$wL^3/24EI$ $19wL^4/2048EI$ $5wL^4/384EI$

(a)

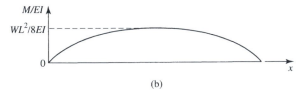

(b)

Figure 9.13.4

Solution: We note that this problem was solved as Example 9.1 by integration of the differential equation where the moment is

$$M(x) = \frac{wLx}{2} - \frac{wx^2}{2}; \tag{9.13.17}$$

the resulting moment diagram shows a parabolic variation with $M_{max} = wL^2/8$ at $x = L/2$ [Fig. (9.13.4b)]. The boundary conditions are $v(0) = v(L) = 0$.

Now, the surface area and centroid of the area bounded by the axis and the M/EI diagram is, in generally, not known *a priori* between, say $x_0 = 0$, and any *arbitrary* $x > 0$ (although these could be calculated). We therefore draw the M/EI diagram 'by parts': that is, we plot each term of Eq. (9.13.17) separately. Thus here we plot $wLx/2EI$ and $-wx^2/2EI$ separately as in Figs. (9.13.5a and b). We recognise that the first term represents the contribution to the moment due to the reaction at A while the second represents the contribution due to the downward uniform load w. [Clearly, the algebraic sum of Figs. (9.13.5a and b) yields the moment diagram of Fig. (9.13.4b).] Moreover, the resulting geometric shapes of the separate M/EI diagrams are now seen to be simple shapes with known properties.

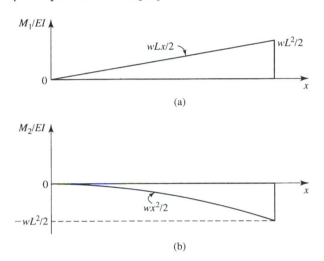

(a)

(b)

Figure 9.13.5

(a) To calculate $v(L/2)$, we note from Eq. (9.13.11) or (9.13.13), that we require θ at some point $x_0 < L/2$. Since this is an unknown quantity we proceed as follows: upon noting the boundary conditions, we let $x_0 = 0$ and $x = L$ in Eq. (9.13.13a), from which we find

$$\theta(0) = \frac{1}{L} \, Q_L|_0^L. \tag{9.13.18a}$$

Now, from Figs. (9.13.5a and b),

$$Q_L|_0^L = \frac{1}{EI} \left\{ \left[\left(\frac{WL^2}{2} \right) \cdot \left(\frac{L}{2} \right) \right] \cdot \left(\frac{L}{3} \right) + \left[\frac{1}{3} \cdot \left(-\frac{wL^2}{2} \right) \cdot L \right] \cdot \left(\frac{L}{4} \right) \right\}$$

$$= \frac{wL^4}{24EI}$$

and therefore

$$\theta(0) = \frac{wL^4}{24EI}, \tag{9.13.18b}$$

which agrees with Eq. (9.4.4b).

Now, setting $x_0 = 0$ and $x = L/2$ and using the B.C. $v(0) = 0$, we write Eq. (9.13.13) explicitly as

$$v(L/2) = \theta(0) \cdot \frac{L}{2} - Q_{L/2}|_0^{L/2}. \tag{9.13.19a}$$

Then

$$Q_{L/2}\big|_0^{L/2} = \frac{1}{EI}\left\{\left[\left(\frac{wL^2}{4}\right)\cdot\left(\frac{L}{4}\right)\right]\cdot\left(\frac{L}{6}\right) + \left[\frac{1}{3}\cdot\left(-\frac{wL^2}{8}\right)\cdot\frac{L}{2}\right]\cdot\left(\frac{L}{8}\right)\right\}$$

$$= \frac{3wL^4}{384EI}.$$

Substituting the calculated values in Eq. (9.13.19a), we obtain

$$v(L/2) = \frac{wL^4}{48EI} - \frac{3wL^4}{384EI} = \frac{5wL^4}{384EI}, \tag{9.13.19b}$$

which is the same result obtained in Example 9.1 [Eq. (9.4.3)].

(b) Having found $\theta(0)$, and setting now $x_0 = 0$ and $x = L/4$ in Eq. (9.13.12), the slope $\theta(L/4)$ is given by

$$\theta(L/4) = \theta(0) - A\big|_0^{L/4}. \tag{9.13.20}$$

From Figs. (9.13.5a and b),

$$A\big|_0^{L/4} = \frac{1}{EI}\left\{\left[\left(\frac{wL^2}{8}\right)\cdot\left(\frac{L}{8}\right)\right] + \left[\frac{1}{3}\cdot\left(-\frac{wL^2}{32}\right)\cdot\frac{L}{4}\right]\right\}$$

$$= \frac{1}{EI}\left[\frac{wL^3}{64} - \frac{wL^3}{384}\right] = \frac{5wL^3}{384EI}. \tag{9.13.21a}$$

Substituting Eqs. (9.13.18b) and (9.13.21a) in Eq. (9.13.20), we obtain

$$\theta(L/4) = \frac{wL^3}{24EI} - \frac{5wL^3}{384EI} = \frac{11wL^3}{384EI}. \tag{9.13.21b}$$

(c) The deflection $v(L/4)$ is similarly found by setting $x_0 = 0$ and $x = L/4$ in Eq. (9.13.13):

$$v(L/4) = \theta(0)\cdot\frac{L}{4} - Q_{L/4}\big|_0^{L/4} \tag{9.13.22a}$$

where[†]

$$Q_{L/4}\big|_0^{L/4} = \frac{1}{EI}\left[\left(\frac{wL^3}{64}\right)\cdot\left(\frac{L}{12}\right) + \left(-\frac{wL^3}{384}\right)\cdot\left(\frac{L}{16}\right)\right] = \frac{7wL^4}{6144EI}.$$

Then, substituting $\theta(0)$ and the above in Eq. (9.13.22a), we obtain

$$v(L/4) = \frac{wL^4}{96EI} - \frac{7wL^4}{6144EI} = \frac{19wL^4}{2048EI}. \tag{9.13.22b}$$

The elastic curve and calculated quantities are shown in Fig. (9.13.4a).   □

The moment–area theorems can also be applied to statically indeterminate beams as illustrated in the following example. They therefore provide another means of determining the unknown reactions for such beams.

[†] Note that the required areas of the component parts, $wL^3/64$ and $-wL^3/384$, respectively, have been calculated previously in Eq. (9.13.21a). To find Q, we therefore need only multiply by the respective distances from $x = L/4$ to their centroid.

Example 9.22: A statically indeterminate beam (having constant *EI*) of length *L*, is fixed at one end A and simply supported at B, as shown in Fig. (9.13.6a). A load *P* is applied at the centre as shown. Determine the unknown reactions.

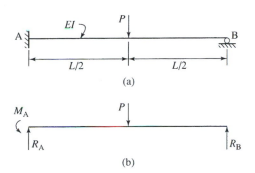

(a)

(b)

Figure 9.13.6

Solution: We denote the unknown reactions by R_A, M_A and R_B, as in Fig. (9.13.6b). Clearly, the two equations of statics,

$$\sum F_y = R_A + R_B - P = 0 \qquad (9.13.23a)$$

and

$$\sum M = M_A - R_B L + \frac{PL}{2} = 0, \qquad (9.13.23b)$$

cannot provide a unique solution for the three unknowns. However, we may write the expression for the moment $M(x)$ in terms of the unknowns; namely

$$M(x) = M_A + R_A x - P\langle x - L/2 \rangle, \quad 0 \le x \le L. \qquad (9.13.24)$$

We now plot the above three terms separately in Figs. (9.13.7a–c); that is, we plot the M/EI diagram 'by parts'.

Noting the boundary conditions $v(0) = v'(0) = v(L) = 0$, $(\theta \equiv v')$, we let $x_0 = 0$ and $x = L$ and writing Eq. (9.13.13) explicitly as

$$v(L) = v(0) + \theta(0) \cdot L - Q_L|_0^L, \qquad (9.13.25a)$$

upon substituting the boundary conditions, we conclude that

$$Q_L|_0^L = 0. \qquad (9.13.25b)$$

Now, from Figs. (9.13.7),

$$Q_L|_0^L = \frac{1}{EI} \left\{ [M_A L] \cdot \left(\frac{L}{2}\right) + \left[R_A \frac{L}{2}\right] \cdot \left(\frac{L}{3}\right) + \left[\frac{L}{4} \cdot \left(-\frac{PL}{2}\right)\right] \cdot \left(\frac{L}{6}\right) \right\}$$

$$= \frac{1}{EI} \left[M_A \frac{L^2}{2} + R_A \frac{L^3}{6} - \frac{PL^3}{48} \right].$$

But, since by Eq. (9.13.25b), this quantity must be zero, we have

$$24 M_A + 8 R_A L = PL. \qquad (9.13.26)$$

Thus, together with Eqs. (9.13.23), this last equation is the additional equation required

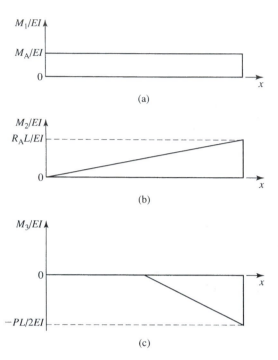

Figure 9.13.7

to solve for the three unknown reactions. Upon solving, we obtain

$$R_A = \frac{11P}{16}, \tag{9.13.27a}$$

$$R_B = \frac{5P}{16}, \tag{9.13.27b}$$

$$M_A = -\frac{3PL}{16}. \tag{9.13.27c}$$

We observe that the result for R_B is in agreement with Eq. (9.10.15b) (for $\zeta = L/2$) of Example 9.13.

Substituting in Eq. (9.13.24), we find the explicit moment expression,

$$M(x) = \frac{P}{16}[-3L + 11x - 16\langle x - L/2 \rangle]. \tag{9.13.28}$$

□

The resulting moment diagram is shown in Fig. (9.13.8a). We note that $M = 0$ at $x = 3L/11 = 0.2727L$. Hence at this point, the elastic curve has zero curvature. This permits us to sketch qualitatively the shape of the elastic curve quite accurately as in Fig. (9.13.8b). Thus, although the exact deflections have not been calculated here, we observe that the direct relation between the deformation and internal resultants (here, the relation between the sign of the curvature and the sign of the moment M) provides an excellent means for understanding the qualitative behaviour of the beam. Thus, knowing the variation of the moment, we were able to sketch the approximate deflection of the beam, as in Fig. (9.13.8).

However, with some experience, one can often easily first visualise the deformation of a structure due to applied loads. Indeed, such an intuitive visualisation of the

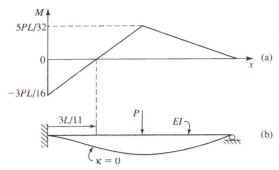

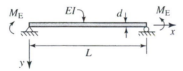

Figure 9.13.8

deformation of a structure often provides the best basic understanding of the behaviour of a structure and leads one to a qualitative estimate of the internal stress resultant.

PROBLEMS

In all problems below, the x-axis is to be taken at the extreme left end of the structure. Assume elastic behaviour.

Sections 2 and 3

9.1: A simply supported prismatic beam, having a rectangular cross-section with depth d, length L and flexural rigidity EI, is subjected to end couples $M_z = M_E$, as shown in Fig. (9P.1), causing maximum flexural stresses $\pm\sigma_0$. (a) Determine the coordinates (x_0, y_0) of the centre of curvature O of the beam according to the (i) exact theory and (ii) linear theory of beams. (b) Evaluate the coordinates (to four significant figures) for typical values $E/\sigma_0 = 10^3$ and $L/d = 100$ according to both the (i) exact and (ii) linear theory.

Figure 9P.1

9.2: A cantilever beam, having a cross section that is symmetric with respect to both the y- and z-axes, with flexural rigidity EI, depth d and length L, is fixed at $x = 0$. The beam is subjected to a vertical load P at the free end, as shown in Fig. (9P.2).
  (a) What is the exact radius of curvature, R_{ex}, at the section, $x = L/2$? Give answer in terms of P, L and EI.
  (b) Due to the load P, the elastic curve, according to the linearised theory of beams, is given as $v(x) = \frac{Px}{6EI}(3Lx - x^2)$. (i) Determine the radius of curvature of the elastic curve, R_L, at $x = L/2$ according to linear beam theory, i.e. using Eq. (9.2.1). (ii) Assuming that P causes a maximum stress σ_0 in the beam, show that the ratio $\frac{R_L}{R_{ex}}$ is

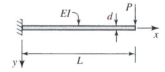

Figure 9P.2

$$\frac{R_L}{R_{ex}} = \left[1 + \frac{9}{16}\left(\frac{\sigma_0}{E}\right)^2\left(\frac{L}{d}\right)^2\right]^{3/2}.$$

(iii) Using the appropriate series expansion, obtain a simplified expression, namely

$$\frac{R_L}{R_{ex}} = 1 + \frac{27}{32}\left(\frac{\sigma_0}{E}\right)^2\left(\frac{L}{d}\right)^2.$$

(c) Evaluate this ratio for a material with $\sigma_0/E = 10^{-3}$ and for two cases, $L/d = 20$ and 100, using the two expressions in (b). Compare the results.

Section 4

9.3: By integrating the Euler–Bernoulli beam equations, $EIv''(x) = -M(x)$, (i) determine the equation of the elastic curve for the beams shown in Figs. (9P.3a–i), (ii) evaluate the maximum deflection and (iii) sketch the elastic curve.

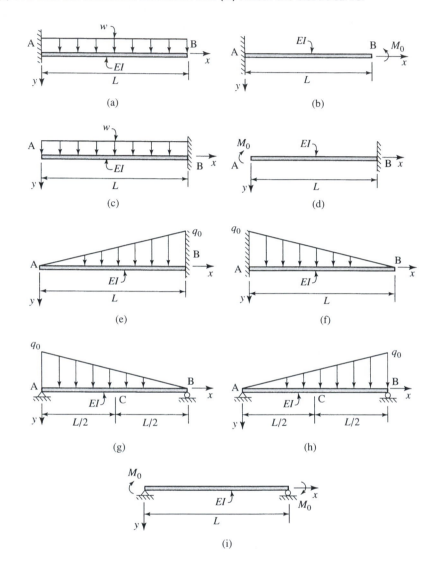

Figure 9P.3

9.4: By integrating the Euler–Bernoulli beam equations, $EIv''(x) = -M(x)$, (i) determine the equation of the elastic curve, namely

$$v(x) = -\frac{q_0 x}{360EIL}(6x^4 - 15Lx^3 + 10L^2x^2 - L^3)$$

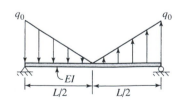

Figure 9P.4

for the beam shown in Fig. (9P.4) and (ii) sketch the elastic curve. (*Note:* See computer-related Problem 9.102.)

9.5: By integrating the Euler–Bernoulli beam equations, $EIv''(x) = -M(x)$, (i) determine the equation of the elastic curve for the prismatic beams shown in Figs. (9P.5a–c), (ii) evaluate the maximum deflection and (iii) sketch the elastic curve.

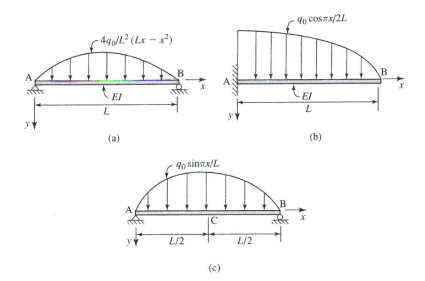

(a)

(b)

(c)

Figure 9P.5

9.6: The cantilever beam shown in Fig. (9P.6) is subjected to an exponentially decaying load $q(x) = q_0 e^{-\alpha x/L}$, where $\alpha \geq 0$ is a constant. By integrating the Euler–Bernoulli beam equations, $EIv''(x) = -M(x)$, (a) Determine the equation of the elastic curve for the cantilever beam, shown in Fig. (9P.6), subjected to the exponentially decaying load. (b) Evaluate the maximum deflection. (c) By taking the limit as $\alpha \to 0$, obtain the maximum deflection for the beam under a load q_0 distributed uniformly over its entire length L, namely $v_{max} = \frac{q_0 L^4}{8EI}$.

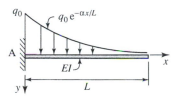

Figure 9P.6

9.7: Determine the deflection Δ at the free end of the cantilevered beam whose cross-section varies as shown in Fig. (9P.7).

9.8: Given a simply supported non-prismatic beam, as shown in Fig. (9P.8), whose whose flexural rigidity is $EI(x)$, where $I(x) = I_0 \sin(\pi x/L)$. The beam is subjected to a symmetrically distributed load $q(x) = \sin(\pi x/L)$. (a) Determine the equation of the elastic curve. (b) Determine the maximum displacement and the slope of the beam at $x = L/2$. (c) Sketch the loading and the elastic curve.

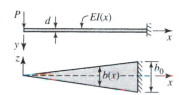

Figure 9P.7

9.9:* The simply supported non-prismatic beam of Problem 9.8 with $I(x) = I_0 \sin(\pi x/L)$ is subjected to an anti-symmetric loading $q(x) = \sin(2\pi x/L)$. (a) Determine the equation of the elastic curve. (b) Determine the displacement and the slope of the beam at $x = L/2$. (c) Determine the maximum displacement and indicate where it occurs. (d) Sketch the loading and the elastic curve.

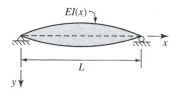

Figure 9P.8

9.10: A beam AB with flexural rigidity EI, simply supported at A and by a linear spring [whose constant is k (N/m)] at B, is subjected to a uniformly distributed load w (N/m), as shown in Fig. (9P.10). (a) By integrating the Euler–Bernoulli relation, determine the equation of the elastic curve $v(x)$. Express the answer in non-dimensional terms x/L and the non-dimensional ratio $\alpha \equiv EI/kL^3$. (b) Show that if $k \to \infty$, i.e. $\alpha = 0$, the mid-span deflection is given by the expression $v(L/2) = \frac{5wL^4}{384EI}$. (c) Express the deflection of the beam for the limiting case of a rigid beam, i.e. $EI \to \infty$ with finite $0 < k$. (d) Sketch the displacement of the beam for the two limiting cases of (b) and (c) above.

9.11: A beam AB having flexural rigidity EI is free at one end B and is anchored at the other end A to a wall. The wall support is assumed to provide only a partial restraint against rotation. The beam is therefore represented by a linear torsional spring that

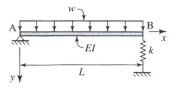

Figure 9P.10

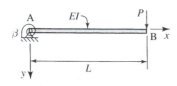

Figure 9P.11

provides a (negative) moment at A, which is linearly proportional to the slope $\theta \approx v'_A$, i.e. $M_A = -\beta\theta$, where β is a constant having dimensions (N-m/rad), as shown in Fig. (9P.11). (a) By integrating the Euler–Bernoulli relation, determine the equation of the elastic curve $v(x)$. Express the answer in non-dimensional terms x/L and the non-dimensional ratio $\gamma \equiv EI/\beta L$. (b) Show that if $\beta \to \infty$, the deflection at the free end is given by the known expression for the rigidly supported cantilever beam, $v(L) = \frac{PL^3}{3EI}$. (c) Express the deflection $v(x)$ of the beam for the limiting case of a rigid beam, i.e. $EI \to \infty$ if $0 < \beta$ is finite. (d) Sketch the displacement of the beam for the two limiting cases of (b) and (c) above.

Section 5

Note: In solving problems in this section, use can be made of results found in Sections 3 and 4 of this chapter or Appendix F.

9.12: The exact axial displacement $(\Delta_x)_{ex}$ of a simply supported beam subjected to end couples $M = M_E$, causing a maximum stress σ_0, is given as [see Eq. (8.6.17b)]

$$(\Delta_x)_{ex} = L\left(1 - \frac{2}{\gamma}\sin\frac{\gamma}{2}\right),$$

where $\gamma = \frac{\sigma_0 L}{Ed}$. (a) Determine Δ_x according to the the linear beam theory. (b) Show that for $\gamma \ll 1$, the ratio $[(\Delta_x) - (\Delta_x)_{ex}]/(\Delta_x)_{ex} = 1 + \frac{\gamma^2}{80}$.

9.13: (a) Determine an expression for the axial displacement $u(x)$ of points lying on the elastic curve of a cantilever beam, with flexural rigidity EI and length $L = 2\,\text{m}$, which is subjected to a concentrated force P acting at the free right end. (b) Evaluate the displacement $\Delta_x/L \equiv u(L)/L$ if $E = 100$ GPa, $I = 10^{-5}\,\text{m}^4$ and $P = 3$ kN.

9.14: A wooden cantilever beam ($E = 15$ GPa) having a 'T-Section', as shown in Fig. (9P.14), is subjected to a uniformly distributed load w (N/m) over its entire length L. If the maximum flexural stress in the beam is 10 MPa, and the length is 2 m, determine the axial displacement Δ_x of the free end and evaluate the ratio Δ_x/L.

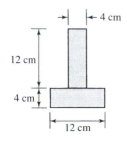

Figure 9P.14

Section 6

9.15: Given a cantilever beam of rectangular cross-section $b \times d$ of length L and whose material properties are E, G and v. (a) Determine an approximation for the deflection v_s at the free end of the elastic curve due to shear deformation if the beam is subjected to a uniformly distributed load w (N/m) over its entire length and (b) evaluate the ratio v_s/v_f, where v_f is the deflection due to flexure. Express the answer in terms of the depth-to-length ratio d/L and the Poisson ratio v.

9.16: Given a simply supported beam of rectangular cross-section $b \times d$ of length L and whose material properties are E, G and v. (a) Determine an approximation for the deflection, v_s, at the centre of the elastic curve due to shear deformation if the beam is subjected to a concentrated load P (N) at the mid-span and (b) evaluate the ratio v_s/v_f where v_f is the deflection due to flexure. Express the answer in terms of the depth-to-length ratio, d/L, and the Poisson ratio, v.

Section 7

9.17: Using singularity functions, (a) determine the equation of the elastic curve for the cantilever beam shown in Fig. (9P.17), (b) evaluate the deflection at the free end and (c) sketch the elastic curve.

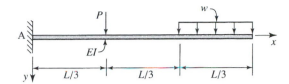

9.18: A steel ($E = 200\,\text{GPa}$) cantilever beam of length L is subjected to a linearly varying load, as shown in Fig. (9P.18). Using singularity functions, (a) determine the equation of the elastic curve, (b) evaluate the deflection at point B, at the free end, point C, and determine the ratio v_C/v_B and, (c) if $L = 4\,\text{m}$ and the cross-section of the beam is given as $W203 \times 22$, evaluate the deflection at the free end if $w=6kN/m$.

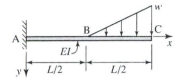

Figure 9P.18

9.19: A simply supported beam having flexural rigidity EI is subjected to two symmetrically placed loads P where $a \leq L/2$, as shown in Fig. (9P.19). (a) Determine the equation of the elastic curve $v(x)$. (b) (i) Evaluate the displacement at the mid-point in terms of P, L, a and EI and (ii) determine the displacement when $a = L/2$.

9.20:* A simply supported prismatic beam of length L with flexural rigidity EI is subjected to an applied couple M_0 at $x = a$, as shown in Fig. (9P.20). (a) Using singularity functions, determine (i) the equation of the elastic curve $v(x)$ and (ii) the deflection at $x = a$. (b) For what values of a/L will the deflection be stationary (maximum or minimum) to the *left* of the applied couple?

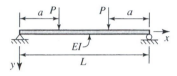

Figure 9P.19

9.21: The overhanging beam ABC, shown in Fig. (9P.21), is subjected to a uniformly distributed load $w\,(N/m)$ over its entire length. (a) Using singularity functions, (i) obtain the equation of the elastic curve $v(x)$, namely

$$v(x) = \frac{wL^4}{24EI}\{(x/L)^4 - 2[1 - (a/L)^2](x/L)^3 + [1 - 2(a/L)^2](x/L) - 2(1 + a/L)^2\langle x/L - 1\rangle^3\}$$

and (ii) the deflection v_C of point C. (b)* For what value of a/L will the deflection $v_C = 0$? (*Note*: See computer-related Problem 9.103.)

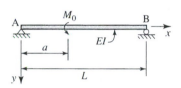

Figure 9P.20

9.22: A concentrated load P acts, as shown in Fig. (9P.22), on an overhanging beam ABC. (a) Determine (i) the equation of the elastic curve $v(x)$ and (ii) the maximum upward displacement in segment AB. (b) Sketch the elastic curve. (c) For what value of a/L will the maximum upward displacement in segment AB be equal to the downward displacement v_C? Determine the value of this displacement $|v|$.

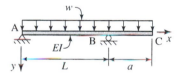

Figure 9P.21

9.23: A simply supported beam ABCD is subjected to a uniformly distributed load over the segment BC, as shown in Fig. (9P.23). Determine the deflection at the centre of the beam.

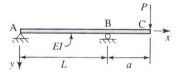

Figure 9P.22

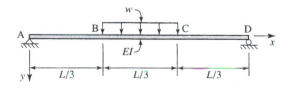

Figure 9P.23

9.24:* Given a beam ABC, containing a hinge at point B and loaded as shown in Fig. (9P.24). (a) Using singularity functions, determine the equation of the elastic curve. (b) Determine the deflection at B and at D, midway under the load w. (c) Determine the discontinuity of the slope at the hinge. (d) Sketch the elastic curve.

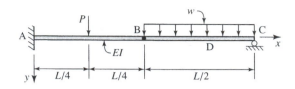

Figure 9P.24

Section 8

Solve Problems 9.25–9.35 by integrating the differential equation, $EIv''(x) = -M(x)$. **Use singularity functions where appropriate.**

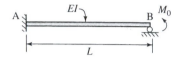

Figure 9P.25

9.25, 9.26: For the indeterminate beams shown in Figs. (9P.25 and 9P.26), (a) determine the reactive forces and moments due to the applied loads, (b) plot the shear and moment diagrams, showing all maximum and minimum values, and (c) sketch the elastic curve and evaluate the maximum deflection.

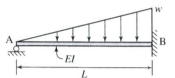

Figure 9P.26

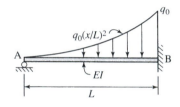

Figure 9P.27

9.27: For the indeterminate beam shown in Fig. (9P.27), subjected to the distributed load $q(x) = q_0(x/L)^2$, (a) determine the reactive forces and moments due to the given loads, (b) plot the shear and moment diagrams, showing all maximum and minimum values, and (c) sketch the elastic curve.

9.28: For the propped cantilever beam shown in Fig. (9P.28), (a) determine the reactions and moment at A due to the load $q(x) = q_0 \cos(\pi x/2L)$, (b) draw the shear and moment diagrams, showing all maximum and minimum values, and (c) obtain the explicit expression for the elastic curve, $v(x)$, namely

$$v(x) = \frac{8q_0 L^4}{\pi^4 EI}\left\{ -\left(\frac{x}{L}\right)^3 + 3\left(\frac{x}{L}\right)^2 + 2\left[\cos\left(\frac{\pi x}{2L}\right) - 1\right]\right\}.$$

(*Note:* See computer-related Problem 9.104.)

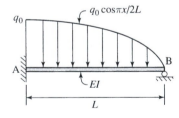

Figure 9P.28

9.29: For the indeterminate beam shown in Fig. (9P.29), (a) determine the reactions due to the applied loads, (b) plot the shear and moment diagrams, showing all maximum and minimum values, and (c) obtain the explicit equation of the elastic curve, namely

$$v(x) = \frac{wL^4}{768EI}\left\{32\left(\frac{x}{L}\right)^4 - 28\left(\frac{x}{L}\right)^3 + 3x - 2\langle 2x - L\rangle^4 - 5\langle 2x - L\rangle^3\right\}$$

and sketch the elastic curve. (*Note:* See computer-related Problem 9.105.)

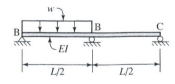

Figure 9P.29

9.30: For the indeterminate beam shown in Fig. (9P.30), (a) determine the reactions due to the applied load, (b) plot the shear and moment diagrams, showing all maximum and minimum values, (c) obtain the explicit equation of the elastic curve, (d) evaluate the maximum deflection $|v|_{max}$ in the segment AB and (e) sketch the elastic curve.

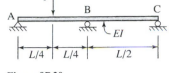

Figure 9P.30

9.31–9.33: The beams shown in Figs. (9P.31–9P.33) are statically indeterminate to degree two, since there exist four unknowns (namely R_A and R_B in the y-direction and M_A and M_B) and only two remaining equations of equilibrium, $\sum F_y = 0$ and $\sum M = 0$. [Note that, according to the comment at the end of Section 5 of this chapter, the equal and opposite axial reactions, $(R_A)_x = (R_B)_x$, are neglected when using linear beam theory.] (a) Determine the reactive forces and moments at A and B. (b) Plot the shear and moment diagrams, showing all maximum and minimum values. (c) Sketch the elastic curve. (*Note*: Use of symmetry due to the symmetrically applied loads simplifies the solution.)

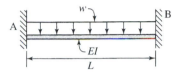

Figure 9P.31

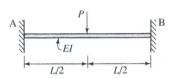

9.34: An elastic beam of flexural rigidity EI is simply supported at the ends A and B and supported, as shown in Fig. (9P.34), at point C by means of a linear spring having stiffness k. (a) Determine the reaction of the spring in terms of the given quantities. (b) Determine the required value of k (in terms of EI and L) to cause the moment at point C to be zero. (c) Sketch the moment diagram using the value of k obtained above and show maximum values of M and indicate where they occur.

Figure 9P.32 and Figure 9P.33

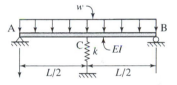

Figure 9P.34

9.35: An elastic beam ABC is fixed at point A and simply supported by a rod BD, as shown in Fig. (9P.35). The flexural rigidity of the beam is given as EI and the axial rigidity of the rod is given as AE. A uniformly distributed load is applied over the span AB. Determine (a) the reaction and moment at point A, (b) the deflection $v(x)$ within the span AB and (c) the vertical displacement of point C. Express answers in terms of w, L and the non-dimensional ratio, $\alpha = hI / AL^3$.

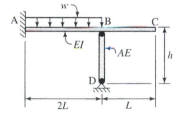

Figure 9P.35

Sections 9 and 10

Problems 9.36–9.48 are to be solved using the force method and/or the principle of superposition. In solving problems in this section, use can be made of results found in Sections 3 to 7 of this chapter or given in Appendix F.

9.36: Determine the required relation between M_0 and w to render $\theta_B = 0$ [see Fig. (9P.36)].

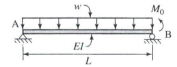

Figure 9P.36

9.37: Determine the required relation between the upward force P and w to render $\theta_B = 0$ [see Fig. (9P.37)].

9.38: Determine the required relation between M_0 and P if $v(L/2) = 0$ [see Fig. (9P.38)].

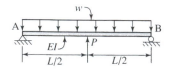

Figure 9P.37

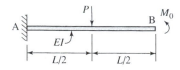

Figure 9P.38

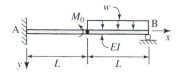

Figure 9P.39

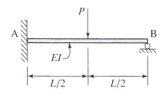

Figure 9P.47

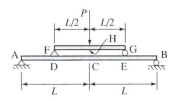

Figure 9P.48

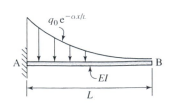

Figure 9P.50

9.39: Determine the required relation between M_0 and w if $v(L) = 0$ [see Fig. (9P.39)].

9.40: Solve for the reactive forces and moment acting on the beam of Fig. (9P.25) if (a) the released structure is a cantilever beam clamped at A and if (b) the released structure is a simply supported beam.

9.41: Solve for the reactions on the beam of Fig. (9P.29) if the released structure is simply supported at points A and B.

9.42: Solve for the reactions on the beam of Fig. (9P.30) if the released structure is simply supported at points A and C.

9.43: Solve for the reactions on the beam of Fig. (9P.34) if the released structure is simply supported at points A and B.

9.44:* Solve for the reactive forces and moments on the beam of Fig. (9P.31) if (a) the released structure is a cantilever beam clamped at A and free at B and if (b) the released structure is simply supported at A and B. (*Note:* This problem is statically indeterminate to degree two.)

9.45:* Solve for the reactive forces and moments on the beam of Fig. (9P.32) if (a) the released structure is a cantilever beam clamped at A and free at B and if (b) the released structure is simply supported at A and B. (*Note:* This problem is statically indeterminate to degree two.)

9.46: (a) Solve for the reactive forces and moments of the beam of Fig. (9P.35) if the released structure is a cantilever beam clamped at A and free at B, (b) determine the displacement at point C (express the answers in terms of w, L and the ratio $\alpha \equiv hI / AL^3$) and (c) determine the above quantities if $\alpha \to 0$. To what, physically, does this correspond?

9.47: Solve for the reactive forces and moment of the beam of Fig. (9P.47) if the released structure is a (a) cantilever beam clamped at A and (b) simply supported beam.

9.48: A beam system, consisting of two beams ACB and FHG, each of flexural rigidity EI and supported as shown in Fig. (9P.48), is subjected to a force P at H. Determine the deflection of points C and H.

Section 11

9.49: Making use of the Green's function for the deflection of a prismatic cantilever beam of length L with flexural rigidity EI [Eq. (9.11.6)], determine the deflection of point B of the beam under a load $q(x) = q_0 \cos(\frac{\pi x}{2L})$, as shown in Fig. (9P.5b).

9.50: Making use of the Green's function for the deflection of a prismatic cantilever beam of length L with flexural rigidity EI [Eq. (9.11.6)], (a) obtain the deflection of point B of the beam due to an exponentially decaying load, as shown in Fig. (9P.50), namely

$$v_B = \frac{q_0 L^4}{6\alpha^4 EI}[6(\alpha - 1) + e^{-\alpha}(6 - 3\alpha^2 - 2\alpha^3)].$$

(b)* Evaluate the deflection at B when $\alpha \to 0$. (*Note:* See computer-related Problem 9.106.)

9.51: Making use of singularity functions (or using the results found in Section 9.7), (a) obtain the influence function (Green's function), $G_d(x, \zeta)$, for the deflection of a simply supported prismatic beam having flexural rigidity EI and length L, namely

$$G_d(x, \zeta)) = \frac{1}{6EI}\left[-\frac{(L-\zeta)x^3}{L} - \frac{(L-\zeta)^3 x}{L} + L(L-\zeta)x + \langle x-\zeta\rangle^3\right].$$

(b) Simplify the expression for $G_d(L, \zeta)$.

9.52: A simply supported prismatic beam with flexural rigidity EI is subjected to a uniformly distributed load w (N/m), as shown in Fig. (9P.52), acting over a length $0 \le x \le a$. (a) Using the Green's function, $G_d(x, \zeta)$, given in Problem 9.51, determine the equation for the elastic curve $v(x)$ and (b) determine $v(x)$ if the uniform load is distributed over the entire length of the beam.

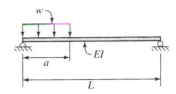

Figure 9P.52

9.53: Making use of the influence function given in Problem 9.51, determine the deflection at the mid-point C of a simply supported beam of length L and flexural rigidity EI if the loading is as shown in (a) Fig. (9P.3h) and (b) Fig. (9P.5c).

9.54: Making use of the influence function given in Problem 9.51, (a) obtain the deflection at point C if the beam is subjected to an exponentially decaying load, as shown in Fig. (9P.54), namely

$$v\left(\frac{L}{2}\right) = \frac{q_0}{48\alpha^4 EI}\left[3e^{-\alpha}(\alpha^2 - 8) + 48e^{-\alpha/2} + 3\alpha^2 - 24\right].$$

(b)* Evaluate v_C when $\alpha \to 0$. (*Note:* See computer-related Problem 9.107.)

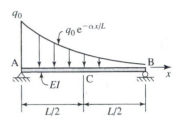

Figure 9P.54

9.55: Based on the influence function given in Problem 9.51 for a simply supported beam of length L, (a) (i) determine the deflection at $x = L/2$ due to a load $P = 1$ acting at $\zeta \le L/2$ and (ii) determine the deflection at any point $x \le L/2$ due to a load $P = 1$ acting at $\zeta = L/2$. (b) What conclusion can be drawn from the results of (i) and (ii) ? Explain by means of a sketch. (*Note:* The same conclusion is valid for a load acting at $\zeta \ge L/2$.)

9.56: Given a simply supported beam of length L, as shown in Fig. (9P.56).
 (a) Obtain the influence functions $G_{V_C} \equiv G_V(x_c, \zeta)$ and $G_{M_C} \equiv G_M(x_c, \zeta)$ for the shear force V_C and moment M_C, respectively. Sketch the influence lines showing all critical ordinates in terms of x_c and L.
 (b) If a uniformly distributed load w (N/m) is applied between points C and B, determine V_C and M_C in terms of w, x_c and L.
 (c) (i) For point C located at $x_c = L/4$, sketch the influence lines and show the critical ordinates in terms of L. (ii) A uniformly distributed load w (N/m), representing a given equipment, acting over a span length $L/4$, is to be applied to the beam. Where must the equipment be placed in order to minimise M_C? What is the minimum value of M_C.
 (d) If the equipment [represented by w (N/m) above] is instead placed within the region $L/8 \le \zeta \le 3L/8$, determine $M_C = M(x_c = L/4)$.
 (e)* Since the equipment is movable, it is necessary to consider the worst case in order to properly design the beam due to flexure. Between what points will the given equipment cause M_C ($x_c = L/4$) to be maximum? What is the maximum moment?

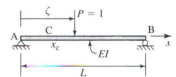

Figure 9P.56

9.57: Given an overhanging beam ABCD containing a hinge at B, as shown in Fig. (9P.57). (a) Determine the Green's functions (influence functions) for (i) $G_{R_A} \equiv G_R(0, \zeta)$, (ii) $G_{R_C} \equiv G_R([a + L], \zeta)$, (iii) $G_{M_A} \equiv G_M(0, \zeta)$ and (iv) $G_{V_{C+}} \equiv G_V([a + L]^+, \zeta)$. (b) Sketch the influence lines for the above quantities as a function of ζ. Indicate values of all critical ordinates. (*Note:* V_{C+} indicates the shear force immediately to the right of point C.)

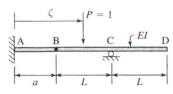

Figure 9P.57

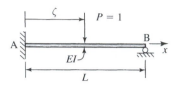

Figure 9P.58

9.58: Given the indeterminate beam shown in Fig. (9P.58). (a) Determine the influence function for the reaction at B and the moment M_A at A (*Note*: The influence functions may be determined either by integration of the differential equation or by using results given in Section 9.7 or given in Appendix F.) (b) Sketch the influence functions $G_{R_B}(\zeta)$ and $G_{M_A}(\zeta)$, showing critical ordinates. (c) If a vertical applied load $P = 1\,kN$ acts on the beam, determine the maximum moment $|M_A|$ if the beam is 2 m in length. Where is P applied?

Section 12

In Problems 9.59–9.65, (a) determine the equation of the elastic curve, (b) evaluate the maximum deflection and (c) sketch the elastic curve. Solve by integration of the fourth-order differential equation $EIv^{iv}(x) = q(x)$.

9.59: Obtain the solution for the cantilever beam loaded, as shown in Fig. (9P.3a).

9.60: Obtain the solution for the cantilever beam loaded, as shown in Fig. (9P.3e).

9.61: Obtain the solution for the simply supported beam loaded, as shown in Fig. (9P.3i).

9.62: Obtain the solution for the simply supported beam loaded, as shown in Fig. (9P.5a).

9.63: Obtain the solution for the cantilever beam loaded, as shown in Fig. (9P.5b).

9.64: Obtain the solution for the cantilever beam loaded, as shown in Fig. (9P.6). Show, in the limit $\alpha \to 0$, that the maximum deflection reduces to $v_{max} = \frac{q_0 L^4}{8EI}$.

9.65: Obtain the solution for the simply supported beam loaded, as shown in Fig. (9P.19).

Problems 9.66–9.70 are to be solved by integration of the fourth-order differential equation $EIv^{iv}(x) = q(x)$.

9.66: Determine the equation of the elastic curve for the beam shown in Fig. (9P.4).

9.67: (a) Obtain the equation for the elastic curve for the simply supported beam loaded, as shown in Fig. (9P.24), and (b) evaluate the deflection at points B and D.

9.68: Given the indeterminate beam, with flexural rigidity EI, subjected to a uniformly distributed load w (N/m), as shown in Fig. (9P.31). (a) Obtain the equation for the elastic curve, (b) evaluate the displacement at the mid-point $x = L/2$, (c) sketch the elastic curve and (d) determine the reactions and moments at the end points, $x = 0, L$.

9.69: Solve Problem 9.68 for a concentrated load P applied at the mid-point $x = L/2$, as shown in Fig. (9P.32).

9.70: An elastic beam with flexural rigidity EI is clamped at A and supported by a linear spring at B, whose constant is k (N/m). The beam is subjected to a uniformly distributed load w (N/m), as shown in Fig. (9P.70). (a) Determine the equation of the elastic curve $v(x)$, namely

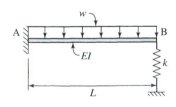

Figure 9P.70

$$v(x) = \frac{wx^2}{48EI}\left[2x^2 - \left(\frac{5\alpha + 24}{\alpha + 3}\right)x + 3\left(\frac{\alpha + 12}{\alpha + 3}\right)L^2\right], \quad \alpha = \frac{kL^3}{EI}.$$

(b) Determine the reactions and end moment at A (express the answer in terms of α). (c) Evaluate the answers to (b) for two limiting cases: (i) $\alpha \to \infty$ and (ii) $\alpha \to 0$. (*Note*: See computer-related Problem 9.108.)

Section 13

Problems 9.71–9.78 are to be solved using the moment–area method.

9.71: A cantilever beam, fixed at $x = 0$ and having flexural rigidity EI, is subjected to a uniformly distributed load w along its entire length L. Determine the slope and displacement at the free end, $x = L$.

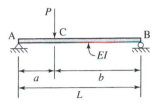

Figure 9P.72

9.72: A concentrated load P acts at point C $(x = a)$ on a simply supported beam with flexural rigidity EI and of length L, as shown in Fig. (9P.72). Determine the displacement under the load.

9.73: A simply supported beam with flexural rigidity EI and length L is subjected to end moments M_A and M_B, as shown in Fig. (9P.73). Determine (a) the slopes at A and B and (b) the deflection at the mid-point of the beam.

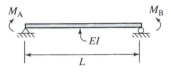

Figure 9P.73

9.74: A linearly varying load acts on a cantilever beam AB with flexural rigidity EI, as shown in Fig. (9P.3f). Determine the deflection at the free end.

9.75:* A propped cantilever beam, having flexural rigidity EI, is fixed at A and simply supported at B. A couple M_0 is applied at C $(x = a)$, as shown in Fig. (9P.75). (a) Determine the reactions and the moment at point A. (b) Determine the deflection v at point C of load application. (c) For what ranges of values of a/L will the moment at A be positive and for which negative? (d) For what range of values of a/L will the beam deflect upward and for which downward? Sketch the elastic curve in both cases.

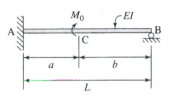

Figure 9P.75

9.76: A couple M_0 is applied at an intermediate point C to a fixed-end beam having flexural rigidity EI, as shown in Fig. (9P.76). (a) Determine the reactions and fixed-end moments M_A and M_B and (b) the deflection of point C.

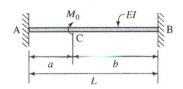

Figure 9P.76

9.77: A simply supported beam ACB has flexural rigidity EI_0 within the segment AC and αEI_0 $(0 < \alpha)$ within segment CB, as shown in Fig. (9P.77). A concentrated force P acts at point C. (a) Determine the displacement v_C. (b) For the case $a = b = L/2$, determine the displacement v_C for $\alpha = 1$ and for the limiting cases (i) $\alpha \to 0$ and (ii) $\alpha \to \infty$. Explain the results of (i) and (ii) in physical terms and sketch the elastic curve where appropriate.

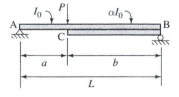

Figure 9P.77

9.78: An indeterminate beam ACB, fixed at both ends A and B, has flexural rigidity EI_0 within the segment AC and αEI_0 $(0 < \alpha)$ within segment CB, as shown in Fig. (9P.78). A concentrated force P acts at point C. (a) Obtain the reactions R_A and M_A. (b) Obtain the displacement v_C in terms of the given parameters, (c) For the case $a = b = L/2$, determine the reactions and the displacement v_C for $\alpha = 1$ and for the limiting cases (i) $\alpha \to 0$ and (ii) $\alpha \to \infty$. Explain the results of (i) and (ii) in physical terms and sketch the elastic curve where appropriate.

Review and comprehensive problems

9.79: The simply supported beam of length L and flexural rigidity EI is subjected to a linearly varying distributed load, as shown in Fig. (9P.79). (a) Determine the

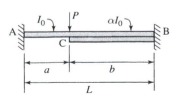

Figure 9P.78

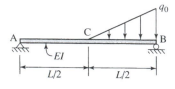

Figure 9P.79

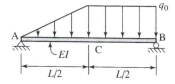

Figure 9P.80

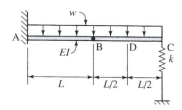

Figure 9P.82

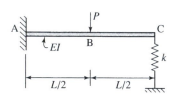

Figure 9P.83

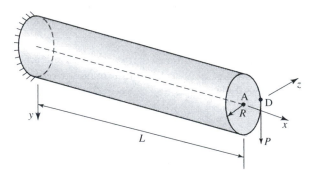

equation of the elastic curve $v(x)$ and evaluate the displacement at $x = L/2$. (b) In which segment AC or CB, does the maximum deflection occur?

9.80: Repeat Problem 9.79 for a loading, as shown in Fig. (9P.80).

9.81: A simply supported beam of length L, having flexural rigidity EI, is subjected to a distributed load $q(x) = q_0 \sin(2\pi x/L)$. (a) Sketch the loading and determine the deflection $v(x)$ and (b) evaluate v at $x = L/4$ and $x = L/2$.

9.82:* The elastic beam of flexural rigidity EI, containing a hinge at B, as shown in Fig. (9P.82), is fixed at A and supported at C by a linear elastic spring of stiffness k. The beam is subjected to a uniformly distributed load w (N/m). (a) Draw the shear and moment diagrams and show the value of all critical ordinates. (b) Determine (i) the deflection $v(x)$ for $0 \le x \le 2L$ and (ii) evaluate v at point D. (c) Determine the value of k (as a function of EI and L) that causes deflections of points B and C to be equal. (d) For the value of k determined in (c), determine the deflection of (i) point B or C and (ii) point D.

9.83: An eccentric force P is applied at point D to the free end of a cylindrical beam of radius R and length L, as shown in Fig. (9P.83). The modulus of elasticity of the rod material is E and the Poisson ratio is $\nu = 0.25$. Assume that all strains and rotations are small. If the ratio of the displacements in the y-direction of points A and D is $v_D/v_A = 1.0125$, what is the ratio R/L of the rod?

9.84:* Given a simply supported non-prismatic beam, as shown in Fig. (9P.8), whose whose flexural rigidity is $EI(x)$, where $I(x)$ is given as $I(x) = I_0 \sin(\pi x/L)$. The beam is subjected to a symmetrically distributed load $q(x) = q_0 \sin(3\pi x/L)$. (a) Sketch the loading and determine the equation of the elastic curve. (b) Determine the maximum displacement and the slope of the beam at $x = L/2$. (c) Sketch the elastic curve.

9.85: (a) Verify, by induction, i.e. by considering several values of n ($n = 0, 1, 2, \ldots$), the validity of Eq. (9.7.2), namely

$$\langle x - k \rangle^n = (x - k)^n - (-1)^n \langle k - x \rangle^n, \quad n \ge 0,$$

and thereby, (b) show that the two expressions for the moment $M(x)$ in the beam of Example 9.7 [see Eq. (9.7.13b)] are equivalent.

9.86: The indeterminate elastic beam of flexural rigidity EI, shown in Fig. (9P.86), fixed at A and supported at C by a linear elastic spring of stiffness k, is subjected to a concentrated force P at B. Using the force method and superposition, determine the force in the spring.

9.87: The indeterminate beam having flexural rigidity EI is supported at a C by means of a spring having constant k and is loaded, as shown in Fig. (9P.87). Determine the reactions on the beam.

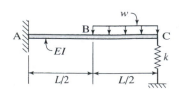

Figure 9P.86

Figure 9P.87

9.88: The overhanging beam ABC of flexural rigidity EI is fixed at A and subjected to a concentrated force P at C, as shown in Fig. (9P.88). (a) Using the force method and superposition, determine the reaction acting on the beam and (b) draw the shear and moment diagrams.

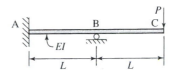

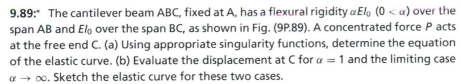

Figure 9P.88

9.89:* The cantilever beam ABC, fixed at A, has a flexural rigidity αEI_0 $(0 < \alpha)$ over the span AB and EI_0 over the span BC, as shown in Fig. (9P.89). A concentrated force P acts at the free end C. (a) Using appropriate singularity functions, determine the equation of the elastic curve. (b) Evaluate the displacement at C for $\alpha = 1$ and the limiting case $\alpha \to \infty$. Sketch the elastic curve for these two cases.

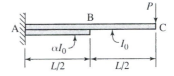

Figure 9P.89

9.90:* A beam AB having flexural rigidity EI, simply supported at the ends and supported at the centre by means of a linear elastic spring having constant k (N/m), is subjected to a uniformly distributed load w (N/m), as shown in Fig. (9P.34). (a) Determine the required value of the spring constant k if the moment at the centre C is (i) $M_C = 0$ and (ii) $M_C = -wL^2/48$. (b) Sketch the shear and moment diagrams for both cases (i) and (ii).

9.91: Given the beam and loading shown in Fig. (9P.25). By means of the fourth-order differential equation, (a) determine the equation of the elastic curve, (b) obtain the reactions on the beam, (c) draw the shear and moment diagrams and (d) sketch the elastic curve.

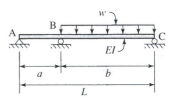

Figure 9P.92

9.92:* Given a continuous beam, simply supported at A, B and C, subjected to a uniformly distributed load w over the span BC, as shown in Fig. (9P.92). Using the moment area theorems, (a) determine the reactions, (b) evaluate M_B and draw the moment diagram and (c) sketch the elastic curve.

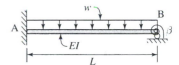

Figure 9P.93

9.93:* A beam AB is fixed at A and simply supported at B. In addition, as shown in Fig. (9P.93), the beam is partially restrained at B by means of a linear torsional spring, whose constant is β (N-m/rad); i.e., the spring tends to restrain the slope θ at B by exerting a restraining moment $M = \beta\theta$. (a) By integrating the Euler–Bernoulli beam equation, $EIv''(x) = -M(x)$, and the equations of equilibrium, determine all the external reactions acting on the beam. (b) Obtain the limiting cases for the above reactions for the two cases: (i) $\beta \to 0$ and (ii) $\beta \to \infty$. (c) Sketch the free body of the beam for cases (i) and (ii), showing all reactions and moments. Sketch the elastic curve for both cases.

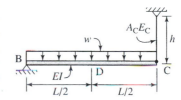

Figure 9P.94

9.94:* A linear elastic beam of length L and flexural rigidity EI is simply supported at one end B and at the other end C by means of an elastic wire of length h as shown in Fig. (9P.94). The axial rigidity of the wire is given as $A_C E_C$ and its coefficient of thermal expansion is α (1/°C). The beam is subjected to a uniformly distributed load q_0. Determine the required change in temperature ΔT such that, under this loading, the vertical deflection of the mid-point of the beam, point D $(x = L/2)$, should be zero. Indicate whether the temperature change represents cooling or heating.

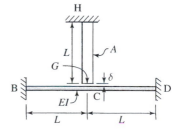

Figure 9P.95

9.95:* An elastic beam BCD, having a flexural rigidity EI, is fixed at its two ends B and D. An elastic rod GH made of the same material and whose cross-sectional area is A, is suspended at H, leaving a gap $\delta \ll L$ between the rod and the beam at C, as shown in Fig. (9P.95). The rod, whose coefficient of thermal expansion is α (°C^{-1}), is heated uniformly and undergoes a change of temperature ΔT. Assuming that ΔT is greater than that necessary to close the gap δ, determine the reactions at B and D as a function of ΔT and the parameters of the problem.

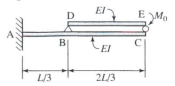

Figure 9P.96

9.96: A beam system consists of two beams of flexural rigidity EI subjected to a couple M_0 at point E, as shown in Fig. (9P.96). Determine the displacement of point C.

9.97: Two cantilever beams, AB and CD, each of flexural rigidity EI, are connected by means of a linear spring having constant k, as shown in Fig. (9P.97). A concentrated force P acts at the mid-point of beam CD. (a) Determine the resulting force R_s in the spring, namely

$$R_s = \frac{5P}{16}\left[\frac{\alpha}{3+2\alpha}\right], \quad \alpha = \frac{kL^3}{EI}.$$

(b) Determine the reactions and resisting moments at A and D and the displacement of C in terms of P, L and the non-dimensional parameter of the system, α. (c) Evaluate R_A, R_D and v_C for the two limiting cases: $\alpha = 0$ and $\alpha \to \infty$. (*Note:* See computer-related Problem 9.109.)

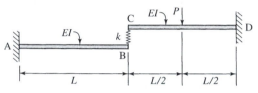

Figure 9P.97

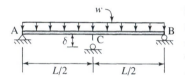

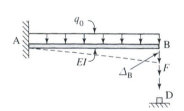

Figure 9P.98

9.98: A beam ACB of flexural rigidity EI is initially supported at the ends and at the mid-point. Due to settlement of the ground, a gap δ now exist between the mid-point of the unloaded beam and the support at the mid-point C. When the beam is subjected to a uniformly distributed load w (N/m), as shown in Fig. (9P.98), the three reactive forces are known to be equal. Determine the gap δ in terms of the given parameters of the system and the load w.

9.99: An initially straight elastic beam with flexural rigidity EI is clamped at A, as shown in Fig. (9P.99). A device exists at point D, which actuates a downward attractive electromagnetic force on point B, which is linearly proportional to Δ_B only if $\Delta_B > 0$ and otherwise is zero, i.e. when $\Delta_B = 0$, $F = 0$ and when $\Delta_B > 0$, $F = \gamma \Delta_B$, where γ (N/m) is a constant. (a) Determine the deflection at B if a uniformly distributed load q_0 is applied over the length of the beam. (b) For what values of γ will the deflection Δ_B be finite? What is the significance if theoretically, $\Delta_B \to \infty$?

Figure 9P.99

9.100: A series of loaded beams (each having the same flexural rigidity EI), (i) to (iv), is shown in Fig. (9P.100). In each case, with $P = wL$, the loadings cause a deflection Δ_B at B. *Without making any calculations*, determine for each series, the sequence of beams (a) to (d) in the order of increasing deflection, Δ_B.

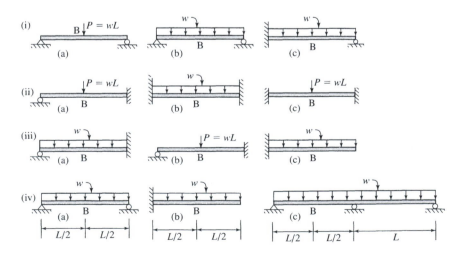

Figure 9P.100

9.101:* Given a 'continuous beam' having flexural rigidity EI, namely one supported along its length by a series of simple supports, as shown in Fig. (9P.101a), and subjected to loadings within each span. Consider now two *adjacent* spans, with supports at $i-1$, i and $i+1$, as shown in Fig. (9P.101b), where the moments at the cross-sections are M_{i-1}, M_i and M_{i+1}, respectively. Due to the linearity of the

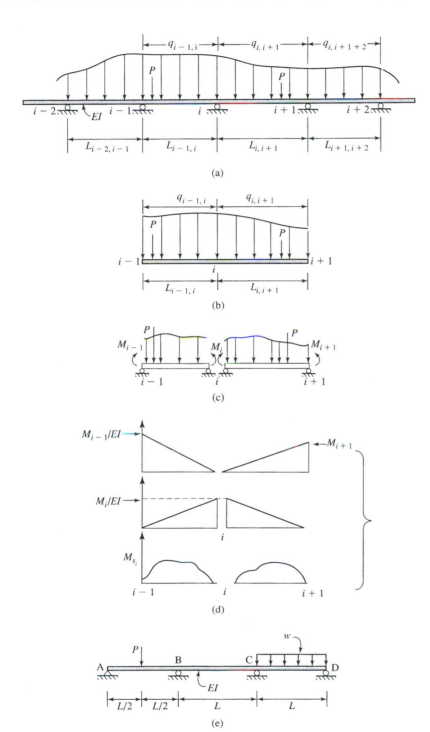

(a)

(b)

(c)

(d)

(e)

Figure 9P.101

problem, we may consider, in the context of superposition, each span to be subjected separately to (a) the end moments and (b) the applied transverse loadings $q_{i-1,i}(x)$ and $q_{i,i+1}(x)$, as shown in Fig. (9P.101c). The resulting 'M/EI' diagram, drawn 'by parts', then appears as in Fig. (9P.101d). (a) Using the moment area theorems, derive a relation between the moments M_{i-1}, M_i and M_{i+1}, namely

$$L_{i-1,i} M_{i-1} + 2(L_{i-1,i} + L_{i,i+1})M_i + L_{i,i+1} M_{i+1} = -6\left[\frac{\tilde{Q}_{i-1}|_{i-1}^{i}}{L_{i-1,i}} + \frac{\tilde{Q}_{i+1}|_{i}^{i+1}}{L_{i,i+1}}\right],$$

where $\tilde{Q}_{i-1}|_{i-1}^{i}$ and $\tilde{Q}_{i+1}|_{i}^{i+1}$ are the first moments, $(M_s)_i$, of the moment diagrams (*not* the 'M/EI' diagram), due to the transverse loading of the equivalent simply supported spans, about $i-1$ and $i+1$ in the spans $(i-1,i)$ and $(i, i+1)$, respectively. The above equation, valid here for the case of prismatic beams with constant EI, is known as the **three-moment equation**, and was first derived by Clapeyron in 1857. (b) Using the three-moment equation, determine the moment M_B in the beam shown in Fig. (9P.92), solve for all reactions and draw the shear and moment diagrams. (c) Using the three moment equation, determine the moments M_B and M_C in the beam shown in Fig. (9P.101e).

The following problems are designed to require the use of a computer. *Note:* In Problems 9.102–9.105, determine the quantities to four significant figures or more.

9.102: Referring to the simply supported beam of Problem 9.4, (a) verify the given expression for $v(x)$ and (b) by means of a computer program, determine numerically the location of the maximum deflection $|v|_{max}$ in the beam and evaluate $|v|_{max}$.

9.103: Referring to the overhanging beam of Problem 9.21, (a) verify the given expression for $v(x)$, (b) by means of a computer program, determine numerically the location of the maximum deflection $|v|_{max}$ in the segment AB as a function of a/L in the range $0 \le a/L \le 1$, (c) evaluate the non-dimensional ratio, $\frac{|v|_{max}}{5q_0 L^4/384EI}$ and (d) using an appropriate plotting routine, plot the resulting ratios as a function of a/L.

9.104: Referring to Problem 9.28, (a) verify the given expression for $v(x)$, (b) by means of a computer program, solve the relevant transcendental equation numerically to determine the location of the maximum deflection $|v|_{max}$ in the beam and (c) evaluate the ratio, $\frac{|v|_{max}}{5q_0 L^4/384EI}$.

9.105: Referring to Problem 9.29, (a) verify the given expression for $v(x)$, (b) by means of a computer program, determine numerically the location of the maximum deflection $|v|_{max}$ in the beam segment AB and evaluate $|v|_{max}$ and (c) determine analytically the location of the maximum deflection $|v|_{max}$ in the beam segment BC and evaluate $|v|_{max}$.

9.106: Referring to Problem 9.50, (a) verify the given expression for $v_C = v(L)$ and (b) by means of a computer program, evaluate the non-dimensional ratio, $\frac{v_C}{q_0 L^4/8EI}$, for several values of α, e.g. 0, 0.25, 0.5, 1.0, 2.0, 5.0, ..., and show these values in a table.

9.107: Referring to Problem 9.54, (a) verify the given expression for $v_C = v(L/2)$ and (b) by means of a computer program, evaluate the non-dimensional ratio, $\frac{v_C}{5q_0 L^4/384EI}$, for several values of α, 0, 0.25, 0.5, 1.0, 2.0, 5.0, ..., and show these values in a table.

9.108: Referring to Problem 9.70, (a) verify the expressions for the given ratios, R_A/wL, R_B/wL and M_A/wL^2, (b) evaluate these ratios as a function of α $(0 \le \alpha \le 20)$ and (c) using an appropriate plotting routine, plot the ratios as a function of α.

9.109: Referring to Problem 9.97, (a) verify the given expression for the force in the spring, (b) determine the non-dimensional ratios of the reactions and moment, R_A/P, R_D/P, M_A/PL, M_D/PL, as a function of α, (c) repeat (b) for the non-dimensional ratio, $v_C = \frac{PL^3}{3EI}$, and (d) using an appropriate plotting routine, plot these quantities as a function of α.

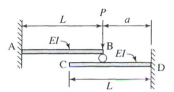

Figure 9P.110

9.110: The beam system shown in Fig. (9P.110) consists of two beams AB and CD, having the same length L and the same flexural rigidity EI. (a) Write a computer program to determine the deflection of point B in terms of a/L and (b) using an appropriate plotting routine, plot the deflection v_B as a function of a/L.

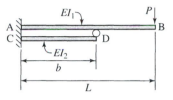

9.111: The beam system shown in Fig. (9P.111) consists of two prismatic beams of different lengths, AB and CD, having flexural rigidities EI_1 and EI_2, respectively. (a) Write a computer program to determine the deflection of point B in terms of the parameters I_1/I_2 and b/L and (b) via a computer, plot a family of curves showing the deflection v_B as a function of b/L for various values of $I_1/I_2 > 0$.

Figure 9P.111

10

Thin-wall pressure vessels: thin shells under pressure

10.1 Introduction

Pressure vessels, which have wide applications, in practice, as containers of fluids or gases under pressure, are of considerable engineering interest. In particular, if the wall thickness of these vessels is small in comparison with the overall dimensions, the vessel is said to be a *thin-wall vessel*. As we shall see, the thin wall characteristic leads to simple analyses particularly for the commonly encountered geometric shapes: cylindrical and spherical vessels. Pressurised vessels represent a special case of what are more generally known as *thin shells*; hence, instead of referring to the containers as *vessels*, we shall refer to them as *cylindrical thin shells* or *spherical thin shells*.

10.2 Thin cylindrical shells

We consider a long thin elastic cylindrical shell of length L with inner and outer radius R_i and R_o, respectively, and wall thickness $t = R_i - R_o$. The mean radius R is then $R = (R_i + R_o)/2$. We construct a polar coordinate system (x, r, θ) where the x-axis denotes the longitudinal axis, i.e. the axial direction, of the cylinder, as shown in Fig. (10.2.1). By a thin shell we mean one for which $t/R \ll 1$. (In practice, a shell is said to be a *thin shell* if $t/R < 0.10$.) The behaviour of the material is elastic with modulus of elasticity E and Poisson ratio v. The closed shell is subjected to an internal hydrostatic pressure p.

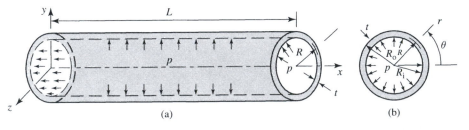

Figure 10.2.1
(a)   (b)

Following the usual *mechanics-of-materials* approach, we consider the deformation of the shell and arrive at specific conclusions based on plausible physical reasoning.

Since the pressure is hydrostatic, the same pressure p evidently acts normally to the internal surface of the wall at all points. Thus we observe that the deformation

392

does not depend on the circumferential coordinate θ; that is, the behaviour is the same at all points of the shell located a distance r from the x-axis; the problem is said to be *axi-symmetric*. Moreover, since the pressure is obviously in the outward direction, it is reasonable to assume that the radius will increase. Let $u = u(R)$ denote the outward displacement of a point on the *mid-circle* C whose initial radius is R [Fig. (10.2.2)].

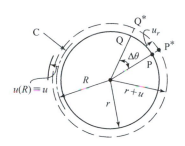

Figure 10.2.2

Using the property that the wall thickness $t \ll R$, we now make a reasonable assumption; namely we assume that any change in the thickness t of the wall is infinitesimal; the thickness t is therefore assumed to remain constant. Consequently, the radial displacements, u_r, at all points $R_i \leq r \leq R_o$ of the shell is the same. Since u_r is constant throughout the thickness, we may write $u_r = u(R) \equiv u$ for all r, $R - t/2 \leq r \leq R + t/2$.

Therefore, due to the outward displacement of the shell, the length of a circumferential segment, $\widehat{PQ} = r\,\Delta\theta$, lying along a circle of any given radius r, will increase to $\widehat{P^*Q^*} = (r + u)\Delta\theta$ [Fig. (10.2.2)]. Hence the strain ϵ_θ in this circumferential direction is

$$\epsilon_\theta(r) = \frac{(r+u)\Delta\theta - r\,\Delta\theta}{r\,\Delta\theta} = \frac{u}{r}. \qquad (10.2.1\text{a})$$

Noting that $r = R \pm \alpha t/2 = R(1 \pm \alpha t/2R)$, where $0 \leq \alpha \leq 1$, and using the binomial expansion, we have

$$\epsilon_\theta = \frac{u}{R(1 \pm \alpha t/2R)} = \frac{u}{R}\left[1 \mp \frac{\alpha t}{R} + \left(\frac{\alpha t}{R}\right)^2 + \cdots\right]. \qquad (10.2.1\text{b})$$

Upon dropping the infinitesimal terms, $t/R \ll 1$,

$$\epsilon_\theta = \frac{u}{R}. \qquad (10.2.2)$$

We thus conclude that for the case of *thin*-wall shells, the strain at all points in the wall can be considered to be independent of the variable radial distance, r, from the x-axis.

We now examine the deformation in the longitudinal direction. Due to the pressure p, it is evident that since this pressure is also acting against the ends of the cylinder, the cylinder will tend to elongate [Fig. (10.2.3)] and all cross-sections will remain plane. It follows that due to the axi-symmetry of this problem, the cross-sections must then necessarily remain perpendicular to the longitudinal x-axis. Moreover, physical reasoning justifies the assumption that the longitudinal displacement throughout the wall thickness is also constant; consequently, we conclude that the displacement in the x-direction is again independent of the radial coordinate and therefore, $\epsilon_x \neq \epsilon_x(r)$.

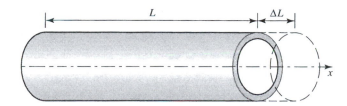

Figure 10.2.3

Due to the uniform pressure p, it is clear that at any point in the cylinder, there exist normal stresses σ_r, σ_θ, σ_x, as shown in Fig. (10.2.4). Furthermore, since the

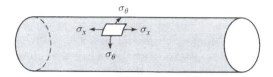

Figure 10.2.4

pressure is uniform, no shear stresses can exist in the cylinder; i.e., $\tau_{r\theta} = \tau_{\theta x} = \tau_{xr} = 0.$[†]

Using the stress–strain relations, i.e. Hooke's law for the elastic shell, we have

$$\epsilon_\theta = \frac{1}{E}[\sigma_\theta - \nu(\sigma_x + \sigma_r)], \qquad (10.2.3a)$$

$$\epsilon_x = \frac{1}{E}[\sigma_x - \nu(\sigma_r + \sigma_\theta)]. \qquad (10.2.3b)$$

From physical considerations, we note that on the inner wall surface, $\sigma_r(r = R - t/2) = -p$, while on the outer wall, $\sigma_r(r = R + t/2) = 0$. Thus it is clear that the stress σ_r at any intermediate point r is

$$-p < \sigma_r(r) < 0. \qquad (10.2.4)$$

We now make a basic assumption; namely, we assume that $|\sigma_r|$ is negligible compared to σ_θ and σ_x. *At this stage, this is merely an assumption that will be justified a posteriori.* Upon neglecting σ_r, we then have, in particular,

$$\epsilon_\theta = \frac{1}{E}(\sigma_\theta - \nu\sigma_x), \qquad (10.2.5a)$$

$$\epsilon_x = \frac{1}{E}(\sigma_x - \nu\sigma_\theta). \qquad (10.2.5b)$$

Since ϵ_x and ϵ_θ are both independent of r, we observe from Eqs. (10.2.5) that the stresses σ_θ and σ_x must also be independent of r; that is, they do not vary across the wall-thickness.

Having established the above, we now examine the equilibrium of a segment of the cylinder using the free body, as shown in Fig. (10.2.5). From symmetry about the vertical axis, it is clear that $\sum F_z = 0$ identically. Taking $\sum F_y = 0$, and noting that the (downward) resultant F_p due to the pressure p is $F_p = p \cdot 2(R - t/2)\Delta x = 2pR\Delta x(1 - t/2R) = 2pR\Delta x,$[‡] we write

$$2\sigma_\theta(t\Delta x) - 2pR\Delta x = 0 \qquad (10.2.6a)$$

and hence

$$\sigma_\theta = \frac{pR}{t}. \qquad (10.2.6b)$$

The circumferential stress σ_θ, as given above, is usually referred to as the *hoop stress.*

[†] Since all cross-sections remain perpendicular to the x-axis, radial and circumferential directions remain perpendicular to the x-direction. Therefore, $\epsilon_{xr} = \epsilon_{x\theta} = 0$ and it follows that $\tau_{xr} = \tau_{x\theta} = 0$.

[‡] From Fig. (10.2.5), the total downward force F_p may alternatively be calculated as

$$F_p = \int_{-\pi/2}^{\pi/2} (p\cos\theta)R(1 - t/R)\,\Delta x\,d\theta = pR(1 - t/R) \cdot \Delta x \int_{-\pi/2}^{\pi/2} \cos\theta\,d\theta = 2pR\Delta x$$

for $t/R \ll 1$.

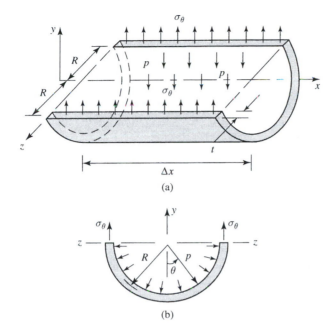

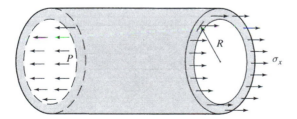

Figure 10.2.5

To satisfy equilibrium in the x-direction, we isolate the free body as shown in Fig. (10.2.6):

$$\sum F_x = \pi (R - t/2)^2 p - \sigma_x (2\pi Rt) = 0$$

or

$$R^2 (1 - t/2R)^2 p - 2Rt\sigma_x = 0,$$

and upon neglecting again $t/R \ll 1$, we obtain for the 'longitudinal stress',

$$\sigma_x = \frac{pR}{2t}. \tag{10.2.7}$$

Note that since $R/t \gg 1$, the stresses $\sigma_\theta \gg p$ and $\sigma_x \gg p$. Recalling Eq. (10.2.4), $|\sigma_r| < p$, it follows that

$$|\sigma_r| \ll \sigma_\theta \quad \text{and} \quad |\sigma_r| \ll \sigma_x.$$

We thus have justified at this stage the previously used assumption, which led to Eqs. (10.2.5), namely that σ_r is negligible with respect to σ_x and σ_θ.

Having found the stresses σ_θ and σ_x, Eq. (10.2.5a) yields

$$\epsilon_\theta = \frac{(2 - \nu)R}{2tE} \cdot p \tag{10.2.8}$$

Figure 10.2.6

and hence, from Eq. (10.2.2), viz. $u = R\epsilon_\theta$,

$$u = \frac{(2-\nu)R^2}{2tE} \cdot p. \qquad (10.2.9)$$

Thus we observe that the outward radial displacement is proportional to the square of the cylinder radius.

Similarly, using Eq. (10.2.5b), the elongation of the cylinder, ΔL, in the x-direction is given by

$$\Delta L = L\epsilon_x = \frac{(1-2\nu)RL}{2Et} \cdot p. \qquad (10.2.10)$$

While we have considered here cylinders that are subjected to a constant pressure p, thus yielding expressions that are independent of x, it is also possible to consider cases where the pressure varies with x, i.e. $p = p(x)$. For example, consider the case of a vertical cylinder (open at the top) filled to a height h with a fluid whose density is ρ, as shown in Fig. (10.2.7), in which case $p = \rho x$. Then, at any level x, equilibrium of a segment of height Δx leads, in lieu of Eq. (10.2.6b), to

$$\sigma_\theta(x) = \frac{p(x)R}{t}, \quad 0 < x < h. \qquad (10.2.11)$$

We note, however, that for this particular case (i.e., for an open cylinder), $\sigma_x = 0$.

Finally, it is worthwhile to recall that in the above analysis we have considered *long* cylindrical shells. Assume, for example, that the cylinders are closed at the ends by plates as shown in Fig. (10.2.8). Then clearly, these plates will tend to resist any deformation at the ends, in particular, any outward displacement of the cylindrical wall. Consequently, as a result of this resistance, near the ends, the shell cannot deform freely and will therefore assume a deformed shape as shown in Fig. (10.2.9). However, we observe from this figure that the deformation due to the end constraints is highly localised. Therefore the stresses and strains obtained by the above analysis are not valid in the end regions. However, by invoking this effect as being similar to that encountered in the principle of de Saint Venant, we may conclude that the stresses and displacements given above are valid away from the ends of the cylindrical shell and that the above analysis is therefore only valid for relatively long cylindrical thin shells.

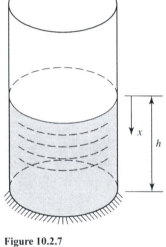

Figure 10.2.7

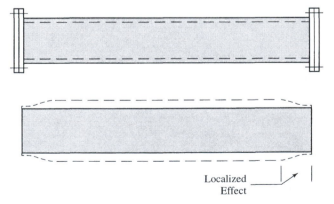

Figure 10.2.8

Figure 10.2.9

Localized
Effect

10.3   Thin spherical shells

The analysis of a thin elastic spherical shell having mean radius R and wall thickness t ($t \ll R$) subjected to an internal pressure p follows closely that of the thin cylindrical shell [Fig. (10.3.1)]. However, we note that for the case of the spherical shell, there exists complete axial symmetry; that is, there are no preferred directions. (For example, as opposed to the cylindrical shell, one cannot refer here to a unique 'axial direction'.)

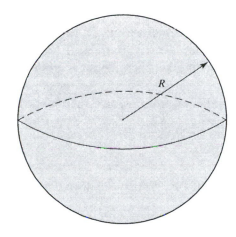

Figure 10.3.1

Furthermore, due to the uniformity of pressure at all interior points of the shell, it is clear that the outward displacement cannot vary with the location of the point. Thus, for a relatively thin shell, the radial displacement can, as in the previous case, be assumed to be constant throughout the thickness of the shell and hence we may write $u \equiv u(R)$.

For convenience, we define a meridian (longitudinal) coordinate θ and a coordinate ϕ, defining the 'latitude' of points on the sphere [Fig. (10.3.2)]. However, due to the perfect radial symmetry the behaviour of the shell is independent of θ and ϕ; moreover, the behaviour is the same in both the meridian and the latitudinal directions. Now, following the same reasoning as for the thin cylindrical shell, due to the outward displacement u, the strain in the shell will be given as u/R [see Eq. (10.2.2)]. But, since as discussed above, the response is the same in the meridian and latitudinal

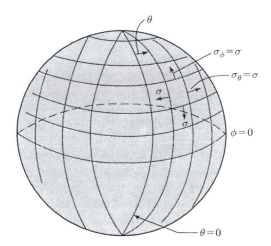

Figure 10.3.2

directions, we denote the strain simply by ϵ; i.e., $\epsilon \equiv \epsilon_\theta = \epsilon_\phi$. Therefore, we write

$$\epsilon = \frac{u}{R}. \tag{10.3.1}$$

It follows also that $\sigma_\theta = \sigma_\phi$ throughout the shell; we therefore denote the stresses by σ [Fig. (10.3.2)]. Furthermore, as with the cylindrical shell, we may neglect σ_r with respect to $\sigma_\theta = \sigma_\phi$. Since these are orthogonal directions, Hooke's law is then simply

$$\epsilon = \frac{(1-v)\sigma}{E}. \tag{10.3.2}$$

We also note that since ϵ is independent of r, so also is the stress σ; that is, σ does not vary across the wall thickness. To determine σ, we isolate, as a free body, a hemisphere obtained by cutting the sphere along any arbitrary great circle as shown in Fig. (10.3.3). Then

$$2\pi R t \cdot \sigma = \pi (R - t/2)^2 \cdot p \simeq \pi R^2 \cdot p \tag{10.3.3a}$$

since $t/R \ll 1$; hence the stress in the shell is given by

$$\sigma = \frac{pR}{2t}. \tag{10.3.3b}$$

Combining Eq. (10.3.1)–(10.3.3), the outward displacement of the shell is

$$u = \frac{(1-v)pR^2}{2Et}. \tag{10.3.4}$$

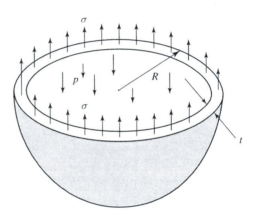

Figure 10.3.3

10.4 Comments and closure

The preceding analyses for pressure vessels are the simplest possible analyses of thin shells. The simplicity is possible due to (a) the simple cylindrical and spherical geometries of the shell, (b) the axi-symmetry or spherical symmetry of the applied normal pressure p at all points of the shell and (c) the validity of the assumptions governing very thin shells. Solutions for shells having more complex geometries, for example, thin shells of revolution (shells whose geometries are defined by the rotation of a curve about an axis), or those subjected to spatially variable pressures can only be obtained using a far more elaborate treatment.

We note also that for the cases of thin shells considered in this chapter, it was possible to assume that all internal stresses act tangentially to the shell geometry; that is, no shear and hence no moments were assumed to exist in the shell. The resulting stresses are referred to, in general, as *membrane stresses*. For shells where the above assumptions are no longer valid or where simple geometries no longer exist, the analysis is far more complex. The behaviour of shells under these conditions is a study in itself and falls within the branch of solid mechanics known as *shell theory*.

PROBLEMS

Sections 2 and 3

10.1: A compressed-air cylindrical tank, 80 cm in diameter, is fabricated by welding a plate 8 mm in thickness along a helix that is at an angle $\alpha = 60°$ with respect to the longitudinal axis, as shown in Fig. (10P.1). If the allowable normal and shear stresses in the weld are 160 and 90 MPa, respectively, determine the allowable internal pressure.

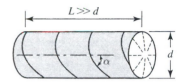

Figure 10P.1

10.2: A number of 10-mm thick square plates are to be welded together to form a pressurised cylinder 1.6 m in diameter. Two options are proposed: one by welding them, as shown in Fig. (10P.2a), and the other, by welding along 45° angles, as shown in Fig. (10P.2b). If the allowable normal and shear stresses in the weld are 120 and 80 MPa, respectively, determine the maximum allowable internal pressure in each case.

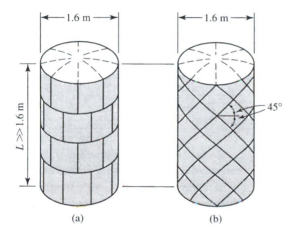

(a)   (b)

Figure 10P.2

10.3: An open thin-wall pipe having length L is to be placed between two walls that are a distance $L - \Delta$ apart. The pipe, with material properties E and Poisson ratio v, having mean radius R and thickness t ($t \ll R$), is therefore compressed by an amount $\Delta \ll L$ and placed between the two rigid frictionless walls (therefore providing no restraint against radial displacement at the two ends), as shown in Fig. (10P.3). The pipe is then subjected to an internal pressure until a critical pressure p_{cr} is reached. (The critical pressure p_{cr} is such that when $p > p_{cr}$, leakage occurs at the smooth walls due to a shortening of the pipe.) (a) Determine p_{cr} in terms of the given parameters of the problem. (b) If the pipe, of initial length $L = 3.5$ m, radius $R = 40$ cm and $t = 1$ cm, is made of aluminum with $E = 70$ GPa and $v = 0.33$, evaluate p_{cr} if the walls are 349.5 cm apart.

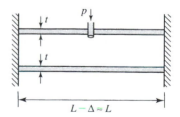

Figure 10P.3

10.4: A cylindrical vessel shown in Fig. (10P.4) is used as a storage tank. Determine the maximum normal stress when the tank is filled half way to the top with water.

10.5: If the storage tank shown in Fig. (10P.4) is filled with oil having a specific density of 0.9, to which height can it be filled if the normal stress is not to exceed 20 MPa?

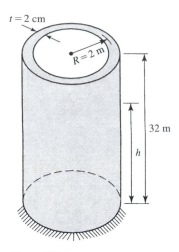

$t = 2$ cm

$R = 2$ m

32 m

h

Figure 10P.4

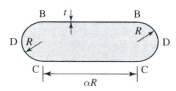

B $t\downarrow$ B

D $\left(R\right.$ R $\left.\right)$ D

C αR C

Figure 10P.7

Figure 10P.8

10.6: A thin-wall cylinder of inner radius R and thickness t ($t/R \ll 1$) is first heated from room temperature in order to force it over another cylinder of the same wall thickness but of outer radius $R + \delta$, where $\delta \ll R$. (Assume that there is sufficient lubrication between the cylinders so that the cylinders are free to expand axially.) Both cylinders are of the same linear elastic material. (a) Determine the circumferential stress in the inner and outer cylinders. (b) Determine the interacting radial stress σ_r acting at the interface between the two cylinders. (c) Verify that the ratio $\left|\frac{\sigma_r}{\sigma_\theta}\right| \ll 1$.

10.7: A compressed-air tank, of constant wall thickness t ($t/R \ll 1$), closed at the ends and having a cross-section as shown in Fig. (10P.7), is subjected to an internal pressure p. Based on the same assumptions used in deriving the relations found in this chapter, (a) determine the expression for the maximum circumferential stress in terms of R, t and p if $\alpha = 4$ and indicate where it occurs in the cross-section; (b) determine the expression for the axial stress in terms of R, t and p if $\alpha = 4$ and, (c) evaluate the maximum circumferential and axial stresses if $R = 0.25$ m, $t = 2$ cm if $p = 1.6$ MPa.

10.8: A thin-wall cylindrical pressure vessel of mean radius R ($t \ll R$) is repaired by welding a helical crack, which makes an angles $0 < \alpha < \pi/2$ with respect to the x-axis, as shown in Fig. (10P.8). The failure stress in the weld is τ_0 in shear and $\sigma_0 = 2\tau_0$ in tension. The cylinder is closed at both ends and subjected to an internal pressure p. Determine, as a function of α and other appropriate parameters, the pressure p at which failure will occur. Indicate whether failure occurs due to tension or shear.

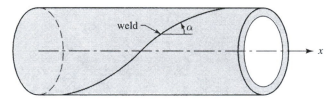

weld α

x

10.9:* In a temperature-controlled laboratory experiment, a closed thin-wall cylindrical specimen of length L, having mean radius R and wall thickness t ($t/R \ll 1$), is placed vertically in a pressure tank of height $H > L + \delta$ such that a gap δ exists between the top of the specimen and the pressure tank. The specimen is made of a material having modulus of elasticity E, Poisson ratio ν and a coefficient of thermal expansion, α ($^\circ C^{-1}$). If the specimen is subjected to an *external* pressure p, what is the maximum allowable temperature change ΔT if no contact is permitted between the top of the specimen and the pressure tank?

10.10:* A cylindrical pressure vessel, of diameter $d = 50$ cm and length $L = 4$ m, is constructed by winding a long thin aluminum sheet, of thickness $t = 6$ mm, whose material properties are $E = 70$ GPa and $\nu = 0.33$. The seams, oriented at an angle $\alpha = 45^\circ$ with respect to the longitudinal axis, are welded together as shown in Fig. (10P.1). The allowable normal and shear stresses in the weld are specified as $\sigma = 90$ MPa and $\tau = 60$ MPa, respectively. The corresponding allowable stresses in the aluminum are given as $\sigma = 100$ MPa and $\tau = 50$ MPa, respectively. (a) Express the (first-order) change in volume dV of the cylinder due to an internal pressure in terms of p, E, t, L *and* ν. (*Note:* The first-order change dV is based on the assumption that dV is small with respect to its original volume V, i.e., $dV/V \ll 1$.) (b) Determine the allowable internal pressure, p_{all}, which can be applied to the cylinder. (c) Based on the expression obtained in (a) evaluate dV numerically when $p = p_{all}$. (d) Determine the exact change in volume of the cylinder, ΔV, due to the internal pressure $p = p_{all}$.

10.11: A thin-wall spherical vessel of mean radius R and thickness t is fabricated by welding together two hemispheres. After filling the sphere with a fluid whose density is ρ (N/m³) and which is under pressure p, the shell is lifted by means of a cable attached at its top. Neglecting the dead weight of the shell itself, determine the normal stress within the weld if the weld lies along the shell's 'equator'.

Review and comprehensive problems

10.12:* A thin-wall cylindrical vessel of mean radius R, thickness t and length L ($t \ll R \ll L$), containing a fluid whose density is ρ (N/m³) and which is under pressure p, is to be lifted by means of cables. Two options that are available are shown in Figs. (10P.12a and b), respectively. Neglecting the dead weight of the shell itself, determine the stresses at points A, B and C. Which is the preferable option?

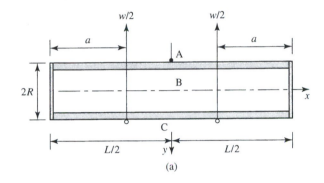

(a)

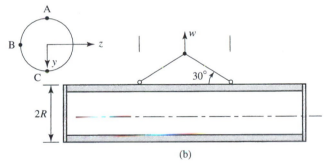

(b)

Figure 10P.12

10.13: Given a long thin-wall cylindrical open pipe, whose mean radius is R and whose thickness is t ($t \ll R$). The ends of the pipe are attached to two rigid walls and the pipe is subjected to an internal pressure p. Noting, for this case, that the axial strain $\epsilon_x = 0$ and assuming a constant radial displacement u, show that

$$\sigma_\theta = \frac{pR}{t}, \qquad \sigma_x = v\frac{pR}{t}, \qquad u = \frac{1-v^2}{E}\frac{pR^2}{t},$$

where E and v are the modulus of elasticity and Poisson ratio of the pipe, respectively.

10.14: Given a closed pressure vessel whose cross-section is as shown in Fig. (10P.7). Based on the same assumptions used in deriving the relations found in this chapter, show that the axial stress σ_x can never equal $(\sigma_\theta)_{max}$.

10.15:* The ends of two open concentric thin-wall cylindrical pipes, having the same material properties (modulus of elasticity E, Poisson ratio v) and the same thickness t, are welded to a rigid support and to a rigid plate, as shown in Fig. (10P.15). The inner pipe is subjected to an internal pressure p. Determine (a) the stresses in the

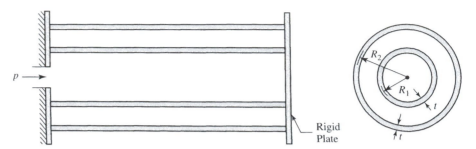

Figure 10P.15

pipes and (b) the axial displacement Δ_x of the rigid plate. (*Note*: Assume that there exists no constraint against radial displacements at the wall and at the plate.)

10.16: Two open thin wall cylindrical pipes, having different radii, R_1 and R_2 ($R_2 > R_1$), but the same material properties, E and v, and thickness t, are connected together by means of welds to a thin rigid plate and attached to rigid walls, as shown in Fig. (10P.16). The pipes are then each subjected to an internal pressure p. Determine (a) the stresses in the pipes and (b) the axial displacement Δ_x of the rigid plate. (*Note*: Neglect the thickness of the rigid plate and assume that there exists no constraint against radial displacements at the wall and at the plate.)

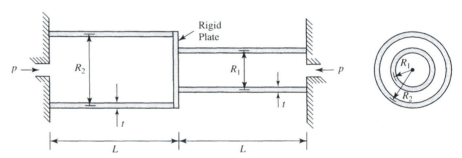

Figure 10P.16

10.17:* A thin-wall cylindrical pressure vessel of mean radius R is fabricated by welding a plate of thickness t along a helix, which is at an angle α with respect to the longitudinal axis, as shown in Fig. (10P.1). The allowable normal and shear stresses in the weld are given as σ_0 and τ_0, respectively, and it is specified that $\sigma_0 = n\tau_0$, where $n > 1$. (a) Determine the minimum value, $n = n_{min}$, for which it is possible that both the the normal stress and the shear stress are equal to their respective allowable values when the cylinder is subjected to an internal pressure p. (b) If $n = n_{min}$, determine the angle α of the weld for which $\sigma_\theta = \sigma_0$ and $\tau = \tau_0$.

10.18:* An open conical pressure vessel with wall thickness t and whose shape is defined by the angle α and varying radius R, is suitably supported at its top, $x = L$. The vessel is filled to a level h, as shown in Fig. (10P.18a), with a fluid whose density is ρ (N/m³). (a) Assuming, as in the case of cylindrical vessels, that the resulting stresses within the wall do not vary across the thickness, using equilibrium considerations, derive an expression for the stress σ_ϕ [see Fig. (10P.18b)], namely

$$\sigma_\phi = \frac{\rho x \left(h - \frac{2}{3}x\right)\tan\alpha}{2t\cos\alpha}, \quad 0 < x \leq h.$$

(b) Determine the location $0 < x < h$ at which the maximum stress σ_ϕ occurs and evaluate $(\sigma_\phi)_{max}$. (c) Determine σ_ϕ at sections above the fluid level. (*Note*: Neglect the dead weight of the vessel.)

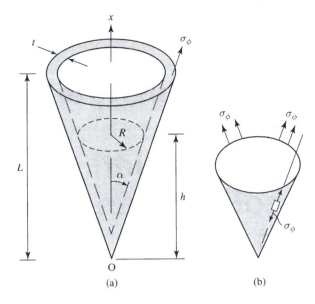

(a)

(b)

Figure 10P.18

10.19:* A hemispherical tank of mean radius R and thickness t, $t \ll R$, is supported by means of a flange, as shown in Fig. (10P.19a). The tank is filled to the top by a fluid whose density is ρ (N/m³). (a) Determine the stress σ_ϕ along the meridian as a function of the angle θ [see Fig. (10P.19b)] and (b) evaluate σ_ϕ at a depth $y = R/2$. (*Note*: Neglect the dead weight of the vessel.)

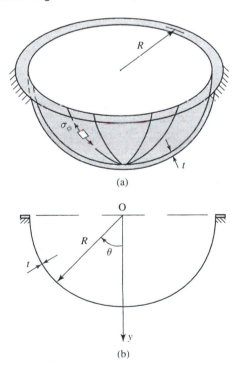

(a)

(b)

Figure 10P.19

11

Stability and instability of rods under axial compression: beam-columns and tie-rods

11.1 Introduction

The analysis of elastic bodies has, as one of its main goals, the determination of internal stresses and strains when the body is subjected to external loads. We often wish to find these quantities in order to determine the maximum loads that will cause the member to fail. In our previous study, the calculation of maximum loads was usually based on a criterion that was dependent on the given material. For example, failure is often assumed to occur when a resulting stress (or strain) in the material exceeds some stress (or strain), which is a characteristic of the material. Under uniaxial loading, this characteristic stress may be taken to be the yield point σ_y. Thus, we have always assumed that we may not exceed some *material property*.

As an example, consider a rod of cross-section A and length L_1 subjected to an axial compressive load P [Fig. (11.1.1a)]. Assume that the rod is made of a given elastic material represented by a stress–strain curve as shown in Fig. (11.1.2), where σ_p is the proportional limit and σ_y the yield point. Clearly, in such a case, the maximum load that can be applied is $P_{max} = \sigma_y A$; we note that this calculation clearly depends on the material property σ_y and is independent of the length.

Now consider a series of rods, as in Figs. (11.1.1b–e), each having the *same* cross-sectional area, A, but different lengths: $L_1 < L_2 < L_3 \ldots$, and assume that in each case an axial load P is applied statically. For rods, say, of length L_2 and L_3, the same maximum load may also be given by $P_{max} = \sigma_y A$. However, if the rod is sufficiently long, say L_5, it is clear that the rod will buckle under some load $P < P_{max} = \sigma_y A$. That is, as the load is increased slowly from zero, the rod will suddenly 'fail' (i.e., buckle) when the load reaches a certain value $P = \sigma A$, where $\sigma < \sigma_y$. Thus, buckling occurs although the stress may be well below σ_y, and so, in this case, the stress σ_y is no longer the governing criteria of failure. Instead, the length (in our case, L_5) now becomes a factor that determines the maximum load that can be applied; thus the criteria for failure is no longer dependent on the material stress property. As we shall see, determination of the load at which buckling occurs requires a completely different type of analysis.

404

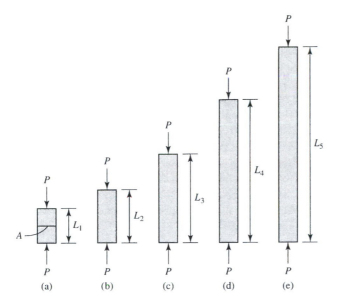

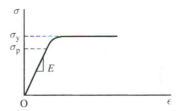

Figure 11.1.1

Let us again consider the long rod, loaded statically by an axial compressive force that increases continuously from zero. For small values of P, the rod will sustain the load and remain straight, but as P reaches a critical value, the rod suddenly buckles – this is a case of instability, i.e., the straight line configuration that was in stable equilibrium has suddenly become unstable. Our goal will be to determine the load at which this instability occurs.

We now pose some questions: If the load P is applied perfectly axially through the centroid, and if the rod is perfectly straight, why then does the rod buckle? Moreover, why does the rod not buckle immediately and why does it buckle only when the load reaches a certain value P (which, we have seen, is independent of σ_y)?

The answer to these questions can only be given after we have studied the phenomenon of instability.

Figure 11.1.2

11.2   Stability and instability of mechanical systems

In order to study stability and instability of mechanical systems, it is necessary to define these terms more precisely. With this in mind, consider a ball that can roll under the force of gravity g on a given track whose shape is as shown in Fig. (11.2.1). We note that at point A, for example, the ball is not in equilibrium; at points B and C, however, the ball is in a state of equilibrium. But clearly

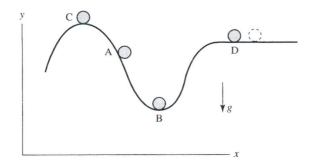

Figure 11.2.1

points B and C represent two different types of equilibrium states: at point B, the ball is in a state of 'stable equilibrium'; on the other hand, at point C, the ball is in 'unstable equilibrium'. Although this agrees with our physical intuition, we must, however, express these ideas in terms that are amenable to a mathematical treatment. To this end, we note the following: (a) if the ball, originally at point B, is given statically a small displacement and then released, it will return to its (original) position at B and (b) if the ball is given a small displacement from its equilibrium position at point C, it will not return; instead it will continue to roll and reach another equilibrium position (namely, point B if the displacement is to the right of point C).

In the study of mechanical systems, we use a more general terminology that is more appropriate to the systematic analysis of stability and instability; i.e., we use the term 'perturbation' instead of 'small displacement'. Thus, rather than referring to the 'small displacement of a mechanical system', we refer to the 'perturbation of a mechanical system'.

We now generalise the above ideas and define stable and unstable equilibrium as follows:

> A mechanical system is said to be in *stable equilibrium* if, when given a perturbation from its equilibrium position, it returns to the same equilibrium position. The system is said to be in *unstable equilibrium* if, when given a perturbation, it does not return to its original equilibrium position but continues to displace.

Now consider the ball on the track at some point D (a general point along the horizontal track). If we give the ball a small displacement, it will neither return to its original position, nor will it continue to displace further. Such an equilibrium state is said to represent *neutral equilibrium*. Thus we may state

> A mechanical system is said to be in *neutral equilibrium* if, when given a perturbation from its equilibrium position, it neither returns to, nor moves further away from, the given equilibrium position. That is, there exist two *adjacent* positions, both of which are in a state of equilibrium.

In order to better understand the phenomenon of instability of rods under compressive loads, we first investigate the simpler cases of a rigid bar as given in the models of the following section. We will observe that in the course of examining such a simple model carefully, we obtain answers to the questions that were posed in the preceding section.

11.3 Stability of rigid rods under compressive loads: the concept of bifurcation

Consider a rigid rod AB of length L, subjected to a vertical force P at B and supported at point A by means of a linear torsional spring having constant β [Fig. (11.3.1a)]; that is, the spring exerts a (restoring) moment, given by

$$M_A = \beta\theta, \qquad (11.3.1)$$

on the rod (which tends to return the rod to its vertical position) whenever the rod rotates by an amount θ.

We wish to examine equilibrium for any position $0 \leq |\theta|$. Taking moments about point A [Fig. (11.3.1b)],

$$\left(\sum M\right)_A = M_A - PL\sin\theta = 0, \qquad (11.3.2)$$

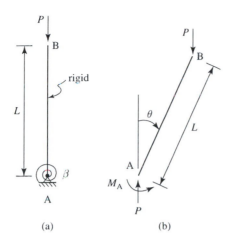

(a)   (b)

Figure 11.3.1

we obtain

$$\beta\theta - PL \sin\theta = 0. \tag{11.3.3}$$

Clearly, for $\theta = 0$, *all values* $P \geq 0$ satisfy this equilibrium equation. If $\theta \neq 0$, then

$$P = \frac{\beta/L}{\sin\theta/\theta}, \quad \theta \neq 0 \tag{11.3.4}$$

Thus, the force $P = P(\theta)$, as calculated by Eq. (11.3.4), is the vertical force that is required to maintain the rod in equilibrium when the latter is inclined at any given angle θ. A plot of this equilibrium force $P(\theta)$ as a function of θ is given by the curve BC of Fig. (11.3.2). We also note, from Eq. (11.3.4), that $\lim_{\theta \to 0} P = \beta/L$.

In addition to the solutions represented by the curve BC, we have observed that for $\theta = 0$, all values of P are solutions to the equilibrium. Thus the heavy line OD of Fig. (11.3.2) also represents solutions to the equilibrium equation.

At this stage of the analysis, it is worthwhile to examine, in the context of the above solutions, the physical behaviour of the rod as it is loaded slowly, from $P = 0$, with increasing values of P. Starting from $P = 0$ [point O of Fig. (11.3.2)], we observe that as P increases, possible solutions can only fall along the line OB, for which $\theta = 0$. Thus below point B, there exists only one equilibrium position, namely $\theta = 0$. Consequently, since there exist no other equilibrium positions for $P < \beta/L$, points along OB necessarily represent a state of stable equilibrium.

As the force P is increased further, it will reach point B; we observe that the solution can then follow two different paths: BC or BD. These paths are called *branches* of the solution. Since, from point B, one can follow either branch BC or BD, point B is called a *bifurcation point*. Note that both branches, BC and BD, represent equilibrium states. We now examine these branches more closely to determine the type of equilibrium.

Consider a generic point along branch BC, e.g. point E, and assume that $\theta = \theta_1$ at this point [Fig. (11.3.3)]. Then, according to Eq. (11.3.4),

$$P = \frac{\beta/L}{\sin\theta_1/\theta_1} = P_1 \tag{11.3.5}$$

is required to satisfy equilibrium. Assume now that P_1 is the actual force acting on the rod. Let us now imagine that the rod, in this equilibrium state, is given a small perturbation $\epsilon > 0$; i.e., let $\theta_1 \to \theta_1 + \epsilon$. (The statement 'the rod is given a small

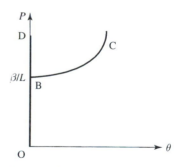

Figure 11.3.2

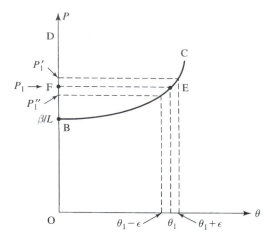

Figure 11.3.3

perturbation ϵ' has the following physical meaning here: the rod is moved by *some external force other than P*, for example, a light wind, which causes the rod to move to the position $\theta_1 + \epsilon$.) Now, the equilibrium force required to maintain the rod in the new position $\theta_1 + \epsilon$ is, again according to Eq. (11.3.4),

$$P = \frac{\beta/L}{\sin(\theta_1 + \epsilon)/(\theta_1 + \epsilon)} = P_1' > P_1 \qquad (11.3.6)$$

However, since P_1 is the actual force acting on the rod, the rod will return to the θ_1 position, i.e., to the equilibrium position from which it was displaced. Similarly, with P_1 acting on the rod, let $\theta_1 \to \theta_1 - \epsilon$; then the required force for equilibrium is $P_1'' < P$. But P_1 is acting on the rod; therefore the rod returns to its θ_1 position.

Thus, since the rod will always return to point E when given a perturbation about this point, we conclude from the previous definitions, that point E represents a stable equilibrium. Since point E is a general point on the branch BC, this branch is said to represent points of stable equilibrium.

Consider now branch BD with $\theta = 0$. Assume, again, that the same force $P_1 > \beta/L$ is acting on the rod, as represented by the generic point F on the branch. Let us give the rod a small perturbation ϵ; i.e., $\theta = 0 \to \theta = 0 + \epsilon = \epsilon$. (Again, as in the case of point E, the cause of this perturbation ϵ can be thought of as being due to some external force acting on the rod.) Then, from Eq. (11.3.4),

$$P = \frac{\beta/L}{\sin \epsilon/\epsilon} \simeq \frac{\beta}{L} \qquad (11.3.7)$$

is the force required to maintain the rod in the position $\theta = \epsilon$, for $|\epsilon| \ll 1$.

Now, as we have seen, $P = \beta/L$ represents the force at the point B of Fig. (11.3.2). But $P_1 > \beta/L$ is actually assumed to be acting; thus, under this load P, the rod cannot stay in the position $\theta = \epsilon$. (Note that there is no point in the neighbourhood of F for which $P = \beta/L$ is the equilibrium load.) Therefore, the rod will snap to a position $\theta = \theta_1$, represented by point E on the branch BC. We thus observe, according to the previous definitions, that the branch BD represents unstable equilibrium positions.[†] Consequently, we finally conclude that when $\theta = 0$, stable

---

[†] We have determined the stable and unstable nature of the branches using physical reasoning. We mention here that the stable and unstable character of the branches can only be established in a rigorous and clear-cut way by energy considerations. Stability analyses using an energy approach will be treated in Chapter 15.

equilibrium is represented only by points along the path OB. In physical terms, if $\theta = 0$, the rod will be in stable equilibrium only if $P < \beta/L$.

Hence, the critical force P, defined as

$$P_{cr} = \frac{\beta}{L}, \tag{11.3.8}$$

is the force such that if a force $P < P_{cr}$ is acting, the rod will remain in stable equilibrium in the position $\theta = 0$; if $P > P_{cr}$, the $\theta = 0$ position is one of unstable equilibrium. Having examined the cases when $P < \beta/L$ and $P > \beta/L$, we now examine the rod when subjected to a force $P = \beta/L$, as represented by the point B of Fig. (11.3.2), with $\theta = 0$. Assume that this is the actual force acting on the rod. If the rod is given a perturbation ϵ such that $\theta = 0 \to \theta = 0 + \epsilon = \epsilon$, then

$$P = \frac{\beta/L}{\sin \epsilon/\epsilon} \simeq \frac{\beta}{L}. \tag{11.3.9}$$

We note that in this perturbed position $\theta = \epsilon$, the required equilibrium force, given by Eq. (11.3.9), is the same as the force existing when the rod is in the position $\theta = 0$. Thus, under the *same* force, two *adjacent* equilibrium positions are possible: $\theta = 0$ and $\theta = \epsilon$. Since these two *adjacent* positions represent equilibrium states, we conclude, from the previous definitions of Section 11.2, that these are *neutral equilibrium* states; hence the rod will neither return to $\theta = 0$, nor will it rotate any further.

We may, therefore, finally conclude that if $P = P_{cr}$, the rod is in neutral equilibrium.

As we now observe, by associating the critical force with a state of neutral equilibrium, we have found a simple means to calculate P_{cr}: namely, we determine the force P required to maintain the rod in equilibrium in a perturbed position $\theta = \epsilon$. Thus we give the rod a small *infinitesimal* rotation $|\theta| > 0$. Because we are now considering only infinitesimal rotations $|\theta| \ll 1$, we may use the known relation $\sin \theta \simeq \theta$. The equilibrium equation, Eq. (11.3.3), then is reduced to the *linearised equation*

$$(PL - \beta)\theta = 0. \tag{11.3.10a}$$

If $\theta \neq 0$, it follows that

$$P = \frac{\beta}{L}, \tag{11.3.10b}$$

which we recognise as the critical load P_{cr}. Thus, by making use of the linearised equation, we obtain the critical load, i.e. the load that causes the system to be in neutral equilibrium: we do so by examining equilibrium of a state $|\theta| > 0$, which is in the neighbourhood of the $\theta = 0$ position.

While the linearised equation provides us with a simple means to determine P_{cr}, we observe that we no longer can find $P = P(\theta)$; i.e., we can no longer find the general force–displacement relation. This, essentially, may be considered as the 'penalty' for using the linear equation. We illustrate the technique by means of the following example.

Example 11.1: A rigid rod of length L is simply supported at point A and by a linear spring of stiffness k at B, which acts as a restoring force. (Note that in this model, the spring is assumed to remain horizontal.) The rod is subjected to a vertical force P at C, as shown in Fig. (11.3.4a). Determine the critical load P_{cr}.

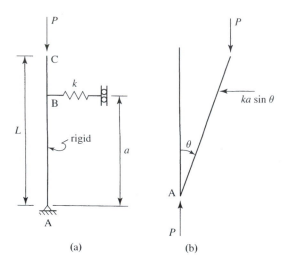

Figure 11.3.4

(a)   (b)

Solution: Consider the rod in a displaced position θ [Fig. (11.3.4b)]. Taking moments about point A,

$$\left(\sum M\right)_{A} = (ka\sin\theta)(a\cos\theta) - PL\sin\theta = 0. \qquad (11.3.11a)$$

Noting that for $|\theta| \ll 1$, $\sin\theta \simeq \theta$ and $\cos\theta \simeq 1$, we obtain the linearised equation

$$(ka^2 - PL)\theta = 0. \qquad (11.3.11b)$$

Hence, for $\theta \neq 0$,

$$P = \frac{ka^2}{L} \qquad (11.3.11c)$$

i.e., the critical load is $P_{cr} = \frac{ka^2}{L}$.   □

The study of the simple models above has provided us with a better understanding of the phenomenon of stability and instability of a mechanical system. It is worthwhile to summarise some of the conclusions derived from this study. In particular, we note the following:

- a critical load P_{cr} has been defined as the maximum load below which a straight rod remains in stable equilibrium or, alternatively, P_{cr} *is the smallest load that can maintain the rod in a perturbed position*,
- the load P_{cr} is associated with the force that causes the system to be in neutral equilibrium; as a result,
- a simple method to determine P_{cr} has been established: namely, we satisfy the equilibrium equations of the system, which is given a small perturbation. The resulting linearised equations then immediately yield P_{cr} (although general force–displacement relations can no longer be determined).

Finally, we should mention that the analysis of the simple model of this section reveals the fundamental ideas of 'bifurcation theory', which govern the study of instability of mechanical systems. We observed that the critical load $P = P_{cr}$ occurs at the bifurcation point. In Fig. (11.3.2), point B is referred to as a *stable bifurcation point* since the path BC (representing the displaced configuration) corresponds to stable equilibrium positions.[†] As we shall see, while more complex systems may perhaps require a more sophisticated mathematical treatment, we shall follow the same line of reasoning; it will not be necessary to introduce any new concepts.

Having established the above ideas, it is now possible to treat the original problem posed in Section 11.1, namely the instability of an elastic rod subjected to an axial force. As will be evident, the restoring force that tends to return the rod to its original straight line position is not a supporting spring (as has been the case in the examples of the present section), but rather the elasticity of the rod itself.

11.4   Stability of an elastic rod subjected to an axial compressive force – Euler buckling load

Consider a rod of cross-sectional area A, length L and made of a material whose stress–strain curve is given by Fig. (11.1.2), such that its flexural rigidity is EI. Assume that the rod is resting on simple supports and that it is subjected to an axial force P acting through the centroid of the cross-section [Fig. (11.4.1a)]. We wish to find the critical load P_{cr} that will cause the rod to buckle. In the context of the previous discussion, we must determine a force P that is required to maintain the rod in a buckled configuration [Fig. (11.4.1b)]. Indeed, as we have seen, the smallest force for which this is possible is the critical force P_{cr} and under this force, the system will be in neutral equilibrium.

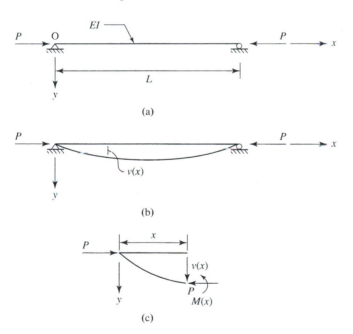

Figure 11.4.1

Let the deflected shape of the rod be given by $v(x)$, where $v(x)$ denotes the lateral displacement of any cross-section. Clearly, when in this configuration, the rod is in

---

[†]   In our subsequent treatment using energy considerations, we shall also determine 'unstable bifurcation points'. Although at this stage of our study the stable and unstable nature of the bifurcation point is irrelevant in obtaining the critical load P_{cr}, we mention here that the critical load P_{cr} of Example 11.1 represents the force at an unstable bifurcation point. This will be shown to be true in Chapter 15 where Example 11.1 is reconsidered in the context of energy considerations.

flexure, and hence its behaviour is governed by the Euler–Bernoulli relation

$$EI\kappa(x) = M(x),\tag{11.4.1}$$

where $\kappa(x)$ is the curvature at any cross-section x. From equilibrium of a free body [Fig. (11.4.1c)], the resulting moment at any cross-section x is

$$M(x) = Pv(x).\tag{11.4.2}$$

Substituting in Eq. (11.4.1) and using the expression for the radius of curvature, Eq. (9.2.1b) of Chapter 9, the explicit equation of the rod is then given by

$$EIv''(x) + \{1 + [v'(x)]^2\}^{3/2}\, v(x)P = 0\tag{11.4.3}$$

Note that this differential equation, which incorporates the equilibrium equations of the rod, is valid for large as well as infinitesimal values of $v(x)$. Solutions to this highly nonlinear equation are given in terms of elliptic integrals that yield a relation between P and $v(x)$. It is worthwhile to note that this nonlinear equation is the analogue to Eq. (11.3.3) for the simple model of the rigid rod studied in Section 11.3 above.

Now, if we are merely interested in obtaining the critical value of the force, P_{cr}, which is required to maintain the rod in a deflected position, this load may be determined, as we have previously seen, by examining equilibrium of the rod in its displaced position as represented by a small perturbation from the original position. Thus, in this case, we shall consider the displacement $v(x)$ to be small (with respect to L), from which it follows that $v'(x)$ must be an infinitesimal, i.e. $|v'(x)| \ll 1$. Hence, following the same reasoning as in Section 2 of Chapter 9, Eq. (11.4.3) yields the linearised equation

$$EIv''(x) = -M(x)\tag{11.4.4a}$$

or

$$EIv''(x) + Pv(x) = 0.\tag{11.4.4b}$$

Note that this linear differential equation is the analogue to the linearised equation, Eq. (11.3.10a), of Section 11.3.

In addition to this equation, one must also satisfy the boundary conditions

$$v(0) = 0,\tag{11.4.5a}$$

$$v(L) = 0.\tag{11.4.5b}$$

It is convenient to divide Eq. (11.4.4b) through by EI; thus we have

$$v''(x) + \lambda^2 v(x) = 0,\tag{11.4.6}$$

where

$$\lambda^2 = \frac{P}{EI}.\tag{11.4.7}$$

In mathematics, the differential equation (11.4.6), together with the associated boundary conditions (11.4.5), is known as a *boundary-value problem*.

The general solution to Eq. (11.4.6) is

$$v(x) = A \sin \lambda x + B \cos \lambda x.\tag{11.4.8}$$

From the boundary condition of (11.4.5a), $v(0) = 0 \rightarrow B = 0$, and hence

$$v(x) = A \sin \lambda x. \qquad (11.4.9)$$

Applying the second boundary condition, Eq. (11.4.5b), one finds

$$A \sin \lambda L = 0. \qquad (11.4.10)$$

Now, if $A = 0$, we obtain the trivial solution $v(x) = 0$ for all x. However, we recall that we are required to investigate the rod *in the displaced configuration*, i.e. $v(x) \neq 0$; therefore A clearly cannot vanish. [It is worthwhile to note that stating that we must satisfy Eq. (11.4.10) with $A \neq 0$ corresponds to satisfying Eq. (11.3.10a) of the preceding section with $\theta \neq 0$.] Therefore, we arrive at the *characteristic equation*

$$\sin \lambda L = 0, \qquad (11.4.11)$$

which, here, is a transcendental equation whose roots are given by

$$\lambda L = n\pi, \quad n = 1, 2, 3, \ldots \qquad (11.4.12a)$$

or

$$\lambda = \frac{n\pi}{L}, \quad n = 1, 2, 3, \ldots \qquad (11.4.12b)$$

Substituting Eq. (11.4.12b) in Eq. (11.4.9),

$$v(x) = A \sin \frac{n\pi x}{L}, \quad n = 1, 2, 3, \ldots \qquad (11.4.13)$$

Hence, we have found that the governing equation, namely Eq. (11.4.6), when subjected to the boundary conditions, Eqs. (11.4.5), can be satisfied only if λ assumes the *discrete* values of Eq. (11.4.12b). Moreover, the buckled configuration must be of the form given by Eq. (11.4.13).

In mathematics, the discrete values of λ are called the *eigenvalues* and the corresponding functions describing $v(x)$ are called the *eigenfunctions*. In terms of mechanics, the functions describing the buckled shape of the rod are called **buckling modes**.

We observe that the constant A remains undetermined. Thus, although we have established the shape of the rod in the buckled configuration, from our analysis, we are unable to determine the magnitude of the displacements (since A, the amplitude of the displacement, remains an unknown). This remark accords with that following Eq. (11.3.10b).

Finally, combining Eq. (11.4.12b) with the definition of λ given by Eq. (11.4.7), and noting that we have examined the neutral equilibrium state, we obtain

$$P_{\text{cr}} = \frac{n^2 \pi^2 E I}{L^2}, \quad n = 1, 2, 3, \ldots \qquad (11.4.14)$$

Thus there exists a set of critical loads $(P_{\text{cr}})_n$, $n = 1, 2, 3, \ldots$. According to our previous discussion, these loads are the axial forces that are required to maintain the rod in the corresponding buckled shapes, namely $v(x) = A \sin(n\pi x/L)$. For example, in order to maintain the rod in the shape $v(x) = A \sin(\pi x/L)$, an axial force not less than $P = (P_{\text{cr}})_1$ must be acting [Fig. (11.4.2a)]. To maintain the rod in the shape $v(x) = A \sin(2\pi x/L)$, a larger axial force, $P = (P_{\text{cr}})_2 = 4(P_{\text{cr}})_1$, must act on the rod [Fig. (11.4.2b)]; a force $P < (P_{\text{cr}})_2$ is insufficient to maintain the rod in the shape $v(x) = A \sin(2\pi x/L)$.

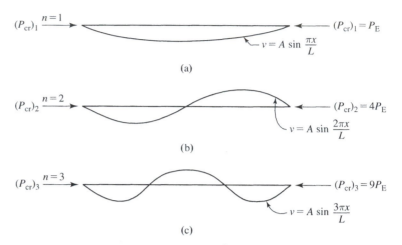

Figure 11.4.2

At this point, several questions can be raised: namely, how does the rod behave if $P > (P_{cr})_1$? Also, if the rod has buckled under the load $P = (P_{cr})_1$ and if the force P is further increased so as to reach $(P_{cr})_2$, will the rod assume the shape $v(x) = A \sin (2\pi x/L)$? To answer these questions, we first note that if the rod buckles in the first mode ($n = 1$), the amplitude of the displacements will increase for loads $P > (P_{cr})_1$, but the rod will *not* assume the form of the second buckling mode, $\sin (2\pi x/L)$. However, from our analysis, we are unable to determine the increased displacements (since the problem has been analysed using the linearised relation which, we recall, precludes the determination of a load–displacement relation). We shall return briefly to this question of 'post buckling' at the end of this section.

In view of the above, how then, can we arrive at the deflection $v(x) = A \sin (2\pi x/L)$ under $(P_{cr})_2$? In order for this equilibrium state to exist, it is initially necessary to prevent the displacement at the mid-point $x = L/2$; this state can be achieved only if a constraint against lateral displacement exists at this point, for example, a brace as in Fig. (11.4.3). Thus, although the solution is also mathematically valid for any $n \geq 2$, it corresponds to a physical situation that can only occur if the rod is braced at points $x = L/n, 2L/n, \ldots, (n-1)L/n$. In the absence of such a bracing, the load at which the straight rod ceases to be stable is given by $P = (P_{cr})_1$.

$P_{cr} = 4P_E$ $4P_E$

Figure 11.4.3

The force $(P_{cr})_1$ – indeed the entire solution to this problem – was first obtained by Euler in 1744. It is therefore customary to define $(P_{cr})_1$ as the Euler buckling load P_E; thus

$$P_E = \frac{\pi^2 E I}{L^2} \qquad (11.4.15)$$

is the smallest critical axial force; when subjected to a force $P \geq P_E$, a straight rod ceases to be in stable equilibrium.

It is clear that $P_E = P_E(L)$. Plotting this as a function of L [Fig. (11.4.4)], we note, as expected, that as $L \to \infty$, $P_E \to 0$; i.e., the critical load decreases to zero as the length increases. On the other hand, according to Eq. (11.4.15), as $L \to 0$, $P_E \to \infty$. Now, while this is mathematically correct, this result is clearly in contradiction with physical reality. Indeed, we know from our previous discussion of Section 11.1 that the maximum load that can be applied to a short rod is a finite force, namely $P = \sigma_y A$. Consequently, it is necessary to investigate this obvious discrepancy between the mathematical solution and physical reality. We first note that corresponding to P_E,

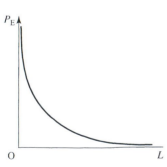

Figure 11.4.4

there exists a compressive stress σ_E,

$$\sigma_E = \frac{P_E}{A} = \frac{\pi^2 E I}{A L^2} = \frac{\pi^2 E}{L^2} \frac{I}{A}. \qquad (11.4.16a)$$

Using the definition of the radius of gyration, r, of the cross-section, $r^2 = I/A$, σ_E is given by

$$\sigma_E = \frac{\pi^2 E}{(L/r)^2}. \qquad (11.4.16b)$$

The ratio L/r is a measure of the slenderness of the rod and is therefore commonly called the **slenderness ratio**. The variation of σ_E with L/r is shown in Fig. (11.4.5).

Now, we recall that we have been investigating the stability of a *linearly elastic* rod for which the stress–strain curve is as given in Fig. (11.1.2). It follows that the stress σ_E as calculated above must fall below σ_p, the proportional limit; otherwise the entire solution is invalid as it contradicts the basic assumption of linear elasticity. Hence the criteria for validity of the elastic solution is $\sigma_E \le \sigma_p$ from which we obtain the domain of validity of the slenderness ratio, namely

$$\frac{L}{r} \ge \pi \sqrt{\frac{E}{\sigma_p}} \equiv \left(\frac{L}{r}\right)_{\text{cr}}. \qquad (11.4.17)$$

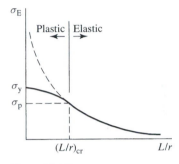

Figure 11.4.5

The stress σ_E for elastic buckling within the domain of physical validity is shown as a solid line in Fig. (11.4.5), while the spurious part of the solution given by Eq. (11.4.16b) is shown by the broken line. Moreover, as $L \to 0$, we have previously established that $\sigma_{\max} = \sigma_y A$. The curve describing the actual behaviour for rods with $L/r \le (L/r)_{\text{cr}}$ is shown by the solid line. For values immediately to the left of the abscissa, $(L/r)_{\text{cr}}$, the rod is said to undergo *plastic buckling*. However, this is beyond the scope of our present study.[†]

While we have been concerned with determining the critical load in the elastic range, the behaviour of a rod under loads $P > P_E$ is often of interest. Such behaviour, as mentioned above, is referred to as *post-buckling behaviour*. Consider, for example, the simple rod in the first buckling mode, $n = 1$, where $\Delta = v(L/2)$ represents the mid-span displacement. From physical reasoning it is clear that Δ will increase with increasing values of P. Although post-buckling behaviour is beyond the scope of our study, we mention here that the load–displacement curve will have the shape as shown in Fig. (11.4.6). From our analysis, where we consider the rod only in the region of the bifurcation point, we have implicitly assumed a behaviour as shown by the dashed line.

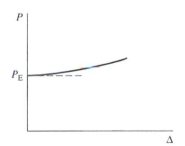

Figure 11.4.6

11.5 Elastic buckling of rods under various boundary conditions

Elastic stability solutions exist for other support conditions; among these are several classical support conditions that are combinations of simple supports, fixed supports and free ends of the rod.

As an example, let us consider the case of a rod, clamped at one end and simply supported at the other [Fig. (11.5.1a)], subjected to an axial compressive load P. From equilibrium, we note that, as opposed to the previous case of a simply supported rod, unknown reactions R exist at A and B [Fig. (11.5.1b)]. The moment

---

[†] While the problem of elastic buckling of a rod was examined by Euler in the 18th century, plastic buckling was only treated in the 20th century, namely by von Karman, Engesser and others.

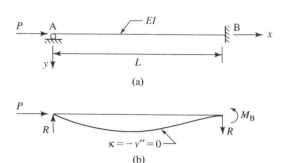

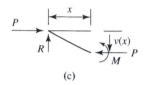

Figure 11.5.1

(c)

$M(x)$ is then [Fig. (11.5.1c)]

$$M(x) = Rx + Pv(x).\qquad(11.5.1)$$

Substituting in Eq. (11.4.4a),

$$EIv''(x) + Pv(x) = -Rx\qquad(11.5.2a)$$

or

$$v''(x) + \lambda^2 v(x) = -\frac{R}{EI}x, \quad \lambda^2 = \frac{P}{EI},\qquad(11.5.2b)$$

with appropriate boundary conditions

$$v(0) = 0\qquad(11.5.3a)$$

$$v(L) = 0\qquad(11.5.3b)$$

$$v'(L) = 0.\qquad(11.5.3c)$$

The general solution to Eq. (11.5.2b) is

$$v(x) = A\sin\lambda x + B\cos\lambda x - \frac{R}{P}x.\qquad(11.5.4)$$

From $v(0) = 0$, $B = 0$ so that

$$v(x) = A\sin\lambda x - \frac{R}{P}x.$$

Differentiating

$$v'(x) = A\lambda\cos\lambda x - \frac{R}{P}$$

and substituting in the boundary condition $v'(L) = 0$, we find $R = A\lambda P\cos\lambda L$. Hence

$$v(x) = A(\sin\lambda x - \lambda x \cdot \cos\lambda L).\qquad(11.5.5)$$

Finally, using the boundary condition $v(L) = 0$, and noting that $A \neq 0$, we obtain

$$\sin\lambda L - \lambda L\cos\lambda L = 0$$

or

$$\tan \lambda L = \lambda L. \tag{11.5.6}$$

Thus, the characteristic equation for the eigenvalue λ, Eq. (11.5.6), is a transcendental equation whose *lowest* root is

$$\lambda L = 4.4934\ldots. \tag{11.5.7}$$

Substituting in the definition of λ^2, $\lambda^2 = P/EI$, we obtain

$$P = (4.4934\ldots)^2 \frac{EI}{L^2}. \tag{11.5.8a}$$

In terms of mathematics, P represents the *lowest* discrete value for which, with boundary conditions given by Eqs. (11.5.3), a non-zero solution $v(x) \neq 0$ exists. In terms of mechanics, it is the smallest force required to maintain the rod in a buckled position; hence, according to our previous discussion, it represents the critical buckling load for this problem.

Upon noting that $4.4934\ldots = 1.4303\pi \simeq \frac{\pi}{0.7}$, we express the load P_{cr} in a convenient form as

$$P_{\mathrm{cr}} = \frac{\pi^2 EI}{(0.7L)^2}. \tag{11.5.8b}$$

We observe again that while the buckling mode is given by Eq. (11.5.5), the amplitude remains unknown, since we cannot solve for the constant A. It is of interest to give an interpretation to the factor '0.7L' appearing in Eq. (11.5.8b). To do so, we investigate the curvature of the rod in the buckled shape. Differentiating Eq. (11.5.5) twice, we obtain

$$v''(x) = -A\lambda^2 \sin \lambda x. \tag{11.5.9}$$

Now, since $A \neq 0$, $v''(x) = 0$ for values $x = 0$ and $\lambda x = \pi$. Hence

$$\frac{x}{L} = \frac{\pi}{\lambda L} = \frac{\pi}{4.4934\ldots} \simeq 0.7.$$

We thus conclude that at $x/L \simeq 0.7$, $v''(x) = 0$; that is, the buckling mode has a point of zero curvature at $x \simeq 0.7L$, as shown in Fig. (11.5.1b). Since the moment is proportional to the curvature, $M(x) = 0$ at this cross-section. This provides us with an immediate interpretation of the '0.7' factor: namely, since the moments at $x = 0$ and $x = 0.7L$ are zero, the value of the critical load P_{cr} for the present case is the same as that of a simply supported rod of length $0.7L$.

Following the same procedure, the critical load of a rod with fixed and free ends, subjected to an axial force P [Fig. (11.5.2a)], is found (upon solving the governing second-order differential equation with the appropriate boundary conditions) to be

$$P_{\mathrm{cr}} = \frac{\pi^2 EI}{4L^2}. \tag{11.5.10}$$

The corresponding buckling mode is

$$v(x) = A\left(1 - \cos\frac{\pi x}{2L}\right), \tag{11.5.11a}$$

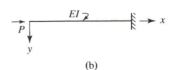

(a)

(b)

Figure 11.5.2

where, here, x is measured from the fixed end.[†] Alternatively, if x is measured from the free end [see Fig. (11.5.2b)], we obtain the same value for P_{cr} and $v(x)$ is expressed as

$$v(x) = A \left(\sin \frac{\pi x}{2L} - 1 \right). \tag{11.5.11b}$$

Note that Eqs. (11.5.11a) and (11.5.11b) are equivalent; i.e., the shape of the buckling mode is the same.

The critical loads for several classical cases are summarised in Figs. (11.5.3) where the shapes of the buckling modes are shown. We note that the expressions for the critical loads can all be written in the same form, namely[‡]

$$P_{cr} = \frac{\pi^2 EI}{(\alpha L)^2}, \tag{11.5.12}$$

where α is given in Figs. (11.5.3) for each case.

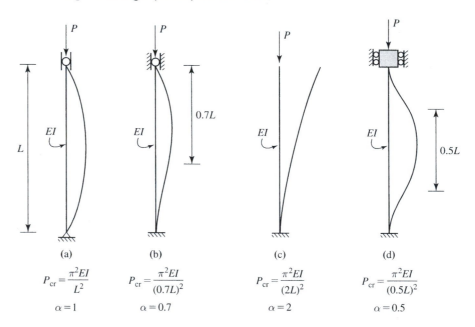

Figure 11.5.3

$$P_{cr} = \frac{\pi^2 EI}{L^2} \qquad P_{cr} = \frac{\pi^2 EI}{(0.7L)^2} \qquad P_{cr} = \frac{\pi^2 EI}{(2L)^2} \qquad P_{cr} = \frac{\pi^2 EI}{(0.5L)^2}$$

$$\alpha = 1 \qquad\qquad \alpha = 0.7 \qquad\qquad \alpha = 2 \qquad\qquad \alpha = 0.5$$

(a)   (b)   (c)   (d)

We observe that the critical loads for all these classical cases are obtained using the same basic method of analysis: equilibrium equations are written in a deflected state under the assumption of infinitesimal displacements thus yielding linear differential equations. These, together with the appropriate boundary conditions then lead to the characteristic equation of an eigenvalue problem. As in the cases previously considered, the characteristic equations are generally transcendental equations whose roots (eigenvalues) correspond to the critical loads.

While the above are the usual classical conditions, evidently other cases may exist. For example, let us consider the case of a simply supported rod ACB containing an interior support at C located at some arbitrary point $a = \gamma L$, as shown in

[†] While this result can be easily established using Eq. (11.4.4a), we shall obtain this solution in Section 11.9 using a fourth-order differential equation.

[‡] Solutions for the case of a rod of length L, fixed at either one or both ends, using Eq. (11.4.4a), are left as problems (see Problems 11.13 and 11.14).

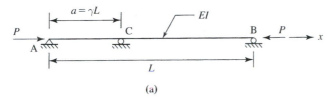

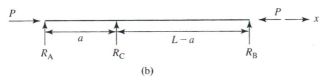

Figure 11.5.4

Fig. (11.5.4a). Clearly, the critical load will depend on the location of C. We recall from our previous discussion that if C is located at the mid-span, $a = L/2$, the critical load is $P_{cr} = 4\pi^2 EI/L^2$ [Fig. (11.4.3)].

The analysis proceeds as before; however, we note that there now exist non-zero reactions R_A, R_B and R_C as in Fig. (11.5.4b).

Taking moments about point C, we have the relation

$$aR_A = (L - a)R_B. \tag{11.5.13}$$

We observe, however, that the structure is statically indeterminate since clearly we cannot solve for the three unknown reactions with two equations of equilibrium.

To determine P_{cr}, we first write, as before, the moment expressions $M(x)$ in terms of P and the unknown reactions, e.g. R_A and R_B. However, note that different expressions for $M(x)$ exist in the two domains $D_1: \{x \mid 0 < x < a\}$ and $D_2: \{x \mid a < x < L\}$; hence we obtain two separate second-order differential equations and therefore two general solutions, $v_1(x)$ and $v_2(x)$, each of which contains two constants of integration. Thus, we have effectively five unknowns: the four constants and an unknown reaction, for example, R_A [since by Eq. (11.5.13), R_B can be expressed in terms of R_A]. However, there exist five boundary conditions, namely

$$v_1(0) = v_1(a) = v_2(a) = v_2(L) = 0 \quad \text{and} \quad v_1'(a) = v_2'(a). \tag{11.5.14}$$

Substituting the general solutions in the above boundary conditions yields the characteristic equation for the eigenvalue $\lambda^2 = P/EI$:

$$\gamma\lambda L(1 - \gamma)\sin(\lambda L) - \sin(\gamma\lambda L) \cdot \sin[(1 - \gamma)\lambda L] = 0, \tag{11.5.15}$$

where $\gamma \equiv a/L$.

When $\gamma = 0.5$ $(a = L/2)$, the above equation is reduced to

$$\sin\left(\frac{\gamma L}{2}\right)\left[\tan\left(\frac{\lambda L}{2}\right) - \frac{\lambda L}{2}\right] = 0, \tag{11.5.16}$$

whose lowest root yields the critical load $P_{cr} = 4\pi^2 EI/L^2$ as found above.

When $\gamma \to 0$ $(a \to 0)$, Eq. (11.5.15) reduces to the characteristic equation, Eq. (11.5.6), i.e.

$$\tan \lambda L = \lambda L, \tag{11.5.17}$$

whose lowest root was seen to yield the critical load $P_{cr} = \pi^2 EI/(0.7L)^2$. We recall that this is the critical load for the structure of Fig. (11.5.3b). A clear physical explanation for this result is shown by means of Fig. (11.5.5): as the interior support C approaches point A, the slope at A necessarily approaches zero. We recognise that the condition of zero displacement and slope at a given point corresponds to a fixed-end support as shown in Fig. (11.5.3b).

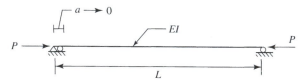

Figure 11.5.5

As another example, consider a rod ACB, fixed at A and free at B, consisting of two segments AC and CB with flexural rigidities EI and αEI ($0 < \alpha$), rigidly attached at B, as shown in Fig. (11.5.6). An axial compressive force acts at B through the centroid. In this case, the governing differential equations are $EIv_1''(x) = -M(x)$ and $\alpha EIv_2''(x) = -M(x)$, respectively, in the two domains $D_1: \{x \mid 0 < x < \gamma L\}$ and $D_2: \{x \mid \gamma L < x < L\}$. Using the appropriate boundary conditions and conditions at B[$v_1(\gamma L) = v_2(\gamma L)$ and $v_1'(\gamma L) = v_2'(\gamma L)$], one obtains the characteristic equation

$$\tan(\gamma \lambda L) = \cot\left[\frac{(1-\gamma)\lambda L}{c}\right], \quad \lambda = \sqrt{P/EI}, \ c = \sqrt{\alpha}. \quad (11.5.18)$$

The development is left as an exercise (see Problem 11.39).

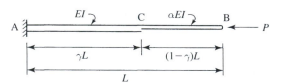

Figure 11.5.6

In the derivation of the above solutions, we have considered rods only for the cases of an axial compressive force passing through the centroid. Now, it is clear that such rods can also be subjected to axial loads that are applied with an eccentricity with respect to the centroidal axes. Moreover, rods may also be subjected simultaneously to both axial and lateral loads. In the following sections, we investigate the behaviour under these loading conditions.

11.6 Rods under eccentric axial loads – the 'secant formula'

In the previous development, it was assumed that the axial compressive force passes through the centroid of a perfectly straight rod. However, in fact, this is an idealisation since no rod is perfectly straight. Moreover, although structures are often designed based on this idealisation, the actual loading may be applied eccentrically with respect to the centroid, either due to imperfections in the rod or misalignments. In fact, very often, it is known *a priori*, that the load is applied eccentrically.

We therefore study the behaviour of a rod under an eccentrically applied axial load. In this case, the ability of a rod or column to withstand an eccentric load can be determined based on an allowable stress. As we shall see, the development also leads to a new mathematical interpretation of instability, which proves to be very useful.

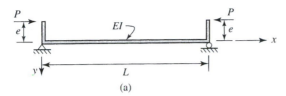

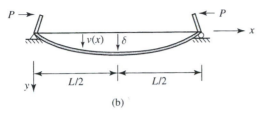

Figure 11.6.1

Consider a simply supported rod of length L and flexural rigidity EI, pinned at the ends and subjected to a load P applied with an eccentricity e, as shown in Fig. (11.6.1a). Due to P, the rod deforms, as shown in Fig. (11.6.1b). Clearly, the moment at any point x is

$$M(x) = P[e + v(x)] \qquad (11.6.1a)$$

and the governing differential equation is

$$EIv''(x) + Pv(x) = -Pe \qquad (11.6.2a)$$

or

$$v''(x) + \lambda^2 v(x) = -\lambda e, \quad \lambda = \sqrt{P/EI} \qquad (11.6.2b)$$

subject to the boundary conditions $v(0) = v(L) = 0$.
The solution to Eq. (11.6.2b) is

$$v(x) = A \sin \lambda x + B \cos \lambda x - e. \qquad (11.6.3)$$

Using B.C. $v(0) = 0$, $B = e$, while $v(L) = 0$ leads to

$$A = \frac{e(1 - \cos \lambda L)}{\sin \lambda L}, \qquad (11.6.4a)$$

which, using standard trigonometric identities, can be rewritten as

$$A = e \tan\left(\frac{\lambda L}{2}\right). \qquad (11.6.4b)$$

Hence

$$v(x) = e\left[\tan\left(\frac{\lambda L}{2}\right)\sin \lambda x + \cos \lambda x - 1\right]. \qquad (11.6.5)$$

It is interesting to note here that the constants of integration are known explicitly in terms of the given eccentricity e and thus, in contradistinction, for example, with Eq. (11.4.13), the *magnitude* of $v(x)$ is known. [This is true since, clearly, as opposed to the case of a non-eccentric load, the rod deforms for *any* load $P > 0$.] By symmetry, the maximum deflection occurs at the centre, $x = L/2$. Hence, letting

$$\delta \equiv v(L/2),$$

$$\delta = e\left[\frac{\sin^2(\lambda L/2)}{\cos(\lambda L/2)} + \cos(\lambda L/2) - 1\right] \qquad (11.6.6a)$$

or

$$\delta = e\left[\sec\left(\frac{\lambda L}{2}\right) - 1\right], \qquad (11.6.6b)$$

where, by definition, $\sec(\lambda L/2) \equiv [\cos(\lambda L/2)]^{-1}$.

Before proceeding with the development, we observe an interesting feature: If $e \neq 0$, then if $\lambda L = \pi$, $\sec(\lambda L/2) \to \infty$, i.e. $\delta \to \infty$.[†] Now, for this value of λ, we have $P = \frac{\pi^2 EI}{L^2}$, which is precisely the Euler buckling load P_E that causes instability. Thus, we arrive at a *mathematical* interpretation of a load causing instability, namely *it is the load that causes the deflection to approach infinity*.[‡] This is readily seen from a $\delta - P$ plot, as shown in Fig. (11.6.2), where we observe that for $e = 0$, the deflection $\delta = 0$ for $P < P_{cr}$ while if $e > 0$, δ increases both with P and with increasing values of e, and approaches infinity as $P \to P_E$.

Now, the maximum moment in the rod is $M_{max} = P(\delta + e)$. Hence, the maximum compressive flexural stress σ_x that occurs in the rod is [see Eq. (8.12.2)]

$$\sigma_{max} = \frac{P}{A} + \frac{M_{max}c}{I} = \frac{P}{A}\left[1 + \frac{P(\delta + e)c}{r^2}\right], \qquad (11.6.7)$$

where A and $r = \sqrt{I/A}$ are the area and radius of gyration of the section, respectively, and c is the largest distance from the neutral axis to a fibre in compression.[§] Substituting δ from Eq. (11.6.6b), we have

$$\sigma_{max} = \frac{P}{A}\left[1 + \frac{ec}{r^2}\sec\left(\frac{\lambda L}{2}\right)\right] \qquad (11.6.8a)$$

and using the definition of λ, $\lambda = \sqrt{P/EI}$, we obtain finally

$$\sigma_{max} = \frac{P}{A}\left[1 + \frac{ec}{r^2}\sec\left(\frac{L}{2r}\sqrt{\frac{P}{EA}}\right)\right]. \qquad (11.6.8b)$$

This expression, which relates the maximum flexural stress (due to an eccentrically applied load) to the average stress, $\sigma_{ave} \equiv P/A$, due to a load applied through the centroid of the section, is known as the **secant formula** for columns. We note, in passing, that the relation between σ_{max} and P is highly nonlinear.

Now, throughout the development, elastic behaviour has been assumed; the derived expressions are therefore valid provided that $\sigma \leq \sigma_p$, where σ_p is the proportional limit of the rod material. Let us consider, for example, the behaviour of a steel rod. In this case, the yield stress σ_y differs but little from σ_p, i.e. $\sigma_y \sim \sigma_p$. In

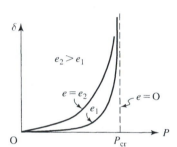

Figure 11.6.2

[†] If $e = 0$, then clearly $\delta = 0$ for all values of $\lambda \neq \pi L$ and if $\lambda = \pi L$, the value of δ, being the product of zero and infinity, is undetermined. This is consistent with the undetermined value of the deflection for a concentric axial loading [see Eq. (11.4.13)].

[‡] We note that, using linearised theory, a deflection $|v(x)| \geq 1$ is clearly not physically relevant and therefore $\delta \to \infty$ is physically meaningless. Nevertheless, the association $\delta \to \infty$ with instability of a system is mathematically valid.

[§] Note that here the radius of gyration does not necessarily correspond to the minimum I of the section (as is the case of a concentrically loaded rod) since bending occurs in a plane perpendicular to the neutral axis, which depends on the given eccentricity of the axial load.

such a case, upon setting $\sigma_{\max} = \sigma_y$, we may write

$$\sigma_y = \frac{P}{A}\left[1 + \frac{ec}{r^2}\sec\left(\frac{L}{2r}\sqrt{\frac{P}{EA}}\right)\right], \qquad (11.6.9a)$$

and therefore

$$\sigma_{\text{ave}} \equiv \frac{P}{A} = \frac{\sigma_y}{\left[1 + \frac{ec}{r^2}\sec\left(\frac{L}{2r}\sqrt{\frac{\sigma_{\text{ave}}}{E}}\right)\right]}. \qquad (11.6.9b)$$

A typical plot showing a family of curves of σ_{ave} vs. L/r is shown in Fig. (11.6.3) for a steel column with $\sigma_y = 240$ MPa and $E = 200$ GPa.[†] Such figures are often used, together with appropriate safety factors, in the design of columns and form the basis of a number of empirical design formulas. Finally, it is worthwhile to observe from the figure that the eccentricity e has a relatively small effect for very long columns and becomes increasing significant for shorter columns.

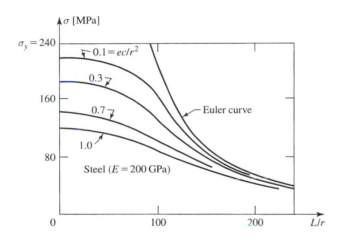

Figure 11.6.3

11.7   Rods under combined axial and lateral loads: preliminary remarks

We now consider structural members, as shown in Fig. (11.7.1a), subjected to both lateral loads and an axial force. We recall that we have treated, in fact, such a case in Chapter 8. However, lateral displacements were neglected when calculating the effect of the axial force on the moments in the member (see e.g. Example 8.18).

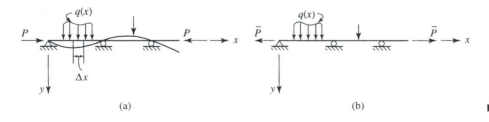

(a)   (b)   Figure 11.7.1

[†] Note that since the quantity 'σ_{ave}' appears on both sides of Eq. (11.6.9b), the curves can only be obtained numerically.

For cases where these displacements are sufficiently small such that the effect of the axial load on the resulting moment is insignificant (compared to the effect of the lateral loads), the solutions thus obtained are reasonably accurate. However, such solutions are only approximate if the axial force has a considerable influence on the moments. In the treatment below, we shall therefore obtain more accurate solutions by taking into account the lateral displacements. Members that are treated in this manner are usually called **beam-columns** [Fig. (11.7.1a)] when the axial load is compressive, and are called **tie-rods** when the axial force is in tension [Fig. (11.7.1b)].

In particular, we shall be interested in determining the effect that the axial forces have on the displacements, moments and internal shearing forces. In order to determine the behaviour of such members, we must first derive the basic governing differential equations of beams under the given loading conditions.

11.8 Differential equations of beams subjected to combined lateral loads and axial forces

We consider an elastic member with flexural rigidity $EI(x)$, located in the x–y plane, where the positive y-direction is taken as downward [Fig. (11.7.1)]. Lateral displacements $v(x)$ are defined as positive in the positive y-direction. We again limit our treatment to beams for which the lateral displacements are small compared to the span lengths so that the slopes $\theta = v'$ are infinitesimals, i.e. $|v'| \ll 1$. Furthermore, we retain the same sign convention for the moments $M(x)$, the shear forces and the lateral loads $q(x)$, as defined in Section 2(b) of Chapter 8. Let us recall that the beam equations, Eqs. (8.3.1) and (8.3.2), for the case where no axial force is present, are

$$\frac{\mathrm{d}V(x)}{\mathrm{d}x} = -q(x), \tag{11.8.1a}$$

$$\frac{\mathrm{d}M(x)}{\mathrm{d}x} = V(x), \tag{11.8.1b}$$

where $M(x)$ and $V(x)$ are the moments and shear forces at a cross-section and the Euler–Bernoulli relation is

$$EI(x)v''(x) = -M(x). \tag{11.8.1c}$$

The governing equation for the member, obtained by combining eqs. (11.8.1a), (11.8.1b) and (11.8.1c) is then

$$\frac{\mathrm{d}^2}{\mathrm{d}x^2}\left[EI(x)v''(x)\right] = q(x). \tag{11.8.2}$$

We now turn our attention to members that are also subjected to axial forces. It is worthwhile first to note that the Euler–Bernoulli relation, $EI(x)v''(x) = -M(x)$, remains valid if axial forces P act on the member. Here, P is taken as a compressive force.

Consider an element Δx of the member of Fig. (11.7.1a) in the deflected state [Fig (11.8.1)]. Let $Q(x)$ denote the *vertical* component of the shear force acting on the cross-section according to the usual sign convention.

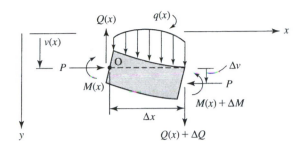

Figure 11.8.1

From equilibrium in the y-direction, $\sum F_y = 0$,

$$-Q(x) + [Q(x) + \Delta Q] + q(\bar{x})\,\Delta x = 0, \qquad (11.8.3)$$

where $x \leq \bar{x} < x + \Delta x$. Therefore, upon dividing through by Δx and taking the limit,

$$\frac{\mathrm{d}Q(x)}{\mathrm{d}x} = -q(x) \qquad (11.8.4)$$

since $\Delta Q \to 0$ and $\bar{x} \to x$ as $\Delta x \to 0$.

Observing that the axial force P is constant, moment equilibrium about point O yields

$$\sum M_0 = -M(x) + [M(x) + \Delta M] - P\Delta v - [Q(x) + \Delta Q]\,\Delta x$$
$$- [q(\bar{x})\,\Delta x]\alpha\,\Delta x = 0, \qquad (11.8.5\mathrm{a})$$

where $0 < \alpha < 1$.

Dividing through by Δx and taking the limit as $\Delta x \to 0$,

$$\lim_{\Delta x \to 0} \frac{\Delta M}{\Delta x} = \lim_{\Delta x \to 0} \left[Q(x) + \Delta Q + P\frac{\Delta v}{\Delta x} + \alpha q(\bar{x})\,\Delta x \right]$$

and therefore

$$\frac{\mathrm{d}M(x)}{\mathrm{d}x} = Q(x) + Pv'(x). \qquad (11.8.5\mathrm{b})$$

Letting

$$V(x) \equiv Q(x) + Pv'(x), \qquad (11.8.5\mathrm{c})$$

we obtain the familiar form

$$\frac{\mathrm{d}M(x)}{\mathrm{d}x} = V(x). \qquad (11.8.5\mathrm{d})$$

From Eqs. (11.8.1c) and (11.8.5d), it follows that

$$\frac{\mathrm{d}}{\mathrm{d}x}[EI(x)v''(x)] = -\frac{\mathrm{d}M(x)}{\mathrm{d}x} = -V(x) \qquad (11.8.6)$$

and therefore

$$\frac{\mathrm{d}^2}{\mathrm{d}x^2}[EI(x)v''(x)] = -\frac{\mathrm{d}}{\mathrm{d}x}[Q(x) + Pv'(x)]$$

or

$$\frac{d^2}{dx^2}[EI(x)v''(x)] = q(x) - Pv''(x) \qquad (11.8.7a)$$

by Eq. (11.8.4).

When $EI = $ constant, this reduces to the simple equation

$$EI\frac{d^4v(x)}{dx^4} + P\frac{d^2v(x)}{dx^2} = q(x). \qquad (11.8.7b)$$

In order to interpret the term V appearing above [defined by Eq. (11.8.5c)], we recall that for infinitesimal slopes v', $\cos(v') \simeq 1$ and $\sin(v') \simeq v'$. Hence, we note that Eq. (11.8.5c) is actually the limiting case of

$$V(x) = Q(x)\cos(v') + P\sin(v'). \qquad (11.8.8)$$

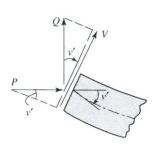

Figure 11.8.2

We observe from Fig. (11.8.2) that V is therefore the resultant shear force acting tangentially on the *deformed* cross-section.

Note that if $P = 0$, then $V(x) = Q\cos(v') \simeq Q$ for *infinitesimal* v'; that is, the vertical shear force Q and the shear force V become (almost) identical. (Thus, in our study of beams in Chapter 8, it was neither necessary nor possible to distinguish, as we now do, between Q and V and therefore we denoted the shear force simply by V. However, based on the present development, it becomes clear that when one refers to the shear force V, one means the force acting tangentially to the cross-section in the deformed state.)

In analysing a member as a beam-column, one has the choice of using either the second-order differential equation, Eq. (11.8.1c), or the fourth-order differential equation, Eq. (11.8.7a) or (11.8.7b). Clearly, the second-order differential equation has a simpler solution and might seem to be preferred. However, if one starts with this equation, it is first necessary to establish an expression for the moment $M(x)$. At times, this is not convenient and rather complicated. In such cases, it may be preferable to use the fourth-order equation since the inhomogeneous term $q(x)$ is a *known* quantity representing the applied loads. Note, however, that in using the second-order equation, proper boundary conditions are only on the displacement v or the slope v'. Additional appropriate boundary conditions taking into account shear and moment conditions must be used when working with the fourth-order equation.[†]

11.9 Stability analysis using the fourth-order differential equation

In Sections 4 and 5 of this chapter, the stability of rods under a compressive axial load was analysed using the second-order differential equation, Eq. (11.4.4a). As mentioned above, we may alternatively use the fourth-order equation, namely Eqs. (11.8.7) by setting the lateral load $q(x) = 0$.

For this case, therefore, Eq. (11.8.7a) reduces to

$$\frac{d^2}{dx^2}[EI(x)v''(x) + Pv(x)] = 0 \qquad (11.9.1a)$$

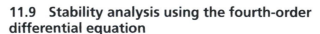

[†] We recall here that the appropriate boundary conditions for an nth-order differential equation can involve only derivatives up to order $(n-1)$.

and for a prismatic rod with $EI = $ constant,

$$EIv^{iv}(x) + Pv''(x) = 0. \tag{11.9.1b}$$

From Eqs. (11.8.1c) and (11.8.6) for the prismatic rod, we have

$$EIv''(x) = -M(x), \tag{11.9.2a}$$

$$EIv'''(x) = -V(x). \tag{11.9.2b}$$

As an example, let us now consider, using the governing equation, Eq. (11.9.1b), the stability of a prismatic rod with free and fixed ends subjected to an axial compressive force P, as shown in Fig. (11.5.2b).

At the free end, $x = 0$, $M(0) = 0$ and the vertical component of the shear force acting on the section in the deformed state is $Q = 0$. Therefore, from Eq. (11.8.5c), $V(0) = Pv'(0)$.

The boundary conditions at the free end $x = 0$ are then

$$v''(0) = 0, \tag{11.9.3a}$$

$$EIv'''(0) = -Pv'(0), \tag{11.9.3b}$$

while at the fixed end, we have

$$v(L) = 0, \tag{11.9.3c}$$

$$v'(L) = 0. \tag{11.9.3d}$$

We therefore must solve the differential equation, Eq. (11.9.1b), subject to the boundary conditions of Eqs. (11.9.3).

As before, dividing Eq. (11.9.1b) through by EI, we have

$$v^{iv}(x) + \lambda^2 v''(x) = 0, \quad \lambda^2 = \frac{P}{EI}, \tag{11.9.4}$$

whose general solution is

$$v(x) = A\sin\lambda x + B\cos\lambda x + Cx + D. \tag{11.9.5}$$

Upon differentiating,

$$v'(x) = A\lambda\cos\lambda x - B\lambda\sin\lambda x + C, \tag{11.9.6a}$$

$$v''(x) = -A\lambda^2\sin\lambda x - B\lambda^2\cos\lambda x, \tag{11.9.6b}$$

$$v'''(x) = -A\lambda^3\cos\lambda x + B\lambda^3\sin\lambda x. \tag{11.9.6c}$$

Note that, from the definition of λ, we may write (11.9.6c) as

$$v'''(x) = \frac{P\lambda}{EI}(-A\cos\lambda x + B\sin\lambda x). \tag{11.9.6c'}$$

Using Eqs. (11.9.3a) and (11.9.6b), $v''(0) = 0$, $B = 0$.

Using Eq. (11.9.3b), $[EIv'''(0) = -Pv'(0)]$,[†] and Eq. (11.9.6a), we obtain $-A\lambda = -(A\lambda + C)$, from which it follows that $C = 0$.

From Eq. (11.9.3d), $v'(L) = 0$ and using Eq. (11.9.6a), we have $A\lambda\cos\lambda L = 0$ and since $\lambda \neq 0$, we obtain the characteristic equation

$$\cos\lambda L = 0, \tag{11.9.7}$$

[†] It is worthwhile to note again that at the free end the shear force $V \neq 0$.

whose roots are

$$\lambda = \frac{n\pi}{2L}, \quad n = 1, 3, 5, \dots. \tag{11.9.8}$$

Then from Eq. (11.9.3c), $v(L) = 0$, we have $A \sin \lambda L + D = 0$ or $D = -A \sin \lambda L = -A \sin (n\pi/2)$. Hence for the lowest root, $n = 1$, $D = -A$.

Using this root, it follows directly that

$$P_{\text{cr}} = EI\lambda^2 = \frac{\pi^2 EI}{4L^2}. \tag{11.9.9}$$

The corresponding buckling mode is then

$$v(x) = A\left(\sin\frac{\pi x}{2L} - 1\right). \tag{11.9.10}$$

We observe that these results were given by Eqs. (11.5.10) and (11.5.11).

11.10 Beam-column subjected to a single lateral force *F* and an axial compressive force *P*

Consider the beam-column with constant EI, as shown in Fig. (11.10.1a). Since the moments in this beam are expressed rather easily, we shall analyse this case using Eq. (11.8.1c) rather than Eq. (11.8.7b). From equilibrium, the moments in the assumed deflected configuration are [Fig. (11.10.1b)]

$$M(x) = \frac{F}{L}(L - a)x + Pv, \quad 0 \le x \le a \tag{11.10.1a}$$

$$M(x) = \frac{Fa}{L}(L - x) + Pv, \quad a \le x \le L. \tag{11.10.1b}$$

Substituting in Eq. (11.8.1c), we obtain separate differential equations in the two domains, namely

$$EIv_1''(x) + Pv_1(x) = -\frac{F}{L}(L - a)x, \quad 0 < x < a, \tag{11.10.2a}$$

$$EIv_2''(x) + Pv_2(x) = -\frac{Fa}{L}(L - x), \quad a < x < L, \tag{11.10.2b}$$

where v_1 and v_2 denote the deflections in the two domains, respectively.

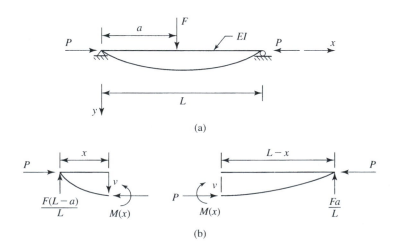

(a)

(b)

Figure 11.10.1

Now, dividing through by EI and setting

$$\lambda^2 = P/EI, \tag{11.10.3}$$

we have

$$v_1''(x) + \lambda^2 v_1(x) = -\frac{F(L-a)x}{EIL}, \quad 0 < x < a, \tag{11.10.4a}$$

$$v_2''(x) + \lambda^2 v_2(x) = -\frac{Fa(L-x)}{EIL}, \quad a < x < L. \tag{11.10.4b}$$

The general solutions to these two equations are then

$$v_1(x) = A_1 \sin \lambda x + B_1 \cos \lambda x - \frac{F(L-a)x}{PL}, \tag{11.10.5a}$$

$$v_2(x) = A_2 \sin \lambda x + B_2 \cos \lambda x - \frac{Fa(L-x)}{PL}, \tag{11.10.5b}$$

where A_1, A_2, B_1 and B_2 are constants of integrations. Appropriate boundary conditions for this problem are

$$v_1(0) = 0, \tag{11.10.6a}$$

$$v_2(L) = 0 \tag{11.10.6b}$$

and

$$v_1(a) = v_2(a), \tag{11.10.6c}$$

$$v_1'(a) = v_2'(a). \tag{11.10.6d}$$

The latter boundary conditions represent continuity of the deflection and slope, respectively, at $x = a$. From Eqs. (11.10.6a) and (11.10.6b) we find, respectively,

$$v_1(0) = B_1 = 0 \tag{11.10.7a}$$

and

$$B_2 = -\frac{\sin \lambda L}{\cos \lambda L} A_2 = -\tan \lambda L \, A_2. \tag{11.10.7b}$$

Hence,

$$v_1(x) = A_1 \sin \lambda x - \frac{F(L-a)x}{PL} \tag{11.10.8a}$$

$$v_2(x) = A_2(\sin \lambda x - \tan \lambda L \cos \lambda x) - \frac{Fa}{PL}(L-x). \tag{11.10.8b}$$

Noting that

$$v_1'(x) = A_1 \lambda \cos \lambda x - \frac{F(L-a)}{PL} \tag{11.10.9a}$$

$$v_2'(x) = A_2 \lambda (\cos \lambda x + \tan \lambda L \sin \lambda x) + \frac{Fa}{PL}, \tag{11.10.9b}$$

substituting in Eqs. (11.10.6c) and (11.10.6d) and using simple trigonometric identities, we obtain, after some algebraic manipulations,

$$A_1 = \frac{F}{P\lambda} \frac{\sin \lambda (L-a)}{\sin \lambda L} \tag{11.10.10a}$$

and

$$A_2 = -\frac{F}{P\lambda} \frac{\sin \lambda a}{\tan \lambda L}. \tag{11.10.10b}$$

Therefore, finally,

$$v_1(x) = \frac{F}{P\lambda} \frac{\sin \lambda (L-a)}{\sin \lambda L} \sin \lambda x - \frac{F(L-a)}{PL} x, \quad 0 \le x \le a \tag{11.10.11a}$$

and

$$v_2(x) = \frac{F}{P\lambda} \frac{\sin \lambda a}{\sin \lambda L} \sin \lambda (L-x) - \frac{Fa}{PL}(L-x), \quad a \le x \le L. \tag{11.10.11b}$$

For convenience, and future use, we note that

$$v_1'(x) = \frac{F}{P} \frac{\sin \lambda (L-a)}{\sin \lambda L} \cos \lambda x - \frac{F(L-a)}{PL} \tag{11.10.12a}$$

$$v_1''(x) = -\frac{F\lambda}{P} \frac{\sin \lambda (L-a)}{\sin \lambda L} \sin \lambda x \tag{11.10.12b}$$

$$v_1'''(x) = -\frac{F\lambda^2}{P} \frac{\sin \lambda (L-a)}{\sin \lambda L} \cos \lambda x. \tag{11.10.12c}$$

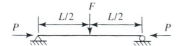

Figure 11.10.2

Similar expressions exist in the region $a \le x \le L$.

Now, it is of interest to examine the behaviour of the beam-column in order to determine the effect of the thrust P. For convenience, we shall consider the case when F acts in the centre, i.e. $a = L/2$ [Fig. (11.10.2)]. Equation (11.10.11a) then becomes

$$v_1(x) = \frac{F}{P\lambda} \frac{\sin \lambda L/2}{\sin \lambda L} \sin \lambda x - \frac{F}{2P} x. \tag{11.10.13}$$

Note that $v_2(x)$ has a similar expressions with x replaced by $(L-x)$. At $x = L/2$ (where $v_1 = v_2$),

$$v(L/2) = \frac{F}{P\lambda} \frac{\sin^2(\lambda L/2)}{\sin \lambda L} - \frac{FL}{4P}. \tag{11.10.14}$$

It is convenient to define the parameter

$$\mu = \frac{\lambda L}{2} = \frac{L}{2}\sqrt{\frac{P}{EI}}. \tag{11.10.15a}$$

Then

$$\mu^2 = \frac{PL^2}{4EI}, \tag{11.10.15b}$$

$$P = \frac{4EI}{L^2}\mu^2, \tag{11.10.15c}$$

$$P\lambda = \frac{8EI\mu^3}{L^3} \tag{11.10.15d}$$

and therefore

$$v\left(\frac{L}{2}\right) = \frac{FL^3}{8EI\mu^3} \frac{\sin^2 \mu}{\sin 2\mu} - \frac{FL^3}{16EI\mu^2}. \qquad (11.10.16a)$$

Since $\sin 2\mu = 2 \sin \mu \cos \mu$, we rewrite this equation as

$$v\left(\frac{L}{2}\right) = \frac{FL^3}{48EI} \left[\frac{3(\tan \mu - \mu)}{\mu^3}\right]. \qquad (11.10.16b)$$

We recognise immediately that $\delta = \frac{FL^3}{48EI}$ represents the mid-span deflection of a simple beam when no thrust P exists [see Eq. (9.4.20)]. Thus

$$v\left(\frac{L}{2}\right) = \delta \left[\frac{3(\tan \mu - \mu)}{\mu^3}\right]. \qquad (11.10.17)$$

The effect of P is then clearly given by the bracketed term $\left[\frac{3(\tan \mu - \mu)}{\mu^3}\right]$.

While the analysis is complete at this point, the expressions in terms of P do not provide any significant insight for a physical interpretation. (For example, does it have any meaning to give a result for, let's say, $P = 1000$ N or $P = 2000$ N?) Clearly, one needs to relate the force P to some known reference axial force. The most appropriate force, in this case, is the Euler buckling load $P_E = \pi^2 EI/L^2$, from which we can construct the ratio P/P_E. With this in mind, we may write, from Eq. (11.10.15a),

$$\mu = \frac{\pi}{2} \sqrt{\frac{P}{P_E}}. \qquad (11.10.18)$$

Using this relation, one may now calculate the effect on $v(L/2)$ for various ratios P/P_E. Thus, for example, if $P/P_E = 0.25$, $\mu = \pi/4$ and $\tan \mu = 1$. Hence, for this ratio of P/P_E, from Eq. (11.10.17), we find $v(L/2) = 1.34\delta$; i.e., an axial force which is 25% of P_E will increase the centre-span deflection by 34%. For $P/P_E = 0.50$, $v(L/2) = 1.99\delta$.

Let us also consider the following two limiting cases:

- $P \to P_E$: Then $\mu \to \pi/2$ and $\tan \mu \to \infty$. Therefore $v(L/2) \to \infty$. We thus observe, as was shown in Section 11.6, that one may associate the critical load leading to instability as the load which causes the deflection to approach infinity.
- $P \to 0$: Then $\mu \to 0$. Noting that the bracketed term is undefined at $\mu = 0$, we make use of the series expansion for $\tan \mu$; thus

$$\frac{3(\tan \mu - \mu)}{\mu^3} = \frac{3}{\mu^3}[(\mu + \mu^3/3 + 2\mu^5/15 + \cdots) - \mu] \qquad (11.10.19)$$

and hence in the limit, as $\mu \to 0$,

$$\frac{3(\tan \mu - \mu)}{\mu^3} = 1,$$

so that we recover $v(L/2) = \delta$.

The effect of P on the displacement $v(L/2)$ is shown as a function of $\frac{P}{P_E}$ in Fig. (11.10.3).

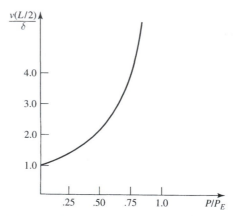

Figure 11.10.3

Substituting for $v_1(x)$ from Eq. (11.10.13) in Eqs. (11.8.1), an expression for the moments (for $0 \leq x \leq L/2$) may be readily obtained:

$$M(x) = -EIv_1''(x) = \frac{EIF\lambda}{P}\frac{\sin \lambda L/2}{\sin \lambda L}\sin \lambda x = \frac{F}{\lambda}\frac{\sin \lambda L/2}{\sin \lambda L}\sin \lambda x.$$

$$(11.10.20a)$$

Similarly, the shear term V becomes, using Eq. (11.8.5d),

$$V = \frac{dM(x)}{dx} = F\frac{\sin \lambda L/2}{\sin \lambda L}\cos \lambda x \qquad (11.10.20b)$$

so that at $x = 0$,

$$V(0) = \frac{F}{2\cos(\lambda L/2)}. \qquad (11.10.20c)$$

This last term is easily seen to be equivalent to

$$V(0) = F/2 + Pv'(0). \qquad (11.10.21a)$$

Finally, using Eq. (11.10.15a), the moment at the centre span becomes according to Eq. (11.10.20a),

$$M\left(\frac{L}{2}\right) = \frac{FL}{4}\frac{\tan \mu}{\mu} \qquad (11.10.21b)$$

from which we obtain the two limiting cases (a) and (b):

(a) As $P \rightarrow P_E$ ($\mu \rightarrow \pi/2$), and therefore $M(L/2) \rightarrow \infty$. This result is consistent with the fact that when $P \rightarrow P_E$, the member becomes unstable.
(b) As $P \rightarrow 0$ ($\mu \rightarrow 0$), and $M(L/2) \rightarrow FL/4$; i.e., we recover the moment existing in a beam with no axial force present.

11.11   Some comments on the solution: use of linear superposition

From the previous example, we make an interesting observation: we observe that the solution of the beam-column does not lead, as in the previous case of the study of instability of compressive members (subjected only to axial loading), to an eigenvalue problem. Indeed, as we have seen, an equilibrium state exists in the deflected state for any value of P, no matter how small; that is, a deflected position can be maintained *for all values of* P. Clearly, this is because the lateral force acting on the member also provides a necessary force to cause a deflection. Mathematically, the difference between the stability analysis and the present problem may be explained by observing that the equations governing beam-columns contain an inhomogeneous term involving lateral loads, as opposed to the homogeneous differential equation for a column. [Compare, e.g., Eq. (11.8.7b) with Eq. (11.9.1b).]

At this point, it is also worthwhile to consider the possibility of the use of linear superposition. Let us first rewrite, for example, Eq. (11.10.11a) explicitly in terms of the given forces F and P. After substituting the definition of λ, we obtain

$$v_1(x) = F \left\{ \frac{\sin[\sqrt{P/EI}(L-a)]}{P\sqrt{P/EI}\sin[\sqrt{P/EI}L]} \sin[\sqrt{P/EI}x] - \frac{L-a}{PL}x \right\}, \quad (11.11.1)$$

which we rewrite symbolically as $v_1(x) = F \cdot f(x, P)$.

We observe immediately that while $v_1(x)$ is a linear function of F, $f(x, P)$ is clearly a nonlinear function of P; hence $v_1(x)$ does not vary linearly with P. Indeed, it is a highly nonlinear function of the axial force P. Thus, linear superposition can be used only with respect to the applied force F and only for a given constant axial force P [Fig. (11.11.1a)]. On the other hand, it is not permissible to construct solutions using linear superposition for a constant force F but different thrusts P [Fig. (11.11.1b)].

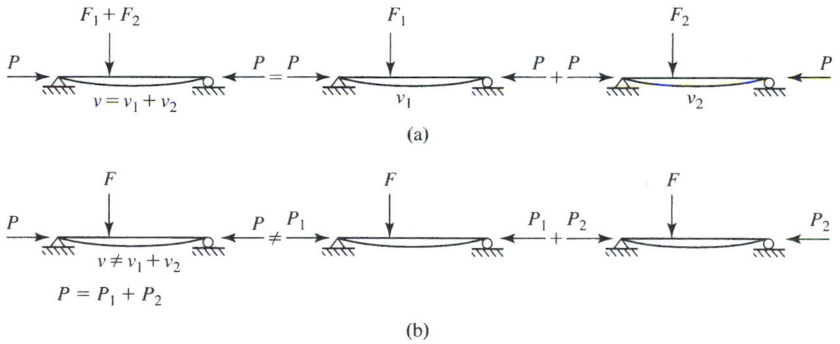

(a)

(b)

P = P₁ + P₂ $P = P_1 + P_2$

Figure 11.11.1

We now pose a significant question: why does such a nonlinear relation exist between the deflection v and the axial force P? To answer this question, we examine the original expression for the moment $M(x)$ given by Eqs. (11.10.1) in the deflected state of the beam. We observe that the product '$Pv(x)$' appears within this expression and that $v(x)$ was, at this stage of the analysis, an unknown quantity. This is in contrast to all previous analyses, where in expressions for moments, shears, etc., such products involving an unknown never appeared. Indeed, it is because of such products, that our present analysis leads to a nonlinear relation, Eq. (11.11.1), between the force P and the deflection.

It now becomes possible to generalise our conclusions. We first recall that equilibrium equations of mechanics of deformable bodies are always written in the deformed state. When, in the case of linearly elastic bodies, the unknown displacements do not appear explicitly within the equilibrium equations, the resulting force–displacement relations will be linear. (This was indeed the case in the treatment of members in Chapters 6–9.) However, when unknown displacements appear *explicitly* in the equilibrium equations as products with a force term, then the resulting force–displacement relation will be nonlinear.

◇11.12   Tie-rods

The analysis of a tie-rod, subjected to a tensile axial force (which we denote as $\overline{P} > 0$), as shown in Fig. (11.12.1), may be treated in a similar manner. Clearly, based on physical intuition, we expect the displacements $v(x)$ and moments $M(x)$ to be reduced with increasing values of $\overline{P}$. As opposed to the moment expressions given by Eqs. (11.10.1), we now have [Fig. (11.12.2)]

$$M(x) = \frac{F(L-a)}{L}x - \overline{P}v, \quad 0 \le x \le a, \qquad (11.12.1a)$$

$$M(x) = \frac{Fa}{L}(L-x) - \overline{P}v, \quad a \le x \le L. \qquad (11.12.1b)$$

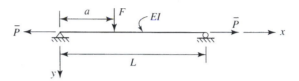

Figure 11.12.1

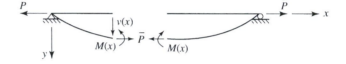

Figure 11.12.2

As in the previous example, we obtain the differential equations

$$v_1''(x) - \overline{\lambda}^2 v_1(x) = -\frac{F(L-a)}{EIL}x, \qquad (11.12.2a)$$

$$v_2''(x) - \overline{\lambda}^2 v_2(x) = -\frac{Fa}{EIL}(L-x), \qquad (11.12.2b)$$

where, here,

$$\overline{\lambda}^2 = \frac{\overline{P}}{EI} > 0. \qquad (11.12.2c)$$

At this point, we may solve the problem analogously as in the preceding case of the beam-column, making use of the same boundary conditions, Eqs. (11.10.6). However, although the analysis is straightforward, we may arrive at the solution in a much simpler manner: we note that the equations of this problem become identical to those of the beam-column if we replace P by $-\overline{P}$, and λ^2 by $-\overline{\lambda}^2$. Thus we set

$$P = -\overline{P}, \qquad (11.12.3a)$$

$$\lambda^2 = -\overline{\lambda}^2 \qquad (11.12.3b)$$

so that

$$\lambda = \imath\overline{\lambda} \qquad (11.12.3c)$$

(where $\imath = \sqrt{-1}$ is imaginary) in all expressions appearing in Section 11.10.

Then, substituting Eqs. (11.12.3) in Eq. (11.10.11a), we obtain, for the displacement v_1,

$$v_1(x) = -\frac{F}{\imath\overline{P}\overline{\lambda}} \frac{\sin[\imath\overline{\lambda}(L-a)]}{\sin(\imath\overline{\lambda})}\sin(\imath\overline{\lambda}) + \frac{F(L-a)}{\overline{P}}x. \qquad (11.12.4)$$

Noting that $\sin \iota\theta = \iota \sinh \theta$, $\cos \iota\theta = \cosh \theta$, $\tan \iota\theta = \iota \tanh \theta$, Eq. (11.12.4) becomes

$$v_1(x) = -\frac{F}{\overline{P}\overline{\lambda}}\frac{\sinh \overline{\lambda}(L-a)}{\sinh \overline{\lambda}L}\sinh \overline{\lambda} + \frac{F(L-a)}{\overline{P}}x. \qquad (11.12.5)$$

Furthermore, upon considering the case for P located at the centre, $a = L/2$, since $\mu = \iota\overline{\mu}$, where $\mu = \lambda L/2$, we replace Eq. (11.10.17) for the deflection $v(x)$ at $x = L/2$ by

$$v\left(\frac{L}{2}\right) = \delta\left[\frac{3(\overline{\mu} - \tanh \overline{\mu})}{\overline{\mu}^3}\right]. \qquad (11.12.6a)$$

Similarly, from Eq. (11.10.21b), the moment $M(L/2)$ becomes

$$M\left(\frac{L}{2}\right) = \frac{FL}{4}\frac{\tanh \overline{\mu}}{\overline{\mu}}. \qquad (11.12.6b)$$

We then observe:

(i) if $\overline{P} \to 0$, $(\overline{\mu} \to 0)$ then $\frac{\tanh \overline{\mu}}{\overline{\mu}} \to 1$

(ii) if $P_E \ll \overline{P}$, $1 \ll \overline{\mu}$ then $\frac{\tanh \overline{\mu}}{\overline{\mu}} \ll 1$.

Upon substituting in Eqs. (11.12.6), we obtain for $\overline{P}/P_E = 0.25$, $v(L/2) = 0.803\delta$ and $M(L/2) = 0.835\frac{FL}{4}$. If $\overline{P}/P_E = 0.5$, $v(L/2) = 0.671\delta$ and $M(L/2) = 0.724\frac{FL}{4}$.

The effect of the axial tension force on the displacement $v(L/2)$ and on the moments, is shown in Fig. (11.12.3). In contrast to the beam-column, we observe, as expected, that the tensile force $\overline{P}$ tends to reduce the lateral displacements, shears and moments.

11.13 General comments and conclusions

In this chapter, we have established the basic ideas required for the analysis of the stability and instability of rods subjected to axial compressive loads. The critical load P_{cr} at which the straight rod ceases to be in stable equilibrium, was found to be the force that leads to a neutral equilibrium state. In other terms, the critical load P_{cr} may also be considered as being the smallest force that can maintain the rod in a deflected position. By studying the simple model of a rigid rod, it was

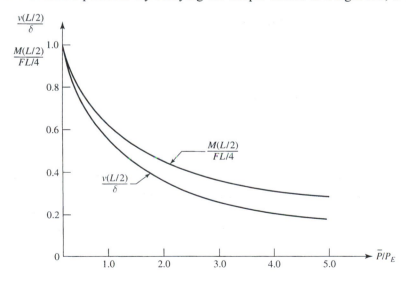

Figure 11.12.3

possible to incorporate all the basic ideas of stability analysis. Indeed the ideas were developed via the simple model as they could be expressed in terms of the most simple mathematics. The stability analysis of elastic rods, while leading to a more complex mathematical treatment, was based on the same essential ideas.

For the case of beam-columns, we observe that the behaviour of the member is quite different: equilibrium states can exist in the deflected position for *all* values of a compressive axial force. This is reflected mathematically by the inhomogeneous nature of the governing differential equation (due to the lateral loading term) and hence we are not led to an eigenvalue problem. Thus, in this case, one cannot physically refer, in the same sense, to instability of a beamcolumn.

Finally, as mentioned previously, we recall that there exists another approach to the analysis of stability; namely an energy approach. By considering stability via such an energy approach, it will be possible to gain further insight into the problem; this approach will also lead to the development of methods that yield approximate solutions in cases where exact solutions are unobtainable. We defer such a study to a later chapter.

PROBLEMS

Section 3

11.1:   Determine the value of the critical load for the rigid rods shown in Fig. (11P.1).

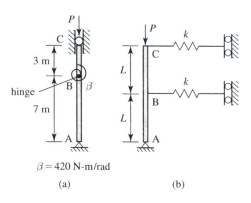

$\beta = 420$ N-m/rad

(a)   (b)

Figure 11P.1

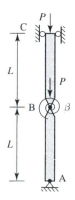

Figure 11P.2

11.2:   Two rigid rods, each of length L, connected by a hinge and a torsional spring with constant β (N-m/rad) at B, are supported at A and C, as shown in Fig. (11P.2). The rods are subjected to two axial forces P acting at B and C. Determine the value of the critical force P_{cr}.

Note: **The systems of Problems 11.3–11.8 are two-degree-of-freedom systems since the displaced configurations are defined by means of two independent parameters (either angles of rotation and/or displacements). The determination of the critical loads is based on the concept of neutral equilibrium, namely that two arbitrary *adjacent* equilibrium states are in equilibrium, i.e. the original (vertical) state and the perturbed state.**

11.3: The system shown in Fig. (11P.3) consists of two rigid rods supported by torsional springs with constant β (N-m/rad). Determine the value of the critical load.

11.4: Given a system consisting of a rigid rod supported by linear springs with constants k_1 and k_2 and by a torsional spring with constant β, as shown in Fig. (11P.4). (a) Determine the critical load P_{cr} in terms of k_1, k_2, β and L. (b) Determine the value of L that leads to a minimum value of P_{cr}. What is $(P_{cr})_{min}$?

11.5:* A system, consisting of two rigid rods, each of length L, is supported by a torsional spring having constant β_1 (N-m/rad) at A and a torsional spring with constant β_2 at the hinge B, as shown in Fig. (11P.5a). (a) Determine the critical axial force P that causes instability. (b) Based on the answer to (a) and letting $\gamma \equiv \beta_2/\beta_1$, show analytically that the critical force P of the system of Fig. (11P.5b) is greater than that of Fig. (11P.5a) for any finite value of β.

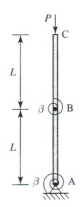

Figure 11P.3

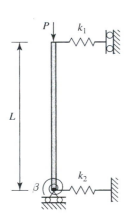

Figure 11P.4

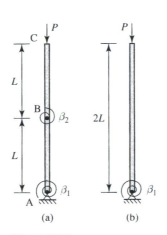

(a)   (b)

Figure 11P.5

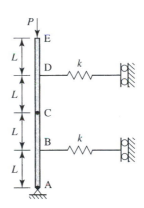

Figure 11P.6

11.6: Two rigid rods, each of length $2L$ are connected at C by means of a hinge and are simply supported at A and by linear springs with constant k (N/m) at B and D, as shown in Fig. (11P.6). Determine the critical axial load P that causes instability of the system.

11.7:* The system shown in Fig. (11P.7) consists of two rigid rods AC and CB, each of length L, which are hinged at C and are supported by a linear spring with constant k at B and a torsional spring with constant β at C. (a) Determine the critical load P_{cr}. (b) If the values for the constants are given as $k = 5$ N/cm and $\beta = 500$ N-cm/rad, determine the critical load when (i) $L = 12$ cm and (ii) $L = 8$ cm. (c) Given the values k and β in (b), for what length L will the system can carry the greatest load P before instability occurs? What is this load?

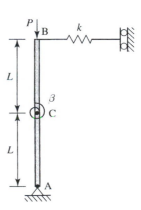

Figure 11P.7

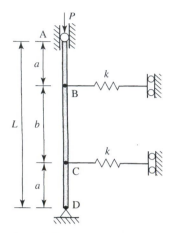

Figure 11P.8

11.8: A system consisting of three rods, AB, BC and CD, whose total length is L, is supported by springs with constant k (N/m) at the hinges B and C, as shown in Fig. (11P.8). An axial load P acts at A. (a) Determine the critical load P_{cr}. (b) What should be the ratio of b/a in order to maximise P_{cr}? What is $(P_{cr})_{max}$?

Sections 4 and 5

11.9: A measuring rod, made of wood, is assumed to behave elastically with $E = 10$ GPa. The length of the rod is 2 m and the dimensions of the cross-section of the rod are 10 mm × 16 mm. The rod is subjected to an axial compressive force P. (a) Under what force will the rod buckle, assuming the rod to be pin-connected at both ends, as shown in Fig. (11P.9a). In what plane will the rod buckle? (b) Determine the buckling load of the rod if, in addition to the two end supports, the rod is braced at point B in such a way that, as shown in Fig. (11P.9b), no displacement in the z-direction can occur at this point. In what plane will the rod buckle?

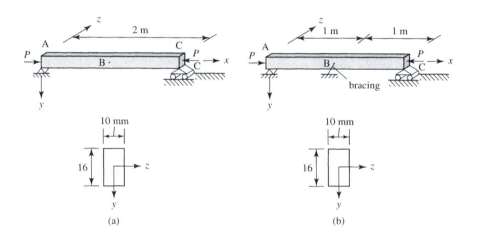

Figure 11P.9   (a)   (b)

11.10: It is required to design a column 3 m in height, which is pinned at both ends. Two options are available: (a) a steel ($E = 200$ GPa) angle 76 × 76 × 12.7 and (b) two steel angles 51 × 51 × 9.5, which are bolted together to form a monolithic section, as shown in Fig. (11P.10). In each case the axial compressive force acts through the centroid of the section. (Note that, from the given tables of Appendix E, the total mass in the two cases is the same.) Determine the Euler buckling load for options (a) and (b).

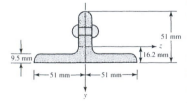

Figure 11P.10

11.11: A rod AB of flexural rigidity EI and length L is fixed at A and free at B. A compressive axial force acts at the free end. Given that the cross-section of the rod is an angle section $a \times 3a$ with thickness t ($t \ll a$). A rigid plate is attached at the free end, as shown in Fig. (11P.11), such that the load P acts through the centroid of the cross-sections. Determine the critical load in terms of E, a, t and L.

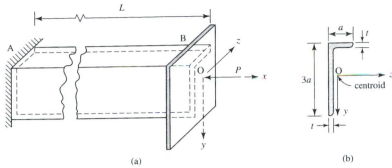

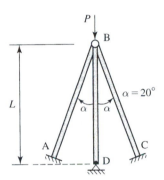

(a)   (b)   **Figure 11P.11**

11.12: A system, which consists of three rods each of length L and flexural rigidity EI, and which are connected by means of a hinge at B, is subjected to a force P, as shown in Fig. (11P.12). The system is assumed to fail due to buckling. (a) Can the rods buckle independently or must they buckle simultaneously? Explain, using sketches. (b) Determine the critical load P_{cr}.

Note: In Problems 11.13–11.21, it is required to solve the problem starting from the basic governing equation, $EIv''(x) = -M(x)$, together with the appropriate boundary conditions.

11.13: A cantilever rod of length L and flexural rigidity EI is subjected to an axial compressive force P acting through the centroid, as shown in Fig. (11P.13a). (a) Determine the loads causing instability and the lowest buckling load P_{cr}. (b) What is the buckling mode (eigenfunction), $v = v(x)$, corresponding to P_{cr}? Sketch the buckling mode. (c) If the cross-section of the rod is a rectangle, as shown in Fig. (11P.13b), express P_{cr} in terms of E, b and L.

Figure 11P.12

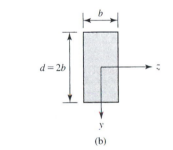

(a)   (b)   **Figure 11P.13**

11.14:* A rod AB having flexural rigidity EI and length L is fixed at A. While the rod is free to move in the longitudinal x-direction at B, there exists a constraint that prevents rotation at B. An axial compressive force P acts through the centroid, as shown in Fig. (11P.14a). (a) Obtain the characteristic equation

$$\sin\left(\frac{\lambda L}{2}\right)\left[\tan\left(\frac{\lambda L}{2}\right) - \frac{\lambda L}{2}\right] = 0, \quad \lambda = \sqrt{\frac{P}{EI}}.$$

(b) Determine the lowest critical load P_{cr} and the corresponding buckling buckling mode $v = v(x)$. Sketch the buckling mode. (c) If a simple support, preventing deflection at the mid-point is attached to the rod as shown in Fig. (11P.14b), determine the critical load P_{cr} for instability. Sketch the corresponding buckling mode.

11.15: A rod AB of length L and flexural rigidity EI is pinned at A and supported by means of a torsional spring having constant β (N-m/rad). An axial compressive force acts at the free end B, as shown in Fig. (11P.15). (a) Show that the characteristic

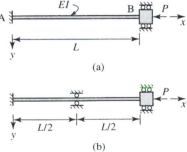

Figure 11P.14

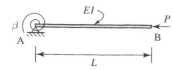

Figure 11P.15

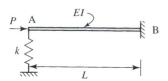

Figure 11P.16

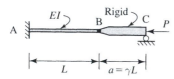

Figure 11P.17

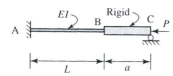

Figure 11P.18

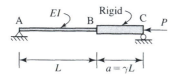

Figure 11P.19

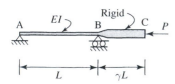

Figure 11P.20

equation whose lowest root yields the critical load P_{cr} is given by

$$\tan(\lambda L) = \frac{\alpha}{\lambda L}, \quad \lambda = \sqrt{\frac{P}{EI}}, \quad \alpha = \frac{\beta L}{EI}.$$

(b) Determine P_{cr} for the limiting case $\beta \to \infty$ and sketch the bucking mode. (c) Determine P_{cr} for the limiting case $EI \to \infty$ and sketch the bucking mode. (*Note:* See computer-related Problem 11.47.)

11.16:* A rod AB of length L and flexural rigidity EI is fixed at B and simply supported by a linear elastic spring having constant k (N/m) ($k > 0$). A force P acts as shown in Fig. (11P.16). (a) Show that the characteristic equation whose lowest root yields the critical load P_{cr} is given by

$$\tan(\lambda L) = \lambda L [1 - (\lambda L)^2/\alpha], \quad \lambda = \sqrt{P/EI}, \quad \alpha = kL^3/EI.$$

(b) What is the eigenvalue λ and P_{cr} for $k \to \infty$? (*Note:* See computer-related Problem 11.48.)

11.17: An elastic rod AB of length L and flexural rigidity EI is fixed at A and connected by means of a hinge to a rigid strut BC of length $a = \gamma L$, which is supported at C, as shown in Fig. (11P.17). An axial compressive force acts at C. Determine the characteristic equation whose lowest root yields the critical load P_{cr}. (*Note:* See computer-related Problem 11.49.)

11.18:* An elastic rod AB of length L and flexural rigidity EI is fixed at A and rigidly connected at B to a rigid strut BC of length a, which is supported at C, as shown in Fig. (11P.18). An axial compressive force acts at C. (a) Determine the characteristic equation whose lowest root yields the critical load P_{cr}. (b) From the characteristic equation obtained above, determine the critical load P_{cr} for the limiting case $a \to \infty$. Sketch the model that represents this limiting case. (c) Obtain the characteristic equation for the limiting case $a \to 0$. Sketch the model that represents this case?

11.19: An elastic rod AB of length L and flexural rigidity EI is simply supported at A and rigidly connected at B to a rigid strut BC of length $a = \gamma L$, which is supported at C, as shown in Fig. (11P.19). An axial compressive force acts through the centroid of the system at C. (a) Determine the characteristic equation whose lowest root yields the critical load P_{cr}. (b) From the characteristic equation obtained above, determine the critical load P_{cr} for the limiting case $a \to 0$. Sketch the model that represents this limiting case. (c) Repeat (b) for the limiting case $a \to \infty$ (*Note:* See computer-related Problem 11.49.)

11.20:* A rigid bar BC of length γL is rigidly attached to a linear elastic rod AB of flexural rigidity EI and length L, as shown in Fig. (11P.20). The system, simply supported at A and B, is subjected to an axial compressive force P acting through the centroid at C. (a) Determine the characteristic equation whose lowest root yields the critical load P_{cr}. (b) Obtain the critical load P_{cr} for the limiting cases $\alpha \to 0$ and $\alpha \to \infty$. Sketch the models that correspond to these cases.

11.21:* An elastic rod ACB, having flexural rigidity EI and length L, is simply supported at A, B and C, as shown in Fig. (11.5.4a), and is subjected to an axial force P at B [see also Fig. (11P.21)]. (a) Obtain the characteristic equation, namely Eq. (11.5.15), for the eigenvalue λ corresponding to the critical load P:

$$\gamma \lambda L (1 - \gamma) \sin(\lambda L) - \sin(\gamma \lambda L) \sin[(1 - \gamma)\lambda L] = 0, \quad \lambda^2 = P/EI.$$

(b) From the above characteristic equation, show that if $\gamma = 0.5$, i.e. when the support C is at the mid-span, the characteristic equation reduces to

$$\sin\left(\frac{\lambda L}{2}\right)\left[\tan\left(\frac{\lambda L}{2}\right) - \frac{\lambda L}{2}\right] = 0,$$

whose lowest root yields the critical value, $P_{cr} = 4EI/L^2$. (c) Show that as $\gamma \to 0$, the characteristic equation reduces to $\tan \gamma L = \gamma L$, namely Eq. (11.5.17). (*Note:* See computer-related Problem 11.50.)

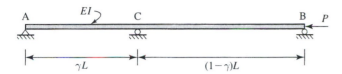

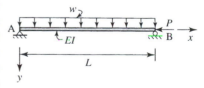

Figure 11P.21

Sections 8 and 9

11.22: The deflection of the rod ACB, shown in Fig. (11P.22), is given by [see Eq. (11.6.5)]

$$v(x) = e\left[\tan\left(\frac{\lambda L}{2}\right)\sin \lambda x + \cos \lambda x - 1\right], \quad \lambda = \sqrt{\frac{P}{EI}}$$

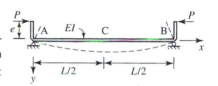

(a) What are the reactions R_A and R_B? What is the shear force $V_C \equiv V(L/2)$? (b) Determine the shear force $V(x)$ and evaluate $V_A = V(x = 0)$, $V_B = V(x = L)$ and V_C. Explain differences with R_A and R_B by means of sketches. Why is there no difference with V_C given in (a)?

Figure 11P.22

Note: **Solve the stability problems of problems 11.23–11.26 using the fourth-order equation, $EIv^{iv}(x) + Pv''(x) = 0$, together with the appropriate boundary conditions.**

11.23: (a) Derive the characteristic equation, (b) determine the buckling mode and (c) the critical buckling load P_{cr} for Problem 11.13.

11.24: (a) Derive the characteristic equation and (b) determine the buckling mode for Problem 11.15.

11.25:* (a) Derive the characteristic equation and (b) determine the buckling mode for Problem 11.16.

11.26: (a) Derive the characteristic equation and (b) determine the buckling mode for Problem 11.17.

Sections 10 and 11

11.27: The beam-column AB of length L and flexural rigidity EI is subjected to a uniform load w (N/m) and a compressive force P acting through the centroid, as shown in Fig. (11P.27). (a) Determine the deflection $v(x)$. (b) Evaluate $v(L/2)$ at the mid-point and express this as $v(L/2) = \delta \cdot f(\mu)$, where δ, the mid-point displacement of a rod subjected to only the uniform load, is given by $\delta = \frac{5wL^4}{584EI}$ and $\mu = (\pi/2)\sqrt{P/P_E}$ [see Eq. (11.10.18)]. (c) Show that as $\mu \to 0$, $v(L/2) \to \delta$. (d) Show that $v(L/2) \to \infty$ as $P \to P_E$, i.e., instability is associated with an infinite deflection. (e) Determine the moment $M(x)$ and shear force $V(x)$.

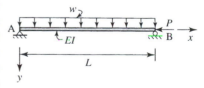

Figure 11P.27

11.28:* A beam-column AB of length L, flexural rigidity EI and fixed at A, is subjected to a uniformly distributed load w and an axial compressive force P, which passes

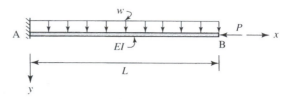

Figure 11P.28

through the centroid at the free end B, as shown in Fig. (11P.28). (a) By means of the governing differential equation $EIv''(x) = -M(x)$, determine the deflection $v(x)$ and evaluate $v(L)$ at the free end. (b) By associating the infinite displacement, $v(L) \to \infty$, with instability, determine the critical load P_{cr} for a cantilevered rod. (c) Determine the ratio $\frac{v(L)}{\delta}$ (where $\delta = wL^4/8EI$ is the deflection at the free end if P is not acting) and express this ratio in terms of P/P_{cr}.

11.29:   Given the beam column AB of length L and flexural rigidity EI subjected to the linearly varying load $q(x)$ and an axial compressive load P, as shown in Fig. (11P.29). (a) Determine the deflection $v(x)$. (b) Determine the shear and moments, $V(x)$ and $M(x)$, and evaluate extreme values. (c) Evaluate $V(x = 0)$ and show by means of a figure why it is not equal to $q_0L/6$.

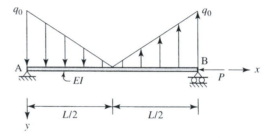

Figure 11P.29

11.30:*   A linearly elastic beam-column having a flexural rigidity EI, is subjected to a thrust P and a moment M_0, as shown in Fig. (11P.30a). (a) Determine the lateral displacement, $v(x)$. (b) From part (a), write the solution for the system subjected to a force P acting at the end C of a rigid bar BC, as shown in Fig. (11P.30b). (c) Determine Δ_C, the horizontal displacement of point C, assuming small rotations. (Assume that the horizontal displacement of point B is negligible.) (d) Rederive the expression for Δ_C if the lateral displacement $v(x)$ is neglected in the moment expression, i.e. if

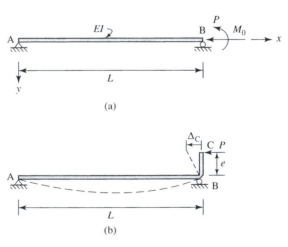

(a)

(b)

Figure 11P.30

$M = M(P, e, L; x)$. Note that this leads to a linearised solution. (e) Show that for small values of the non-dimensional quantity PL^2/EI, the linearised approximation of the solution obtained in (c) above, approaches that of (d). (*Hint*: Expand the solution of (c) in a Taylor series.) (f) Explain why, although the beam is made of a linearly elastic material, the results of part (c) are nonlinear. (*Note*: See computer-related Problem 11.51.)

Review and comprehensive problems

11.31: Two rigid rods, AB and BC, having length a and b, respectively, are connected by means of a hinge at B to which is attached a torsional spring with constant β (N-m/rad), as shown in Fig. (11P.31). An axial compressive load is applied at C. (a) Determine the critical load P_{cr} in terms of L, β and c, where $c \equiv a/b$. (b) What is the ratio a/b for which P_{cr} will be a minimum? What is $(P_{cr})_{min}$?

11.32: A rigid strut BC of length a is attached to a rod AB of length L and flexural rigidity EI. A compressive axial force P acts at C, as shown in Fig. (11P.32). (a) Determine the characteristic equation whose lowest root yields the critical value P_{cr} for instability. (b) Evaluate P_{cr} for the limiting cases $a \to 0$ and $a \to \infty$.

11.33: For the beam-column AB, shown in Fig. (11P.33), subjected to a force F at the mid-span and a compressive thrust P, the deflection $v(x)$ is given by Eq. (11.10.13) and hence the slope $\theta \simeq v'(x)$ is given by the expression

$$v'(x) = \frac{F}{P}\left(\frac{\sin \lambda L/2}{\sin \lambda L}\right)\cos \lambda x - \frac{F}{2P}, \quad \lambda = \sqrt{P/EI}, \quad 0 \le x \le L/2$$

(a) Show, for infinitesimal v', that the shear V at $x = 0$, given by $EIv'''(x) = -V(x)$, is

$$V(0) = \frac{F}{2\cos \lambda L/2}, \tag{1}$$

and that this is equivalent to

$$V(0) = F/2 + Pv'(0). \tag{2}$$

(b) Show that if P/P_E is sufficiently small, the expressions above reduce to the approximate relation

$$V(0) = \frac{F}{2}\left(1 + \frac{P}{P_E}\frac{\pi^2}{8}\right), \tag{3}$$

where $P_E = \pi^2 EI/L^2$ is the Euler buckling load. (*Note*: See computer-related Problem 11.52.)

11.34: A simply supported rod AB of length L and flexural rigidity EI is subjected to two eccentrically applied loads, as shown in Fig. (11P.34). Determine the deflection $v(x)$.

11.35: * A linear elastic circular rod of radius R and length L, fixed at A and simply supported at B, rests against a linear elastic spring of stiffness k (N/m), as shown in Fig. (11P.35). The modulus of elasticity of the rod is E and the coefficient of thermal expansion is α. The rod is slowly subjected to an increasing temperature ΔT such that both the rod and the spring come under compression. (a) Determine the critical temperature increase, ΔT_{cr}, under which the rod will buckle. (b) What is the elongation of the rod, e, immediately before buckling takes place? (c) What is ΔT_{cr} if $k \to \infty$?

11.36: 'Due to a compressive axial load passing through the centroid, a rod will always buckle in a plane that contains a principal centroidal axis of the cross-section'. Comment on the validity of this statement, and give reasons.

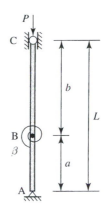

Figure 11P.31

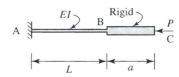

Figure 11P.32

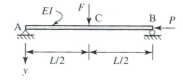

Figure 11P.33

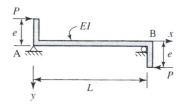

Figure 11P.34

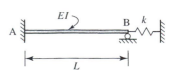

Figure 11P.35

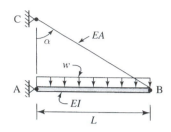

Figure 11P.37

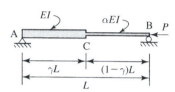

Figure 11P.38

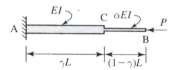

Figure 11P.39

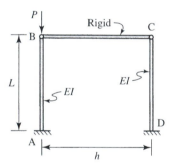

Figure 11P.40

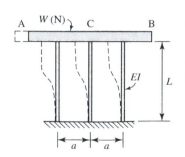

Figure 11P.41

11.37: A rod AB of length L and having flexural rigidity EI is simply supported at A and at B by a wire BC whose axial rigidity is EA, as shown in Fig. (11P.37). Determine the deflection $v(x)$ of the rod in terms of E, I, A, L and α. (*Note:* Assume rotations are infinitesimal.)

11.38: An axial compressive force P acts on a simply supported elastic rod consisting of two rigidly attached segments AC and CB with flexural rigidity EI and αEI, $0 < \alpha$, respectively, as shown in Fig. (11P.38). (a) Obtain the characteristic equation for the eigenvalue $\lambda = \sqrt{P/EI}$, namely

$$c\left[\tan\left(\frac{\gamma\lambda L}{c}\right) - \tan\left(\frac{\lambda L}{c}\right)\right] = \tan(\gamma\lambda L)\left[1 + \tan\left(\frac{\lambda L}{c}\right)\tan\left(\frac{\gamma\lambda L}{c}\right)\right],$$

where $c = \sqrt{\alpha}$. (b) By examining the limiting cases $\gamma = 1$ and $\gamma \to 0$, verify that the roots of the characteristic equation correspond to the known Euler buckling loads. (*Note:* See computer-related Problem 11.53.)

11.39:* An axial compressive force P acts on an elastic rod fixed at A and free at B. The rod consists of two rigidly attached segments AC and CB with flexural rigidity EI and αEI $(0 < \alpha)$, respectively, as shown in Fig. (11P.39). (a) Show that the characteristic equation for the eigenvalue $\lambda = \sqrt{P/EI}$, is given by Eq. (11.5.18), namely

$$\tan(\gamma\lambda L) = \cot\left[\frac{(1-\gamma)\lambda L}{c}\right],$$

where $c = \sqrt{\alpha}$. (b) By examining the limiting cases $\gamma = 1$ and $\gamma \to 0$, verify that the roots of the characteristic equation correspond to the known Euler buckling loads. (*Note:* See computer-related Problem 11.53.)

11.40:* Two rods AB and CD having flexural rigidity EI and fixed at A and D, are connected by means of hinges at B and C to a bar BC, as shown in Fig. (11P.40). A vertical force P acts at B. Determine the characteristic equation whose lowest root yields the critical load P_{cr}.

11.41: A rigid bar ACB is placed horizontally on three identical (massless) rods, which are fixed at one end and are rigidly attached to the bar, as shown in Fig. (11P.41). The rods are of length L and flexural rigidity EI. Determine the maximum permissible weight W of the horizontal bar if bucking is not to occur.

11.42: A spring of length L_s, having an unknown constant k (N/m), is situated in a frictionless slot of radius R [see Fig. (11P.42a)]. In order to determine k, an elastic rod of length $L < L_s$, having the same radius R, is inserted in the slot, as shown in Fig. (11P.42b). The rod is found to buckle under a critical load P_{cr}. (a) Determine k in terms of E, R, L and P_{cr}. (b) Determine the value of k if the rod is made of steel $(E = 200\,\text{GPa})$, $R = 2\,\text{cm}$, $L = 2\,\text{m}$, and if $P_{cr} = 50{,}000\,\text{N}$.

11.43:* A force P acts on a horizontal bar BC, which is connected by means of hinges, to two rods, AB and CD, as shown in Fig. (11P.43). Rod AB is fixed at A and rod CD is hinged at D. Both rods have the same flexural rigidity EI. Failure is assumed to occur due to buckling of the rods when P reaches P_{cr}. Determine (a) the position of the load P, namely a, for which P_{cr} is a maximum and (b) $(P_{cr})_{max}$.

11.44:* A linear elastic circular rod having flexural rigidity EI and length L is inserted in a smooth (frictionless) circular slot of length $L_s > L$ and the same radius as the rod. The rod is inserted by applying a force P, as shown in Fig. (11P.44). If the slot contains

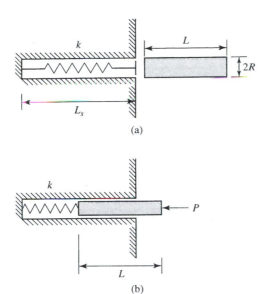

(a)

(b)

Figure 11P.42

a linear spring of initial length L_s whose constant is k, determine the maximum value of k if buckling is not to occur as P increases until $y = L$.

11.45:* A system consists of two rods AB and BC, which are connected by means of a hinge at B, as shown in Fig. (11P.45). The two rods have the same flexural rigidity EI. Failure is assumed to occur due to buckling when a force P is applied at B. (a) Can each rod buckle independently or must they buckle simultaneously? Explain. (b) At what angle ϕ with respect to BC should the force be applied in order to maximise the critical force P_{cr}? (c) Determine $(P_{cr})_{max}$ in terms of EI, L and α.

11.46:* A rod ACB having flexural rigidity EI and length L is simply supported at the two ends A and B and by a spring of constant k (N/m), at the mid-point C, as shown in Fig. (11P.46). For which values of k will the rod buckle such that $v_C = 0$. (*Note*: See computer-related Problem 11.54.)

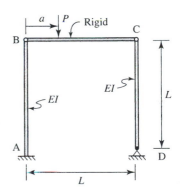

Figure 11P.43

The following problems, which require the use of a computer, are to be solved in conjunction with the referenced problems.

11.47: Referring to Problem 11.15, express the given characteristic equation as $\tan(\eta) = \alpha/\eta$, where $\eta \equiv \lambda L$ is a non-dimensional variable, and (a) Determine, by means of a computer, the roots of the transcendental equation. (b) Using a plotting routine, plot the non-dimensional ratio $\frac{P_{cr}}{P_{cr}|_{\beta \to \infty}}$ as a function of α, where $P_{cr}|_{\beta \to \infty} = \pi^2 EI/4L^2$.

11.48: Referring to Problem 11.16, (a) rewrite the given characteristic equation as $\tan \eta = \eta(1 - \eta^2/\alpha)$, where $\eta \equiv \lambda L$ and $\alpha \equiv kL^3/EI$. (b) By means of a computer, solve the above transcendental equation for η, determine the critical load as a function of η and using an appropriate plotting routine, plot P_{cr}/P_E as a function of α, where P_E is the Euler buckling load.

11.49: Referring to Problem 11.17 or 11.19, (a) rewrite the relevant characteristic equation in terms of the non-dimensional variable, $\eta = \lambda L$, and determine, by means of a computer, the roots of the transcendental equation for $\gamma > 0$. (b) Using a plotting routine, plot η as a function of γ.

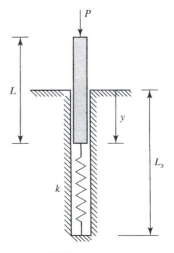

Figure 11P.44

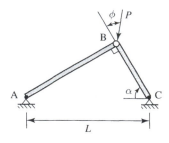

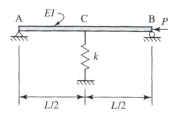

Figure 11P.45

Figure 11P.46

11.50: Referring to Problem 11.21, (a) determine the roots of the characteristic equation for values $0 < \gamma \leq 0.5$ and (b) using a plotting routine, plot the non-dimensional quantity P_{cr}/P_E (where $P_E = \pi^2 EI/L^2$), as a function of γ in this range.

11.51: Referring to Problem 11.30, (i) prepare a table of numerical results of the non-dimensional displacement $L\Delta_C/ea$ as a function of PL^2/EI $(0 \leq PL^2/EI \leq 3)$, as obtained in parts (c) and (d) of Problem 11.30, and (ii) using a plotting routine, plot the two curves on the same graph. Determine the percent error of the approximate solution of (d) and interpret the behaviour of the curves. For what values of PL^2/EI is the percent error less than 10%?

11.52: Referring to Problem 11.33, (a) by means of Eq. (1), determine the effect of P on $V(0)$ as a function of $P/P_E (0 \leq P/P_E < 3)$ and (b) using a plotting routine, plot the results of (a) and the approximate relation [Eq. (3) of Problem 11.33] on the same graph. For what values of P/P_E is the approximate relation accurate to within 10%?

11.53: Referring to Problem 11.38 or 11.39, rewrite the relevant characteristic equation in terms of the non-dimensional variable $\eta = \lambda L$ and choose one or more of the following options:

    (i) By means of a computer, determine the roots (eigenvalues), η, of the relevant transcendental equation for discrete values of $\alpha \equiv \sqrt{c}$ $(0 < \alpha \leq 1)$ and, using a plotting routine, plot the resulting family of curves as a function of $\gamma (0 \leq \gamma \leq 1)$.

    (ii) By means of a computer, determine the roots (eigenvalues), η, of the relevant transcendental equation for discrete values of γ $(0 \leq \gamma \leq 1)$ and, using a plotting routine, plot the resulting family of curves as a function of $\alpha \equiv \sqrt{c} (0 < \alpha \leq 1)$.

11.54: Referring to Problem 11.46, (a) verify that the characteristic equation for buckling in the symmetric mode is given by

$$\tan\zeta - \zeta + \zeta^3/\alpha = 0,$$

where $\zeta = \lambda L/2 = \frac{L}{2}\sqrt{P/EI}$ and $\alpha = kL^3/16EI$. (b) Using a plotting routine, plot a graph of P/P_E vs. α (where P_E is the Euler buckling load) for the case of symmetric buckling and show that if $\alpha \geq \pi^2$, buckling occurs in the anti-symmetric mode with $v_C = 0$.

12

Torsion of elastic members of arbitrary cross-section: de Saint Venant torsion

12.1 Introduction

In the development below, we consider the stresses and deformation of relatively long prismatic elastic members of arbitrary cross-section subjected to torsion. We first establish a Cartesian coordinate system (x, y, z), where the y- and z-axes lie in the plane of the cross-section and the x-axis is the longitudinal axis [Fig. (12.1.1)]. A torque $T \equiv M_x$ is applied about the x-axis as shown, and all sections rotate about this axis without changing their shape; we refer to points on the x-axis as the *centre of twist*. As we shall see, the analysis of torsion of members of arbitrary cross-section is much more complex than that of members whose cross-sections consist of circles. Let us therefore first recall the solution for members of circular cross-section, as developed in Chapter 7. This solution was based upon the following conclusions (which were deduced from symmetry considerations):

- all plane cross-sections remain plane after deformation,
- all radial lines remain radial straight lines,
- circular cross-sections remain circular.

Figure 12.1.1

From the above, using strain–displacement and elastic stress–strain relations, it was found that the resultant shear stresses vary linearly with r, the radial distance measured from the x-axis, and more specifically the shear stress component in the circumferential direction θ was found to be given by the expression

$$\tau_{x\theta} = \frac{Tr}{J}, \tag{12.1.1a}$$

where J is the polar moment of the cross-sectional area about the x-axis. Further, the unit angle of twist Θ (i.e., the relative rotation of two cross-sections a unit

447

distance apart) was given by

$$\Theta = \frac{T}{GJ}, \tag{12.1.1b}$$

where GJ is the **torsional rigidity**.

Now, the above assumptions and results are valid *only* for members of circular cross-section, the solution of which was obtained by Coulomb in 1784. The more general solution for the torsion of members of arbitrary cross-section was given in 1855 by de Saint Venant. As we shall see, the results are rather startling when we compare them with the circular case. We list here three interesting results (which will be shown to be true for elliptic and rectangular sections):

- all cross-sections (except circular cross-sections) warp, i.e., points displace non-uniformly in the longitudinal x-direction;
- the maximum torsional shear stresses at the edge generally occur at points that are *closest* to the centre of twist (the x-axis) [compare with Eq. (12.1.1a)].
- if two members of the same material and same cross-sectional area (but different shapes) are subjected to the same torque, the one having an area with a smaller polar moment will have a greater torsional rigidity [compare with Eq. (12.1.1b)] provided that the cross-sections are 'simply connected'.[†]

We now wish to consider the solution of the general torsion problem, which is usually referred to as the *de Saint Venant torsion solution*. In developing this solution, it is no longer possible to use the traditional *mechanics-of-materials* approach as was the case with the Coulomb solution for circular members. Instead we require a more exact approach; namely, we must use the basic equations of the *theory of elasticity*. However, before doing so, we first discuss and indicate the aspects of the approach that will be used.

12.2 Semi-inverse methods: uniqueness of solutions

Let us recall that in solving a problem in solid mechanics where known forces (tractions) are applied to a body, we are interested in determining (i) the state of stress at all points (six independent quantities) and (ii) the displacements at all points (three independent quantities) within the body. Then, from the resulting displacements, we can also obtain (iii) the strains within the body.

Thus, in general, recalling our discussion in Chapter 5, we must solve for 15 unknowns: 6 stress components, 6 strain components and 3 displacement components. (See Table 5.1 of Chapter 5.)

On the other hand, as we have also seen in Chapter 5, for an elastic body there exist, in general, 15 equations: (a) 3 stress equations of equilibrium, (b) 6 strain–displacement relations, and (c) 6 elastic stress–strain relations. In addition, (d) the stress state at the boundary of the elastic body must correspond to the known tractions that exist at the surface.

It is quite understandable that to solve in a straightforward manner for all the unknowns, using the 15 equations, is a very difficult, if not impossible, task.

Now, instead of assuming that absolutely nothing is known about the stresses or displacements in the body, it is reasonable to use some physical knowledge and

---

[†] Mathematically, a cross-section is said to be *simply connected* if the cross-section contains no 'holes'. If the cross-section contains n holes, it is said to be $n + 1$ *connected*. Thus a pipe is said to be *doubly connected*. In general, one refers to a cross-section as being either *simply connected* or *multi-connected*.

intuition, based on plausible reasoning and assume, for example, that the displacements behave according to a certain pattern.

In making assumptions on the displacements it will also be reasonable to assume that the displacements are continuous throughout the body and that all displacements are 'single-valued', i.e., the displacement of any point is described by a single vector.

Let us say that we have made some physically plausible assumption on the displacement pattern (satisfying the above conditions), and that from these we can satisfy the relevant equations within the body and the boundary tractions [i.e., (a)–(d) above]. We will then have *a* solution to the problem. Such a solution, obtained by making certain 'guesses' in the unknowns and then satisfying all relevant equations, is known as a **semi-inverse method**.

Assume now that we are able to obtain a solution that satisfies all the relevant equations; we then pose the following question: is it possible that there exists some other solution that also satisfies these equations and the surface traction? For example, if we were to start with a different assumption on the basic displacement pattern, could we then obtain a new and different solution? Fortunately, there exists a *uniqueness theorem*, which may be stated as follows: If internal stresses and compatible displacements satisfying the relevant equations (equilibrium, stress–strain, strain–displacement) are found and if the prescribed (known) boundary conditions on the body are satisfied, then the solution obtained is the *only possible* solution, i.e., the solution is *unique*.

Thus, a solution obtained by the semi-inverse method is *the* solution of the problem.

The above uniqueness theorem, which was stated without proof, is a general theorem valid for all linear elastic media subject to small strains. We shall accept this uniqueness theorem without proof, based on physical intuition. (A rigorous mathematical proof, which is beyond the scope of our study, may be found in any text on the *theory of elasticity*, and is based on a consideration of the total energy of a system.)

Having accepted the uniqueness theorem as given above, we can now proceed confidently with the de Saint Venant torsion problem.

12.3 The general de Saint Venant torsion solution

Consider a linearly elastic prismatic member of arbitrary cross-section, as shown in Fig. (12.3.1a), subjected to a torque $T \equiv M_x$.

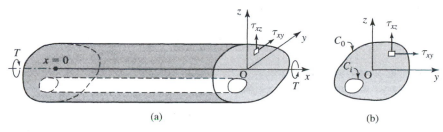

(a)   (b)   **Figure 12.3.1**

In general, the cross-section may have any number of holes say n, as shown in Fig. (12.3.1b), i.e., it may be *multi-connected* or may have no holes, i.e., it may be *simply connected*.

We denote the external boundary curve by C_0 and all interior boundary curves by C_i, $i = 1, 2, \ldots, n$.

Let the continuous displacement of a generic point P be given by

$$\boldsymbol{u} = u\boldsymbol{i} + v\boldsymbol{j} + w\boldsymbol{k}. \tag{12.3.1}$$

Without loss of generality, we assume that the displacement v and w of all points in the plane $x = 0$ are zero; i.e., the cross-section at this plane is assumed to be fixed against rotation. Note that we have *not* assumed u to be zero.

Since we expect shear stresses to exist on the cross-section, we denote the shear stress components on any x-plane by τ_{xy} and τ_{xz}, as shown.

The solution to the problem is restricted to those cases for which strains and relative rotations are small.

According to the semi-inverse method, we shall now make a basic assumption on the displacement pattern, based on a physically plausible argument.

Due to an applied torque, all cross-sections clearly will undergo rotations (without changing shape); i.e., all cross-sections are assumed to rotate about the x-axis, namely the 'centre of twist' with respect to the section $x = 0$, which is assumed fixed. Denoting the unit angle of twist by Θ, the rotation of a plane located at any x will then be Θx. Thus, the relative rotation of two cross-sections, a distance Δx apart, is $\Theta \Delta x$.

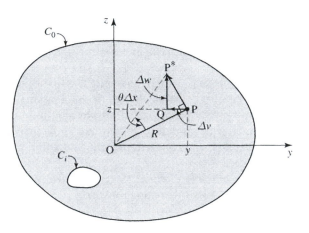

Figure 12.3.2

Let the distance from the centre of twist to the generic point $P(x, y, z)$ be denoted by R [Fig. (12.3.2)]. Then, if the unit angle of twist Θ is small, the relative displacements $\overrightarrow{PP^*}$ in a cross-section, located at $x + \Delta x$ (with respect to those at the cross-section located at x), are perpendicular to the radial line OP (i.e., $\overline{PP^*} \perp \overline{OP}$). These displacements are therefore given by

$$\overline{PP^*} = R \cdot (\Theta \cdot \Delta x).$$

The displacement components in the y- and z-directions are then

$$\Delta v = -\overline{PP}^* \cdot \sin(\angle PP^*Q) = -R\Theta\Delta x \cdot \sin(\angle POy)$$
$$= -R\Theta\Delta x(z/R) = -\Theta\Delta x \cdot z \tag{12.3.2a}$$

and

$$\Delta w = (\overline{PP}^*) \cdot \cos(\angle PP^*Q) = R\Theta\Delta x \cdot \cos(\angle POy)$$
$$= R\Theta\Delta x(y/R) = \Theta\Delta x \cdot y. \tag{12.3.2b}$$

Furthermore, we shall *assume* that the displacement components of points in the x-direction (and all its partial derivatives) are continuous, but are *independent of x*; i.e., *all planes warp identically*.

Thus, we start the solution with the following basic assumption on the displacement pattern:[†]

$$u = u(y, z), \quad \text{(continuous, class } C^2\text{)} \tag{12.3.3a}$$

$$\Delta v = -\Theta \Delta x \cdot z \tag{12.3.3b}$$

$$\Delta w = \Theta \Delta x \cdot y, \tag{12.3.3c}$$

where we note that in the limit $\Delta x \to 0$,

$$\frac{\partial v}{\partial x} = -\Theta z, \tag{12.3.4a}$$

$$\frac{\partial w}{\partial x} = \Theta y. \tag{12.3.4b}$$

Now, observing that Δv and Δw are not functions of y and z, respectively, and that all cross-sections are assumed to rotate with respect to the fixed cross-section at $x = 0$, it follows that v and w are not functions of y and z respectively.[‡] Therefore, from the strain–displacement relations, Eq. (3.7.20), we have

$$\epsilon_{xx} = \frac{\partial u}{\partial x} = 0, \qquad \epsilon_{yy} = \frac{\partial v}{\partial y} = 0, \qquad \epsilon_{zz} = \frac{\partial w}{\partial z} = 0. \tag{12.3.5a}$$

Furthermore, since all cross-sections rotate without changing shape, lines that were initially orthogonal remain mutually perpendicular; specifically, lines originally parallel to the y- and z-axes remain orthogonal after deformation. It follows that

$$\epsilon_{yz} = \frac{1}{2}\left(\frac{\partial w}{\partial y} + \frac{\partial v}{\partial z}\right) = 0. \tag{12.3.5b}$$

However,

$$\epsilon_{xy} = \frac{1}{2}\left(\frac{\partial u}{\partial y} + \frac{\partial v}{\partial x}\right) = \frac{1}{2}\left(\frac{\partial u}{\partial y} - \Theta z\right), \tag{12.3.6a}$$

$$\epsilon_{xz} = \frac{1}{2}\left(\frac{\partial u}{\partial z} + \frac{\partial w}{\partial x}\right) = \frac{1}{2}\left(\frac{\partial u}{\partial z} + \Theta y\right). \tag{12.3.6b}$$

Then, from the stress–strain relations, Eqs. (4.4.10), we obtain

$$\tau_{xx} = \tau_{yy} = \tau_{zz} = \tau_{yz} = 0, \tag{12.3.7a}$$

[†] By class C^2 we mean that $u(y, z)$ as well as its partial derivatives up to the second order with respect to y and z is assumed continuous.

[‡] To demonstrate this assertion, we first note that integration of Eqs. (12.3.4) yields

$$v = -\Theta xz + \tilde{v}(y, z), \tag{a}$$

$$w = \Theta xy + \tilde{w}(y, z). \tag{b}$$

Since the cross-section at $x = 0$ is assumed to be fixed against rotation, it follows that $\tilde{v} = \tilde{w} = 0$ identically and therefore v and w are not functions of y or z, respectively. Thus, we have

$$v = -\Theta xz, \tag{c}$$

$$w = \Theta xy. \tag{d}$$

It is worthwhile to observe that Eqs. (c) and (d) are valid only if $|\Theta x| \ll 1$ and are otherwise spurious. [We note here, in passing, that Eqs. (c) and (d) will not be used in the analysis.]

and the only remaining non-zero stresses are

$$\tau_{xy} = 2G\epsilon_{xy} = G\left(\frac{\partial u}{\partial y} - \Theta z\right), \tag{12.3.7b}$$

$$\tau_{xz} = 2G\epsilon_{xz} = G\left(\frac{\partial u}{\partial z} + \Theta y\right). \tag{12.3.7c}$$

In passing, we observe from the above that the only existing stresses τ_{xy} and τ_{xz} are continuous functions of y and z (and do not depend on x).

Up to this point, we have made use of the strain–displacement and the stress–strain relations. Let us now examine the stress equations of equilibrium, Eqs. (2.4.4). From Eq. (2.4.4b),

$$\frac{\partial \tau_{xy}}{\partial x} + \frac{\partial \tau_{yy}}{\partial y} + \frac{\partial \tau_{zy}}{\partial z} = 0,$$

we note that the second and last terms vanish by virtue of Eq. (12.3.7a) while the first term vanishes since τ_{xy} is not a function of x. Thus this stress equation of equilibrium is *satisfied identically*. Repeating the same process, we see that Eq. (2.4.4c),

$$\frac{\partial \tau_{xz}}{\partial x} + \frac{\partial \tau_{yz}}{\partial y} + \frac{\partial \tau_{zz}}{\partial z} = 0,$$

is also satisfied identically.

Consider now the first equation of equilibrium, Eq. (2.4.4a),

$$\frac{\partial \tau_{xx}}{\partial x} + \frac{\partial \tau_{yx}}{\partial y} + \frac{\partial \tau_{zx}}{\partial z} = 0. \tag{12.3.8a}$$

Since $\tau_{xx} = 0$, this reduces to

$$\frac{\partial \tau_{xy}}{\partial y} + \frac{\partial \tau_{xz}}{\partial z} = 0, \tag{12.3.8b}$$

which is the remaining equation that must be satisfied throughout the body.

At this point of the development we have resolved the problem down to two unknown stress components, which are continuous in y and z. Now, upon examining Eq. (12.3.8b), we observe that if we could express the shear stress components in terms of a single function $\phi(y, z)$, we might simplify the problem even further.

Let us therefore *assume* that there exists an unknown *continuous* function $\phi(y, z)$ of class C^2 such that the stress components are given by

$$\tau_{xy} = \frac{\partial \phi}{\partial z}, \tag{12.3.9a}$$

$$\tau_{xz} = -\frac{\partial \phi}{\partial y}. \tag{12.3.9b}$$

Then, Eq. (12.3.8b) is satisfied identically since

$$\frac{\partial^2 \phi}{\partial y \partial z} - \frac{\partial^2 \phi}{\partial z \partial y} = 0 \tag{12.3.10}$$

for such a continuous function of (y, z). Such a function, whose partial derivatives yield stress components, is called a **stress function**.

We now seek the appropriate equation that the stress function $\phi(y, z)$ must satisfy. From Eqs. (12.3.7b) and (12.3.7c), we have

$$\frac{\partial \phi}{\partial z} = G\left(\frac{\partial u}{\partial y} - \Theta z\right), \qquad (12.3.11\text{a})$$

$$\frac{\partial \phi}{\partial y} = -G\left(\frac{\partial u}{\partial z} + \Theta y\right). \qquad (12.3.11\text{b})$$

Operating on Eqs. (12.3.11a) and (12.3.11b) by $\partial/\partial z$ and $\partial/\partial y$ respectively,

$$\frac{\partial^2 \phi}{\partial z^2} = G\left(\frac{\partial^2 u}{\partial z \partial y} - \Theta\right), \qquad (12.3.11\text{c})$$

$$\frac{\partial^2 \phi}{\partial y^2} = -G\left(\frac{\partial^2 u}{\partial y \partial z} + \Theta\right), \qquad (12.3.11\text{d})$$

and adding these last two equations, we obtain

$$\frac{\partial^2 \phi}{\partial y^2} + \frac{\partial^2 \phi}{\partial z^2} = -2G\Theta + G\left(\frac{\partial^2 u}{\partial z \partial y} - \frac{\partial^2 u}{\partial y \partial z}\right).$$

However, we have assumed $u(y, z)$ and its partial derivatives to be continuous. Therefore the expression in the parentheses vanishes, and we remain with

$$\frac{\partial^2 \phi}{\partial y^2} + \frac{\partial^2 \phi}{\partial z^2} = -2G\Theta. \qquad (12.3.12)$$

Equation (12.3.12) is thus the equation that the stress function must satisfy. We observe that this equation corresponds to the physical condition that all displacements u are continuous and single-valued.

We note that Eq. (12.3.12) can be written more concisely as

$$\nabla^2 \phi = -2G\Theta,$$

where

$$\nabla^2 \equiv \frac{\partial^2}{\partial y^2} + \frac{\partial^2}{\partial z^2} \qquad (12.3.13)$$

is the (two-dimensional) Laplacian operator. Equations of the type of Eq. (12.3.12) are known in mathematics as *Poisson equations*.

We may also determine the equation governing the displacement $u(y, z)$. Operating on Eqs. (12.3.11a) and (12.3.11b) by $\frac{\partial}{\partial y}$ and $\frac{\partial}{\partial z}$ respectively, we obtain

$$\frac{\partial^2 \phi}{\partial y \partial z} = G\frac{\partial^2 u}{\partial y^2}, \qquad \frac{\partial^2 \phi}{\partial z \partial y} = -G\frac{\partial^2 u}{\partial z^2}. \qquad (12.3.14)$$

Subtracting the second of Eq. (12.3.14) from the first, we have by Eq. (12.3.10),

$$\frac{\partial^2 u}{\partial y^2} + \frac{\partial^2 u}{\partial z^2} = 0 \qquad (12.3.15)$$

or

$$\nabla^2 u = 0.$$

Equations of the type of Eq. (12.3.15) are known as *Laplace equations*.

Summarising at this point, the problem has been reduced to obtaining the stress function $\phi(y, z)$ satisfying the equation $\nabla^2 \phi = -2G\Theta$. Once this function has been determined, the two non-zero stress components are determined from Eqs. (12.3.9).

Furthermore, noting that τ_{xy} and τ_{xz} are the scalar components of the traction (vector) $\boldsymbol{T}_x$ acting on the x-plane [see Eq. (2.3.7a)], the resultant shear stress acting on this plane can be obtained simply from the relation [Fig. (12.3.3)][†]

$$|\tau_{xR}| = \left(\tau_{xy}^2 + \tau_{xz}^2\right)^{1/2} = \left[\left(\frac{\partial \phi}{\partial y}\right)^2 + \left(\frac{\partial \phi}{\partial z}\right)^2\right]^{1/2}. \qquad (12.3.16)$$

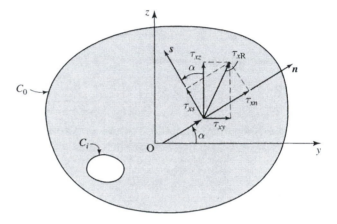

Figure 12.3.3

Now, this resultant shear stress that acts on the x-plane can be expressed in terms of components in *any* two perpendicular directions lying in this plane. Let us therefore resolve the resultant shear stress τ_{xR} in the two perpendicular directions, e.g., into τ_{xn} and τ_{xs}, as shown in Fig. (12.3.3). We wish to obtain an expression for these components in terms of ϕ. Let $\boldsymbol{n}$ and $\boldsymbol{s}$ be orthogonal unit vectors lying in the cross-sectional plane, which specify the n- and s-directions. Further, we denote by α the angle that $\boldsymbol{n}$ makes with the y-axis. Then

$$\tau_{xn} = \tau_{xy} \cos \alpha + \tau_{xz} \sin \alpha.$$
$$\tau_{xs} = -\tau_{xy} \sin \alpha + \tau_{xz} \cos \alpha. \qquad (12.3.17)$$

But, from geometry [Fig. (12.3.4)]

$$\cos \alpha = \frac{\partial y}{\partial n} = \frac{\partial z}{\partial s} \quad \text{and} \quad \sin \alpha = \frac{\partial z}{\partial n} = -\frac{\partial y}{\partial s}.$$

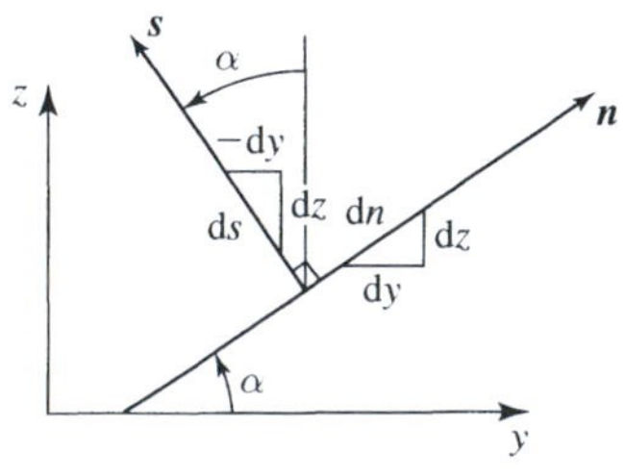

Figure 12.3.4

Therefore substituting Eq. (12.3.9) and the above in Eq. (12.3.17), there results

$$\tau_{xn} = \left(\frac{\partial \phi}{\partial z}\right)\left(\frac{\partial z}{\partial s}\right) - \left(\frac{\partial \phi}{\partial y}\right)\left(-\frac{\partial y}{\partial s}\right) = \frac{\partial \phi}{\partial s}, \qquad (12.3.18a)$$

$$\tau_{xs} = \left(-\frac{\partial \phi}{\partial z}\right)\left(\frac{\partial z}{\partial n}\right) - \left(\frac{\partial \phi}{\partial y}\right)\left(\frac{\partial y}{\partial n}\right) = -\frac{\partial \phi}{\partial n}. \qquad (12.3.18b)$$

[†] Here the upper case R is used simply to signify the 'resultant' shear stress rather than a component of the stress tensor.

We then have

$$|\tau_{xR}| = \left(\tau_{xn}^2 + \tau_{xs}^2\right)^{1/2} = \left[\left(\frac{\partial\phi}{\partial n}\right)^2 + \left(\frac{\partial\phi}{\partial s}\right)^2\right]^{1/2}. \qquad (12.3.18c)$$

From Eq. (12.3.18a) we note that the shear stress component τ_{xn} acting in the n-direction is equal to the 'slope' of ϕ in the s-direction, while τ_{xs} is given by the negative of the 'slope' of ϕ in the n-direction. We shall have reason to return to this remark subsequently.

Let us now return to our main problem: determining the stress function ϕ. Although we know the equation that this function must satisfy [viz. Eq. (12.3.12)] at all points in the interior of the body, this is not sufficient to determine ϕ. Additional information is required, namely, we must know the boundary conditions that ϕ must satisfy; i.e., we must determine conditions that ϕ must satisfy at all points on the external boundary C_0 and on any internal boundaries C_i (if such boundaries exist).[†]

However, we notice that we have not yet specified the tractions existing on the lateral surfaces of the element, implying that these tractions must vanish. It is this condition that provides us with a mathematical statement of the boundary condition on ϕ.

Consider any boundary C_i ($i = 0, 1, 2, \ldots, n$) of the body. Let n now be defined such that *at all points on the boundary*, it represents a direction normal to the boundary [Fig. (12.3.5)]. Then it follows that s is always tangential to the boundary line.

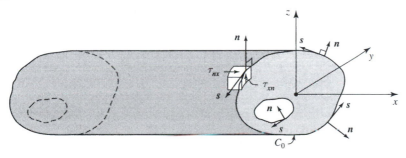

Figure 12.3.5

Now, as mentioned above, physically, any lateral boundary (interior as well as exterior) is *traction-free*, i.e. the surface traction $T_n = 0$; specifically $\tau_{nx} = 0$.

But, since $\tau_{xn} = \tau_{nx}$ it follows that $\tau_{xn} = 0$ at any boundary, i.e., at the edge of the cross-section. Thus we conclude that *the shear stresses existing at all points lying on the boundaries C_i must act tangentially to C_i*. Hence, using Eq. (12.3.18a), we have on any boundary C_i, the boundary condition

$$\left.\frac{\partial\phi}{\partial s}\right|_{C_i} = 0, \quad i = 0, 1, \ldots, n. \qquad (12.3.19)$$

Since s represents the parameter tangential to the boundary, Eq. (12.3.19) states that the function ϕ does not change as we proceed along any boundary; thus, on any boundary C_i, the stress function has a constant value, i.e.

$$\phi|_{C_i} = k_i, \quad i = 0, 1, \ldots, n, \qquad (12.3.20)$$

[†] According to the mathematical theory of differential equations, it is proved that to obtain a solution for a function governed by a Poisson equation in a given domain, one must specify either the function (or its normal derivative) everywhere on the boundary of the domain. Such a problem is known in mathematics as a *Dirichlet Problem* (or a *Neumann Problem*).

where k_i are constants. It should be emphasised that, while ϕ must be constant on any single boundary, the value of the constant is usually different from one boundary C_i to another. We note, however, that the values of the constants k_i are as yet unknown.

We have thus determined, from the physical condition of traction-free surfaces, the mathematical boundary condition that ϕ must satisfy. This condition together with Eq. (12.3.12) is sufficient to properly state the problem on ϕ. We observe too, at this stage, that Θ, appearing in Eq. (12.3.12), is still an unknown quantity.

We have yet one remaining task before completing the general solution: viz. we must find the relation between the applied torque T and the function ϕ.

To this end, consider the increment of torque $\mathrm{d}T$ about the x-axis produced by τ_{xy} and τ_{xz} acting over an infinitesimal area at point P [Fig. (12.3.6)]. Using a right-handed sign convention,

$$\mathrm{d}T = (-\tau_{xy}\,\mathrm{d}A) \cdot z + (\tau_{xz}\,\mathrm{d}A) \cdot y$$

or

$$\mathrm{d}T = (-\tau_{xy}z + \tau_{xz}y)\,\mathrm{d}A. \tag{12.3.21a}$$

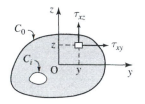

Figure 12.3.6

Integrating over the net area A,

$$T = \iint_A (-\tau_{xy}z + \tau_{xz}y)\,\mathrm{d}A. \tag{12.3.21b}$$

Substituting from Eqs. (12.3.9),

$$T = \iint_A \left[-\left(\frac{\partial \phi}{\partial z}\right)z - \left(\frac{\partial \phi}{\partial y}\right)y \right]\mathrm{d}A. \tag{12.3.21c}$$

Then, noting that

$$\frac{\partial (\phi y)}{\partial y} = y\frac{\partial \phi}{\partial y} + \phi, \tag{12.3.22a}$$

$$\frac{\partial (\phi z)}{\partial z} = z\frac{\partial \phi}{\partial z} + \phi, \tag{12.3.22b}$$

we may rewrite Eq. (12.3.21c) as

$$T = -\iint_A \left[\frac{\partial (\phi y)}{\partial y} + \frac{\partial (\phi z)}{\partial z} \right]\mathrm{d}A + 2\iint_A \phi\,\mathrm{d}A. \tag{12.3.23}$$

We now wish to transform the first surface integral over the area A appearing above into a line integral along the boundaries C_i. To this end, we recall Green's Theorem in a plane (for a multi-connected region) [see Eq. (B.2.3) of Appendix B.2]:

$$\iint_A \left(\frac{\partial Q}{\partial y} - \frac{\partial P}{\partial z} \right)\mathrm{d}y\,\mathrm{d}z = \oint_{C_0}(P\,\mathrm{d}y + Q\,\mathrm{d}z) - \sum_{i=1}^{n}\oint_{C_i}(P\,\mathrm{d}y + Q\,\mathrm{d}z),$$
$$\tag{12.3.24}$$

where P and Q are continuous functions of y and z with continuous first partial derivatives within the domain and on the boundaries C_i.

Letting $P \equiv -\phi z$, $Q \equiv \phi y$, Eq. (12.3.23) becomes, according to Green's theorem,

$$T = -\oint_{C_0}(y\phi\,\mathrm{d}z - z\phi\,\mathrm{d}y) + \sum_{i=1}^{n}(y\phi\,\mathrm{d}z - z\phi\,\mathrm{d}y) + 2\iint_A \phi\,\mathrm{d}A. \quad (12.3.25a)$$

But, according to Eq. (12.3.20), ϕ is constant along any boundary; i.e., $\phi = k_i$ on any C_i. Taking the constants outside the integral, we have

$$T = -k_0\oint_{C_0}(y\,\mathrm{d}z - z\,\mathrm{d}y) + \sum_{i=1}^{n}k_i\oint_{C_i}(y\,\mathrm{d}z - z\,\mathrm{d}y) + 2\iint_A \phi\,\mathrm{d}A. \quad (12.3.25b)$$

We now wish to express these simpler line integrals as surface integrals. Thus, again using Green's theorem, but now letting $Q \equiv y$, $P \equiv -z$, we transform back to surface integrals:

$$T = 2\iint_A \phi\,\mathrm{d}A - 2k_0 A_0 + 2\sum_{i=1}^{n}k_i A_i, \quad (12.3.26a)$$

where A denotes the *net area* of the cross-section, A_0 the *total area* within the external boundary C_0 and A_i $(i = 1, 2, \ldots)$ denotes the area of the ith hole within C_i.

For a simply connected cross-section, i.e. if no interior holes exist, the relation naturally becomes

$$T = 2\iint_A \phi\,\mathrm{d}A - 2k_0 A, \quad (12.3.26b)$$

where A is the actual cross-sectional area.

Having completed the derivation of the general solution of the de Saint Venant torsion problem, we summarise the results as follows:

Summary of general de Saint Venant torsion solution

The problem is resolved to one of obtaining a stress function $\phi(y, z)$ satisfying the partial differentiale quation

$$\nabla^2\phi = -2G\Theta$$

and subject to the boundary condition

$$\phi|_{C_i} = k_i, \quad (i = 0, 1, 2, \ldots, n)$$

on any boundary C_i.

The stresses within the body are then given by

$$\tau_{xy} = \frac{\partial\phi}{\partial z}, \qquad \tau_{xz} = -\frac{\partial\phi}{\partial y}.$$

The relation between the applied torque T and ϕ is

$$T = 2\iint_A \phi\,\mathrm{d}A - 2k_0 A_0 + 2\sum_{i=1}^{n}k_i A_i.$$

It is worthwhile to recall that we originally defined the x-axis to be the 'centre of twist'. For cross-sections with axes of symmetry, it is clear that the centre of twist lies on such axes of symmetry. However, if no axis of symmetry exists, the location of the centre of twist remains, at this stage, unknown and indeed, its location cannot be determined here. (The location of the centre of twist for elastic members will be established later after consideration of energy principles.)

Finally, we make an additional but important comment. We observe that in obtaining the general solution, we have not stipulated any constraints on the displacements $u(y, z)$ in the axial direction other than that they be continuous. (Specifically, we also note that τ_{xx} is zero.) Consequently, the displacement $u(y, z)$ at any section of the member can displace freely, and not necessarily uniformly in the x-direction. Thus, in the de Saint Venant torsion solution, we implicitly assume *free warping of any cross-section*.

We now apply this general solution to the specific problem of the torsion of a member having an elliptic cross-section.

12.4 Torsion of a member of elliptic cross-section

Let the semi-major and semi-minor axes of the cross-section be denoted by a and b respectively, $b \le a$ [Fig. (12.4.1)].

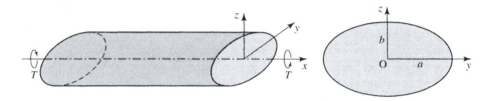

Figure 12.4.1

Recall that the equation of an ellipse is

$$\frac{y^2}{a^2} + \frac{z^2}{b^2} = 1 \tag{12.4.1a}$$

or

$$\frac{y^2}{a^2} + \frac{z^2}{b^2} - 1 = 0. \tag{12.4.1b}$$

Now, we observe that if we assume a stress function $\phi(y, z)$ of the form

$$\phi = \kappa \left(\frac{y^2}{a^2} + \frac{z^2}{b^2} - 1 \right), \tag{12.4.2}$$

where κ is an unknown constant, then on the boundary C_0, $\phi|_{C_0} = 0$. Thus we have already satisfied the boundary condition that the stress function ϕ is a constant on the boundary, namely $k_0 = 0$.

Now, we have *assumed* that Eq. (12.4.2) is of a form that satisfies Eq. (12.3.12). To see if this is possible, we substitute in this latter equation:

$$\kappa \left(\frac{2}{a^2} + \frac{2}{b^2} \right) = 2\kappa \left(\frac{b^2 + a^2}{a^2 b^2} \right) = -2G\Theta. \tag{12.4.3a}$$

Hence, the equation is satisfied if

$$\kappa = -\left(\frac{a^2 b^2}{a^2 + b^2}\right) G\Theta. \tag{12.4.3b}$$

Therefore, we conclude that we have indeed found a stress function

$$\phi(y, z) = -\left(\frac{a^2 b^2}{a^2 + b^2}\right)\left(\frac{y^2}{a^2} + \frac{z^2}{b^2} - 1\right) G\Theta, \tag{12.4.4}$$

which satisfies both the partial differential equation and the required boundary condition.

From Eq. (12.3.26b), we obtain with $k_0 = 0$,

$$T = 2\kappa \iint_A \left(\frac{y^2}{a^2} + \frac{z^2}{b^2} - 1\right) dA. \tag{12.4.5}$$

This integral may be evaluated directly. However, its value can be found very simply if we recall that

$$\iint_A y^2 \, dA = I_{zz} = \frac{\pi a^3 b}{4}, \quad \iint_A z^2 \, dA = I_{yy} = \frac{\pi a b^3}{4}, \quad \iint_A dA = A = \pi a b, \tag{12.4.6}$$

where I_{yy} and I_{zz} are the known second moments of area of an ellipse about the y- and z-axes, respectively.

Therefore, Eq. (12.4.5) becomes

$$T = 2\kappa \left[\frac{\pi a b}{4} + \frac{\pi a b}{4} - \pi a b\right] = -\kappa (\pi a b)$$

and hence

$$\kappa = -\frac{T}{\pi a b}. \tag{12.4.7}$$

Substituting in Eq. (12.4.3b),

$$T = \pi \left(\frac{a^3 b^3}{a^2 + b^2}\right) G\Theta, \tag{12.4.8}$$

which is of the form $T = CG \cdot \Theta$.

Now, recalling the Coulomb solution of Chapter 7, the ratio $\frac{T}{\Theta}$ clearly is the *torsional rigidity*; thus, this quantity is given by

$$CG = \pi \left(\frac{a^3 b^3}{a^2 + b^2}\right) G. \tag{12.4.9}$$

We observe again that the torsional rigidity is a function of the geometry, represented by the torsional rigidity constant C, as well as of the shear stiffness of the material, G.

We emphasise here that the expression

$$\Theta = \frac{T}{GC}, \tag{12.4.10}$$

relating the unit angle of twist Θ to the torsional moment T, is a general relation that is valid for all members subject to torsion. Thus, according to the linear theory developed above, the unit angle of twist Θ is always inversely proportional to the torsional rigidity GC.

Having determined the stress function $\phi(y, z)$, we find the shear stresses by taking partial derivatives according to Eqs. (12.3.9) using Eqs. (12.4.2) and (12.4.7). Thus

$$\tau_{xy} = \frac{\partial \phi}{\partial z} = \kappa\left(\frac{2z}{b^2}\right) = -\left(\frac{2T}{\pi a b^3}\right)z, \qquad (12.4.11a)$$

$$\tau_{xz} = -\frac{\partial \phi}{\partial y} = -\kappa\left(\frac{2y}{a^2}\right) = \left(\frac{2T}{\pi a^3 b}\right)y. \qquad (12.4.11b)$$

The resultant shear stress τ_{xR} is, by Eq. (12.3.16),

$$|\tau_{xR}| = \frac{2T}{\pi ab}\left(\frac{y^2}{a^4} + \frac{z^2}{b^4}\right)^{1/2}. \qquad (12.4.11c)$$

For $b < a$, the maximum shear stress occurs at point ($y = 0, z = \pm b$), that is, at points D of Fig. (12.4.2):

$$|\tau_{xR}|_D = |\tau_{xy}|_D = \frac{2T}{\pi ab^2}. \qquad (12.4.11d)$$

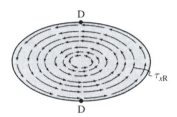

Figure 12.4.2

Note that points D are the points on the boundary which are *closest* to the centre of twist (a confirmation of the statement of Section 12.1). The general distribution of the resultant shear stress over the cross-section is shown in Fig. (12.4.2), where the directions of the resultant stresses τ_{xR} are seen to lie along concentric ellipses. We notice, in particular, that the resultant shear stress always acts tangentially along the boundary. (We recall that this pattern is necessary at the boundary in order to have a traction-free surface.)

Finally, it is clear that the centre of twist of an elliptic cross-section lies at point O, the intersection of the axes of symmetry.

In closing it is worth noticing that the case of a circular cross-section is a degenerate case of the elliptic section. Setting $a = b = R$, the radius of the circle, we obtain from Eq. (12.4.9),

$$C = \frac{\pi R^4}{2},$$

which is recognised as the polar moment J of a circle about the x-axis.

Furthermore

$$|\tau_{xR}| = \frac{2T}{\pi R^2} \cdot \left(\frac{y^2}{R^4} + \frac{z^2}{R^4}\right)^{1/2} = \frac{T}{\pi R^4/2}(y^2 + z^2)^{1/2}$$

or

$$|\tau_{xR}| = \frac{Tr}{J} \qquad (12.4.12)$$

since $r = (y^2 + z^2)^{1/2}$ is the radial distance from the centre of twist to a point P

[Fig. (12.4.3)]. Thus, we have recovered the Coulomb solution for torsion of a circular member, as given in Chapter 7.

We now recall that in Section 1 of this chapter, it was stated that, in general, warping takes place in a member subjected to torsion; only if its cross-section is a circle does warping vanish (i.e., plane cross-sections remain plane). In this latter case, the displacement in the x-direction is $u = $ const. (e.g., zero). For all other geometries, cross-sections will warp; that is, there exist displacements $u = u(y, z)$, which are functions of y and z. We now verify this statement by investigating the displacement u for the elliptic section.

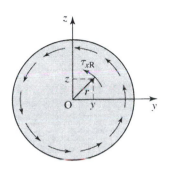

Figure 12.4.3

From Eqs. (12.3.7b) and (12.3.7c), we have

$$\frac{\partial u}{\partial y} = \frac{\tau_{xy}}{G} + \Theta z, \qquad (12.4.13a)$$

$$\frac{\partial u}{\partial z} = \frac{\tau_{xz}}{G} - \Theta y. \qquad (12.4.13b)$$

Substituting for τ_{xy} and τ_{xz} from Eqs. (12.4.11a) and (12.4.11b), we obtain

$$\frac{\partial u}{\partial y} = \left(\frac{2\kappa}{Gb^2} + \Theta \right) z, \qquad (12.4.14a)$$

$$\frac{\partial u}{\partial z} = -\left(\frac{2\kappa}{Ga^2} + \Theta \right) y. \qquad (12.4.14b)$$

Integrating each of the above,

$$u(y, z) = \left(\frac{2\kappa}{Gb^2} + \Theta \right) yz + f_1(z), \qquad (12.4.15a)$$

$$u(y, z) = -\left(\frac{2\kappa}{Ga^2} + \Theta \right) yz + f_2(y), \qquad (12.4.15b)$$

where $f_1(z)$ and $f_2(y)$ are functions of z and y respectively. Clearly, since we assume that the displacement u is unique, both expressions must therefore describe the same displacement.

Subtracting Eq. (12.4.15b) from Eq. (12.4.15a), we find

$$\left[\frac{2\kappa}{G} \left(\frac{1}{b^2} + \frac{1}{a^2} \right) + 2\Theta \right] yz + f_1(z) - f_2(y) = 0. \qquad (12.4.16a)$$

However, from Eq. (12.4.3a), we note that the bracketed term

$$\frac{2\kappa}{G} \left(\frac{1}{b^2} + \frac{1}{a^2} \right) + 2\Theta = 0. \qquad (12.4.16b)$$

Therefore, we arrive at the condition $f_1(z) = f_2(y)$. Now a function of y can be equal to a function of z at *all* points only if both functions are, in fact, the same constant, say B. Thus we find

$$f_1(z) = f_2(y) = B. \qquad (12.4.16c)$$

Adding Eqs. (12.4.15a) and (12.4.15b), and using Eq. (12.4.16c), we obtain,

$$u(y, z) = \frac{\kappa}{G} \left(\frac{1}{b^2} - \frac{1}{a^2} \right) yz + B. \tag{12.4.17a}$$

To evaluate the constant B we observe, using an argument of symmetry, that at point O, the displacement must necessarily be zero, i.e., $u(y = 0, z = 0) = 0$. Consequently, $B = 0$. It follows that

$$u(y, z) = \frac{\kappa}{G} \left(\frac{1}{b^2} - \frac{1}{a^2} \right) yz \tag{12.4.17b}$$

or, using Eq. (12.4.7),

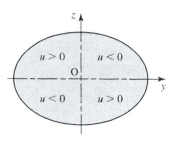

$$u(y, z) = -\frac{T}{G} \frac{a^2 - b^2}{\pi a^3 b^3} yz. \tag{12.4.17c}$$

We also note here that Eq. (12.3.15), $\nabla^2 u(y, z) = 0$, is identically satisfied.

From the expression for $u(y, z)$, we observe that $u < 0$ in the first and third quadrants; in the second and fourth quadrants, $u > 0$ [see Fig. (12.4.4)]. Thus we find that the cross-sections do indeed warp. Moreover, we observe that for the case of a circular cross-section with $a = b = R$, $u(y, z) = 0$ identically. Hence, we have verified that plane cross-sections will remain plane for members having a circular cross-section.

Figure 12.4.4

12.5 Torsion of a member of rectangular cross-section

We consider here a rectangular cross-section ($a \times b$) shown in Fig. (12.5.1). As discussed in the previous sections, we seek a stress function $\phi(y, z)$ that satisfies the Poisson equation, Eq. (12.3.12), i.e.

$$\frac{\partial^2 \phi}{\partial y^2} + \frac{\partial^2 \phi}{\partial z^2} = -2G\Theta, \tag{12.5.1a}$$

which is subject to the boundary condition

$$\phi|_{C_0} = k_0. \tag{12.5.1b}$$

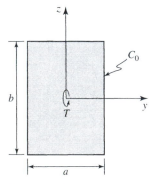

Figure 12.5.1

Without loss of generality, we set $k_0 = 0$.[†]

Now, in general, it is much easier to solve Laplace equations than Poisson equations. We therefore seek to transform the mathematical problem consisting of Eqs. (12.5.1) to a system with a Laplace equation and associated boundary conditions. We thus proceed as follows: noting that the Poisson equation is a linear equation, we assume that $\phi(y, z)$ can be decomposed into two parts; i.e., we let

$$\phi(y, z) = \phi_0(y) + \phi_1(y, z) \tag{12.5.2}$$

such that $\phi_0(y)$ satisfies the simple ordinary differential equation

$$\nabla^2 \phi_0(y) \equiv \frac{d^2 \phi_0}{dy^2} = -2G\Theta \tag{12.5.3a}$$

[†] Since the stress components τ_{xy} and τ_{xz} depend only on the *derivatives* of $\phi(y, z)$, we can always choose one of the constants k_i arbitrarily without affecting the stresses. We therefore choose $k_0 = 0$.

and such that, on the boundary $y = \pm a/2$,

$$\phi_0(y = \pm a/2) = 0. \tag{12.5.3b}$$

Note that $\phi_0(y)$, being a function only of y, is *not* a constant on the $z = \pm b/2$ boundary.

Using the linearity property of the differential equation, $\phi_1(y, z)$ must then satisfy the Laplace equation

$$\frac{\partial^2 \phi_1}{\partial y^2} + \frac{\partial^2 \phi_1}{\partial z^2} = 0. \tag{12.5.4}$$

Furthermore, on the boundary $y = \pm a/2$, we require that

$$\phi_1(\pm a/2, z) = 0. \tag{12.5.5a}$$

In addition, since $k_0 = 0$, $\phi_1(y, z)$ must assume values so as to make the $\phi = \phi_0 + \phi_1 = 0$ everywhere on C_0; hence ϕ_1 must also satisfy the condition

$$\phi_0(y) + \phi_1(y, z = \pm b/2) = 0. \tag{12.5.5b}$$

Thus we observe that instead of having to solve the original Poisson equation, we must now solve two equations: the simpler Laplace equation with the associated boundary conditions [Eqs. (12.5.5)] and the ordinary differential equation with boundary conditions given by Eq. (12.5.3b).

Integrating Eq. (12.5.3a) and using the condition of Eq. (12.5.3b), we readily find $\phi_0(y)$:

$$\phi_0(y) = -G\Theta(y^2 - a^2/4). \tag{12.5.6}$$

We now turn our attention to $\phi_1(y, z)$, which must satisfy Eq. (12.5.4). Let us suppose that $\phi_1(y, z)$ can be represented in the form[†]

$$\phi_1(y, z) = Y(y) \cdot Z(z), \tag{12.5.7}$$

where $Y(y)$ and $Z(z)$ are each a function of a single variable.

Substituting in Eq. (12.5.4), we obtain

$$\frac{d^2 Y}{dy^2} Z + \frac{d^2 Z}{dz^2} Y = 0,$$

which, upon dividing through by $Y \cdot Z$, yields

$$\frac{d^2 Y/dy^2}{Y} = -\frac{d^2 Z/dz^2}{Z} = -\alpha^2, \quad \text{a constant,} \tag{12.5.8}$$

where $\alpha^2 > 0$.

Note that we have set both terms of Eq. (12.5.8) equal to a constant. We justify this as follows: we note that the first ratio, $\frac{d^2 Y/dy^2}{Y}$ is a function only of y while the second ratio is a function only of z. Clearly, for this to be true at all points in the body, these two ratios can be equal to each other only if they are equal to a (same) constant (which we have called $-\alpha^2$). The justification for the minus sign appearing above will become apparent later.

[†] This form of solution for a partial differential equation is known, in mathematics, as *separation of variables*.

Multiplying out, from Eq. (12.5.8), we obtain

$$\frac{d^2 Y}{dy^2} + \alpha^2 Y(y) = 0,$$ (12.5.9a)

$$\frac{d^2 Z}{dz^2} - \alpha^2 Z(z) = 0.$$ (12.5.9b)

Solutions to Eqs. (12.5.9) are

$$Y(y) = A \cos \alpha y + B \sin \alpha y,$$ (12.5.10a)

$$Z(z) = C \cosh \alpha z + D \sinh \alpha z,$$ (12.5.10b)

where cosh and sinh are the hyperbolic functions. Hence, by Eq. (12.5.7),

$$\phi_1(y, z) = (A \cos \alpha y + B \sin \alpha y)(C \cosh \alpha z + D \sinh \alpha z),$$ (12.5.11)

where A, B, C and D are arbitrary constants. Note that this form of $\phi_1(y, z)$ satisfies Eq. (12.5.4) for all values of these constants and for any value of α.

We now show that two of these constants are zero. To do so we consider the anticipated stress distribution on the cross-section, using arguments of symmetry and anti-symmetry. Due to a torque acting on the cross-section, we conclude by symmetry that for a given y and z [Fig. (12.5.2)],

$$\tau_{xy}(y, z) = -\tau_{xy}(y, -z),$$ (12.5.12a)

$$\tau_{xz}(y, z) = -\tau_{xz}(-y, z).$$ (12.5.12b)

We first recall Eqs. (12.3.9),

$$\tau_{xy} = \frac{\partial \phi}{\partial z},$$ (12.5.13a)

$$\tau_{xz} = -\frac{\partial \phi}{\partial y}.$$ (12.5.13b)

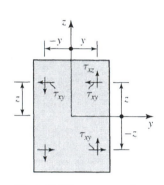

Figure 12.5.2

Then, from Eqs. (12.5.12a) and (12.5.13a) it follows, using Eq. (12.5.10b), that

$$C \sinh(\alpha z) + D \cosh(\alpha z) = -C \sinh(-\alpha z) - D \cosh(-\alpha z).$$

Recalling that the cosh and sinh hyperbolic functions are even and odd functions, respectively, we have

$$C \sinh(\alpha z) + D \cosh(\alpha z) = C \sinh(\alpha z) - D \cosh(\alpha z),$$ (12.5.14a)

from which we find $D = 0$.

Similarly, using Eqs. (12.5.12b), (12.5.13b) and (12.5.10a), and noting the even and odd properties of the cos and sin functions, respectively, we obtain the relation

$$A \sin(\alpha y) + B \cos(\alpha y) = A \sin(\alpha y) - B \cos(\alpha y)$$ (12.5.14b)

from which we conclude that $B = 0$.[†]

[†] We note that from the anti-symmetry of the shear stresses, $\tau_{xy}(y, z) = \tau_{xy}(-y, z)$, $\tau_{xz}(y, z) = \tau_{xz}(y, -z)$, appearing in Fig. (12.5.2), we would also obtain the same result, namely $B = D = 0$.

Equation (12.5.11) therefore reduces to

$$\phi_1(y, z) = \kappa \cos \alpha y \cdot \cosh \alpha z, \qquad (12.5.15)$$

where $\kappa \equiv AC$ is an unknown constant.[†]

Now, from Eq. (12.5.5a), we must satisfy the condition $\phi_1(\pm a/2, z) = 0$. Since $\kappa \neq 0$ and $\cosh(\alpha z) \geq 1$, we therefore require that

$$\cos(\alpha a/2) = 0$$

from which

$$\alpha = \frac{n\pi}{a}, \qquad n = 1, 3, 5, \ldots. \qquad (12.5.16)$$

Therefore, for any odd value of n, solutions for ϕ_1, given by

$$\phi_{1n}(y, z) = \kappa_n \cos \frac{n\pi y}{a} \cosh \frac{n\pi z}{a}, \qquad n = 1, 3, 5, \ldots, \qquad (12.5.17)$$

will satisfy the condition $\phi_1(\pm a/2, z) = 0$. Note that we have attached a subscript n in the above expressions to indicate that there are many (indeed an infinity of) solutions,

$$\phi_{11}, \phi_{13}, \phi_{15}, \phi_{17}, \ldots,$$

which satisfy the above boundary condition as well as Eq. (12.5.4). Therefore, the sum of all these solutions also satisfies the equation and boundary condition Eq. (12.5.5a). Thus we let

$$\phi_1(y, z) = \sum_{\substack{n=1 \\ n \text{ odd}}}^{\infty} \kappa_n \cos \frac{n\pi y}{a} \cosh \frac{n\pi z}{a}. \qquad (12.5.18)$$

Then from Eqs. (12.5.2) and (12.5.6), we have

$$\phi(y, z) = -G\Theta(y^2 - a^2/4) + \sum_{\substack{n=1 \\ n \text{ odd}}}^{\infty} \kappa_n \cos \frac{n\pi y}{a} \cosh \frac{n\pi z}{a}. \qquad (12.5.19)$$

We note that Eq. (12.5.19) satisfies the original equation, Eq. (12.5.1a), and $\phi(y = \pm a/2, z) = 0$. It remains for us to satisfy the boundary condition $\phi(y, z = \pm b/2) = 0$, i.e., Eq. (12.5.5b). Therefore, we set

$$G\Theta(y^2 - a^2/4) = \sum_{\substack{n=1 \\ n \text{ odd}}}^{\infty} \left(\kappa_n \cosh \frac{n\pi b}{2a} \right) \cos \frac{n\pi y}{a}. \qquad (12.5.20)$$

We recognise that the right-hand side of Eq. (12.5.20) is actually the Fourier series representation (in the y-variable) of the function on the left side where the Fourier coefficient is given by the term within parentheses $(\cdots)$.

To solve for this coefficient (effectively, to obtain κ_n) we multiply both sides of the equation by $\cos \frac{m\pi y}{a}$, integrate over the interval $y = -a/2$ to $y = a/2$ and

---

[†] It is worthwhile to note that we started with four unknown arbitrary constants A, B, C and D and have arrived at a single constant κ, using arguments based on physical reasoning of the problem. The above discussion is a very good illustration of applied mathematics in solving an engineering problem: we use the physics of the problem to 'help' us solve a rather complex mathematical problem!

sum over the odd values of n. Upon interchanging the summation and integration processes, we obtain

$$\sum_{n\ \text{odd}}^{\infty} \kappa_n \cosh \frac{n\pi b}{2a} \int_{-\frac{a}{2}}^{\frac{a}{2}} \cos \frac{n\pi y}{a} \cos \frac{m\pi y}{a}\, \mathrm{d}y = G\Theta \int_{-\frac{a}{2}}^{\frac{a}{2}} (y^2 - a^2/4) \cos \frac{m\pi y}{a}\, \mathrm{d}y.$$

(12.5.21)

Then from the orthogonality condition

$$\int_{-\frac{a}{2}}^{\frac{a}{2}} \cos \frac{n\pi y}{a} \cos \frac{m\pi y}{a}\, \mathrm{d}y = \begin{cases} a/2, & m = n \\ 0, & m \neq n, \end{cases}$$

(12.5.22)

we find

$$\kappa_n \frac{a}{2} \cosh \frac{n\pi b}{2a} = G\Theta \int_{-\frac{a}{2}}^{\frac{a}{2}} (y^2 - a^2/4) \cos \frac{n\pi y}{a}\, \mathrm{d}y.$$

(12.5.23a)

Upon integrating, the expression for κ_n becomes

$$\kappa_n = -\frac{8a^2 \Theta G (-1)^{(n-1)/2}}{n^3 \pi^3 \cosh \frac{n\pi b}{2a}}.$$

(12.5.23b)

Substitution in Eq. (12.5.19), yields the final expression for $\phi(y, z)$:

$$\phi(y, z) = -G\Theta(y^2 - a^2/4) - \frac{8a^2 \Theta G}{\pi^3} \sum_{n\ \text{odd}}^{\infty} \frac{(-1)^{(n-1)/2} \cos \frac{n\pi y}{a}}{n^3 \cosh \frac{n\pi b}{2a}} \cosh \frac{n\pi z}{a}.$$

(12.5.24)

Having established the stress function, the stresses are then readily obtained; upon differentiating term by term,

$$\tau_{xy} = \frac{\partial \phi}{\partial z} = -\frac{8a\Theta G}{\pi^2} \sum_{n\ \text{odd}}^{\infty} \frac{(-1)^{(n-1)/2} \cos \frac{n\pi y}{a}}{n^2 \cosh \frac{n\pi b}{2a}} \sinh \frac{n\pi z}{a}$$

(12.5.25a)

$$\tau_{xz} = -\frac{\partial \phi}{\partial y} = 2G\Theta y - \frac{8a\Theta G}{\pi^2} \sum_{n\ \text{odd}}^{\infty} \frac{(-1)^{(n-1)/2} \sin \frac{n\pi y}{a}}{n^2 \cosh \frac{n\pi b}{2a}} \cosh \frac{n\pi z}{a}.$$

(12.5.25b)

The resulting shear stress distribution is then as shown in Fig. (12.5.3).

Examining these expressions, the maximum value of τ_{xz} is seen to occur at points $D(y = \pm a/2, z = 0)$ of Fig. (12.5.3). Thus

$$\tau_{xz}\big|_{\substack{y=\pm a/2 \\ z=0}} = G\Theta a \cdot Q,$$

(12.5.26a)

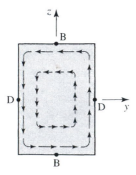

Figure 12.5.3

where

$$Q = 1 - \frac{8}{\pi^2} \sum_{n \text{ odd}}^{\infty} \frac{1}{n^2 \cosh \frac{n\pi b}{2a}}. \tag{12.5.26b}$$

The maximum value of τ_{xy} occurs at the points B $(y = 0, z = \pm b/2)$ of Fig. (12.5.3) and is given by

$$\tau_{xy}\Big|_{\substack{y=0 \\ z=\pm b/2}} = -\frac{8G\Theta a}{\pi^2} \sum_{n \text{ odd}}^{\infty} \frac{(-1)^{(n-1)/2} \tanh \frac{n\pi b}{2a}}{n^2}. \tag{12.5.27}$$

By evaluating Eqs. (12.5.26) and (12.5.27), it can be shown that for $b > a$,

$$|\tau_{xz}(\pm a/2, 0)| > |\tau_{xy}(0, \pm b/2)|. \tag{12.5.28}$$

Thus the maximum shear stresses $|\tau|_{\max}$, as given by Eq. (12.5.26a), occur at points D, the points on the boundary, that are *closest* to the centre of twist.

We observe now that Θ, the unit angle of twist, has, at this stage, not yet been obtained. However, we have not made use of the remaining relation [Eq. (12.3.25a)]

$$T = 2 \iint_A \phi(y, z) \, dA. \tag{12.5.29}$$

Substituting Eq. (12.5.24) in Eq. (12.5.29) and integrating, we find

$$T = \beta b a^3 G \Theta, \tag{12.5.30a}$$

where

$$\beta = \frac{1}{3} \left[1 - \frac{192a}{\pi^5 b} \sum_{n \text{ odd}}^{\infty} \frac{\tanh \frac{n\pi b}{2a}}{n^5} \right]. \tag{12.5.30b}$$

Hence, we have again obtained the $T - \Theta$ relation

$$\Theta = \frac{T}{GC}, \tag{12.5.31a}$$

where

$$C = \beta b a^3. \tag{12.5.31b}$$

The maximum shear stress $\tau_D = |\tau|_{\max}$ can now be evaluated in terms of the torque. Substitution of Eqs. (12.5.31) in Eq. (12.5.26a) yields

$$\tau_{\max} = \frac{TaQ}{C}. \tag{12.5.32}$$

If we define the non-dimensional constant

$$\gamma = \frac{C}{ba^3 Q} = \frac{\beta}{Q}, \tag{12.5.33a}$$

we may then write

$$\tau_{\max} = \frac{T}{\gamma ba^3} \cdot a. \tag{12.5.33b}$$

A summary of the final results and a numerical evaluation of the constants β, Q and γ for various values of b/a are presented in the accompanying table.

TORSION OF RECTANGULAR CROSS-SECTION:
Summary of Results

where

$$\theta = \frac{T}{GC}, \qquad C = \beta a^3 b, \qquad \tau_{max_D} = \frac{TaQ}{C} = \frac{T}{a^2 b\gamma},$$

$$\beta = \frac{1}{3}\left[1 - \frac{192a}{\pi^5 b}\sum_{n\,odd}^{\infty}\frac{\tanh\frac{n\pi b}{2a}}{n^5}\right],$$

$$Q = 1 - \frac{8}{\pi^2}\sum_{n\,odd}^{\infty}\frac{1}{n^2\cosh\frac{n\pi b}{2a}}$$

and

$$\gamma = \frac{\beta}{Q}.$$

b/a	γ	Q	β
1.0	0.208	0.675	0.141
1.2	0.219	0.759	0.166
1.5	0.231	0.848	0.196
2.0	0.246	0.950	0.229
2.5	0.258	0.968	0.249
3.0	0.267	0.985	0.263
4.0	0.282	0.997	0.281
5.0	0.291	0.999	0.291
10.0	0.312	1.000	0.312
∞	0.333	1.000	0.333

12.6   The membrane analogy

The de Saint Venant torsion solution, as developed in Section 12.3, is a *general* solution under the assumption of free warping. However, there exist but very few cross-sectional shapes for which one may obtain *specific* exact and closed-form solutions. Indeed the elliptic section and the equilateral triangular section are among the few shapes that yield such solutions, while that for a rectangular cross-section is expressed as a series solution.

For more geometrically complex shapes, an exact solution is difficult, if not impossible, to find. However, the general de Saint Venant solution does serve a very important purpose. We are fortunate that there exists a very useful analogy, called the **membrane analogy**, which will permit us to obtain approximate stress functions ϕ for a variety of sections. Moreover, the membrane analogy also permits one to determine the stress distribution and torsional rigidity experimentally.

To develop the idea behind this analogy, we now concentrate our discussion on the behaviour of a membrane and we momentarily digress from our discussion of torsion.

Let us therefore start by considering a thin flexible membrane stretched over an arbitrary area. (With a view to our subsequent analogy with torsion, we assume that the membrane has the same shape as the cross-section of a member under torsion). For simplicity we consider first a simply connected member (i.e., having no internal boundaries).

Let the membrane be under an initial tension force F (N/m), attached at the boundary C_0, and let the membrane be subjected to a uniform pressure p. The membrane in the undeflected and deflected position is shown in Figs. (12.6.1a) and (12.6.1b), respectively. Note that in the deflected position, the tensile force F

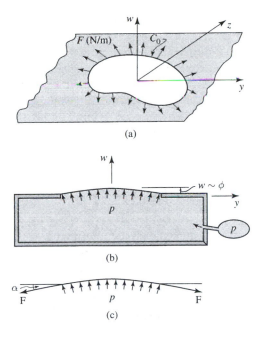

(a)

(b)

(c)

Figure 12.6.1

acts, at all points, tangentially to the membrane; at the boundary C_0, the angle of inclination is $\alpha(s)$.

We assume that the membrane is stretched sufficiently such that, due to the pressure, it undergoes only small deflections w from the undeflected position in the y–z plane of a Cartesian coordinate system, as shown in Figs. (12.6.1b and c). (Note that following the assumption of small deflections, $|\alpha| \ll 1$.) The equation governing the deflection w of the membrane, Eq. (C.5), as derived in Appendix C, is then

$$\frac{\partial^2 w}{\partial y^2} + \frac{\partial^2 w}{\partial z^2} = -\frac{p}{F}. \tag{12.6.1}$$

We notice immediately an analogy with the Poisson equation governing the stress function ϕ, Eq. (12.3.12), namely,

Torsion	Membrane
ϕ	w
$2G\Theta$	p/F

Further, since the membrane is attached at the external boundary C_0, the deflection along C_0 is constant (and is, in fact, zero).

Thus, although the two physical phenomena are quite unrelated, they are governed by similar equations. Because the two phenomena are described mathematically by analogous equations, the deflection of the membrane is said to be analogous to the stress function of a member under torsion. Therefore, as given above, the displacement w of the membrane is a direct analogy to the stress function ϕ.

Before proceeding with a discussion and the consequences of this analogy, it is worthwhile to consider the use and importance of any analogy in the analysis of physical problems.

In the case with which we are concerned, where ϕ and w are analogous, it is evident that the displacement of a stretched membrane due to a uniform pressure is rather simple to visualise since, from experience with the physical world, we have some physical intuition of the shape a membrane will take under such a pressure loading. On the other hand, the form of the stress function ϕ is an abstract concept with which we have no experience. An analogy is therefore always useful when it permits an understanding of one phenomenon by considering another phenomenon with which one is more familiar (in this case the membrane).

Let us now return specifically to the membrane analogy. Due to the analogous relation $\phi \sim w$, we have seen that we may now think of the stress function in terms of the deflection of a membrane having the same shape as the cross-section of the rod undergoing torsion [Fig. (12.6.1)]. With this interpretation in mind, and choosing arbitrarily $k_0 = 0$,[†] the expression for the torque is

$$T = 2 \iint\limits_A \phi \, dA \sim 2 \iint\limits_A w \, dA .$$

The second integral may immediately be interpreted as the volume under the deflected membrane. Thus the torque T is proportional to twice the volume under the displaced membrane [Fig. (12.6.2a)]. The analogy also gives us a very accurate description of the stress distribution.

We first note that it is possible to describe the deflection of the membrane by drawing 'contour lines' representing points of constant deflection, on the projected y–z plane [Fig. (12.6.2b)].

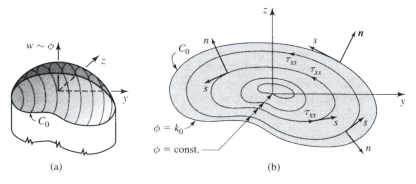

Figure 12.6.2   (a)   (b)

Now, let the parameter s be taken tangentially along any such contour (w or ϕ constant). Then, according to the definition of a 'contour line', it follows that $\frac{\partial w}{\partial s} = 0$ and therefore by analogy

$$\frac{\partial \phi}{\partial s} = 0. \qquad (12.6.2)$$

Hence, by Eq. (12.3.18a), $\tau_{xn} = 0$ on all contours of $\phi = $ const. Thus, by Eq. (12.3.18c), the resultant shear stress τ_{xR} is

$$|\tau_{xR}| = |\tau_{xs}| = \left| \frac{\partial \phi}{\partial n} \right| \sim \left| \frac{\partial w}{\partial n} \right| ; \qquad (12.6.3)$$

i.e., the resultant shear stress at any point acts *in the direction of the contour* and is proportional, at the point, to the slope of the membrane *normal* to the contour. Hence, the largest shear stress occurs at those points where the contours are closest to each other, i.e. where the membrane is steepest.

This may be illustrated best by considering two known solutions: the circular and elliptic members. We note that for the circular member [Fig. (12.6.3)], the magnitude

[†] See footnote p. 462.

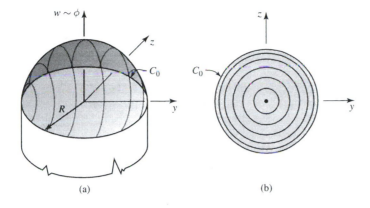

Figure 12.6.3

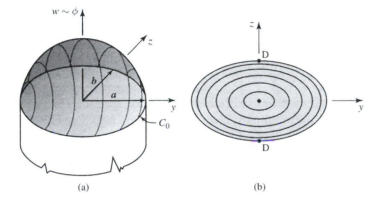

Figure 12.6.4

of the slope of the membrane at the centre is zero and increases continuously as we approach the edge. Hence the shear stress is greatest at the edge. Further, since the resultant shear stress is always in the direction tangential to the contour, the shear stress acts along concentric circles.

Similarly, the shear stress distribution of an elliptic cross-section is shown in Fig. (12.6.4). We note that the resultant shear stress is greatest at points D where the contours are closest to each other, that is, again where the slope of the membrane is steepest. Furthermore, we note that the resultant shear stress at the edge always acts tangentially to the boundary.

Having discussed the membrane analogy for simply connected members, the analysis for multi-connected members is very simple. We merely require that at all boundaries C_i, the analogous membrane have a constant value $w = \gamma_i$, where γ_i may be a different constant on every C_i. For example, for the triply connected area (containing two internal boundaries), as shown in Fig. (12.6.5), the values at the boundary are γ_1 and γ_2. Without loss of generality, γ_0 is set at zero.

By analogy with Eq. (12.3.26a), with $\gamma_0 = 0$, the torque T is proportional to twice the volume under the deflected membrane as follows:

$$T \sim 2 \iint_A w \, dA + 2 \sum_{i=1}^{n} \gamma_i A_i, \qquad (12.6.4)$$

where again A is the net area under the actual membrane and A_i is the area within C_i.

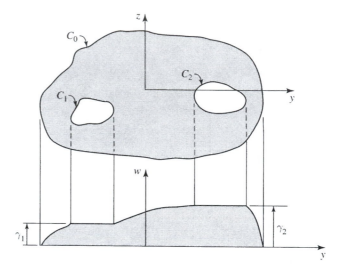

Figure 12.7.5

12.7 Torsion of a member having a narrow rectangular cross-section

(a) Derivation of membrane analogy solution

As an example of the usefulness of the membrane analogy in obtaining solutions, we shall solve the problem of a member having a *narrow* rectangular cross-section, shown in Fig. (12.7.1a).

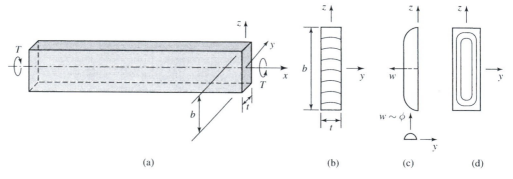

Figure 12.7.1   (a)   (b)   (c)   (d)

We denote the thickness by t and the height by b, where $t \ll b$. A Cartesian coordinate system is established with origin at the centre, as shown.

If we consider a rectangular area to be covered by a stretched membrane, based on our physical intuition, the deflection of the membrane due to a pressure will be as in Figs. (12.7.1b and c). Contour lines of equal deflection are shown in Fig. (12.7.1d).

Thus if b is much greater than the thickness t, it is evident that over an overwhelming part of the area the membrane will have a cylindrical shape, i.e. the curvature in the z-direction vanishes ($\partial^2 w/\partial z^2 = 0$). (The only parts of the cross-section where this is not true are near the ends, but those regions are not significant for a *narrow* section. Therefore if we assume that the membrane takes on a purely cylindrical shape, that is, if we neglect the end effects, our solution will not be valid near these ends.) With this approximation, the equation of the membrane becomes

$$\frac{\partial^2 w}{\partial y^2} = -\frac{p}{F}.$$

(12.7.1)

Note that w now is a function of y alone. Therefore the analogous equation for the stress function $\phi = \phi(y)$ becomes

$$\frac{d^2\phi}{dy^2} = -2G\Theta. \tag{12.7.2}$$

Letting $k_0 = 0$, the boundary condition on ϕ is then

$$\phi(y = \pm t/2) = 0.$$

We note, however, that the boundary condition at $y = -t/2$ may be replaced by a simpler condition, since by symmetry, the slope $\frac{d\phi}{dy}$ vanishes along $y = 0$. Hence, we shall use equivalent boundary conditions

$$\phi|_{y=t/2} = 0, \qquad \frac{d\phi}{dy}\bigg|_{y=0} = 0. \tag{12.7.3}$$

Integrating Eq. (12.7.2), and substituting the above boundary conditions, we obtain

$$\phi(y) = G\Theta(t^2/4 - y^2), \tag{12.7.4}$$

which is the equation of a second degree parabola.

The torsional rigidity is obtained as follows. Noting that $\phi(y = t/2) = k_0 = 0$, we have from Eq. (12.3.26b),

$$T = 2 \iint\limits_A \phi(y)\, dA$$

$$= 2G\Theta \int_{-b/2}^{b/2} \int_{-t/2}^{t/2} [(t^2/4 - y^2)\, dy]\, dz = 2G\Theta \int_{-b/2}^{b/2} (t^3/4 - t^3/12)\, dz$$

or

$$T = G\Theta bt^3/3. \tag{12.7.5a}$$

Therefore the torsion rigidity T/Θ is given by

$$\frac{T}{\Theta} = G\frac{bt^3}{3}. \tag{12.7.5b}$$

Defining the torsional rigidity constant C as before, such that $T/\Theta = GC$,

$$C = \frac{bt^3}{3}. \tag{12.7.5c}$$

The shear stresses may be immediately determined from the partial derivatives of ϕ.

First, we note that

$$\tau_{xy} = \frac{\partial\phi}{\partial z} = 0. \tag{12.7.6}$$

Then, by Eq. (12.7.4),

$$\tau_{xz} = -\frac{\partial\phi}{\partial y} = 2G\Theta y \tag{12.7.7}$$

or, substituting from Eqs. (12.7.5b) and (12.7.5c),

$$\tau_{xz} = \frac{2Ty}{bt^3/3} = \frac{2Ty}{C}. \tag{12.7.8a}$$

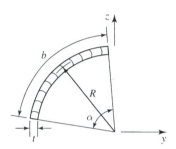

(a)   (b)

Figure 12.7.2

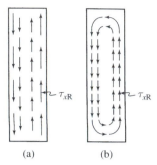

Figure 12.7.3

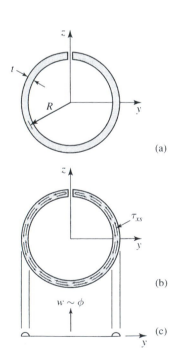

(a)

(b)

(c)

Figure 12.7.4

The maximum resultant shear stress τ_{xR} (which, here, is equal to τ_{xz}) occurs at the edges $y = \pm t/2$; i.e.,

$$\tau_{xR} = \pm \frac{Tt}{C}. \qquad (12.7.8b)$$

Note that the resultant shear stress acts along the contours and is directed in the z-direction [Fig. (12.7.2a)].

It will be noticed that the shear stress, as obtained in our solution, does not vary with z. This is consistent with our initial assumption that ϕ describes a cylindrical surface. However, near the ends $z = \pm b/2$, the membrane deflection, and hence the stress function is certainly a function of both y and z. Therefore, as mentioned above, our solution may be said to be valid at all points except near the ends $z = \pm b/2$. Away from these regions, which for $t \ll b$ are relatively very small, the results obtained are accurate. The true directions of the shear stresses are shown in Fig. (12.7.2b).[†]

Thus we see again that it is only possible to obtain results that are consistent with initial assumptions. Had we not been satisfied with such an approximation, we could not have solved the problem so easily. The inaccuracy near the ends is the penalty we have to expect from our simplification.

Before closing, we notice that the above solution may be readily extended to the torsion of a member having a cross-section consisting of a narrow curved arc, as shown in Fig. (12.7.3). If R is the radius of curvature to the centreline and α the subtended angle, then $b \to \alpha R$. Hence, from Eq. (12.7.5c),

$$C = \alpha R t^3 / 3$$

while $(\tau_{xR})_{max}$ is still given by Eq. (12.7.8b).

If α approaches 2π, the cross-section consists of a circular tube with a slit, as shown in Fig. (12.7.4a). (Note that this is still a simply connected member, since there is only a single boundary.) For this section,

$$C = \frac{2\pi R t^3}{3}. \qquad (12.7.9)$$

The shear stresses acting in the section are shown in Fig. (12.7.4b). A section through the analogous membrane is shown in Fig. (12.7.4c).

Based on our qualitative discussion, it is clear that the solution obtained here using the membrane analogy is approximate. Having found the exact solution for a general rectangular cross-section, we may now determine the error incurred by this approximation.

(b) Comparison of exact solution with membrane analogy solution for narrow rectangular sections

The numerical results presented in the table of Section 12.5 permit us to compare the exact values of τ_{max} and C for torsion of a narrow rectangular section with the approximate values as obtained using the membrane analogy. Denoting the narrow width of the rectangle by t, the exact solution given by Eqs. (12.5.31b) and

---

[†] Recall that we have determined in Section 12.3 that the shear stress existing at all points on a boundary must be tangential to the boundary.

(12.5.33b), upon letting $a \equiv t$, becomes

$$C_{(ex)} = \beta b t^3, \qquad \tau_{max(ex)} = \frac{Tt}{[(3\gamma)(bt^3/3)]},$$

while the membrane analogy solution [Eqs. (12.7.5c) and (12.7.8b)] is

$$C_{(ma)} = 0.333 b t^3, \qquad \tau_{max(ma)} = \frac{Tt}{bt^3/3}.$$

From the tabulated values of Section 12.5, we observe, as expected, that the membrane analogy solution fails to provide a good approximation for ratios of b/t which are not sufficiently large; the solution becomes exact only in the limit as $b/t \to \infty$. To obtain a quantitative measure of the error, we calculate the relative errors as follows:

$$C_{err} = \frac{[C_{(ma)} - C_{(ex)}]}{C_{(ex)}} = \frac{0.333 - \beta}{\beta} = 1/3\beta - 1,$$

$$\tau_{err} = \frac{[\tau_{max(ma)} - \tau_{max(ex)}]}{\tau_{max(ex)}} = \frac{3 - 1/\gamma}{1/\gamma} = 3\gamma - 1.$$

From the table we find the following:

For $b/t = 5$

$$C_{err} = 1/(3 \cdot 0.291) - 1 = 0.145 = 14.5\%,$$
$$\tau_{err} = 3 \cdot (0.291) - 1 = -0.127 = -12.7\%.$$

For $b/t = 10$

$$C_{err} = 1/(3 \cdot 0.312) - 1 = 0.068 = 6.8\%,$$
$$\tau_{err} = 3 \cdot (0.312) - 1 = -0.064 = -6.4\%.$$

First we note that the torsional rigidity as calculated from the membrane analogy solution is always greater than the true torsional rigidity, while the approximate maximum shear stresses are less than the true values. However, for $b/t > 10$ we observe that the approximate solution gives an error of less than 6.8%. Such an error is usually acceptable in many engineering applications, and consequently, the approximate expressions are generally used in practice for relatively narrow members.

12.8   Torsion of thin-wall open-section members

The results of Section 12.7 are now used in considering the problem of an 'open-section' prismatic member subjected to an applied torque T.

By an 'open section', we mean a cross-section that is defined by a single boundary; mathematically we have described such a section as being 'simply connected'. In particular we shall consider open sections consisting of a series of thin rectangular components, as shown in Fig. (12.8.1). Such sections are found largely in engineering practice, e.g., as with wide-flange beams, channels, angles, etc.

Consider now a torque applied to a section, as shown in Fig. (12.8.1), consisting of n component parts. Since, as we have seen, all sections other than circular cross-sections, warp due to torsion, we naturally expect the open-section to warp. We recall that the de Saint Venant solution for torsion is derived under the assumption

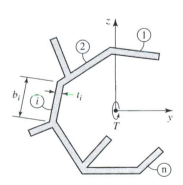

Figure 12.8.1

that any *warping can take place freely*. Therefore, in the development below, we must assume too that the entire section is free to warp with no restraint.

We continue with the basic assumption that all sections rotate with respect to one another without changing shape; i.e., all components of the entire cross-section, undergo the same rotation given by the unit angle of twist Θ.

Now, we have shown that for any cross-section subjected to a torque, the relation between the unit angle of twist and the existing torque is given by $T = GC\Theta$, or $\Theta = \frac{T}{GC}$, where GC is the torsional rigidity.

Let us denote the resisting torque of the ith component part of a section by T_i. The total resisting torque (which is equal to the applied torque T) is then the sum of the resisting torques T_i, that is

$$T = \sum_{i=1}^{n} T_i. \tag{12.8.1}$$

Note, too, from Eqs. (12.7.5), that the rotation of the ith component is given by

$$\Theta_i = \frac{T_i}{GC_i}, \tag{12.8.2a}$$

where the torsional rigidity of the ith component is

$$C_i = \frac{b_i t_i^3}{3} \tag{12.8.2b}$$

and where b_i and t_i are the height and thickness of the ith component, respectively.

Then, if the rotation of each component is the same, i.e., if

$$\Theta_1 = \Theta_2 = \cdots = \Theta_i = \cdots = \Theta_n = \Theta, \tag{12.8.3}$$

from Eq. (12.8.2a),

$$\frac{T_1}{GC_1} = \frac{T_2}{GC_2} = \cdots = \frac{T_i}{GC_i} = \cdots = \frac{T_n}{GC_n} = \Theta. \tag{12.8.4a}$$

In the above, Θ is the unit angle of twist of the entire section. Thus, from this last relation, we find that the resisting torque of the ith component is

$$T_i = GC_i\Theta. \tag{12.8.4b}$$

Substituting in Eq. (12.8.1),

$$T = G\Theta \sum_{i=1}^{n} C_i = GC\Theta, \tag{12.8.5}$$

where C, the torsional rigidity of the entire cross-section, is

$$C = \sum_{i=1}^{n} C_i. \tag{12.8.6a}$$

Hence

$$T_i = \left(\frac{C_i}{C}\right) T, \tag{12.8.6b}$$

i.e., each component carries a part of the total torque proportional to its own torsional rigidity.

Once the quantities T_i have been established, the largest shear stress in the ith component can be obtained from Eq. (12.7.8b); that is

$$[\tau_{xR}]_i = \frac{T_i t_i}{C_i}$$

and hence

$$[\tau_{xR}]_i = \frac{T \cdot t_i}{C}. \qquad (12.8.7)$$

From Eq. (12.8.7) we observe that the *maximum shear stress will occur in the thickest component*.

We emphasise here that the expressions derived above are approximate. The approximations are due to (a) the approximations inherent in the results of Section 12.7, (b) neglecting the effects at points where the ends are joined and (c) the basic assumption that the shape of the cross-section is not altered during rotation.

Specifically, we recall that the stresses, as given by Eq. (12.7.8b) or (12.8.7), are not valid near the ends of a narrow rectangular cross-section. Moreover, it is clear that for a cross-section consisting of several such components, the shear stress distribution existing at the junctures of the components is very complex. In the following section we obtain an approximate expression for the shear stresses in the region of the junctures.

Finally, as was mentioned at the beginning of this section, the above analysis is valid under the assumption that there is no restraint against warping. If permitted to warp freely ('free warping'), it is found that thin-wall open sections (e.g., I-sections, open tubular sections, etc.) can undergo considerable warping. In practice, however, actual supports for such members usually provide restraints against warping thereby inducing longitudinal stresses τ_{xx}. The analysis for torsion that takes into account such restraints is much more complex and is beyond the scope of our present study.

Example 12.1:   A structural steel angle, L76×76×6.4, is subjected to a torque $T = 200$ N-m acting at the end of a cantilever beam of length $L = 1.2$ m, as shown in Fig. (12.8.2). Determine the angle of twist α of the free end and the maximum shear stress. (*Note: G = 76 GPa.*)

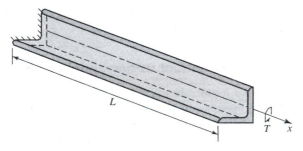

Figure 12.8.2

Solution: From tables of structural sections, the length and thickness of the legs of the angle are 76 mm and 6.4 mm respectively. The true cross-section of the angle is shown in Fig. (12.8.3a). We note that at the juncture of the two legs, there exists a rounded 'fillet'. (These fillets are a result of the rolling process used in the manufacture of structural sections. As we shall see in the following section, these fillets are useful in eliminating high stresses at the juncture.) However, to obtain the torsional rigidity

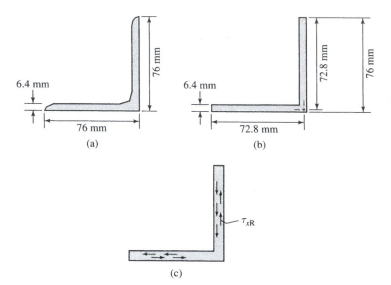

Figure 12.8.3

and shear stress in the section, one neglects the fillets and models the section as shown in Fig. (12.8.3b).

Using Eqs. (12.8.2b) and (12.8.6a),

$$C = \sum_{i=1}^{2} C_i = \sum_{i=1}^{2} \frac{b_i t_i^3}{3} = 2 \frac{72.8 \cdot 6.4^3}{3} = 1.27 \times 10^4 \text{ mm}^4.$$

Therefore (noting that $G = 76$ GPa $= 76 \times 10^3$ N/mm²),

$$\alpha = \Theta L = \frac{TL}{GC} = \frac{(200 \times 10^3)(1200)}{(76 \times 10^3)(1.27 \times 10^4)} = 0.249 \text{ rad} = 14.3°$$

and from Eq. (12.8.7),

$$\tau_{xR} = \frac{T \cdot t}{C} = \frac{(200 \times 10^3)(6.4)}{1.27 \times 10^4} = 101 \text{ N/mm}^2 = 101 \text{ MPa}.$$

Note that since the thickness of the angle is constant, the same shear stress will occur in both legs. The shear stress distribution is shown in Fig. (12.8.3c). We emphasise again that these calculated values are valid only in the region sufficiently far away from the ends and far from the juncture.

12.9 Shear stress at a re-entrant corner: approximate solution

We obtain here a solution for the shear stress in the region of the juncture of two (narrow rectangular) components of an open section. Let us, in particular, consider a two-component section that forms a right angle.

We consider a section of constant thickness t, containing a fillet, as described in Example 12.1 above. Such a fillet is described by a portion of a circle with centre O and radius a, as shown in Fig. (12.9.1). Since the thickness is constant, the shear stresses in both rectangular elements away from both the ends and the juncture is, by Eq. (12.8.7),

$$\tau = \frac{Tt}{C}.$$

(12.9.1a)

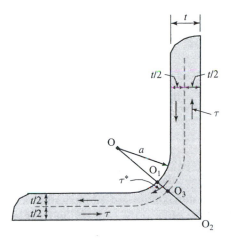

Figure 12.9.1

Noting that $\Theta = \frac{T}{GC}$, we observe the relationship

$$G\Theta = \frac{\tau}{t}.$$ (12.9.1b)

For future reference, we define points O_1 and O_2 as shown in the figure, and further define point O_3 such that the distance $\overline{O_1\text{–}O_3} = t/2$. We denote the shear stress at points in the neighbourhood of the line $O_1\text{–}O_2$ in the region of the juncture by τ^*.

Our analysis below is again based on the membrane analogy. Now, it is clear that the deflected membrane, which is analogous to the stress function ϕ and which is governed by Eq. (12.6.1), $\nabla^2 w = -p/F$, has the form of a cylindrical surface, as shown in Fig. (12.9.2). In the region of the corner, it is clearly more appropriate to use a cylindrical coordinate system (r, θ), where θ is an angle as shown in the figure. We now make a reasonable and simplifying, but important, assumption: we assume that the deflected membrane surface in the region of the line $O_1\text{–}O_3$ is independent of θ; that is, $w = w(r)$. The Laplacian operator ∇^2 in this coordinate system is then given by

$$\nabla^2 \equiv \frac{d^2}{dr^2} + \frac{1}{r}\frac{d}{dr}$$ (12.9.2)

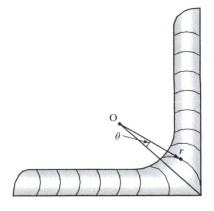

Figure 12.9.2

and hence the governing equation of the membrane becomes

$$\frac{d^2 w}{dr^2} + \frac{1}{r}\frac{dw}{dr} = -\frac{p}{F}.$$ (12.9.3)

From the membrane analogy, we then write

$$\frac{d^2 \phi}{dr^2} + \frac{1}{r}\frac{d\phi}{dr} = -2G\Theta.$$ (12.9.4a)

Substituting Eq. (12.9.1b),

$$\frac{d^2 \phi}{dr^2} + \frac{1}{r}\frac{d\phi}{dr} = -\frac{2\tau}{t}.$$ (12.9.4b)

Now, we recall that the slope of the stress function ϕ represents the shear stress; more specifically, the slope of ϕ in the r-direction represents the shear stress in a direction perpendicular to r, i.e. in the circumferential direction. Hence for the desired shear stress, which we denote here as $\tau^*(r)$ [see Fig. (12.9.1)], we have

$$\tau^*(r) = -\frac{d\phi(r)}{dr}.$$ (12.9.5)

Using this relation, Eq. (12.9.4b) becomes

$$\frac{d\tau^*}{dr} + \frac{\tau^*}{r} = \frac{2\tau}{t}.$$ (12.9.6)

Now Eq. (12.9.6) can be solved as a first-order inhomogeneous differential equation with variable coefficients. However, noting the identity

$$\frac{d(\cdots)}{dr} + \frac{1}{r}(\cdots) = \frac{1}{r}\frac{d\,[r\,(\cdots)]}{dr},$$ (12.9.7)

we may write Eq. (12.9.6) in a more simple form as

$$\frac{1}{r}\frac{d\,(r\tau^*)}{dr} = \frac{2\tau}{t}.$$ (12.9.8a)

Integrating, we find

$$\tau^*(r) = \frac{\tau r}{t} + \frac{B}{r},$$ (12.9.8b)

where B is a constant of integration. Now, from the membrane analogy, it is clear that $\tau^* = 0$ at some point along the line O_1–O_2. Let us *assume* that $\tau^* = 0$ at O_3; i.e., $\tau^*(r = a + t/2) = 0$. Using this condition, we obtain $B = -\frac{\tau}{t}(a + t/2)^2$. Hence

$$\tau^*(r) = \frac{\tau}{t}\left[r - \frac{(a + t/2)^2}{r}\right].$$ (12.9.9)

We note that this last equation yields the shear stress along the line O_1–O_2 in terms of the approximate shear stress τ as calculated by Eq. (12.9.1a). Specifically, at point O_1, $r = a$,

$$|\tau^*(a)| = \tau(1 + t/4a).$$ (12.9.10)

From this last expression we observe that as $a \to 0$, $\tau^* \to \infty$; that is, the shear stress tends to infinity if no fillet exists. It is therefore evident that by introducing a fillet in an open section, the shear stress at the juncture is significantly reduced.

The variation of the ratio τ^*/τ at $r = a$ is plotted in Fig. (12.9.3) as a function of a/t.

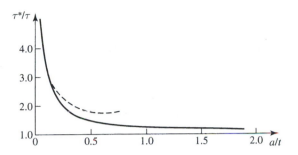

Figure 12.9.3

Finally, it should be mentioned that although these results prove to be very useful, the above analysis is only approximate. (Recall that two basic simplifying assumptions were made: (i) that ϕ is independent of the angle θ and (ii) that $\tau^* = 0$ at the point O_3.) A comparison with the more exact solution, shown by the broken lines in Fig. (12.9.3), reveals that the error of the approximation is insignificant for relatively small values of a/t.

12.10   Torsion of closed-section members: thin-wall sections

The analysis of members having closed thin-walled sections (i.e., multiply connected areas) is also simplified considerably by use of the membrane analogy. In particular, consider the case of a tubular section (*doubly connected*) containing a single interior boundary [Fig. (12.10.1a)]. We denote the (tangential) parameter along the centreline of the wall by s (*positive in a counter-clockwise direction*) and assume that the variable thickness of the wall $t(s)$ is much smaller than $R(s)$, which is measured from the centre of twist [Fig. (12.10.1b)].

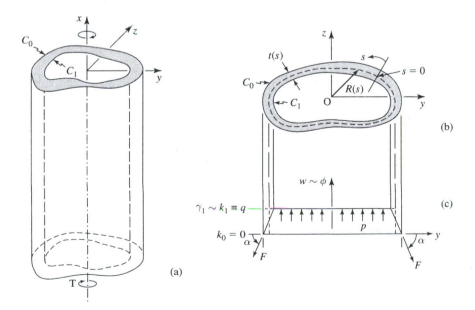

Figure 12.10.1

Now, for a doubly connected section, from Eq. (12.3.26a),

$$T = 2 \iint\limits_{A} \phi \, \mathrm{d}A - 2k_0 A_0 + 2k_1 A_1, \qquad (12.10.1a)$$

where k_0, k_1 are constant values of ϕ on C_0, C_1, respectively.

Using the membrane analogy, we recall that we may interpret the stress function ϕ as the deflection of a membrane. Therefore, if $t \ll R$, the membrane deflection may readily be assumed to have the approximate shape, as shown in Fig. (12.10.1c); i.e., we *assume* a linear variation of ϕ through the wall thickness. Then, without loss of generality we may set $k_0 = 0$ and, letting $k_1 \equiv q$, a constant, we have

$$T = 2 \iint\limits_{A} \phi \, \mathrm{d}A + 2q \, A_1. \qquad (12.10.1b)$$

Again, letting s represent the parameter (tangential to the centreline) along a contour describing a constant value of ϕ, the shear stress τ_{xs} is given by Eq. (12.3.18b), namely

$$\tau_{xs} = -\frac{\partial \phi}{\partial n}. \qquad (12.10.2)$$

However, since s is measured along the contour line, n is always the outward directed normal to the centreline. Therefore, assuming a linear variation of ϕ, at all points

$$\frac{\partial \phi}{\partial n} = -q/t \qquad (12.10.3a)$$

and hence

$$\tau_{xs}(s) = \frac{q}{t(s)}. \qquad (12.10.3b)$$

Note that as a result of the assumption that ϕ varies linearly (and consequently that the slope of ϕ is constant) through the thickness t at any fixed value of s, the shear stress τ does not vary through the thickness.

The quantity

$$q = \tau_{xs} t \qquad (12.10.3c)$$

is usually called the **shear flow**. We thus see that the shear flow is constant throughout the section.

Now, recall that the torque T, as given by Eq. (12.10.1b), may be interpreted as being represented by twice the volume under the deflected membrane. From Fig. (12.10.1), it is evident that the volume may be approximated by Aq, where A is now the *cross-sectional area within the centreline*.[†] Thus

$$T = 2Aq. \qquad (12.10.4)$$

Substituting from Eq. (12.10.3c)

$$\tau_{xs}(s) = \frac{T}{2At(s)}. \qquad (12.10.5)$$

[†] The volume of the rectangle represented by the broken lines is clearly the same as that represented by the trapezoid.

From Eq. (12.10.5), we observe that the shear stress is *maximum at those points where the wall thickness is smallest*. The shear flow is shown in Fig. (12.10.2). Note, as mentioned above, that across any line segment (e.g., BB of the figure), the shear stress is constant; i.e., it does not vary across the thickness of the wall. Essentially, we have computed the average value across BB, a result that is inherent to the assumption of a linear variation of ϕ throughout the wall thickness.

In passing, we notice that the shear stress $\tau_{xn} = 0$, since $\frac{\partial \phi}{\partial s} = 0$. Thus the resultant shear stress is

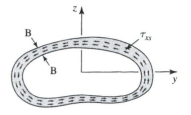

Figure 12.10.2

$$\tau_{xR} = \tau_{xs} = \frac{T}{2At(s)} \qquad (12.10.6)$$

and is always directed, at all points, tangentially to the centreline, as shown in Fig. (12.10.2).

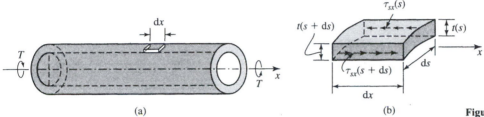

(a)   (b)   **Figure 12.10.3**

It is appropriate to point out that, alternatively, the above relations can be easily derived directly, as follows, without invoking the membrane analogy. To this end, consider a small element of the thin-wall section subjected to the torque T, as shown in Fig. (12.10.3). Then, from the equilibrium condition $\Sigma F_x = 0$, we find

$$- [\tau_{sx}(s) \cdot t(s)]\, dx + [\tau_{sx}(s + ds) \cdot t(s + ds)]\, dx = 0. \qquad (12.10.7a)$$

Noting that

$$\tau_{sx}(s + ds) \cdot t(s + ds) = \tau_{sx}(s) \cdot t(s) + \frac{d}{ds}[\tau_{sx}(s) \cdot t(s)]\, ds,$$

we obtain

$$\frac{d}{ds}[\tau_{sx}(s) \cdot t(s)] = 0, \qquad (12.10.7b)$$

that is, the bracketed term does not vary with s. Therefore

$$\tau_{sx}(s)t(s) = q \quad \text{(constant)} \qquad (12.10.7c)$$

or

$$\tau_{xs}(s) = \frac{q}{t(s)},$$

which agrees with Eq. (12.10.3b).

Consider now the resulting torque due to the stresses τ_{xs}, which act in the tangential direction s of the wall. Let $\mathbf{r} = R(s)\mathbf{g}$, (where $|\mathbf{g}| = 1$) denote the vector from the centre of twist to point P [Fig. (12.10.4)]. The torque about the x-axis due to the stresses acting on a small element $t \cdot ds$ is then given by the vector product

$$d\mathbf{T} = \mathbf{r} \times [\tau_{xs} \cdot t\, ds]\mathbf{s} = [\tau_{xs}t\, ds]\mathbf{r} \times \mathbf{s}, \qquad (12.10.8)$$

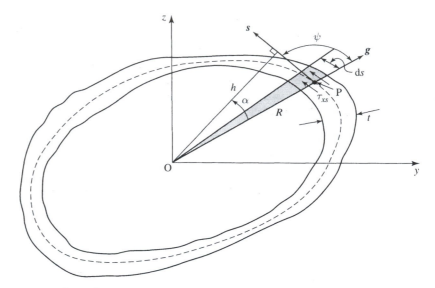

Figure 12.10.4

where s is a unit vector ($|s| = 1$) acting tangentially at all points of the centreline. We note that

$$|r \times s| = R|g||s| \sin(g, s) = R|g||s| \sin \psi. \tag{12.10.9a}$$

But $|g| = |s| = 1$ and $\sin \psi = \cos \alpha$, as shown in Fig. (12.10.4). Hence

$$|r \times s| = R \cos \alpha = h, \tag{12.10.9b}$$

where h is the perpendicular distance from point O to s. Substituting in Eq. (12.10.8),

$$dT = |dT| = \tau_{xs} t(s) h \, ds = 2\tau_{xs} t(s) \, dA, \tag{12.10.10a}$$

where $dA = h \cdot ds/2$ is the incremental triangular shaded area of Fig. (12.10.4). Using Eq. (12.10.3b), we obtain

$$T = 2q \iint\limits_A dA = 2q A \tag{12.10.10b}$$

or $q = \frac{T}{2A}$, from which we again recover the relation $\tau_{xs} = \frac{T}{2At(s)}$.

To obtain an expression relating the unit angle of twist Θ to the applied torque T, we again invoke the membrane analogy. From Fig. (12.10.1c) we obtain, using the equilibrium condition $\Sigma F_w = 0$ (w, denoting here the direction of the analogous membrane displacement),

$$pA - \oint\limits_{C_0} F \sin \alpha \, ds = 0, \tag{12.10.11a}$$

where we recall from Section 12.6 that p represents the pressure under the membrane, and F is the tensile force (N/m) acting tangentially to the membrane at an angle $\alpha(s) \ll 1$ along the contour C_0. (Here $\oint_{C_0}$ indicates integration around the closed contour C_0.) Since $F = $ const., we have

$$\frac{p}{F} = \frac{1}{A} \oint\limits_{C_0} \sin \alpha \, ds. \tag{12.10.11b}$$

Now, according to the membrane analogy, all membrane displacements are assumed to be small and consequently $\alpha \ll 1$. Therefore $\sin\alpha \sim \tan\alpha = \gamma/t(s)$ [see Fig. (12.10.1c)[†]]. Then, recalling the analogy,

$$p/F \sim 2G\Theta, \qquad \gamma \equiv q, \qquad (12.10.13)$$

we have

$$2G\Theta = \frac{1}{A} \oint_{C_0} \frac{q}{t(s)} \, ds = \frac{1}{A} \oint_{C_0} \tau_{xs} \, ds$$

or

$$\Theta = \frac{1}{2GA} \oint_{C_0} \tau_{xs} \, ds. \qquad (12.10.14)$$

This relation, known as *Bredt's formula*, relates the unit angle of twist to the shear stress existing in a thin-wall closed-section member [see Fig. (12.10.5)]. It is emphasised here that this relation is valid in general; that is, for any τ_{xs} existing in the section (irrespective of the cause), the angle of twist is given by Eq. (12.10.14).[‡]

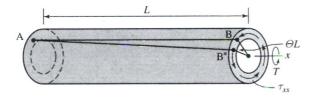

Figure 12.10.5

If the shear stress is due to an applied torque, we obtain, after substituting Eq. (12.10.6),

$$\Theta = \frac{T}{4A^2 G} \oint_{C_0} \frac{ds}{t(s)}, \qquad (12.10.15a)$$

where, as before, A is the area within the centreline of the thin wall.

For a constant wall thickness t, Eq. (12.10.15a) becomes

$$\Theta = \frac{TS}{4A^2 Gt}, \qquad (12.10.15b)$$

where S is the perimeter along the centreline.

Example 12.2: A torque $T = 9,000$ N-m is applied to an aluminium member ($G = 26$ GPa) having a cross-section as shown in Fig. (12.10.6). Determine (a) the shear stress distribution in the cross-section and (b) the unit angle of twist.

Solution:

$A = 50 \times 100 = 5 \times 10^3$ mm^2.
$q = \frac{T}{2A} = \frac{9 \times 10^6}{10 \times 10^3} = 900$ N/mm.

[†] Note that for pictorial clarity, the height of the deflected membrane γ has been shown exaggerated.
[‡] The validity of this statement will be more clearly apparent following the rederivation, in Chapter 14, of Bredt's formula, using the principle of complementary virtual work.

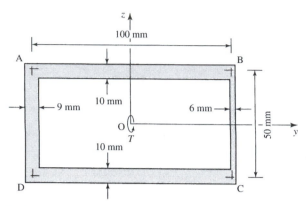

Figure 12.10.6

Using the relation, $\tau_{xR} = \frac{q}{t}$ we obtain:

From A to B: $\tau_{xR} = \frac{900}{10} = 90 \text{ N/mm}^2 = 90 \text{ MPa}$;

From B to C: $\tau_{xR} = \frac{900}{6} = 150 \text{ MPa}$;

From C to D: $\tau_{xR} = 90 \text{ MPa}$;

From D to A: $\tau_{xR} = \frac{900}{9} = 100 \text{ MPa}$.

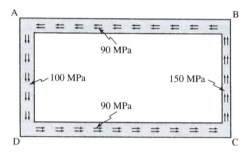

Figure 12.10.7

The stress distribution is as shown in Fig. (12.10.7). From Eqs. (12.10.15),

$$\Theta = \frac{T}{4A^2 G} \oint \frac{ds}{t} = \frac{T}{4A^2 G} \sum_{i=1}^{4} \frac{S_i}{t_i},$$

i.e.

$$\Theta = \frac{9 \times 10^6}{4(5 \times 10^3)^2 (26 \times 10^3)} \left(\frac{100}{10} + \frac{50}{6} + \frac{100}{10} + \frac{50}{9} \right)$$

$$= 1.173 \times 10^{-4} \text{ rad/mm} = 6.72°/\text{m}.$$

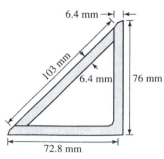

Figure 12.10.8

Example 12.3: It is decided to strengthen the section of Example 12.1 by welding a steel plate, 6.4-mm thick, to the ends of the legs of the L76×76×6.4 structural steel angle so as to form a closed section over the entire length $L = 1.2$ m of the cantilevered member [Fig. (12.10.8)]. Determine (a) the total angle of twist α of the free end and (b) the maximum shear stress in the member due to the applied torque $T = 200$ Nm.

Solution: The total angle of twist, from Eq. (12.10.15b), for this closed section, is

$$\alpha = \Theta L = \frac{TSL}{4A^2 Gt}. \tag{12.10.16}$$

Recalling that A and S relate to the centre lines of the thin wall, we have $A = 72.8 \times 72.8/2 = 2650 \text{ mm}^2$ and $S = 2(72.8) + 103 = 249$ mm.

Therefore, substituting in Eq. (12.10.16), with $G = 76$ GPa,

$$\alpha = \frac{(200 \times 10^3)(249)(1200)}{4(2650)^2(76 \times 10^3)(6.4)} = 4.37 \times 10^{-3} \text{ rad} = 0.25°.$$

The maximum shear stress is $\tau = \frac{T}{2At} = \frac{200 \times 10^3}{(2)(2650)(6.4)} = 5.90$ N/mm$^2 = 5.90$ MPa.

$\square$

In comparing the rotation of the strengthened (closed) section with the calculations of the open section of Example 12.1, we observe that the rotation has been reduced from $\alpha = 14.3°$ to $\alpha = 0.25°$, while the maximum shear stress has been reduced from 101 to 5.90 MPa. Thus we note that a closed thin-wall section is much more efficient and very much stiffer than a corresponding open section. This point will be discussed further in the context of the membrane analogy.

12.11 Torsion of multi-cell closed thin-wall sections

The membrane analogy lends itself very well to analysing the problem of a member having a cross-section consisting of more than one 'hole' (but for which all walls are relatively thin). Specifically, we shall consider a section that is triply connected, i.e., one having two 'holes'. The section is conventionally referred to as having two 'cells' [Fig. (12.11.1a)].

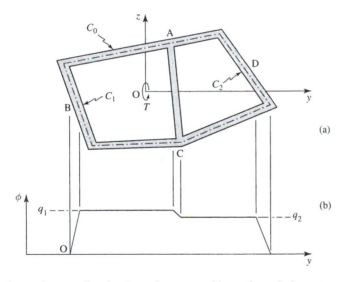

Figure 12.11.1

Without loss of generality, let ϕ on the external boundary C_0 be zero, and further let ϕ have (unknown) constant values q_1 and q_2 on the interior boundary C_1 and C_2, respectively [see Fig. (12.11.1b)]. Again, we assume a linear variation of ϕ across the wall thicknesses.

From the membrane analogy, the resisting torque of the left and right cells is given, respectively, by

$$T_1 = 2q_1 A_1, \qquad T_2 = 2q_2 A_2, \tag{12.11.1}$$

where A_1 and A_2 are the areas of the respective cells within the centerline of the walls.

The total torque is then given by

$$T = T_1 + T_2 = 2(q_1 A_1 + q_2 A_2). \tag{12.11.2}$$

As in the case of open sections, we assume that the entire section rotates without changing its shape; it then follows that $\Theta_1 = \Theta_2 = \Theta$, where Θ_i is the unit angle of rotation of the ith cell.

Hence, by Eq. (12.10.14),

$$\frac{1}{A_1} \oint_{C_1} (\tau_{xR})_1 \, ds = \frac{1}{A_2} \oint_{C_2} (\tau_{xR})_2 \, ds, \qquad (12.11.3)$$

where the parameter s is always taken in a counter-clockwise direction and where $(\tau_{xR})_i$ is the resultant shear stress along the closed path C_i ($i = 1$, ABCA; $i = 2$, ACDA).

Now, using Eq. (12.10.3b), along the path ABC

$$(\tau_{xR})_1 = q_1/t. \qquad (12.11.4a)$$

Similarly, along the path CDA,

$$(\tau_{xR})_2 = q_2/t. \qquad (12.11.4b)$$

However, since the contour integrals of Eq. (12.11.3) are for closed paths, we must also obtain the shear stress along AC. We note, however, from Fig. (12.11.1b) that the shear flow in this segment is given by

$$q = q_1 - q_2. \qquad (12.11.5)$$

Hence, the shear stress in A to C is given by

$$(\tau_{xR})_{1 \, or \, 2} = (q_1 - q_2)/t. \qquad (12.11.6)$$

Substituting Eqs. (12.11.4) and (12.11.6) in Eq. (12.11.3) yields an equation in the unknowns q_1 and q_2. Equations (12.11.2) and (12.11.3) may then be regarded as two equations from which we may solve for these unknowns.

The solution is best illustrated by means of an example.

Example 12.4: A member whose cross-section consists of two cells having walls of constant thickness t, as shown in Fig. (12.11.2), is subjected to a torque T. Determine (a) the shear stresses and (b) the unit angle of twist.

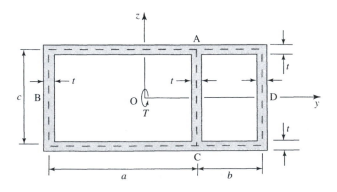

Figure 12.11.2

Solution: From Eq. (12.11.2),

$$2(q_1 \cdot ac + q_2 \cdot bc) = T. \qquad (12.11.7a)$$

From Eq. (12.11.3),

$$\frac{1}{ac}\frac{1}{t}[q_1(a + c + a) + (q_1 - q_2)c] = \frac{1}{bc}\frac{1}{t}[q_2(b + c + b) - (q_1 - q_2)c]$$

or

$$[2b(a+c)+ac]q_1 - [2a(b+c)+bc]q_2 = 0. \qquad (12.11.7b)$$

Solving Eqs. (12.11.7) for the two unknowns, we obtain

$$q_1 = \frac{T}{4c}\left[\frac{bc+2a(b+c)}{a^2(b+c)+b^2(a+c)+abc}\right],$$

$$q_2 = \frac{T}{4c}\left[\frac{ac+2b(a+c)}{a^2(b+c)+b^2(a+c)+abc}\right]. \qquad (12.11.8)$$

Then, along ABC, $\tau_{xR} = q_1/t$; along ADC, $\tau_{xR} = q_2/t$ and along AC, $\tau_{xR} = (q_1 - q_2)/t$.

Using Bredt's formula, the unit angle of twist is given, for example, by

$$\Theta = \frac{1}{2A_1 G}\oint_{C_1}(\tau_{xR})_1\,ds = \frac{1}{2actG}[q_1(2a+c)+(q_1-q_2)c]. \quad (12.11.9)$$

If the two cells are the same ($a = b$),

$$q_1 = q_2 = \frac{T}{4ac}. \qquad (12.11.10)$$

Thus, in this case, the shear flow vanishes in the centre wall; hence the section acts as though the centre wall did not exist. The unit angle of twist, Eq. (12.11.9), is then given by

$$\Theta = \frac{1}{G}\left(\frac{2a+c}{8a^2c^2t}\right)T. \qquad (12.11.11)$$

The analysis of multi-connected cells having a larger number of cells proceeds similarly as above.

12.12 Closure

The analysis of prismatic members subjected to torsion was based entirely, as we have seen, upon the general de Saint Venant torsion solution. Indeed, the solution for the torsion of members of arbitrary cross-section is much more complex than that for members having a circular cross-section. The distinct difference in the solutions is based upon the fact that, in general, warping in the longitudinal direction takes place for cross-sections other than circles. At this point, it may be worthwhile to re-emphasise a basic assumption in the de Saint Venant solution, namely, that there is no restraint against such warping.

We have noticed, too, that it is only possible to obtain *exact* solutions from the general solution for a small number of relatively simple geometric shapes. If this were the only use of the de Saint Venant solution, it would be of limited value. However, we have seen that the general solution, together with the membrane analogy, permits us to obtain approximate solutions for members having many particular shapes. Using this analogy, we can often 'guess' intuitively at the general shape of the stress function and thus obtain approximate solutions. From the consideration of the analogy, we can deduce the general shear stress distribution and the relative torsional rigidity. For example, recalling that the torque is proportional to the volume under the deflected membrane, and hence under the surface of the stress function

ϕ, we may immediately deduce qualitatively that the torque T required to produce a given unit angle of twist Θ in a closed section is much greater than that required for the corresponding open section [compare Figs. (12.10.1c) and (12.7.4c)]. Hence the torsional rigidity of a closed section is always greater than that of the corresponding open section.

A second use of the analogy, which was not considered here, is its application in analysing members of great geometric complexity by experimental means. Indeed for particularly complex cross-sectional shapes, even approximate solutions are not possible. Experimental measurements of the displacement of thin membranes subject to pressures, then permit us to determine the stress distribution and torsional rigidities to a desired degree of accuracy.

PROBLEMS

Sections 3 and 4

12.1: Using the representation for the shear stresses τ_{xy} and τ_{xz} expressed in terms of the stress function $\phi(y, z)$ [Eqs. (12.3.8)], show that the resulting shear forces $V_y=0$ and $V_z=0$ on any arbitrary cross-section.

12.2: A torsional moment is applied to two different bars of the same material. One is a bar of circular cross-section and the other of elliptical cross-section, both of which have the same cross-sectional area. The torsional rigidity constants of the two cross-sections are C_0 and C, respectively. The ratio of the semi-major axis a to the semi-minor axis b of the elliptic cross-section is defined as k ($k > 1$). Show that the ratio of the torsional rigidities of the two sections is $\frac{C}{C_0} = \frac{J_0}{J}$; i.e., prove that the ratio of the torsional rigidities of the two bars is *inversely* proportional to the ratio of their polar moments of cross-sectional area.

12.3:* A prismatic member, having an equilateral triangular cross-section whose sides are of length a as shown in Fig. (12P.3), is subjected to a torsional moment T.

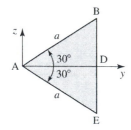

Figure 12P.3

(a) Analogous to the torsion solution for an elliptic section, construct a stress function $\phi(y, z) = \kappa \cdot f(y, z)$ such that $\phi = 0$ identically along the outer edges of the section.

(b) Derive the expression for the torsional rigidity, namely $GC = G\frac{\sqrt{3}a^4}{80}$.

(c) (i) Assuming that the maximum shear stress occurs along the edge of the cross-section, show (analytically) that along the edge BE it occurs at point D. (ii) Indicate (by means of a figure) other points in the cross-section where the maximum shear stress occurs. What conclusion can be drawn? (iii) Determine the maximum shear stress.

12.4:* Using the general de Saint Venant torsion solution, derive directly, by integrating the equation $\nabla^2\phi=-2G\Theta$, the Coulomb solution for a member having a solid circular cross-section of radius R; that is, determine the shear stress $\tau_{x\theta}$ and the torsional stiffness.

Note: Using polar coordinates (r,θ), the Laplacian for the axisymmetric case is given by

$$\nabla^2 = \frac{d^2}{dr^2} + \frac{1}{r}\frac{d}{dr} = \frac{1}{r}\frac{d}{dr}\left(r\frac{d}{dr}\right).$$

12.5:* Show that the resultant shear stresses τ_{xR} due to a torsional moment acting on a member of elliptic cross-section, as shown in Fig. (12.4.2), always lie along elliptic curves concentric with the ellipse defining the cross-section. (*Hint*: Show that the slope of τ_{xR} at any point $P(y, z)$ is that of a concentric ellipse.)

Section 5

12.6: An engineer naively calculates (wrongly) the maximum shear stress due to torsion of a member having a rectangular cross-section $a \times b$ (where $b=2a$) by using the expression $\tau = TR/J$ (where J is the polar moment of area of the section and R is the largest distance from the centre of twist) and assumes that GJ represents the torsional rigidity. Making use of the results given in the table of p. 468, determine the percentage error in the calculations.

12.7: A steel rod ($G = 76$ GPa) of rectangular cross-section 6 mm $\times$ 18 mm and 2 m in length is subjected to a torsional moment. The two ends rotate with respect to each other by $4.5°$. Determine the maximum shear stress in the member.

Sections 6 and 7

12.8: The shear stresses due to a torsional moment T acting on a member having a narrow rectangular cross-section $b \times t$ ($t \ll b$) are given by Eqs. (12.6.6) and (12.6.7). (a) By integrating appropriately the stresses over the area, determine the torsional moment that is produced by these stresses and (b) explain the discrepancy with the applied torque T.

12.9:* (a) In deriving the equation, $\nabla^2\phi(y, z) = -2G\Theta$, which the stress function $\phi(y, z)$ must satisfy, the origin O of the (y, z) coordinates was taken to coincide with the centre of twist. Show that all equations remain valid irrespective of the location of the origin O. (b) The torsional stress function ϕ of a rod, having a narrow rectangular cross-section $b \times t$ ($t \ll b$), is given by Eq. (12.6.4), namely $\phi = G\Theta(t^2/4 - y^2)$. Invoking the membrane analogy solution, construct analogously a stress function ϕ for a member whose cross-section is a narrow triangular shape, as shown in Fig. (12P.9), where $t_0 \ll b$. Express ϕ in terms of a (y, z) coordinate system (whose origin, O is as shown in the figure), and determine an approximate expression for the resulting torsional rigidity GC.

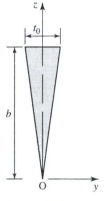

Figure 12P.9

12.10: A channel member of length L, whose cross-section is as shown in Fig. (12P.10) (where $t \ll c$, $t \ll h$), is fixed at one end and subjected to a torsional moment T at the free end. (a) Determine the horizontal component of displacement Δ_y of point A at the free end in terms of c, h, t, L, G and T. (b) From the given data, it is not possible (from the development in this chapter) to determine the vertical component of displacement Δ_z of A? Explain. (c) Determine Δ_y and Δ_z at A for the limiting case $c \to 0$. Why is it possible to determine Δ_z for this limiting case?

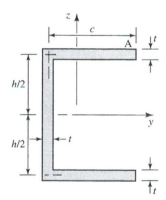

Figure 12P.10

Section 8

12.11: A thin-wall closed equilateral triangular cross-section ($t \ll a$), as shown in Fig. (12P.11), is subjected to a torsional moment T. Determine (a) the shear stress and (b) the torsional rigidity constant C of the section, in terms of a and t.

12.12: The shear stress existing at points on the centreline of a closed thin-wall circular tube subjected to a torque T is given by the exact expression $\tau_{ex} = \frac{TR}{J}$, where J is the polar moment of area and R is the radius of the centreline. The average stress in the tube, as calculated from the membrane analogy solution, is $\tau_{ma} = \frac{T}{2At} = \frac{T}{2\pi R^2 t}$, where $t \ll R$ is the thickness of the tube. Show that the relative error is

$$\frac{\tau_{ma} - \tau_{ex}}{\tau_{ex}} = \frac{\eta^2}{4},$$

where $\eta = t/R$.

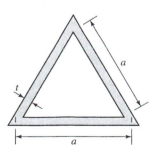

Figure 12P.11

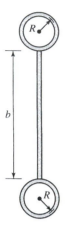

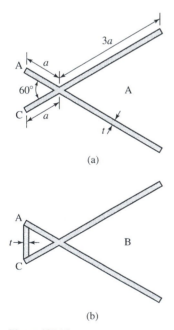

Figure 12P.13

Figure 12P.14

12.13:   The cross-section of a member, as shown in Fig. (12P.13), consists of a thin plate of width b and thickness t $(t \ll b)$ and two thin-wall hollow cylinders of thickness t and mean radius R. The torsional stiffness of the cross-section is given as 10 times the torsional stiffness of the plate alone. Determine the ratio b/R.

12.14:   Two shafts, consisting of thin tubular sections, one seamless and the other with a slit [Figs. (12P.14a and b), respectively], are each subjected to the same torque T. From membrane analogy considerations and from the derived expressions, (a) evaluate the ratio of shear stresses developed in each shaft. Indicate carefully the shear flow in a figure for each case. (b) Determine the ratio of the unit angle of twist of the two sections when subjected to the same torque.

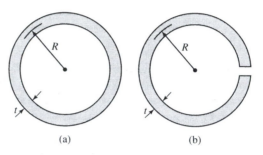

(a)   (b)

Review and comprehensive problems

12.15:   Consider two solid sections, a circular cylinder of radius R and an equilateral triangle with sides a whose torsional rigidity constant C is given as $C = \sqrt{3}a^4/80$. The two sections have the same cross-sectional area A. Show, for these two cross-sections, that the one with the greater polar moment of area (about the centre of twist) has a smaller torsional rigidity than the other.

12.16:   A member having a cross-section 'A', which consists of two steel ($E = 200$ GPa) plates as shown in Fig. (12P.16a), is subjected to a torque T. In Fig. (12P.16b) a steel plate AC has been welded, as shown, 'B.' (a) What is the torsional rigidity of each section? Express answers in terms of G and the dimensions of the section. (b) What is the ratio of the torsional rigidities of section B to section A if $a = 80$ mm and $t = 4$ mm. (c) Determine the maximum shear stress and the unit angle of twist of each section if $T = 150,000$ N-mm.

12.17:   A wide plate made of a material whose shear modulus is G, having dimensions $t \times \ell \times L$, is cut into 12 plates each $t \times a$, where $t \ll a$ ($a = \ell/12$), as shown in Fig. (12P.17). The 12 plates can then be welded to form members (of length L) whose cross-sections are as shown in Fig. (12P.17b, c or d). For each case, determine (a) the torsional stiffness CG and (b) the maximum shear stress if the member is subjected to a torque T. (c) If the original plate is, instead, cut into n strips of width ℓ/n, which are welded to form a closed polygonal section, what is the limiting value of C as $n \to \infty$?

12.18:*   A hollow member, having an elliptic cross-section with outer semi-axes a and b and inner semi-axes αa and αb, as shown in Fig. (12P.18), is subjected to a torsional moment T. Determine (a) the torsional rigidity CG of the cross-section, (b) the shear stresses at points A, B, D and E in terms of a, b, α and T and (c) determine $|\tau|_{max}$.

12.19:   Two members, whose cross-sections are as shown in Figs. (12P.19a and b), are subjected to the same torsional moment T. The cross-section of Fig (12P.19a) consists

(a)

(b)

Figure 12P.16

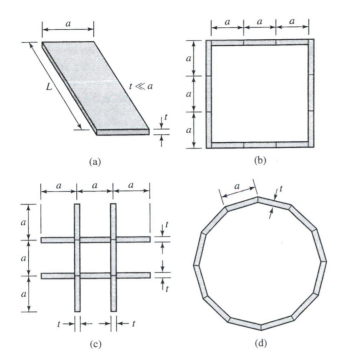

Figure 12P.17

of a T-section whose horizontal and vertical legs have thickness t and $2t$, respectively, where $t \ll a$. The cross-section of Fig. (12P.19b) consists of two angles (each leg having thickness t), which are welded together over the length of the members at two points c and d. In each case, the cross-sections are assumed to rotate without changing shape when subjected to the torques. Draw a sketch of the shear flow in each cross-section. Discuss *qualitatively* the differences in the shear flow. Are the torsional rigidities of the two sections the same or different?

12.20: A thin-wall member of length L is constructed by means of n plates $a \times t \times L$ ($t \ll a$), which are welded together to form a closed section whose shape is an n-sided regular polygon, as shown in Fig. (12P.20). (a) Derive an expression for the torsion

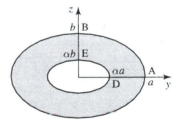

Figure 12P.18

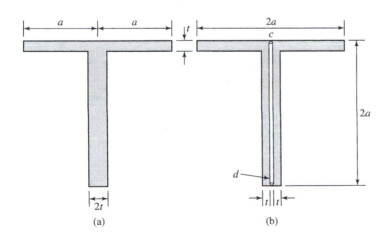

Figure 12P.19

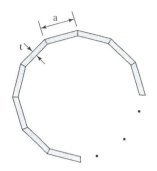

rigidity constant $C = T/G\Theta$ in terms of a, t and n. (b) What is C for (i) a triangular section, (ii) a square section and (iii) a hexagonal section. (c) If the sections are subjected to a torsional moment T, determine a general expression for the shear stress in an n-sided regular polygon. (d) Repeat (b) for the stress τ_{xs}. (e) Show that the general expressions derived in (a) and (c) above lead to C and $\tau_{x\theta}$ of a circle when $n \to \infty$, $a \to 0$ such that in the limit, the product $na \to 2\pi R$, where R is the mean radius of the circle.

12.21: Determine the torsional stiffness CG for a member whose thin-wall two-cell cross-section is as shown in Fig. (12P.21). What is the maximum shear stress and indicate where it occurs by means of a figure?

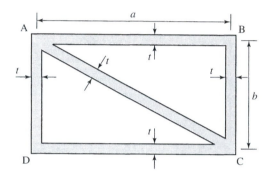

Figure 12P.21

12.22: Determine the torsional stiffness CG for a member whose thin-wall three-cell cross-section is as shown in Fig. (12P.22). What is the maximum shear stress and indicate where it occurs by means of a figure.

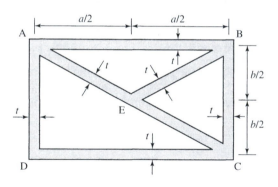

Figure 12P.22

The following problem is designed to require the use of a computer

12.23: A torsional moment T is applied to two members each having a rectangular cross-section but the same cross-sectional area A. The first member has a square cross-section ($a_s \times a_s$) as shown in Fig. (12P.23a). The second member has a general rectangular cross-section shape ($a_r \times b$), where $b/a_r = \alpha$ ($\alpha > 0$), as shown in Fig. (12P.23b).

Let the torsional stiffness of the rectangular and square cross-sections be C_r and C_s, respectively, and let the maximum shear stress in the two cross-sections be τ_r and τ_s, respectively. (a) Using the results obtained for the solution of a general rectangular cross-section, (i) write a computer program to evaluate the relevant coefficients β [Eq. (12.7.30b)] (using the first five terms of the series), (ii) determine the ratio of the torsional stiffnesses, C_r/C_s, as a function of α and (iii) present the results in tabular

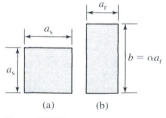

Figure 12P.23

form and, using a plotting routine, plot this ratio as a function of α. [*Note: Alternatively*, determine the ratio C_r/C_s, as a function of α using values of β given in the table on p. 468 and evaluate this ratio for values of $\alpha = 1.0$, 1.2, 1.5, 2.0, 2.5, 3.0, 4.0, 5.0 and 10.0. Present the results in tabular form and using a plotting routine, plot this ratio as a function of α. (b) Repeat (a) above (using the coefficient γ) for the ratio τ_r/τ_s and plot as a function of α. (c) What conclusions can be drawn from the above calculations ?]

13

General bending theory of beams

13.1. Introduction

The study of the bending of elastic beams aims at the determination of the internal stresses and deflections of beams due to flexure. In our previous treatment, expressions for these quantities were developed for a limited class of beams, namely for beams whose cross-sections possess an axis of symmetry (which was taken as the y-axis). It was seen that if, in addition, the beam is subjected only to lateral loads lying in the plane of symmetry [Fig. (13.1.1)] (such that moments about the y-axis, $M_y = 0$), it is possible to conclude that (a) the cross-sections do not rotate about the longitudinal x-axis, (b) the neutral axis coincides with the z-axis (which was shown to be a centroidal axis) and (c) the beam deflects in the plane of symmetry, i.e. in the y-direction.

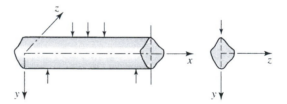

Figure 13.1.1

Based on these conclusions, an expression for the flexural stress at any point of the beam was derived for slender beams in a state of pure bending: namely Eq. (8.6.9b),

$$\sigma_x = \frac{M_z y}{I_{zz}}, \tag{13.1.1}$$

where y is the perpendicular distance from the neutral axis and I_{zz} is the second moment of the cross-section about the z-axis.

It is important to re-emphasise that according to the derivation, the second moment of area is, in fact, about the neutral axis and it is only because the neutral axis coincides with the z-axis that we write I_{zz}.

In this chapter, we extend our study to the bending of beams having arbitrary cross-sections. Moreover, loads and couples are assumed to be applied in arbitrary planes. As lack of symmetry precludes prior knowledge of the location of the neutral axis, the direction of deflection of such beams is not known *a priori*; moreover, the beam may undergo rotations about the x-axis. In considering the flexure of such beams in this chapter, our development will follow the same reasoning as in Chapter 8, with the exception that no symmetry is assumed.

13.2. Moment–curvature relation for elastic beams in flexure

Consider a beam having an arbitrary cross-section of area A, which is initially straight with a longitudinal axis lying along the x-axis [Fig. (13.2.1a)]. Note that at this stage of the analysis the x-axis is *not* necessarily a centroidal axis. Let the beam be subjected to end couples. As a result, the beam will be in a state of pure bending (since the moments do not vary with x) and thus the beam will bend with constant curvature, as shown in Fig. (13.2.1b). Because of the arbitrary nature of the cross-section and the arbitrary plane of the end couples, we cannot conclude that the beam deflects in the x–y plane but must assume that the beam deflects in some unknown direction. The deflected position is therefore as shown in Fig. (13.2.1b).

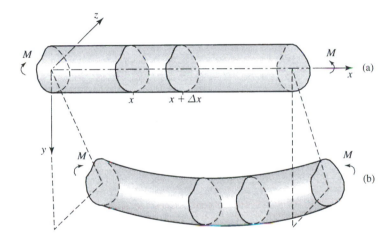

Figure 13.2.1

Let Δx represent an element of the beam in its undeformed state [Fig. (13.2.2a)]. Since the beam is in a state of pure bending after deformation, plane cross-sections *necessarily* remain plane; i.e., cross-sections rotate about some unknown axis N that lies in the y–z plane. It is important to observe that the deflection of the beam is in a direction perpendicular to the line N.

We denote by $\Delta\theta$ the subtended angle between the two end cross-sections of the element after deformation. Clearly, in the deformed state, some fibres (bb′) will elongate and others (tt′) will shorten [Figs. (13.2.2a and b)]. Therefore, there must exist some fibres, originally lying in some plane P of the undeformed element, that neither shorten nor elongate. (The x-axis is taken to lie along this plane.) Let N denote the intersection of the plane P with the cross-section and let nn′ denote fibres in this plane. After deformation, the plane P becomes a curved surface P′, with typical fibres lying along the arc $\overset{\frown}{nn}{}'$ of Fig. (13.2.2b); for these fibres, the extensional strain is clearly $\epsilon_x = 0$.

Let R denote the radius of curvature to the deformed nn′ fibres. Because of the arbitrary nature of the cross-section and loading, the line N in the undeformed

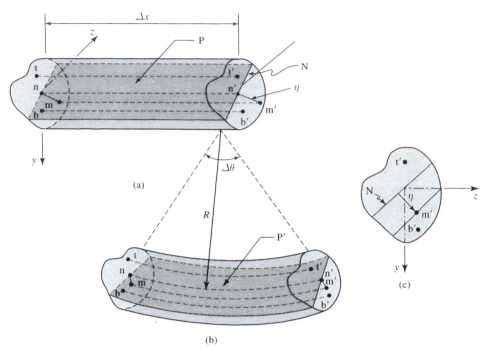

Figure 13.2.2

element, which represents points of $\epsilon_x = 0$, is in general, inclined as shown in Fig. (13.2.2c).

From geometry, the arc length $\widehat{nn'} = R\Delta\theta$. But since $\epsilon_x = 0$ for the nn' fibres, $R\Delta\theta = \Delta x$. Consider now the fibres mm' located at some distance η of the undeformed element, *measured perpendicularly from the line* N [Figs. (13.2.2a and c)]. After deformation, the arc length $\widehat{mm'} = (R + \eta)\Delta\theta$. But since all fibres in the element were initially of length Δx, the strain ϵ_x in the mm' fibres is given by

$$\epsilon_x = \frac{\widehat{mm'} - \Delta x}{\Delta x} = \frac{(R + \eta)\Delta\theta - R\Delta\theta}{R\Delta\theta}, \tag{13.2.1a}$$

i.e.,

$$\epsilon_x = \frac{\eta}{R}. \tag{13.2.1b}$$

As in Chapter 8, we assume that for a slender beam, $\sigma_y = \sigma_z = 0$ throughout the body. It follows that the stress–strain relation is simply $\sigma_x = E\epsilon_x$. Hence

$$\sigma_x = \frac{E\eta}{R}. \tag{13.2.2}$$

Thus along the line N representing the nn' fibres ($\eta = 0$), the stress $\sigma_x = 0$. This line is called the *neutral axis*. From Eqs. (13.2.1) and (13.2.2), we note that the strain and stress vary linearly with the perpendicular distance from the neutral axis.

The moment about the neutral axis, denoted by M_n, is given by

$$M_n = \iint\limits_A \eta\sigma_x \, dA. \tag{13.2.3a}$$

Substituting Eq. (13.2.2),

$$M_n = \iint_A \eta \left(\frac{E\eta}{R} \right) dA = \frac{E}{R} \iint_A \eta^2 dA. \qquad (13.2.3b)$$

Clearly, by definition, $I_n = \iint_A \eta^2 \, dA$ is the second moment of the cross-sectional area *about the neutral axis*. Hence we obtain the moment–curvature relation

$$M_n = \frac{EI_n}{R}. \qquad (13.2.4)$$

Note that the same relation was obtained for beams of symmetric cross-section in Chapter 8 [Eq. (8.6.6a) or (8.6.9a)], but because of the symmetry of the case considered, the discrepancy between the neutral axis and the z-axis was masked in the final expression. Here, however, we observe clearly that the relation refers to the neutral axis.

Consider now the resultant force F_x acting normal to the cross-section in the x-direction, which is given by

$$F_x = \iint_A \sigma_x \, dA. \qquad (13.2.5a)$$

Substituting for σ_x,

$$F_x = \iint_A \frac{E\eta}{R} \, dA = \frac{E}{R} \iint_A \eta \, dA. \qquad (13.2.5b)$$

Since for pure bending, the resultant normal force on the cross-section must vanish, i.e. $F_x = 0$, it follows that $\iint_A \eta \, dA = 0$. *Thus the neutral axis must always pass through the centroid of the cross-section.*

Combining the moment–curvature relation, Eq. (13.2.4), and Eq. (13.2.2), the expression for the flexure stress becomes

$$\sigma_x(\eta) = \frac{M_n \eta}{I_n}. \qquad (13.2.6)$$

Hence, if one knows the moment about the neutral axis, M_n, and the second moment I_n, one can always obtain the stress σ_x at any point η measured perpendicularly from the neutral axis.

In the case of symmetric bending, with all loads applied, for example, in the x–y plane of symmetry, we know, *a priori*, that the neutral axis coincides with the z-axis. However, if the cross-section is arbitrary, then for any loading system, the orientation of the neutral axis is not known. Therefore, although Eq. (13.2.6) is conceptually correct, it proves not to be very useful in practice.

We therefore seek to develop a useful expression for the flexural stress σ_x which is valid for all arbitrary cross-sections and for loads acting in an arbitrary plane. Since loads are applied arbitrarily, we must, contrary to Chapter 8, consider moments about both the y- and z-axes, as well as applied forces and shear forces in the y- and z-coordinate directions.

13.3. Sign convention and beam equations for bending about two axes

Consider a beam of arbitrary cross-section, located in the x, y, z coordinate system, as shown in Fig. (13.3.1). Because of the asymmetry, it is necessary to establish a sign convention that takes into account forces in both the y- and z-directions as well as moments about these axes.

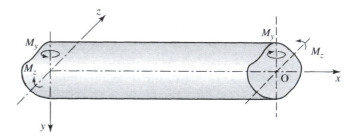

Figure 13.3.1

(a) Sign convention

We first recall the standard sign convention that defines the faces of the cross-section: a positive (negative) face is one for which the outward normal is acting in the positive (negative) x-direction.

The sign convention for the shear forces is as follows:

■ Positive shear forces V_y and V_z act on a positive (negative) face in the positive (negative) y- and z-directions, respectively [Fig. (13.3.2a)].

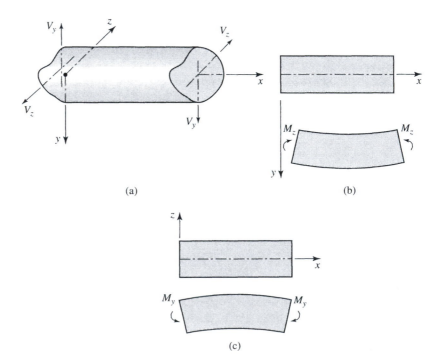

Figure 13.3.2

Consequently, negative V_y and V_z forces act, respectively, on a positive (negative) face in the negative (positive) y- and z-directions.

Denoting the moments acting about the z- and y-axes as M_z and M_y, respectively, the following sign convention is adopted:

- A moment M_z is said to be positive (negative) if it tends to cause tension in the fibres having positive (negative) y-coordinates [Fig. (13.3.2b)].
- A moment M_y is said to be positive (negative) if it tends to cause tension in the fibres having positive (negative) z-coordinates [Fig. (13.3.2c)].

The vector representation of the positive moments acting on a positive face is, according to the right-hand rule, shown in Fig. (13.3.3). Note that, while the vector representing positive M_y (on a positive x-face) points in the positive y-direction, *the vector representing positive M_z points in the negative z-direction.*

The following sign convention is adopted for applied loads:

- Positive forces $q_y(x)$ and $q_z(x)$ act in the positive y- and z-directions, respectively [Fig. (13.3.4)].

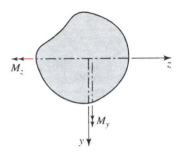

Figure 13.3.3

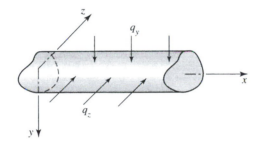

Figure 13.3.4

(b) Differential beam equations

The differential beam equations relating V_y, M_z and q_y, derived previously in Chapter 8, are

$$\frac{dV_y(x)}{dx} = -q_y(x), \tag{13.3.1a}$$

$$\frac{dM_z(x)}{dx} = V_y(x). \tag{13.3.1b}$$

so that

$$\frac{d^2 M_z(x)}{dx^2} = -q_y(x). \tag{13.3.2}$$

Similarly, using the above sign convention, the analogous relations between V_z, M_y and q_z, obtained from the equilibrium equations, $\sum F_z = 0$ and $\sum M_y = 0$, are

$$\frac{dV_z(x)}{dx} = -q_z(x), \tag{13.3.3a}$$

$$\frac{dM_y(x)}{dx} = V_z(x). \tag{13.3.3b}$$

so that

$$\frac{d^2 M_y(x)}{dx^2} = -q_z(x). \tag{13.3.4}$$

Having established the basic equations of equilibrium for the beam, we are now in a position to derive the general expressions for the flexural stress.

13.4. General expression for stresses due to flexure

(a) Derivation: stresses in beams under pure bending

Consider a prismatic beam made of a homogeneous, isotropic linear elastic material in an x, y, z coordinate system as shown, such that the longitudinal x-axis passes through the *centroids* O of the cross-sections. The beam is assumed to have an arbitrary cross-sectional area A. Let the beam be subjected to end couples such that the beam is in a state of pure bending [Fig. (13.4.1a)]. We therefore assume that known positive internal resultant moments M_y and M_z are acting on the cross-section. These moments are shown according to their vectorial representation in Fig. (13.3.3).

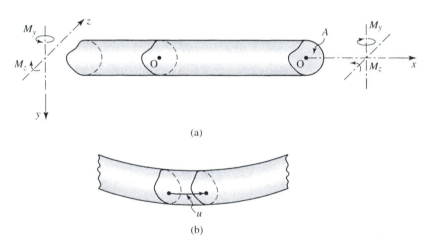

(a)

(b)

Figure 13.4.1

Now, internal stress resultants F_x, M_y and M_z are given by

$$F_x = \iint_A \sigma_x \, dA, \tag{13.4.1a}$$

$$M_y = \iint_A z\sigma_x \, dA, \tag{13.4.1b}$$

$$M_z = \iint_A y\sigma_x \, dA. \tag{13.4.1c}$$

As was seen in Chapter 8, since the beam is in pure bending, the cross-sections must necessarily remain plane; i.e., no warping of the cross-section takes place during bending. Therefore, the displacement in the x-direction, $u(x, y, z)$, of any point of the cross-section is given by [Fig. (13.4.1b)]

$$u(x, y, z) = C_1(x) + C_2(x)y + C_3(x)z, \tag{13.4.2a}$$

where, in the case of pure bending, $C_i(x)$ are all linear functions of x. Note that Eq. (13.4.2a) represents the equation of a plane.

The strains at any point, $\epsilon_x = \frac{\partial u}{\partial x}$, are therefore given by

$$\epsilon_x(x, y, z) = C_1' + C_2'y + C_3'z, \qquad (13.4.2b)$$

where C_i' are constants. Hence the stress, $\sigma_x = E\epsilon_x$, becomes[†]

$$\sigma_x = a + by + cz, \qquad (13.4.3)$$

where $a \equiv EC_1'$, $b \equiv EC_2'$ and $c \equiv EC_3'$ are constants (for this case of pure bending).

Substituting in Eq. (13.4.1a),

$$F_x = \iint_A (a + by + cz)\,\mathrm{d}A = a \iint_A \mathrm{d}A + b \iint_A y\,\mathrm{d}A + c \iint_A z\,\mathrm{d}A. \qquad (13.4.4)$$

Since the y- and z-axes are centroidal axes, it follows that $\iint_A y\,\mathrm{d}A = \iint_A z\,\mathrm{d}A = 0$. Now, for a beam in simple flexure, the normal force resultant F acting on the cross-section vanishes; therefore $a = 0$.

Substituting now in Eqs. (13.4.1b) and (13.4.1c), and noting that b and c are constants with respect to the integral,

$$\iint_A (by + cz)z\,\mathrm{d}A = b \iint_A yz\,\mathrm{d}A + c \iint_A z^2\,\mathrm{d}A = M_y, \qquad (13.4.5a)$$

$$\iint_A (by + cz)y\,\mathrm{d}A = b \iint_A y^2\,\mathrm{d}A + c \iint_A yz\,\mathrm{d}A = M_z. \qquad (13.4.5b)$$

But

$$I_{yy} = \iint_A z^2\,\mathrm{d}A, \qquad (13.4.6a)$$

$$I_{zz} = \iint_A y^2\,\mathrm{d}A, \qquad (13.4.6b)$$

$$I_{yz} = \iint_A yz\,\mathrm{d}A \qquad (13.4.6c)$$

are the moments of the area about the y- and z-centroidal axes of the cross-section. Hence Eqs. (13.4.5) become

$$I_{yz}b + I_{yy}c = M_y, \qquad (13.4.7a)$$

$$I_{zz}b + I_{yz}c = M_z, \qquad (13.4.7b)$$

which we recognise as two simultaneous equations in the two unknown constants

[†] Note that, as in the derivation for symmetric bending of long beams (see Chapter 8), we ignore σ_y and σ_z with respect to σ_x.

b and c, whose solutions are

$$b = \frac{I_{yy}M_z - I_{yz}M_y}{I_{yy}I_{zz} - I_{yz}^2},$$ (13.4.8a)

$$c = \frac{I_{zz}M_y - I_{yz}M_z}{I_{yy}I_{zz} - I_{yz}^2}.$$ (13.4.8b)

Finally, substituting in Eq. (13.4.3), we obtain the general expression for the flexure stress:

$$\sigma_x = \left(\frac{I_{yy}M_z - I_{yz}M_y}{I_{yy}I_{zz} - I_{yz}^2}\right) y + \left(\frac{I_{zz}M_y - I_{yz}M_z}{I_{yy}I_{zz} - I_{yz}^2}\right) z.$$ (13.4.9)

Since $\sigma_x(y=0, z=0)=0$, we conclude that *the neutral axis always passes through the centroid* of the cross-section. Note that this same result was previously observed from Eq. (13.2.5b).

(b) Extension of expression for flexural stress in beams due to applied lateral loads

The above expression for the flexural stress, derived for beams that were assumed to be in a state of pure bending ($M_y =$ const., $M_z =$ const.), is an 'exact' expression for slender linear elastic beams. Following the explanation given in Section 9 of Chapter 8, this expression is also correct if the moments are *linear functions* of x. For moments $M_y = M_y(x)$, $M_z = M_z(x)$ that are general functions of x, the stresses, calculated according to Eq. (13.4.9), are not 'exact'. However, recalling our discussion in Chapter 8 (Section 7), we may, on the basis of Navier's hypothesis, use Eq. (13.4.9) to calculate σ_x for all loading conditions irrespective of the resulting moment variation with the x-coordinate. Such calculations yield excellent approximations particularly for very long beams.

(c) Some particular cases

Case (i): *y- and z-axes, as principal axes*
If the y- and z-axes are principal axes, then $I_{yz}=0$ and Eq. (13.4.9) reduces to

$$\sigma_x = \frac{M_z y}{I_{zz}} + \frac{M_y z}{I_{yy}},$$ (13.4.10)

which coincides with Eq. (8.12.1), as obtained by linear superposition in Chapter 8, for beams having cross-sections for which both the y- and the z-axes were assumed to be axes of symmetry.

Case (ii): $M_y = 0$
Suppose that the beam of arbitrary cross-section is subjected only to loads acting in the x–y plane so that $M_y = 0$. It follows that

$$\sigma_x = \left(\frac{I_{yy}M_z}{I_{yy}I_{zz} - I_{yz}^2}\right) y - \left(\frac{I_{yz}M_z}{I_{yy}I_{zz} - I_{yz}^2}\right) z$$ (13.4.11)

or

$$\sigma_x = \left(\frac{I_{yy}y - I_{yz}z}{I_{yy}I_{zz} - I_{yz}^2}\right) M_z. \qquad (13.4.12)$$

Under this loading condition, the stress at the z-centroidal axis, i.e. the $y = 0$ axis, is

$$\sigma_x = \left(-\frac{I_{yz}z}{I_{yy}I_{zz} - I_{yz}^2}\right) M_z. \qquad (13.4.13)$$

We observe that if all loads are acting in the y-direction, $\sigma_x = 0$ at the z-axis *only if* $I_{yz} = 0$. Thus, under such loads, *the neutral axis will lie on the z-axis if, and only if, the z-axis is a principal axis.* (Recall that if either the y- or the z-axes are axes of symmetry, they are, in fact, also principal axes.)[†] Hence in our previous treatment of beams in Chapter 8, the limitation to beams with cross-sections that possessed an axis of symmetry was unduly restrictive: it would have been only necessary to require that the axes be principal axes. From the present discussion, we know this only *a posteriori*.

Now, having established that under this loading condition, the neutral axis does not, in general, coincide with the z-axis, we wish to determine its orientation. Since the neutral axis is *defined* by the condition $\sigma_x = 0$, it follows from Eq. (13.4.12) that, for this $M_y = 0$ case, the equation representing the neutral axis is

$$I_{yy}y - I_{yz}z = 0, \qquad (13.4.14)$$

which we note is the equation of a straight line passing through the centroid [Fig. (13.4.2)].

We denote the slope of the neutral axis by β (where β is measured in the positive clock-wise direction with respect to the z-axis). Note then that β is given by $\beta = \tan^{-1}(y/z)$. Hence, from Eq. (13.4.14) for this $M_y = 0$ case, we have established the orientation of the neutral axis, namely [Fig. (13.4.2)]

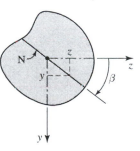

Figure 13.4.2

$$\tan \beta = \frac{I_{yz}}{I_{yy}}. \qquad (13.4.15)$$

We observe here that $\beta = 0$ if, and only if, $I_{yz} = 0$, i.e. if the y- and z-axes are principal axes. This corroborates the statement made in the previous paragraphs.

Example 13.1: A 2-m long cantilever beam is subjected to a load $P = 300$ N acting in the y-direction at the free end. The cross-section of the beam is a Z-section with dimensions as shown in Fig. (13.4.3). Determine (a) the stress σ_x existing at the fixed support ($x = 0$) at points A and B of the section, (b) the orientation of the neutral axis and (c) the direction of the deflection δ of point O at the free end.

Solution: The properties of this section are $I_{zz} = 26.3$ cm^4, $I_{yy} = 23.9$ cm^4, $I_{yz} = -19.6$ cm^4. The moments at the fixed end are $M_y = 0$, $M_z = (-2 \times 300) = -600$ N-m $= -60{,}000$ N-cm.

[†] Note that if the y-axis is an axis of symmetry, it is also a principal axis. However, as seen in Appendix A.1, it is clear that not all principal axes are necessarily axes of symmetry.

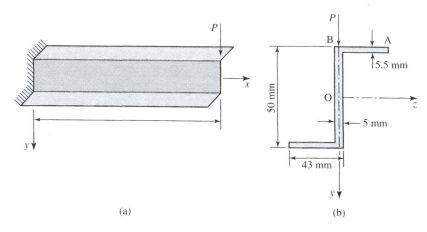

Figure 13.4.3   (a)   (b)

(a) Using Eq. (13.4.12), the stresses σ_x are

$$\sigma_{xA} = \frac{(23.9)(-2.5) - (-19.6)(4.05)}{(23.9)(26.3) - (19.6)^2}(-60{,}000)$$

$$= -4820 \text{ N/cm}^2 = -48.2 \text{ MPa (compression)}$$

$$\sigma_{xB} = \frac{(23.9)(-2.5) - (-19.6)(-0.25)}{(23.9)(26.3) - (19.6)^2}(-60{,}000)$$

$$= 15{,}900 \text{ N/cm}^2 = 159 \text{ MPa (tension)}$$

(b) The slope β of the neutral axis is given, according to Eq. (13.4.15), by
$\beta = \tan^{-1}(-19.6/23.9) = \tan^{-1}(-0.820) \rightarrow\rightarrow \beta = -39.36°$.

(c) Since the deflection is always perpendicular to the neutral axis, the deflection δ will be in a direction inclined at an angle of 50.64° with respect to the z-axis.

The resulting stress distribution on the flange AB of the section as well as the orientation of the neutral axis and direction of the deflection are shown in Fig. (13.4.4).

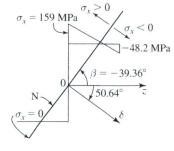

Figure 13.4.4

(d) General case

We now consider the general case of a beam loaded by a set of inclined lateral loads such that both moments M_y and M_z are acting on a cross-section of the beam. In this case, according to Eq. (13.4.9), the equation of the neutral axis, defined by $\sigma_x = 0$, is

$$(I_{yy}M_z - I_{yz}M_y)y + (I_{zz}M_y - I_{yz}M_z)z = 0. \qquad (13.4.16)$$

Assume now that all applied loads act at a given angle θ to the y-axis [see Fig. (13.4.5)]. Note that the moments M_z are then due to the components of loads $\pm P\cos\theta$ and that M_y moments are due to $\mp P\sin\theta$ (according to the adopted sign convention). (Note also that θ is measured clockwise positive with respect to the y-axis.) Then

$$\frac{M_z}{|M|} = \pm\cos\theta, \qquad (13.4.17a)$$

$$\frac{M_y}{|M|} = \mp\sin\theta, \qquad (13.4.17b)$$

where $|M| \equiv M$, the magnitude of the moment resultant due to the load, is perpendicular to P [Fig. (13.4.5)]. Substituting in Eq. (13.4.16),

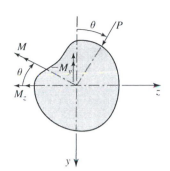

Figure 13.4.5

$$[(I_{yy} \cos\theta)\, M + (I_{yz} \sin\theta)\, M]\, y + [(-I_{zz} \sin\theta)\, M - (I_{yz} \cos\theta)\, M]\, z = 0\,.$$

$$(13.4.18)$$

Hence the orientation of the neutral axis is given by

$$\tan\beta = \frac{I_{zz} \sin\theta + I_{yz} \cos\theta}{I_{yy} \cos\theta + I_{yz} \sin\theta}, \qquad (13.4.19)$$

where again β is measured clockwise from the z-axis. Note that if y and z are principal axes, then

$$\tan\beta = \left(\frac{I_{zz}}{I_{yy}}\right) \tan\theta. \qquad (13.4.20)$$

We illustrate these results by means of the following example.

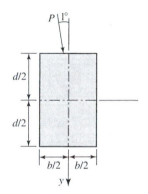

Figure 13.4.6

Example 13.2:   A set of given forces is applied to a beam of rectangular cross-section of width b and depth d. If all the forces are applied at an angle of $1°$, as shown in Fig. (13.4.6), determine the location of the neutral axis.

Solution: Since the cross-section is rectangular, the y- and z-axes, being axes of symmetry, are also principal axes, i.e. $I_{yz} = 0$ and therefore Eq. (13.4.20) is valid here. Then with $I_{zz} = bd^3/12$, $I_{yy} = db^3/12$,

$$\frac{I_{zz}}{I_{yy}} = \left(\frac{d}{b}\right)^2.$$

Hence, by Eq. (13.4.20),

$$\tan\beta = \left(\frac{d}{b}\right)^2 \tan\theta. \qquad (13.4.21)$$

We note here that $\theta = -1°$ and $\tan(-1°) = -0.01746$. We consider three separate cases:

(a) $b = d$: then, from Eq. (13.4.21), $\beta = \theta$, i.e. the neutral axis lies in a direction perpendicular to the applied loads. The beam will therefore deflect in the same direction as the applied loads.
(b) $d = 2b$: then $\tan\beta = 4\tan(-1°) = -0.0698 \rightarrow\rightarrow \beta = -4.0°$.
(c) $d = 5b$: then $\tan\beta = 25(-0.01746) = -0.4365 \rightarrow\rightarrow \beta = -23.6°$.

The location of the neutral axis for this loading, in the above three cases, is shown in Figs. (13.4.7a–c). Recalling that the deflection δ of the beam is always perpendicular to the neutral axis, we observe that the deviation of the direction of the deflection with respect to the direction of the applied load becomes very sensitive (even for a slightly inclined load), as the ratio d/b of the beam increases.

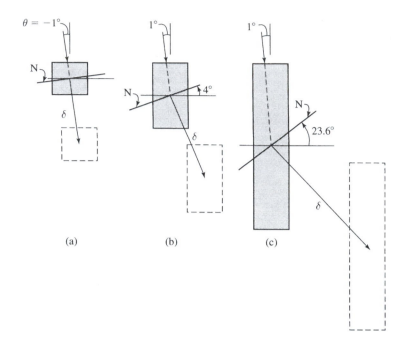

$\theta = -1°$

$1°$

$1°$

N

N

$4°$

N

$23.6°$

δ

δ

δ

(a)   (b)   (c)

Figure 13.4.7

13.5.   Shear stresses due to bending of beams

(a) Derivation

If the moments M_y and M_z are functions of x, i.e. not constants, then according to Eqs. (13.3.1b) and (13.3.3b), the shear forces at a cross-section V_y and V_z will not vanish. Due to these non-zero shear forces, shear stresses τ_{xy} and τ_{xz} will exist at the cross-section. Consider now an element Δx of the beam, as shown in Fig. (13.5.1a), and let the left and right cross-sections be located at x and $x + \Delta x$, respectively. Furthermore, $M_y(x + \Delta x) = M_y(x) + \Delta M_y$ and $M_z(x + \Delta x) = M_z(x) + \Delta M_z$. As a result, according to Eq. (13.4.9), the flexural stresses σ_x are also functions of x. Let $\sigma_x(x)$ and $\sigma_x(x + \Delta x) = \sigma_x + \Delta \sigma_x$ denote the flexural stresses on the two faces of the element, respectively [Fig. (13.5.1a)]. Clearly, the entire element Δx is in equilibrium in the x-direction.

As in Chapter 8, we consider an arbitrary portion of the element. We therefore 'cut' the element along some arbitrarily inclined plane H that is parallel to the x-axis (i.e., one whose normal n is perpendicular to the x-axis). Note that the intersection of this plane with the cross-section is a line K whose length we denote by b [Fig. (13.5.1b)]. Let us now isolate a portion of the element, for example, the bottom portion.

We now consider this portion of the element [Fig. (13.5.1c)] as a free body and examine it for equilibrium. Let us denote the area of this portion of the cross-section by $\overline{A}$. Now, the flexural stresses acting in the x-direction on the areas $\overline{A}$ of the left and right faces are, as before, σ_x and $\sigma_x + \Delta \sigma_x$, respectively. Clearly, since $\sigma_x(x) \neq \sigma_x(x + \Delta x)$, these stresses alone cannot maintain this free body in equilibrium in the x-direction. Therefore, shear stresses that act in the x-direction must necessarily exist on the plane H. Note that the area A' of this plane is $A' = b \times \Delta x$. Let us denote the *average* shear stress that acts on the plane H by $\tau \equiv \tau_{nx}$ [Fig. (13.5.1c)]. Note also that we have taken τ to be *positive when acting to the left* on the plane H of this isolated free body.

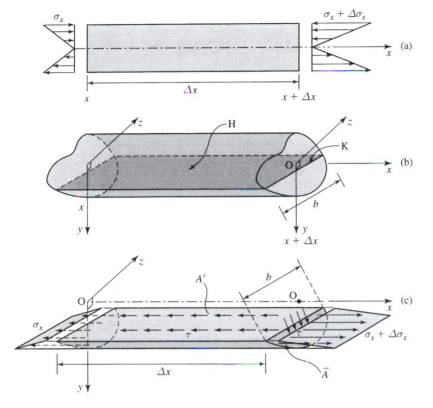

Figure 13.5.1

For equilibrium in the x-direction, $\sum F_x = 0$, we have

$$-\tau(b\Delta x) - \iint\limits_{\overline{A}} \sigma_x(x)\,\mathrm{d}A + \iint\limits_{\overline{A}} [\sigma_x(x) + \Delta\sigma_x]\,\mathrm{d}A = 0 \qquad (13.5.1a)$$

or

$$\tau(b\Delta x) = \iint\limits_{\overline{A}} \Delta\sigma_x\,\mathrm{d}A . \qquad (13.5.1b)$$

Dividing through by $b\Delta x$, we have

$$\tau = \frac{1}{b} \iint\limits_{\overline{A}} \frac{\Delta\sigma_x(x)}{\Delta x}\,\mathrm{d}A \qquad (13.5.1c)$$

and taking the limit as $\Delta x \to 0$, we arrive at[†]

$$\tau = \frac{1}{b} \iint\limits_{\overline{A}} \frac{\mathrm{d}\sigma_x(x)}{\mathrm{d}x}\,\mathrm{d}A . \qquad (13.5.2)$$

[†] As in the derivation of shear stresses for symmetrical cross-sections given in Chapter 8, we note that upon taking the limit as $\Delta x \to 0$, the plane H is reduced to the line K. Consequently, the average shear stress $\tau \equiv |\tau_{nx}|$ acting on H becomes, in fact, the average shear stress acting at points along the length b of this line.

Now, the flexural stress σ_x is given by Eq. (13.4.9). Hence, upon taking its derivative,

$$\frac{d\sigma_x}{dx} = \left(\frac{I_{yy}\frac{dM_z}{dx} - I_{yz}\frac{dM_y}{dx}}{I_{yy}I_{zz} - I_{yz}^2} \right) y + \left(\frac{I_{zz}\frac{dM_y}{dx} - I_{yz}\frac{dM_z}{dx}}{I_{yy}I_{zz} - I_{yz}^2} \right) z. \qquad (13.5.3a)$$

But, according to Eqs. (13.3.1b) and (13.3.3b),

$$\frac{dM_z(x)}{dx} = V_y(x), \qquad \frac{dM_y(x)}{dx} = V_z(x).$$

Hence

$$\frac{d\sigma_x}{dx} = \left(\frac{I_{yy}V_y - I_{yz}V_z}{I_{yy}I_{zz} - I_{yz}^2} \right) y + \left(\frac{I_{zz}V_z - I_{yz}V_y}{I_{yy}I_{zz} - I_{yz}^2} \right) z. \qquad (13.5.3b)$$

Substituting in Eq. (13.5.2), and noting that V_y and V_z are not functions of dA, we obtain

$$\tau = \frac{1}{b} \left[\left(\frac{I_{yy}V_y - I_{yz}V_z}{I_{yy}I_{zz} - I_{yz}^2} \right) \iint_{\overline{A}} y \, dA + \left(\frac{I_{zz}V_z - I_{yz}V_y}{I_{yy}I_{zz} - I_{yz}^2} \right) \iint_{\overline{A}} z \, dA \right].$$

$$(13.5.4)$$

As in Chapter 8, we observe that $\iint y \, dA$ integrated over the area $\overline{A}$ [shown shaded in Fig. (13.5.2)] represents the moment of this area *with respect to the z-centroidal axis*, which we denote by Q_z; thus

$$Q_z = \iint_{\overline{A}} y \, dA = \overline{y}\,\overline{A}, \qquad (13.5.5a)$$

where $\overline{y}$ represents the distance from the z-centroidal axis to the centroid C of $\overline{A}$. Similarly, let

$$Q_y = \iint_{\overline{A}} z \, dA = \overline{z}\,\overline{A} \qquad (13.5.5b)$$

be the moment of $\overline{A}$ with respect to the y-centroidal axis, where $\overline{z}$ is the distance from the y-axis to the centroid of $\overline{A}$.

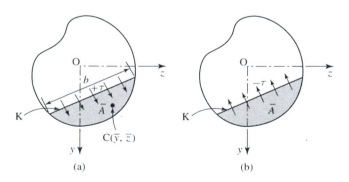

Figure 13.5.2   (a)   (b)

Thus we may finally write

$$\tau = \frac{1}{b}\left[\left(\frac{I_{yy}V_y - I_{yz}V_z}{I_{yy}I_{zz} - I_{yz}^2}\right)Q_z + \left(\frac{I_{zz}V_z - I_{yz}V_y}{I_{yy}I_{zz} - I_{yz}^2}\right)Q_y\right]. \qquad (13.5.6)$$

If y and z are principal axes, then Eq. (13.5.6) becomes

$$\tau = \frac{1}{b}\left(\frac{V_y Q_z}{I_{zz}} + \frac{V_z Q_y}{I_{yy}}\right). \qquad (13.5.7)$$

Furthermore, if no applied forces are acting in the z-direction (so that $V_z = 0$), we recover the simple expression for $\tau \equiv \tau_{nx}$ as derived in Chapter 8, namely

$$\tau = \frac{V_y Q_z}{I_{zz}b}. \qquad (13.5.8)$$

Finally, recalling the equality of the conjugate shear stresses, $\tau_{nx} = \tau_{xn}$, we note that Eqs. (13.5.6)–(13.5.8) represent the average of the shear stresses acting in the cross-section along the line K.

(b) Comments on the expressions

We observe that the derivation of the above expressions for τ follows closely that given in Chapter 8 [Eqs. (8.8.1)–(8.8.4)], the difference being that, here, the more general expression for the flexural stress σ_x [Eq. (13.4.9)] enters in the derivation. Thus the comments and conclusions stated in Chapter 8 remain valid. It is worthwhile to recall these features, namely

 (i) The final expression for τ represents an average value of the shear stress acting in the cross-section at points along the cut of length b.
 (ii) Positive values of τ, as calculated from Eqs. (13.5.4)–(13.5.8), signify that τ acting on a cross-section having a positive face, is pointing inward into the surface area $\overline{A}$; a negative τ indicates a stress pointing out of the area $\overline{A}$ [Figs. (13.5.2a and b) respectively].
 (iii) Since the calculated values are average values over a cut of length b, these average values are better approximations to the true value of the shear stress at points along a cut whose length b is relatively small.

Thus, finally, it should again be clear that the values for τ given by the expressions derived in this section are approximate and not exact values.

Example 13.3: A horizontal load P is applied to a beam whose cross-section is a structural angle (L102 × 102 × 9.5), as shown in Fig. (13.5.3a). The resulting

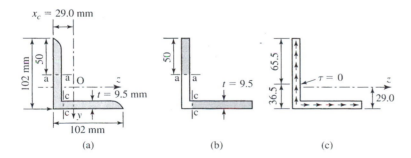

(a)   (b)   (c)   **Figure 13.5.3**

shear forces at a cross-section of the beam are given by $V_z = 100,000$ N, $V_y = 0$. Determine the average shear stresses τ along the lines a–a and c–c.

Solution: From Eq. (13.5.6), τ is given by

$$\tau = \frac{V_z}{b} \frac{\left(I_{zz}Q_y - I_{yz}Q_z\right)}{\left(I_{yy}I_{zz} - I_{yz}^2\right)}.$$

I_{yy}, I_{zz} and z_c (defining the location of the centroid O [Fig. (13.5.3a)]) are found in standard tables of structural sections:

$$I_{yy} = I_{zz} = 1.81 \times 10^6 \text{ mm}^4, \qquad t = 9.5 \text{ mm}, \qquad z_c = 29.0 \text{ mm}.$$

To calculate the 'mixed second moment' (loosely referred to as the 'product of inertia')[†] of this thin-wall section (which does not appear in the standard tables), we consider the component parts to be thin rectangles, thus ignoring the fillets [Fig. (13.5.3b)].

Hence, using this representation, I_{yz} is found as follows:[‡]

$$I_{yz} = (102 \times 9.5)(24.25)(22) + (92.5 \times 9.5)(-24.25)(-26.75)$$

$$= 1.09 \times 10^6 \text{ mm}^4$$

With these values, the above expression for τ becomes

$$\tau = \frac{100,000}{9.5} \frac{[1.81 \times 10^6 \, Q_y - 1.09 \times 10^6 \, Q_z]}{[(1.81 \times 10^6)^2 - (1.09 \times 10^6)^2]}$$

$$= 5.04 \times 10^{-9} \, (1.81 \times 10^6 \, Q_y - 1.09 \times 10^6 Q_z) \text{ MPa}.$$

To determine Q_y and Q_z for this thin-wall sections, we again make use of the representation of Fig. (13.5.3b). Thus, for the stress along a–a, we calculate the moments of the area $\overline{A}$ about the y- and z-axes, located above the line a–a as follows:

$$Q_y = (50 \times 9.5) \times (-24.25) = -11.52 \times 10^3 \text{ mm}^3,$$

$$Q_z = (50 \times 9.5) \times (-48) = -22.8 \times 10^3 \text{ mm}^3.$$

Substituting in the last expression,

$$\tau = 20.2 \text{ MPa along a–a}$$

Similarly, to obtain the stress along c–c, the moments of the area $\overline{A}$, calculated for an area to the right of line c–c, are

$$Q_y = (92.5 \times 9.5)(55.75 - 29) = 23.51 \text{ mm}^3,$$

$$Q_z = (92.5 \times 9.5)(24.25) = 21.3 \times 10^3 \text{ mm}^3.$$

Substituting in the final expression above, we find

$$\tau = 93.7 \text{ MPa along c–c}.$$

[†] Analogously to the second moment of an area, which is loosely called *moment of inertia*, the mixed second moment is sometimes called the *product of inertia* (see footnote p. 688).

[‡] Alternatively, from the table of Appendix E, the area of the angle section is given as $A = 1845$ mm^2, and its minimum radius of gyration is $r = 20$ mm; hence $I_{\min} = 0.738 \times 10^6$ mm^4. Using Eq. (A.1.14) of Appendix A and noting that $I_2 \equiv I_{\min}$, we obtain $I_{yz} = 1.07 \times 10^6$ mm^4. The descrepancy is due to the fillets and slight taper of the angle legs.

The resulting shear stresses τ acting on a positive face are shown in Fig. (13.5.3c). Note that the shear stresses change direction in the vertical leg of the angle.

13.6. Distribution of shear stresses in a thin-wall section: shear centres

As we have seen in Chapter 8, expressions for the shear stress τ can be used to determine the shear stress distribution throughout a cross-section. We now examine the shear stress distribution for a channel section. In doing so, we will find that this leads to a new property of a section, the 'shear centre'. It is convenient to develop our ideas via a specific case.

We examine a cantilever beam of length L having a channel cross-section [Fig. (13.6.1a)]. Such a section is called a **thin-wall section** if $t/c \ll 1$ and $t'/h \ll 1$, as shown in Fig. (13.6.1b). (Using the thin-wall property, we may consider dimensions c and h to be between the centres of the flange and web, as shown in the figure.) The beam is subjected to a load P acting in the y-direction at the free end. Note that the point of application of the force P within the cross-section at the free end is *not specified*. However, irrespective of the y-line of action of P, the shear force at any cross-section is $V_y = P > 0$ (with $V_z = 0$). We first determine the shear stress distribution throughout any cross-section at any arbitrary location along the longitudinal x-axis. Observing that the z-axis is an axis of symmetry and therefore a principal axis, we make use of Eq. (13.5.8),

$$\tau = \frac{V_y Q_z}{I_{zz} b}. \tag{13.6.1}$$

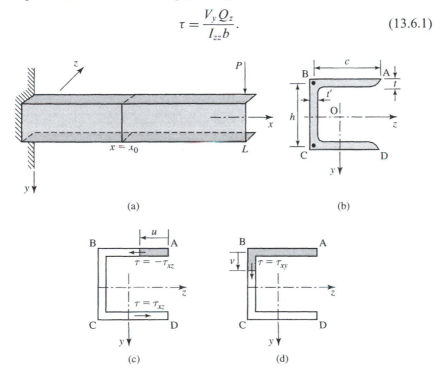

(a)

(b)

(c)

(d)

Figure 13.6.1

(a) Shear stress distribution

To determine the shear stress in the flange AB of the channel, we might make a *horizontal* cut somewhere between the top and bottom faces of the flange. We note that if we do so, the length b here is equal to the dimension 'c', as given in Fig. (13.6.1b). The resulting shear stresses τ would then be acting in the vertical direction. On the other hand, we may also make a *vertical* cut at some variable distance u from A. Then the b appearing in the expression is equal to the thickness t of the flange. Noting that $t/c \ll 1$, and that the b term appears in the denominator of the expression, it is clear that the horizontal component of τ is much greater than the vertical component. (In fact, we know that at the top and bottom free faces of the flange, vertical components of τ must vanish.) Thus if $t/c \ll 1$, we may then neglect the vertical component and calculate τ in the flange by making a vertical cut at some variable distance u from point A [Fig. (13.6.1c)]. For $0 \le u \le c$,

$$Q_z = \overline{A} \cdot \overline{y} = (tu) \cdot (-h/2). \tag{13.6.2}$$

Substituting in Eq. (13.6.1) and noting that with $b \equiv t$, $\tau(u) = -V_y hu/2I_{zz}$. We observe that τ varies linearly with u in the flange AB and that it is negative for $V_y > 0$; thus, according to comments (ii) of Section 13.5b, τ is acting to the left on a positive face. Note also that in the coordinate system of Fig. (13.6.1c), $\tau_{xz} = \tau = -V_y hu/2I_{zz}$ in the flange AB. At point B $(u = c)$, $\tau = -V_y hc/2I_{zz}$ (is maximum in absolute value).

The shear stress in the flange CD is calculated similarly upon defining u as the distance from point D: we obtain the same result but with opposite sign; i.e., $\tau(u) = \tau_{xz} = V_y hu/2I_{zz}$. Therefore, for $V_y > 0$, τ is acting to the right in this flange.

Consider now the web BC. Here, it is clear that the major stress component will be acting in the y-direction. Defining the 'cut' by means of the variable v (measured from the intersection of the centrelines), the moment Q_z of the area $\overline{A}$ above the cut [Fig. (13.6.1d)] is, by Eq. (13.5.5a),

$$Q_z = \overline{A} \cdot \overline{y} = -\frac{tch}{2} + (vt') \cdot [-(h/2 - v/2)] = -\frac{tch}{2} - \frac{vt'(h - v)}{2}. \tag{13.6.3}$$

Substituting in the above expression, and noting that here $b \equiv t'$,

$$\tau(v) = -\frac{V_y}{2I_{zz}t'}[cth + vt'(h - v)]. \tag{13.6.4a}$$

We observe that τ has a parabolic distribution in the web and is maximum when $v = h/2$, i.e. at the z-axis. Note also that for $V_y > 0$, $\tau(v) < 0$. Hence according to the previous comment (ii) of Section 13.5b, τ acts downward on the positive x-face. In the established coordinate system, $\tau_{xy} = -\tau$, i.e.

$$\tau_{xy}(v) = \frac{V_y}{2I_{zz}t'}[cth + vt'(h - v)]. \tag{13.6.4b}$$

The physical shear stresses and their distribution on the positive x-face are shown in Fig. (13.6.2a).

(b) The shear centre

Let us now calculate the resultant forces acting in the flanges, R_H, and in the web, R_V, due to the τ above. Upon observing that the τ distribution in the flange AB is linear, we obtain

$$R_H = \frac{1}{2}\left(-V_y \frac{hc}{2I_{zz}}\right)(ct) = -V_y \frac{htc^2}{4I_{zz}}. \tag{13.6.5}$$

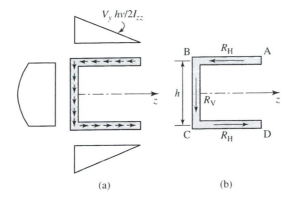

(a)   (b)   **Figure 13.6.2**

Note again that the minus sign here indicates that R_H is acting to the left in the flange AB. In the flange CD, we will clearly obtain the same R_H with the opposite sign; thus R_H acts to the right on the positive face of the flange CD.

The vertical resultant force R_V in the web BC, calculated from τ of Eq. (13.6.4b), is

$$R_V = \int_0^h \tau_{xy}(vt)t' \, dv = V_y \frac{t'}{2I_{zz}t'} \int_0^h (cth + t'hv - t'v^2) \, dv \quad (13.6.6)$$

or

$$R_V = \frac{V_y}{I_{zz}} \left[\frac{cth}{2}v + \frac{t'hv^2}{4} - \frac{t'v^3}{6} \right]_0^h. \quad (13.6.6a)$$

Therefore

$$R_V = \frac{V_y}{I_{zz}} \left(\frac{cth^2}{2} + \frac{t'h^3}{12} \right). \quad (13.6.6b)$$

To calculate I_{zz}, we note that we have assumed $t/c \ll 1$ and $t'/h \ll 1$. We therefore neglect these quantities when they appear as third-order infinitesimals in the calculation [i.e., we set $(t/c)^3 = 0$, etc.]. The second moment I_{zz}, calculated on this basis, becomes

$$I_{zz} = \frac{cth^2}{2} + \frac{t'h^3}{12}, \quad (13.6.7)$$

from which we find $R_V = V_y$. [The resultant forces R_H and R_V acting on the positive x-face are shown in Fig. (13.6.2b).] Thus, we observe that the internal stress resultant of the *average* shear stresses τ_{xy} satisfies the given vertical shear force condition on this positive face of the cross-section.

Consider now a segment $x_0 - L$ of arbitrary length of the beam where we have made a cut at some section $x = x_0$ [Fig. (13.6.3a)]. We note that three forces exist on the *negative* face of the cross-section at x_0: namely, an *upward* force in the web, $R_V = V_y$, and the two forces in the flanges: a force R_H acting to the right in flange AB, and to the left in flange CD. We observe that the forces R_H constitute, in effect, a couple $R_H h$.

Now, we recall that one can always replace *any* force system in terms of an 'equivalent force system' consisting of a single resultant (or a couple). We thus wish to replace the three-force system by a single resultant. Since the two flange forces are equivalent to a *clockwise* couple, it is clear that the resultant is a vertical

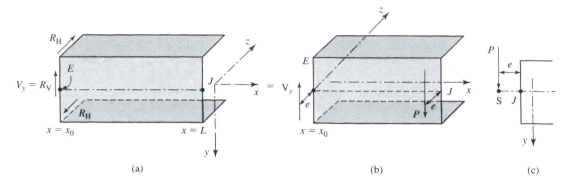

Figure 13.6.3   (a)   (b)   (c)

force $V_y = P$ that acts in the upward y-direction. To determine the location of this resultant such that it represents an equivalent force system, the moment M_x about *any* axis due to this resultant must be identical with that due to the original existing force system. (Here, it is convenient to take moments about an axis EJ of Fig. (13.6.3b), parallel to the centroidal x-axis of the section.) Since the flange forces represent a clockwise couple, we therefore require that

$$R_H h = V_y e, \qquad (13.6.8a)$$

that is,

$$e = \frac{R_H h}{V_y}. \qquad (13.6.8b)$$

Substituting Eq. (13.6.5),

$$e = \frac{c^2 h^2 t}{4 I_{zz}}. \qquad (13.6.9)$$

Thus, the segment of the member, $x_0 - L$, can be considered to be subjected at the section x_0 to a force system consisting of a single upward force $V_y = P$ whose line of action is located at a distance e from the web, as shown in Fig. (13.6.3b).

Treating the segment $x_0 - L$ as a free body, we observe that it is, in effect, subjected to two forces: the upward force $V_y = P$ at x_0 and the applied downward load P at $x = L$. By inspection, the equilibrium conditions $\sum F_y = 0$, $\sum F_z = 0$ are seen to be satisfied. Examining the moment condition, $(\sum M_x)_{EJ} = 0$, we find that this condition is satisfied only if the applied vertical load P acts along a line of action at the same distance e from the web [see Fig. (13.6.3b)].

Thus we conclude that if the beam is to bend without twisting when loaded by a force acting in the y-direction, it is necessary that its line of action (in the y-direction) be at a distance e as measured from the centre of the web.

If P does not lie along the line of action given by e above, the beam will rotate (i.e., twist) due to an external torque $T = Pe$.

If we consider now a force acting in the horizontal z-direction on the channel beam, it is clear from arguments of symmetry, that the beam will bend about the y-axis without twisting only if the line of action of the applied horizontal force coincides with the z-axis.

The intersection of these two lines of action, point S, is called the **shear centre** of the channel [Fig. (13.6.3c)]. We therefore define the shear centre as follows:

■ The shear centre of a cross-section is that point through which all external loads must be applied such that the beam bend without twisting.

It is instructive to consider the location of the shear centre for the channel section investigated above for the case where the thickness of the wall is constant, i.e. $t' = t$. From Eq. (13.6.7), the second moment I_{zz} becomes

$$I_{zz} = \frac{th^2}{12}(6c + h). \qquad (13.6.10a)$$

Substituting in Eq. (13.6.9),

$$e = \frac{3\gamma}{1 + 6\gamma} \cdot c, \quad \gamma = \frac{c}{h}. \qquad (13.6.10b)$$

The position of the shear centre is seen to depend on the ratio c/h and falls in the range $0 \le e \le 0.5c$. Thus, if this ratio is relatively large, the shear centre for this section will lie at a distance $e \simeq 0.5c$ from the web. Note also that if $\gamma = 0$, i.e. if the section consists of a thin rectangle, the shear centre lies along its y-axis of symmetry.

(c) Some remarks and comments

 (i) The shear centre S of a cross-section is a geometric property of the section alone, and is not a property of the loading. All cross-sections possess a shear centre.
 (ii) If a load is applied through any other point of the cross-section other than the shear centre, the beam will twist due to a resulting torque.
 (iii) It will later be shown (in Chapter 14) that the shear centre of a linear elastic beam coincides with the 'centre of twist' of a member when subjected to a torsional moment T.
 (iv) From standard arguments of symmetry, we may conclude that the shear centre will always lie on the axis of symmetry if a cross-section possesses an axis of symmetry. Thus, for example, the shear centre S of a rectangular cross-section, etc. or of an I-section will be as located as shown in Figs. (13.6.4a and b).
 (v) Let us consider cross-sections composed of thin rectangular components, as shown in Figs. (13.6.4c and d). Clearly, irrespective of the applied loads, the resulting shear stresses must be parallel to the 'long' dimension of these rectangular components. Thus, for the sections shown, for example, in Figs. (13.6.4c and d), the resultant forces of these shear stresses must pass through the point S and consequently produce no moments about this point. Therefore, the external forces themselves must be such that they too contribute no moments about this point. Hence point S, as shown in the figure, must represent the location of the shear centre.
 (vi) Finally, we should address the following question: if the shear centre lies outside the physical section (as, e.g., in the case of the channel) how then can one load

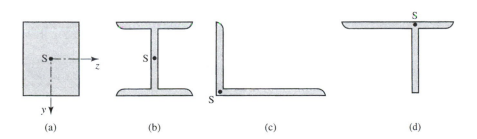

(a)   (b)   (c)   (d)   **Figure 13.6.4**

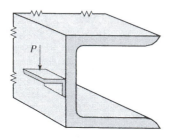

Figure 13.6.5

the beam through S? In engineering practice, this is achieved by adding a bracket to the beam, as shown in Fig. (13.6.5). However, the importance of establishing the location of the shear centre lies mainly in determining the rotation (twist) of a beam when loads do *not* pass through this point. This problem is considered in the section below.

13.7. Deflections and rotations of a beam under applied loads

We have seen in our previous study, that (a) the deflection of a beam is always perpendicular to the neutral axis and (b) applied forces that do not act through the shear centre of a cross-section will cause the beam to rotate.

Let us return to the problem of the channel cross-section. Assume that a vertical force P is acting, for example, through the centroid O of the section, as shown in Fig. (13.7.1a). Using the principle of superposition, one may consider this loading case as a superposition of loads given by Figs. (13.7.1b and c), where S is the shear centre of the cross-section. For the load of Fig. (13.7.1b), we note that P passes through S and acts in the direction of a principal axis. Hence from Eq. (13.4.15), the neutral axis coincides with the z-axis and therefore the deflection δ in this case will be in the direction of P.

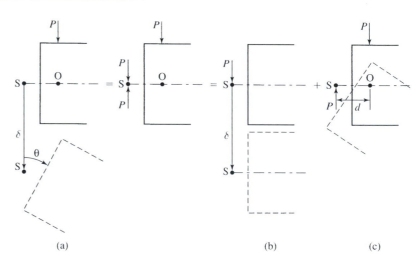

Figure 13.7.1   (a)   (b)   (c)

Examining the loading case of Fig. (13.7.1c), we recognise that this loading system consists of a couple producing a torque $T \equiv M_x = Pd$. Hence this force system will cause a rotation about the centre of twist which, according to comment (iii) of Section 13.6 above, coincides with the shear centre S. Thus, due to this loading case, the beam may be treated as a member subjected to a torsional moment T. The rotation θ and the additional shear stresses τ in the beam due to the torque may be found from the expressions of Chapter 12.

Combining the two effects, we observe that the cross-section of the beam will both deflect and rotate, as shown in Fig. (13.7.1a), if P is applied through the centroid.

Let us now examine another case, namely a beam whose cross-section is an angle (whose legs have equal length) situated in the y–z system, as shown in Fig. (13.7.2a). We observe that here the z-axis is a principal axis. Assume now that a vertical load is applied as shown. Note that the shear centre S is located at the intersection of the two legs [as was observed in remark (v) of Section 13.6b]. Then clearly, the cross-section will deflect in the vertical y-direction and will not rotate, since, as we have seen, the load passes through the shear centre.

If the beam is subjected to a force acting parallel to the principal axis but not through the shear centre, then the shear centre S will deflect in the direction of the load but some rotation will take place [Fig. (13.7.2b)].

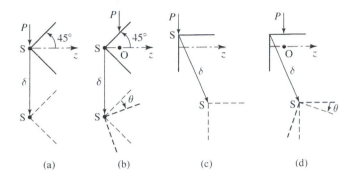

| (a) | (b) | (c) | (d) | **Figure 13.7.2** |

On the other hand, assume that the angle is situated as in Fig. (13.7.2c) and that a vertical load is applied. Here, the z-axis is not a principal axis. Since P is not acting parallel to a principal axis, the cross-sections will not deflect in the direction of P. But, again, because P is acting through the shear centre S, no rotation will take place; the beam will therefore assume the position shown in Fig. (13.7.2c).

Assume now that the beam is situated as in Fig. (13.7.2d) and that a force P is acting through a point other than S. Then, as in the previous case, the beam will deflect in a direction not parallel to P. Moreover, since the force is not acting through the shear centre S, the beam will also rotate, as shown in the figure.

We summarise our results by means of the following table [where (a) to (d) refer to Fig. (13.7.2)].

Beam loading	Through shear centre	Not through shear centre
Parallel to principal axis	Shear centre deflects in direction of loading with no rotation (a)	Shear centre deflects in direction of loading and cross-section rotates about the shear centre (b)
Not parallel to principal axis	Deflects with no rotation but not in direction of loading (c)	Shear centre deflects in some arbitrary direction and cross-section rotates about the shear centre (d)

Finally, it is important to note that the expressions for τ derived in Section 13.5 are due to bending and do not take into account twisting of the beam. These expressions are therefore valid only for the case when the applied loads pass through the shear centre of the section. The effect of torsional moments, which occur when the applied loads do not pass through the shear centre, must then be analysed according to the expressions developed in Chapter 12. Since all the derived relations are linear, the total effect, due to bending and torsion, is then obtained by simple superposition.

13.8.   Shear stresses in closed thin-wall sections

Thin-wall sections appear in many engineering structures, in particular, in aircraft structures. While the determination of the flexural stresses σ_x in such cross-sections does not require any special treatment, it is necessary to develop expressions for the shear stresses.

We examine a beam with a thin-wall cross-section of arbitrary shape and varying thickness t, as shown in Fig. (13.8.1a). For simplicity, we consider the case where the y- and z-axes are principal centroidal axes. The beam is subjected to a known shear force V_y acting through a given point J as shown. We wish to determine the shear stresses existing throughout the section. As we have seen from our previous discussions, shear stresses existing in a thin wall must always act tangentially to the lateral surfaces; it is these shear stresses that we wish to obtain.

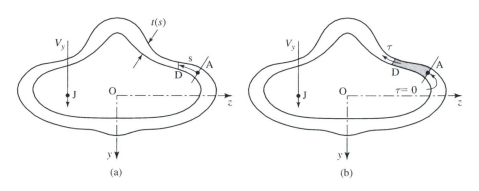

Figure 13.8.1   (a)   (b)

We first define a parameter s as the variable tangential distance along the centre-line of the wall measured from some arbitrary point A in the section [Fig. (13.8.1)]. Then, from the derived expressions of Section 13.5, one would expect τ at any point D, at a distance s from A, to be given by Eq. (13.5.8), namely

$$\tau = \frac{V_y Q_z(s)}{I_{zz} t(s)}.$$

We recall, however, that Q_z appearing in this expression represents the moment about the z-axis of the area $\overline{A}$ of an *isolated portion* of the cross-section. Now, in order to isolate such a portion and consider it as a free body, we note that here it is necessary to introduce two 'cuts', say at A and D. On each one of these cuts, unknown shear stresses must be assumed to be acting. Thus clearly, it is not possible to calculate the shear stresses at D since the shear stresses at A are not known. Indeed, in deriving Eqs. (13.5.4)–(13.5.8), it was clearly implied that only a single cut had to be made. If, however, τ were known to be zero, for example at A, then the above expression could readily be used to calculate τ at D by isolating the segment AD. We therefore make the *assumption* that at A, the stress $\tau = 0$ and, on this basis, calculate Q_z for the area $\overline{A}$ of the segment AD [Fig. (13.8.1b)]. As a result, in using the above expression, we do not obtain the true average physical stress τ, but rather a fictitious stress τ_f; i.e.,

$$\tau_f = \frac{V_y Q_z(s)}{I_{zz} t(s)}, \qquad (13.8.1)$$

which must later be corrected in the analysis. Note that the resultant of these internal stresses will always be equal to V_y.

Having calculated τ_f from Eq. (13.8.1), it is then possible to determine the torsional moment M' about the x-axis (passing through point O), which is caused by these stresses. Setting this moment equal to $V_y a$, we find $a = M'/V_y$, the distance from O to the point C, as shown in Fig (13.8.2). Thus, the stresses τ_f are the stresses *which would exist* if the resultant shear force V_y were to pass through point C. However, V_y has been specified to pass through point J.

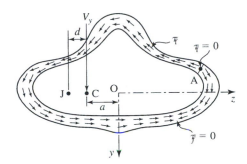

Figure 13.8.2

Now, we know that a force system consisting of a single force passing through a given point can always be replaced by an equivalent force system consisting of a force (having the same resultant) and a moment. Thus, here, the force system V_y acting through point J [Fig. (13.8.3a)] can be replaced by the force V_y acting through point C and a couple $M = V_y d$ acting about the x-axis, where d is the distance between J and C [Figs. (13.8.3b and c)].

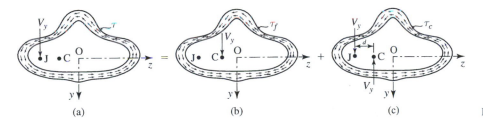

(a)   (b)   (c)   Figure 13.8.3

Let τ denote the actual stresses due to the given shear force appearing in Fig. (13.8.3a). Since all the stress–force relations are linear, we may use the principle of superposition; thus

$$\tau = \tau_f + \tau_c, \qquad (13.8.2a)$$

where τ_f are the fictitious stresses due to the shear force acting through point C [Fig. (13.8.3b)] and given by Eq. (13.8.1). The stresses τ_c, produced by the couple $T = V_y d$ [Fig. (13.8.3c)] may be considered as corrective stresses [which essentially are corrections to our initial assumption that $\tau_A \equiv \tau(s=0)=0$]. From the expressions derived in Chapter 12, these stresses, due to a torsional moment acting on a closed thin-wall section, are given by [see Eq. (12.10.6)]

$$\tau_c(s) = \frac{T}{2At(s)} = \frac{V_y d}{2At(s)}, \qquad (13.8.2b)$$

where A represents the area within the centre-line of the thin wall. Combining these expressions, we arrive at an explicit expression for the true average stress,

$$\tau(s) = \tau_f(s) + \tau_c(s) = \frac{V_y Q_z(s)}{I_{zz}t(s)} + \frac{V_y d}{2At(s)}. \qquad (13.8.3)$$

Example 13.4: A rectangular thin-wall section, as shown in Fig. (13.8.4), is subjected to a shear force $V_y = 100,000$ N whose line of action passes through the point J, 50 mm from the y-axis. (a) Determine the average shear stresses τ at points along the lines a–a and b–b and along an arbitrary line c–c located at a distance v, as shown in the figure. [Note that here v is measured differently than in Fig. (13.6.1d).] (b) Draw the distribution of the shear stresses τ in the section. (c) Determine the unit angle of rotation Θ due to the given applied load if the cross-section is made of steel ($G = 79$ GPa).

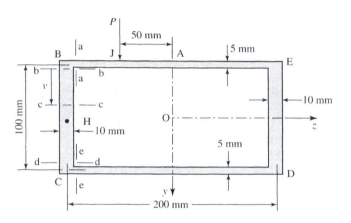

Figure 13.8.4

Solution:

(a) Following the development above, we assume that the stress $\tau = 0$ at point A. The resulting stresses τ_f are then calculated according to Eq. (13.8.1). For the section,

$$I_{zz} = (210 \times 105^3)/12 - (190 \times 95^3)/12 = 668.33 \times 10^4 \text{ mm}^4.$$

At a–a: $Q_z = (95 \times 5.0)(-50.0) = -23.75 \times 10^3 \text{ mm}^3$
At b–b: $Q_z = (105 \times 5.0)(-50.0) = -26.25 \times 10^3 \text{ mm}^3$
At c–c (defined by the variable v): $Q_z = -26.25 \times 10^3 - 475v + 5v^2$

Note: Along the segment B-E (or C-D) Q_z has a linear variation. Along the segment B-C, Q_z is a quadratic function of v; the shear stresses τ_f therefore vary parabolically along this segment and will have a maximum value at point H where $Q_z(v = 47.5) = -37.53 \times 10^3 \text{ mm}^3$.

Substituting in Eq. (13.8.1):

At a–a: $\tau_f = \frac{100,000(-23.75 \times 10^3)}{668.33 \times 10^4 \times 5.0} = -71.1$ MPa

At b–b: $\tau_f = \frac{100,000(-26.25 \times 10^3)}{668.33 \times 10^4 \times 10.0} = -39.3$ MPa

At H: $\tau_f = \frac{100,000(-37.53 \times 10^3)}{668.33 \times 10^4 \times 10.0} = -56.15$ MPa

Note:

(i) Clearly, Q_z at d–d must be the same as Q_z at b–b and Q_z at e–e must be the same as Q_z at a–a. Hence τ_f at d–d and e–e are equal to τ_f at b–b and a–a, respectively.

(ii) Upon examining the expressions for Q_z, we immediately deduce, by inspection, that the fictitious shear stresses τ_f are symmetric with respect to the y-axis.

The τ_f stress distribution as well as the directions of the stresses are shown in Fig. (13.8.5a).

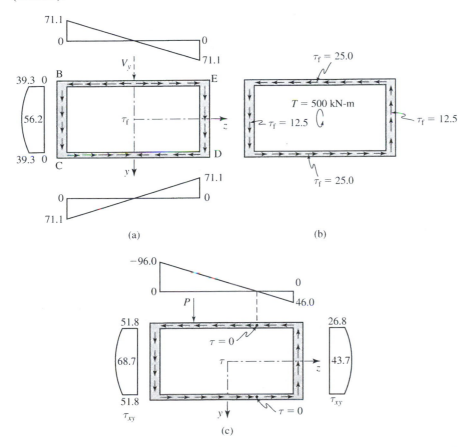

Figure 13.8.5

From this stress distribution, it is evident that the resulting moment about the x-axis passing through point O vanishes. Therefore, we may conclude that the calculated stresses τ_f are those that would be due to a shear force V_y that acts along the y-axis. (Thus, *if* V_y had been specified as passing through point A, the above stresses would then indeed be the true average shear stresses τ and no correction would be necessary.)

However, since V_y has been stated to pass through the point J, there exist corrective stresses τ_c due to a torque $T = 50 \times 100,000 = 5.0 \times 10^6$ N-mm. Hence from

Eq. (13.8.2b),

$$\tau_c = \frac{T}{2At} = \frac{5 \times 10^6}{2(100 \times 200)t} = \frac{125.0}{t} \text{ MPa.}$$

Along the top and bottom segments (B-E and C-D), $t = 5.0$ mm and therefore $\tau_c = 25.0$ MPa.

Along the vertical segments (B-C and D-E), $t = 10.0$ mm and therefore $\tau_c = 12.50$ MPa.

The shear distribution of the corrective stresses τ_c is shown in Fig. (13.8.5b).

Finally, the actual stresses, obtained by superposition, $\tau = \tau_f + \tau_c$, can be immediately calculated.

(b) The distribution of shear stresses τ is shown in Fig. (13.8.5c). Note that there exist two points in the section where $\tau = 0$. Since the shear flow is always continuous, and since, in order to satisfy the equilibrium conditions $\sum F_y = \sum F_z = 0$ in the cross-section, it must 'change directions', it follows that the shear flow must necessarily vanish (at least) at two points within a closed section.

(c) To obtain the rotation of the section due to the applied loads, we use Bredt's formula as derived in Chapter 12 [Eq. (12.10.14)]:

$$\Theta = \frac{1}{2AG} \oint \tau(s)\, ds. \tag{13.8.4a}$$

Substituting from Eq. (13.8.2a),

$$\Theta = \frac{1}{2AG} \oint [\tau_f(s) + \tau_c(s)]\, ds, \tag{13.8.4b}$$

where $\oint$ is over the closed path of the centre lines. Due to the symmetry of the cross-section and of τ_f, it is evident, for this case, that the integral of τ_f over the closed path vanishes. Recalling that τ_c is given by Eq. (13.8.2b), the above equation reduces to

$$\Theta = \frac{T}{4A^2G} \oint \left[\frac{ds}{t(s)} \right] = \frac{T}{4A^2G} \sum \left[\frac{S_i}{t_i} \right],$$

where t_i and S_i are the thickness and length of each component segment, respectively. Hence we obtain (upon using the symmetric properties), for $G = 79$ GPa $= 79 \times 10^3$ N/mm^2,

$$\Theta = \frac{5 \times 10^6}{[4 \times (2.00 \times 10^4)^2 \times (79 \times 10^3)]} \left(\frac{200}{5} + \frac{100}{10} \right) \times 2$$

$$= 0.00396 \text{ rad/m} = 0.227°/\text{m.}$$

Thus for a member 10 m long, the ends will rotate $2.1°$ with respect to each other.

It should be noted that the above example was considerably simplified due to the symmetry of the cross-section. In addition, by assuming A to be the point of zero τ in the solution, we obtained a symmetric distribution of the stresses τ_f with respect to the y-axis. The analysis of sections of arbitrary shape requires the use of Eq. (13.5.6) to calculate τ_f in lieu of Eq. (13.5.8) used here. Although this leads to some more complex calculations, the basic ideas remain the same and the analysis proceeds as in the given example.

PROBLEMS

Sections 2–4

13.1: (a) Derive the parallel axis theorem for the product of inertia of an area. Is the product of inertia about the centroidal axes necessarily a minimum? Why? (b) Let (y, z) be centroidal axes of a cross-section and let (y', z) be another set of axes, as shown in Fig. (13P.1). What statement can be made concerning the relation between I_{yz} and $I_{y'z}$?

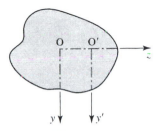

13.2: A vertical force $P = 2000$ N acting in the y-direction is applied at the centre of a simply supported steel beam of length $L = 5$ m whose cross-section is an angle with unequal legs, L89 × 76 × 12.7, as shown in Fig. (13P.2). Determine the maximum flexural stress at the mid-span at points A, B and C of the cross-section. (*Note:* I_{yz} can be calculated from the properties of the cross-section, I_{yy}, I_{zz} and the principal moment of inertia $I_{min} = Ar^2$ (and its principal direction), which are given in the tables of Appendix E.)

13.3:* Given a beam of arbitrary cross-section, subjected to bending moments about any two orthogonal axes. Starting from the general expression for the flexural stress, Eq. (13.4.9), show that this leads to Eq. (13.2.6), $\sigma_x = \frac{M_n \eta}{I_n}$, where I_n is about the neutral axis and η is the perpendicular distance to any point of the cross-section from the neutral axis.

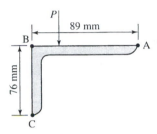

Figure 13P.2

13.4:* The flexural and shear stresses in a beam are given by Eqs. (13.4.9) and (13.5.6), respectively. Clearly, for finite stresses, the denominator, $I_{yy}I_{zz} - I_{yz}^2 \neq 0$. Prove that $I_{yy}I_{zz} - I_{yz}^2 > 0$ for any arbitrary cross-sectional area A. (*Hint:* $\iint_A (y - \lambda z)^2 dA > 0$ for any real λ.)

13.5: A rectangular beam 15 cm wide and 20 cm deep is simply supported on a span of 5 m. Two loads of $P = 3500$ N each are applied to the beam, each load being 1.25 m from the support. The loads are located in a plane that makes an angle of 30° with the y-axis, as shown in Fig. (13P.5).

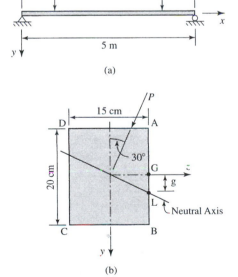

(a)

(b)

Figure 13P.5

(a) Determine the angle of inclination of the neutral axis with respect to the z-axis, namely $\beta = 45.75°$.

(b) Using the general expression for the bending stresses, determine the stresses at points A, B, C, D and G that exist at the centre of the span.

(c) Show the stress distribution for points along the face A-B by means of a figure.

(d) Evaluate g, the point of zero bending stress. Show that this point lies on the neutral axis as found from (a).

(e) Knowing the orientation of the neutral axis for the given loads obtained in (a) above, verify that one obtains the same value for the stress at point G [as in part (b) above] from the expression $\sigma_x = M_n \eta / I_n$, where η is the perpendicular distance from the neutral axis to point G and M_n is the moment of the loads about the neutral axis.

13.6: For the Z-section shown in Fig. (13P.6), determine (a) the centroidal moments of area I_{zz} and I_{yy} and the mixed product of area I_{yz}, (b) the directions of the principal centroidal axes (indicate their orientation by means of a figure) and (c) the principal moments of area.

13.7: A cantilever beam fixed at one end is subjected to a moment M_z and a force P at the free end, as shown in Fig. (13P.7a). The moments and product of area of the cross-section are given as $I_{yy} = 4I_0$, $I_{zz} = 2I_0$ and $I_{yz} = -I_0$. In addition, it is known that at a cross-section A located a distance $2a$ from the free end, the flexural stress at point D ($y = -2b$, $z = b$) of the cross-section [see Fig. (13P.7b)] is $\sigma_x = \sigma_0$. Determine (a) M_z and (b) the location of the neutral axis with respect to the z-axis at cross-section B, located a distance a from the free end. Show the location of the neutral axis by means of a figure.

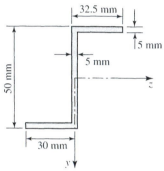

Figure 13P.6

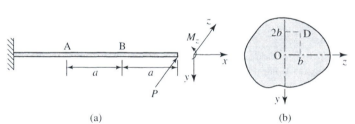

(a)   (b)

Figure 13P.7

Sections 5 and 6

13.8: A steel ($G = 77$ GPa) channel, C102 × 11, used as a cantilever 2 m long, is subjected to a load $P = 4$ kN at the free end passing through the centroid, as shown in Fig. (13P.8). (a) Determine the distance e from the web to the shear centre S. (b) Determine the angle of twist at the free end of the beam. (c) Find the maximum shear stress at the fixed end, including the effects of both torsion and flexure.

13.9: Determine e, the location of the shear centre S with respect to point F of the thin-wall cross-section having thickness t ($t \ll a$, $t \ll h$), as shown in Fig. (13P.9). Indicate its location by mean of a figure.

13.10: (a) Calculate the shear stress distribution acting on the Z-section, shown in Fig. (13P.6), when subjected to a vertical shear force V_y that passes through the shear centre. (b) Plot the variation of the shear stress τ / V_y along the flanges and the web. Indicate the shear flow by means of a figure. (c) What is the location of the shear centre of this section? *Explain* the reasoning that justifies your answer. (*Given:* $I_{zz} = 19.2$ cm^4, $I_{yy} = 9.1$ cm^4, $I_{yz} = -10.1$ cm^4.)

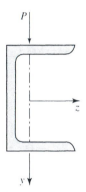

Figure 13P.8

13.11:   (a) Determine the distance *e* from point O to the shear centre S of the thin-wall ($t \ll d$, $t \ll h$) symmetrical section shown in Fig. (13P.11). Express the answer as the ratio e/d.

13.12:   A shear force $V_y > 0$ is applied at the free end of a cantilever beam whose open thin-wall ($t \ll R$) cross-section is as shown in Fig. (13P.12). (a) Determine the shear stress distribution in segments AB and CD of the cross-section and sketch the shear stresses in the cross-section. (b) Determine the distance *e* from point G to the shear centre. Show the location of the shear centre by means of a figure.

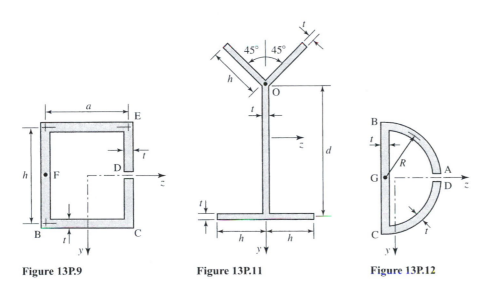

Figure 13P.9   **Figure 13P.11**   **Figure 13P.12**

13.13:   (a) Determine the shear stress distribution $\tau = \tau(\phi)$ for a semi-circular cross-section of radius *R* and thickness *t* ($t/R \ll 1$) [Fig. (13P.13a)] that is subjected to a vertical shear force V_y. (b) Show that the location of the shear centre of this section is given by $e = 4R/\pi$. (c) Determine the distance *e* from point O to the shear centre for the section shown in Fig. (13P.13b). Show the location by means of a figure. (*Note*: The results of part (b) may be used to obtain the solution to part (c).)

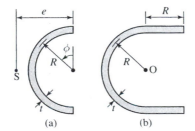

(a)   (b)   **Figure 13P.13**

13.14:   Two plates are welded along the entire length of a channel section to form a monolithic cross-section as shown in Fig. (13P.14). Determine the value of α if the shear centre lies at point O.

13.15:   For the given symmetric section with constant wall thickness *t*, as shown in Fig. (13P.15), where $t \ll h_1, h_2$ and *d*, determine the location of the shear centre, namely e measured from point O in terms of h_1, h_2 and d and show the location by means of a figure.

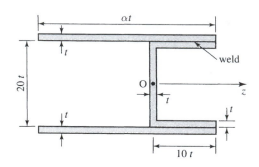

Figure 13P.14

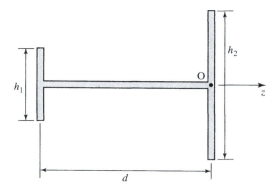

Figure 13P.15

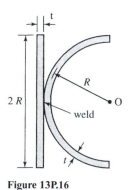

Figure 13P.16

13.16: Determine the location of the shear centre S with respect to point O for the thin-wall cross-section $t \ll R$ shown in Fig. (13P.16). Indicate the location of S by means of a figure.

13.17:* A vertical force P acts at point D at the free end of a cantilever beam, as shown in Fig. (13P.17). Determine the value of α if the beam does not twist.

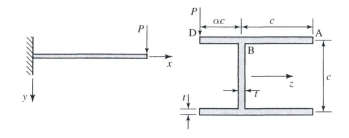

Figure 13P.17

13.18:* The cross-section of a beam is described as a crescent-shaped section whose width $t(\theta) = t_0 \sin\theta$ varies, as shown in Fig. (13P.18). Determine the distance e to the shear centre with respect to point O.

Section 7

13.19:* Given an elastic beam loaded by a single force P acting at an angle θ with respect to the y-axis such that the displacement of the shear centre, Δ, is in the same direction as the force *P*. *Prove* that the load is acting in a direction parallel to the principal axis of the cross-section.

13.20:   A load *P* acts at an angle θ with respect to the *y*-axis on a beam having a thin-wall ($t \ll a$) Z-section, as shown in Fig. (13P.20). If $d = 5a$, determine the possible values of θ if the shear centre deflects in a direction parallel to the load *P*.

13.21:   A thin-wall cross-section consists of three plates ($a \times L$) welded together to form a beam of length *L* having an equilateral triangular cross-section, as shown in Fig. (13P.21). At point B, the weld fails along the entire length of the beam. Determine the location of the shear centre S with respect to point D.

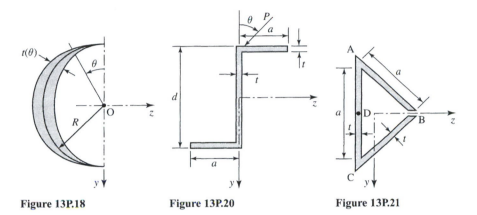

Figure 13P.18	Figure 13P.20	Figure 13P.21

13.22:   A prismatic beam of rectangular cross-section is subjected to a load acting along one of its diagonal planes, as shown in Fig. (13P.22). *Prove* that the neutral axis coincides with the second diagonal of the cross-section.

13.23:*   The cross-section of a beam consists of two solid square elements ($2a \times 2a$) welded together with an overlap 2*e*, as shown in Fig. (13P.23). (a) Determine the ratio *e/a* if point O deflects in a direction 30° with respect to the *y*-axis when the beam is subjected only to vertical loads. (b) What is the largest possible value of the angle of inclination of the displacement with respect to the *y*-axis under this loading; for what value of *e* ($0 \le e \le a$) does this occur?

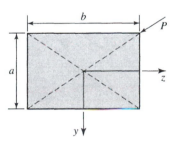

Figure 13P.22

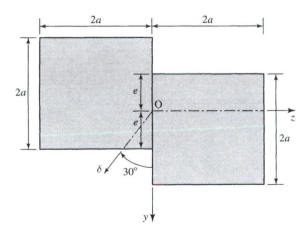

Figure 13P.23

13.24: A cantilever beam of length L, fixed at one end and having a thin-wall cross-section of thickness t ($t \ll a$), as shown in Fig. (13P.24), is subjected to a load P at the free end. The load is inclined at an angle θ with respect to the y-axis and passes through the centroid C. (a) Determine the angle θ if the resultant displacement of point B at the free end is in the z-direction. (b) For θ found in (a) above, determine Δ_y, the component of the displacement of the centroid in the y-direction. Express Δ_y in terms of a, t, L, P and G, the shear modulus.

13.25:* Given an elastic beam of arbitrary cross-section [see Fig. (13P.25)] loaded by a single force P acting at some angle θ with respect to the y-axis. Prove that the displacement of the shear centre, Δ, can never be perpendicular to the direction of the applied force. (*Hint*: Note the property of Problem 13.4.)

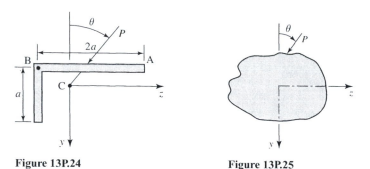

Figure 13P.24   **Figure 13P.25**

Section 8

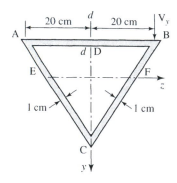

Figure 13P.26

13.26: A shear force $V_y = 200{,}000$ N acts on a thin-wall equilateral triangular cross-section of a beam having constant thickness of 1 cm, as shown in Fig. (13P.26). (a) Determine the resulting shear stress distribution in the section. (b) Draw the shear flow in the section indicating values at points A to F. (*Suggestion*: Assume initially that $\tau = 0$ along the line d–d of the cross-section.)

13.27: A weight W is suspended at point B of a thin-wall pipe ($t \ll R$), as shown in Fig. (13P.27). (a) Determine the shear stress in the pipe at points A, B, C and D. (b) Sketch the shear flow within the pipe and determine the points of zero shear stress and indicate their location by means of a figure. (*Hint*: The shear flow due to a vertical force acting along the y-axis is symmetric with respect to this axis.)

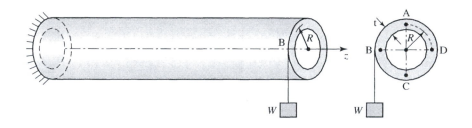

Figure 13P.27

13.28: Determine the distance e from point E to the shear centre for the closed thin-wall cross-section, shown in Fig. (13P.28), and indicate its location by means of a figure. (*Suggestion*: Assume initially that $\tau = 0$ along the line a–a of the cross-section.)

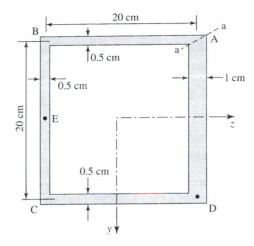

Figure 13P.28

Review and comprehensive problems

13.29: (a) Determine the location of the shear centre S with respect to point C of the thin-wall open cross-section, shown in Fig. (13P.29a), and show its location by means of a figure. (b) The following statement is made: 'If a force P is applied to the section in any arbitrary direction α, as shown in Fig. (13P.29b), the deflection of the shear centre will always be parallel to P.' Is this statement true? Give precise reasons for the answer.

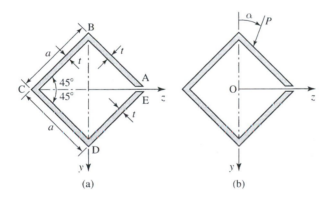

(a)   (b)

Figure 13P.29

13.30: A structural steel angle L203 × 203 × 19.1 is used as a 2.5-m long cantilever beam, and is subjected to a vertical concentrated load of 1400 N acting at the free end through the shear centre of the cross-section. (a) Calculate the flexural stress σ_x at the points A, B and C at the fixed end of the beam, as well as the average shear stress τ acting along the line d–d of this cross-section [see Fig. (13P.30)]. (b) Find the position of the neutral axis and the direction of the deflection and indicate on a sketch of the cross-section with respect to the centroidal axes. (*Note:* I_{yy} and I_{zz} are given in the tables of Appendix E. I_{yz} must be calculated.)

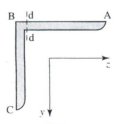

Figure 13P.30

13.31: A load P is applied at the free end of a cantilever beam at an angle α with respect to the y-axis. The thin-wall cross-section ($t \ll a$) of the beam is as shown in Fig. (13P.31b). (a) Determine the angle α if the resultant displacement of the shear centre is in the y-direction. (b) For α, as found in (a) above, determine the component of displacement in the z-direction of point B of the cross-section of the free end in terms of P, the geometry of the problem and G, the shear modulus of the beam material.

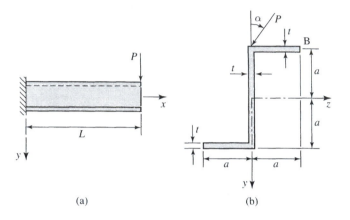

Figure 13P.31   (a)   (b)

13.32:* A channel member of length L whose cross-section of constant wall thickness t ($t \ll c$, $t \ll h$) is as shown in Fig. (13P.32), is fixed at one end and subjected to a vertical load P, passing through the web of the channel at the free end. (a) Determine the horizontal and vertical components of displacement, Δ_y and Δ_z, of point A at the free end in terms of c, h, t, L, P and the moduli of the beam material, E and G. (b) Determine Δ_y and Δ_z at A for the limiting case $c \to 0$.

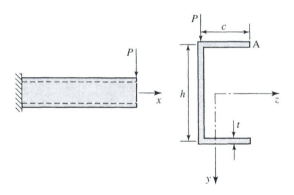

Figure 13P.32

13.33:* Determine the value of α such that the shear centre of the thin-wall cross-section shown in Fig. (13P.33) coincides with point O.

13.34: A beam whose cross-section is a thin wall ($t \ll a$), as shown in Fig. (13P.34b), is subjected to two vertical loads, as shown in Fig. (13P.34a). Determine the vertical component of displacement of point O at the free end in terms of P, L, a, t and the moduli of the beam material, E and G.

13.35: (a) Determine the ratio b/a if the shear centre S of the thin-wall symmetric cross-section, shown in Fig. (13P.35a), is to coincide with point D. (b) Determine the horizontal and vertical displacement components of point B at the free end of the beam, shown in Fig. (13P.35b), if a torsional moment T is applied. Express the answer in terms of a, t, L, T and G, the shear modulus of the beam material.

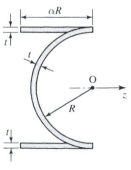

Figure 13P.33

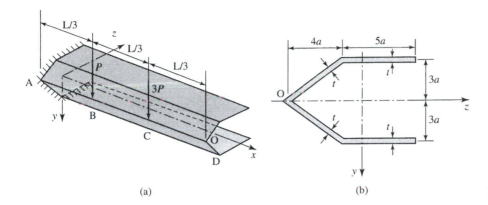

(a)   (b)   **Figure 13P.34**

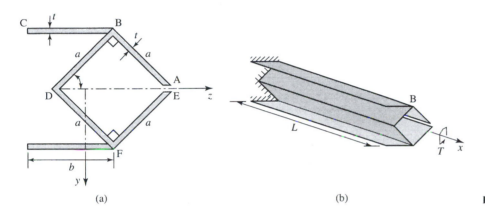

(a)   (b)   **Figure 13P.35**

13.36:   The cross-section of a beam consists of two solid rectangular elements ($a \times b$) welded together along its entire length at point O, as shown in Fig. (13P.36). The beam is subjected to a load acting in the y-direction. If the displacement is at an angle $\alpha = 45^\circ$ with respect to the y-axis, determine the ratio b/a.

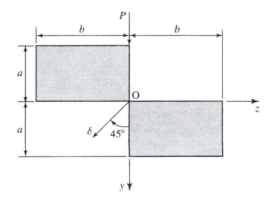

Figure 13P.36

13.37:   (a) Determine the location of the shear centre S with respect to point O of the thin-wall open section shown in Fig. (13P.37a). (b) Determine the displacement of point D at the free end if a cantilever beam of length L is subjected to a torsional moment at the free end, as shown in Fig. (13P.37b). Express the answer in terms of the given geometry and G, the shear modulus of the beam material.

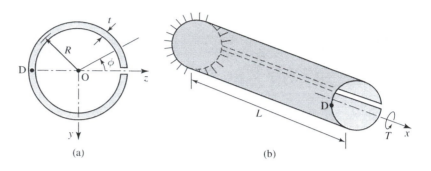

Figure 13P.37

(a)   (b)

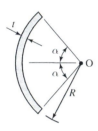

Figure 13P.38

13.38:* Show that the distance e from point O to the shear centre S of a cross-section consisting of an arc of a circle having thickness t with mean radius R ($t \ll R$) [see Fig. (13P.38)], is given by

$$e = 4R\frac{\sin\alpha - \alpha\,\cos\alpha}{2\alpha - \sin 2\alpha}.$$

The following problems are designed to require the use of a computer.

13.39: (a) Determine the distance e from point O to the shear centre S of the symmetric thin-wall cross-section shown in Fig. (13P.39). Express the answer as the ratio e/d as a function of θ and (b) using a plotting routine, plot e/d as an function of θ for $0 \le \theta \le 90°$.

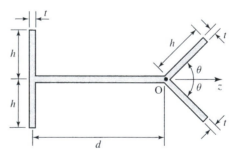

Figure 13P.39

13.40: A load P acts at any arbitrary angle θ with respect to the y-axis on a beam having a thin-wall ($t \ll a$) Z-section, as shown in Fig. (13P.20). (a) Derive the equation for the ratio d/a as a function of θ if the shear centre is to deflect in a direction parallel to the load P. (b) Using a plotting routine, plot θ vs. $\alpha \equiv d/a$ for $0 \le \alpha < \infty$.

Energy methods and virtual work

14

Basic energy theorems, principles of virtual work and their applications to structural mechanics

14.1 Introduction

In the study of mechanics, there exist two approaches that complement each other: (a) a vectorial approach in which the laws of mechanics are written in vectorial (or tensor) form and (b) an energy approach, expressed in scalar form. Although each approach has its respective merits, it is sometimes preferable to use one or the other in analysing any specific problem. As will be seen, the energy and virtual work principles often lead to very simple and elegant solutions to many problems, which, if approached vectorially, would prove to be quite complicated. Moreover, as we shall show in this and the next chapter, these principles can provide a means of obtaining approximate solutions to problems for which no exact analytical solution exists.

Such practical reasons should be sufficient motivation for the study of energy methods. However, aside from practical considerations, energy principles, together with the principles of virtual work, prove to be of great importance on the theoretical level: namely they afford a different viewpoint in mechanics and deepen our conceptual understanding of the behaviour of mechanical systems.

14.2 Elastic strain energy

From our previous discussion, we recall that when an elastic body undergoes deformation, internal energy in the form of strain energy is stored in the body. This concept was developed in Chapter 4 and a basic principle, the Principle of Conservation of Energy, was derived for the one-dimensional case, i.e. the uniaxial case. This principle states essentially that when an elastic body is subjected to an external force system, the work done by this force system is equal to the resulting internal strain energy.

We first review some of the basic results and expressions as developed in Chapter 4.

(a) Review of results for the uniaxial state of stress[†]

We first recall that an elastic material has been defined as one for which the stress state τ at any given point is a function of the state of strain ϵ at the point, and that a unique one-to-one inverse relation exists; thus

$$\tau = f(\epsilon), \qquad \epsilon = f^{-1}(\tau) \tag{14.2.1}$$

where f^{-1} is the symbolic representation of the inverse function.

If a *slender* elastic rod of length L and cross-sectional area $A(x)$ is subjected to a statically applied uniaxial load P passing through the centroid, as shown in Fig. (14.2.1), the axial stress is $\tau_{xx} = P/A(x)$ for any value of P; all other stress components are taken as zero. For this uniaxial state of stress, the material properties of the elastic rod may be represented by the stress–strain curve shown in Fig. (14.2.2); i.e., under the given loading, $\tau_{xx} = \tau_{xx}(\epsilon_{xx})$. The resulting strain energy density is then given by [Eq. (4.4.23)]

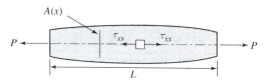

$$U_0 = \int_0^{\epsilon_{xx}^{\mathrm{f}}} \tau_{xx}(\epsilon_{xx}) \, \mathrm{d}\epsilon_{xx}, \tag{14.2.2a}$$

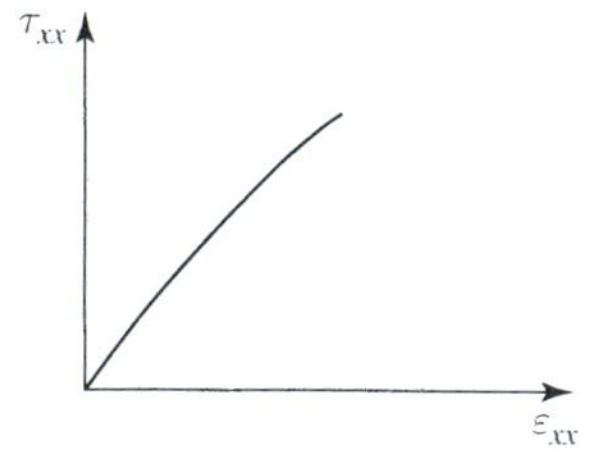

where $\epsilon_{xx}^{\mathrm{f}}$ represents the final strain state.

The total strain energy stored in the body, U, is therefore [Eq. (4.4.24)]

$$U = \iiint_V U_0 \, \mathrm{d}\Omega, \tag{14.2.2b}$$

Figure 14.2.2

where $\mathrm{d}\Omega$ is a volume element; that is,

$$U = \iiint_V \left[\int_0^{\epsilon_{xx}^{\mathrm{f}}} \tau_{xx}(\epsilon_{xx}) \, \mathrm{d}\epsilon_{xx} \right] \mathrm{d}\Omega. \tag{14.2.3}$$

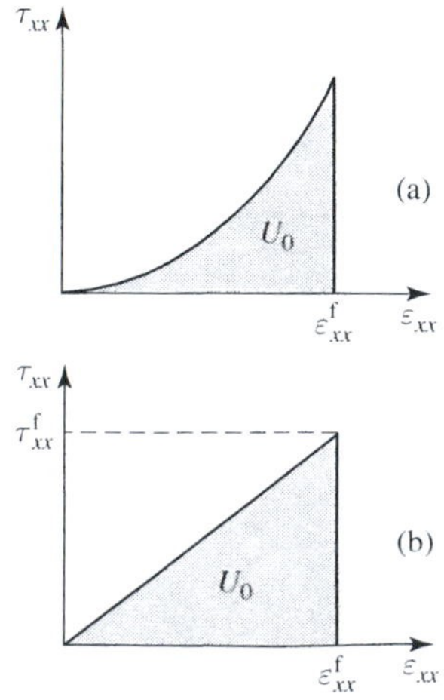

From the known geometric representation of the integral of Eq. (14.2.2a), we recall that the strain energy density U_0 may be represented by the area under the stress–strain curve, as in Fig. (14.2.3a).

For the special case of the linear isotropic elastic material, $\tau_{xx} = E\epsilon_{xx}$; the strain energy density U_0 is then

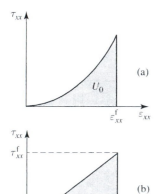

$$U_0 = E \int_0^{\epsilon_{xx}^{\mathrm{f}}} \epsilon_{xx} \, \mathrm{d}\epsilon_{xx} = \frac{E}{2} \left(\epsilon_{xx}^{\mathrm{f}} \right)^2. \tag{14.2.4}$$

Figure 14.2.3

[†] A detailed development of the derived expressions of this sub-section is given in Chapter 4, Section 4b.

Using again the stress–strain relation, we may write

$$U_0 = \frac{E}{2}\left(\epsilon_{xx}^{\mathrm{f}}\right)^2 = \frac{\tau_{xx}^{\mathrm{f}}\epsilon_{xx}^{\mathrm{f}}}{2} = \frac{\left(\tau_{xx}^{\mathrm{f}}\right)^2}{2E}. \tag{14.2.5}$$

We observe that the strain energy density for a linear elastic material is represented by the triangular area in Fig. (14.2.3b).

For convenience we drop the superscript 'f' and, using the simplified notation, we write Eq. (14.2.2a) for the general uniaxial case as

$$U_0 = \int_0^{\epsilon_{xx}} \tau_{xx}(\epsilon_{xx})\,\mathrm{d}\epsilon_{xx}. \tag{14.2.6}$$

Similarly, we rewrite Eq. (14.2.5) as

$$U_0 = \frac{E\epsilon_{xx}^2}{2} = \frac{\tau_{xx}\epsilon_{xx}}{2} = \frac{\tau_{xx}^2}{2E}. \tag{14.2.7}$$

In the above it is to be understood that the quantities represent the *final actual values* of the stress and strain components.

Recalling that $E > 0$, from the first or third terms of Eq. (14.2.7), we observe that U_0 is never negative; i.e., $U_0 \geq 0$. The strain energy density is then said to be a **positive definite** quantity.[†]

(b) General stress state

We now generalise our results to bodies in a three-dimensional state of stress. Consider first the simple case for linear isotropic elastic materials. From the second term of Eq. (14.2.7), we observe that the strain energy density is given by half the product of the stress times the corresponding strain component; i.e., $U_0 = \frac{\tau_{xx}\epsilon_{xx}}{2}$. This represents the work done on an element of *unit area* and *unit length* by the stress τ_{xx} [see Fig. (14.2.4)]. Since the material is isotropic, the same behaviour must exist in the y- and z-directions. Thus, the contribution to U_0 due to τ_{yy} and τ_{zz} will be $\frac{1}{2}[\tau_{yy}\,\epsilon_{yy}]$ and $\frac{1}{2}[\tau_{zz}\,\epsilon_{zz}]$, respectively. Consider now the work done by the shear stress τ_{yx} on a unit element [Fig. (14.2.5)]. Recalling that the strain ϵ_{xy} represents half the angle change, the work done by τ_{yx} is $\frac{1}{2}[2\epsilon_{yx}\,\tau_{yx}]$. Similar terms appear for the contribution due to τ_{yz} and τ_{zx}. We therefore conclude that the linear elastic strain energy density for a general state of stress is

$$U_0 = \frac{1}{2}[\tau_{xx}\epsilon_{xx} + \tau_{yy}\epsilon_{yy} + \tau_{zz}\epsilon_{zz} + 2(\tau_{xy}\epsilon_{xy} + \tau_{yz}\epsilon_{yz} + \tau_{zx}\epsilon_{zx})]. \tag{14.2.8a}$$

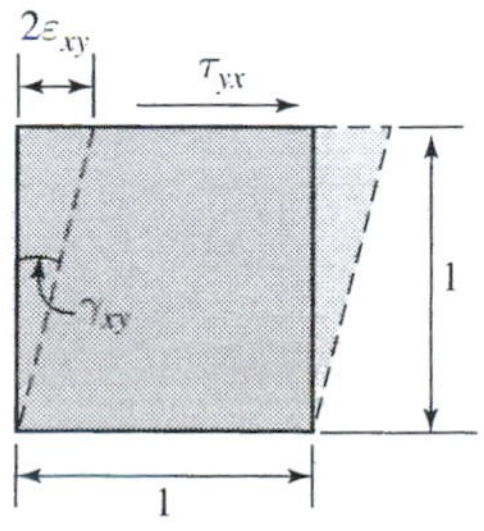

The total strain energy is then

$$U = \frac{1}{2}\iiint_V [\tau_{xx}\epsilon_{xx} + \tau_{yy}\epsilon_{yy} + \tau_{zz}\epsilon_{zz} + 2(\tau_{xy}\epsilon_{xy} + \tau_{yz}\epsilon_{yz} + \tau_{zx}\epsilon_{zx})]\,\mathrm{d}\Omega. \tag{14.2.8b}$$

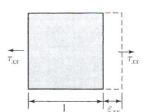

Figure 14.2.4

Figure 14.2.5

[†] While this result is correct, it is based on a practical engineering approach. We mention here that the strain energy density U_0 can be shown to be positive definite using basic thermodynamic considerations. It then follows that $E > 0$. Moreover, based on similar considerations, the strain energy density U_0 for *any elastic material* is shown to be positive definite. Such an approach is beyond the scope of our study.

For a two-dimensional case ($\tau_{zz} = \tau_{zx} = \tau_{zy} = 0$), we have

$$U = \frac{1}{2} \int\!\!\int\!\!\int_V [\tau_{xx}\epsilon_{xx} + \tau_{yy}\epsilon_{yy} + 2\tau_{xy}\epsilon_{xy}] \, d\Omega. \qquad (14.2.8c)$$

Alternative forms for U_0, expressed only in terms of stress or strain, can be obtained by substituting the appropriate stress–strain relations. Again, it is emphasised that U_0 will always be a positive definite quantity.

We consider here the general case of a nonlinear elastic material (not necessarily isotropic) under a three-dimensional state of stress and strain, i.e. a generalisation of the relation given by Eq. (14.2.6). In this last equation, we note that U_0 was obtained by integrating τ_{xx} over the corresponding strain component (as it was for the linear case). If, however, a general state of stress and strain exists, then τ_{xx} will be a function not only of ϵ_{xx} but also of all the strain components. Let us denote this symbolically as $\tau_{xx}(\epsilon)$. Contributions to the strain energy density due to other stress components will then be represented by $\tau_{yy}(\epsilon) \, d\epsilon_{yy}$, $\tau_{zz}(\epsilon) \, d\epsilon_{zz}$, $\tau_{xy}(\epsilon) \, d\epsilon_{xy}$, etc.

Noting that $\tau_{xy} = \tau_{yx}$, $\epsilon_{xy} = \epsilon_{yx}$, etc., the total strain energy density is, therefore, given by

$$U_0 = \int_0^{\epsilon_{xx}} \tau_{xx}(\epsilon) \, d\epsilon_{xx} + \int_0^{\epsilon_{yy}} \tau_{yy}(\epsilon) \, d\epsilon_{yy} + \int_0^{\epsilon_{zz}} \tau_{zz}(\epsilon) \, d\epsilon_{zz}$$
$$+ 2 \left[\int_0^{\epsilon_{xy}} \tau_{xy}(\epsilon) \, d\epsilon_{xy} + \int_0^{\epsilon_{yz}} \tau_{yz}(\epsilon) \, d\epsilon_{yz} + \int_0^{\epsilon_{zx}} \tau_{zx}(\epsilon) \, d\epsilon_{zx} \right]. \qquad (14.2.9)$$

From this last expression, we may deduce an interesting result. Since the strain energy here is a function of strain only, for a material under an existing state of strain ϵ, the increment of strain energy density due to changes in the various strain components is

$$dU_0 = \tau_{xx}(\epsilon) \, d\epsilon_{xx} + \tau_{yy}(\epsilon) \, d\epsilon_{yy} + \tau_{zz}(\epsilon) \, d\epsilon_{zz} +$$
$$+ 2[\tau_{xy}(\epsilon) \, d\epsilon_{xy} + \tau_{yz}(\epsilon) \, d\epsilon_{yz} + \tau_{zx}(\epsilon) \, d\epsilon_{zx}]. \qquad (14.2.10a)$$

We note that the strains ϵ appearing in Eq. (14.2.10a) denote the *final* strain components of a body in its deformed state. Furthermore, recalling that an elastic material is, by definition, one for which its final state is independent of the 'deformation history' or 'loading path' (i.e., from its initial to its final deformed state) (see Chapter 4, Section 4), it follows that dU_0, given by[†]

$$dU_0 = \frac{\partial U_0}{\partial \epsilon_{xx}} d\epsilon_{xx} + \frac{\partial U_0}{\partial \epsilon_{yy}} d\epsilon_{yy} + \frac{\partial U_0}{\partial \epsilon_{zz}} d\epsilon_{zz} + \frac{\partial U_0}{\partial \epsilon_{xy}} d\epsilon_{xy} + \frac{\partial U_0}{\partial \epsilon_{yz}} d\epsilon_{yz}$$
$$+ \frac{\partial U_0}{\partial \epsilon_{zx}} d\epsilon_{zx} + \frac{\partial U_0}{\partial \epsilon_{yx}} d\epsilon_{yx} + \frac{\partial U_0}{\partial \epsilon_{zy}} d\epsilon_{zy} + \frac{\partial U_0}{\partial \epsilon_{xz}} d\epsilon_{xz}, \qquad (14.2.10b)$$

is a perfect differential. Hence, comparing Eqs. (14.2.10a) and (14.2.10b), we find

$$\tau_{xx} = \frac{\partial U_0}{\partial \epsilon_{xx}}, \qquad \tau_{yy} = \frac{\partial U_0}{\partial \epsilon_{yy}}, \qquad \tau_{xy} = \frac{\partial U_0}{\partial \epsilon_{xy}}, \qquad \text{(etc.)} \quad (14.2.11)$$

[†] Note that here we treat mathematically ϵ_{xy}, ϵ_{yx}, etc. as independent variables of a function.

Thus, if the strain energy is expressed in terms of strain only, each stress component is then given by the partial derivative of U_0 with respect to the corresponding strain component.[†]

(c) Examples of strain energy for linear elastic bodies

We derive below expressions for the linear elastic strain energy for two specific cases that are often encountered in engineering analysis: the case of an axially loaded prismatic bar and the case of a member subjected to flexure.

(i) *Linear elastic prismatic rod subjected to a uniform axial load P.* Consider a prismatic bar of cross-sectional area A and length L subjected to a load P passing through the centroid of each section, as shown in Fig. (14.2.6). The resulting stress $\tau_{xx} = P/A$. Substituting in Eq. (14.2.7), $U_0 = P^2/2A^2E$ and hence by Eq. (14.2.2b), the total elastic strain energy in the bar is

$$U = \iiint_V U_0 \, d\Omega = \int_0^L \frac{P^2}{2A^2E} A \, dx = \frac{P^2 L}{2AE}. \qquad (14.2.12)$$

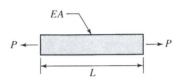

Figure 14.2.6

(ii) *Flexural strain energy in a linear elastic beam.* Consider a beam of varying cross-section with $A = A(x)$, $I = I(x)$ and length L [Fig. (14.2.7a)].[‡] Assume that the moment about the z-axis, $M = M(x)$, is known [Fig. (14.2.7b)]. The flexural stress is then [Eq. (8.7.1b)]

$$\tau_{xx}(x, y) = \frac{M(x)y}{I(x)}.$$

Substituting in Eq. (14.2.7)

$$U_0 = \frac{M^2(x)y^2}{2EI^2(x)}. \qquad (14.2.13)$$

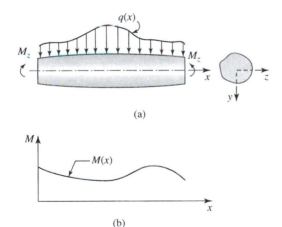

(a)

(b)

Figure 14.2.7

[†] We mention here that, contrary to the more physical definition of an elastic material as given by Eq. (14.2.1), an elastic material may alternatively be *defined* more abstractly as a material that possesses a potential U_0 such that the stresses are given by the derivatives with respect to the corresponding strain component.

[‡] The second moment of the area, I, is taken about the z-axis, which is also assumed to be a principal axis.

Note that here the strain energy density varies throughout the beam in contrast to the previous case. The total strain energy is then given by

$$U = \iiint_V U_0 \, d\Omega = \int_0^L \left(\iint_A U_0 \, dA \right) dx. \qquad (14.2.14)$$

Substituting U_0 from Eq. (14.2.13),

$$U = \int_0^L \frac{M^2(x)}{2EI^2(x)} \left(\iint_A y^2 \, dA \right) dx. \qquad (14.2.15)$$

But $\iint_A y^2 \, dA = I$ by definition. Hence we obtain

$$U = \frac{1}{2} \int_0^L \frac{M^2(x)}{EI(x)} \, dx. \qquad (14.2.16a)$$

For the case where $I(x) = I$, a constant,

$$U = \frac{1}{2EI} \int_0^L M^2(x) \, dx. \qquad (14.2.16b)$$

14.3 The principle of conservation of energy for linear elastic bodies

(a) Derivation of the principle

We now derive a basic and important principle relating to strain energy, viz. the principle of conservation of energy. This principle, which was developed in Chapter 4 under a uniaxial state of stress, states that if an elastic body is subjected to applied forces, then the external work done by the applied forces is equal to the elastic strain energy stored in the body.

For mathematical simplicity, we first prove this principle for the two-dimensional state of stress and strain only and then generalise the results to the three-dimensional case. We further restrict our proof to the case of linear elastic bodies subject to small strains and small relative rotations.

Consider a linear elastic body in an x, y, z coordinate system, having a volume V enclosed by a surface S. Furthermore, let the outward normal to S at any point be a unit vector $\boldsymbol{n}$, as shown in Fig. (14.3.1), which is given here for the two-dimensional case:

$$\boldsymbol{n} = \ell_x \boldsymbol{i} + \ell_y \boldsymbol{j}, \qquad (14.3.1)$$

where ℓ_x and ℓ_y are the direction cosines with respect to the x- and y-axes, respectively. Let the body be in a state of equilibrium under a set of external tractions $\boldsymbol{T}_n$ acting on S (see Section 8 of Chapter 2) and body forces $\boldsymbol{B}$ acting within V, as shown in the figure.

We denote the two-dimensional displacements at any point due to deformation of the body by

$$\boldsymbol{u}(x, y) = u \boldsymbol{i} + v \boldsymbol{j}. \qquad (14.3.2)$$

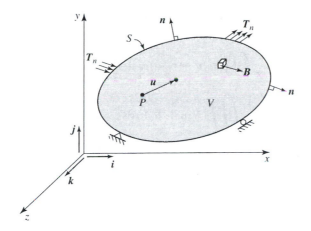

Figure 14.3.1

Due to the externally applied forces, stresses will exist at points in the body and consequently for the two-dimensional case considered, the elastic strain energy is, by Eq. (14.2.8c),

$$U = \frac{1}{2} \iiint_V (\tau_{xx}\epsilon_{xx} + \tau_{yy}\epsilon_{yy} + 2\tau_{xy}\epsilon_{xy})\, d\Omega, \qquad (14.3.3)$$

where $d\Omega$ again represents a volume element.

Assuming that strains and rigid-body rotations are infinitesimal, from Eq. (3.7.20),

$$\epsilon_{xx} = \frac{\partial u}{\partial x}, \qquad \epsilon_{yy} = \frac{\partial v}{\partial y}, \qquad \epsilon_{xy} = \frac{1}{2}\left(\frac{\partial u}{\partial y} + \frac{\partial v}{\partial x}\right).$$

Substituting in Eq. (14.3.3),

$$U = \frac{1}{2} \iiint_V \left[\tau_{xx}\frac{\partial u}{\partial x} + \tau_{yy}\frac{\partial v}{\partial y} + \tau_{xy}\left(\frac{\partial u}{\partial y} + \frac{\partial v}{\partial x}\right) \right] d\Omega. \qquad (14.3.4)$$

Now note that

$$\tau_{xx}\frac{\partial u}{\partial x} = \frac{\partial}{\partial x}(\tau_{xx}u) - u\frac{\partial \tau_{xx}}{\partial x}, \qquad \tau_{xy}\frac{\partial u}{\partial y} = \frac{\partial}{\partial y}(\tau_{xy}u) - u\frac{\partial \tau_{xy}}{\partial y}, \qquad \text{(etc.)}$$

Using this relation and the corresponding relation for other components, we have, upon rearranging the terms,

$$U = \frac{1}{2} \iiint_V \left[\frac{\partial}{\partial x}(\tau_{xx}u + \tau_{xy}v) + \frac{\partial}{\partial y}(\tau_{yy}v + \tau_{xy}u) \right] d\Omega$$

$$+ \frac{1}{2} \iiint_V \left[-\left(\frac{\partial \tau_{xx}}{\partial x} + \frac{\partial \tau_{yx}}{\partial y}\right)u - \left(\frac{\partial \tau_{yy}}{\partial y} + \frac{\partial \tau_{xy}}{\partial x}\right)v \right] d\Omega. \qquad (14.3.5)$$

From the stress equations of equilibrium, Eqs. (2.4.4),

$$-\left(\frac{\partial \tau_{xx}}{\partial x} + \frac{\partial \tau_{yx}}{\partial y}\right) = B_x, \qquad -\left(\frac{\partial \tau_{yy}}{\partial y} + \frac{\partial \tau_{xy}}{\partial x}\right) = B_y,$$

where B_x and B_y are the components of the body forces in the x- and y-directions, respectively.

Applying now the divergence theorem [see Eq. (B.3.4) of Appendix B.3] to the first integral of Eq. (14.3.5), we obtain

$$U = \frac{1}{2} \iint_S [(\tau_{xx} u + \tau_{xy} v)\ell_x + (\tau_{xy} u + \tau_{yy} v)\ell_y] \, ds$$

$$+ \frac{1}{2} \iiint_V (B_x u + B_y v) \, d\Omega, \tag{14.3.6}$$

where ℓ_x and ℓ_y are the components of the unit normal vector $\boldsymbol{n}$, as defined by Eq. (14.3.1). Rearranging the terms,

$$U = \frac{1}{2} \iint_S [\ell_x \tau_{xx} + \ell_y \tau_{yx})u + (\ell_y \tau_{yy} + \ell_x \tau_{xy})v] \, ds$$

$$+ \frac{1}{2} \iiint_V (B_x u + B_y v) \, d\Omega. \tag{14.3.7}$$

From Eqs. (2.8.8), we observe that

$$\ell_x \tau_{xx} + \ell_y \tau_{yx} = X_n, \qquad \ell_x \tau_{xy} + \ell_y \tau_{yy} = Y_n, \tag{14.3.8}$$

where, according to Eq. (2.8.2b), X_n and Y_n are the components of traction $\boldsymbol{T}_n$ on the surface S. Thus,

$$U = \frac{1}{2} \iint_S (X_n u + Y_n v) \, ds + \frac{1}{2} \iiint_V (B_x u + B_y v) \, d\Omega. \tag{14.3.9}$$

Upon noting that the above integrands are the scalar products of the tractions and body forces $\boldsymbol{B}$, respectively, with the displacement vector $\boldsymbol{u}$ of Eq. (14.3.2), the above may be written concisely in vector notation as

$$U = \frac{1}{2} \iint_S \boldsymbol{T}_n \cdot \boldsymbol{u} \, ds + \frac{1}{2} \iiint_V \boldsymbol{B} \cdot \boldsymbol{u} \, d\Omega, \tag{14.3.10}$$

where again $\boldsymbol{T}_n$ and $\boldsymbol{B}$ are the applied surface tractions acting on S and the applied body forces within V, respectively. Now, the scalar product of forces with displacements is the basic definition of work. Hence, the right-hand side of Eq. (14.3.10) clearly represents the work W done by the statically applied external forces. (An explanation of the 1/2 factor will be given below.) Thus we write

$$W = \frac{1}{2} \iint_S \boldsymbol{T}_n \cdot \boldsymbol{u} \, ds + \frac{1}{2} \iiint_V \boldsymbol{B} \cdot \boldsymbol{u} \, d\Omega \tag{14.3.11}$$

and hence we have the relation

$$U = W. \tag{14.3.12}$$

This last relation leads to the statement of the theorem:

If a linear elastic body is subjected to external forces producing a state of equilibrium, the external work done by the applied forces is equal to the internal strain energy.

This theorem is generally known as Clapeyron's theorem.

(b) Application of the principle

Consider a linear elastic cantilever beam of length L and flexural rigidity EI, to which a load P is applied statically at the free end [Fig. (14.3.2a)]. Determine the deflection Δ under the load due to flexural deformation.

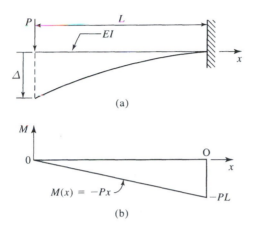

(a)

(b)

$M(x) = -Px$

$-PL$

Figure 14.3.2

Noting that the load is applied statically, the external work done by P is given by $W = \frac{P}{2}\Delta$.[†] The internal strain energy, by Eq. (14.2.16b), is

$$U = \frac{1}{2EI} \int_0^L M^2(x)\, dx.$$

Noting that $M(x) = -Px$ [Fig. (14.3.2b)],

$$U = \frac{P^2}{2EI} \int_0^L x^2\, dx = \frac{P^2 L^3}{6EI}. \qquad (14.3.13)$$

Equating $U = W$, we have

$$\Delta = \frac{PL^3}{3EI}, \qquad (14.3.14)$$

which is the same result obtained by integration of the differential equation of the beam [Eq. (9.4.7b)].

The principle of conservation of energy for elastic bodies may be considered as a first extension of the general energy conservation principles encountered in the study of rigid bodies and in mechanics, in general. Thus, this principle is an important one

---

[†] For problems (such as considered in this section) where the relations are linear, the 1/2 term can be proven directly as follows: If we assume that the deflection is linearly proportional to P, then $P = K\Delta$, where K is a constant of proportionality. It follows that the work done by the load is

$$W = \int_0^{\Delta^f} P(\Delta)\, d\Delta = K \int_0^{\Delta^f} \Delta\, d\Delta = \frac{K(\Delta^f)^2}{2} = \frac{P^f \Delta^f}{2}.$$

Upon dropping the superscript 'f', we have $W = \frac{P\Delta}{2}$, where Δ now represents the actual final displacement. Another, heuristic, proof is as follows. If a load is applied statically to a linear body, then the 'average' force applied is equal to the sum of one-half the initial (zero) force and the final force P. The work done is then the product of the 'average' force and the displacement through which it acts.

conceptually as it re-emphasises that no dissipation of energy occurs in an elastic body. Yet, despite its importance, this theorem has only limited use in the practical solution of problems in solid and structural mechanics. This limitation becomes evident when one attempts to determine displacements for bodies subjected to several concentrated loads or to distributed applied forces. However, the principle is of great importance as it provides a basis for the development of powerful theorems and relations that may be used in practice.

14.4 Betti's law and Maxwell's reciprocal relation: flexibility coefficients

Both Betti's law and Maxwell's reciprocal relation are applicable to a class of elastic bodies that are called **linear elastic bodies**.

We define a **linear body** or **linear structure** here as one for which the displacements of any point of the body are linearly proportional to the applied loads.[†] If a body satisfies this condition, then it follows that the principle of linear superposition is valid.

Consider now a 'linear body' supported in such a way that no rigid-body motion occurs. Without loss of generality, consider the static application of two applied independent forces P_1 and P_2 applied, say, at points 1 and 2, respectively [Fig. (14.4.1a)].

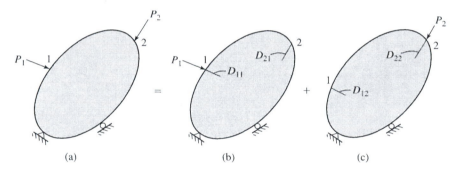

Figure 14.4.1   (a)   (b)   (c)

Further, assume that the body is in a state of equilibrium. As a result of the linearity, superposition is valid; we therefore consider the effect of each load separately [Fig. (14.4.1b and c)]. Let

D_{11} be the component of displacement of point 1 in the direction of P_1 due to P_1
D_{12} be the component of displacement of point 1 in the direction of P_1 due to P_2
D_{21} be the component of displacement of point 2 in the direction of P_2 due to P_1
D_{22} be the component of displacement of point 2 in the direction of P_2 due to P_2

To study the effect of each load separately, we consider that the loads are applied in turn rather than simultaneously. Thus, assume that P_1 is first applied statically. Then, the external work done by P_1 is (see footnote, p. 545)

$$W_1 = \frac{1}{2}P_1 D_{11}. \tag{14.4.1a}$$

[†] It is emphasised here that the *material* properties of a linear elastic body must necessarily be linear elastic. However, this is not sufficient for it does not necessarily follow that a body made of such material will have a linear load–displacement relation. For example, if a linear elastic beam-column is subjected to a lateral load F and an axial thrust P, the lateral displacement is not linearly proportional to P [see, e.g., Eq. (11.11.1)].

Now apply P_2 statically, remembering that P_1 is already acting on the body. The work done by the load P_2 will then be $\frac{1}{2}P_2 D_{22}$. However, since P_1 is already acting on the body, it too will do work since it displaces when P_2 is applied. The work done by P_1 due to the application of P_2 is then $P_1 D_{12}$. Thus, the total work done on the system is

$$W_1' = \frac{1}{2}P_1 D_{11} + \frac{1}{2}P_2 D_{22} + P_1 D_{12}. \qquad (14.4.1b)$$

Now we imagine that we reverse the order in which the loads are applied. Thus, assume that P_2 is first applied. The resulting external work is then

$$W_2 = \frac{1}{2}P_2 D_{22}. \qquad (14.4.2a)$$

Keeping P_2 on the body, we now apply P_1 statically. The additional work done is then $\frac{1}{2}P_1 D_{11} + P_2 D_{21}$, and hence the total work done on the system is

$$W_2' = \frac{1}{2}P_2 D_{22} + \frac{1}{2}P_1 D_{11} + P_2 D_{21}. \qquad (14.4.2b)$$

Because the final loading condition in both cases is identical, the strains and therefore the internal strain energy in both cases must also be identical. Furthermore, since the internal strain energy is equal to the work done by the external forces, then necessarily $W_1' = W_2'$, from which it follows that

$$P_1 D_{12} = P_2 D_{21}. \qquad (14.4.3)$$

This relation is known as **Betti's law** and is valid for all linear structures. We may state this law more generally as follows:

Let two separate systems of loads act on a linear body, such that in each case the body is in equilibrium. Then the work done by the forces of the first system in going through the displacements caused by the second system is equal to the work of the second system in going through the displacements caused by the first system.

At this point, it is convenient to define a new quantity: the displacement at a point due to a unit load. We denote this quantity by f_{ij} and define it as the displacement of point i in a specified direction due to a *unit load*, $P_j = 1$, applied at point j (in a specified direction). In structural mechanics, the quantities f_{ij} are called *flexibility coefficients*. For example, in the above, f_{12} represents the displacement of point 1 in the direction of P_1 due to a unit force $P_2 = 1$. (This definition also holds true if the point j at which P_j is applied coincides with point i at which P_i is applied. However, the direction of the 'j-force', P_j, need not be the same as the direction of the 'i-force', P_i. Thus, for example, for the *same given point*, f_{ij} might represent the displacement in the y-direction due to a unit force in the x-direction at the same point.)

Since, for a linear body, displacements are proportional to the loads,

$$D_{11} = P_1 f_{11}, \qquad D_{22} = P_2 f_{22}, \qquad D_{12} = P_2 f_{12}, \qquad D_{21} = P_1 f_{21}$$

or, in general,

$$D_{ij} = P_j f_{ij}. \qquad (14.4.4a)$$

Thus we note that, consistent with the definition of the flexibility coefficient, for a given force P_i acting at point i, the displacement of this force, using the linearity property, is given by

$$D_{ii} = P_i f_{ii} \,. \tag{14.4.4b}$$

Substituting Eq. (14.4.4a) in Eq. (14.4.3), $P_1 P_2 f_{12} = P_2 P_1 f_{21}$ and thus we establish that

$$f_{12} = f_{21}. \tag{14.4.5a}$$

In general, if we have two points i and j on a body, and if P_i and P_j are applied at these two points, then we may write

$$f_{ij} = f_{ji}, \quad i \neq j. \tag{14.4.5b}$$

This result is known as **Maxwell's reciprocal relation**[†] and may be stated formally as follows:

> In a linear body, the displacement of point i (in the direction of a force at point i) due to a unit force applied at point j is equal to the displacement of point j (in the direction of a force at point j) due to a unit force applied at point i.

We observe that Maxwell's reciprocal relation is a direct consequence of Betti's Law, which was derived from a basic work–energy concept. Since work is also the product of moment and rotation, the above relations may be generalised immediately if for 'force' we substitute 'moment' and for 'displacement' we substitute 'rotation'.

An illustration of Maxwell's reciprocal relation is shown in Fig. (14.4.2) for a cantilever beam ABC subjected to three separate unit 'forces' as shown. We define the lateral displacement of the end A of the beam by f_{1j}, the displacement of the centre B of the beam by f_{2j} and the rotation of the end A by f_{3j}. (Note that, as previously mentioned, for a given point, we may define more than one 'direction'. For example, here, we have defined directions '1' and '3' as the vertical displacement and rotation at point A, respectively.) The corresponding forces are shown in Figs. (14.4.2a–c), respectively. The resulting deformations are also shown in the figure where the equal quantities are connected by means of dashed lines.

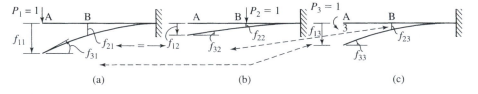

Figure 14.4.2

(a)   (b)   (c)

At this point, we re-emphasise that the property of linearity was required in the derivations and definitions given above, and that therefore the principles are only applicable to linear structures.

To conclude, we observe that as a result of the validity of the principle of superposition for linear bodies or structures, the displacement at any point in a body due to a combination of applied forces can be expressed simply in terms of the flexibility coefficients f_{ij}.

[†] It follows from Maxwell's relations that influence functions (viz. Green's functions) for *displacements* of a linear elastic system, G_v, as developed in Chapter 9, are symmetric; that is, $G_v(x, \zeta) = G_v(\zeta, x)$.

Consider a body subjected to n forces, $P_1, P_2, \ldots, P_i, P_j, \ldots, P_n$, applied at points $1, 2, \ldots, i, j, \ldots, n$, as shown in Fig. (14.4.3). Using the linearity property, the displacement of point i, Δ_i, may then be expressed as

$$\Delta_i = P_1 f_{i1} + P_2 f_{i2} + \cdots + P_j f_{ij} + \cdots + P_n f_{in}$$

or

$$\Delta_i = \sum_{j=1}^{n} P_j f_{ij}. \qquad (14.4.6)$$

Again, in Eq. (14.4.6), the loads P can be considered either as concentrated forces or moments in the generalised sense, provided the corresponding flexibility coefficients f_{ij} are considered as displacement or rotations.

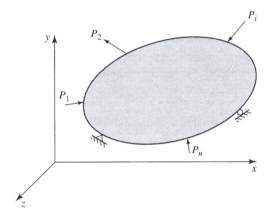

Figure 14.4.3

Example 14.1: In our study of general bending of elastic beams in Chapter 13, the location of the shear centre was established for a variety of cross-sections. Moreover, in the development of the theory of torsion of an element in Chapter 12, it was postulated that there exists an axis about which all cross-sections must rotate; the cross-sections were said to rotate about the 'centre of twist'. Prove that the centre of twist coincides with the shear centre for any arbitrary cross-section.

Solution: Let the elastic prismatic beam be subjected to a torsional moment T and a lateral force P passing through the shear centre, point S [Fig. (14.4.4)]. The two 'forces' are therefore T and P, respectively. Let θ_P denote the rotation of the section due to P and let Δ_T denote the displacement of point S due to the torsional moment

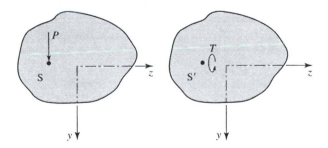

Figure 14.4.4

T. Now, by Betti's law,

$$P\Delta_T = T\theta_P.$$

Since point S is the shear centre, the load P causes no rotation, that is, $\theta_P = 0$. It follows, since $P \neq 0$, that $\Delta_T = 0$, i.e., point S does not displace due to the torsional moment. Now the only point that does not displace due to a torque is the centre of twist; it follows that point S must, in fact, be the centre of twist.

Hence, the shear centre and centre of twist coincide for any cross-section of a linear elastic beam.

14.5   Castigliano's second theorem

Castigliano's second theorem,[†] which we develop below, is an important theorem that permits us to find displacements for a linear structure subjected to a set of concentrated loads.

Consider a linear body subjected to n statically applied *independent* loads P_1, $P_2, \ldots, P_n$ such that equilibrium is maintained [Fig. (14.4.3)]. Let the components of displacement at the points of application in the direction of the applied loads be given by $\Delta_1, \Delta_2, \ldots, \Delta_i, \ldots, \Delta_n$.

Then[‡]

$$\Delta_j = \sum_{k=1}^{n} P_k f_{jk} \tag{14.5.1}$$

and the external work is

$$W = \frac{1}{2} P_1 \Delta_1 + \frac{1}{2} P_2 \Delta_2 + \cdots + \frac{1}{2} P_i \Delta_i + \cdots + \frac{1}{2} P_n \Delta_n \tag{14.5.2}$$

or

$$W = \frac{1}{2} \sum_{j=1}^{n} P_j \Delta_j.$$

By the principle of conservation of energy [Eq. (14.3.12)], the external work is equal to the internal strain energy; thus

$$U = \frac{1}{2} \sum_{j=1}^{n} P_j \Delta_j. \tag{14.5.3}$$

Substituting Δ_j from Eq. (14.5.1),

$$U = \frac{1}{2} \sum_{j=1}^{n} \sum_{k=1}^{n} P_j P_k f_{jk}. \tag{14.5.4}$$

[†]   Castigliano, in a thesis, presented several important theorems governing the behaviour of elastic bodies. The theorem given here was presented by Castigliano as his second theorem. For historic reasons, it is therefore currently referred to as *Castigliano's second theorem*.

[‡]   Note that the index k, appearing as subscripts, is immaterial since we sum up on the k here. Such a subscript is called a '*dummy subscript*', in the same sense as a '*dummy variable*' appearing in an integral expression. Thus, whenever a subscript is summed, we may change its 'name' without affecting the results.

From this last equation, we observe that the strain energy can be expressed as a function of the applied loads, i.e., $U = (P_1, P_2, \ldots, P_n)$, and that for linear bodies it is a quadratic function of the n loads.

Now let us take the partial derivative of U with respect to any particular force, say P_i. From Eq. (14.5.4), we have

$$\frac{\partial U}{\partial P_i} = \frac{1}{2} \sum_{j=1}^{n} \sum_{k=1}^{n} \frac{\partial P_j}{\partial P_i} P_k f_{kj} + \frac{1}{2} \sum_{j=1}^{n} \sum_{k=1}^{n} P_j \frac{\partial P_k}{\partial P_i} f_{kj}. \qquad (14.5.5)$$

Since the applied loads P are independent of each other,

$$\frac{\partial P_j}{\partial P_i} = \begin{cases} 1, & i = j \\ 0, & i \neq j \end{cases} \qquad (14.5.6)$$

and therefore terms appearing in the summations of Eq. (14.5.5) vanish except when the subscripts in the partial derivatives are identical.

Thus, we obtain

$$\frac{\partial U}{\partial P_i} = \frac{1}{2} \sum_{k=1}^{n} P_k f_{ki} + \frac{1}{2} \sum_{j=1}^{n} P_j f_{ij} \qquad (14.5.7a)$$

and hence[†]

$$\frac{\partial U}{\partial P_i} = \frac{1}{2} \sum_{k=1}^{n} P_k (f_{ki} + f_{ik}). \qquad (14.5.7b)$$

But by Maxwell's reciprocal relation [Eq. (14.4.5b)], $f_{ik} = f_{ki}$, and therefore

$$\frac{\partial U}{\partial P_i} = \sum_{k=1}^{n} P_k f_{ik}.$$

Comparing with Eq. (14.5.1), we establish the final result:

$$\Delta_i = \frac{\partial U}{\partial P_i}. \qquad (14.5.8)$$

The above relation is known as **Castigliano's second theorem** and may be stated as follows:

> If a linear body is subjected to n independent loads $P_1, P_2, \ldots, P_n$ such that equilibrium is maintained, then the first partial derivative of the elastic strain energy with respect to any particular load is equal to the component of displacement of the point under the load (in the direction of the load).

Again we emphasise that the terms 'load' and 'displacements' are used in the generalised sense and signify force or moment and displacement or rotation, respectively.

Application of Castigliano's second theorem to problems in structural mechanics is illustrated by means of several examples.

Example 14.2: Consider a linear elastic prismatic bar of cross-sectional area A and length L, subjected to an applied longitudinal load (passing through the

---

[†] See the previous footnote.

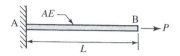

Figure 14.5.1

centroid of each section) as shown in Fig. (14.5.1). Determine the horizontal displacement Δ at the end B.

Solution: From Eq. (14.2.12),

$$U = \frac{P^2 L}{2AE}.$$ (14.5.9)

Applying Castigliano's (second) theorem,

$$\Delta = \frac{\partial U}{\partial P} = \frac{PL}{AE}.$$ (14.5.10)

In the following example we illustrate a technique that simplifies the calculation in the applications of Castigliano's theorem.

Example 14.3:   An elastic cantilever beam AD of flexural rigidity *EI* and length *L* is subjected to a uniform load *w*, a concentrated load *P* at the end and a couple *C*, as shown in Fig. (14.5.2). Determine (a) the vertical displacement Δ_A of point A due to flexure and (b) the rotation θ_A of the beam at point A due to flexure.

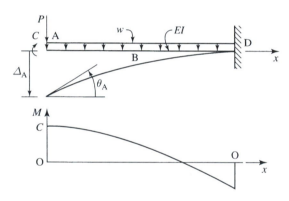

Figure 14.5.2

Solution: From Eq. (14.2.16b),

$$U = \frac{1}{2EI} \int_0^L M^2 \, \mathrm{d}x$$ (14.5.11)

where

$$M = M(P, C, w; x) = -Px + C - wx^2/2.$$ (14.5.12)

From Castigliano's (second) theorem,

$$\Delta_A = \frac{\partial U}{\partial P} = \frac{1}{2EI} \frac{\partial}{\partial P} \int_0^L M^2(x) \, \mathrm{d}x.$$ (14.5.13)

We might proceed by substituting Eq. (14.5.12) in (14.5.13), first integrating and then performing the required differentiation on the integral. Simplification generally follows if we reverse the order of integration and differentiation; thus[†]

$$\Delta_A = \frac{1}{2EI}\int_0^L \frac{\partial M^2(x)}{\partial P}\,dx = \frac{1}{EI}\int_0^L M\frac{\partial M}{\partial P}\,dx. \qquad (14.5.14)$$

From Eq. (14.5.12), we have

$$\frac{\partial M}{\partial P} = -x, \qquad \frac{\partial M}{\partial C} = 1.$$

Hence

$$\Delta_A = \frac{\partial U}{\partial P} = \frac{1}{EI}\int_0^L (-x)(-Px + C - wx^2/2)\,dx$$

or

$$\Delta_A = \frac{1}{EI}[PL^3/3 - CL^2/2 + wL^4/8]. \qquad (14.5.15a)$$

Similarly,

$$\theta_A = \frac{\partial U}{\partial C} = \frac{1}{EI}\int_0^L (-Px + C - wx^2/2)\,dx$$

or

$$\theta_A = \frac{1}{EI}[-PL^2/2 + CL - wL^3/6]. \qquad (14.5.15b)$$

$\square$

It is worth noting, from Eq. (14.5.15a), that

$$\Delta_A]_{\substack{\text{due to}\\P=1}} = \frac{L^3}{3EI}, \qquad \Delta_A]_{\substack{\text{due to}\\C=1}} = -\frac{L^2}{2EI}$$

while, on the other hand, from Eq. (14.5.15b),

$$\theta_A]_{\substack{\text{due to}\\P=1}} = -\frac{L^2}{2EI}, \qquad \theta_A]_{\substack{\text{due to}\\C=1}} = \frac{L}{EI}$$

These last quantities are flexibility coefficients as defined previously. We observe that $\theta_A]_{P=1} = \Delta_A]_{C=1}$ as we should expect according to Maxwell's reciprocal relation. Note that negative signs appearing above indicate displacements or rotations in a sense opposite to the assumed sense of the applied loads.

In the following example, we introduce a 'trick' in the use of Castigliano's theorem.

Example 14.4: For the structure and loads of the previous example [Fig. (14.5.2)], determine the vertical displacement of the mid-point B.

Solution: To find the displacement of a point in a body using Castigliano's theorem, we must take the derivative with respect to a load acting at that point. However, no

[†] If M as well as its derivative $\partial M/\partial P$ exist and are piece-wise continuous within the range of integration of x, then the partial derivative can be taken inside the integral.

applied load at B has been specified here. We overcome this problem by the following technique.

Let us *assume* that there is a load, say F, applied to point B, as shown [Fig. (14.5.3)]. Then, after performing the required differentiation, we will merely set $F = 0$. For $F \neq 0$,

$$M_1(x) = -Px + C - wx^2/2, \quad 0 \leq x \leq L/2$$

$$M_2(x) = -Px + C - wx^2/2 - F(x - L/2), \quad L/2 \leq x \leq L$$

(14.5.16)

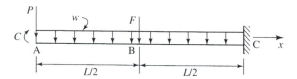

Figure 14.5.3

Applying Castigliano's (second) theorem, we have

$$\Delta_B = \frac{\partial U}{\partial F} = \frac{1}{EI} \int_0^L M \frac{\partial M}{\partial F} \, dx$$

$$= \frac{1}{EI} \left[\int_0^{L/2} M_1 \frac{\partial M_1}{\partial F} \, dx + \int_{L/2}^L M_2 \frac{\partial M_2}{\partial F} \, dx \right].$$

(14.5.17)

Note that it is necessary to split the integration into two regions since different expressions for $M(x)$ exist to the left and right of point B.

From Eqs. (14.5.16),

$$\frac{\partial M_1}{\partial F} = 0, \qquad \frac{\partial M_2}{\partial F} = -(x - L/2).$$

(14.5.18)

At this point in the problem, we may set $F = 0$ (since it has 'performed its function'), thus obtaining

$$\Delta_B = \frac{1}{EI} \int_{L/2}^L [-Px + C - wx^2/2][-(x - L/2)] \, dx.$$

(14.5.19a)

Simple integration gives finally

$$\Delta_B = \frac{1}{EI} \left[\frac{5PL^3}{48} - \frac{CL^2}{8} + \frac{17wL^4}{384} \right].$$

(14.5.19b)

$\square$

The application of Castigliano's theorem to a more complex structure is shown below.

Example 14.5: Compute the horizontal component of deflection at point E of the pin-connected elastic truss due to the applied load P, as shown in Fig. (14.5.4). (The cross-sectional area of each member is given in parenthesis.) Let the modulus of elasticity of all members be E.

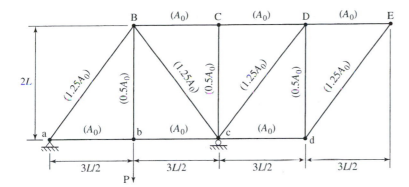

Figure 14.5.4

Solution: To determine the horizontal deflection at E (following the technique of Example 14.4), we assume a force Q acts as shown. Since the truss is pin-connected, it follows that the internal forces must be colinear with the members. The resulting axial forces F_i (in the ith member) are shown in Fig. (14.5.5) where '+' indicates tension and '−' indicates compression.

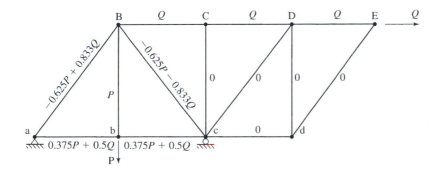

Figure 14.5.5

By Castigliano's (second) theorem,

$$\Delta_E = \frac{\partial U}{\partial Q}. \qquad (14.5.20)$$

Now the strain energy for the ith member is, by Eq. (14.2.12),

$$U_i = \frac{F_i^2 L_i}{2 A_i E}, \qquad (14.5.21a)$$

where A_i and L_i are respectively the cross-sectional area and length of the ith member. Hence the total strain energy in the entire truss is

$$U = \sum_{i=1}^{n} U_i = \frac{1}{2} \sum_{i=1}^{n} \frac{F_i^2 L_i}{A_i E}. \qquad (14.5.21b)$$

Using Eq. (14.5.20), we find

$$\Delta_E = \sum_{i=1}^{n} \left[\frac{F_i L_i}{A_i E} \cdot \frac{\partial F_i}{\partial Q} \right]_{Q=0}. \qquad (14.5.22)$$

The calculations are illustrated in the following table.

Bar	L_i	A_i	L_i/A_i	F_i	$\partial F_i/\partial Q$	$F_i \dfrac{\partial F_i}{\partial Q}\Big|_{Q=0}$	$\dfrac{F_i L_i}{A_i} \cdot \dfrac{\partial F_i}{\partial Q}\Big|_{Q=0}$
ab	$1.5L$	A_0	$1.5L/A_0$	$0.375P + 0.5Q$	0.5	$0.1875P$	$0.2813PL/A_0$
bc	$1.5L$	A_0	$1.5L/A_0$	$0.375P + 0.5Q$	0.5	$0.1875P$	$0.2813PL/A_0$
aB	$2.5L$	$1.25A_0$	$2L/A_0$	$-0.625P + 0.833Q$	0.833	$-0.522P$	$-1.044PL/A_0$
Bc	$2.5L$	$1.25A_0$	$2L/A_0$	$-0.625P - 0.833Q$	-0.833	$0.522P$	$1.044PL/A_0$
							$\sum = 0.5626PL/A_0$

Thus

$$\Delta_{\mathrm{E}} = 0.5626\frac{PL}{A_0 E}. \qquad (14.5.23)$$

14.6 Geometric representation (complementary strain energy and Castigliano's first theorem)

By considering the case of the simple uniaxial stress state, it is possible to interpret some of the previously developed relations from a geometric point of view.

To this end, consider an elastic rod (not necessarily linear) subjected to a load P, which is statically applied from its initial value $P = 0$ to its final value P^{f} [Fig. (14.6.1)]. For each value of P, we may then measure and plot the resulting displacement Δ. Let us assume that the load–displacement curve is as given in Fig. (14.6.2) where the coordinates of point A are Δ^{f} and P^{f}. For an elastic body, the curve can be represented by the *unique* relation $P = P(\Delta)$ or by its *unique inverse* $\Delta = \Delta(P)$.

The work done by the force P is given by

$$W = \int_0^B P(\Delta)\, \mathrm{d}\Delta. \qquad (14.6.1)$$

Since the rod is elastic, the principle of conservation of energy, Eq. (14.3.12), is valid, i.e. $U = W$, and thus

$$U = \int_0^B P(\Delta)\, \mathrm{d}\Delta. \qquad (14.6.2)$$

Clearly then, the total elastic strain energy in the rod is represented by the area OAB.

Consider now the complementary area OAC. Although this area does not represent strain energy, its dimensions have units of energy. Hence we define the area as representing '**complementary strain energy**' and denote it by U^*. Thus we write

$$U^* = \int_0^C \Delta(P)\, \mathrm{d}P. \qquad (14.6.3)$$

Now, assume that the applied force is increased by a small amount $\mathrm{d}P$ such that we

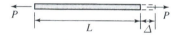

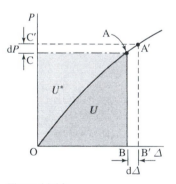

P L P

Figure 14.6.1

Figure 14.6.2

arrive at point A′ on the curve. Then we observe that

$$dU = P \cdot d\Delta \qquad (14.6.4a)$$

and hence

$$\frac{dU(\Delta)}{d\Delta} = P. \qquad (14.6.4b)$$

We may, therefore, state the following:

Let an elastic body be in equilibrium under a load P, and let the strain energy U be expressed in terms of the displacement Δ of the load. Then the first derivative of U with respect to the displacement of P is equal to the applied load.

This theorem is referred to as **Castigliano's first theorem**. Note that it is valid for *general elastic behaviour, not necessarily linear.*

When proceeding from A to A′ on the curve, we observe also that

$$dU^* = \Delta \cdot dP \qquad (14.6.5a)$$

or

$$\Delta = \frac{dU^*(P)}{dP}. \qquad (14.6.5b)$$

Thus, for an elastic body, linear or nonlinear, the derivative of the complementary energy with respect to a load is equal to the displacement of the point under the load (in the direction of the load).

In general, $U^* \neq U$. However, for a *linear* elastic body, $U = U^*$ [Fig. (14.6.3)]. It follows that for a *linear* elastic body,

$$\Delta = \frac{dU(P)}{dP}, \qquad (14.6.6)$$

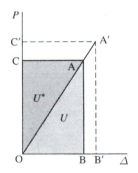

Figure 14.6.3

which is Castigliano's second theorem as previously given by Eq. (14.5.8). Thus, we have again demonstrated that this theorem is valid only if a linear load–displacement relation exists.

It is worth noting that if we consider, for any elastic body, the quantity

$$d(P\Delta) = P\,d\Delta + \Delta\,dP, \qquad (14.6.7a)$$

we have, from Eqs. (14.6.4a) and (14.6.5a),

$$d(P\Delta) = dU + dU^* \qquad (14.6.7b)$$

and hence

$$P\Delta = \int dU + \int dU^* = U + U^*. \qquad (14.6.7c)$$

We therefore observe that the two areas represented by U and U^* together yield the rectangular area $P\Delta$, as in Fig. (14.6.2).

Finally, since

$$dU = \frac{dU(\Delta)}{d\Delta}\,d\Delta \qquad (14.6.8a)$$

and

$$dU^* = \frac{dU^*(P)}{dP}\,dP, \qquad (14.6.8b)$$

upon noting that for a *linear* elastic body $U = U^*$ and therefore $dU = dU^*$, it follows that

$$\frac{dU(\Delta)}{d\Delta} \, d\Delta = \frac{dU(P)}{dP} \, dP. \tag{14.6.9a}$$

Hence

$$\frac{dU(\Delta)}{d\Delta} = \frac{dU(P)}{dP} \cdot \frac{dP}{d\Delta}. \tag{14.6.9b}$$

14.7   The principle of virtual work

(a) Introduction

The principle of virtual work, which is derived in this section, is perhaps the most general principle encountered in mechanics since it may be applied to any body, be it elastic or dissipative, plastic, viscoelastic, etc. As we shall see, the principle of virtual work, as applied to rigid bodies, is but a degenerate case of the principle derived for deformable bodies. The versatility and generality of the principle of virtual work thus make it one of the most powerful tools in the study of mechanics. The principle is the basis for a wide class of methods, both analytic and numerical, known as **variational methods**.

In the following section, Section 14.8, we will derive a related principle, the principle of complementary virtual work. As will be seen, the two principles are parallel, each being a counterpart of the other. Moreover, the two principles provide the basis for effective methods in the solution of problems in mechanics.

(b) Definitions of external and internal virtual work: virtual displacements

(i) Virtual displacements

Consider a body in an x, y, z coordinate system, occupying a space V enclosed by a surface S and supported at points A and B [Fig. (14.7.1a)]. Let the body be subjected to a set of forces that may, as we have seen, be of two kinds: (a) surface forces acting on the surface S and (b) body forces acting within V. If the surface forces are distributed over an area, then their intensities are specified in terms of the traction T_n on the surface. In addition, concentrated forces P and couples C may also act on the surface as shown in the figure.[†] The body forces that act on elements within the body (e.g., gravity) are denoted by B. It is necessary to emphasise here that, since S denotes the *entire* surface, in stating 'the body is subjected to a set of forces acting on the surface S', we include as the surface tractions T_n *both the known applied tractions as well as the unknown reactive tractions* [Fig. (14.7.1b)].

In general, the idea of work implies the product of force and displacements. To define virtual work, we must first define the term 'virtual displacements'.

[†] We recall from Chapter 1 that concentrated forces are, in fact, idealisations represented by the limiting case of a distribution of load of infinite intensity acting over an infinitesimal area as the area tends to zero. Using now the terminology developed in solid mechanics, we consider a concentrated force P as an idealisation representing infinite tractions acting over an infinitesimal area Δs; the force is therefore defined as

$$P = \lim_{\substack{\Delta S \to 0 \\ |T_n| \to \infty}} \iint_{\Delta s} T_n \, ds$$

[cf. Eq. (1.2.1)]. Thus concentrated forces and couples can exist *only if we include the possibility of discontinuous tractions* on the surface S.

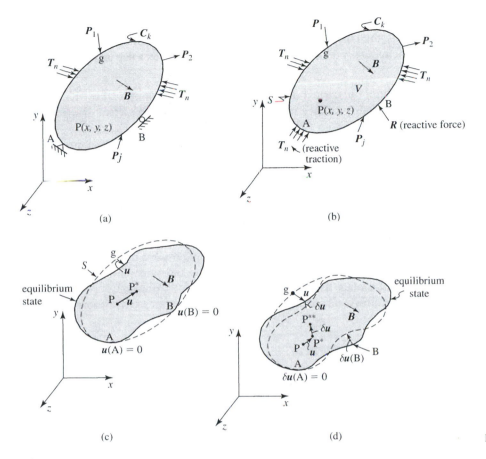

(a)

(b)

(c)

(d)

Figure 14.7.1

Now, if a deformable body is subjected to a set of external forces, points in the body undergo displacements $u(x, y, z)$ and the body assumes a new deformed configuration in its equilibrium state [Fig. (14.7.1c)]. The displacement $u(P)$ of a typical point $P(x, y, z)$ describing $P \rightarrow P^*$ is as shown in the figure. Clearly, under static conditions, the body will no longer move once it has reached the equilibrium position. Let us imagine however that, from this equilibrium position, points in the body were to undergo some additional (imaginary) displacements $P^* \rightarrow P^{**}$, which we denote by δu [Fig. (14.7.1d)]. We thus say that we 'give all points in the body some displacements'; we refer to such imaginary displacements as **virtual displacements**. However, we impose a restriction on these displacements: the virtual displacements must be *geometrically possible*. By this we shall mean that the virtual displacements within the body must, in general, be continuous except, perhaps, at a finite number of points. Using a more mathematical terminology, we define virtual displacements of a body to be any *piece-wise continuous displacement field* and denote these virtual displacements by $\delta u(x, y, z)$. The 'δ' is used to remind us that these displacements are not real but are imaginary displacements that points of the body undergo. (For example, if the body is rigid, the imaginary displacements must be such that the body or separate parts of the body remain rigid.) Thus, with u representing the real (actual) displacement of a point to its deformed position, we may regard the 'δ' of δu as denoting an (imaginary) *variation* and therefore also refer to δu as the *variation of the displacement* about its equilibrium position.

If, in addition, the virtual displacement field also satisfies all the *geometric boundary conditions*, then the field is said to be **kinematically admissible**. Thus, we may consider kinematically admissible virtual displacements as a subset of $\delta \boldsymbol{u}$. Note the following:

- We do not limit our virtual displacements to small quantities (i.e., the 'δ' does not signify 'small'). In fact, at this stage of our study, the virtual displacements $\delta \boldsymbol{u}$ may, according to our definition, be as large as we wish.
- According to our definition, the virtual displacements need not be kinematically admissible. However, although the virtual displacements do not *necessarily* have to satisfy the geometric boundary conditions, one may, of course, choose a set of virtual displacements that satisfy some or all the geometric boundary conditions of the body. For example, if a body is 'pinned' to supports A and B in Fig. (14.7.1a), we may imagine the body to displace in such a manner that the constraint is violated at B but not at A, as shown in Fig. (14.7.1d). Thus, we may imagine the virtual displacements to be any possible displacements. In certain problems, we find it convenient to adhere to the constraints; in other cases, we purposely violate the constraint. Since, by definition, the virtual displacements are arbitrary, our choice of virtual displacements will depend upon our purpose.

(ii) External virtual work

Having defined virtual displacements, we now define 'external virtual work', which we denote by δW_{ext}, as follows: the **external virtual work** is the 'work' done by a set of *actual existing external forces* (already acting on the body) in 'riding' through the virtual displacements. Thus, as real external work is defined as the (scalar) product of the external forces (traction $\boldsymbol{T}_n$, body forces $\boldsymbol{B}$, concentrated forces $\boldsymbol{P}$ and couples $\boldsymbol{C}$) with the displacements [see Eq. (14.3.11)], we define 'external virtual work' in a similar way; namely

$$\delta W_{\text{ext}} = \iint_S \boldsymbol{T}_n \cdot \delta \boldsymbol{u} \, ds + \iiint_V \boldsymbol{B} \cdot \delta \boldsymbol{u} \, d\Omega + \sum_{j=1}^n \boldsymbol{P}_j \cdot \delta \boldsymbol{u}_j + \sum_{k=1}^m \boldsymbol{C}_k \cdot \delta \boldsymbol{\theta}_k,$$

(14.7.1)

where $\delta \boldsymbol{u}_j$ and $\delta \boldsymbol{\theta}_k$, respectively, are the virtual displacement and rotation at the points at which $\boldsymbol{P}_j$ and $\boldsymbol{C}_k$ act.

Note that here, in contradistinction to Eq. (14.3.11), we also have included explicitly n concentrated forces and m concentrated couples that may be acting on the body. We also observe that, in contradistinction to Eq. (14.3.11), no $1/2$ factor appears in the expression for the external virtual work. This is because, according to our definition, the virtual work represents the work of an *existing system* of external forces in 'riding' through the virtual displacements.[†] Hence, the prefix 'δ' appearing in δW_{ext} is used to denote that the work is 'virtual' (and not 'real').

(iii) Internal virtual work

We now turn our attention to internal virtual work. For mathematical simplicity, we shall confine our discussion and proof below assuming a two-dimensional state

---

[†] Note that the existing set of external forces is considered to be *constant* both in magnitude and direction. In particular, we emphasise that the directions of the forces do not change as they 'ride' through the virtual displacements.

of stress exists everywhere. The results may then be generalised to bodies where three-dimensional states exist.

Consider a deformable body in a *state of equilibrium* subjected to a set of external forces such that a two-dimensional state of stress $(\tau_{xx}, \tau_{yy}, \tau_{xy})$ exists [Fig. (14.7.2)]. Due to the existing stresses, elements in the body undergo displacements u and v in the x- and y-directions, respectively. (For simplicity, we shall also assume here that all displacements take place in the x–y plane.) Let the strains measuring the deformation be $\epsilon_{xx}, \epsilon_{yy}, \epsilon_{xy}$. Note that since the body is not necessarily elastic, the strains are not, in general, uniquely determined by the stress. (However, this is irrelevant, for here we are *not at all concerned with these real strains and, moreover, we shall never calculate them*.) The only fact that is of interest to us is the statement that the element is in equilibrium.

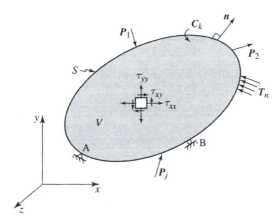

Figure 14.7.2

Now, as before, we imagine that while under the existing *constant* state of stress,[†] points in the body undergo virtual displacements from their equilibrium position, i.e. $u \to u + \delta u, v \to v + \delta v$. Then, due to the virtual displacements δu, corresponding 'virtual strains' will occur; we denote these by $\delta\epsilon_{xx}, \delta\epsilon_{yy}, \delta\epsilon_{xy}$. It is important to note, too, that the virtual strains are independent of, and completely unrelated to, the actual state of stress existing in the body. Furthermore, we observe that, up to now, no restrictions have been imposed on the virtual displacements or strains other than the statement that they must be geometrically possible, i.e. be piece-wise continuous with respect to x and y.

We now wish to calculate the work done by the original stress state (which is in equilibrium) in going through the virtual strains.

For simplicity, consider an element, as shown in Fig. (14.7.3), where only the stress τ_{xx} is acting. Due to τ_{xx}, the original length of the element, dx, first undergoes a (real) elongation $\epsilon_{xx} dx$. If we imagine that the element now undergoes a virtual strain $\delta\epsilon_{xx}$, the work done by τ_{xx} (which is already acting) on the element is $(\tau_{xx} dy dz)\delta\epsilon_{xx} dx$, where $dy dz$ is the area over which τ_{xx} acts. We call this work, which is done by (actual) stresses 'riding' through the virtual strains, **internal virtual work**.

Similarly, if other stresses act through the virtual strains, the total internal virtual

Figure 14.7.3

[†] By a 'constant state of stress', we mean stresses that do not change, for example, with time. However, they may vary in space, i.e. with x and y.

work done on the element is

$$d(\delta W_{\text{int}}) = (\tau_{xx}\delta\epsilon_{xx} + \tau_{yy}\delta\epsilon_{yy} + 2\tau_{xy}\delta\epsilon_{xy}) \, dx \, dy \, dz \qquad (14.7.2a)$$

so that the total internal virtual work done on the entire body becomes[†]

$$\delta W_{\text{int}} = \iiint_V (\tau_{xx}\delta\epsilon_{xx} + \tau_{yy}\delta\epsilon_{yy} + 2\tau_{xy}\delta\epsilon_{xy}) \, d\Omega, \qquad (14.7.2b)$$

where $d\Omega = dx \, dy \, dz$ denotes the elementary volume.

This last equation is taken as the definition of internal virtual work. It represents the 'work' done by (real) stresses when 'riding' through the virtual strains.

For the three-dimensional case, it is clear that

$$\delta W_{\text{int}} = \iiint_V [\tau_{xx}\delta\epsilon_{xx} + \tau_{yy}\delta\epsilon_{yy} + \tau_{zz}\delta\epsilon_{zz} + 2(\tau_{xy}\delta\epsilon_{xy} + \tau_{yz}\delta\epsilon_{yz} + \tau_{zx}\delta\epsilon_{zx})] \, d\Omega$$

$$(14.7.2c)$$

(c) Proof of the principle of virtual work: comments on the principle
(i) Derivation of the principle

Consider a deformable body in a state of equilibrium under a set of forces, and let the resulting stresses at each point be τ_{xx}, τ_{yy}, τ_{xy}. Since each element is in equilibrium, it follows that, for the two-dimensional case considered, the stress components must, at each point in the body, satisfy the stress equations of equilibrium [see Eqs. (2.4.4)]

$$\frac{\partial \tau_{xx}}{\partial x} + \frac{\partial \tau_{yx}}{\partial y} + B_x = 0 \qquad (14.7.3a)$$

$$\frac{\partial \tau_{xy}}{\partial x} + \frac{\partial \tau_{yy}}{\partial y} + B_y = 0. \qquad (14.7.3b)$$

Now, let us first multiply the above two equations by arbitrary scalar quantities $\delta u(x, y)$ and $\delta v(x, y)$, respectively, and then add these equations. Clearly, we then have

$$\left(\frac{\partial \tau_{xx}}{\partial x} + \frac{\partial \tau_{yx}}{\partial y} + B_x\right)\delta u + \left(\frac{\partial \tau_{xy}}{\partial x} + \frac{\partial \tau_{yy}}{\partial y} + B_y\right)\delta v = 0. \qquad (14.7.4)$$

The arbitrary character of δu and δv should be quite clear, for we could have called these variables, say, $\delta f_1(x, y)$ and $\delta f_2(x, y)$ and Eq. (14.7.4) would be equally valid. If we integrate this last equation over the volume V of the body, then

$$\iiint_V \left[\left(\frac{\partial \tau_{xx}}{\partial x} + \frac{\partial \tau_{yx}}{\partial y} + B_x\right)\delta u + \left(\frac{\partial \tau_{xy}}{\partial x} + \frac{\partial \tau_{yy}}{\partial y} + B_y\right)\delta v\right] d\Omega = 0.$$

$$(14.7.5)$$

Considering a typical term, we can write

$$\frac{\partial \tau_{xx}}{\partial x}\delta u = \frac{\partial(\tau_{xx}\delta u)}{\partial x} - \tau_{xx}\frac{\partial(\delta u)}{\partial x}. \qquad (14.7.6)$$

[†] It is worthwhile comparing Eq. (14.7.2b) with Eq. (14.2.8c). In the former, the stress components are *constant* (i.e., 'fixed') at any point whereas in the latter they are functions of the strain.

Operating similarly on the remaining terms, we obtain

$$\iiint_V \left[\frac{\partial(\tau_{xx}\delta u)}{\partial x} - \tau_{xx}\frac{\partial(\delta u)}{\partial x} + \frac{\partial(\tau_{xy}\delta u)}{\partial y} - \tau_{xy}\frac{\partial(\delta u)}{\partial y} \right.$$

$$\left. + \frac{\partial(\tau_{yy}\delta v)}{\partial y} - \tau_{yy}\frac{\partial(\delta v)}{\partial y} + \frac{\partial(\tau_{xy}\delta v)}{\partial x} - \tau_{xy}\frac{\partial(\delta v)}{\partial x} \right] d\Omega$$

$$+ \iiint_V \left(B_x\delta u + B_y\delta v \right) d\Omega = 0. \tag{14.7.7a}$$

Rearranging the terms leads to

$$\iiint_V \left[\frac{\partial}{\partial x}(\tau_{xx}\delta u + \tau_{xy}\delta v) + \frac{\partial}{\partial y}(\tau_{yy}\delta v + \tau_{xy}\delta u) \right] d\Omega$$

$$+ \iiint_V \left(B_x\delta u + B_y\delta v \right) d\Omega$$

$$= \iiint_V \tau_{xx}\frac{\partial(\delta u)}{\partial x} + \tau_{xy}\left[\frac{\partial(\delta v)}{\partial x} + \frac{\partial(\delta u)}{\partial y} \right] + \tau_{yy}\frac{\partial(\delta v)}{\partial y} \, d\Omega. \tag{14.7.7b}$$

Note that in the integral appearing on the right side of the last equation, terms of the form $\frac{\partial}{\partial x}(\delta u)$ appear. How are these to be interpreted? It is at this stage, and only now, that we interpret δu and δv as virtual displacements. Moreover, *we restrict our discussion to small, infinitesimal values of these virtual displacements such that the resulting virtual strains and relative rotations are small.* We recall too that, under this restriction, strains are given by the expressions [see Eq. (3.7.20)]

$$\epsilon_{xx} = \frac{\partial u}{\partial x}, \tag{14.7.8a}$$

$$\epsilon_{yy} = \frac{\partial v}{\partial y}, \tag{14.7.8b}$$

$$\epsilon_{xy} = \frac{1}{2}\left(\frac{\partial u}{\partial y} + \frac{\partial v}{\partial x} \right). \tag{14.7.8c}$$

Consider, for example, a variation $\delta\epsilon_{xx}$ of the strain ϵ_{xx} [Fig. (14.7.4)]. Since, for small strains, the extensional strain is the change of length of an element divided by its original length, we can write

$$\epsilon_{xx} + \delta\epsilon_{xx} = \frac{\left\{ \left[u + \frac{\partial u}{\partial x}dx + \delta u + \frac{\partial(\delta u)}{\partial x}dx \right] + dx - (u + \delta u) \right\} - dx}{dx}$$

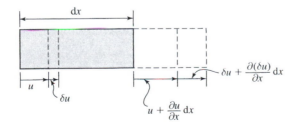

Figure 14.7.4

or

$$\epsilon_{xx} + \delta\epsilon_{xx} = \frac{\partial u}{\partial x} + \frac{\partial(\delta u)}{\partial x}. \tag{14.7.9a}$$

Using Eq. (14.7.8a), it follows that

$$\delta\epsilon_{xx} = \frac{\partial(\delta u)}{\partial x}. \tag{14.7.9b}$$

Therefore

$$\delta\left(\frac{\partial u}{\partial x}\right) = \frac{\partial(\delta u)}{\partial x}. \tag{14.7.9c}$$

In general, for any function $f(x, y)$, which is continuous and has continuous partial derivatives,

$$\delta\left(\frac{\partial f}{\partial x}\right) = \frac{\partial}{\partial x}(\delta f), \qquad \delta\left(\frac{\partial f}{\partial y}\right) = \frac{\partial}{\partial y}(\delta f). \tag{14.7.9d}$$

Using the relation of Eq. (14.7.9d), we now return to Eq. (14.7.7b) and write

$$\iiint\limits_{V} \left[\frac{\partial}{\partial x}(\tau_{xx}\delta u + \tau_{xy}\delta v) + \frac{\partial}{\partial y}(\tau_{yy}\delta v + \tau_{xy}\delta u)\right] d\Omega$$

$$+ \iiint\limits_{V} (B_x\delta u + B_y\delta v) d\Omega$$

$$= \iiint\limits_{V} (\tau_{xx}\delta\epsilon_{xx} + 2\tau_{xy}\delta\epsilon_{xy} + \tau_{yy}\delta\epsilon_{yy}) d\Omega. \tag{14.7.10}$$

Application of the divergence theorem (Appendix B.3) to the first integral appearing above yields[†]

$$\iiint\limits_{V} (\tau_{xx}\delta\epsilon_{xx} + 2\tau_{xy}\delta\epsilon_{xy} + \tau_{yy}\delta\epsilon_{yy}) d\Omega$$

$$= \iint\limits_{S} [\ell_x(\tau_{xx}\delta u + \tau_{xy}\delta v) + \ell_y(\tau_{yy}\delta v + \tau_{xy}\delta u)] ds + \iiint\limits_{V} (B_x\delta u + B_y\delta v) d\Omega \tag{14.7.11}$$

where ℓ_x and ℓ_y are the components (in the x- and y-directions, respectively) of the unit normal vector $\boldsymbol{n}$ to the surface S enclosing the volume V [see Eq. (2.8.7)].

Rearranging the terms, the surface integral above becomes

$$\iint\limits_{S} [(\ell_x\tau_{xx} + \ell_y\tau_{xy})\delta u + (\ell_x\tau_{xy} + \ell_y\tau_{yy})\delta v] ds,$$

which by Eqs. (2.8.8) is equal to

$$\iint\limits_{S} (X_n\delta u + Y_n\delta v) ds,$$

[†] For mathematical simplicity, we assume here that all functions are continuous. Hence, in using the divergence theorem, concentrated forces and couples, which can be represented only by discontinuous functions, will not appear.

where again X_n and Y_n denote, respectively, the x- and y-components of the traction T_n acting on S. Hence we recognise the integrand of this last expression as the scalar product $T_n \cdot \delta u$. Similarly, the integrand $B_x \delta u + B_y \delta v = B \cdot \delta u$. Thus the right-hand integrals of Eq. (14.7.11) are

$$\iint_S (T_n \cdot \delta u)\, ds \quad \text{and} \quad \iiint_V (B \cdot \delta u)\, d\Omega,$$

respectively, and Eq. (14.7.11) can finally be written as

$$\iiint_V (\tau_{xx}\delta\epsilon_{xx} + 2\tau_{xy}\delta\epsilon_{xy} + \tau_{yy}\delta\epsilon_{yy})\, d\Omega$$

$$= \iint_S (T_n \cdot \delta u)\, ds + \iiint_V (B \cdot \delta u)\, d\Omega. \tag{14.7.12}$$

Now we recognize the left-hand side of Eq. (14.7.12) as being the internal virtual work δW_{int}. Similarly, the first integral on the right side is seen to be the external virtual work of the applied tractions on the surface of the body, while the second integral of Eq. (14.7.12) represents the external virtual work of the body forces. We conclude that the right side of Eq. (14.7.12) represents the total virtual work δW_{ext} of all external forces acting on the body.[†] Hence, Eq. (14.7.12) is a statement that

$$\delta W_{\text{int}} = \delta W_{\text{ext}}. \tag{14.7.13}$$

Equation (14.7.13) expresses the *principle of virtual work* for deformable bodies and may be stated as follows:

Let a body under a set of forces produce an equilibrium state of stress at all points. Then, the internal virtual work of the stresses (acting through the virtual strains) is equal to the external virtual work of the applied forces (acting through the virtual displacements).

Writing the stresses and strains symbolically as the tensors τ and ϵ, we may rewrite Eq. (14.7.12) as

$$\iiint_V \tau \cdot \delta\epsilon\, d\Omega = \iint_S T_n \cdot \delta u\, ds + \iiint_V B \cdot \delta u\, d\Omega, \tag{14.7.14}$$

where, as indicated by the corresponding arrows, we note that the (actual) equilibrium stress state τ at each point is due to the external tractions and body forces and the virtual strains $\delta\epsilon$ are compatible with the virtual displacements. (Note that here we have for convenience again omitted, in Eq. (14.7.14), the concentrated forces and couples.)

(ii) Some comments

The following remarks are now in order:

(1) The principle of virtual work was derived without reference to any energy expression. Consequently, at this stage, the principle should be considered as separate from energy considerations.

[†] In comparing the definition of δW_{ext} given in Eq. (14.7.1) with the right side of Eq. (14.7.12), we note that the summations due to concentrated forces and couples do not appear in the latter equation. If traction discontinuities had also been taken into account, the two equations would be identical. (See previous footnotes on pages 558 and 564.)

(2) The *only* condition imposed on the external forces and on the internal stress state is that they must be in equilibrium.

(3) In the derivation of the principle, we imposed only two conditions on the virtual displacements: namely (a) the virtual displacements within the body must be piece-wise continuous and (b) all their partial derivatives, $\frac{\partial(\delta u)}{\partial x}$, $\frac{\partial(\delta v)}{\partial x}$, $\frac{\partial(\delta u)}{\partial y}$, $\frac{\partial(\delta v)}{\partial y}$, must be small. *Hence, the principle, as derived, is limited to bodies undergoing small virtual strains and relative rotations* [see Eq. (3.5.16)]. Conversely, the principle remains valid if the body undergoes *finite virtual translations*. Aside from this, no other conditions have been imposed on the virtual displacements. For emphasis, we reiterate that nowhere have we stated or required that the virtual displacements δu, δv satisfy constraints at the boundary. Indeed, if we wish, we may actually violate the constraints. And, in fact, we often do violate the constraints in order, for example, to evaluate reactions.

(4) It is of interest to compare the principle of conservation of energy (Section 14.3) with the principle of virtual work. A comparison of the principles is presented in the following table.

Energy	Virtual work
Real work is done by stresses (forces) acting through displacements that are a result of the stresses (forces) themselves.	Virtual work is done by stresses (forces) acting through arbitrary virtual strains (displacements).
Stress components are acting statically from zero to their final value.	Stresses are assumed to be already acting on the body. The virtual work is done by constant stress components.
Conservation of energy implies that there is no dissipation of energy.	No implication has been made concerning energy dissipation. The principle is valid for any type of body provided it is in equilibrium.

(5) If we give a body virtual displacements $\delta \boldsymbol{u}$ describing rigid-body motion (such that the virtual strains are zero), it follows from Eq. (14.7.12) that if the body is in equilibrium, then

$$\iint_S \boldsymbol{T}_n \cdot \delta \boldsymbol{u} \, \mathrm{d}s + \iiint_V \boldsymbol{B} \cdot \delta \boldsymbol{u} \, \mathrm{d}\Omega = 0 \qquad (14.7.15a)$$

or

$$\delta W_{\text{ext}} = 0, \qquad (14.7.15b)$$

which is the principle of virtual work for rigid bodies. We note that the principle of virtual work for rigid bodies is but a special case of the general principle developed here for deformable bodies.

(6) In the above derivation, we started from the equations of equilibrium [Eqs. (14.7.3)] and arrived at the principle [Eq. (14.7.13)]. Thus we state that if a body is in equilibrium, then it must satisfy the principle of virtual work. Now, if we were to start from Eq. (14.7.12) or (14.7.13) and work backwards, we would arrive at Eqs. (14.7.3). Hence, we may state that if a body satisfies the principle of virtual work when given a virtual displacement, then it must be in

equilibrium. Thus, the equilibrium conditions are both necessary and sufficient. We therefore make the following statement:

A body subjected to a set of external forces satisfies the principle of virtual work if, and only if, it is in a state of equilibrium.

(7) Although, as has been pointed out, the internal virtual work δW_{int} of a body is *not* equal to the internal strain energy U of the body, for a given existing constant state of stress within the body, the *change* of strain energy $\delta U = \delta W_{int}$ due to any variation of strains (virtual strains) $\delta \epsilon$. (We show this explicitly in Section 14.9 below.)

In the above development of the principle, it was possible, at each stage of the derivation, to clearly define the required terms and to state the specific limitations and conditions under which the principle is valid. Thus the principle, as stated above, is quite general and may be applied to any given problem.

Now, in structural mechanics, we often encounter specific structural components such as rods, beams, bars, etc. (Such components, as we previously observed, are often referred to as *one-dimensional elements* since their cross-sections are defined by means of a single coordinate, say x.) In such cases, it is more useful to express the internal virtual work for these structural elements in terms of internal stress resultants and deformation patterns (which describe globally the deformation of the cross-sections) rather than in terms of stresses and strains. In the following section, we therefore derive the principle specifically for beams undergoing flexure in the x–y plane. As we shall observe, the derivation follows closely the more general development given above.

(d) The principle of virtual work for flexure of beams

We derive the principle of virtual work explicitly for a beam, starting from the basic governing equilibrium equation, namely Eq. (8.3.3),

$$\frac{d^2 M(x)}{dx^2} + q(x) = 0, \tag{14.7.16}$$

where $M \equiv M_z$ and the external loads $q(x)$ lie in the x–y plane [Fig. (14.7.5a)].[†] We note too that this equation is the analogue to Eqs. (14.7.3); as we shall see, the

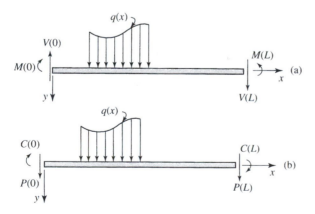

Figure 14.7.5

[†] Note that in the derivation of this equation, both equilibrium conditions, $\sum F_y = 0$ and $\sum M = 0$, were satisfied.

present derivation will follow analogously the general derivation in the preceding section.

We now assume that we give the beam a virtual lateral displacement $\delta v(x)$. Then clearly,

$$\left[\frac{d^2 M(x)}{dx^2} + q(x)\right]\delta v = 0 \qquad (14.7.17a)$$

and hence

$$\int_0^L \left\{\left[\frac{d^2 M(x)}{dx^2} + q(x)\right]\delta v\right\} dx = 0 \qquad (14.7.17b)$$

or

$$\int_0^L \frac{d^2 M(x)}{dx^2}\delta v\, dx = -\int_0^L q(x)\delta v\, dx. \qquad (14.7.17c)$$

Integrating the left-hand side by parts,

$$\int_0^L \frac{d^2 M(x)}{dx^2}\delta v\, dx = \frac{dM(x)}{dx}\delta v\Big|_0^L - \int_0^L \frac{dM(x)}{dx}\frac{d(\delta v)}{dx}\, dx. \qquad (14.7.18a)$$

Now, by Eq. (8.3.2), $\frac{dM(x)}{dx} = V(x)$. Hence

$$\int_0^L \frac{d^2 M(x)}{dx^2}\delta v\, dx = V(x)\delta v\Big|_0^L - \int_0^L \frac{dM(x)}{dx}\frac{d(\delta v)}{dx}\, dx. \qquad (14.7.18b)$$

Integrating the last integral of Eq. (14.7.18b) again by parts, we have[†]

$$\int_0^L \frac{d^2 M(x)}{dx^2}\delta v\, dx = V(x)\delta v\Big|_0^L - M(x)\frac{d(\delta v)}{dx}\Big|_0^L + \int_0^L M(x)\frac{d^2(\delta v)}{dx^2}\, dx. \qquad (14.7.18c)$$

We now note from Eq. (14.7.9d) that $\frac{d(\delta v)}{dx} = \delta\left(\frac{dv}{dx}\right) \equiv \delta v'(x)$, etc. We recall too that for small rotations (slopes), $v''(x)$ represents the curvature $\kappa \equiv \kappa_z$ of the beam in the x–y plane [see Eq. (9.2.3)]; i.e., $v''(x) = -\kappa(x)$. Equation (14.7.18c) then becomes

$$\int_0^L \frac{d^2 M(x)}{dx^2}\delta v\, dx = V(x)\delta v\Big|_0^L - M(x)\delta v'\Big|_0^L - \int_0^L M(x)\delta\kappa(x)\, dx. \qquad (14.7.18d)$$

[†] For mathematical simplicity, we assume no discontinuities in $M(x)$ or $V(x)$ within $0 < x < L$. As a result, concentrated forces and couples *within* the beam will not appear in the integrated expressions (see p. 564).

Finally, substituting Eq. (14.7.18d) in Eq. (14.7.17c), we have

$$\int_0^L M(x)\delta\kappa(x)\,dx = \int_0^L q(x)\delta v\,dx + V(x)\delta v\Big|_0^L - M(x)\delta v'\Big|_0^L ; \quad (14.7.19a)$$

that is,

$$\int_0^L M(x)\delta\kappa(x)\,dx = \int_0^L q(x)\delta v\,dx + V(L)\delta v(L) - V(0)\delta v(0)$$

$$- M(L)\delta v'(L) + M(0)\delta v'(0). \quad (14.7.19b)$$

The left side of the above equation represents the internal virtual work of the moments due to the *variation of curvature*, while the right-hand side represents the external virtual work of external forces.

Equations (14.7.19) thus express the principle of virtual work for beams undergoing flexure. It is important to note that since the right-hand side of Eq. (14.7.19b) represents the virtual work of external forces acting on the beam, it specifically also includes the virtual work done by shear forces and moments acting at the ends $x = 0$ and $x = L$ [Fig. (14.7.5a)].[†] Thus the right-hand side of Eq. (14.7.19b) can be written formally as

$$\delta W_{\text{ext}} = \int_0^L q(x)\delta v\,dx + \sum P\delta v\Big|_{x=0,L} + \sum C\delta\theta\Big|_{x=0,L}, \quad (14.7.20a)$$

where $\delta\theta \equiv \delta v'$ denotes the slope, positive P here denotes *downward* shear forces and positive C denotes *clock-wise* moments acting at the two ends [Fig. (14.7.5b)].

In the above, $q(x)$ has been implicitly assumed to be a continuous function of x. However, if we assume that concentrated transverse loads and applied moments (couples) are also acting at x_j, and x_k, respectively, within the span $0 < x < L$ [Fig. (14.7.6)], the function $q(x)$ will contain discontinuities at these points [see Section 11b of Chapter 9]. Hence, in place of the simple

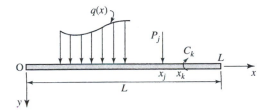

Figure 14.7.6

[†] The appearance of the minus signs in Eq. (14.7.19b) is a result of our adopted sign convention: positive $V(L)$ acts downward, positive $V(0)$ acts upward at the left end $x = 0$ and positive $M(L)$ acts counterclock-wise at the right end, $x = L$.

integral $\int q(x)\delta v \, dx$, we have

$$\int_0^L q(x)\delta v \, dx + \sum P_j \, \delta v_j + \sum C_k \delta \theta_k, \quad 0 < x_j, \ x_k < L.$$

In this case, the expression for the external virtual work becomes

$$\delta W_{\text{ext}} = \int_0^L q(x)\,\delta v \, dx + \sum_{j=1}^n P_j \, \delta v_j + \sum_{k=1}^m C_k \, \delta \theta_k, \quad 0 \leq x_j, \ x_k \leq L .$$

(14.7.20b)

The principle of virtual work for beams under the combined forces [Fig. (14.7.7)] can therefore be written as

$$\int_0^L M(x)\,\delta\kappa(x)\,dx = \int_0^L q(x)\,\delta v \, dx + \sum_{j=1}^n P_j \, \delta v_j + \sum_{k=1}^m C_k \, \delta \theta_k . \quad (14.7.21)$$

It is clear that this specific derivation for a beam in flexure follows in parallel steps the derivation of the general proof. [While Eq. (14.7.21) is the analogue of Eqs. (14.7.12), integration by parts – in this one-dimensional case – is analogous to the use of the divergence theorem in the more general derivation.]

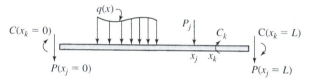

Figure 14.7.7

Finally, it is worthwhile to mention here again that material constants do not enter into the derivation; thus the expressions of Eqs. (14.7.19)–(14.7.21) are valid for beams of any arbitrary material.

(e) Application of the principle of virtual work to evaluate reactions and internal stress resultants: the 'method of virtual displacements'

The principle of virtual work has many applications. In particular, the principle can be used directly to evaluate reactions and internal stress resultants (i.e., axial forces, moments and shear forces) in a body. However, in calculating the virtual work (both internal or external), we must conform to the limitations imposed in the derivation of the principle: namely the virtual displacements must be piece-wise continuous [see Section 14.7b (i)] and the body must undergo a virtual motion consistent with small strains and small rotations.

In the use of the principle to evaluate reactions or internal force resultants, we shall essentially release all (or some) of the constraints. The virtual work of the unknown external forces corresponding to the releases (when acting through an appropriate virtual displacement of the 'released structure') is then calculated. By satisfying the principle of virtual work, we require the unknown forces to be, in fact, equilibrium forces. Thus it will be noticed, in the applications which follow, that we shall actually choose virtual displacements which violate the constraints, i.e. the geometric boundary conditions.

The following example illustrates explicitly the *arbitrary nature of the virtual displacements*.

Example 14.6: A cantilever beam AB of length L (having arbitrary material properties) is subjected to a uniformly distributed load $q(x) = w$ [Fig. (14.7.8a)]. Determine the reactions acting on the beam.

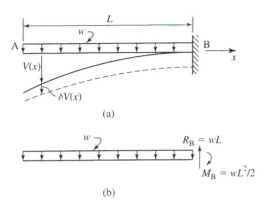

(a)

(b)

Figure 14.7.8

Solution: For this simple problem, the moment is $M(x) = -wx^2/2$ and it is clear that the reactions at $x = L$ consist of an *upward* force $R_B = wL$ and a *clock-wise* moment $M_B = wL^2/2$ [Fig. (14.7.8b)]. We verify here that the principle of virtual work leads to this same result *for any arbitrary variation* $\delta v(x)$ [Fig. (14.7.8a)].

Although the material of the beam has not been specified, it is evident that the beam will undergo some real lateral deflection $v(x)$ and reach an equilibrium state under the given loading. We are not concerned with this actual (true) displacement. Irrespective of what $v(x)$ may be, let us assume that this displacement $v(x)$ is varied from its equilibrium position by $\delta v(x)$, which we prescribe to be of the form

$$\delta v(x) = \delta A \left[\left(\frac{x}{L} \right)^n + 1 \right], \quad n \geq 0, \qquad (14.7.22a)$$

where δA (assumed positive) is a coefficient and $n \geq 0$ is arbitrary.

It is worthwhile to sketch the virtual displacements δv for several values of $n, n \geq 0$; these are shown in Figs. (14.7.9a and b). Note that these virtual displacement patterns are quite arbitrary due to the arbitrary nature of n and do not bear any resemblance to the possible real deflection of the beam. We also observe that the chosen $\delta v(x)$ *violates the boundary conditions at* $x = L$ for all $n \geq 0$.

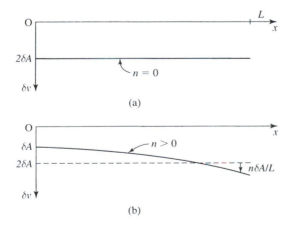

(a)

(b)

Figure 14.7.9

From Eq. (14.7.22a),

$$\delta v'(x) = \delta A \left[n \frac{x^{n-1}}{L^n} \right] \qquad (14.7.22b)$$

$$\delta v''(x) = \delta A \left[n(n-1) \frac{x^{n-2}}{L^n} \right] \qquad (14.7.22c)$$

Recalling that $\kappa(x) = -v''(x)$ for $|v'| \ll 1$, substitution in the left-hand side of Eq. (14.7.19) representing the internal virtual work, yields

$$\int_0^L M(x) \delta \kappa(x)\, dx = \frac{\delta A}{L^n} \frac{w}{2} n(n-1) \int_0^L x^n\, dx = \delta A \frac{wL}{2} \frac{n(n-1)}{n+1}. \qquad (14.7.23a)$$

Consider now the right-hand side of Eq. (14.7.19b). For the problem at hand, taking into account our adopted sign convention,

$$V_A \equiv V(0) = 0; \quad V_B \equiv V(L) = -R_B; \quad M_A \equiv M(0) = 0; \quad M(L) = -M_B.$$

Moreover, from our assumed variation $\delta v(x)$, Eqs. (14.7.22),

$$\delta v(L) = 2\delta A; \qquad \delta v'(L) = n\frac{\delta A}{L} \geq 0,$$

Therefore, the right-hand side of Eq. (14.7.19b), representing the external virtual work, becomes

$$\int_0^L q(x) \delta v\, dx + V(L) \delta v(L) - V(0) \delta v(0) - M(L) \delta v'(L) + M(0) \delta v'(0)$$

$$= w\delta A \int_0^L \left[\frac{x^n}{L^n} + 1 \right] dx - 2R_B \delta A + n M_B \frac{\delta A}{L}$$

$$= \delta A \left[wL \left(\frac{n+2}{n+1} \right) - 2R_B + n\frac{M_B}{L} \right]. \qquad (14.7.23b)$$

Applying the principle of virtual work, i.e. equating Eq. (14.7.23a) to Eq. (14.7.23b), and noting that this is true for any arbitrary δA, we obtain

$$\frac{wL}{2}(n-4) + 2R_B - \frac{nM_B}{L} = 0, \quad n \geq 0. \qquad (14.7.24a)$$

For $n = 0$, $R_B = wL$. Substituting this value back in Eq. (14.7.24a) leads to

$$M_B \equiv M(L) = \frac{wL^2}{2} \quad \text{for all } n > 0. \qquad (14.7.24b)$$

$\square$

We observe from this example that one may obtain reactions and solve for unknown forces using *any appropriate set of arbitrary virtual displacements* [see, e.g., Fig. (14.7.10a)]. Indeed, one could obtain the same results as above by choosing, for example, a virtual displacement of the form

$$\delta v(x) = \delta A \sin \frac{n\pi x}{2L}, \quad n > 0. \qquad (14.7.25)$$

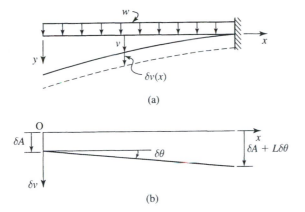

(a)

(b)

Figure 14.7.10

Since the principle of virtual work is indeed valid for *any* arbitrary virtual displacement, in applying the principle of virtual work to obtain relations between internal stress resultants and reactions, we find it useful to prescribe virtual displacements which describe rigid-body motion so that $\delta W_{int} = 0$ *identically*. Hence, when prescribing such rigid-body virtual displacements, the principle will be satisfied by Eq. (14.7.15b), namely $\delta W_{ext} = 0$. As we shall now show, by prescribing virtual rigid-body motions, we simplify considerably the required calculations since we need not calculate δW_{int}!

For the given problem, instead of the virtual displacement given by Eq. (14.7.22a), let us now assume a virtual displacement representing rigid-body motion defined by

$$\delta v(x) = \delta A + \delta \theta \cdot x, \qquad (14.7.26)$$

where δA and $\delta \theta$ are shown in Fig. (14.7.10b).

Using Eq. (14.7.20), the external virtual work is given by

$$\delta W_{ext} = -R_B(\delta A + L\delta \theta) + M_B \cdot \delta \theta + w \int_0^L (\delta A + \delta \theta \cdot x)\,dx. \qquad (14.7.27)$$

Integrating and setting $\delta W_{ext} = 0$,

$$(-R_B + wL) \cdot \delta A + \left(-R_B L + M_B + \frac{wL^2}{2}\right) \cdot \delta \theta = 0. \qquad (14.7.28a)$$

Since δA and $\delta \theta$ are arbitrary, each of the coefficients of δA and $\delta \theta$ must vanish: thus setting the coefficient of δA to zero, we have $R_B = wL$; setting the coefficient of $\delta \theta$ to zero and replacing with the calculated value of R_B, we find

$$R_B = wL, \qquad M_B = \frac{wL^2}{2}, \qquad (14.7.28b)$$

which agrees with our previous results. We observe that by prescribing virtual displacements representing rigid-body motion, the required calculations have become much simpler.

(i) Applications of the principle of virtual work to statically determinate structures

Some further applications of the principle of virtual work to statically determinate structures are given in the following illustrative examples.

Example 14.7: Determine the reactions R_{Ax} and R_{By} of a simply supported beam of length L, subjected to a uniform load w and a concentrated force P, as shown in Fig. (14.7.11a).

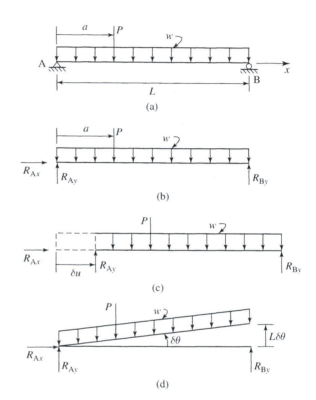

Figure 14.7.11

Solution: First we assume that the beam is in equilibrium under the set of forces given in Fig. (14.7.11b). According to the discussion above, we may choose to give the beam virtual displacements that represent rigid-body motion.

To find R_{Ax}, we let the virtual displacement be δu in the x-direction. Then the virtual work done is [Fig. (14.7.11c)]

$$\delta W_{\text{ext}} = R_{Ax}\delta u = 0. \tag{14.7.29a}$$

Since Eq. (14.7.29a) must be valid for any δu, it follows that $R_{Ax} = 0$ must be true. To find R_{By}, we let the virtual displacement be a rotation $\delta\theta$ about point A [Fig. (14.7.11d)[†]]. Then

$$\delta W_{\text{ext}} = R_{By}(L\delta\theta) - P(a\delta\theta) - wL\left(\frac{L\delta\theta}{2}\right) = 0. \tag{14.7.29b}$$

[†] It is important to note that all forces in this figure remain vertical. See footnote p. 560.

Hence, since this must be valid for all $\delta\theta$,

$$R_{By} = \frac{Pa}{L} + \frac{wL}{2}.$$

□

We again note that in obtaining the above solutions, we released the constraints and chose virtual displacements that, in each case, violated the boundary conditions. Moreover, we observe that to obtain the solution for any desired unknown, we have chosen an appropriate virtual displacements in such a way that *only the single unknown* appears in the expression for the external virtual work. As we shall see in the following examples, we can apply this simple procedure for any statically determinate structure.

Clearly, in the simple problem considered above, the results could have been obtained just as well from the equations of equilibrium. Indeed, the principle of virtual work as applied here does not yield any results that could not be obtained from equilibrium equations. However, the principle provides us with a different method of solution, and in certain problems, use of the virtual work method proves to be much simpler and efficient than solving a set of equilibrium equations. This is demonstrated in the following example, where we determine internal force resultants.

Example 14.8:   A simply supported beam of length L is subjected to a force P acting at a *variable* point $x = \zeta$, as shown in Fig. (14.7.12a). Determine[†] (a) the moment M at section $B(x = b)$ and (b) the shear force V at section $B(x = b)$.

Solution: Since we are now interested in determining stress resultants within the structure, we choose to give the structure appropriate virtual displacements such that these unknowns appear in the expressions for the virtual work. Now, the unknowns M and V can appear as terms in the virtual work expressions only if we choose virtual displacements containing discontinuities within the structure. *It is to be recalled that since virtual displacements can be piece-wise continuous, discontinuities are permitted in the virtual displacement field* (see p. 559). The discontinuities can be either in the displacements themselves or in the rotations.

Having made the above comments, we proceed with the solution.

(a) **Moment at B**

In choosing a virtual displacement, we seek to have the moment appear as the *only unknown* in the virtual work expression [see Fig. (14.7.12b)]. We therefore choose as the virtual displacement, rotations $\delta\theta_1$, $\delta\theta_2$ such that the bar remains rigid except for a discontinuity in the slope at point B, i.e., we introduce a 'kink' at B but permit no separation at B, as shown in Fig. (14.7.12c).[‡] In doing so, we effectively decompose the beam into two rigid segments, $0 < x < b$ and $b < x < L$. In effect, we have transformed the beam into a 'mechanism'.

The moment M at the cross-section B is then considered as an external moment acting on each segment of the beam. However, we must consider two separate cases: $\zeta \leq b$ and $\zeta \geq b$, i.e., when the load is applied to the left and to the right of the cross-section at B.

[†]   In this example, and all subsequent examples in this section, positive moments and shear forces follow the adopted sign convention as shown in Fig. (8.2.8).

[‡]   Note that in prescribing virtual displacements such that no separation exists at B, the virtual work of the downward shear force V_B acting on the right face of segment AB cancels with the upward shear force V_B acting on the left face of segment BC.

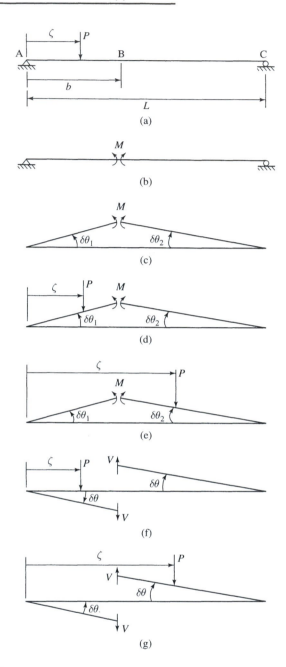

Figure 14.7.12

Case (i): $\zeta \leq b$ [Fig. (14.7.12d)]

The external virtual work done is

$$\delta W_{\text{ext}} = -P(\zeta\,\delta\theta_1) + M(\delta\theta_1 + \delta\theta_2). \qquad (14.7.30)$$

Noting that the virtual displacements define a rigid-body motion (hence, $\delta W_{\text{int}} = 0$), it follows that $\delta W_{\text{ext}} = 0$, and therefore

$$M(\delta\theta_1 + \delta\theta_2) = P(\zeta\,\delta\theta_1). \qquad (14.7.31)$$

Now, if no 'virtual separation' exists between the two segments, for $|\theta_1| \ll 1$, $|\theta_2| \ll 1$, we require that

$$\delta\theta_1 b = (L - b)\,\delta\theta_2. \tag{14.7.32}$$

Substituting the above Eq. (14.7.31),

$$M\left(1 + \frac{b}{L - b}\right)\delta\theta_1 = P\zeta\,\delta\theta_1. \tag{14.7.33a}$$

Since this must be true for all arbitrary $\delta\theta_1$,

$$M = \frac{P\zeta}{1 + \frac{b}{L-b}} = \frac{P(L - b)}{L}\zeta, \quad \zeta \le b. \tag{14.7.33b}$$

Case (ii): $b \le \zeta$ [Fig. (14.7.10e)]

Proceeding as above, we find, instead of Eq. (14.7.30),

$$\delta W_{\text{ext}} = -P[(L - \zeta)\delta\theta_2] + M(\delta\theta_1 + \delta\theta_2). \tag{14.7.34a}$$

Hence, since again $\delta W_{\text{ext}} = 0$,

$$M(\delta\theta_1 + \delta\theta_2) = P[(L - \zeta)\,\delta\theta_2]. \tag{14.7.34b}$$

Using Eq. (14.7.32), we obtain

$$M\left(1 + \frac{L - b}{b}\right)\delta\theta_2 = P(L - \zeta)\,\delta\theta_2. \tag{14.7.34c}$$

Since this must be valid for all $\delta\theta_2$,

$$M = \frac{P(L - \zeta)}{1 + \frac{L-b}{b}} = \frac{Pb}{L}(L - \zeta), \quad b \le \zeta. \tag{14.7.34d}$$

(b) Shear Force V at B

Here, we choose an appropriate virtual displacement such that only the unknown V appears in the virtual work expression. To this end, we introduce a discontinuity in the virtual displacement δv at point B while at the same time we permit no discontinuity in the slope $\delta\theta_B$. (By not permitting a discontinuity in $\delta\theta$, the total virtual work of the moment M at B will be zero since the counter-clock-wise moment M_B acting on the segment AB then cancels out with the clock-wise moment M_B acting on BC.) Hence, the virtual displacement is, as shown in Fig. (14.7.12f or g), where $\delta\theta_1 = \delta\theta_2 = \delta\theta$.

In doing so, as in the previous case for the moment, we effectively have decomposed the beam into two rigid segments, $0 < x < b$ and $b < x < L$, thus again creating a 'mechanism'. The shear force V at the cross-section B is then considered as an external force acting on each segment of the beam.

Again, we consider two separate cases

Case (i): $\zeta < b$ [Fig. (14.7.12f)]

The external work is

$$\delta W_{\text{ext}} = P\zeta\,\delta\theta + V[b\delta\theta + (L - b)\delta\theta] = P\zeta\,\delta\theta + VL\delta\theta. \tag{14.7.35a}$$

Applying the principle, $\delta W_{\text{ext}} = 0$,

$$V = \frac{-P\zeta}{L}, \qquad 0 < \zeta < b. \tag{14.7.35b}$$

Case (ii): $b < \zeta < L$ [Fig. (14.7.12g)]

Here

$$\delta W_{\text{ext}} = -P[(L - \zeta)\delta\theta] + VL\delta\theta. \tag{14.7.35c}$$

Again, applying the principle, $\delta W_{\text{ext}} = 0$,

$$V = \frac{P(L - \zeta)}{L}, \quad b < \zeta < L. \tag{14.7.35d}$$

$\square$

Example 14.9: The statically determinate structure of Fig. (14.7.13) is subjected to a load P, which acts at a variable position $x = \zeta$. Determine (a) the reaction at A, (b) the reaction at C, (c) the moment at C, (d) the moment at E as a function of the variable load position ζ, $0 \leq \zeta \leq 4L$.

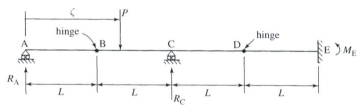

Figure 14.7.13

In each of the solutions, we choose an appropriate virtual displacement consistent with small rotations.

Solution:

(a) **Reaction at A**
We choose a virtual displacement $\delta\Delta$ as shown by the mechanism in Fig. (14.7.14a). Then, for $\zeta \leq L$,

$$\delta W_{\text{ext}} = R_A \delta\Delta - P\left(\frac{L - \zeta}{L}\right)\delta\Delta. \tag{14.7.36a}$$

Note that a hinge exists at B. Therefore with this imposed virtual displacement, the moment M_B does not contribute to the virtual work since $M_B = 0$. Then, setting $\delta W_{\text{ext}} = 0$,

$$R_A = \left(\frac{L - \zeta}{L}\right)P. \tag{14.7.36b}$$

For $L \leq \zeta \leq 4L$, it is clear that $R_A = 0$. The resulting reaction R_A is plotted as a function of the position ζ of the applied load in Fig. (14.7.15a).

(b) **Reaction at C**
We choose a virtual displacement $\delta\Delta$ as shown by the mechanism in Fig. (14.7.14b). Then, for $\zeta \leq L$,

$$\delta W_{\text{ext}} = R_C \delta\Delta - 2\frac{P\zeta}{L}\delta\Delta. \tag{14.7.37a}$$

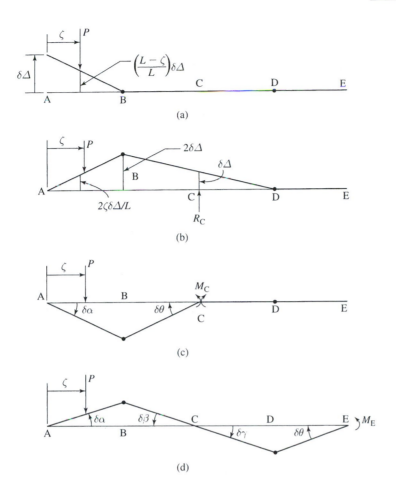

Figure 14.7.14

Again, since $\delta W_{\text{ext}} = 0$,

$$R_C = \frac{2P\zeta}{L}, \quad \zeta \le L. \tag{14.7.37b}$$

For $L \le \zeta \le 3L$,

$$\delta W_{\text{ext}} = R_C \delta\Delta - \left[\frac{P(3L - \zeta)}{2L}\right] 2\delta\Delta \tag{14.7.37c}$$

from which

$$R_C = \frac{P}{L}(3L - \zeta), \quad L \le \zeta \le 3L. \tag{14.7.37d}$$

For $3L \le \zeta \le 4L$, it is clear that $R_C = 0$. The resulting reaction R_C is plotted as a function of the position ζ of the applied load in Fig. (14.7.15b).

(c) Moment at C
We choose a virtual displacement as in Fig. (14.7.14c) such that a discontinuity $\delta\theta$ occurs in the slope at C. By choosing this as the virtual displacement, the only unknown appearing in the virtual work expressions is the unknown moment M_C. Note that $\delta\alpha = \delta\theta$. Hence, for $\zeta \le L$, $\delta W_{\text{ext}} = M_C\delta\theta + P\zeta\delta\theta$ and therefore, following the

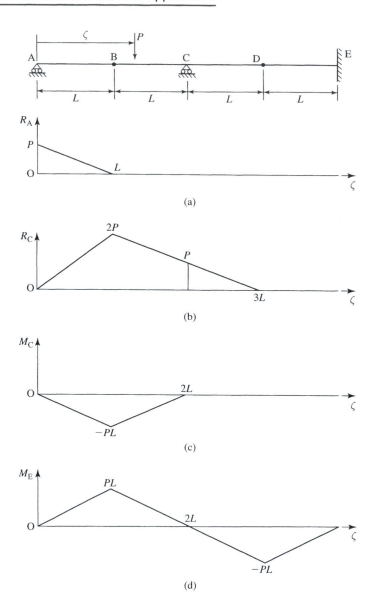

Figure 14.7.15

previous reasoning,

$$M_C = -P\zeta. \tag{14.7.38a}$$

Similarly, for $L \le \zeta \le 2L$, $\delta W_{ext} = M_C \delta\theta + P(2L - \zeta)\delta\theta$, from which

$$M_C = -P(2L - \zeta). \tag{14.7.38b}$$

For $2L \le \zeta \le 4L$, $\delta W_{ext} = 0$ identically, from which it follows that

$$M_C = 0. \tag{14.7.38c}$$

In Fig. (14.7.15c), the resulting moment M_C is plotted as a function of the position ζ of the applied load.

(d) **Moment at E**

In order to have only M_E appear as the unknown, we choose a virtual displacement $\delta\theta$, as shown in Fig. (14.7.14d). Note that the virtual rigid-body displacement requires that $\delta\alpha = \delta\beta = \delta\gamma = \delta\theta$.

The final results, obtained by setting $\delta W_{ext} = 0$ are as follows:

$$\text{For } 0 \le \zeta \le L, \qquad \delta W_{ext} = M_E \delta\theta - P\zeta\delta\theta \longrightarrow\longrightarrow M_E = P\zeta. \qquad (14.7.39a)$$

$$\text{For } L \le \zeta \le 2L, \qquad \delta W_{ext} = M_E \delta\theta - P(2L - \zeta)\delta\theta \longrightarrow\longrightarrow M_E = P(2L - \zeta).$$
$$(14.7.39b)$$

$$\text{For } 2L \le \zeta \le 3L, \qquad \delta W_{ext} = M_E \delta\theta + P(\zeta - 2L)\delta\theta \longrightarrow\longrightarrow M_E = -P(\zeta - 2L).$$
$$(14.7.39c)$$

$$\text{For } 3L \le \zeta \le 4L, \qquad \delta W_{ext} = M_E \delta\theta + P(4L - \zeta)\delta\theta \longrightarrow\longrightarrow M_E = -P(4L - \zeta).$$
$$(14.7.39d)$$

The resulting moment M_E is plotted as a function of the position ζ of the applied load in Fig. (14.7.15d).

We observe that each of the graphs of Figs. (14.7.15a–d) has the same shape as the virtually displaced structure, namely (corresponding to this statically determinate structure) the mechanisms of Figs. (14.7.14a–d), respectively. [In the section on *Influence Lines*, we shall find that this interesting feature has useful and important applications.] □

In examining each of the cases in the above solutions, it is worthwhile to recall a general comment: when solving for a given unknown force in a statically determinate structure by the principle of virtual work, it is always possible to choose a virtual displacement describing a mechanism such that *only the desired unknown* appears in the virtual work expression. In principle, this comment is valid for any structure, no matter how complex, provided the structure is statically determinate. Indeed, the principle of virtual work is used extensively in the field of structural mechanics where it has a wide variety of applications not only for beams but also in the analysis of more elaborate structures, for example, frames and trusses. We illustrate the application of the principle to more complex structures by means of the following examples.

Example 14.10: The statically determinate frame structure shown in Fig. (14.7.16) consists of elements rigidly connected at points C, F and G and connected by means of hinges at points A, B, D, E and H. The structure is subjected to a horizontal force P.

Using the principle of virtual work, determine (a) the horizontal component R_{xA} of the reaction at A, (b) the horizontal component R_{xH} of the reaction at H, (c) the moment at point C, (d) the shear force V in member BC at point B and (e) the axial force N in member BC at point B.

Solution: As in the case of beams of the preceding examples, we solve for the unknowns by specifying virtual displacements such that the component parts of the structure move as rigid bodies; i.e., the structure becomes a 'mechanism'. Hence

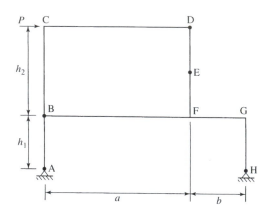

Figure 14.7.16

the principle reduces to $\delta W_{\text{ext}} = 0$. (Since the given frame is a statically determinate structure, as has been previously mentioned, we shall always find it possible to prescribe an appropriate mechanism such that only a single unknown appears in the virtual work expression.)

(a) Reaction at A
We give the structure a virtual displacement $\delta_{x\text{A}}$, as shown in Fig. (14.7.17a).[†]

$$\delta W_{\text{ext}} = -R_{x\text{A}}\delta_{x\text{A}} = 0$$

and hence $R_{x\text{A}} = 0$.[‡]

(b) Reaction at H
We give the structure a virtual displacement $\delta_{x\text{H}}$, as shown in Fig. (14.7.17b). Then

$$\delta W_{\text{ext}} = P\delta_{x\text{C}} - R_{x\text{H}}\delta_{x\text{H}} = 0,$$

and since $\delta_{x\text{C}} = \delta_{x\text{H}}$, we have $(P - R_{x\text{H}})\delta_{x\text{C}} = 0$ from which $R_{x\text{H}} = P$. Thus the total horizontal resisting force of the structure to the applied load P is due to R_{H}.

(c) Moment at C
We give the structure a virtual displacement $\delta\theta$, as shown in Fig. (14.7.17c). Then, since $\delta_{x\text{C}} = h_2\,\delta\theta$,[§]

$$\delta W_{\text{ext}} = -M_{\text{C}}\delta\theta + Ph_2\delta\theta = 0,$$

from which $M_{\text{C}} = Ph_2$.

(d) Shear V at B
We imagine that we 'cut' member BC at B so as to give the structure a virtual displacement $\delta_{x\text{B}}$, as shown in Fig. (14.7.17d). Then

$$\delta W_{\text{ext}} = P\delta_{x\text{B}} - V\delta_{x\text{B}} = 0,$$

from which $V = P$. Observe that since the moments $M_{\text{D}} = M_{\text{E}} = 0$, no work is done by these moments due to the angle changes at D and E.

[†] In Figs. (14.7.17, 19a and 21b), the original structure and the virtually displaced structure are represented by broken and solid lines, respectively.

[‡] We might have anticipated this result by noting that, for the load P specified as acting on the structure, member AB is a 'two-force member'; it can therefore only carry an axial load and hence can have no shear component. It follows necessarily that $R_{x\text{A}} = 0$.

[§] We emphasise here that since the rotation $|\delta\theta| \ll 1$ is an infinitesimal, the resulting displacement (in this case point C) is *perpendicular* to the (original) radial line from the centre of rotation (here, point B).

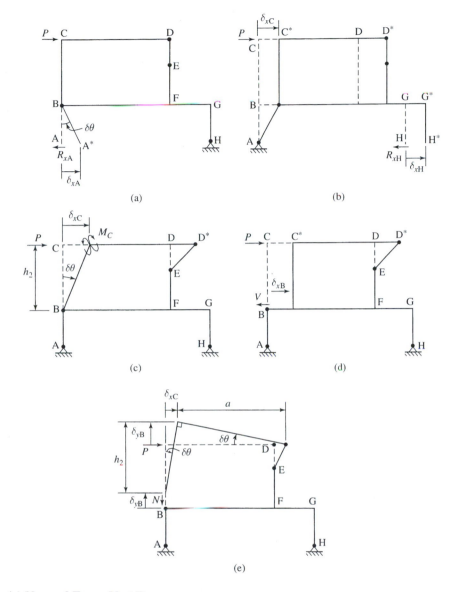

(a)   (b)   (c)   (d)   (e)

Figure 14.7.17

(e) Normal Force N at B

We give the structure a virtual displacement δ_{yB} as shown in Fig. (14.7.17e) (noting that element CD rotates with respect to DE, and DE with respect to EF. This is permissible since hinges exist at D and E.) We denote the rotation of element BC by $\delta\theta$. However, since the connection at C is rigid, we require that the angle BCD remain a right angle. It follows that CD also rotates by $\delta\theta$ as shown. From geometry, $\delta\theta = (\delta_{yB})/a$. Hence the virtual displacement $\delta_{xC} = h_2\,\delta\theta = h_2(\delta_{yB})/a$. Therefore

$$\delta W_{\text{ext}} = P\delta_{xC} - N\delta_{yB} = \left(\frac{h_2 P}{a} - N\right)\delta_{yB} = 0,$$

from which $N = \frac{h_2}{a}P$.   □

The principle of virtual work can also be applied to the analysis of trusses. In general, the axial forces existing in a (statically determinate) hinged truss subjected to specified loads can be obtained from the equilibrium equations using, for example, equations of equilibrium at each joint or the 'method of sections'. We note that when using either method, one must usually solve simultaneous equations for two or more unknowns. However, since, as we have seen, it is always possible to write external virtual work expressions for statically determinate structures in which only one unknown appears, the principle of virtual work permits finding the axial force in any particular member of such a truss, thus eliminating the need to calculate the forces in other members. That is, we need not make an analysis of the entire truss if it is necessary to determinate the force only in a specified member. We demonstrate this feature below for the statically determinate truss structure of Example 14.5.

Example 14.11:   The truss shown in Fig. (14.7.18) is subjected to a downward force P at point b and a horizontal force Q at point E. Determine the axial force existing in member aB.

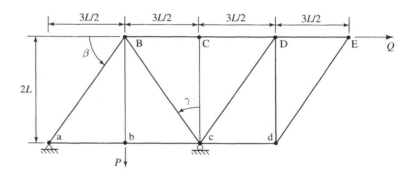

Figure 14.7.18

Solution: We imagine that we 'cut' members aB and ab (just to the left of element bB). The truss may then be considered to consist of two rigid segments: one to the left of bB and the other to the right (and including bB). We now give the right-hand truss segment a rigid-body rotation $\delta\theta$ about c, as shown in Fig. (14.7.19a). Due to this virtual rotation, the horizontal virtual displacement of point E is $\delta_{xE} = 2L\delta\theta$ so that the virtual work of the force Q is $2LQ\,\delta\theta$. Similarly, since the vertical virtual displacement component of b is given by $\delta_{yb} = (3L/2)\cdot\delta\theta$, the virtual work of the force P is $(-3PL/2)\cdot\delta\theta$.

We now consider the virtual work of the force F_{aB}. We first note that $L_{Bc} = 2.5L$. Furthermore, let β be the angle of inclination of aB with respect to the x-axis, and let γ be the angle of inclination of Bc with respect to the y-axis. Note that

$$\beta = \cos^{-1}(0.6) = \sin^{-1}(0.8), \qquad \gamma = \cos^{-1}(0.8) = \sin^{-1}(0.6).$$

Due to the virtual rotation $\delta\theta$, the virtual displacement $\delta(\overline{BB}^*) = L_{Bc}\delta\theta = 2.5L\cdot\delta\theta$ where we note that $\delta(\overline{BB}^*)$ is perpendicular to the line L_{Bc}. To obtain the virtual work of the force F_{aB}, we require the component of the virtual displacement, δ_{aB}, in the (original) direction of member aB [BG as shown in Fig. (14.7.19b)].[†] From geometry,

$$\delta_{aB} = (2.5L\delta\theta)\,[\cos(\beta - \gamma)] = (2.5L\delta\theta)\,[\cos\beta\cos\gamma + \sin\beta\sin\gamma]$$

and using the values given above,

$$\delta_{aB} = (2.5L)\,[2(0.6)(0.8)] = 2.5L\cdot 0.96\delta\theta = 2.4L\delta\theta\,,$$

[†] Note that the direction of the force F_{aB} is considered to be constant; i.e., it acts in the *original* direction of member aB (see footnote, p. 560).

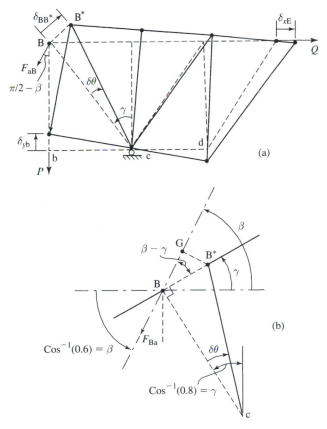

Figure 14.7.19

the virtual work of the force is $\delta W = F_{aB} \cdot \delta_{aB} = -2.4 F_{aB} L\delta\theta$. Setting the external virtual work of all forces acting on the truss segment to zero,

$$\delta W_{ext} = \left[-2.4L \cdot F_{aB} - \frac{3L}{2} \cdot P + 2L \cdot Q \right] \delta\theta = 0,$$

we obtain

$$F_{aB} = -\frac{1.5}{2.4}P + \frac{2.0}{2.4}Q = -0.625P + 0.833Q,$$

which agrees with the results as shown in Fig. (14.5.5). □

We observe that in using the principle of virtual work, it is necessary to first obtain the geometric relations for the given virtual displacement and rotations. The complexity of the problem thus depends on its geometry. Indeed, the major effort in using the principle for this class of problems consists of obtaining these geometric relationships.

(ii) Applications of the principle of virtual work to statically indeterminate structures

In the application of the principle of virtual work for statically determinate structures, it was emphasised that the choice of an appropriate virtual displacement led to a mechanism where *only a single desired unknown* appeared in the virtual work expression; consequently, it was possible to immediately determine the unknown force. In treating statically indeterminate structures, it is again possible to transform the structure into a 'mechanism'. However, for such structures, more than a single unknown force will always appear in the virtual work expression. As a result, one can never solve for the unknowns explicitly. Instead, as we shall see in the following examples, we are only able to obtain *relations between the various unknown quantities* existing in the structure.

Example 14.12: For the statically indeterminate structure shown [Fig. (14.7.20a)], obtain (a) a relation between M_A and R_C, (b) a relation between M_A, M_B and V_A.

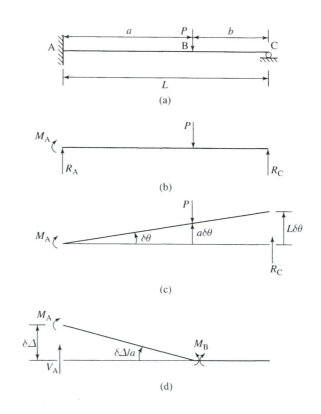

Figure 14.7.20

Solution: We let the unknown forces and moments be as shown in Fig. (14.7.20b).

(a) Choose a virtual displacement $\delta\theta$ as shown in Fig. (14.7.20c). Then

$$\delta W_{\text{ext}} = R_C L\delta\theta - M_A\delta\theta - Pa\delta\theta.$$

Applying the principle, $M_A\delta\theta = R_C L\delta\theta - Pa\delta\theta$, and hence

$$M_A = R_C L - Pa. \tag{14.7.40a}$$

(b) Choose a virtual displacement $\delta\Delta$ as shown in Fig. (14.7.20d). Then

$$\delta W_{\text{ext}} = V_A\delta\Delta + M_A(\delta\Delta/a) - M_B(\delta\Delta/a).$$

Hence, since $\delta W_{\text{ext}} = 0$,

$$M_B = M_A + V_A a. \tag{14.7.40b}$$

$\square$

We observe that both of Eqs. (14.7.40) could easily be obtained from an appropriate free-body diagram and the equations of equilibrium $(\sum M)_A = 0$ and $(\sum M)_B = 0$, respectively. We note too that, in this present example, in choosing rigid-body virtual displacements more than a single unknown always appears in the virtual work expression. This corroborates with the feature, indeed the definition, of a statically indeterminate structure; namely there exist more unknowns than independent equations of equilibrium.

Example 14.13: The statically indeterminate frame structure shown in Fig. (14.7.21a) is subjected to a horizontal load P. Determine a relation between M_A at A, the axial force N and the shear force V in member BC at point D.

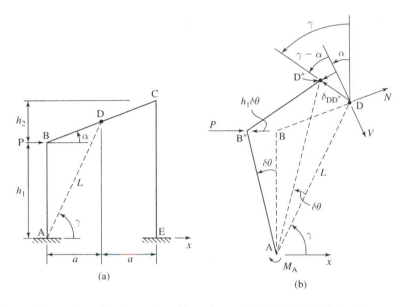

(a)

(b)

Figure 14.7.21

Solution: We apply a virtual rotation $\delta\theta$ as shown in Fig. (14.7.21b). Let L represent the length of line AD, and let the angles of inclination of lines BC and AD (with respect to the x-axis) be α and γ, respectively.

Hence for the given virtual rotation, the virtual displacement $\delta(DD^*) = L\delta\theta$. Note that since $|\delta\theta| \ll 1$, $\delta(DD^*)$ is perpendicular to the line L.[†] From simple geometry, applying the principle of virtual work,

$$\delta W_{\text{ext}} = -M_A\delta\theta - Ph_1\delta\theta - N[L\delta\theta \cdot \sin(\gamma - \alpha)] - V[L\delta\theta \cdot \cos(\gamma - \alpha)] = 0$$

or

$$\frac{M_A}{L} + N\sin(\gamma - \alpha) + V\cos(\gamma - \alpha) = -\frac{Ph_1}{L},$$

where $L = [a^2 + (h_1 + a\tan\alpha)^2]^{1/2}$.   □

We note that, as in Example 14.12, it is *not* possible to solve explicitly for unknown forces in a statically indeterminate structure; here again, the principle of virtual work can provide us only with *relations* between the unknowns. However, this in itself can

[†] See footnote, p. 582.

prove to be quite useful. Indeed, we should not expect the principle to yield explicit solutions for a statically indeterminate structure since the principle is merely an alternative means of stating equilibrium conditions.

(f) Influence lines for reactions, shears and moments in beams by the principle of virtual work

(i) Influence lines for statically indeterminate structures

We recall from Chapter 9 (Section 10) that the expression for the moment M (or shear V) at any arbitrary point x in a beam due to a unit load $P = 1$ acting at a variable point $x = \zeta$ is called the **fundamental solution** for the moment (or shear) in the beam. As was seen in Chapter 9, these quantities are expressed by means of Green's functions (or, alternatively called **influence functions**), which we denote by $G_M(x, \zeta)$ and $G_V(x, \zeta)$, respectively. Having determined the appropriate Green's function, we showed, using linear superposition, that the resulting moment and shear for *any given distributed load* $q(\zeta)$ acting on the beam over, say, a length $0 < x < L$ is given by [see Eqs. (9.11.18) and (9.11.20a)]

$$M(x) = \int_0^L q(\zeta) G_M(x, \zeta) \, d\zeta \tag{14.7.41a}$$

and

$$V(x) = \int_0^L q(\zeta) G_V(x, \zeta) \, d\zeta. \tag{14.7.41b}$$

In particular, if we wish to know these quantities at a specific point $x = x_0$, we have

$$M(x_0) = \int_0^L q(\zeta) G_M(x_0, \zeta) \, d\zeta \tag{14.7.42a}$$

and

$$V(x_0) = \int_0^L q(\zeta) G_V(x_0, \zeta) \, d\zeta. \tag{14.7.42b}$$

Similarly, using the principle of superposition, the reaction R existing at a given point x_0 is

$$R(x_0) = \int_0^L q(\zeta) G_R(x_0, \zeta) \, d\zeta. \tag{14.7.42c}$$

Note that, using the standard notation given in Chapter 9, the function $G(x_0, \zeta)$ represents the required quantity (e.g., M, V or R) at $x = x_0$ due to the unit load $P = 1$ acting at the variable point $x = \zeta$ [Fig. (14.7.22)].

In Chapter 9, influence functions were determined for statically determinate beams by means of the equations of equilibrium. While, for a simply supported beam or one with cantilever supports, such calculations are rather simple, these calculations can become rather cumbersome for a beam having more complex supports. As we shall now show, the principle of virtual work simplifies considerably

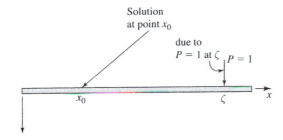

Solution
at point x_0

due to
$P = 1$ at ζ $P = 1$

x_0 ζ x

Figure 14.7.22

the calculation for these functions. In fact, the principle permits us to obtain directly a *plot* of G as a function of the position of the unit load acting at ζ. Such a plot is called an **influence line**. Thus the influence line for G of any given quantity (say, M, V or R) at a specified point is a graph that shows the variation of the solution as a function of the position ζ of an applied unit load.

To illustrate the application of the principle, let us reconsider the structure discussed in Example 14.9 and shown again in Fig. (14.7.23a). Assume that we wish to find the influence function $G_M(2L, \zeta)$ for the moment M_C, that is, M at $x = 2L$ due to a load $P = 1$ acting at some variable point $x = \zeta$. As discussed in the example, for a given load P applied at $x = \zeta$, we chose virtual displacements that cause a discontinuity $\delta\theta$ at point C and thus obtained a mechanism as shown in Fig. (14.7.14c) and repeated here as Fig. (14.7.23b). The resulting moment $M(x = 2L)$ was found to be [see Eqs. (14.7.38)]

$$M(2L, \zeta) = -P\zeta, \quad \zeta \leq L, \tag{14.7.43a}$$

$$M(2L, \zeta) = -P(2L - \zeta), \quad L \leq \zeta \leq 2L, \tag{14.7.43b}$$

$$M(2L, \zeta) = 0, \quad 2L \leq \zeta, \tag{14.7.43c}$$

whose plot, given in Fig. (14.7.15c), is repeated in Fig. (14.7.23c). We note that this figure has the identical shape as the chosen mechanism of Fig. (14.7.23b), i.e., the

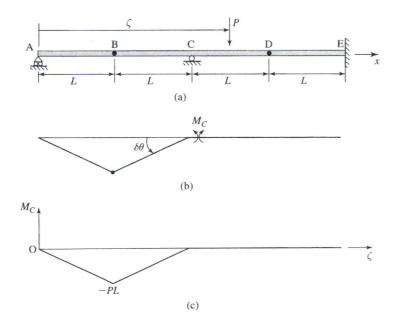

(a)

(b)

(c)

Figure 14.7.23

plot of $M(2L, \zeta)$ is the same shape as that of the mechanism. Moreover, we note that if we set $P = 1$ and *artificially* define $\delta\theta$ to be $\delta\theta = 1$,[†] the ordinates of the virtually deformed structure (i.e. the mechanism) are given by

$$M(2L, \zeta) = -\zeta, \quad \zeta \leq L, \tag{14.7.44a}$$

$$M(2L, \zeta) = -(2L - \zeta), \quad L \leq \zeta \leq 2L, \tag{14.7.44b}$$

$$M(2L, \zeta) = 0, \quad 2L \leq \zeta \tag{14.7.44c}$$

Now, by definition, this is precisely the required Green's function $G_M(2L, \zeta)$.[‡] We therefore observe that by choosing a virtual displacement (which corresponds to the desired force quantity [here, $M(2L)$]) equal to unity, the shape of the virtually deformed structure yields the influence line *directly*. That is, if the structure is given a virtual displacement such that the virtual displacement of the desired force quantity is unity, the influence line of the desired quantity coincides identically with the virtually deformed structure. Hence, letting $\delta\theta = 1$, the influence line for $G_M(2L, \zeta)$ is as shown in Fig. (14.7.24a). Thus the principle of virtual work provides us with a powerful means of immediately drawing the influence line without having to derive an analytic expression for the influence function. We summarise here the practical results as follows:

> The influence line for a specific desired reactive force quantity (reaction, shear force or moment) of a statically determinate structure is obtained by first violating the associated geometric constraint such that the 'released structure' is transformed into a mechanism. Upon giving the desired force quantity a virtual displacement of unity, the influence line corresponds identically to the displaced mechanism.[§]

The influence lines for other quantities of the structure of Example 14.9, viz. R_A, R_C, M_E, drawn immediately using this procedure, are presented in Figs. (14.7.24b–d). [Note that these graphs are the same as those of Figs. (14.7.15a, b and d), respectively, when $P = 1$].

To further illustrate the application of the principle, we draw directly the influence line for the shear V existing immediately to the right of point C (which we denote as V_C^+) of the same beam, as shown in Fig. (14.7.23a). Here we require that V_C^+ be the only internal resultant that does virtual work. We therefore choose a virtual deformation containing a virtual discontinuity δv at $x = 2L^+$ while, at the same time, the virtual displacement of the structure must be such that no discontinuity appears in the slope at point C (since we wish M_C to do no virtual work). The resulting mechanism is as shown in Fig. (14.7.25a). Upon setting $\delta v = 1$, we obtain the influence line for $G_V(2L^+, \zeta)$ of Fig. (14.7.25b), where we show all critical ordinates.

In the following example, we obtain the influence line for a reaction and demonstrate a particularly useful application to structural mechanics.

[†]　We recall that the principle of virtual work, as developed, is valid only for small rotations. Here, we merely *assign* a value $\delta\theta = 1$.

[‡]　Note that this Green's function was found in Chapter 9, using the equations of equilibrium for this structure [see Eq. (9.11.24)].

[§]　It is worthwhile to also point out that the influence lines for reactions, moments and shear forces existing in *statically determinate structures* always consist of straight lines. For statically indeterminate structures, this is not the case.

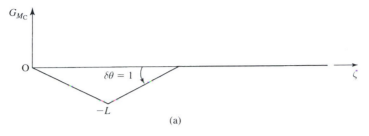

(a)

(b)

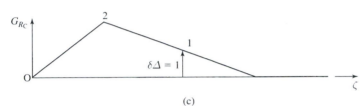

(c)

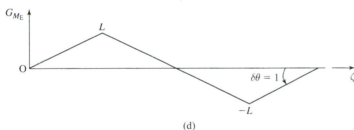

(d)

Figure 14.7.24

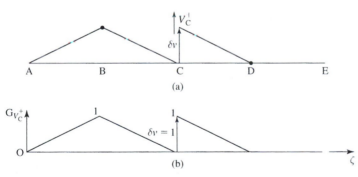

(a)

(b)

Figure 14.7.25

Example 14.14:

(a) Draw the influence line for the upward vertical reaction R at point H ($x_0 = 7L$) of the structure shown in Fig. (14.7.26a).

(b) A piece of machinery, represented by a uniform load w (N/m) of length $2L$ is to be placed on this structure.
   (i) Determine R if the load is placed between C and E.
   (ii) Where must the load be placed for the reaction $|R|$ to have a minimum value? What will this value be?

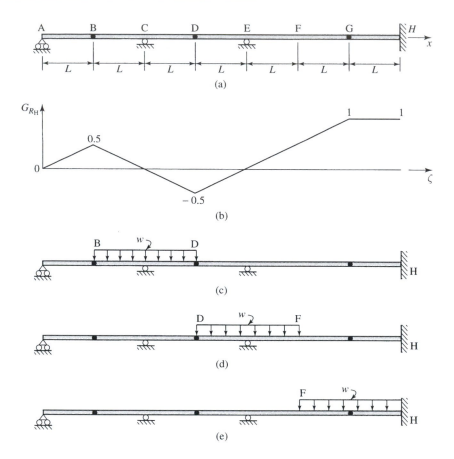

Figure 14.7.26

(iii) Where must the load be placed for the reaction $|R|$ to have a maximum value? What will this value be?

Solution: (a) We give the structure a virtual displacement $\delta v = 1$ at H and obtain the mechanism, and hence the influence line for R as shown in Fig. (14.7.26b). [Note that since hinges exist at points B, D and G, i.e., $M_B = M_D = M_G = 0$ (as opposed to the previous example), slope discontinuities at these points are permissible since no virtual work is done by the zero internal moments.]

(b) (i) Using Eq. (14.7.42c), with $q(\zeta) = w$, R is given by

$$R = w \int_C^E G_R(7L, \zeta)\, d\zeta . \tag{14.7.45a}$$

Noting that the above integral represents the area under G_R between points C and E, we find

$$R = w\left[\frac{1}{2}(2L)(-0.5)\right] = -0.5wL \text{ (down)} . \tag{14.7.45b}$$

(ii) To determine the placement of the uniform load (acting over a span $2L$) required to yield a minimum value of $|R|$, we first note that the reaction is given by the integral expression

$$R = w \int_a^b G_R(7L, \zeta) \, d\zeta, \qquad (14.7.46)$$

where the upper and lower limits must be such that $b - a = 2L$. We now observe that the total area under the influence line for G_R between points B and D (or between D and F) cancel out algebraically; hence we conclude that if the load is placed either between B and D (or between D and F), $|R|$ will have a minimum value, $R = 0$ [see Figs. (14.7.26c and d)].

(iii) To determine the placement of the uniform load (acting over a span $2L$) required to yield a maximum value of $|R|$, we again note that the reaction is given by the integral expression of Eq. (14.7.46). By inspection, we observe that the greatest area under the influence line (over a span $2L$) lies between F and H and that this area has a value $1.75L$. Therefore, $R = 1.75wL$ is the greatest possible reaction for the loading and will occur if the load is placed between F and H, as shown in Fig. (14.7.26e).   □

We point out that for $q(\zeta)$ not constant, the required quantities $R(x_0)$, $M(x_0)$ and $V(x_0)$ at any point x_0 are found from Eq. (14.7.42). The analytic expressions for the Green's functions can then be readily obtained from the analytic equations of the influence lines.

(ii) Influence lines for statically indeterminate linearly elastic structures: the Müller–Breslau principle

The principle of virtual work can also be applied to obtain influence lines for reactions, shear forces and moments of statically indeterminate structures. However, application of the principle is not as direct as for the case of determinate structures. We develop the method by means of the following example.

Consider a beam AB fixed at point A, which is simply supported at B, as shown in Fig. (14.7.27a). We wish to obtain the Green's function $G_R(L, \zeta)$ and more specifically, the influence line for the reaction R at B due to a unit downward force $P = 1$ acting at $x = \zeta$. Hence we remove the constraint against the y-displacement of point B and give the released beam (here, a cantilever beam) a virtual displacement such that R and P do external virtual work; specifically, we choose virtual displacements

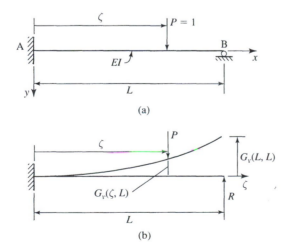

Figure 14.7.27

that correspond (and, in fact, are identical) to the elastic curve of the released beam due to a unit upward force, $R = 1$, at the free end [Fig. (14.7.27b)], namely $G_v(\zeta, L)$. Thus, the virtual displacement δv at any point ζ is $\delta v(\zeta) = G_v(\zeta, L)$. The external virtual work is therefore

$$\delta W_{\text{ext}} = R\, G_v(L, L) - P\, G_v(\zeta, L), \tag{14.7.47}$$

where, from Eq. (9.7.15c) (upon noting that here we have assumed displacements to be upward, i.e. in the negative y-direction),

$$G_v(\zeta, L) = -\frac{1}{6EI}[3L\zeta^2 - \zeta^3], \quad \zeta \leq L; \tag{14.7.48a}$$

in particular,

$$G_v(L, L) = -\frac{L^3}{3EI}. \tag{14.7.48b}$$

Considering the *linear elastic* beam to be subjected to two forces P and R, from Castigliano's second theorem, the displacement at point B is given by

$$\Delta_{\text{B}} = \frac{\partial U}{\partial R}. \tag{14.7.49}$$

Since $\Delta_{\text{B}} = 0$ is the required compatibility condition for the given beam of Fig. (14.7.27a), $\frac{\partial U}{\partial R} = 0$ and therefore

$$\delta U = \frac{\partial U}{\partial R}\, \delta R = 0. \tag{14.7.50}$$

Now, for an elastic body, $\delta W_{\text{int}} = \delta U$, where δU is the variation of strain energy. (See remark 7 of Subsection c(ii) of this section, p. 567). It follows that $\delta W_{\text{int}} = 0.$[†] Applying the principle of virtual work, we then have $\delta W_{\text{ext}} = 0$.

Upon setting Eq. (14.7.47) equal to zero, from the basic definition of the Green's function for the reaction R, $G_R(L, \zeta) \equiv R|_{P=1}$, we obtain for $P = 1$[‡]

$$G_R(L, \zeta) = \frac{G_v(\zeta, L)}{G_v(L, L)} \tag{14.7.51}$$

[†] We note that δU appearing in Eq. (14.7.50) represents the change in strain energy due to a variation of a force while δW_{int} is the internal virtual work due to a variation of displacements (i.e., virtual displacements). However, from Section 14.6 [see Eq. (14.6.9b)], we recall that for a *linear* elastic body,

$$\frac{dU(\Delta)}{d\Delta} = \frac{dU(R)}{dR} \cdot \frac{dR}{d\Delta},$$

where Δ denotes the displacement. Therefore, if $\frac{dU}{dR} = 0$, it follows that $\frac{dU(\Delta)}{d\Delta} = 0$ and hence $\delta U = (\frac{dU}{d\Delta}) \cdot \delta \Delta = 0$. This last expression represents the change in strain energy due to a variation of displacement (i.e., due to a virtual displacement). Thus, since $\delta U = 0$, Eq. (14.7.50) leads, for this linear case, to $\delta W_{\text{int}} = 0$.

[‡] An alternative derivation for the Green's function $G_R(L, \zeta)$ is as follows: For the given beam of Fig. (14.7.27a), we have, using the principle of superposition,

$$\Delta_{\text{B}} = PG_v(L, \zeta) - RG_v(L, L). \tag{a}$$

But, from the condition of geometric compatibility, $\Delta_{\text{B}} = 0$. Hence

$$R = P\frac{G_v(L, \zeta)}{G_v(L, L)}. \tag{b}$$

However, by Maxwell's reciprocal theorem,

$$G_v(L, \zeta) = G_v(\zeta, L). \tag{c}$$

where $G_v(\zeta, L)$ is the upward deflection at any point ζ of the beam due to an upward unit force acting at point B. Substituting Eq. (c) in Eq. (b) and using the definition of the Green's function $G_R(L, \zeta) \equiv R|_{P=1}$, we recover Eq. (14.7.51).

or, using Eqs. (14.7.48), we have explicitly

$$G_R(L, \zeta) = \frac{3L\zeta^2 - \zeta^3}{2L^3} = \frac{\zeta^2}{2L^2}(3 - \zeta/L). \qquad (14.7.52)$$

A plot of $G_R(L, \zeta)$ is shown in Fig. (14.7.28). We observe from the figure as well as from Eq. (14.7.52a) that $G_R(L, L) = 1$; i.e., the value of G_R at the point of application of R, $\zeta = L$, is equal to unity.

Figure 14.7.28

In the above development, $G_v(L, \zeta) = G_v(\zeta, L)$ were known functions as they were originally found in Chapter 9; consequently, the explicit expression for $G_R(L, \zeta)$, Eq. (14.7.52), was readily determined. Let us instead assume momentarily that $G_v(\zeta, L)$ and $G_v(L, L)$ are unknown. We now show that application of the principle of virtual work nevertheless provides us with an immediate *qualitative description of the shape* of the desired function $G_R(L, \zeta)$. Noting that $G_v(L, L)$ is a constant, it follows from Eq. (14.7.51) that $G_R(L, \zeta)$ and $G_v(\zeta, L)$ have the same functional relation in ζ. Hence, the influence line for $G_R(L, \zeta)$ must have the same shape as the structure deformed by application of a force at B. This provides us with a means to estimate *qualitatively* the shape of the influence line since, most often, the deformed shape of a structure to a unit concentrated force is intuitively known. [For example, for the problem at hand, it required but little experience to assume that the structure when subjected to a force at B deforms as in Fig. (14.7.27b).] Thus, for the case considered here, we observe that the influence line for $G_R(L, \zeta)$ [Fig. (14.7.28)] has indeed the same scaled shape as the deflected beam shown in Fig. (14.7.27b).

As a further example, assume we wish to determine qualitatively the influence line represented by $G_M(0, \zeta)$, for the moment at A ($x = 0$) of the given beam, [Fig. (14.7.27)], due to a downward unit load acting at any point, $x = \zeta$. We therefore release the constraint against rotation at A and choose for virtual displacements δv, the elastic curve due to an applied moment $M_A = 1$ [Fig. (14.7.29)];[†] i.e., at all points ζ, we let $\delta v(\zeta) = G_v(\zeta, 0)$ represent the displacement due to a unit applied moment at A. Furthermore, letting $\delta\theta = G_\theta(\zeta, 0)$ represent the (virtual) slope at ζ due to $M_A = 1$ such that at A $\delta\theta_A = G_\theta(0, 0)$, the external virtual work is

$$\delta W_{\text{ext}} = M_A G_\theta(0, 0) + P G_v(\zeta, 0). \qquad (14.7.53)$$

Figure 14.7.29

From geometric compatibility, we require that $\theta_A = 0$ for the unreleased beam and hence by Castigliano's second theorem, we have $\frac{dU}{dM_A} = 0$. As in the previous case

---

[†] Note that by releasing the constraint against rotation at A, the fixed-end condition becomes a simple support, i.e., the 'released structure' is a simply supported beam.

for $G_R(L, \zeta)$, this leads to $\delta U = \delta W_{\text{int}} = 0$. Then, by the principle of virtual work, $\delta W_{\text{ext}} = 0$ and we obtain directly, for the case $P = 1$,[†]

$$G_{M_A} \equiv G_M(0, \zeta) = -\frac{G_v(\zeta, 0)}{G_\theta(0, 0)}. \qquad (14.7.54)$$

Since here $G_\theta(0, 0)$ is a constant, it is clear that the shape of the resulting influence line is again the same as that given by $G_v(\zeta, 0)$, as shown in Fig. (14.7.29). In other words, we may immediately sketch the shape of the influence line for G_{M_A} as the deformed structure due to the applied moment M_A. However, for this problem, we are again able not only to sketch the influence line but also to evaluate it at all points, ζ, since $G_v(\zeta, 0)$ and $G_\theta(0, 0)$ were determined effectively from the solution of Example 9.4 [see Eqs. (9.4.13c) and (9.4.14)]:

$$G_v(\zeta, 0) = \frac{1}{6EIL}[\zeta^3 - 3L\zeta^2 + 2L^2\zeta] \qquad (14.7.55a)$$

$$G_\theta(0, 0) = \frac{L}{3EI}. \qquad (14.7.55b)$$

Hence, from Eq. (14.7.54),

$$G_M(0, \zeta) = -\frac{1}{2L}[\zeta^3 - 3L\zeta^2 + 2L^2\zeta]. \qquad (14.7.56a)$$

Note that since $G_v'(\zeta, 0) \equiv G_\theta(\zeta, 0)$, from Eq. (14.7.54) [or more directly from Eq. (14.7.56a)], we find $G_M'(0, 0) = -1$; i.e., the absolute value of the *slope*, $|dG_M(0, \zeta)/d\zeta|$, of G_M at the point of application of M_A, $\zeta = 0$, is equal to unity. (Note that both the virtual rotation at A and the applied moment M_A are clock-wise.) Moreover, from Eq. (14.7.56a), we find

$$|G_M(0, \zeta)|_{\text{max}} = \frac{\sqrt{3}\, L}{9}. \qquad (14.7.56b)$$

A plot of $G_{M_A} \equiv G_M(0, \zeta)$ is shown in Fig. (14.7.30). Note that $G_M(0, \zeta)$ has precisely of the same shape as that appearing in Fig. (9.4.6c) for the deformed beam of Example 9.4 when subjected to an end moment at A.

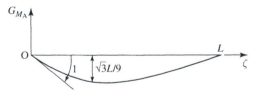

Figure 14.7.30

From the above cases, we may summarize the practical results as follows:

[†] Alternatively, as in the case for $G_R(L, \zeta)$, by superposition,

$$\theta_A = M_A\, G_\theta(0, 0) + P\, G_\theta(0, \zeta), \qquad (a)$$

where $G_\theta(0, \zeta)$ represents the slope at A due to a unit force acting at ζ. However, by Maxwell's reciprocal theorem,

$$G_v(\zeta, 0) = G_\theta(0, \zeta). \qquad (b)$$

Therefore

$$\theta_A = M_A\, G_\theta(0, 0) + P\, G_v(\zeta, 0). \qquad (c)$$

Since $\theta_A = 0$ for the given beam, we obtain, upon setting $P = 1$, G_{M_A} as given by Eq. (14.7.54).

The influence line for a specific reactive force quantity (reaction, shear force or moment) of an elastic indeterminate structure is obtained by first violating the associated geometric constraint. The structure is then given virtual displacements (corresponding to the elastic curve of the 'released structure') such that the virtual displacement associated with the desired force quantity is unity. The resulting shape of the influence line then corresponds identically to the scaled shape of the deformed released structure due to the reactive force quantity.[†]

The above is often referred to as the Müller–Breslau principle for statically indeterminate structures.

As illustrated in the previous example, we remark that violation of a constraint in a statically indeterminate structure does not lead to a 'mechanism' as in the case of statically determinate structures. Thus, influence lines for reactions, shears forces and moments of a statically indeterminate structures are not straight lines but instead are always curves.

Finally, it is worthwhile to mention that the method, as developed above, can be used to obtain the shape of influence functions for statically indeterminate structures having higher degrees of indeterminacy, i.e., for structures having more than one redundant, namely more than one 'extra reaction'.

We illustrate the method further by means of the following examples.

Example 14.15: Sketch the shape of the influence line for the Green's function of the structure shown in Fig. (14.7.31a) for R_C and M_B due to a unit load $P = 1$ acting at any point ζ.

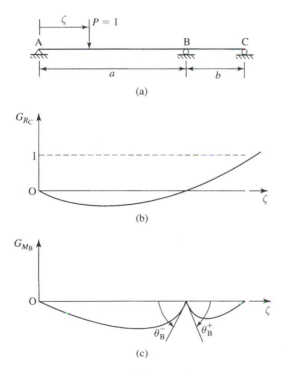

(a)

(b)

(c)

Figure 14.7.31

[†] Considering moments as '*generalised forces*', for influence lines for a given moment, the terms 'force' and 'displacement' must be replaced by 'moment' and 'slope', respectively.

Solution: We follow the procedure given above: for G_{R_C}, upon releasing the constraint against translation at C, we apply a unit virtual displacement at A; for G_{M_B}, we release the continuous slope constraint at B (which evidently results in a 'kink') and apply unit virtual rotations at this point. The resulting influence lines represented by the deformation of the released structure due to applied forces R_C and M_B are shown in Figs. (14.7.31b and c), respectively.

Note that although $G_{R_C}(\zeta = a + b) = 1$ and the slopes of G_{M_B} at B, $\theta_B^- = a/3EI$, $\theta_B^+ = b/3EI$ are known [see Eq. (14.7.55b)], the values of the influence lines at all points, ζ, have not been determined; i.e. only the general shapes are known. To obtain values of $G_{R_C}(\zeta)$ and $G_{M_B}(\zeta)$, an analysis of the statically indeterminate beam is required.

Example 14.16: Sketch the shape of the influence lines for R_A and M_C of the statically indeterminate beam of Fig. (14.7.32a). Note that this beam is statically indeterminate to the second degree.

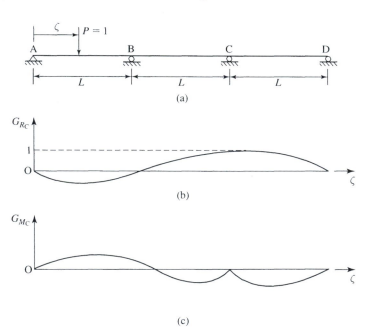

Figure 14.7.32

(c)

Solution: Following the above procedure, the shape of the influence lines, shown in Figs. (14.7.32b and c) respectively, are immediately obtained.

Example 14.17: Sketch the shape of the influence lines for R_A, M_C, R_C, M_E, V_C^-, V_C^+ and R_E for the statically indeterminate structure shown in Fig. (14.7.33). Note that this is the same structure as that of Fig. (14.7.23a), with the exception that no hinge exists at B; the present structure is therefore indeterminate to degree one.

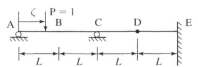

Figure 14.7.33

Solution: We apply the procedure as developed above. (For V_C^- and V_C^+, the shear immediately to the left and right of point C, we release the constraint against relative displacement in the vertical direction of the structure at $\zeta = 2L^-$ and $\zeta = 2L^+$, respectively.) The resulting influence lines are shown in Figs. (14.7.34a–g).[†]

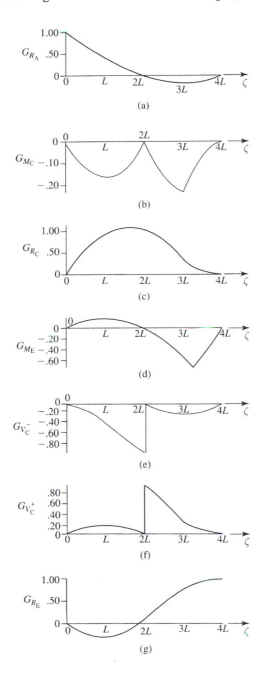

Figure 14.7.34

[†] Note that here the actual plots show values that were calculated using expressions of the form of Eqs. (14.7.51) and (14.7.54) together with an analysis of the beam as an indeterminate structure. Nevertheless, it should be quite evident that if only the *shapes* are required, they conform with the shape of the appropriately released deformed structure.

It is instructive to compare Fig. (14.7.34) with the influence lines of Figs. (14.7.24) and (14.7.25) as found for the corresponding statically determinate structure.

14.8   The principle of complementary virtual work

(a) Introduction

In this section, we develop the principle of complementary virtual work, which is the counterpart of the previously derived principle of virtual work. As we shall see, although the basic idea is quite different, the derivation follows essentially the same operative steps. Application of the principle of complementary virtual work leads to a simple but effective method for determining displacements of specific points in a body.

(b) Development and derivation of the principle
(i) Some preliminaries

Before proceeding with the derivation of the principle, we review some ideas and define some necessary terms. Let us therefore consider again a deformable body (not necessarily elastic) occupying a space V enclosed by a surface S that is in equilibrium. As we have seen, if external forces are acting on the body, these consist, in general, of (a) tractions T_n acting on the surface S of the body and (b) body forces B acting at points within the body V [Fig. (14.8.1a)].[†] As a result, since the body is deformable, a state of strain will exist throughout the body. However, here we shall consider the case where deformations may also be due to causes other than external forces, for example, temperature changes. For mathematical simplicity, we consider a two-dimensional plane case.

Thus, *irrespective of the cause of the deformation*, we may assume that since the body is deformable, all points undergo displacements; we denote these, for the two-dimensional case under consideration, as $u(x, y) = ui + vj$ and assume that $u(x, y)$ and its partial derivatives are continuous throughout V. Clearly, at

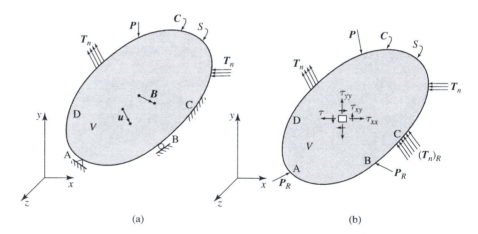

Figure 14.8.35   (a)   (b)

supports (i.e., at particular points or areas of the surface S, the displacements are

---

[†] As in Section 14.7, in referring to the tractions T_n acting on the surface S, we include not only the applied known tractions but also the resisting tractions; that is, we do not differentiate between applied and reactive tractions on S. In anthropomorphic terms, 'the body knows only that it is acted upon by forces on its surface'; it does not distinguish between applied and reactive forces. These reactive forces are labelled with a subscript R in Fig. (14.8.1b).

subject to constraints, i.e. geometric boundary conditions. (For example, at point A of Fig. (14.8.1a), the displacement components are $u = v = 0$.) Thus the real displacements are represented by a *kinematically admissible* displacement field, as defined previously under in Section 14.7b. We also mention here that the external force system, consisting of the applied as well as reactive forces (which are required to prevent violation of the constraints), is said to represent a *statically admissible force system.*[†]

We limit our discussion here to displacements such that all strains and rotations are small. It is clear that corresponding to any displacement field $u(x, y)$, a strain field $(\epsilon_{xx}, \epsilon_{yy}, \epsilon_{xy})$ exists, which is related to the displacements through Eqs. (14.7.8). As with the real displacements, the corresponding strain field is kinematically admissible throughout the body. We denote these strains symbolically by $\epsilon = \epsilon(x, y)$. Moreover, due to the external force system, or whatever the cause of the displacements, real (actual) stresses $(\tau_{xx}, \tau_{yy}, \tau_{xy})$ exist at each point. Since the body is assumed to be in equilibrium, the stress components (for a two-dimensional case) must satisfy the stress equations of equilibrium [Eqs. (14.7.3)]. However, while the actual strain field is of concern to us, we shall not be interested in these stresses, per se; in fact, they will be considered irrelevant in the development below since we shall never use them explicitly.[‡]

(ii) Derivation of the principle

Having established the above ideas, we proceed with the derivation as follows: Let us imagine that instead of actual external forces that may be acting on the body (i.e., the applied tractions, the reactive tractions T_n and the body forces B), we apply a set of imaginary equilibrium forces denoted by δT_n and δB and referred to as **virtual external forces** [Fig. (14.8.2)]. (Here, the symbol δ is used to indicate that these are not actual forces and that they are simply forces which we 'imagine' to be acting on the body.) We emphasise here that the external virtual forces must satisfy equilibrium conditions; otherwise they are *completely arbitrary*.

Now, due to the application of the virtual external force system, it is clear that stresses will be induced at all points in the body. We refer to these stresses by **virtual stresses** and denote them by $\delta\tau_{xx}, \delta\tau_{yy}, \delta\tau_{xy}$. These stresses, which we denote symbolically by $\delta\tau(x, y)$, are then referred to as defining a **virtual stress field** in equilibrium. We note that since we have imposed no other conditions, aside from equilibrium conditions on the virtual external force system, the same is true of the resulting virtual stress field $\delta\tau$. Thus the stress field $\delta\tau$ is completely arbitrary and need satisfy only the equilibrium equations; namely

$$\frac{\partial(\delta\tau_{xx})}{\partial x} + \frac{\partial(\delta\tau_{yx})}{\partial y} + \delta B_x = 0 \qquad (14.8.1a)$$

$$\frac{\partial(\delta\tau_{xy})}{\partial x} + \frac{\partial(\delta\tau_{yy})}{\partial y} + \delta B_y = 0. \qquad (14.8.1b)$$

The important feature of Eqs. (14.8.1) is that they represent, at all points in the body,

[†]   In our study of mechanics of deformed bodies as discussed previously throughout this book, all external force systems in equilibrium have, in fact, been statically admissible since they satisfy *both* the equilibrium equations and the support conditions. Hence it has not been necessary to use the term 'statically admissible' explicitly. We do so here because we shall have need to discuss systems that are *not* statically admissible.

[‡]   Note that, while real stresses were considered in the development of the principle of virtual work, the real strain field was considered to be totally irrelevant (see p. 561).

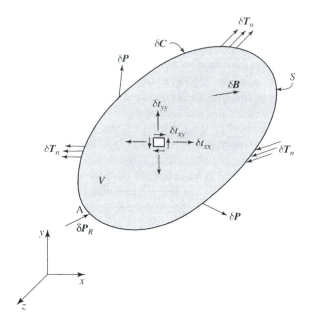

Figure 14.8.1

equations for *any arbitrary* equilibrium stress state $\delta\tau(x, y)$, which is compatible with the δT_n–δB external force system.

Now let us multiply Eqs. (14.8.1a) and (14.8.1b), respectively, by the *actual* displacements u and v, which occur in the body, and add the resulting expressions; we then obtain

$$\left[\frac{\partial(\delta\tau_{xx})}{\partial x} + \frac{\partial(\delta\tau_{yx})}{\partial y} + \delta B_x\right]u + \left[\frac{\partial(\delta\tau_{xy})}{\partial x} + \frac{\partial(\delta\tau_{yy})}{\partial y} + \delta B_y\right]v = 0. \quad (14.8.2)$$

The analogy with Eq. (14.7.4) is immediately observed. However, there is a significant difference, for u and v here are actual displacements while in Eq. (14.7.4) the displacements δu and δv were arbitrary. It is emphasised again that when referring to the 'actual displacements', they may be due to any cause; for example, an external force system T_n–B or temperature changes, etc. In this sense, whatever the cause of the actual displacements, Eq. (14.8.2) remains valid provided the virtual stress field $\delta\tau$ is in equilibrium.

Although the expressions (14.7.4) and (14.8.2) are formally similar, they clearly are not identical; the difference is that the δ now appears with the stress terms (i.e., the $\delta\tau$ stress components are associated with the δT_n–δB external force system), while in Eq. (14.7.4), the δ was associated with an arbitrary variation of the displacements.

Now, having assumed that u and v satisfy the conditions that all partial derivatives are small such that the strains are expressed by means of Eqs. (14.7.8), then, upon performing the same subsequent operations as in Eq. (14.7.6), Eqs. (14.7.7) and Eqs. (14.7.10)–(14.7.12) (namely expressing strains in terms of displacements and using the divergence theorem), we are lead to

$$\iiint\limits_V [\delta\tau_{xx}\epsilon_{xx} + 2\delta\tau_{xy}\epsilon_{xy} + \delta\tau_{yy}\epsilon_{yy}]\,d\Omega = \iint\limits_S \delta T_n \cdot u\,ds + \iiint\limits_V \delta B \cdot u\,d\Omega.$$

$$(14.8.3a)$$

Finally, we note that in obtaining the right-hand side of Eq. (14.8.3a), the divergence theorem is used under the assumption that all variations of the traction are continuous on S. If discontinuous traction variations, such as n concentrated forces δP_j or m couples δC_k exist on the surface, then these terms must be added and we rewrite Eq. (14.8.3a) as

$$\iiint\limits_V [\delta\tau_{xx}\epsilon_{xx} + 2\delta\tau_{xy}\epsilon_{xy} + \delta\tau_{yy}\epsilon_{yy}]\, d\Omega$$

$$= \iint\limits_S \delta T_n \cdot u \, ds + \iiint\limits_V \delta B \cdot u \, d\Omega + \sum_{j=1}^{n} \delta P_j \cdot u_j + \sum_{k=1}^{m} \delta C_k \cdot \theta_k, \quad (14.8.3b)$$

where u_j and θ_k, respectively, are the actual displacement and rotation at the points at which δP_j and δC_k act.

Analogous with the definition of virtual work, the left-hand side of Eqs. (14.8.3), denoted by δW_{int}^*, is called the **internal complementary virtual work**:

$$\delta W_{int}^* = \iiint\limits_V [\delta\tau_{xx}\epsilon_{xx} + 2\delta\tau_{xy}\epsilon_{xy} + \delta\tau_{yy}\epsilon_{yy}]\, d\Omega. \quad (14.8.4)$$

Similarly, the right-hand side of the equations represents **external complementary virtual work**, δW_{ext}^*. Thus we have arrived at the principle of complementary virtual work, namely

$$\delta W_{int}^* = \delta W_{ext}^*. \quad (14.8.5)$$

The *principle of complementary virtual work* may therefore be stated as follows:

Let a body subjected to arbitrary virtual external forces produce an equilibrium state of virtual stress at all points. Then, the internal complementary virtual work of the stresses (acting through the real strains in the body) is equal to the complementary external virtual work of the arbitrary external forces (acting through the actual displacements).

Writing the arbitrary stresses and real strains, symbolically, as the tensors $\delta\tau$ and ϵ, we may rewrite Eq. (14.8.3a) as

$$\iiint\limits_V \delta\tau \cdot \epsilon \, d\Omega = \iint\limits_S \delta T_n \cdot u \, ds + \iiint\limits_V \delta B \cdot u \, d\Omega,$$
$$(14.8.6)$$

where, as indicated by the corresponding arrows, the (arbitrary) equilibrium stress state $\delta\tau$ at each point is due to the virtual external tractions and body forces and the actual strains ϵ are compatible with the actual displacements. [Note that here we have, for convenience, again omitted the concentrated forces and couples in Eq. (14.8.6).]

(iii) Comments on mechanical boundary conditions, static admissibility and inadmissibility

In applying the principle of complementary virtual work, it is important to understand the meaning of static admissibility and inadmissibility of an external force system and the corresponding internal stress fields within a body. We therefore dwell upon this concept here.

As we have previously noted, an external force system acting upon a body consists, in general, of known applied forces (or tractions), as well as the reactive forces (or tractions), which are necessary to prevent violation of the support conditions (or using a more general terminology, the geometric boundary conditions). If a body is in equilibrium, the external force system must necessarily satisfy the equations of equilibrium. If the reactive components of the force system are such that they prevent violation of the geometric boundary conditions, then the force system is said to satisfy the required 'mechanical boundary conditions'. Thus, an external force system that satisfies *both* the equilibrium conditions and the mechanical boundary conditions is said to be statically admissible. However, if there exists an external force system that is in equilibrium but whose forces do *not* prevent violation of the existing geometric boundary conditions, then we say that the force system does not satisfy the mechanical boundary conditions and therefore it is *statically inadmissible*.

Now, corresponding to any external force system in equilibrium there exists an equilibrium stress field that satisfies the stress equations of equilibrium at all points within the body. (We assume here that it is always possible to obtain such a field.) If the external force system is statically admissible or inadmissible, then the same will be true of the stress field.

In developing the principle of complementary virtual work, we have emphasised that both the virtual external forces system $\delta T_n - \delta B$ and the corresponding virtual stress field $\delta\tau$ within a body are completely arbitrary and need only satisfy the equations of equilibrium.[†] Using the above terminology, we therefore observe that the principle of complementary virtual work is valid for both statically admissible as well as statically inadmissible systems.

It is instructive to illustrate the above ideas by means of the following example and counter-example: we consider a cantilever beam subjected to a given (not virtual) load P, as shown in Fig. (14.8.3a). For simplicity, we assume a rectangular cross-section, having a cross-section area A, as shown.

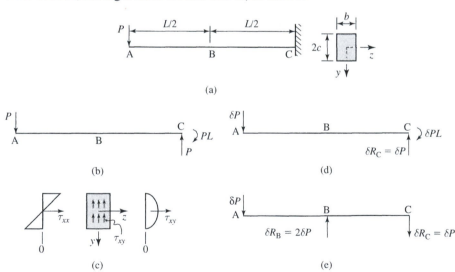

Figure 14.8.2

[†] Note that in deriving the principle, we required only the equilibrium equations [Eqs. (14.8.1)]; no reference was made to mechanical boundary conditions. Thus static admissibility was not a necessary assumption.

For equilibrium to exist, we require a reactive force $R_C = P$ and a clock-wise moment, $M_C = PL$ at C [Fig. (14.8.3b)]. The force R_C and moment M_C are thus those reactions that are required to satisfy the geometric boundary conditions at C; namely $\Delta = 0$, $\theta = 0$. If, for example, the beam is linearly elastic, using the expressions of Chapter 8, the actual stress field existing throughout the body will be such that at the boundary C [see Eqs. (8.6.8) and (8.8.6b), respectively],

$$\tau_{xx} = \frac{My}{I} = -\frac{PLy}{I}, \qquad \tau_{xy} = \frac{VQ}{Ib} = \frac{3P}{2A}\left(1 - \frac{y^2}{c^2}\right) \qquad (14.8.7a)$$

whose distributions over the cross-section are shown in Fig. (14.8.3c).

Since the resultant of the (real) stress field satisfies the mechanical boundary conditions at C, it is said to be statically admissible. Thus, by definition, a real stress field is *always* statically admissible.

Now, let us consider that we apply a virtual force δP at A. Clearly, the reactions δR_C and δM_C, as shown in Fig. (14.8.3d), satisfy equilibrium conditions. Moreover, due to this external virtual force system $(\delta P, \ \delta R_C, \ \delta M)$ the geometric boundary conditions at C ($\Delta = 0, \theta = 0$) are again satisfied. Thus, using the same reasoning as above, the virtual stress field due to this virtual external load system is also statically admissible.

Having shown an example of a statically admissible virtual stress field, we now illustrate the case of a virtual stress field that is *not* statically admissible.

To this end, let us again consider that a virtual force δP is acting on the body ABC, as shown in Fig. (14.8.3e), together with an upward force $\delta R_B = 2\delta P$ at B, a downward force $\delta R_C = \delta P$ acting at C (and with $\delta M_C = 0$). We recognize immediately that this virtual external force system $(\delta P, \delta R_B, \delta R_C)$ is, in fact, in equilibrium. However, the external virtual force system is not statically admissible since it does not satisfy the mechanical boundary conditions of the given structure. Note that the equilibrium virtual stress field due to this system is such that at the cross-section C, the stresses $\delta\tau_{xx}$ and $\delta\tau_{xy}$ are

$$\delta\tau_{xx} = 0, \qquad \delta\tau_{xy} = \frac{\delta VQ}{Ib} = -\frac{3\delta P}{2A}\left(1 - \frac{y^2}{c^2}\right). \qquad (14.8.7b)$$

Therefore, their resultants over the cross-section at C are not those required to satisfy the geometric boundary conditions; for example, they do not prevent rotation at C. Thus we have here a virtual stress field that does not satisfy the mechanical boundary conditions and hence is not statically admissible.

As we shall see later, virtual force systems that are not statically admissible will prove to be very useful in simplifying the application of the principle of complementary virtual work to structural problems.

(c) Comparison and analogues between the two principles

The two principles, the principle of virtual work and the principle of complementary virtual work, may now be compared. In both principles, there exists a common requirement: the body must be in equilibrium under a set of externally applied forces.

For convenience we repeat here the two principles; namely the statement of the principle of virtual work:

$$\iiint\limits_V \tau \cdot \delta\epsilon\, \mathrm{d}\Omega = \iint\limits_S T_n \cdot \delta u\, \mathrm{d}s + \iiint\limits_V B \cdot \delta u\, \mathrm{d}\Omega$$

and the statement of the principle of complementary virtual work:

$$\iiint\limits_V \delta\tau \cdot \epsilon\, \mathrm{d}\Omega = \iint\limits_S \delta T_n \cdot u\, \mathrm{d}s + \iiint\limits_V \delta B \cdot u\, \mathrm{d}\Omega.$$

In the principle of virtual work, the stress field is the *actual* equilibrium state due to an actual external force system; hence this actual stress state must satisfy the mechanical boundary conditions; i.e., the stress field is statically admissible. However, the virtual displacements are arbitrary; namely they must merely be geometrically possible (i.e. must be piece-wise continuous); they are *arbitrary and need not necessarily satisfy the geometric constraints* at the boundary. The principle of virtual work then states that the total internal work done by the stresses in 'riding' through these virtual strains is equal to the work done by the external forces in 'riding' through the corresponding virtual displacements if (and only if) the body is in equilibrium.

In the principle of complementary virtual work, we require *any arbitrary* stress state that is in equilibrium with arbitrary external forces. In analogy to the principle of virtual work, we may give a physical interpretation to the principle of complementary virtual work: if a body is in a state of equilibrium, the total internal complementary virtual work done by arbitrary equilibrium stresses in 'riding' through the *actual strains* is equal to the external complementary virtual work done by the arbitrary external forces in 'riding' through the *actual displacements* of the body.

Finally, using the terminology defined above, we restate the two principles concisely, side by side, and thus observe that they are, in fact, exact counterparts.

Virtual work	Complementary virtual work
A body is in a state of equilibrium if, and only if, the work done by a *statically admissible stress field* acting through the *arbitrary virtual strains* is equal to the work of the (real) external equilibrium forces acting through the arbitrary virtual displacements.	A body is in a state of equilibrium if, and only if, the work done by an *arbitrary stress field acting* through *kinematically admissible strains* is equal to the work of the arbitrary external equilibrium forces acting through the (real) displacements.

It should, of course, always be remembered that although we have repeatedly used the terms *arbitrary*, the real or varied stresses must always be associated with the real or varied external force system respectively, and the real or virtual strains must be compatible with the real or virtual displacements of the body, respectively.

(d) Expressions for internal complementary virtual work in terms of internal stress resultants: generalised forces and displacements

(i) General expressions

In deriving the principle of complementary virtual work for deformable bodies, the internal complementary virtual work was defined in terms of virtual stresses and (real) strains as the integral [Eq. (14.8.4)]

$$\delta W_{\text{int}}^* = \iiint\limits_V (\delta\tau_{xx}\epsilon_{xx} + \delta\tau_{yy}\epsilon_{yy} + 2\delta\tau_{xy}\epsilon_{xy})\, d\Omega.$$

We have previously observed that one often encounters structural components, such as rods, beams, bars, etc., which undergo specific types of deformations. (These components are often referred to as *one-dimensional elements* since the properties existing at their cross-sections are defined by means of a single coordinate, say x.) Although one can, in principle, always use Eq. (14.8.4), it is more convenient to express the internal complementary virtual work for these structural elements in terms of internal stress resultants and deformation patterns (which describe globally the deformation of the cross-sections) rather than in terms of stresses and strains. In this section, we derive such expressions by means of a physical approach.

Without loss of generality, we consider a typical structural element of a coplanar structure, whose longitudinal axis lies along the centroidal x-axis and which is subjected to axial and lateral loads as shown in Fig. (14.8.4).

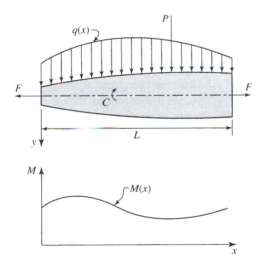

Figure 14.8.3

To establish the expression for complementary internal virtual work in terms of virtual stress resultants, let us first consider the various typical types of deformation that occur in one-dimensional structural elements due to such external loads. These include the following:

- Simple axial extension where all points in the cross-section undergo the same displacement in the x-direction, i.e. $\epsilon_{xx} = \epsilon_{xx}(x)$ [Fig. (14.8.5a)].
- Bending of a member under the assumption that plane sections remain plane. The strain at any point in the cross-section is then given by $\epsilon_{xx}(x, y) = y\kappa(x)$, where

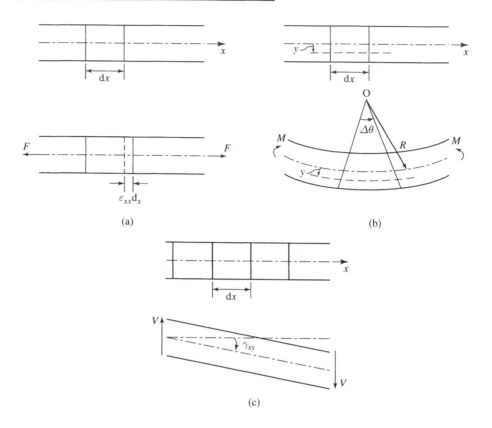

Figure 14.8.4

$\kappa = 1/R$ is the curvature of the deformed longitudinal axis [Fig. (14.8.5b)]. Note $\kappa \equiv \kappa_z$, i.e. the curvature is in the x–y plane.[†]

■ Shear deformation of a beam. Assuming that a uniform shear deformation γ_{xy} exists, the member will undergo distortions as shown in Fig. (14.8.5c). We note that γ_{xy} represents the change of angle of a fibre lying on the x-axis and is given by $2\epsilon_{xy}$, where ϵ_{xy} is the shear strain component [see Eq. (3.7.19) of Chapter 3].

Now, for a coplanar system, the real internal force resultants at any cross-section are: an axial force F, a moment and a shear force, given respectively by Eqs. (2.5.4a), (8.6.5a) and (2.5.4b):[‡]

$$F = \iint_A \tau_{xx} \, dA, \tag{14.8.8a}$$

$$M = \iint_A \tau_{xx} \, y \, dA, \tag{14.8.8b}$$

$$V = \iint_A \tau_{xy} \, dA. \tag{14.8.8c}$$

[†] Recall that this relation is purely geometric and does not depend on the material of the beam. Since the x-axis lies on the neutral surface, $R \, d\theta = dx$ or $\kappa \equiv 1/R = d\theta/dx$.
[‡] The moment M is about the z-axis, i.e. $M \equiv M_z$, and the shear force V is in the y-direction, $V \equiv V_y$.

Similarly, if an external virtual force system is applied, stresses $\delta\tau_{xx}$ and $\delta\tau_{xy}$ will exist throughout the element; the corresponding virtual stress resultants are then

$$\delta F = \int\int_A \delta\tau_{xx}\, \mathrm{d}A, \qquad (14.8.8\mathrm{d})$$

$$\delta M = \int\int_A \delta\tau_{xx}\, y\, \mathrm{d}A, \qquad (14.8.8\mathrm{e})$$

$$\delta V = \int\int_A \delta\tau_{xy}\, \mathrm{d}A, \qquad (14.8.8\mathrm{f})$$

respectively.

We now consider the complementary internal virtual work due to displacements conforming to the above deformation patterns.

(i) *Axial Extension*: $\epsilon_{xx} = \epsilon_{xx}(x)$. From the basic definition

$$\delta W_{\mathrm{int}}^* = \int\int\int_V \delta\tau_{xx}\, \epsilon_{xx}\, \mathrm{d}\Omega = \int_0^L \int\int_A \delta\tau_{xx}\, \epsilon_{xx}(x)\, \mathrm{d}A\, \mathrm{d}x$$

$$= \int_0^L \epsilon_{xx}(x) \left(\int\int_A \delta\tau_{xx}\, \mathrm{d}A \right) \mathrm{d}x.$$

Using Eq. (14.8.8d), we obtain

$$\delta W_{\mathrm{int}}^* = \int_0^L \delta F\, \epsilon_{xx}(x)\, \mathrm{d}x. \qquad (14.8.9\mathrm{a})$$

(ii) *Bending Deformation*: $\epsilon_{xx}(x, y) = y\kappa(x)$. Then

$$\delta W_{\mathrm{int}}^* = \int\int\int_V \delta\tau_{xx}\, \epsilon_{xx}\, \mathrm{d}A\, \mathrm{d}x = \int_0^L \kappa(x) \left(\int\int_A y \cdot \delta\tau_{xx}\, \mathrm{d}A \right) \mathrm{d}x$$

and by Eq. (14.8.8e),

$$\delta W_{\mathrm{int}}^* = \int_0^L \kappa(x)\, \delta M(x)\, \mathrm{d}x. \qquad (14.8.9\mathrm{b})$$

(iii) *Shear Deformation*: $\bar{\epsilon}_{xy}(x, y) = \frac{1}{2}\bar{\gamma}_{xy}.$[†] Then

$$\delta W_{\mathrm{int}}^* = \int\int\int_V 2\delta\tau_{xy}\, \bar{\epsilon}_{xy}\, \mathrm{d}A\, \mathrm{d}x = \int_0^L \bar{\gamma}_{xy} \int\int_A \delta\tau_{xy}\, \mathrm{d}A\, \mathrm{d}x$$

[†] The shear strain ϵ_{xy}, which varies throughout the cross-section of the beam, necessarily vanishes at the traction-free lateral surfaces. Hence, $\bar{\epsilon}_{xy}(x, y) = \frac{1}{2}\bar{\gamma}_{xy}$ is used here to express this 'average' shear strain. For a linear elastic beam, a 'weighted average' of the shear strain is expressed as $\bar{\gamma}_{xy} = \alpha V/AG$, where α is a 'shape factor', which depends on the shape of the cross-section (recall footnote on p. 265 of Chapter 8). For example, for a rectangular cross-section $\alpha = 1.2$, while for a structural I-section, $\alpha = A/A_{\mathrm{w}} \simeq 1$, where A_{w} is the area of the 'web'.

and by Eq. (14.8.8f),

$$\delta W_{\text{int}}^* = \int_0^L \delta V(x)\, \overline{\gamma}_{xy}(x)\, \mathrm{d}x. \qquad (14.8.9c)$$

We observe that each of the expressions for the complementary internal virtual work given by Eqs. (14.8.9) is the product of a (virtual) internal stress resultant and a corresponding global deformation that represents the deformation that of the entire cross-section. It is often customary to refer to the internal (virtual) stress resultants as (virtual) *generalised forces* and to the deformation as *generalised displacements*.

There exists another type of deformation that is also of importance in structural mechanics, namely that of twisting due to a torsional moment (torque) $T \equiv M_x$. Analogously to the above development, we may consider the torque $T(x)$ as a 'generalised force' acting at a cross-section and Θ, the unit angle of twist, as the generalised displacement. Then, as above, the internal virtual work becomes

$$\delta W_{\text{int}}^* = \int_0^L \delta T(x)\, \Theta\, \mathrm{d}x. \qquad (14.8.9d)$$

The total internal complementary virtual work is then given by[†]

$$\delta W_{\text{int}}^* = \int_0^L [\delta F \cdot \epsilon_{xx} + \delta M \cdot \kappa + \delta V \cdot \overline{\gamma}_{xy} + \delta T \cdot \Theta]\, \mathrm{d}x. \qquad (14.8.10)$$

It is again emphasised that the 'generalised displacements' ($\epsilon_{xx}, \kappa, \overline{\gamma}_{xy}, \Theta$) here are real quantities and are in no way related to the virtual internal stress resultants δF, δM, δV, δT appearing above. Equation (14.8.10), for a one-dimensional structural element, is the analogue of the more general equation [Eq. (14.8.4)]. Both expressions are valid for any material.

In analogy to the physical interpretation given for the internal virtual work δW_{int} in Section 14.7b, we may interpret δW_{int}^* as the product of virtual generalised internal forces 'riding' through the corresponding generalised displacements.

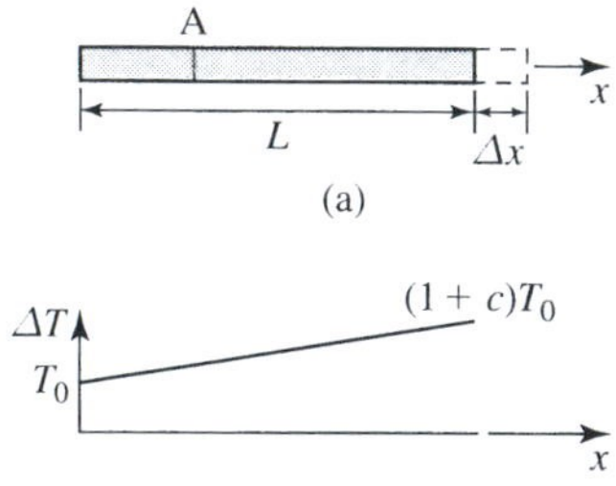

Example 14.18: Consider a prismatic rod of uniform cross-sectional area A and length L [Fig. (14.8.6a)] and made of some material for which only the coefficient of thermal expansion, α, is known. The rod is heated in such a way that the *increase* in temperature ΔT at any point x is given by [Fig. (14.8.6b)]

$$\Delta T(x) = (1 + cx/L)\, T_0, \qquad (14.8.11a)$$

where T_0 and c are constants. Determine the elongation Δ_x of the bar.

Solution: Let ϵ_{xx} denote the strain due to the increase in temperature. Then

$$\epsilon_{xx}(x) = \alpha \Delta T(x) = \alpha(1 + cx/L)\, T_0. \qquad (14.8.11b)$$

To determine the elongation of the bar, we apply a virtual axial force δF acting through the centroid, as shown in Fig. (14.8.7). Then, as we have seen in Chapter 6,

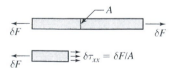

Figure 14.8.5

Figure 14.8.6

[†] In a three-dimensional structure, where, in addition, a shear force V_z and a moment M_y can exist, additional terms, $\delta V_z \overline{\gamma}_{xz}$ and $\delta M_y \kappa_y$ respectively, will appear in the integral.

the resulting stresses $\delta\tau_{xx}$ will be in equilibrium with the external virtual force δF if $\delta\tau_{xx} = \delta F/A$. (Although a trivial observation, we note that we have thus established a virtual stress state that is in equilibrium with the applied virtual force δF.)

Furthermore, the internal complementary virtual work is, from Eq. (14.8.9a),

$$\delta W^*_{int} = \delta F \int_0^L \epsilon_{xx}(x)\,dx = \alpha\delta F\, T_0 \int_0^L \left(1 + \frac{cx}{L}\right) dx$$

or

$$\delta W^*_{int} = \alpha\delta F\, T_0 \cdot (1 + c/2)L. \tag{14.8.12a}$$

Moreover, the external complementary virtual work is

$$\delta W^*_{ext} = \delta F \Delta_x. \tag{14.8.12b}$$

Now, since the given virtual force and stress systems are in equilibrium, we apply the principle of complementary virtual work,

$$\delta W^*_{ext} = \delta W^*_{int},$$

and obtain

$$\Delta_x = \alpha\, T_0(1 + c/2)L. \tag{14.8.13}$$

Note that in this problem, no elastic constant appears. The solution is therefore valid for a rod of any material having a coefficient of thermal expansion, α.   □

The principle of complementary virtual work also proves to be useful in deriving general expressions in mechanics. For example, in Chapter 12, we derived Bredt's formula [see Eq. (12.10.14)]:

$$\Theta = \frac{1}{2AG} \oint_C \tau_{xs}\,ds$$

for the unit angle of twist, Θ, of an elastic member having a thin-wall closed cross-section, as shown in Fig. (14.8.8). Although it was emphasised that this formula is valid *irrespective of the cause* of τ_{xs} existing in the section, this statement was not justified at that stage. By rederiving the formula via the principle of complementary virtual work, it will become clear why this statement is true.

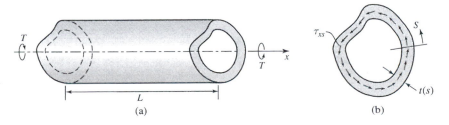

(a)   (b)   **Figure 14.8.7**

For the case of the thin-wall member under consideration, the only shear stress and shear strain components are τ_{xs} and ϵ_{xs}; hence, the internal complementary virtual work is given by [see Eq. (14.8.4)]

$$\iiint_V 2\delta\tau_{xs}\epsilon_{xs}\,d\Omega.$$

Noting that $d\Omega = [t(st) \cdot ds]L$, we may therefore write

$$\delta W_{\text{int}}^* = 2L \oint_C \delta\tau_{xs}\epsilon_{xs}t(s)\,ds, \tag{14.8.14a}$$

and, using Hooke's law, $\epsilon_{xs} = \tau_{xs}/2G$, we have

$$\delta W_{\text{int}}^* = \frac{L}{G} \oint_C \delta\tau_{xs}\tau_{xs}t(s)\,ds. \tag{14.8.14b}$$

Let us now apply a virtual torsional moment δT to the member. The external complementary virtual work is then $\delta W_{\text{ext}}^* = \delta T \cdot \Theta L$. Furthermore, from Eq. (12.10.5), the virtual stresses, in equilibrium with δT, are

$$\delta\tau_{xs} = \frac{\delta T}{2At(s)}. \tag{14.8.14c}$$

Substituting in the above,

$$\delta W_{\text{int}}^* = \frac{\delta T L}{2AG} \oint_C \tau_{xs}\,ds. \tag{14.8.14d}$$

Equating the external work to the internal work, we again obtain Bredt's formula, namely

$$\Theta = \frac{1}{2AG} \oint_C \tau_{xs}\,ds.$$

We note that in this rederivation of Bredt's formula, the cause of the existing τ_{xs} (or ϵ_{xs}) has not been specified. Bredt's formula is therefore seen to be valid for *any* τ_{xs} existing in a thin-wall closed section irrespective of its cause.

(ii) Expressions for elastic elements

For the case of linear elastic members, the (real) generalised displacements $(\epsilon_{xx}, \kappa, \gamma_{xy}, \Theta)$ can, according to previously derived expressions, be expressed in terms of (real) internal stress resultants F, M, V and T existing at any section [see Eqs. (6.2.10), (8.6.6b), (13.4.10) and footnote, p. 609]:

$$\epsilon_{xx} = \frac{F}{AE}, \tag{14.8.15a}$$

$$\kappa = \frac{M}{EI}, \tag{14.8.15b}$$

$$\overline{\gamma}_{xy} = \alpha\frac{V}{AG}, \tag{14.8.15c}$$

$$\Theta = \frac{T}{CG}, \tag{14.8.15d}$$

where AE, EI, AG, CG are the axial, flexural, shear and torsional rigidities, respectively. The coefficient α is called the **shape factor** and depends on the geometry of the given cross-section of the member.

Thus, for a linear elastic member, upon substituting in the general expression for the internal complementary virtual work, Eq. (14.8.10), we obtain

$$\delta W_{\text{int}}^* = \int\limits_0^L \left[\frac{F \cdot \delta F}{AE} + \frac{M \cdot \delta M}{EI} + \alpha \frac{V \cdot \delta V}{AG} + \frac{T \cdot \delta T}{CG} \right] dx. \quad (14.8.16)$$

(e) Internal complementary virtual work in linear elastic rods and beams: explicit expressions (some generalisations)

The expression for the internal complementary virtual work for linear elastic one-dimensional elements, given by Eq. (14.8.16), is generally applicable to an element subjected to (real) external forces from which the (real) internal stress resultants (F, V, M and T) can be obtained. However, it is desirable to obtain an expression that includes also deformation due to causes other than applied loads. In the previous development, using a physical approach for the derivation, it is not immediately evident that such other causes have been included. We therefore rederive the expression for internal complementary virtual work due to axial and flexural deformation directly from its basic definition [Eq. (14.8.4)], using a more direct and formal approach. As we shall see, the final expression will include the relevant terms appearing in Eq. (14.8.16).

We consider an elastic bar of cross-sectional area $A(x)$ and length L, which is subjected to an axial load F passing through the centroid and lateral loads $q(x)$ such that moments $M(x)$ exist at any cross-section [Fig. (14.8.4)].[†] We wish to calculate the internal complementary virtual work due to the axial and flexural effects. (Therefore we neglect here any deformation due to shear strains ϵ_{xy}.) The strain ϵ_{xx} due to F and to the flexure of the element is given by [Eqs. (6.2.10) and (8.6.9c)]

$$\epsilon_{xx} = \frac{F}{EA(x)} + \frac{M(x)y}{EI(x)},$$

where the x-axis denotes the centroidal longitudinal axis.

Let us assume, too, that the element undergoes additional real strains $\epsilon_{xx0}(x)$ due to some other cause (e.g., temperature increases). Then

$$\epsilon_{xx}(x, y) = \epsilon_{xx0}(x) + \frac{F}{EA(x)} + \frac{My}{EI(x)}. \quad (14.8.17)$$

We note that the strains ϵ_{xx}, being due to real forces acting on the actual element, are kinematically admissible.

In general, we might wish to calculate (a) the displacement Δ_x, which the bar undergoes in the x-direction or (b) the lateral displacement Δ_y of the bar at some particular point x_p or perhaps (c) the slope θ of the bar at some point x_c in the deformed state.

Let us, therefore, imagine that there exists an arbitrary system of virtual forces consisting of an axial load δF as well as some arbitrary lateral load δP at x_p and/or couples at x_c, which cause moments $\delta M(x)$ in the beam [see Fig. (14.8.9)]. Then, it is clear that the virtual stresses, given by

$$\delta \tau_{xx} = \frac{\delta F}{A(x)} + \frac{\delta M(x)y}{I(x)}, \quad (14.8.18)$$

---

† The moment is assumed to be acting about a principal centroidal axis of the cross-section having a second moment of the area, I.

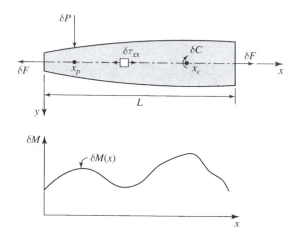

Figure 14.8.8

are in an *equilibrium state* with the applied virtual forces and, according to Eq. (14.8.4), the internal complementary virtual work due to axial and bending effects then is

$$\delta W^*_{\text{int}} = \iiint_V \delta\tau_{xx}\epsilon_{xx}\,dv = \int_L \left(\iint_A \delta\tau_{xx}\epsilon_{xx}\,dA \right) dx. \quad (14.8.19a)$$

Substituting Eqs. (14.8.17) and (14.8.18) in Eq. (14.8.19a),

$$\delta W^*_{\text{int}} = \int_0^L \left\{ \iint_A \left[\frac{\delta F}{A(x)} + \frac{\delta M(x)y}{I(x)} \right] \left[\epsilon_{xx0}(x) + \frac{F}{EA(x)} + \frac{M(x)y}{EI(x)} \right] dA \right\} dx$$

or

$$\delta W^*_{\text{int}} = \int_0^L \left\{ \iint_A \left[\frac{\delta F\epsilon_{xx0}(x)}{A(x)} + \frac{\delta F \cdot F}{EA^2(x)} + \frac{\delta F \cdot M(x)y}{EA(x)I(x)} + \epsilon_{xx0}(x)\frac{\delta M(x)y}{I(x)} \right. \right.$$

$$\left. \left. + \frac{F \cdot \delta M(x)y}{EA(x)I(x)} + \frac{\delta M(x) \cdot M(x)}{EI^2(x)}y^2 \right] \right\} dA\,dx. \quad (14.8.15)$$

But

$$\iint_A dA = A, \qquad \iint_A y\,dA = 0, \qquad \iint_A y^2\,dA = I. \quad (14.8.20)$$

Hence

$$\delta W^*_{\text{int}} = \int_0^L \left[\delta F \cdot \epsilon_{xx0}(x) + \frac{\delta F \cdot F}{EA(x)} + \frac{\delta M(x) \cdot M(x)}{EI(x)} \right] dx. \quad (14.8.21)$$

We note that the last two terms appearing in the above integral, which are due to axial and flexural deformation respectively, coincide with the corresponding terms of Eq. (14.8.16).

(f) Application of the principle of complementary virtual work to evaluate displacements of linear elastic bodies: the method of 'virtual forces'

We now apply the principle of complementary virtual work to elastic bars, and will show its application in the determination of displacements by means of several illustrative examples. Since, as we have seen, the principle is applicable for *any arbitrary stress state, not necessarily statically admissible*, we shall choose to apply, at our convenience, appropriate arbitrary external forces that are in equilibrium with a stress state. Because these forces do not, in fact, actually act on the body (i.e., since the forces are only imagined to exist), the method is sometimes called the **virtual force method**.

Example 14.19:   Determine the elongation of the rod Δ_x of Example 14.15 due to an applied axial force F [Fig. (14.8.10a)] if the material is linearly elastic with modulus of elasticity E.

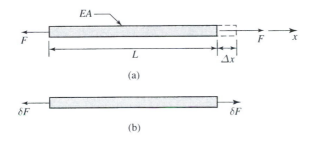

(a)

(b)

Figure 14.8.9

Solution: Due to F, the strain is $\epsilon_{xx} = F/AE$ at all points of the rod. We imagine again that a separate axial force δF acts at the ends, as shown in Fig. (14.8.10b). The resulting virtual equilibrium stresses are $\delta \tau_{xx} = \delta F/A$. Then $\delta W^*_{\text{ext}} = \delta F \Delta_x$ and using Eq. (14.8.21),

$$\delta W^*_{\text{int}} = \frac{\delta F \cdot FL}{AE}.$$

Applying the principle of virtual work, $\delta W^*_{\text{ext}} = \delta W^*_{\text{int}}$, we have

$$\delta F \Delta_x = \frac{\delta F \cdot FL}{AE} \qquad (14.8.22a)$$

or

$$\Delta_x = \frac{FL}{AE}. \qquad (14.8.22b)$$

□

Example 14.20:   A linear elastic prismatic beam AC of flexural rigidity EI and length L is subjected to a uniformly distributed load w, as shown in Fig. (14.8.11a). Determine the vertical displacement Δ_A at the free end A due to flexure.

Solution: Let the moment due to w be given by $M(x)$. Then

$$M(x) = -wx^2/2, \qquad 0 \leq x \leq L. \qquad (14.8.23a)$$

The moment diagram $M(x)$ is shown in Fig. (14.8.11b). Now since we wish to calculate the displacement at point A, let us apply a virtual force δP at A, as shown in

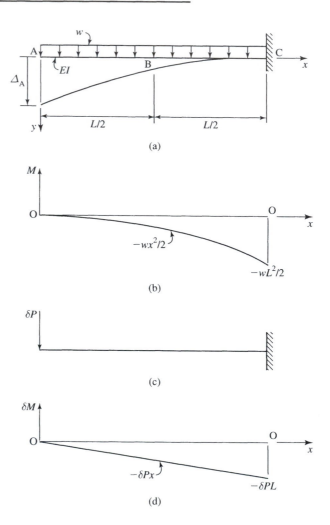

Figure 14.8.10

Fig. (14.8.11c). The resulting moments in the beam are then [Fig. (14.8.11d)]

$$\delta M(x) = -\delta P x, \quad 0 \le x \le L. \tag{14.8.23b}$$

From Eq. (14.8.16) or (14.8.21), the internal complementary virtual work is

$$\delta W^*_{\text{int}} = \int_0^L \frac{\delta M \cdot M}{EI}\, \mathrm{d}x,$$

and the external complementary virtual work is

$$\delta W^*_{\text{ext}} = \delta P \Delta_{\text{A}},$$

so that, applying the principle of complementary virtual work,

$$\delta P \Delta_{\text{A}} = \int_0^L \frac{\delta M \cdot M}{EI}\, \mathrm{d}x. \tag{14.8.24a}$$

Substituting the expressions for $M(x)$ and $\delta M(x)$, we have

$$\delta P \Delta_A = \frac{\delta P w}{EI} \int_0^L \left(\frac{x^2}{2}\right) x \, dx \qquad (14.8.24b)$$

or

$$\Delta_A = \frac{wL^4}{8EI}, \qquad (14.8.25)$$

which coincides with Eq. (9.7.22b).

Note that in the above the force δP cancels out. One may therefore apply a unit load, i.e. $\delta P = 1$, and obtain the same result. It is for this reason that the method is also often referred to as the *unit dummy load method*.

Example 14.21:   Determine the slope θ_B at point B$(x = L/2)$ of the beam of Example 14.20 due to the given applied load w.

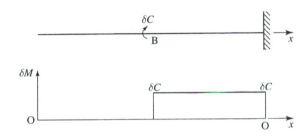

Figure 14.8.11

Solution: Since we are interested in calculating the slope at B, we apply a (virtual) couple δC at B, as shown in Fig. (14.8.12a). The resulting external complementary virtual work is then

$$\delta W_{\text{ext}}^* = \delta C \, \theta_B. \qquad (14.8.26)$$

The moments produced by δC are [Fig. (14.8.12b)]

$$\begin{aligned}
\delta M(x) &= 0, \quad 0 \le x \le L/2 \\
\delta M(x) &= \delta C, \quad L/2 \le x \le L.
\end{aligned} \qquad (14.8.27)$$

Hence the internal complementary virtual work is, by Eq. (14.8.16) or (14.8.21),

$$\delta W_{\text{int}}^* = \frac{1}{EI} \int_0^L \delta M \cdot M \, dx = \frac{\delta C}{EI} \int_{L/2}^L (-wx^2/2) \, dx = -\frac{7\delta C w L^3}{48EI}. \qquad (14.8.28a)$$

Equating $\delta W_{\text{ext}}^* = \delta W_{\text{int}}^*$, we obtain

$$\theta_B = -\frac{7wL^3}{48EI}. \qquad (14.8.28b)$$

We note here that the negative sign appearing in Eq. (14.8.28b) indicates that the slope is opposite to the assumed sense of δC.

In the following example, we determine the deflection of an indeterminate beam. The problem will be solved in two ways to expose the advantage of introducing a statically inadmissible system when applying the principle of complementary virtual work to such structures.

Example 14.22:   A linear elastic prismatic beam AB, of flexural rigidity $E\,I$ and length L, is fixed at both ends. A uniformly distributed load w is applied as shown in Fig. (14.8.13a). Determine the lateral displacement of point C due to flexural deformation.

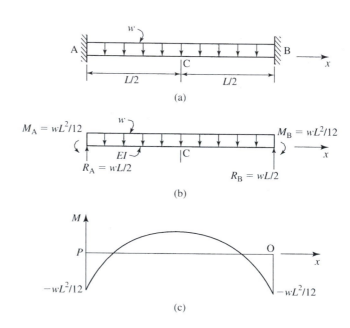

Figure 14.8.12

First Solution: We first observe that since the given beam is statically indeterminate, a static analysis must first be performed to find the moment (e.g., using the force method as discussed in Chapter 9). For the example at hand, the end reactions R_A and R_B and end moments M_A and M_B are given in Appendix F and are shown in Fig. (14.8.13b).[†] Knowing these, we then have

$$M(x) = -\frac{wL^2}{12} + \frac{wLx}{2} - \frac{wx^2}{2}, \qquad (14.8.29)$$

whose moment diagram is shown in Fig. (14.8.13c).

Now, to determine the displacement of point C, we apply a virtual load δP as shown in Fig. (14.8.14a). For this case, the end moments and reactions are again given in Appendix F and are shown in Fig. (14.8.14b).[‡] (Note that, here, it is required to know the reactions and moments for *two* loading cases).

The resulting moment $\delta M(x)$ is

$$\delta M = -\frac{\delta PL}{8} + \frac{\delta P}{2}x, \qquad 0 \le x \le L/2 \qquad (14.8.30a)$$

$$\delta M = -\frac{\delta PL}{8} + \frac{\delta P}{2}(L - x), \qquad L/2 \le x \le L. \qquad (14.8.30b)$$

[†]   End reactions and moments for a large number of 'standard' indeterminate beams can be found in tables and texts of the engineering literature. For example, see R.J. Roark, *Formulas for Stress and Strain*.

[‡]   In passing, we note that for this linear analysis we neglect the horizontal reactions at A and B since these are second-order effects (see Section 5 of Chapter 9).

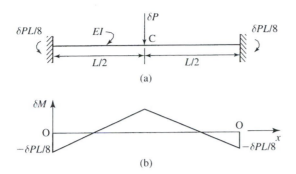

Figure 14.8.13

The internal complementary virtual work is then [Eq. (14.8.16) or Eq. (14.8.21)],

$$\delta W_{\text{int}}^* = \frac{1}{EI} \int_0^L \delta M(x) M(x) \, dx.$$

Substituting Eqs. (14.8.29) and (14.8.30) and performing the simple integrations, we obtain

$$\delta W_{\text{int}}^* = \frac{\delta P w L^4}{384 EI}. \tag{14.8.31}$$

Now, since the external complementary virtual work is $\delta W_{\text{ext}}^* = \delta P \Delta_c$, applying the principle yields

$$\Delta_c = \frac{w L^4}{384 EI}. \tag{14.8.32}$$

It is emphasised here that the moments resulting from the virtual loading system of Fig. (14.8.14a) produce a (virtual) stress field that is *statically admissible*; that is, the reactions $\delta R_A = \delta R_B = \delta P/2$, $\delta M_A = \delta M_B = -\delta P L/8$ are those required to maintain zero displacement and rotation of the given elastic beam at A and B.

We recall, however, that the principle of complementary virtual work does *not* require that the virtual stress field be statically admissible; the stress field need only be one that is in equilibrium with the external virtual loading system. We shall see that this leads to the much simplified alternative solution given below.

Alternative Solution: Since the principle of complementary virtual work is valid for *any* equilibrium state imposed on the body, we shall choose the following as our external virtual force system: a virtual force δP acting at C, a reacting force at B and a reacting moment at B. In addition, we *set the reacting force and moment at A equal to zero* [Fig. (14.8.15a)]. Note that this external virtual force system is then the same as one that acts on an equivalent 'released' structure shown in Fig. (14.8.15b).[†] The system will be in equilibrium provided the upward force $\delta R_B = \delta P$ and $\delta M_B = -\delta P L/2$. Note that the moment throughout the beam,

$$\delta M(x) = 0, \quad 0 \le x \le L/2,$$
$$\delta M(x) = -\delta P(x - L/2), \quad L/2 \le x \le L, \tag{14.8.33}$$

is then in equilibrium with the virtual external force system. We observe that the δ-force system acts on the released structure, which is statically determinate. Thus, in

---

[†] We observe from the figure that the constraints against displacement and rotation at point A of the given original structure have been 'released'.

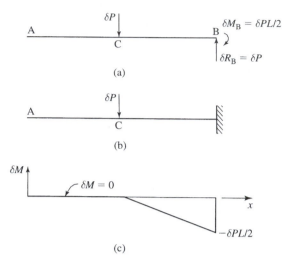

Figure 14.8.14

this alternative solution, we require to know only *one* statically indeterminate solution, namely that due to the real applied loading.

Since equilibrium is the *only* criterion that is required, we may apply this statically inadmissible virtual system in the principle of complementary virtual work.

Substituting Eqs. (14.8.29) and (14.8.33) in Eq. (14.8.16) or (14.8.21), the internal complementary virtual work becomes

$$\delta W_{\text{int}}^* = \frac{1}{EI} \int_0^L \delta M \cdot M \, dx = -\delta P w \int_{L/2}^L \left(-\frac{L^2}{12} + \frac{Lx}{2} - \frac{x^2}{2} \right) \cdot \left(x - \frac{L}{2} \right) dx,$$

which, upon performing the necessary integration, yields

$$\delta W_{\text{int}}^* = \frac{\delta P w L^4}{384 EI}. \tag{14.8.34a}$$

Now the boundary conditions at $x = L$ for the *actual* displacements at B($x = L$) are $v(L) = 0$ and $v'(L) \equiv \theta(L) = 0$. Hence for this loading system, the only contribution to the external complementary virtual work is that of the force δP; therefore

$$\delta W_{\text{ext}}^* = \delta P \Delta_c. \tag{14.8.34b}$$

Application of the principle yields

$$\Delta_c = \frac{w L^4}{384 EI}; \tag{14.8.35}$$

that is, we obtain the same displacement as before.   □

In the solution of the above problem, we have seen that it is possible to obtain more than one equilibrium state for the statically indeterminate element AB. (In fact, we could have chosen other equilibrium states in order to satisfy the conditions required for the principle of complementary virtual work.) It is therefore worthwhile here to make a more general comment on the use of statically admissible and inadmissible force systems.

We first recall that a statically indeterminate structure is, by definition, one for which the equations of statics are not sufficient to provide a *unique* solution for the unknown reactions. Indeed there exist an infinite number of possible equilibrium states.[†] (However, only *one* of these states will be statically admissible.) In physical terms, there exist more reactions than are necessary to give stability to the structure and hence the structure is said to be **overly constrained**, i.e. to have 'redundant constraints'. Therefore, it is always possible to 'release' some of the constraints and thereby obtain an equivalent released (statically determinate) structure corresponding to the given statically indeterminate stable structure.[‡] Since the remaining constraints are sufficient to maintain a stable structure, it is *not necessary to introduce new constraints* to ensure stability of the corresponding equivalent structure. Now, in releasing the redundant constraints, we effectively set the (virtual) reactive forces to zero and thus obtain a resulting virtual external force system, which is, by definition, statically inadmissible (since the geometric constraints of the original structure are violated).

Since one may usually designate any of the $n - m$ reactions as 'redundant reactions', one has the choice of deciding which constraint is to be released. The resulting statically inadmissible external force system will therefore depend on the choice of the redundant(s).[§] Note, however, that one must make a judicious choice to avoid a contribution in the external work expression due to virtual forces 'riding' through *unknown* (real) displacements. Since there exists a choice of several possible statically inadmissible systems, we therefore usually choose a virtual force system that leads to the simplest calculations.

However, in the case of a statically determinate structure, the number of constraints (i.e. reactions) is just sufficient to maintain the structure in equilibrium: there exist no 'redundant reactions'. Thus, if we were to choose a statically inadmissible virtual external force system (thereby releasing a constraint), it would be necessary to introduce a *new* constraint to maintain a stable structure. This new constraint would then appear in the expression for the complementary virtual work as a product with an additional unknown, namely a (real) displacement.

Consequently, it follows from the above remarks that *it is advantageous to apply a statically inadmissible external force system* (as in the alternate solution above), *only if the original structure is statically indeterminate.*

As a practical comment, we observe that in the first solution given above, one theoretically must solve for the reactions of an indeterminate structure under two different loading systems: the real and the virtual loading systems. In the alternative solution, it is only necessary to solve a single indeterminate problem; namely for the reactions due to the actual (real) loading system. It follows that choosing a statically inadmissible virtual loading system leading to an equivalent statically determinate structure (as was done in the alternative solution) will always considerably simplify the analysis.

[†] In mathematical terms, there exist more unknown reactions than independent equations of equilibrium in a statically indeterminate structure. Consequently, there are an *infinite* number of solutions to the equations of equilibrium. If there exist n unknowns and only m ($m < n$) independent equations of equilibrium, the 'degree of indeterminacy' of the structure is said to be $n - m$. According to the laws of linear algebra, $n - m$ unknowns may then be assigned arbitrary values.

[‡] The mathematical equivalence of 'releasing' the redundant constraints is to set the $n - m$ reactions to zero.

[§] For example, in the structure of the preceding problem, there exist four reactive components: R_A, R_B, M_A and M_B. In the alternative solution, we released the constraints against vertical displacement and rotation at A; that is, we designated R_A and M_A to be redundants. The inadmissible virtual external force system (δP, δR_B, δM_B) acting on the resulting equivalent structure was as shown in Fig. (14.8.15a). We might instead have chosen to release the constraints against rotation at the two ends (i.e., chosen M_A and M_B to be the redundants forces) therefore leading to a different inadmissible virtual force system, consisting only of δP, δR_A and δR_B. The resulting equivalent released structure would then be a simply supported beam.

Example 14.23: (a) Compute the horizontal component of deflection Δ_E of point E of the pin-connected elastic truss shown in Fig. (14.5.4) if member Bc is lengthened by an amount 'e'. (For example, this might occur if the member suddenly undergoes an expansion due to heat. However, the cause of the expansion is irrelevant here.)

(b) Compute the horizontal component of deflection of point E, Δ_E, of the truss due to an applied load P acting as shown [Fig. (14.8.16b)]. [Note that the cross-sectional area of each member is given in parentheses in Fig. (14.5.4).]

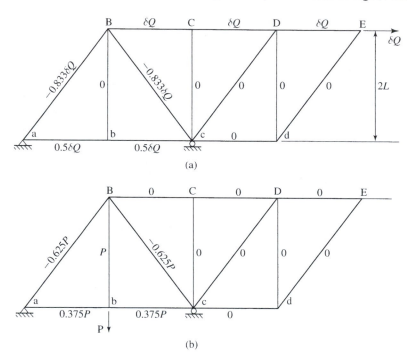

Figure 14.8.15

[We recall that part (b) of this example was solved previously in Example 14.5 by means of Castigliano's (second) theorem.]

Solution:

(a) To calculate the horizontal displacement component Δ_E due to a lengthening 'e' of member Bc, we apply a virtual force δQ at E, as shown in Fig. (14.8.16a). (The resulting internal axial forces δF_i in the ith member, previously calculated in Example 14.5, are given in the figure.) The external work is then $\delta W_{ext}^* = \delta Q \Delta_E$, while the internal complementary virtual work is, using the first term of Eq. (14.8.21),

$$\delta W_{int}^* = \int_0^L \delta F_{Bc}\, \epsilon_{xx0}(x)\, dx = \delta F_{Bc} \int_0^L \epsilon_{xx0}(x)\, dx = \delta F_{Bc}\, e, \quad (14.8.36a)$$

since, by Eq. (3.3.6), the total elongation $e = \int_0^L \epsilon_{xx0}(x)\, dx$. Therefore, since $\delta F_{Bc} = -0.833\delta Q$, we have

$$\delta W_{int}^* = \delta F_{Bc}\, e = -0.833\, \delta Q\, e. \quad (14.8.36b)$$

Hence, applying the principle of complementary virtual work,

$$\Delta_{\mathrm{E}} = -0.833e. \qquad (14.8.37)$$

Note that the minus sign indicates that point E moves in the opposite direction to the assumed direction of δQ; i.e., it moves to the left. Note too that modulus of elasticity E and the cross-sectional areas of the bars are irrelevant here.

(b) In Example 14.5, the forces F_i in the members of the truss due to P were calculated; these are shown again in Fig. (14.8.16b).

We again apply a horizontal virtual force δQ at point E [Fig. (14.8.16a)]. From Eq. (14.8.21), the internal virtual work in the ith member is $F_i \delta F_i L_i / A_i E$. Therefore, the internal virtual work of the entire truss is

$$\delta W_{\mathrm{int}}^* = \sum_{i=1}^{n} \frac{F_i \delta F_i L_i}{A_i E}, \qquad (14.8.38)$$

where n is the number of truss members. Noting that the external complementary virtual work is $\delta W_{\mathrm{ext}}^* = \delta Q \Delta_{\mathrm{E}}$, application of the principle yields

$$\delta Q \Delta_{\mathrm{E}} = \sum_{i=1}^{n} \frac{F_i \delta F_i L_i}{E A_i}. \qquad (14.8.39)$$

The calculations are best shown by means of the following table:

Bar	L_i	A_i	L_i / A_i	F_i	δF_i	$\frac{F_i \delta F_i L_i}{A_i}$
ab	$1.5L$	A_0	$1.5L/A_0$	$0.375P$	$0.5\delta Q$	$0.2813\,P\delta QL/A_0$
bc	$1.5L$	A_0	$1.5L/A_0$	$0.375P$	$0.5\delta Q$	$0.2813\,P\delta QL/A_0$
aB	$2.5L$	$1.25A_0$	$2L/A_0$	$-0.625P$	$0.833\delta Q$	$-1.044\,P\delta QL/A_0$
Bc	$2.5L$	$1.25A_0$	$2L/A_0$	$-0.625P$	$-0.833\delta Q$	$1.044\,P\delta QL/A_0$
						$\sum = 0.5626\,P\delta QL/A_0$

Hence

$$\Delta_{\mathrm{E}} = 0.5626 PL/A_0 E, \qquad (14.8.40)$$

which is identical with the solution of Example 14.5.

14.9   The principle of stationary potential energy

(a) Derivation of the principle and some applications

We consider an elastic body (not necessarily linear) subjected to a set of external forces [see, e.g., Fig. (14.7.1a)] such that a state of stress (τ_{xx}, τ_{yy}, τ_{xy}, ...) as well as a corresponding state of strain (ϵ_{xx}, ϵ_{yy}, ϵ_{xy}, ...) exists at all points.[†] It is then possible to compute the total strain energy, e.g., by Eq. (14.2.9).

We now pose the following question: Suppose the actual strains in the body were changed arbitrarily by a small amount, how would the strain energy change? Or, in other words, if we vary the strain components, how would the strain energy vary? We again denote, e.g., the variation of ϵ_{xx} by $\delta\epsilon_{xx}$. Thus, we consider a variation of

[†] At this stage, the body need not necessarily be in equilibrium.

the strains such that

$$\epsilon_{xx} \to \epsilon_{xx} + \delta\epsilon_{xx}, \qquad \epsilon_{yy} \to \epsilon_{yy} + \delta\epsilon_{yy}, \qquad \epsilon_{xy} \to \epsilon_{xy} + \delta\epsilon_{xy}, \qquad \text{(etc.)}.$$

From Eq. (14.2.10a), we have, with $U \to U + \delta U$,

$$\delta U = \iiint_V [\tau_{xx}\delta\epsilon_{xx} + \tau_{yy}\delta\epsilon_{yy} + \tau_{zz}\delta\epsilon_{zz} + 2(\tau_{xy}\delta\epsilon_{xy} + \tau_{yz}\delta\epsilon_{yz} + \tau_{zx}\delta\epsilon_{zx})]\,d\Omega.$$

$$(14.9.1)$$

Note that Eq. (14.9.1) represents the *change* in the strain energy due to a variation of strains under an existing *constant stress state*.[†] The expression is thus valid for any *given* stress field 'moving' through any arbitrary virtual strain field. Note, too, that although the elastic strain energy is positive definite, i.e. $U \geq 0$, δU can be either positive or negative.

Now, comparing this last expression with the internal virtual work expression given by Eq. (14.7.2c), we observe that the two expressions are identical; i.e.,

$$\delta U = \delta W_{\text{int}}.$$

$$(14.9.2)$$

This last equation may be interpreted as follows: Given an elastic body in which there exists, at all points, a stress state corresponding to a state of strain. Then, if the strain components are varied, the internal virtual work is equal to the change in the strain energy.

At this stage, we impose the condition of equilibrium. If the state of stress is one for which the body is in equilibrium, then from the principle of virtual work,

$$\delta W_{\text{int}} = \delta W_{\text{ext}}.$$

It therefore follows that

$$\delta U = \delta W_{\text{ext}}.$$

$$(14.9.3)$$

We now turn our attention to the external forces acting upon the system. We shall exclude from our consideration all non-conservative forces and consider only the application of conservative forces, i.e. those forces F that, by definition, can be expressed by a relation[‡]

$$F = -\nabla V,$$

$$(14.9.4a)$$

where

$$\nabla \equiv \frac{\partial}{\partial x}i + \frac{\partial}{\partial y}j + \frac{\partial}{\partial z}k,$$

$$(14.9.4b)$$

i.e., where ∇ is the gradient operator and $V = V(x, y, z)$ is a scalar function called a **potential function**.

The virtual work δW_{ext} done by a force F located at a point defined by the position vector r in going through a displacement δu is then [Fig. (14.9.1)]

$$\delta W_{\text{ext}} = F \cdot \delta u,$$

$$(14.9.5a)$$

where $\delta u \equiv \delta r = \delta x i + \delta y j + \delta z k$. Substituting Eq. (14.9.4a),

$$\delta W_{\text{ext}} = -\nabla V \cdot \delta u \equiv -\nabla V \cdot \delta r = -\delta V.$$

$$(14.9.5b)$$

[†] Note that the stress components can be functions of x and y. By 'constant', we mean that at any given point they do not change in magnitude or direction.

[‡] The minus sign, though arbitrary, is universally adopted.

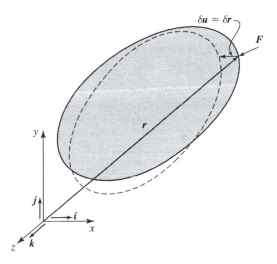

Figure 14.9.1

In mechanics, the function V is said to represent the *potential energy* of the force. Thus, from Eq. (14.9.5b), we see that work done by a conservative force is equal to the negative change of its potential energy, i.e. the loss of potential energy.[†]

Now, we wish here to examine the change in strain energy of the body *from its equilibrium position*. Furthermore, in considering the change in potential energy of the external forces, we wish to consider only the change in potential energy of the *applied forces and not of the reactive forces*. From Eqs. (14.9.5), it is clear that δV of the *reactive* forces can be made to vanish identically only if all the virtual displacements δu are identically zero at points on the surface S where there exist non-zero (reactive) constraints [see Section 14.7b]. We therefore impose an additional restriction on these variations: namely *the variations δu must be such that they satisfy the geometric boundary conditions of the body. Thus the variations must now be kinematically admissible.*

Substituting Eq. (14.9.5b) in Eq. (14.9.3),

$$\delta U = -\delta V$$

or

$$\delta(U + V) = 0. \tag{14.9.6}$$

We now define the **total potential energy of the system**, Π, by

$$\Pi = U + V \tag{14.9.7}$$

and therefore finally write

$$\delta \Pi = 0. \tag{14.9.8}$$

Thus, the variation of the total potential vanishes, i.e. no change occurs in Π when the system is displaced *from its equilibrium position*. Equation (14.9.8) is a mathematical statement of the principle of stationary potential energy and may be stated as follows:

An elastic body in equilibrium under a set of externally applied forces will deform in such a way so as to render the total potential of the system stationary.

[†] Note that the force here is a given *constant* force and hence the work done by F in going through any arbitrary displacement δu is the same as the virtual work.

Alternatively, we may state

Of all possible ways that an elastic body deforms in reaching its equilibrium position, it will 'choose' that deformation which causes Π to have a stationary value.[†]

The principle of stationary potential energy has many useful applications in the analysis of elastic bodies. In particular, it is often used in the study of elastic stability of structural members. This subject will be studied in the following chapter.

The following examples will serve to illustrate the application of the principle.

Example 14.24:   A number n of linearly elastic rods, each of cross-sectional area A and arranged symmetrically about the y-axis, are hinged at one end, point B, where a vertical force P acts, as shown in Fig. (14.9.2a). The angle of inclination of the ith rod with respect to the y-axis is given as α_i and the length of each rod is denoted by L_i. Determine the displacement of point B and the axial force in each rod.

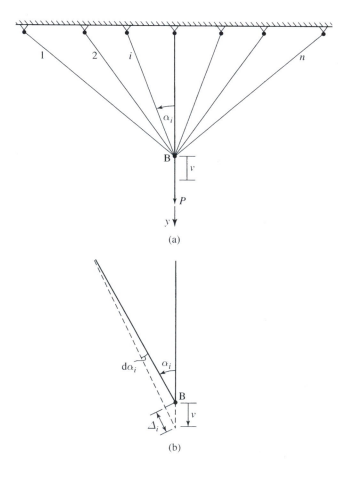

Figure 14.9.2   (b)

[†] By stationary we mean that Π takes on an extreme value, i.e., maximum or minimum. If the body is in stable equilibrium, then it can be shown that Π has a minimum value. It is for this reason that the principle is sometimes called the **Principle of minimum potential energy**.

Solution: Because of the symmetry of the problem, point B will displace only in the y-direction. Denote this displacement by v. Let the axial force, stress and strain in the ith rod be denoted by F_i, τ_i, ϵ_i, respectively. The strain energy U_i existing in the ith rod is, by Eqs. (14.2.7) and (14.2.2b),

$$U_i = \frac{E\epsilon_i^2}{2}AL_i = \frac{EA}{2L_i}\Delta_i^2, \tag{14.9.9}$$

where $\Delta_i = \epsilon_i L_i$ denotes the extension of the ith bar. From geometry [Fig. (14.9.2b)], we obtain

$$\Delta_i = v\cos(\alpha_i - d\alpha_i),$$

where $d\alpha_i$ represents the change of angle of inclination. Since we are limited to small rotations [$|d\alpha_i| \ll 1$], $\cos(\alpha_i - d\alpha_i) \simeq \cos\alpha_i$ and therefore we have

$$\Delta_i = v\cos\alpha_i. \tag{14.9.10}$$

Substituting in Eq. (14.9.9)

$$U_i = \left[\frac{EA}{2L_i}\cos^2\alpha_i\right]v^2. \tag{14.9.11a}$$

Hence the strain energy in the entire system is

$$U = \sum_{i=1}^{n}\left[\frac{EA}{2L_i}\cos^2\alpha_i\right]v^2. \tag{14.9.11b}$$

Since the potential energy of the externally applied force is simply $V = -Pv$, it follows that the total potential of the system Π is given by

$$\Pi = U + V = \sum_{i=1}^{n}\left[\frac{EA}{2L_i}\cos^2\alpha_i\right]v^2 - Pv. \tag{14.9.12}$$

We now apply the principle of stationary potential energy, $\delta\Pi = 0$, noting that $\Pi = \Pi(v)$.[†] Hence

$$\delta\Pi = \frac{d\Pi(v)}{dv}\delta v = 0. \tag{14.9.13a}$$

Since this must be true for *any arbitrary* δv about the equilibrium position, we conclude that

$$\frac{d\Pi(v)}{dv} = 0. \tag{14.9.13b}$$

Upon carrying out the above differentiation, we obtain

$$\sum_{i=1}^{n}\left(\frac{EA}{L_i}\cos^2\alpha_i\right)v - P = 0$$

from which

$$v = \frac{P}{EA\sum_{i=1}^{n}\frac{\cos^2\alpha_i}{L_i}}. \tag{14.9.14a}$$

[†] In general, Π can be a function of several displacement quantities. In this problem, it is a function of only one variable, v.

Using Eq. (14.9.10), the strain in member i is then

$$\epsilon_i = \frac{v \cos \alpha_i}{L_i} \qquad (14.9.14\text{b})$$

and thus, since $F_i = EA\epsilon_i$,

$$F_i = \frac{P}{L_i} \frac{\cos \alpha_i}{\sum_{i=1}^{n} \frac{\cos^2 \alpha_i}{L_i}}. \qquad (14.9.14\text{c})$$

The following example illustrates the application of the principle to a problem where a geometric nonlinearity occurs.

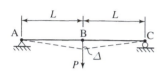

Figure 14.9.3

Example 14.25: Consider a structure consisting of two linearly elastic members (each having cross-sectional area A) hinged at the three points A, B, C. A load P is applied at B, as shown [Fig. (14.9.3a)]. Determine the load–displacement relation.

Solution: By symmetry, point B will displace in the y-direction only. We denote this displacement by Δ. We need to first find Π as a function of Δ.

For any Δ, the strain ϵ in each of the elastic bars is given by

$$\epsilon = \frac{\sqrt{L^2 + \Delta^2} - L}{L} = \sqrt{1 + \left(\frac{\Delta}{L}\right)^2} - 1, \qquad (14.9.15\text{a})$$

which by the binomial theorem becomes

$$\epsilon = \left[1 + \frac{1}{2}\left(\frac{\Delta}{L}\right)^2 - \frac{1}{8}\left(\frac{\Delta}{L}\right)^4 + \cdots\right] - 1.$$

For small rotations, $|\Delta/L| \ll 1$ and hence, if we drop all higher order terms,

$$\epsilon = \frac{1}{2}\left(\frac{\Delta}{L}\right)^2. \qquad (14.9.15\text{b})$$

Since the state of stress is uniaxial, the strain energy is, by Eqs. (14.2.2b) and (14.2.7),

$$U = \frac{E}{2} \iiint_V \epsilon^2 \, d\Omega = \frac{E}{8}\left(\frac{\Delta}{L}\right)^4 \cdot 2AL = \frac{EA}{4L^3} \cdot \Delta^4. \qquad (14.9.16)$$

Noting that the potential energy of the applied force P is $V = -P\Delta$, the total potential Π is

$$\Pi = U + V = \frac{AE}{4L^3}\Delta^4 - P\Delta. \qquad (14.9.17)$$

Now, if Δ denotes the displacement of point B in the *equilibrium state* then, by the principle of stationary potential energy, $\delta\Pi(\Delta) = 0$; therefore

$$\frac{d\Pi(\Delta)}{d\Delta}\delta\Delta = 0. \qquad (14.9.18\text{a})$$

Again, since $\delta\Delta$, the variation of the displacement from the equilibrium state, is arbitrary, it follows that

$$\frac{d\Pi(\Delta)}{d\Delta} = 0. \qquad (14.9.18\text{b})$$

From Eq. (14.9.17b),

$$\frac{AE}{L^3}\Delta^3 - P = 0 \tag{14.9.19a}$$

and therefore

$$\Delta = \left(\frac{P}{AE}\right)^{1/3} \cdot L. \tag{14.9.19b}$$

$\square$

Note that although the *material* behaviour of the elastic bars is linear, the force–displacement relation of the system is *nonlinear*. Such a nonlinearity is called a **geometric nonlinearity**, since it arises due to the particular geometry of the problem.

◇(b) Approximate solutions – the Rayleigh–Ritz method

The principle of stationary potential energy lends itself to obtaining approximate solutions to problems consisting of bodies for which there is no dissipation of energy, namely for elastic bodies.

We introduce the ideas by means of an example. Let us consider a simply supported beam subjected to a uniformly distributed lateral load w, as shown in Fig. (14.9.4). We recall from Chapter 9, that integration of the differential equation,

$$EIv''(x) = -M(x), \tag{14.9.20a}$$

satisfying the geometric boundary conditions,

$$v(0) = v(L) = 0, \tag{14.9.20b}$$

yielded the deflection [see Eq. (9.4.2b)]

$$v(x) = \frac{w}{24EI}(x^4 - 2Lx^3 + L^3x) \tag{14.9.21a}$$

and, specifically at the centre,

$$v\left(\frac{L}{2}\right) = \frac{5wL^4}{384EI} = 0.0103021\frac{wL^4}{EI}. \tag{14.9.21b}$$

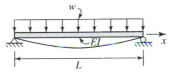

Figure 14.9.4

Let us assume that this solution is not known to us and that we wish to obtain the deflection $v(x)$ using the principle of stationary potential energy. From Eq. (14.2.16), the flexural energy is given by $U = \frac{1}{2EI}\int_0^L M^2(x)\,dx$. Upon substituting Eq. (14.9.20a), we may write the strain energy as

$$U = \frac{EI}{2}\int_0^L [v''(x)]^2\,dx. \tag{14.9.22}$$

The potential energy of the applied external load $q(x)$ with respect to the undeflected position is, in general,

$$V = -\int_0^L q(x)v(x)\,dx \tag{14.9.23a}$$

and for the specific case, $q(x) = w$,

$$V = -w\int_0^L v(x)\,dx. \tag{14.9.23b}$$

The Rayleigh–Ritz method consists essentially of *assuming* a form for the deflection $v(x)$ of the beam, which *satisfies all geometric boundary conditions*. (Such functions are called **admissible functions**.) For the problem at hand, let us assume, for example, that

$$v(x) = A\,x(L - x), \tag{14.9.24}$$

whose derivatives are $v'(x) = A(L - 2x)$ and $v''(x) = -2A$. Substituting in Eqs. (14.9.22) and (14.9.23b) respectively, we obtain after simple integration,

$$U = 2A^2 EIL \quad \text{and} \quad V = -wAL^3/6. \tag{14.9.25}$$

The total potential $\Pi = U + V$ is then

$$\Pi = 2A^2 EIL - wAL^3/6. \tag{14.9.26}$$

We observe that Π is a function of the unknown parameter A. According to the principle of stationary potential energy, for equilibrium we require that $\delta\Pi = 0$ and hence $\frac{d\Pi}{dA} = 0$. Taking the derivative, we obtain

$$4A\,EIL - \frac{wL^3}{6} = 0$$

or

$$A = \frac{wL^2}{24EI}. \tag{14.9.27}$$

Substituting back in Eq. (14.9.24), yields an approximate expression for the deflection

$$v(x) = \frac{wL^2}{24EI}x(L - x). \tag{14.9.28a}$$

The deflection at the mid-point $x = L/2$ is

$$v\left(\frac{L}{2}\right) = \frac{wL^4}{96EI}. \tag{14.9.28b}$$

Upon comparing the approximate deflection $v_{\text{app}}(x)$ with the exact deflection $v_{\text{ex}}(x)$ given by Eq. (14.9.21b) (i.e., 'exact' in the framework of linear beam theory), we find that the assumed deflection yields, at the mid-point, a percentage error of 25%, which is relatively large.[†]

We also observe that the approximate deflection is smaller that the 'exact' deflection given by Eq. (14.9.21b). We may explain this feature physically by considering physically the beam in an anthropomorphic sense.

When subjected to a given load, a beam 'chooses' to deflect into a 'natural' shape – here, that described by the function, Eq. (14.9.21a) – for which the total potential of the system is stationary (and in fact, for a stable system, a minimum). However, in specifying an arbitrary function such as Eq. (14.9.24), the beam is essentially forced to deflect in an unnatural shape, i.e., an additional unnatural constraint is imposed on the beam. This constraint thus tends to stiffen the beam and hence the

---

[†] A more global measure of the error that takes into account the differences, $v_{\text{ex}} - v_{\text{app}}$, of the deflections of the two solutions over the entire beam is given by the root-mean-square expression

$$\text{Error} = \sqrt{\frac{\int_0^L [v_{\text{ex}}(x) - v_{\text{app}}(x)]^2\,dx}{\int_0^L v_{\text{ex}}^2(x)\,dx}}.$$

Substituting Eqs. (14.9.21a) and (14.9.28a) for v_{ex} and v_{app}, respectively, and integrating yields a percentage error of 18%.

resulting deflections are generally smaller than the 'true' deflections. This can be proven rigorously for the deflection under a concentrated load.

With this in mind, we note that the deflection given by Eq. (14.9.24) is dependent on a single parameter, namely A. Now, it is clear that if one assumes a function that depends on several parameters, say a_n ($n > 1$), then one introduces some flexibility into the system since several degrees of freedom can be adjusted to render the potential Π of the system a minimum.

Considering again the simply supported beam, we note that the function $v(x) = a_n \sin(n\pi x/L)$ satisfies the geometric boundary conditions, Eq. (14.9.20b); that is, it is an admissible function for all $n \geq 1$. Let us therefore assume a deflection for the beam in the form

$$v(x) = \sum_{n=1}^{\infty} a_n \sin\left(\frac{n\pi x}{L}\right), \tag{14.9.29a}$$

which we observe, in passing, is a Fourier sine series with Fourier coefficients a_n.

Taking derivatives, we have

$$v''(x) = -\sum_{n=1}^{\infty} a_n \left(\frac{n\pi}{L}\right)^2 \sin\left(\frac{n\pi x}{L}\right) \tag{14.9.29b}$$

and upon substituting in Eq. (14.9.22),

$$U = \frac{EI}{2} \sum_{n=1}^{\infty} \sum_{m=1}^{\infty} a_n \left(\frac{n\pi}{L}\right)^2 a_m \left(\frac{m\pi}{L}\right)^2 \int_0^L \sin\left(\frac{n\pi x}{L}\right) \sin\left(\frac{m\pi x}{L}\right) dx$$

$$= \frac{EI}{2} \left(\frac{\pi}{L}\right)^4 \sum_{n=1}^{\infty} \sum_{m=1}^{\infty} a_n a_m n^2 m^2 \int_0^L \sin\left(\frac{n\pi x}{L}\right) \sin\left(\frac{m\pi x}{L}\right) dx. \tag{14.9.30}$$

Now, we recall again the orthogonality condition

$$\int_0^L \sin\left(\frac{n\pi x}{L}\right) \sin\left(\frac{m\pi x}{L}\right) dx = \begin{cases} L/2, & n = m \\ 0, & n \neq m. \end{cases} \tag{14.9.31}$$

Hence, the strain energy becomes

$$U = \frac{EIL}{4} \left(\frac{\pi}{L}\right)^4 \sum_{n=1}^{\infty} a_n^2 n^4. \tag{14.9.32}$$

Similarly, the potential energy of the applied external load [Eq. (14.9.23b)] is

$$V = -w \sum_{n=1}^{\infty} a_n \int_0^L \sin\left(\frac{n\pi x}{L}\right) dx,$$

which, upon integrating, yields

$$V = -w \sum_{n=1}^{\infty} a_n \left[-\frac{L}{n\pi} \cos\left(\frac{n\pi x}{L}\right)\right]_0^L$$

$$= w \sum_{n=1}^{\infty} a_n \frac{L}{n\pi} [\cos n\pi - 1].$$

Noting that $[\cos(n\pi) - 1]$ vanishes for even values of n, we have

$$V = -\frac{2wL}{\pi} \sum_{n=1,3,5,\ldots}^{\infty} \frac{a_n}{n}.$$

The total potential $\Pi = U + V$ therefore is

$$\Pi = \frac{EIL}{4}\left(\frac{\pi}{L}\right)^4 \sum_{n=1}^{\infty} a_n^2 n^4 - \frac{2wL}{\pi} \sum_{n=1,3,5,\ldots}^{\infty} \frac{a_n}{n}. \tag{14.9.33}$$

We observe that $\Pi = \Pi(a_n)$. Now, for equilibrium, $\delta\Pi = 0$; hence, for all $n \geq 1$, we require that

$$\frac{\partial\Pi}{\partial a_n} = 0. \tag{14.9.34}$$

Upon taking the derivative with respect to any a_i (noting that the coefficients are independent of each other; i.e. $\frac{\partial a_n}{\partial a_i} = 1$ if $n = i$ and zero if $n \neq i$), we have

$$\frac{EIL}{2}\left(\frac{\pi}{L}\right)^4 a_i i^4 = \frac{2wL}{\pi i}$$

or

$$a_n = \frac{4wL^4}{\pi^5 n^5 EI}, \quad n \text{ odd.} \tag{14.9.35}$$

Substituting in Eq. (14.9.29a), we then obtain

$$v(x) = \frac{4wL^4}{\pi^5 EI} \sum_{n=1,3,5,\ldots}^{\infty} \frac{1}{n^5} \sin\left(\frac{n\pi x}{L}\right). \tag{14.9.36}$$

At $x = L/2$, the deflection becomes

$$v\left(\frac{L}{2}\right) = \frac{4wL^4}{\pi^5 EI} \sum_{n \text{ odd}}^{\infty} \frac{1}{n^5} \sin\left(\frac{n\pi}{2}\right)$$

$$= 0.013071 \frac{wL^4}{EI} \sum_{n \text{ odd}}^{\infty} \left(1 - \frac{1}{3^5} + \frac{1}{5^5} - \frac{1}{7^5} + \cdots\right).$$

Note that the first term yields $v(L/2) = 0.013071\frac{wL^4}{EI}$ with a percentage error, when compared to Eq. (14.9.21b) of 0.386%. However, upon taking the first three terms, i.e.,

$$v(L/2) = 0.013071\frac{wL^4}{EI}\left(1 - \frac{1}{3^5} + \frac{1}{5^5}\right),$$

we obtain $v(L/2) = 0.013021\frac{wL^4}{EI}$ with an error of only 0.0047%.[†] Having found the deflection, we can immediately obtain internal stress resultants in the beam. For

---

[†] Note that the convergent series,

$$1 - \frac{1}{3^5} + \frac{1}{5^5} - \frac{1}{7^5} + \cdots = \frac{5\pi^5}{1536}.$$

Therefore, if an infinite number of terms were taken into account, one would obtain identically the deflection given by Eq. (14.9.21b).

example, using the relation $M(x) = -EIv''(x)$, and Eqs. (14.9.29b) and (14.9.35), the moment at the centre $x = L/2$ is found to be

$$M\left(\frac{L}{2}\right) = \frac{4wL^2}{\pi^3} \sum_{n=1,3,5,\ldots} \frac{1}{n^3} \sin\left(\frac{n\pi}{2}\right) = \frac{4wL^2}{\pi^3}\left(1 - \frac{1}{3^3} + \frac{1}{5^3} - \frac{1}{7^3} + \cdots\right).$$
(14.9.37)

Taking only the first term yields $M(L/2) = 0.12901wL^2$ while, using the first three terms we obtain $M(L/2) = 0.12526wL^2$, with relatively small percent errors of 3.2 and 0.21%, respectively, compared to the exact result, $M(L/2) = wL^2/8$.[†] It is interesting to observe that the series for the deflection, Eq. (14.9.36), converges as $\frac{1}{n^5}$ while that for the moment, Eq. (14.9.37), converges only as $\frac{1}{n^3}$. Although both series yield excellent results, it is a characteristic of the method that convergence is always more rapid for deflections in a body than for internal forces or stresses.

The Rayleigh–Ritz method can be applied to both statically determinate and indeterminate structures and is particularly useful in obtaining approximate solutions in the latter case. For example, if one assumes for a beam clamped at both ends and subjected to a uniform load w [Fig. (14.9.5)], an admissible function of the form

$$v(x) = \sum_{n=2,4,6,\ldots} a_n\left[1 - \cos\left(\frac{n\pi x}{L}\right)\right]$$
(14.9.38)

and applies the same method as above, one finds the following for the deflection at $x = L/2$ and the moment at $x = 0$ and $x = L$:

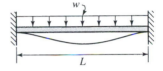

Figure 14.9.5

$$v\left(\frac{L}{2}\right) = \frac{wL^4}{4\pi^4 EI}\left(1 + \frac{1}{3^4} + \frac{1}{5^4} + \frac{1}{7^4} + \cdots\right)$$
(14.9.39a)

and

$$M(0) = M(L) = -\frac{wL^2}{2\pi^2}\left(1 + \frac{1}{2^2} + \frac{1}{3^2} + \frac{1}{4^2} + \cdots\right).$$
(14.9.39b)

The infinite series in brackets,

$$1 + \frac{1}{2^2} + \frac{1}{3^2} + \frac{1}{4^2} + \cdots = \frac{\pi^2}{6}.$$
(14.9.40)

Thus, upon taking the infinite terms into account, we obtain $M(0) = M(L) = -wL^2/12$. [Note that this value was used in Eq. (14.8.29).]

Finally, it is worthwhile to mention that the Rayleigh–Ritz method as presented here in the context of the principle of stationary potential energy forms a fundamental basis for numerical methods of solution such as the finite element method. This method yields approximate solutions to important practical problems in engineering that are not amenable to a tractable analytic treatment. However, a treatment of this method is beyond the scope of our discussion.

[†] Note that the convergent series

$$1 - \tfrac{1}{3^3} + \tfrac{1}{5^3} - \tfrac{1}{7^3} + \cdots = \tfrac{\pi^3}{32}.$$

Therefore, if an infinite number of terms were taken into account, one would obtain identically the exact value for the moment, $M(L/2) = wL^2/8$.

14.10   Summary and conclusions

In this chapter, the fundamental principles of energy and virtual work have been established. These have a wide application in the field of solid and structural mechanics. As we have seen, they lead to various methods, either to determine equilibrium conditions or to find deflections of deformable bodies under applied loading systems.

We conclude with a summary of the above-mentioned principles according to their validity of application.

General (elastic and dissipative)

■ Principle of virtual work
■ Principle of complementary virtual work

Elastic: linear or nonlinear

■ Conservation of energy
■ Theorem of stationary potential energy
■ Castigliano's first theorem

Elastic: linear

■ Betti's law and Maxwell's reciprocal theorem
■ Castigliano's second theorem

PROBLEMS

Sections 2–3

14.1:   For a linear elastic material, with $\tau_{yz} = \tau_{zx} = 0$, the strain energy density U_0 due to shear is given by

$$U_0 = \tau_{xy}\epsilon_{xy} = \frac{\tau_{xy}^2}{2G}. \tag{1}$$

Consider a linear elastic prismatic beam of length L having a rectangular cross-sectional area A, which is subjected to lateral loads in the y-direction such that the shear $V(x) = V_y(x)$, $V_z = 0$. Starting from the expression for U_0 above, show that

$$U = \frac{\alpha}{2} \int\limits_0^L \frac{V^2(x)}{AG}\, dx \tag{2}$$

where $\alpha = 1.2$. (*Note*: The coefficient α is called a **shape factor** since it depends on the shape of the cross-section.)

14.2:   An elastic prismatic cantilever beam of rectangular cross-section is subjected to a force P at the free end, as shown in Fig. (14P.2).

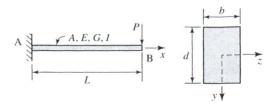

Figure 14P.2

(a) Using Eq. (2) of Problem 14.1 above, apply the principle of conservation of energy to determine the vertical deflection Δ_s of point B due to shear deformation.

(b) Determine the ratio $\frac{\Delta_s}{\Delta_f}$ of point B in terms of d, L, E and G, where Δ_f, the displacement due to flexural deformation, is $\Delta_f = PL^3/3EI$.

(c) Based on the results of (b) above, what conclusion may be established concerning the displacements due to shear deformation of long slender beams.

14.3:* The strain energy density of an elastic material under a three-dimensional state of stress and deformation is given by the quadratic expression

$$U_0 = \frac{1}{2}\left[\lambda\Delta^2 + 2\mu\left(\epsilon_{xx}^2 + \epsilon_{yy}^2 + \epsilon_{zz}^2 + \epsilon_{xy}^2 + \epsilon_{yz}^2 + \epsilon_{zx}^2 + \epsilon_{yx}^2 + +\epsilon_{zy}^2 + \epsilon_{xz}^2\right)\right],$$

where $\Delta \equiv \epsilon_{xx} + \epsilon_{yy} + \epsilon_{zz}$ is the dilatation and where λ and μ are material constants (known as the *Lamé constants*).

(a) Using the relations $\tau_{xx} = \frac{\partial U_0}{\partial \epsilon_{xx}}$, $\tau_{xy} = \frac{\partial U_0}{\partial \epsilon_{xy}}$, etc., express τ_{xx}, τ_{yy}, τ_{zz}, τ_{xy} in terms of the strain components.

(b) Is the material *linearly* elastic? If U_0 were expressed in terms of strains to the fourth power instead of quadratics, what could be said about the stress–strain relations?

(c) Can U_0 be expressed in terms of third powers of the strain components for all values of strain? Explain why/why not.

(d) From the results of part (a), solve for the extensional strain components in terms of the normal stress components τ_{xx}, τ_{yy}, τ_{zz}. Write the shear stress components in terms of the relevant strain components.

(e) What combination of λ and μ lead to Hooke's law in the form

$$\epsilon_{xx} = \frac{1}{E}[\tau_{xx} - \nu(\tau_{yy} + \tau_{zz})], \qquad \epsilon_{xy} = \frac{\tau_{xy}}{2G}, \qquad \text{(etc.)},$$

where E and ν are the modulus of elasticity and Poisson ratio of the material, respectively, and G is the shear modulus.

Sections 4–6

14.4: Consider a linear body subjected to two forces P_1 and P_2 applied statically to its surface S, as shown in Fig. (14P.4). Let Δ_i denote the component of displacement of the point of application of the force P_i ($i = 1, 2$) in the direction of the force P_i. The external work done by the two forces is then

$$W = \frac{1}{2}P_1\Delta_1 + \frac{1}{2}P_2\Delta_2 = \frac{1}{2}\sum_{i=1}^{2}P_i\Delta_i. \qquad (1)$$

Assume now that the forces P_1 and P_2 are applied sequentially; namely P_1 is first applied and then P_2. The work done is then

$$W = \frac{1}{2}P_1D_{11} + P_1D_{12} + \frac{1}{2}P_2D_{22}, \qquad (2)$$

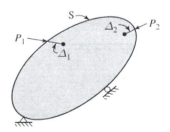

Figure 14P.4

where D_{ij} is the component of displacement of P_i (in the direction of P_i) due to the force P_j. (a) Prove that the two expressions for the work of the external forces, Eqs. (1) and (2), are equivalent. (b) Show that if the order of applications is reversed (i.e., if P_2 is applied first and then P_1), the external work is the same.

Note: In Problems 14.5–14.13, determine the deflections and rotations due to flexural deformation unless otherwise specified.

14.5: Given the linear elastic beam shown in Fig. (14P.5). By means of Castigliano's second theorem, determine (a) the vertical displacement of point B, (b) the vertical displacement of point A and (c) the rotation of point A.

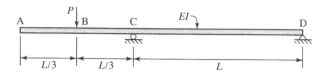

Figure 14P.5

14.6:   An elastic structural member ABCD, simply supported at C and D, is subjected to an applied moment M_0 at A and a horizontal force P, as shown in Fig. (14P.6). (a) Using Castigliano's second theorem, determine the horizontal component of displacement of point A in terms of M_0, P, E, I, L and h. (b) From the answer to part (a) above, determine the angle that member AB makes with the vertical at point A due to a horizontal force $P = 1$ applied at A.

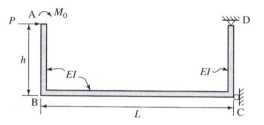

Figure 14P.6

14.7:   Beam A–D of flexural rigidity $E\,I$, simply supported at A, B and D and containing a hinge at point C, is subjected to a uniformly distributed load w, as shown in Fig. (14P.7). (a) Using Castigliano's second theorem, determine the deflection of point G and (b) sketch the deflected shape of the beam.

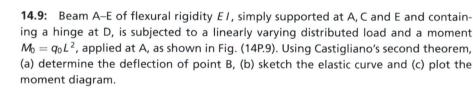

Figure 14P.7

14.8:   By means of Castigliano's second theorem, determine (a) the rotation of point A and (b) the vertical displacement of point A of the structure shown in Fig. (14P.8).

14.9:   Beam A–E of flexural rigidity $E\,I$, simply supported at A, C and E and containing a hinge at D, is subjected to a linearly varying distributed load and a moment $M_0 = q_0 L^2$, applied at A, as shown in Fig. (14P.9). Using Castigliano's second theorem, (a) determine the deflection of point B, (b) sketch the elastic curve and (c) plot the moment diagram.

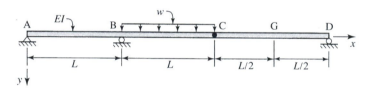

Figure 14P.8

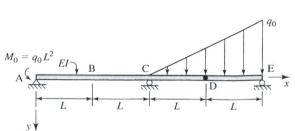

Figure 14P.9

14.10: The simply supported beam ABC of flexural rigidity EI, as shown in Fig. (14P.10), is subjected to a uniformly distributed load w within span BC. What moment M_0 must be applied at C in order to cause the slope at A, θ_A, to be zero?

14.11: (a) Using Castigliano's second theorem, determine the vertical displacement $v(x)$, $0 \le x \le 2L$, of the beam of flexural rigidity EI, containing a hinge at B, when loaded as shown in Fig. (14P.11a). Sketch the deflection $v(x)$. (b) What is the deflection v_B of point B if the load is applied at any point $x = a$ $(0 \le a \le 2L)$, as shown in Fig. (14P.11b)? Write the expression for v_B and sketch the resulting deflection of the beam.

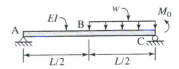

Figure 14P.10

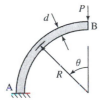

(a) (b)

Figure 14P.11

14.12: For a curved beam whose lateral dimensions are small with respect to the radius of curvature, the flexural and normal strain energy may be expressed as

$$U = \frac{1}{2} \int \frac{M^2}{EI}\, ds + \frac{1}{2} \int \frac{N^2}{AE}\, ds,$$

where A and I are the area and moment of inertia of the cross-section and $M = M(\theta)$, $N = N(\theta)$ are the moments and axial forces at a cross-section. (Note that the integrals represent the elastic strain energy due to flexural and axial deformation, respectively.)

(a) Using Castigliano's second theorem, determine the horizontal component of deflection of point B of Fig. (14P.12) due to a vertical load. Consider both axial and flexural deformation.

Figure 14P.12

(b) Let $\Delta_{B,axial}$ and $\Delta_{B,flex}$ represent the displacement due to axial and flexural deformation, respectively. If the cross-section of the beam is rectangular ($b \times d$), $d \ll R$, show that

$$\frac{\Delta_{B,axial}}{\Delta_{B,flex}} = \frac{1}{12}\left(\frac{d}{R}\right)^2.$$

(c) What conclusion can be stated regarding the importance of the two kinds of deformation?

14.13: Using Castigliano's second theorem, determine the deflection in the y-direction of the free end of the elastic cantilever beam whose cross-section varies as shown in Fig. (14P.13).

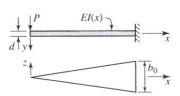

Figure 14P.13

14.14: For the pin-connected truss loaded as shown in Fig. (14P.14) (with axial rigidity AE the same for all members), determine, by means of Castigliano's second theorem, (a) the vertical displacement of point B, (b) the horizontal displacement of point C and (c) the vertical displacement of point C.

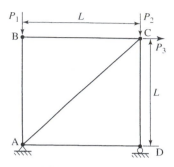

Figure 14P.14

14.15:* A solid cylindrical rod of length L, consisting of a linear isotropic elastic material (with modulus of elasticity E and Poisson ratio v), is subjected, at some arbitrary cross-section, to two equal and opposite colinear radial forces P, as shown in Fig. (14P.15). Determine the change in length of the cylinder, ΔL, in terms of P, the material constants E and v and the geometry.

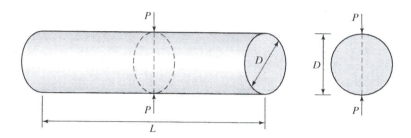

Figure 14P.15

Section 7

14.16: Given the cantilever beam of arbitrary material behaviour, subjected to a force at A, as shown in Fig. (14P.16). Show that an arbitrary kinematically *inadmissible* virtual displacement (i.e., one which violates the boundary conditions) of the form of Eq. (14.7.25), namely

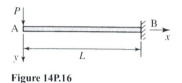

Figure 14P.16

$$\delta v = \delta A \sin \frac{n\pi x}{2L}, \quad n > 0,$$

leads to the correct reactions at B.

14.17: Beam A–E, containing hinges at B and D, is subjected to a uniformly distributed load w over the span CDE and a couple $M_0 = wL^2$ at point E, as shown in Fig. (14P.17). Using the principle of virtual work, determine (a) the moment M_A, (b) the reaction at C and (c) the moment M_C.

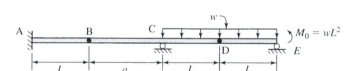

Figure 14P.17

Figure 14P.18

14.18: Using the principle of virtual work, determine the forces F_{AB} and F_{BC} of the statically determinate pin-connected truss shown in Fig. (14P.18).

14.19:* The statically determinate structure shown in Fig. (14P.19) consists of elements rigidly connected at B and F and connected by hinges at A, C, D, E and G. Loads P_1 and P_2 are applied at D. (a) Using the principle of virtual work, determine (i) the reactions R_{Ax} and R_{Ay}, (ii) the moment M_B in member AB at point B and (iii) the shear force V_F in member FG at point F. [*Note:* For each case show, by means of a sketch, the virtually displaced structure (with respect to the original structure) which is used such that only the desired unknown appears in the virtual work expression.] (b) Check the answers using the equations of equilibrium.

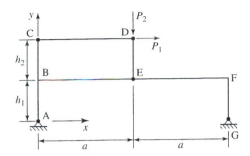

Figure 14P.19

Note: In Problems 14.20–14.22, use the principle of virtual work to draw the influence lines (due to applied downward loads) for the structure without first determining analytical expressions for the relevant influence functions. Use the adopted sign convention for moments and shear and indicate all critical ordinates of the influence lines.

14.20: The statically determinate structure A–E shown in Fig. (14P.20) contains hinges at points B and D. Draw the influence lines due to a downward force for (a) the upward reaction R_C, (b) the moment M_E at point E, (c) the shear force V_D at D, (d) the shear force V_C^+ immediately to the right of point C and (e) the shear force V_C^- immediately to the left of point C.

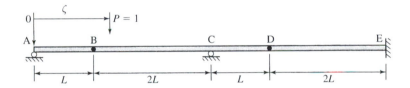

Figure 14P.20

14.21: The statically determinate structure A–I shown in Fig. (14P.21), containing hinges at C, E and H, is simply supported at A, B and F and is fixed at I.

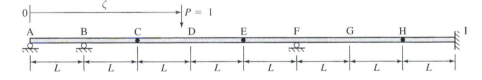

Figure 14P.21

(a) Sketch directly the influence lines due to a downward force for the following quantities: (i) the upward reaction at B, (ii) shear V_G at point G, (iii) the moment M_G at G, (iv) the upward reaction at I and (v) the shear V_B^+ immediately to the right of point B.

(b) Given a downward uniformly distributed load w acting over a span of $2L$. (i) Between which points [nL and $(n+2)L$, (n is an integer)] should the loading be applied to cause $|M_G|$ to be a minimum? What is the resulting minimum value of $|M_G|$? (ii) Repeat (i) for maximum $|M_G|$.

(c) Repeat (b)(i) and (ii) above for $|R_I|$.

14.22: The statically determinate structure A–H shown in Fig. (14P.22), containing hinges at E and G, is simply supported at A and C and fixed at H.

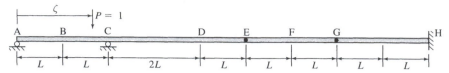

Figure 14P.22

(a) Sketch directly the influence lines for the following quantities: (i) the upward reaction at A, (ii) the moment M_D at point D, (iii) the upward reaction R_H at H, (iv) the shear V_C^- immediately to the left of point C and (v) the shear V_C^+ immediately to the right of point C.

(b) Given a downward uniformly distributed load w acting over a span of L. (i) Between which points [nL and $(n+1)L$, (n an integer)] will the loading cause $|M_D|$ to be maximum? What is the resulting value of $|M_D|_{max}$? (ii) Repeat (i) for maximum V_C^-.

(c)* Repeat (b) if the given downward load w can be placed between any two points ζ_0 and $\zeta_0 + L$.

14.23: Based on the Müller–Breslau principle, sketch qualitatively the general shape of the influence lines of the indeterminate structure A–E, containing a hinge at B, as shown in Fig. (14P.23), for the following quantities: (a) R_A, (b) M_C, (c) M_D, (d) V_B, (e) R_D and (f) V_D^-.

Figure 14P.23

Section 8

14.24: Solve Problem 14.5 using the principle of complementary virtual work (method of virtual forces).

14.25: Repeat Problem 14.11 using the principle of complementary virtual work (method of virtual forces).

14.26: The linear elastic structure shown in Fig. (14P.26) consists of a beam ABCD supported by members AE and CE. The properties of the members are shown in the figure.

(a) Using the principle of complementary virtual work, determine the vertical displacement Δ_B of point B taking into account both axial and flexural deformation.

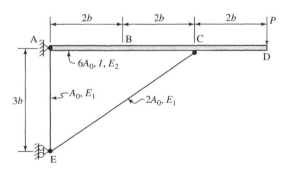

Figure 14P.26

(b) The following values are given for the properties and the load : $A_0 = 10$ cm^2, $b = 2$ m, $I = 50 \times 10^2$ cm^4, $P = 2000$ N, $E_1 = E_2 = 200$ GPa (steel). (i) Evaluate the displacement Δ_B numerically. (ii) What is the displacement due to axial deformation and flexural deformation? What percentage of the total displacement does each deformation contribute? (iii) If bars AE and CE are rigid (i.e., $E_1 \to \infty$), what is the displacement Δ_B?

(c) What conclusion can be reached concerning the relative importance of the axial deformation in the various members, based on the results of part (b) above?

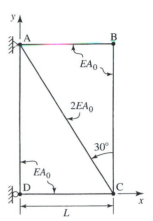

Figure 14P.28

14.27: Repeat Problem 14.9 using the principle of complementary virtual work.

14.28:* A pin-connected truss consists of five elastic members whose axial rigidity is as indicated in Fig. (14P.28). Member AC undergoes a temperature change $\Delta T > 0$ and member CD undergoes a temperature change $2\Delta T > 0$. (The remaining members undergo no temperature change.) The coefficient of thermal expansion of all members is given as α ($^\circ C^{-1}$). Determine the component of displacement of point C in the y-direction by means of the principle of complementary virtual Work.

14.29: Solve Problem 14.13 using the principle of complementary virtual work.

14.30: Repeat Problem 14.7 using the principle of complementary virtual work.

14.31: The statically indeterminate beam A–E of flexural rigidity EI [Fig. (14P.31a)] is subjected to a uniformly distributed load w over its entire length. (a) By means of the principle of complementary virtual work (method of virtual forces), determine the deflection of points B and D due to flexure using the statically admissible virtual force system. (*Note*: Statically admissible reactions for the indeterminate structure are given in Appendix F.) (b) Obtain the solution using the (statically inadmissible) symmetric virtual force system shown in Fig. (14P.31b).

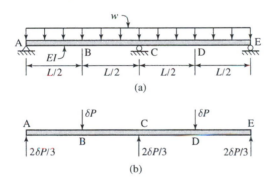

(a)

(b)

Figure 14P.31

14.32:* A statically indeterminate beam of uniform cross-section is subjected to a uniform load, as shown in Fig. (14P.32a). (The statically admissible reactions are given in Appendix F.)

(a) Using the principle of complementary virtual work, determine the vertical displacement of point F. For the internal complementary virtual work of the entire beam, use the statically admissible virtual force system as shown in Fig. (14P.32b). (External reactions due to a load P acting at F are given in Appendix F.) (*Note*: Take advantage of symmetry.)

(b) Using the principle of complementary virtual work, determine the vertical displacement of point F if the equilibrium state of stress is $\sigma_{xx} = 0$ between A and B and between C and D, i.e., if the state of virtual equilibrium stresses is statically inadmissible, namely that of a simply supported beam, as shown in Fig. (14P.32c).

(c) Explain precisely in terms of the principle of complementary virtual work, the reason the scheme of Fig. (14P.32c) will give the correct solution. Explain further why using a scheme, as shown in Fig. (14P.32d) (with known end reactions and

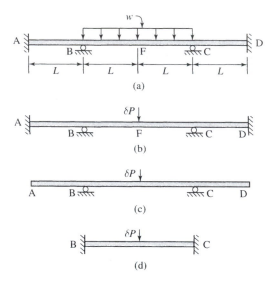

Figure 14P.32

moments) one nevertheless requires additional information. What is this information?

14.33: A cantilever beam having a cross-section with uniform depth d, originally at a reference temperature T_0, is subjected to a linear temperature variation ΔT, as shown in Fig. (14P.33), causing the beam to bend. The beam material is linear elastic with a stress–strain relation given by $\sigma_{xx} = E\epsilon_{xx}$. Using the principle of complementary virtual work for deformable bodies and assuming that the only resulting stresses are flexural stresses and that plane cross-sections remain plane, show that the vertical deflection of point B is given by

$$\Delta = (\alpha \Delta T L^2)/d,$$

where α is the coefficient of thermal expansion.

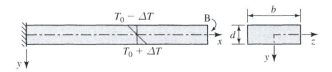

Figure 14P.33

Section 9

14.34: The structure shown in Fig. (14P.34) consists of two rigid rods connected by a hinge at B. Using the principle of stationary potential energy, determine the equilibrium position of the structure.

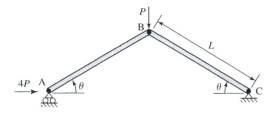

Figure 14P.34

14.35:* A rigid rod ABC of length $2L$ is simply supported at A and by wires having axial rigidity AE, as shown in Fig. (14P.35). The rod is subjected to a force P at C. Using the principle of stationary potential energy, determine (a) the vertical component of displacement of point C and (b) the forces in the wires. Assume that the system undergoes small rotations.

14.36:* Two rigid rods AB and BC each of length L are connected by a hinge at B. The two degree-of-freedom system is simply supported at A and by wires each having axial rigidity AE, as shown in Fig. (14P.36). The rods are loaded over their entire length by a uniformly distributed load w (N/m). Using the principle of stationary potential energy, determine (a) the vertical component of displacement of points B and C and (b) the forces in the wires. (*Note:* Assume all displacements to be small with respect to L.)

14.37: (a) Using the Rayleigh–Ritz method and assuming a kinematically admissible function of the form

$$v(x) = \sum_{n=1,3,5,\dots}^{\infty} a_n\left(1 - \cos\frac{n\pi x}{2L}\right),$$

obtain an approximate solution for the deflection $v(x)$, moments and shear forces in a cantilever beam of length L subjected to a uniformly distributed load w, as shown in Fig. (14P.37). Consider only flexural deformation. (b) Evaluate $v(L)$ and $M(0)$ using several terms of the series and compare with the exact linear solution, $v(L) = wL^4/8EI$, $M(0) = -wL^2/2$.

14.38: By means of the Rayleigh–Ritz method, determine the displacement and moment at the mid-point of the statically indeterminate beam shown in Fig. (14P.38), using the admissible function $v(x)$ given by Eq. (14.9.38). Evaluate these quantities using the first four terms of the series and calculate the percent error when compared with the exact values given in Appendix F.

Review and comprehensive problems

14.39: Given an arbitrary elastic body, suitably supported, to which a single concentrated load P is applied. Can the the point of application of P displace in a direction perpendicular to the load? Explain the reasoning.

14.40:* The strain energy due to shear deformation of an elastic beam is given by $U = \frac{\alpha}{2}\int_0^L \frac{V^2(x)}{AG}\,dx$, where α is the shape factor (see Problem 14.1). Determine the shape factor for a beam having a circular cross-section of radius R.

14.41: The statically determinate structure A–I of Fig. (14P.21) contains hinges at C, E and H. Using the principle of virtual work, sketch directly the influence lines due to a downward concentrated force for the following quantities: (i) the upward reaction at A, (ii) the upward reaction at F (iii) the moment at I, (iv) the shear force V_F^- (i.e., immediately to the left of point F) and (v) the shear force V_F^+. (See note immediately preceding Problem 14.20.)

14.42: The statically determinate structure, A–J of Fig. (14P.42) contains hinges at C, F and I.

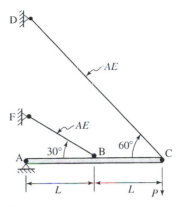
Figure 14P.35

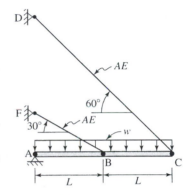
Figure 14P.36

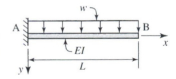
Figure 14P.37

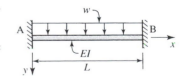
Figure 14P.38

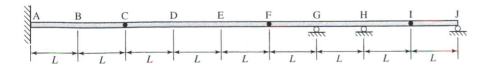
Figure 14P.42

(a) Using the principle of virtual work, sketch directly the influence lines due to a downward concentrated force for the following quantities: (i) M_A, (ii) V_H^+, (iii) V_E, (iv) the upward reaction at H and (v) M_E. (See note immediately preceding Problem 14.20.)

(b) The structure is to be designed to supported a piece of machinery modelled as a uniformly distributed load w (N/m) over a span $2L$. (i) Between which two points, nL and $(n+2)L$, (n is an integer), will the load cause the absolute value of the moment $|M_A|$ to be a maximum? What is the resulting value of $|M_A|_{max}$. (ii) Between which two points, ζ_0 and $\zeta_0 + 2L$ (ζ_0 not necessarily an integer), will the load cause the absolute value of the moment $|M_A|$ to be a maximum? What is the resulting value of $|M_A|_{max}$?

(c) Repeat (b) for $(R_H)_{max}$.

14.43:   The statically determinate structure A–E shown in Fig. (14P.43a) containing hinges at B and D, is fixed at A and E. (a) Using the principle of virtual work, sketch directly the influence lines for the following quantities: (i) the upward reaction at A, (ii) the moment M_A and (iii) the the shear V_C at C. Show values of all critical ordinates. (b) Given a downward linearly varying distributed load $q(x)$ acting over the span BD, as shown in Fig. (14P.43b). Based on (a), determine (i) R_A, (ii) M_A and (iii) V_C due to the given loading.

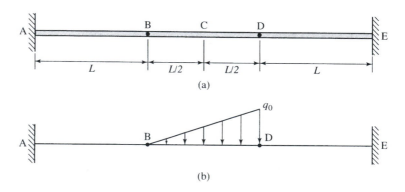

(a)

(b)

Figure 14P.43

14.44:* Two identical members AB and BC having axial and flexural rigidities AE and EI, respectively, are rigidly connected at B, as shown in Fig. (14P.44). Determine the change of distance $\Delta_{|AC|}$ due to the load P. Consider both flexural and axial deformation.

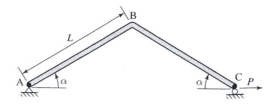

Figure 14P.44

14.45:* From the results of Problem 14.33, it is observed that the displacement of point B is dependent only on the coefficient of thermal expansion, α, and not on the material constant E. Repeat Problem 14.33 using the following conditions: (a) The beam is made of some *arbitrary material* whose only known property is α. (b) *Do not assume* that plane sections remain plane.

14.46: Based on the Müller–Breslau principle, sketch qualitatively the shape of the indeterminate structure ABCD of Fig. (14P.46) for the following quantities: (i) M_C, (ii) V_B^-, (iii) V_B^+ and (iv) V_C.

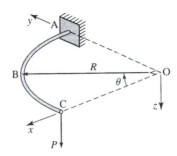

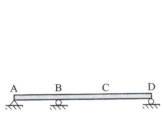

Figure 14P.46   **Figure 14P.47**

14.47:* A thin elastic member ABC in the shape of a quarter circle of radius R lies in the x–y plane, as shown in Fig. (14P.47). The member is subjected to a force P acting in the z-direction. The flexural and torsional rigidities of the member are given as $E\,I$ and GC, respectively. Determine the component of deflection in the z-direction of point C due to flexural and torsional deformation.

14.48:* Given a three-dimensional linear isotropic elastic body (with modulus of elasticity E and Poisson ratio v) having an arbitrary shape and subjected to two colinear forces P acting at the surface S of the body, as shown in Fig. (14P.48). Derive an expression for the change in volume, ΔV, of the body in terms of P, E, v and b, the distance between the two forces.

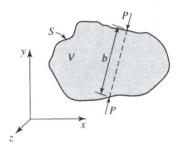

Figure 14P.48

14.49: A force P acts at point A of an elastic frame ABC at an angle α with respect to the x-axis, as shown in Fig. (14P.49). (a) Determine the value(s) of α if the resultant displacement of point A due to flexural deformation is in the direction of the force P. (b) (i) Show, for the given geometry of the frame, that the resultant displacement of point A due to axial deformation is *always* parallel to P. (ii) If AB and BC were not of the same length, would this be true?

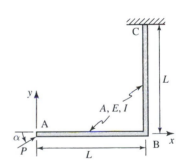

Figure 14P.49

14.50:* (a) Show, by means of Castigliano's second theorem, that the resultant displacement of point A of the linear elastic frame of Fig. (14P.49) due to flexural deformation can never be perpendicular to P. (b) Show that this statement is valid also for the case of axial deformation.

14.51: Using the Rayleigh–Ritz method and the admissible function given in Problem 14.37, obtain an approximate solution for the deformation, moment and shear in a cantilever beam of length L subjected to a concentrated force P at the free end, as shown in Fig. (14P.51). Evaluate the quantities using the first four terms of the series and determine the percentage error with the known solution.

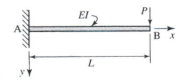

Figure 14P.51

15

Stability of mechanical systems by energy considerations: approximate methods

15.1 Introduction

In the previous chapter, we have found that a system is in equilibrium if, and only if, the total potential of the system, $\Pi = U + V$, is stationary; that is, if the value of Π is an 'extremum'. Moreover, we recall (from Chapter 11) that there exist several types of equilibrium: stable, unstable and neutral equilibrium. As we shall see, the character of the equilibrium state is intimately related to the type of extremum of the total potential energy of the system. Indeed, it will become evident that the nature of the equilibrium state can be clearly and rigorously established only by energy considerations.

Although we shall mainly investigate the stability of simple members subjected to axial forces, the ideas that will be developed are quite general. However, in addition to their theoretical importance, energy considerations are also of great practical value as they lead to approximate methods for determining critical forces of systems. Such approximate solutions are of importance particularly in complex structural systems where, very often, exact analytic solutions cannot be obtained.

15.2 Classification of equilibrium states according to energy criteria

We first recall (from Chapter 11) that a system is in a stable equilibrium position if, when given a small perturbation, it returns to the original position. Conversely, if the system does not return to the original position when given a small perturbation (but moves instead to a new equilibrium state), the position is said to be in unstable equilibrium. Finally, if the system neither returns to the original equilibrium position nor moves further away, it is said to be in neutral equilibrium.

We now wish to relate the type of equilibrium state to the energy of the system. We first approach this study by considering the same intuitive case as used in Chapter 11, namely, a model consisting of a *rigid* sphere of mass M, which, under the effect of a gravitation force, can roll on a track in the x–y plane, defined by a prescribed function $y = f(x)$ [see Fig. (15.2.1)]. From our previous study, stable,

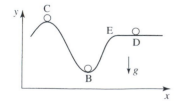

Figure 15.2.1

646

unstable and neutral equilibrium positions are known to exist at points B, C and D respectively.

Now, the forces acting on the sphere are gravity g and the reactive force of the track. In using the principle of stationary potential energy, the perturbation of the system is given in terms of *virtual displacements*. However, we recall that in deriving this principle, the virtual displacements are limited to those that are *kinematically admissible*; that is, they must satisfy the constraints of the system. For the given system here, the sphere is constrained to roll along the track; therefore, virtual displacements must be taken as displacements *along the track*. Consequently, there can be no contribution of the normal force (which the track exerts on the sphere) to the potential energy of the system. The potential energy is therefore only that of the gravity force which (if measured arbitrarily from the x-axis, $y = 0$) is given by $V = Mg \cdot y(x)$. Furthermore, since the sphere is rigid, no strains exist, and therefore the strain energy $U = 0$. Hence, for this special case, $\Pi = V = Mg \cdot y$. Note that $\Pi = \Pi[y(x)]$, that is, Π is a function only of the single variable, x through $y = y(x)$.

We note that for all points to the left of E, except at points B and C,

$$\delta \Pi = \left(\frac{d\Pi}{dy} \right) \delta y = Mg \left(\frac{dy}{dx} \right) \delta x \neq 0. \qquad (15.2.1)$$

Thus the principle of stationary potential energy leads us to the obvious but trivial conclusion that equilibrium can exist only at points B, C and at all points to the right of point E, where $\frac{dy}{dx} = 0$.

We note that at B, a point of stable equilibrium, Π is a relative minimum; any change in Π from its original equilibrium position is positive, i.e. $\Delta \Pi > 0$.

Conversely, at C, a point of unstable equilibrium, Π is a maximum and any change in Π from its original equilibrium position is negative, i.e. $\Delta \Pi < 0$.

Consider now the sphere at some point D along the horizontal track, which we recall is a neutral equilibrium state. It is clear that here Π is neither a relative maximum or minimum, for there is no change in the potential due to a small perturbation, that is $\Delta \Pi = 0$. In fact, we note here that according to our model, neutral equilibrium implies that *all* derivatives

$$\frac{d^n \Pi}{dx^n} = 0, \quad n \geq 1. \qquad (15.2.2)$$

Thus, we may summarise the results as follows:

- $\Delta \Pi > 0$ corresponds to a stable equilibrium state,
- $\Delta \Pi < 0$ corresponds to an unstable equilibrium state,
- $\Delta \Pi = 0$ corresponds to a neutral equilibrium state.

We emphasise here that $\Delta \Pi$ is the difference in the potential energy from an equilibrium position. Using this simple model, we have established a direct relation between the stability and instability of a system in terms of its total potential Π. In determining the critical loads for stability of various systems, we shall find, as in Chapter 11, that the concept of neutral (and what will later be defined as *pseudo-neutral*) equilibrium plays an important role.

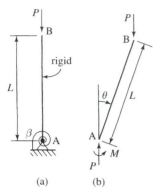

Figure 15.3.1

15.3   Stability of a rigid rod subjected to a compressive axial force

Using the definition of stability/instability in terms of energy, let us reconsider the same system as was studied in Section 3 of Chapter 11: namely, a rigid rod AB of length L supported at point A by a linear torsional spring having a stiffness β [Fig. (15.3.1a)]. The rod is subjected to a vertical axial force P (which always acts in the downward y-direction). We recall that if the rod is given a rotation θ from its original vertical position, as in Fig. (15.3.1b), the spring exerts a moment $M = \beta\theta$ that tends to bring the rod back to its original position.

Since the position of this system is described by a single variable θ, the system is said to have a *single degree-of-freedom* (1d.o.f). In *mechanics*, the variable θ, which may vary arbitrarily, is called a **generalised coordinate** of the system. Furthermore, it is important to note that here we consider the spring to be part of the mechanical system.

We seek (a) to establish the magnitude of P that is required to maintain the rod in equilibrium positions $|\theta| > 0$, and (b) to determine for which forces P the original position $\theta = 0$ is a stable or unstable equilibrium position.

For any position θ, the strain energy U of the spring is

$$U = \frac{1}{2} M_A \theta = \frac{\beta}{2} \theta^2, \tag{15.3.1a}$$

while the potential energy of the force P is

$$V = -PL(1 - \cos\theta). \tag{15.3.1b}$$

Hence

$$\Pi(P, \theta) = U + V = \frac{\beta\theta^2}{2} - PL(1 - \cos\theta). \tag{15.3.2}$$

According to the principle of stationary potential energy, for equilibrium we require that

$$\delta\Pi = \frac{d\Pi(\theta)}{d\theta} \delta\theta = 0 \tag{15.3.3a}$$

and since this must be true for *any* variation $\delta\theta$, the equilibrium condition becomes

$$\frac{d\Pi(\theta)}{d\theta} = 0, \tag{15.3.3b}$$

that is,

$$\beta\theta - PL\sin\theta = 0. \tag{15.3.4}$$

First note that for $\theta = 0$, *all* finite values of P satisfy this equilibrium condition; i.e., the rod is in equilibrium in its original vertical position for all forces P.

For $\theta \neq 0$, the equilibrium force is

$$P = \frac{\beta/L}{\left(\frac{\sin\theta}{\theta}\right)}, \qquad 0 < |\theta| < \pi. \tag{15.3.5}$$

This solution represents the force P required to maintain the rod in equilibrium for any position θ. From the P–θ relation, shown by the heavy line in Fig. (15.3.2), we note that as $\theta \to 0$, $P \to \beta/L$. Solutions of Eq. (15.3.4) are given by points lying

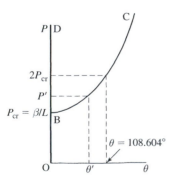

Figure 15.3.2

along the branches OB, BC and BD, where we recall that point B is referred to as a *bifurcation point*.[†] Note that these same results were obtained in Chapter 11.

For a given applied force $P < \beta/L$, the only possible equilibrium position is $\theta = 0$. Thus, for such loads, if the rod is given a rotation θ, it will always return to the $\theta = 0$ position since no other equilibrium position exists for these values of P. Hence $\theta = 0$ necessarily represents a stable equilibrium position for any $P < \beta/L$.[‡] Consider now that P is slowly increased from $P = 0$ such that it reaches $P = \beta/L$ [point B in Fig. (15.3.2)]. If we increase P further, say to P', then we note that θ may assume two different values: $\theta = 0$ or $\theta = \theta'$ for the given P'. Thus two different equilibrium branches emanate from the bifurcation point B: path BC and path BD. To investigate the nature of the equilibrium positions represented by points on the various equilibrium paths, using the energy criteria established in Section 15.2, we examine the change $\Delta\Pi$ of the total potential due to a given virtual displacement $\delta\theta$. To this end, let us first expand Π in a power series about any equilibrium position, θ_{eq}.[§]

$$\Pi(\theta_{eq} + \delta\theta) = \Pi(\theta_{eq}) + \left.\frac{d\Pi(\theta)}{d\theta}\right|_{\theta_{eq}} \delta\theta + \frac{1}{2!}\left.\frac{d^2\Pi(\theta)}{d\theta^2}\right|_{\theta_{eq}} (\delta\theta)^2$$

$$+ \frac{1}{3!}\left.\frac{d^3\Pi(\theta)}{d\theta^3}\right|_{\theta_{eq}} (\delta\theta)^3 + \cdots. \qquad (15.3.6a)$$

Therefore

$$\Delta\Pi \equiv \Pi(\theta_{eq} + \delta\theta) - \Pi(\theta_{eq}) = \left.\frac{d\Pi(\theta)}{d\theta}\right|_{\theta_{eq}} \delta\theta + \frac{1}{2!}\left.\frac{d^2\Pi(\theta)}{d\theta^2}\right|_{\theta_{eq}} (\delta\theta)^2$$

$$+ \frac{1}{3!}\left.\frac{d^3\Pi(\theta)}{d\theta^3}\right|_{\theta_{eq}} (\delta\theta)^3 + \cdots. \qquad (15.3.6b)$$

Since $\theta = \theta_{eq}$ represents an equilibrium position, by Eq. (15.3.3b), the first derivative vanishes, and therefore

$$\Delta\Pi = \frac{1}{2!}\left.\frac{d^2\Pi(\theta)}{d\theta^2}\right|_{\theta_{eq}} (\delta\theta)^2 + \frac{1}{3!}\left.\frac{d^3\Pi(\theta)}{d\theta^3}\right|_{\theta_{eq}} (\delta\theta)^3 + \frac{1}{4!}\left.\frac{d^4\Pi(\theta)}{d\theta^4}\right|_{\theta_{eq}} (\delta\theta)^4 + \cdots.$$

$$(15.3.7)$$

Now if we take $\delta\theta$ as a small variation from the equilibrium position θ_{eq}, it is evident that *the sign of $\Delta\Pi$ is given by the sign of the lowest non-vanishing derivative of even order*. This then is the explicit criterion to determine the stability or instability of any mechanical system. For the problem at hand, from Eq. (15.3.2),

$$\frac{d^2\Pi}{d\theta^2} = \beta - PL\cos\theta. \qquad (15.3.8)$$

[†] The heavy line of the figure may be considered as a 'loading path' starting from point O ($P = 0$) to point B and then proceeding along either branch with increasing P.

[‡] This conclusion, derived by a simple logical argument, will be shown to be rigorously correct using the energy criteria of the preceding section.

[§] We assume that Π is an *analytical function*, which can be expanded in a Taylor series. By this we mean here that Π is continuous and possesses all derivatives at $\theta = \theta_{eq}$.

We first examine the derivative along OB and BD (with $\theta = \theta_{eq} = 0$):

$$\left.\frac{d^2\Pi}{d\theta^2}\right|_{\theta=0} = \beta - PL. \tag{15.3.9}$$

If $P < \beta/L$, $\frac{d^2\Pi}{d\theta^2} > 0$, and therefore $\Delta\Pi > 0$. It follows that along OB, the equilibrium is stable (which confirms our previous conclusion). If $P > \beta/L$, $\Delta\Pi < 0$ and hence, for such values of P, the position $\theta = 0$ represents unstable equilibrium positions. (Note that the second derivative is multiplied by $(\delta\theta)^2$ and therefore the sign of $\Delta\Pi$ is necessarily the same as $d^2\Pi/d\theta^2$.)

It is of interest to establish the stable/unstable character of equilibrium at point B ($\theta = 0$), i.e. where $P = \beta/L$. From Eq. (15.3.8), $d^2\Pi/d\theta^2 = 0$ and hence provides no information on the sign of $\Delta\Pi$. We must therefore examine higher order derivatives. To this end, note that

$$\frac{d^3\Pi}{d\theta^3} = PL\sin\theta, \tag{15.3.10a}$$

which also vanishes at $\theta = 0$; we therefore must examine the next derivative,

$$\frac{d^4\Pi}{d\theta^4} = PL\cos\theta. \tag{15.3.10b}$$

At $\theta = 0$,

$$\left.\frac{d^4\Pi}{d\theta^4}\right|_{\theta=0} = PL > 0. \tag{15.3.10c}$$

Hence, since this derivative is multiplied by $(\delta\theta)^4$, we conclude that when $P = \beta/L$, $\Delta\Pi > 0$. Thus, the $\theta = 0$ position is stable under this load.

In summary, the vertical position is in stable equilibrium for all $P \leq \beta/L$ and is unstable for $P > \beta/L$; i.e., if the rod is subjected to a force $P \leq \beta/L$ and is given a small displacement from $\theta = 0$, it will return to its original position.[†]

We now examine the equilibrium position along the branch BC. Noting that the required equilibrium force P along this branch is given by Eq. (15.3.5), upon substituting this value for P in Eq (15.3.8),

$$\frac{d^2\Pi}{d\theta^2} = \beta\left(1 - \frac{\theta}{\tan\theta}\right). \tag{15.3.11}$$

Now, using the Taylor expansion for the tan function,

$$\frac{\tan\theta}{\theta} = \frac{1}{\theta}\left(\theta + \theta^3/3 + 2\theta^5/15 + 17\theta^7/515 + \cdots\right)$$

$$= 1 + \theta^2/3 + 2\theta^4/15 + 17\theta^6/515 + \cdots > 1. \tag{15.3.12}$$

Therefore, we observe that along BC, $d^2\Pi/d\theta^2 > 0$ and hence $\Delta\Pi > 0$. Thus, the branch BC represents stable equilibrium positions since along this path Π is a minimum.

We have observed that point B, where $P = \beta/L$, represents the greatest force for which $\theta = 0$ is a stable equilibrium position. We therefore define

$$P_{cr} = \beta/L \tag{15.3.13}$$

[†] We recall that in examining this system in Chapter 11, we found that stable equilibrium exists only if $P < \beta/L$. The discrepancy with the present results is explained in Section 15.4.

as the critical load P of the system: for all forces $P \leq P_{cr}$ the system is stable for $\theta = 0$ and for all $P > P_{cr}$ the original position $\theta = 0$ becomes unstable.

Further insight into this problem can be gained by examining the total energy Π of the system for various values of $|\theta| < \pi$. Let us assume that a given force $P = \alpha P_{cr}$ (where $\alpha \geq 0$ is a constant) is acting upon the system. Then, substituting in Eq. (15.3.2),

$$\Pi = \beta\theta^2/2 - \alpha\beta(1 - \cos\theta). \qquad (15.3.14)$$

In Fig. (15.3.3), Π/β is plotted as a function of θ for typical values $\alpha < 1$, $\alpha = 1$, $\alpha > 1$. We observe again that if $P > P_{cr}$ (e.g., $\alpha = 2$), $\theta = 0$ represents a position of unstable equilibrium since for any $|\theta| > 0$ in the neighbourhood of $\theta = 0$, the total energy Π is less than $\Pi(\theta = 0)$ of the original vertical position. Hence the system will jump to a position of lower energy as represented by the path BC of Fig. (15.3.2). We thus note that the system 'seeks a position of stable equilibrium'.[†]

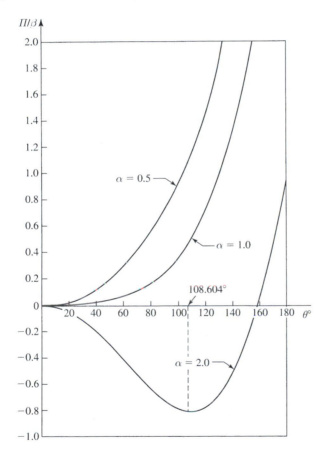

Figure 15.3.3

On the other hand, for $P \leq P_{cr}$ (e.g., $\alpha = 0.5$ or 1), the $\theta = 0$ position represents one for which the energy is a relative minimum. Hence, under such loads, the rod will always return to its original position, $\theta = 0$, which represents a stable position.

[†] We note, for example, that for the curve $\alpha = 2 (P = 2\beta/L)$ the energy has a minimum value at $\theta = 108.604°$; hence the rod is in stable equilibrium at this position when $P = 2P_{cr}$. Note that this value of P is in agreement with the $P-\theta$ relation of Eq. (15.3.5) [see Fig. (15.3.2)].

(It is worthwhile to note that for $P = P_{\text{cr}}$, the Π curve in the neighbourhood of $\theta = 0$ is considerably flatter than for $P < P_{\text{cr}}$.)

Finally, we mention that since the branch BC represents displaced *stable* equilibrium positions in the vicinity of point B, we refer to this point as a *stable bifurcation point*. Now, critical loads may also occur in systems for which there exists an *unstable bifurcation point*. We recall that in Example 11.1 of Chapter 11, it was asserted (without proof) that the critical load for that system occurs at an unstable bifurcation point. We now reconsider this problem using the energy criterion as developed above and will show that the bifurcation point is indeed unstable.

Example 15.1:   Consider the system consisting of a rigid rod of length L and supported by a linear elastic spring with constant k, as shown in Fig. (15.3.4a). The spring is attached to the wall in such a way that it always remains horizontal. Determine the critical load and the stable and unstable character of the equilibrium positions.

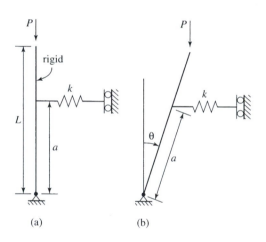

Figure 15.3.4   (a)   (b)

Solution: We examine the rod in a displaced position, defined by a rotation θ, as shown in Fig. (15.3.4b). (Note again that this is a 1-d.o.f. system since its position is defined by one generalised coordinate).

Since the shortening of the spring $\Delta_s = a \sin \theta$, the internal strain energy of the system is

$$U = \frac{k}{2} a^2 \sin^2 \theta. \tag{15.3.15a}$$

The potential energy of the force P, measured from the vertical position, is

$$V = -PL(1 - \cos \theta) \tag{15.3.15b}$$

and hence

$$\Pi(\theta) = U + V = \frac{k}{2} a^2 \sin^2 \theta - PL(1 - \cos \theta). \tag{15.3.15c}$$

The first derivative is then

$$\frac{d\Pi(\theta)}{d\theta} = (ka^2 \cos \theta - PL) \sin \theta. \tag{15.3.16}$$

As before, the required condition for equilibrium is $\frac{d\Pi}{d\theta} = 0$. Clearly, for $\theta = 0$ and $\theta = \pi$, *all values* of P satisfy this condition. If $\theta \neq 0, \theta \neq \pi, \frac{d\Pi}{d\theta} = 0$ yields

$$P = \frac{ka^2}{L} \cos\theta, \quad \theta \neq 0, \pi. \tag{15.3.17}$$

These equilibrium values of P are shown in Fig. (15.3.5). If we now imagine that the force P, applied to the vertical rod, is increased slowly from zero along the loading path OB, we first reach the bifurcation point B from which two branches emanate: BD and BCE. To determine the character of these branches, we examine the sign of the second derivative:

$$\frac{d^2\Pi}{d\theta^2} = ka^2(\cos^2\theta - \sin^2\theta) - PL\cos\theta. \tag{15.3.18}$$

At $\theta = 0$,

$$\left.\frac{d^2\Pi}{d\theta^2}\right|_{\theta=0} = ka^2 - PL. \tag{15.3.19}$$

Hence, according to the criterion, we conclude that

■ $\theta = 0$ is a stable equilibrium position if $P < \frac{ka^2}{L}$,
■ $\theta = 0$ is an unstable equilibrium position if $P > \frac{ka^2}{L}$.

We therefore conclude that the critical load is $P_{cr} = ka^2/L$.

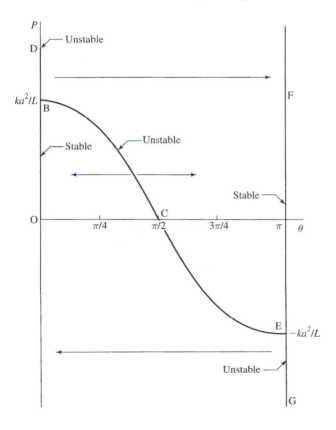

Figure 15.3.5

At $\theta = \pi$,

$$\left.\frac{d^2\Pi}{d\theta^2}\right|_{\theta=\pi} = ka^2 + PL. \tag{15.3.20}$$

Hence,

- $\theta = \pi$ represents stable equilibrium positions for all values $P > -ka^2/L$. For $P < -ka^2/L$, equilibrium will be unstable at $\theta = \pi$.[†]

Thus, we have established that the paths OB and EF represent stable equilibrium while BD and EG represent unstable equilibrium.

We now examine the branch BCE for any equilibrium position $0 < \theta = \theta_{eq} < \pi$. Substituting the required equilibrium force, Eq. (15.3.17), in Eq. (15.3.18),

$$\left.\frac{d^2\Pi}{d\theta^2}\right|_{\theta_{eq}} = -ka^2 \sin^2 \theta_{eq}. \tag{15.3.21}$$

Therefore $\Delta\Pi < 0$ and hence BCE represents unstable equilibrium.

We observe that the stable or unstable character of equilibrium at the bifurcation point B (when approaching this point along the branch BCE) cannot be established from the second derivative since $d^2\Pi/d\theta^2|_{\theta\to0} = 0$ as $P \to ka^2/L$. According to the above discussion, we must therefore examine higher order derivatives. From Eq. (15.3.18),

$$\frac{d^3\Pi}{d\theta^3} = -2ka^2 \sin 2\theta + PL \sin\theta. \tag{15.3.22a}$$

Note that this derivative also vanishes identically as $\theta \to 0$; therefore differentiating once more,

$$\frac{d^4\Pi}{d\theta^4} = -4ka^2 \cos 2\theta + PL \cos\theta, \tag{15.3.22b}$$

that is,

$$\left.\frac{d^4\Pi}{d\theta^4}\right|_{\theta\to0} = -4ka^2 + PL. \tag{15.3.22c}$$

Substituting $P = ka^2/L$, we find that at point B, $d^4\Pi/d\theta^4|_B = -3ka^2 < 0$, and hence $\Delta\Pi < 0$. Therefore we conclude that point B is an unstable bifurcation point.[‡]

Several comments and remarks related to Fig. (15.3.5) are now in order:

- Since more than one equilibrium position exists for values $P < ka^2/L$, it is not possible to immediately conclude from Fig. (15.3.5) [as was concluded from Fig. (15.3.2) for the case of the system considered in Fig. (15.3.1)] that under such loads the vertical position $\theta = 0$ is in a state of stable equilibrium.
- For loads $P > ka^2/L$, two possible equilibrium positions exist: the unstable vertical position $\theta = 0$ and the stable $\theta = \pi$ position. Thus if the rod, subjected to such loads, is given a small perturbation from $\theta = 0$, it will 'jump' to the position $\theta = \pi$. This is indicated by the horizontal arrow in Fig. (15.3.5).

[†] Negative values, $P < 0$, clearly represent an upward force acting on the rod. We note that to maintain the rod in a position $\theta = \pi$, a downward force $P > ka^2/L$ or an *upward* force having magnitude $|P| < ka^2/L$ is required.

[‡] An unstable bifurcation point, such as B in this problem, is often referred to as a *limit point* in the stability analysis of structures.

■ If the rod is in an unstable position represented by the branch BCE, it will jump to the $\theta = \pi$ position if it is given a small perturbation $\theta \to \theta + \epsilon$; if given a perturbation $\theta \to \theta - \epsilon$, it will jump to the $\theta = 0$ position. Thus, we again observe that a system always 'seeks' a stable equilibrium position.   □

In investigating the above problems, it was possible to determine the critical force P_{cr} and establish the stable or unstable character of the equilibrium positions for all possible values of $0 \le \theta \le \pi$. However, we often are interested merely in determining the critical load, for in engineering problems this is often the only quantity that is required. If indeed this is our sole concern, we need only consider the system in the neighbourhood of the original position; that is, the system may be investigated using a small (infinitesimal) displacement analysis in the neighbourhood of the original position. Such an analysis is mathematically much simpler. As we shall see, it will lead us to define an equilibrium state that we shall call *pseudo-neutral*. In fact, pseudo-neutral equilibrium will become our principal criterion for determining critical loads. In the following section, we determine the critical load to the previous problems, using this procedure.

15.4 Determination of critical loads using a small deflection analysis – pseudo-neutral equilibrium

We again consider the rod shown in Fig. (15.3.1). For this system, it was found that [Eq. (15.3.2)]

$$\Pi = \frac{\beta\theta^2}{2} - PL(1 - \cos\theta).$$

Let us now consider that the system undergoes only small displacements $0 < |\theta| \ll 1$. Then from the Taylor series expansion

$$\cos\theta = 1 - \frac{\theta^2}{2!} + \frac{\theta^4}{4!} \cdots, \tag{15.4.1}$$

Eq. (15.3.2) becomes, upon neglecting terms of order higher than θ^2,

$$\Pi(\theta) = (\beta - PL)\theta^2/2. \tag{15.4.2}$$

Note that following the small displacement assumption, Π is now a quadratic function of θ and since all higher order terms have been dropped, for all $n > 2$, the derivatives $\frac{d^n\Pi}{d\theta^n} = 0$, *identically.*[†]
For equilibrium

$$\frac{d\Pi}{d\theta} \delta\theta = (\beta - PL)\theta \, \delta\theta = 0. \tag{15.4.3}$$

For any arbitrary $\delta\theta$, the equilibrium condition is satisfied under the following conditions:

■ If $\theta = 0$, all values of P satisfy the condition,
■ If $\theta \ne 0$, $P = \beta/L$.

We first examine the equilibrium states for the solution, by examining $\frac{d^2\Pi}{d\theta^2}(\delta\theta)^2$ at $\theta = 0$.

[†] Clearly, having examined the general case in which higher non-zero derivatives were obtained [see Eqs. (15.3.22)], we note that the result here is due to truncation of the power series of Eq. (15.3.6a).

From Eq. (15.4.2),

$$\frac{d^2 \Pi}{d\theta^2} (\delta\theta)^2 = (\beta - PL)(\delta\theta)^2. \tag{15.4.4}$$

Hence, since the first derivative vanishes at $\theta = 0$, we note, from Eq. (15.3.6b), that

$$\Delta\Pi > 0 \quad \text{if } P < \beta/L, \tag{15.4.5a}$$

$$\Delta\Pi = 0 \quad \text{if } P = \beta/L, \tag{15.4.5b}$$

$$\Delta\Pi < 0 \quad \text{if } P > \beta/L. \tag{15.4.5c}$$

Therefore we again conclude that for $\theta = 0$, the rod is in stable equilibrium if $P < \beta/L$, while if $P > \beta/L$, the rod is in unstable equilibrium. When $P = \beta/L$, according to the criterion of Section 15.2, the rod *appears* to be in neutral equilibrium. However, from the analysis of the previous section, we have established that for a value $P = \beta/L$, the $\theta = 0$ position is actually a stable equilibrium position. *The discrepancy in the results is due to the assumption of small displacements in which we neglected terms of order higher than that of the quadratic.* Hence, since the system under this load is *not truly* in neutral equilibrium, we refer to the rod as being in pseudo-neutral equilibrium with the clear understanding that this is merely a result obtained by the dropping of the higher order terms.[†]

However, in practical terms the difference between neutral and pseudo-neutral equilibrium states is immaterial if our goal is only to determine the critical load, for, in physical terms, we have established that for any load $P > \beta/L$ the rod is in unstable equilibrium. It follows that, *using the small displacement analysis, we associate the critical load*

$$P_{\text{cr}} = \beta/L \tag{15.4.6}$$

with a pseudo-neutral equilibrium state (i.e., when $d^2\Pi/d\theta^2 = 0$). In other words, according to the small displacement analysis, when $P < \beta/L$ the original position is in stable equilibrium; as P increases and reaches $P = \beta/L$ the rod assumes a pseudo-neutral equilibrium state for $\theta = 0$; as P increases further, the equilibrium position $\theta = 0$ becomes unstable.

From the finite analysis of the preceding section, we conclude that a rod in the $\theta = 0$ position, under a force P_{cr}, will return to its original state when given any disturbance; however, according to the small displacement analysis, $\Delta\Pi = 0$ when $P = P_{\text{cr}}$. Upon recalling the definition of neutral equilibrium as given in Chapter 11 (Section 2), this latter analysis lead to the conclusion that if the rod is given a small displacement from $\theta = 0$ with $P = P_{\text{cr}}$, it will neither return to $\theta = 0$ nor will the displacement increase. Thus, when using a small displacement analysis, $P = P_{\text{cr}}$ is said to be the *smallest force*, which can maintain the rod in a slightly deflected position.

It is necessary, however, to emphasise here an important limitation of the small displacement analysis: namely, while the small displacement approximation with Π expressed as a quadratic in the displacements (or rotations) can be used for a large variety of problems, such an approximation clearly cannot yield solutions if

[†] Pseudo-neutral equilibrium is clearly a result of an imposed *mathematical* restriction and hence does not have a distinct physical meaning. Therefore, many authors do not make a distinction between pseudo-neutral and neutral equilibrium states and refer to both of these states as 'neutral' equilibrium. For clarity, we shall continue to use the term 'pseudo-neutral' where appropriate.

all derivatives up to the second order vanish *identically for all P* when evaluated at the original equilibrium position. In such cases, we must expand beyond the quadratic term, i.e. use the large displacement analysis. Furthermore, we observe that, as opposed to the finite displacement analysis, the small displacement analysis does not permit us to study equilibrium states on branches representing displaced configurations of a system.

Example 15.2:   Determine the critical axial load P for a force acting on a rigid rod AB of length L, hinged at point A and restrained by a nonlinear elastic spring as shown in Fig. (15.4.1a). The spring is governed by the relation

$$F = c(1 - e^{-\gamma\Delta}), \quad 0 < \gamma, \ 0 < c, \tag{15.4.7}$$

where F is the force in the spring, Δ is the change in length of the spring and c and γ are constants. [We observe that γ and c have units of (length)$^{-1}$ and Newtons respectively.]

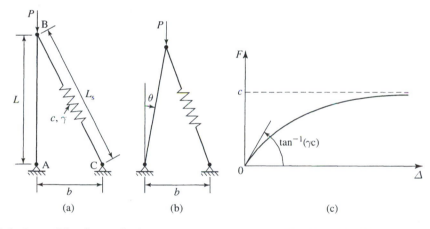

(a)   (b)   (c)   **Figure 15.4.1**

Solution: The force–displacement curve represented by Eq. (15.4.7) is shown in Fig. (15.4.1c). We note that the derivative $\frac{dF}{d\Delta} = \gamma c\, e^{-\gamma\Delta}$; at $\Delta = 0$, $\frac{dF}{d\Delta} = \gamma c$. (Thus the slope of the force–displacement curve increases with γ, and as $\gamma \to \infty$, the spring becomes infinitely stiff at $\Delta = 0$.)

The strain energy in the spring is given by[†]

$$U = \int_0^{\Delta^f} F(\Delta)\, d\Delta = c \int_0^{\Delta^f} (1 - e^{-\gamma\Delta})\, d\Delta$$

$$= c\left[\Delta + \frac{1}{\gamma}e^{-\gamma\Delta}\right]_0^{\Delta^f} = c\left[\Delta - \frac{1}{\gamma}(1 - e^{-\gamma\Delta})\right]. \tag{15.4.8}$$

Assuming small rotations, $|\theta| \ll 1$, $|\Delta| \ll 1$ [Fig. (15.4.1b)] and using the representation for the exponential function

$$e^x = 1 + x + \frac{x^2}{2!} + \frac{x^3}{3!} + \cdots, \tag{15.4.9}$$

[†] The superscript f is dropped in the last expression of Eq. (15.4.8), with the understanding that Δ represents the final displacement.

upon dropping all terms higher than the quadratic,

$$U = \frac{c\gamma}{2}\Delta^2. \tag{15.4.10}$$

If θ is taken as the generalised coordinate, then from simple geometry of the displaced system [Fig. (15.4.1b)],

$$\Delta = L_s - (L^2 + b^2 - 2bL\sin\theta)^{1/2} = L_s\left[1 - \left(1 - \frac{2bL}{L_s^2}\sin\theta\right)^{1/2}\right], \tag{15.4.11}$$

where $L_s = \sqrt{L^2 + b^2}$ is the original spring length.

Using the Taylor expansion

$$\sin\theta = \theta - \frac{\theta^3}{3!} + \frac{\theta^5}{5!} + \cdots \tag{15.4.12a}$$

and the binomial theorem

$$(1 \pm x)^{1/2} = 1 \pm \frac{x}{2} - \frac{x^2}{8} + \cdots \tag{15.4.12b}$$

and again dropping all terms higher than the quadratic, we have

$$\Delta = L_s\left[1 - \left(1 - \frac{bL}{L_s^2}\theta + \cdots\right)\right]$$

or

$$\Delta = \frac{bL\theta}{L_s}. \tag{15.4.13}$$

The strain energy in the spring is therefore

$$U = \frac{c\gamma b^2 L^2}{2L_s^2}\theta^2. \tag{15.4.14}$$

Noting that the potential energy of P, taken with respect to the original $\theta = 0$ position, is $V = -PL(1 - \cos\theta)$; upon using the Taylor expansion, Eq. (15.4.1), we have $V = -PL\theta^2/2$. Therefore

$$\Pi = \frac{c\gamma b^2 L^2}{2L_s^2}\theta^2 - \frac{PL\theta^2}{2}$$

or

$$\Pi = \frac{1}{2}\left(\frac{\gamma c b^2 L^2}{L_s^2} - PL\right)\theta^2. \tag{15.4.15}$$

From the derivative

$$\frac{d\Pi(\theta)}{d\theta} = \left(\frac{\gamma c b^2 L^2}{L_s^2} - PL\right)\theta, \tag{15.4.16}$$

we note that according to the condition $d\Pi/d\theta|_{\theta=0} = 0$, the original position is satisfied for all values of P.

Setting the second derivative to zero (such that $\Delta\Pi = 0$), we then find the critical load corresponding to pseudo-neutral equilibrium:

$$P_{cr} = \frac{\gamma c b^2 L}{L_s^2} = \frac{\gamma c b^2 L}{L^2 + b^2}. \tag{15.4.17}$$

Several special cases are of interest:

- If $b = 0$, $P_{cr} = 0$. This corresponds to a physical situation where, if the system is given a small disturbance, the spring is unable to exert a restoring moment about point A in order to bring the rod back to its original $\theta = 0$ position.
- If $b \to \infty$, $P_{cr} \to \gamma cL$. This limiting case corresponds to a system with a horizontal spring. The term γc corresponds to a linear spring constant k.[†] (See Example 11.1 with $a = L$.)
- As $\gamma \to \infty$, $P_{cr} \to \infty$. As mentioned previously, this corresponds to a spring that is initially infinitely stiff and therefore does not permit any motion of the rod. □

In anticipation of the discussion in the next section, we observe the following feature in the small displacement analysis (which will prove to have a practical value in a stability analysis): the critical load P_{cr}, which was obtained from the stability criterion, Eq. (15.4.17), could also be obtained from the equilibrium criterion, Eq. (15.4.16), if $\theta \neq 0$. Indeed, we obtain the same value of P_{cr} if we set $\Pi(\theta) = 0$ in Eq. (15.4.15) for all $\theta \neq 0$ ($|\theta| \ll 1$).

To explain this, we examine the expressions for the total potential in greater generality, using small displacements in the following section. From the conclusions, we will arrive at a simple procedure for determining critical loads.

15.5 The total potential for small displacements: reconsideration of the stability criterion

In Section 15.3, using a finite displacement analysis, the total potential was represented by an infinite series [Eq. (15.3.6a)]. If we wish only to obtain the critical load, we have seen that it is often possible to consider small displacements and drop all terms higher than the quadratic term. Moreover, we are usually interested in the variation from the initial position where the coordinate defining the position is taken as zero. Let us therefore expand the total potential about this point. We consider here a 1-d.o.f. system and let q denote the generalised coordinate (as defined in Section 15.3) that specifies the position of the system. Then, using Eq. (15.3.6a), and dropping higher order terms, the total potential expanded about $q = q_{eq} = 0$ becomes

$$\Pi(\delta q) = \Pi(q)\big|_{q=0} + \frac{d\Pi(q)}{dq}\bigg|_{q=0} \delta q + \frac{1}{2!} \frac{d^2\Pi(q)}{dq^2}\bigg|_{q=0} (\delta q)^2 \quad (15.5.1a)$$

or

$$\Delta\Pi(\delta q) = \frac{d\Pi(q)}{dq}\bigg|_{q=0} \delta q + \frac{1}{2!} \frac{\partial^2\Pi(q)}{\partial q^2}\bigg|_{q=0} (\delta q)^2. \quad (15.5.1b)$$

If $q = 0$ is an equilibrium position, then

$$\frac{d\Pi(q)}{dq}\bigg|_{q=0} = 0 \quad (15.5.2)$$

and hence

$$\Delta\Pi(\delta q) = \frac{1}{2} \frac{d^2\Pi(q)}{dq^2}\bigg|_{q=0} (\delta q)^2. \quad (15.5.3)$$

[†] Note that if $c\gamma \to k$, the strain energy U, given by Eq. (15.4.10), becomes $U = k\Delta^2/2$ as in the case of a linear elastic spring.

Thus we observe that, when measured from an equilibrium position, $\Delta \Pi$ contains only quadratic terms. We note further that $\Pi(q = 0)$ may always be taken as zero since it depends on an arbitrary datum [i.e., the datum may always be chosen in such a way as to render $\Pi(q = 0) = 0$]. Hence if $\Pi(q = 0) = 0$,

$$\Delta \Pi(\delta q) = \Pi(\delta q) \tag{15.5.4}$$

and Eq. (15.5.3) may be written as

$$\Pi(\delta q) = \frac{1}{2} \frac{d^2 \Pi(q)}{dq^2} \bigg|_{q=0} (\delta q)^2. \tag{15.5.5}$$

Letting

$$a \equiv \frac{1}{2} \frac{d^2 \Pi(q)}{dq^2} \bigg|_{q=0}, \tag{15.5.6}$$

we have

$$\Pi(\delta q) = a(\delta q)^2. \tag{15.5.7}$$

This indeed was the case in the two previous examples considered in Section 15.4, where (with $q = 0$ and $\delta q \equiv q - q_{eq} = q - 0 = q$) Π was of the form

$$\Pi(q) = aq^2, \tag{15.5.8}$$

that is, purely quadratic. In the first problem, that of the rod supported by a torsion spring, $a = \frac{1}{2}(\beta - PL)$ [see Eq. (15.4.2)] while in the second problem we had $a = \frac{1}{2}(\frac{\gamma cb^2 L^2}{L_s^2} - PL)$ [see Eq. (15.4.15)].

We recall now that using the small displacement analysis, the critical load was found to be associated with pseudo-neutral equilibrium. In effect *this means that for $P = P_{cr}$, the original $q = 0$ position, as well as neighbouring positions $0 < |q| \ll 1$, is an equilibrium position.* Therefore, we set

$$\frac{d\Pi}{dq} \bigg|_{q \neq 0} = 2aq \bigg|_{q \neq 0} = 0 \tag{15.5.9}$$

and, hence, we require that $a = 0$. For pseudo-neutral equilibrium, we require that the second derivative vanish:

$$\frac{d^2 \Pi}{dq^2} = 2a = 0. \tag{15.5.10}$$

Thus we observe that from either of Eqs. (15.5.9) or (15.5.10), we obtain the same result, viz., $a = 0$ (which, e.g., in the two previous examples cited, led to critical loads P_{cr}). We note too that for 1-d.o.f. systems, as examined here, we can obtain the same critical loads by setting $\Pi(q \neq 0) = 0$ in Eq. (15.5.8). However, as we shall see, this last procedure cannot be used so simply for n-d.o.f. systems.

The above discussion may be summarised as follows:

To obtain the critical loads, which cause instability of a system in its original configuration, we may consider only small displacements, and expand Π in a power series, dropping higher order terms; as a result the total potential Π is represented by a pure quadratic expression.

The critical load is then taken as that force which causes the system to be in pseudo-neutral equilibrium, i.e., the system is assumed to be in equilibrium in a position adjacent to its original position (as well as in its original position). The critical load is thus determined by setting the first derivative of Π equal to zero for non-zero displacements. The second derivative will then automatically vanish, confirming the state of pseudo-neutral equilibrium.

As was mentioned in Section 15.4, such a procedure is permissible only if neither the first nor the second derivative of Π vanish identically for *all* values of P when the generalised coordinate is different than zero.

15.6 Systems having several degrees-of-freedom – small displacement analysis

In our previous study, we have considered only 1-d.o.f. systems, i.e. systems whose configurations are completely defined by a single generalised coordinate, q. We now consider systems having several degrees of freedom. For example, the system given in Fig. (15.6.1a) has 2 d.o.f. since two coordinates q_1, q_2 are required to define its position, while the system shown in Fig. (15.6.1b) has 3 d.o.f. with generalised coordinates q_1, q_2, q_3.

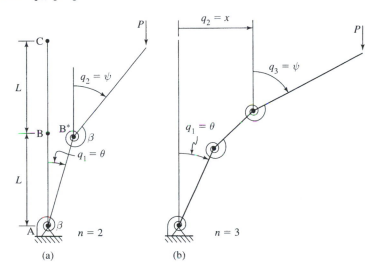

(a)　　　(b)　　　　　　　　　　　　　　**Figure 15.6.1**

In general, a system may have n d.o.f. with generalised coordinates $q_1, q_2, \ldots, q_n$. The important feature of the generalised coordinates is that they are independent coordinates; i.e., they can be varied independently. Note that the generalised coordinates may be displacements or angles [or even combinations of those as in Fig. (15.6.1b)].

Two-degree-of-freedom system

For mathematical simplicity, we first consider a 2-d.o.f. system; we will generalise the results at a later stage. We follow essentially the discussion given in the previous section, and consider generalised coordinates q_1 and q_2, $|q_i| \ll 1$. If we neglect higher order terms, the expansion of the total potential Π about q_1 and q_2 becomes

$$\Pi(q_1 + \delta q_1, q_2 + \delta q_2) = \Pi(q_1, q_2) + \frac{\partial \Pi(q_1, q_2)}{\partial q_1} \delta q_1 + \frac{\partial \Pi(q_1, q_2)}{\partial q_1} \delta q_2$$

$$+ \frac{1}{2}\left[\frac{\partial^2 \Pi}{\partial q_1^2}(\delta q_1)^2 + 2\frac{\partial^2 \Pi}{\partial q_1 \partial q_2}(\delta q_1)(\delta q_2) + \frac{\partial^2 \Pi}{\partial q_2^2}(\delta q_2)^2 \right].$$

$$(15.6.1)$$

As in the previous section, we expand about $q_1 = q_2 = 0$. Then

$$\Delta\Pi \equiv \Pi(\delta q_1, \delta q_2) - \Pi(0,0) = \frac{\partial\Pi}{\partial q_1}\bigg|_{q_i=0}(\delta q_1) + \frac{\partial\Pi}{\partial q_2}\bigg|_{q_i=0}(\delta q_2)$$

$$+ \frac{1}{2}\left[\frac{\partial^2\Pi}{\partial q_1^2}\bigg|_{q_i=0}(\delta q_1)^2 + 2\frac{\partial^2\Pi}{\partial q_1 \partial q_2}\bigg|_{q_i=0}(\delta q_1)(\delta q_2) + \frac{\partial^2\Pi}{\partial q_2^2}\bigg|_{q_i=0}(\delta q_2)^2\right].$$

$$(15.6.2)$$

If $q_1 = q_2 = 0$ represents an equilibrium position, the variation of Π is

$$\delta\Pi = \frac{\partial\Pi}{\partial q_1}\bigg|_{q_i=0}(\delta q_1) + \frac{\partial\Pi}{\partial q_2}\bigg|_{q_i=0}(\delta q_2), \quad i = 1, 2, \qquad (15.6.3a)$$

and since this must be true for any arbitrary variation, δq_i, about the equilibrium position, we require that

$$\frac{\partial\Pi}{\partial q_1}\bigg|_{q_i=0} = \frac{\partial\Pi}{\partial q_2}\bigg|_{q_i=0} = 0, \quad i = 1, 2. \qquad (15.6.3b)$$

For convenience, we use the following notation:

$$a_{11} = \frac{1}{2}\frac{\partial^2\Pi}{\partial q_1^2}\bigg|_{q_i=0}, \qquad a_{12} = a_{21} = \frac{1}{2}\frac{\partial^2\Pi}{\partial q_1 \partial q_2}\bigg|_{q_i=0}, \qquad a_{22} = \frac{1}{2}\frac{\partial^2\Pi}{\partial q_2^2}\bigg|_{q_i=0},$$

$$(15.6.4)$$

where the a's are constants. Note that these constants are dependent upon the force P.[†] Using Eqs. (15.6.3) and (15.6.4), we then may rewrite Eq. (15.6.2) as

$$\Delta\Pi(\delta q_1, \delta q_2) = a_{11}(\delta q_1)^2 + 2a_{12}(\delta q_1)(\delta q_2) + a_{22}(\delta q_2)^2. \qquad (15.6.5)$$

Furthermore, since the total potential can always be referred to any arbitrary datum, it is *always* possible to set $\Pi(0,0) = 0$. It follows that

$$\Pi(\delta q_1, \delta q_2) = \Delta\Pi(\delta q_1, \delta q_2), \qquad (15.6.6)$$

and since $\delta q_i = q_i - 0 = q_i$ we have

$$\Pi(q_1, q_2) = a_{11}q_1^2 + 2a_{12}q_1 q_2 + a_{22}q_2^2, \qquad (15.6.7)$$

which is observed to be a pure quadratic.

We now apply our criterion for pseudo-neutral equilibrium, which states that in addition to the equilibrium state $q_i = 0$, equilibrium exists at neighbouring positions, $0 < |q_i| \ll 1$ when $P = P_{\text{cr}}$. Hence, the stationary condition

$$\frac{\partial\Pi}{\partial q_1}\bigg|_{q_i\neq0} = \frac{\partial\Pi}{\partial q_2}\bigg|_{q_i\neq0} = 0, \quad i = 1, 2, \qquad (15.6.8)$$

must be satisfied. From Eq. (15.6.7), we have

$$2a_{11}q_1 + 2a_{12}q_2 = 0, \qquad (15.6.9a)$$

$$2a_{12}q_1 + 2a_{22}q_2 = 0. \qquad (15.6.9b)$$

For a non-trivial solution of Eqs. (15.6.9), the determinant

$$D \equiv \begin{vmatrix} a_{11} & a_{12} \\ a_{21} & a_{22} \end{vmatrix} = 0 \qquad (15.6.10a)$$

[†] In general, the coefficients a_{11} and a_{22} will be positive for small values of P.

or

$$a_{11}a_{22} - a_{12}^2 = 0. \qquad (15.6.10b)$$

Equation (15.6.10b) is thus the required condition for equilibrium. Let us now verify from the basic definition of pseudo-neutral equilibrium (namely, $\Delta\Pi = 0$) that Eq. (15.6.10b) corresponds to such an equilibrium state.

To this end, we digress here to recall a mathematical property of quadratic functions. Let

$$f(x, y) = Ax^2 + 2Bxy + Cy^2 \qquad (15.6.11)$$

be a continuous quadratic function of the two independent variables x and y. Then, if and only if

$$A > 0 \qquad (15.6.12a)$$

and

$$AC - B^2 > 0, \qquad (15.6.12b)$$

it follows that $f(x, y) > 0$ for all $x \neq 0$, $y \neq 0$. The function $f(x, y)$ is then said to be **positive definite**.[†] Thus $f(x, y)$, given by Eq. (15.6.11) and satisfying the conditions of Eqs. (15.6.12), has a relative minimum at $x = y = 0$.

Now, we observe that Π appearing in Eq. (15.6.7) is precisely of the same form as $f(x, y)$ of Eq. (15.6.11). Therefore, since $\Delta\Pi = \Pi$, $\Delta\Pi > 0$ for all $\delta q_i \equiv q_i$, provided that

$$a_{11} > 0 \qquad (15.6.13a)$$

and that the determinant

$$D \equiv \begin{vmatrix} a_{11} & a_{12} \\ a_{21} & a_{22} \end{vmatrix} > 0. \qquad (15.6.13b)$$

We recall that for all $P < P_{cr}$ the system will be in a state of stable equilibrium, i.e., $\Delta\Pi > 0$ if the system is given a displacement when subjected to $P < P_{cr}$. Therefore, for stable equilibrium, we require that $D > 0$. When D ceases to be positive, it is possible to give the system displacements δq_1, δq_2 that will cause $\Delta\Pi$ to vanish, i.e., the state $q_1 = 0$, $q_2 = 0$ ceases to be in stable equilibrium and becomes instead one of pseudo-neutral equilibrium. Now we observe that the vanishing of the determinant of Eq. (15.6.13b) is exactly the same as that obtained from the equilibrium conditions, Eq. (15.6.10a). In practice we need therefore write only the equilibrium conditions in the displaced configuration and these will *automatically* correspond to a pseudo-neutral equilibrium state, thus leading to the determination of the critical load. (This is the same conclusion as was reached in the previous section for systems having 1 d.o.f.).

Example 15.3: Determine the critical buckling load P for the system shown in Fig. (15.6.1a), where AB and BC are rigid bars. The linear torsional springs at A and B, each have a constant β (N-m/rad).

Solution: The given system has 2 d.o.f. We choose as general coordinates $q_1 \equiv \theta$, $q_2 \equiv \psi$, and consider the system in its displaced configuration [Fig. (15.6.1a)].

[†] From Chapter 14, we recall that elastic strain energy possesses this same property.

For any given displacement, the strain energy U of the springs and the potential energy of P are given by

$$U = \frac{\beta}{2}[\theta^2 + (\psi - \theta)^2] \qquad (15.6.14a)$$

and

$$V = -PL(2 - \cos\theta - \cos\psi) \qquad (15.6.14b)$$

respectively; hence,

$$\Pi = \frac{\beta}{2}(2\theta^2 - 2\theta\psi + \psi^2) - PL(2 - \cos\theta - \cos\psi). \qquad (15.6.15a)$$

Assuming small displacements,

$$\Pi(\theta, \psi) = \beta/2(2\theta^2 - 2\theta\psi + \psi^2) - \frac{PL}{2}(\theta^2 + \psi^2). \qquad (15.6.15b)$$

For equilibrium, we require that

$$\frac{\partial\Pi}{\partial\theta} = (2\beta - PL)\theta - \beta\psi = 0, \qquad (15.6.16a)$$

$$\frac{\partial\Pi}{\partial\psi} = -\beta\theta + (\beta - PL)\psi = 0. \qquad (15.6.16b)$$

Non-trivial solutions for θ and ψ exist provided

$$\begin{vmatrix} 2\beta - PL & -\beta \\ -\beta & \beta - PL \end{vmatrix} = 0 \qquad (15.6.17a)$$

or

$$(2\beta - PL)(\beta - PL) - \beta^2 = 0, \qquad (15.6.17b)$$

that is,

$$(PL)^2 - 3\beta(PL) + \beta^2 = 0. \qquad (15.6.17c)$$

Solving for P,

$$P = \frac{\beta}{2L}(3 \pm \sqrt{5}). \qquad (15.6.18a)$$

Note that two values have been obtained:

$$P_1 = \frac{\beta}{2L}(3 - \sqrt{5}), \qquad (15.6.18b)$$

$$P_2 = \frac{\beta}{2L}(3 + \sqrt{5}). \qquad (15.6.18c)$$

Corresponding to P_1, we substitute in either of the Eqs. (15.6.16) and find that

$$\psi = \frac{1}{2}(\sqrt{5} + 1)\theta = 1.618\theta \qquad (15.6.19a)$$

while for P_2,

$$\psi = \frac{1}{2}(1 - \sqrt{5})\theta = -0.618\theta. \qquad (15.6.19b)$$

Equations (15.6.19) define two *modes of buckling* and are shown in Fig. (15.6.2). Thus, e.g., we observe that if P_1 acts on the system, it will buckle in the shape shown in Fig. (15.6.2a).

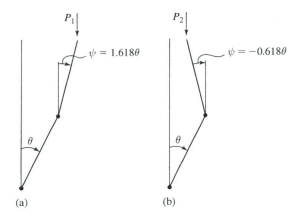

$$\psi = 1.618\theta$$

$$\psi = -0.618\theta$$

(a)   (b)   **Figure 15.6.2**

Both of these configurations represent pseudo-neutral equilibrium positions. However, since we are interested in the lowest buckling load, we define the critical load as

$$P_{cr} = P_1 = \frac{\beta}{2L}(3 - \sqrt{5}) = 0.3820\,\frac{\beta}{L}. \tag{15.6.20}$$

We observe that while the general shape of the buckled configuration has been established, the amplitudes of the displacements remain unknown. This confirms our initial comments that by assuming small displacements, we lose the ability to establish exact equilibrium positions.

It is worthwhile to mention that if we substitute Eqs. (15.6.19a) and (15.6.20) in Eq. (15.6.15b), we find that $\Pi(\theta, \psi) = 0$ identically. This confirms the state of pseudo-neutral equilibrium since $\Delta\Pi \equiv \Pi(\theta, \psi) - \Pi(0, 0) = 0$.

Finally, we observe that mathematically this problem is an eigenvalue problem, where the discrete forces P_1, P_2 are the eigenvalues and the buckling modes $\{q_1 q_2\}^{\mathrm{T}}$ are the eigenvectors corresponding to these eigenvalues.

◇n-Degree-of-freedom systems

The basic ideas developed for the 2-d.o.f system carry over to the n-d.o.f system, where the total potential is given by

$$\Pi = \Pi(q_1, q_2, q_3, \ldots, q_n). \tag{15.6.21}$$

Specifically, an n-d.o.f system has the quadratic function

$$\begin{aligned}
\Pi = &\left(a_{11}q_1^2 + a_{12}q_1q_2 + a_{13}q_1q_3 + \cdots + a_{1n}q_1q_n\right) \\
&+ \left(a_{21}q_2q_1 + a_{22}q_2^2 + a_{23}q_2q_3 + \cdots + a_{2n}q_2q_n\right) \\
&+ \left(a_{31}q_3q_1 + a_{32}q_3q_2 + a_{33}q_3^2 + \cdots + a_{3n}q_3q_n\right) \\
&\quad\vdots \\
&+ \left(a_{n_1}q_nq_1 + a_{n2}q_nq_{n2} + \cdots + a_{nn}q_n^2\right),
\end{aligned}$$

that is,

$$\Pi = \sum_{i=1}^{n}\sum_{j=1}^{n} a_{ij}q_iq_j, \tag{15.6.22a}$$

where a_{ij} are coefficients. In matrix form

$$\Pi = \{q\}^{\mathrm{T}}[A]\{q\}, \tag{15.6.22b}$$

where $\{q\}$ represents the vector having components q_i, $\{q\}^{\mathrm{T}}$ is its transpose and $[A]$ is the matrix of the coefficients,

$$A = \begin{bmatrix} a_{11} & a_{12} & a_{13} & \cdots & a_{1n} \\ a_{21} & a_{22} & a_{23} & \cdots & a_{2n} \\ a_{31} & a_{32} & a_{33} & \cdots & a_{3n} \\ & & \vdots & & \\ a_{n1} & a_{n2} & a_{n3} & \cdots & a_{nn} \end{bmatrix}.$$

The necessary and sufficient conditions for Π to be positive definite are that all the principal minors,

$$a_{11} > 0, \qquad \begin{vmatrix} a_{11} & a_{12} \\ a_{21} & a_{22} \end{vmatrix} > 0, \qquad \begin{vmatrix} a_{11} & a_{12} & a_{13} \\ a_{21} & a_{22} & a_{23} \\ a_{31} & a_{32} & a_{33} \end{vmatrix} > 0, \ldots$$

and that D, the determinant of $[A]$, be positive, i.e.

$$D \equiv \begin{vmatrix} a_{11} & a_{12} & a_{13} & \cdots & a_{1n} \\ a_{21} & a_{22} & a_{23} & \cdots & a_{2n} \\ a_{31} & a_{32} & a_{33} & \cdots & a_{3n} \\ & & \vdots & & \\ a_{n1} & a_{n2} & a_{n3} & \cdots & a_{nn} \end{vmatrix} > 0. \tag{15.6.23}$$

As with the 2-d.o.f system, a pseudo-neutral equilibrium state will exist if it is possible to give the system a set of variations $\delta q_1, \delta q_2, \ldots, \delta q_n$ such that $\Delta \Pi = 0$. Hence (since the determinant D involves all the variables q_i) the corresponding condition for pseudo-neutral equilibrium is obtained by setting $D = 0$.

Having investigated the stability of systems consisting of rigid elements supported by elastic springs, we now study the stability of elastic elements. We shall see that the ideas and concepts developed in the preceding sections also apply to elastic bodies.

15.7   Stability of an elastic rod: the Rayleigh quotient

We consider the stability of a linear elastic rod AB of length L, having flexural rigidity EI. The rod is subjected to a compressive axial force P and is assumed to be simply supported at the two ends, as shown in Fig. (15.7.1a).

In Chapter 11, the critical load was found to be

$$P_{\mathrm{cr}} = \frac{\pi^2 EI}{L^2}, \tag{15.7.1}$$

which is known as the *Euler buckling load*. When $P = P_{\mathrm{cr}}$ the rod assumes a buckled configuration as shown in Fig. (15.7.1b). We now investigate the stability of the rod from energy considerations.

Let us assume that the axial force is applied statically starting from a zero value. Clearly, for low values of P, $P < P_{\mathrm{cr}}$, the rod will remain straight, although it will

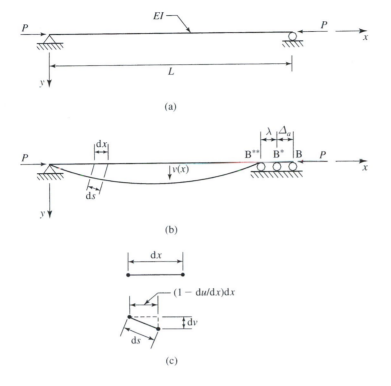

Figure 15.7.1

undergo axial strains. Point B will undergo a small axial displacement Δ_a to say B*
and the resulting strain energy, given by Eq. (14.2.12), is

$$U_a = \frac{P^2 L}{2AE}, \qquad P < P_{cr}. \qquad (15.7.2)$$

Upon reaching $P = P_{cr}$ the rod can assume a buckled configuration. We are therefore interested in investigating the *difference* in the total potential energy of the
system *in the buckled configuration with that which exists immediately prior to
buckling.* Thus, we are concerned only with the energy that exists due to the bending
of the rod.[†] This strain energy, given by Eq. (14.2.16a), is

$$U = \frac{1}{2} \int_0^L \frac{M^2}{EI} \, dx. \qquad (15.7.3)$$

Now, we shall be interested here only in finding the critical buckling load, i.e.,
the load above which the original straight-line configuration ceases to be stable.
Following our previous discussion, we therefore use a small displacement analysis.
We recall, from Chapter 9, that using the assumption of small displacements and
rotations, the Euler–Bernoulli relation for flexural deformation is [Eq. (9.2.4)]

$$EI v''(x) = -M(x), \qquad (15.7.4)$$

[†] Effectively, we set the arbitrary datum of Π equal to zero, immediately prior to buckling, i.e. when the
right end of the rod is at point B*.

where $v(x)$ is the displacement in the positive y-direction. Substituting in Eq. (15.7.3),

$$U = \frac{1}{2} \int_0^L EI(v'')^2 \, dx.$$ (15.7.5)

Now in addition to the displacement $v(x)$ in the y-direction, points also displace in the x-direction due to the change from the straight-line position to the buckled configuration. In particular $B^* \to B^{**}$ [Fig. (15.7.1b)]; we denote this distance by λ. The potential energy of the force P is then

$$V = -P\lambda,$$ (15.7.6)

where $V = 0$, *immediately prior to buckling*.

The calculation of λ is merely a problem of geometry.[†]

To this end, let $u(x)$ represent the displacement *to the left* of any point of the longitudinal axis; i.e., we take here positive $u(x)$ in the *negative x-direction*. Consider now a typical element of length dx, as shown in Fig. (15.7.1c). In the buckled state, the length is given by

$$ds^2 = \left[\left(1 - \frac{du}{dx} \right) dx \right]^2 + dv^2,$$ (15.7.7a)

that is

$$ds^2 = \left[\left(1 - \frac{du}{dx} \right)^2 + \left(\frac{dv}{dx} \right)^2 \right] dx^2.$$ (15.7.7b)

We now recall that flexural deformation is *inextensional*; that is, there is no change in the length of the element lying on the longitudinal centroidal axis [see Section 6b of Chapter 8]. Therefore setting $ds^2 = dx^2$,

$$\left(1 - \frac{du}{dx} \right)^2 + \left(\frac{dv}{dx} \right)^2 = 1.$$ (15.7.8a)

Solving for du/dx,

$$\frac{du}{dx} = 1 - \sqrt{1 - \left(\frac{dv}{dx} \right)^2}.$$ (15.7.8b)

Now, by the binomial theorem [Eq. (15.4.12b)],

$$[1 - (v')^2]^{1/2} = 1 - \frac{1}{2}(v')^2 - \frac{1}{8}(v')^4 + \cdots.$$ (15.7.9a)

Consistent with the small displacement analysis, recalling that $|v'| \ll 1$ we neglect all terms higher than the quadratic:[‡]

$$[1 - (v')^2]^{1/2} \simeq 1 - \frac{1}{2}(v')^2,$$ (15.7.9b)

[†]   An expression for λ was derived in Section 5 of Chapter 9, where the notation Δ_x was used, i.e., $\Delta_x \equiv \lambda$. We repeat it here for continuity and convenience.

[‡]   This approximation is analogous to the small displacement approximations used in Section 15.4.

and therefore, from Eq. (15.7.8b),

$$\frac{du}{dx} = 1 - \left[1 - \frac{1}{2}(v')^2\right] = \frac{1}{2}(v')^2. \tag{15.7.10}$$

Integrating,

$$u(x) - u(0) = \frac{1}{2} \int_0^x (v')^2 \, dx. \tag{15.7.11}$$

Since $u(0) = 0$, we finally obtain for $\lambda \equiv u(L)$,

$$\lambda = \frac{1}{2} \int_0^L (v')^2 \, dx. \tag{15.7.12}$$

Substituting in Eq. (15.7.6), the potential energy V of the force P is

$$V = -\frac{P}{2} \int_0^L (v')^2 \, dx, \tag{15.7.13}$$

and therefore upon combining with Eq. (15.7.5), we obtain

$$\Pi = \frac{1}{2} \int_0^L EI(v'')^2 \, dx - \frac{P}{2} \int_0^L (v')^2 \, dx. \tag{15.7.14}$$

Alternatively, using Eq. (15.7.3) for the strain energy, we have

$$\Pi = \frac{1}{2} \int_0^L \frac{M^2(x)}{EI} \, dx - \frac{P}{2} \int_0^L (v')^2 \, dx. \tag{15.7.15}$$

We observe that the potential Π, given by Eq. (15.7.14), is expressed in terms of pure quadratic terms, as was the case in Section 15.5 for the rigid rods, using a small displacement analysis. We note too that the total potential

$$\Pi = \Pi(P, EI, L) \tag{15.7.16}$$

here is actually a set of numbers defined by the three given parameters. However, Π also depends on the form of $v(x)$, which in fact is *unknown*. Such a quantity is called a **functional**.

Proceeding as in the previous sections, we investigate possible equilibrium positions when $v(x) \neq 0$, and thus seek stationary values of Π. We therefore require that

$$\delta\Pi = \delta\left[\frac{1}{2} \int_0^L EI(v'')^2 \, dx - \frac{P}{2} \int_0^L (v')^2 \, dx\right] = 0. \tag{15.7.17}$$

Finding extreme values of Π, given by an expression such as that appearing in Eq. (15.7.17), requires the use of the calculus of variations and falls beyond the scope of our treatment. [However, we state that by use of this calculus, a differential equation for $v = v(x)$ is obtained together with appropriate boundary conditions at $x = 0$ and $x = L$. The solution of the resulting boundary value problem then yields the shape of the function $v(x)$ for which Π as well as the critical force P_{cr} will be stationary.]

Having precluded the use of the calculus of variations in our treatment, we seek a different approach that, while not yielding, in general, an exact value of the critical load, will provide us with approximate solutions to the instability problem.

First, we recall that if we assume infinitesimal displacements, the critical load corresponds to the force that causes the system to be in a neutral (or more precisely a pseudo-neutral) equilibrium state.[†] This occurs when the system (described by a potential that is quadratic in the displacements) undergoes displacements for which the change $\Delta \Pi = 0$.

Now, let us consider Π as given by Eq. (15.7.14), where we have noted that Π depends on the quadratic terms v' and v''. We first observe that in the original position $v(x) \equiv 0$, $\Pi = 0$. Moreover, it is important to note that the $v = 0$ position is an equilibrium position. Hence Π as given by Eq. (15.7.14) is also, in fact, the *change* $\Delta \Pi$ in the total potential, which occurs due to buckling.[‡] Using the neutral equilibrium condition, $\Delta \Pi = 0$, and substituting Eq. (15.7.14), we may therefore write

$$\int_0^L E I(x)(v'')^2 \, dx - P \int_0^L (v')^2 \, dx = 0. \qquad (15.7.18)$$

Hence P is given by the quotient

$$\frac{\int_0^L E I(x)(v'')^2 \, dx}{\int_0^L (v')^2 \, dx},$$

called *Rayleigh's quotient* (named after Lord Rayleigh); the calculated value of P is referred to as the *Rayleigh buckling load* and is denoted by P_R, i.e.,

$$P_R = \frac{\int_0^L E I(x)(v'')^2 \, dx}{\int_0^L (v')^2 \, dx}. \qquad (15.7.19)$$

Now, clearly, the value of P_R depends on the function $v = v(x)$, which, as mentioned previously, is generally unknown. Thus P_R will assume various numerical values depending on the function $v(x)$. The question here, then, is what is the relation between P_R and the critical load P_{cr}. To answer this question, we must recall that the critical load is the smallest possible load that can maintain the system in a deflected position, $v(x) \neq 0$, since if $P < P_{cr}$ the rod will return to its original position $v(x) = 0$. Hence, we conclude that the critical load is the smallest value of P_R that can be obtained from the Rayleigh quotient; that is,

$$P_{cr} = \min P_R = \min \frac{\int_0^L E I(x)(v'')^2 \, dx}{\int_0^L (v')^2 \, dx}. \qquad (15.7.20)$$

Indeed, there is a unique function $v = v(x)$ that will cause P_R to be a minimum. (This is the function that we would have obtained, had we proceeded with the calculus of variations previously mentioned!).

In the next sections we shall develop a method that gives approximate values of P_{cr} and shall also give a mathematical proof that $P_{cr} \leq P_R$. Since P_R is never less than P_{cr} (and, in fact, it is generally larger), P_R is said to be an *upper bound* to P_{cr}.

However, before concluding, we note that the numerator appearing in Eq. (15.7.20) represents the strain energy. Consequently if U is given by Eq. (15.7.3) instead of

---

[†] At this stage, we will use the conventional term 'neutral' equilibrium with the understanding that we are referring to 'pseudo-neutral' equilibrium.

[‡] Note that a similar case existed for the previous systems considered. [See, e.g., Eq. (15.5.4).]

Eq. (15.7.5), the Rayleigh load may be expressed as

$$P_R = \frac{\int_0^L \frac{M^2}{EI(x)}\, dx}{\int_0^L (v')^2\, dx}.$$

(15.7.21)

At times we shall find it advantageous to use this form in obtaining critical loads. Note, however, that in practice, Eq. (15.7.21) proves to be useful only if $M(x)$ can be written explicitly, i.e. only for statically determinate systems.

15.8   The Rayleigh method for critical loads

(a) Development of the method

In the previous section, the Rayleigh load was defined as

$$P_R = \frac{\int_0^L EI(x)(v'')^2\, dx}{\int_0^L (v')^2\, dx}$$

and it was observed that the value of P_R depends on the form $v = v(x)$, which, in general, is unknown.

We develop here the Rayleigh method to obtain approximate values of P_{cr}. Essentially, recalling the development of Section 9 of Chapter 14, we *assume* a form $v = f(x)$ that satisfies all the *geometric* boundary conditions of the problem. Thus, as defined in Chapter 14, these are *admissible functions*. (Note that there exist many admissible functions – actually an infinite number of such functions – which satisfy the geometric boundary conditions of a given problem.) Therefore, in general, we choose the simplest forms of the admissible functions and substitute in Eq. (15.7.19). Upon performing the integration, a value of P_R is obtained, which will always be greater than P_{cr}. However, we may repeat the procedure any number of times, each time using a different admissible function and thus obtain a different P_R. Since $P_R \geq P_{cr}$, the smallest value of P_R will be the best approximation to P_{cr}. We illustrate the method in the following example.

Example 15.4:   Determine the Rayleigh buckling load for the simply supported elastic rod shown in Fig. (15.7.1), assuming as an admissible function, the parabola

$$v(x) = Ax(L - x).$$

(15.8.1)

Solution: Taking derivatives,

$$v'(x) = A(L - 2x),$$

(15.8.2a)

$$v''(x) = -2A,$$

(15.8.2b)

the numerator of Rayleigh's Quotient is

$$EI \int_0^L (v'')^2\, dx = 4A^2 EI \int_0^L dx = 4A^2 EIL$$

(15.8.3a)

while the denominator is

$$\int_0^L (v')^2\, dx = A^2 \int_0^L (L - 2x)^2\, dx = \frac{A^2 L^3}{3}.$$

(15.8.3b)

Hence, substituting in Eq. (15.7.19),

$$P_R = \frac{4A^2 EIL}{A^2 L^3/3} = \frac{12EI}{L^2}. \tag{15.8.4}$$

If we compare with the exact value of $P_{cr} = \frac{\pi^2 EI}{L^2} = 9.8696 \frac{EI}{L^2}$ given by Eq. (15.7.1), we note that the percentage error is

$$\frac{P_R - P_{cr}}{P_{cr}} = 21.5\%, \tag{15.8.5}$$

which is, in fact, a large error.

We may explain this large relative error by noting, from Eq. (15.8.2b), that for our assumed buckled configuration, $v'' = -2A \neq 0$. Now we know that the moment $M = 0$ at $x = 0$, $x = L$. Hence since $M = -EIv''(x)$, the non-vanishing second derivative is a clear contradiction with the actual conditions that exist at the two ends A and B. In other words, our assumed admissible function, Eq. (15.8.1), yields a second derivative which clearly does not approximate that which occurs for the actual elastic curve in the true buckled position.

However, let us solve the problem again, with the same admissible function, Eq. (15.8.1), using instead the form for the Rayleigh load given by Eq. (15.7.21). Then

$$M(x) = Pv(x) = PAx(L - x). \tag{15.8.6}$$

The denominator of Eq. (15.7.21) remains the same as before, while the numerator is

$$\frac{1}{EI} \int_0^L M^2(x)\,dx = \frac{P^2 A^2}{EI} \int_0^L x^2(L - x)^2\,dx = \frac{P^2 A^2 L^5}{30EI}. \tag{15.8.7}$$

Therefore

$$P_R = \frac{P_R^2 A^2 L^5/30EI}{A^2 L^3/3}$$

from which

$$P_R = \frac{10EI}{L^2}. \tag{15.8.8}$$

The resulting percentage error is

$$\frac{P_R - P_{cr}}{P_{cr}} = \frac{10 - \pi^2}{\pi^2} = 1.32\%,$$

which is a relatively small error. $\qquad\qquad\qquad\qquad\qquad\qquad \square$

The question may well be raised as to why, if the *same* admissible function is assumed, do we get such a poor approximation in the first case, while in the second case the approximation is very good.

We observe that in the second case, the approximation for the second derivative does not appear *explicitly* in the expression for the strain energy. Thus, while it is often possible to choose an admissible function that approximates the true buckled shape, choosing a function whose second derivative approximates that of the true buckled curve proves to be much more difficult. This, in fact, was the case for the admissible function given by Eq. (15.8.1).

We thus conclude that in general, the Rayleigh load will approximate the true critical load, depending on how well the chosen admissible function approximates the true shape of the buckled curve.[†] The latter is referred to as the *buckling mode*, as we have seen in Section 15.6. It follows that if by chance, we choose a curve corresponding to the natural buckling mode, then we shall find that $P_R = P_{cr}$. For the problem at hand we know that the true buckling mode corresponding to P_{cr} (with $n = 1$) is given by [Eq. (11.4.13)]

$$v(x) = A \sin \frac{\pi x}{L} \qquad (15.8.9)$$

for which

$$v'(x) = \frac{A\pi}{L} \cos \frac{\pi x}{L}, \qquad (15.8.10a)$$

$$v''(x) = -A \left(\frac{\pi}{L}\right)^2 \sin \frac{\pi x}{L}. \qquad (15.8.10b)$$

Note that here

$$v(0) = v(L) = v''(0) = v''(L) = 0, \qquad (15.8.11)$$

i.e., *all* boundary condition, both geometric and mechanical, are satisfied.
Substituting in the numerator and denominator of Eq. (15.7.19),

$$EI \int_0^L [v''(x)]^2 \, dx = EI A^2 \left(\frac{\pi}{L}\right)^4 \int_0^L \sin^2 \frac{\pi x}{L} \, dx = \frac{EI A^2 \pi^4}{2L^3}, \quad (15.8.12a)$$

$$\int_0^L [v'(x)]^2 \, dx = A^2 \left(\frac{\pi}{L}\right)^2 \int_0^L \cos^2 \frac{\pi x}{L} \, dx = \frac{A^2 \pi^2}{2L} \qquad (15.8.12b)$$

from which

$$P_R = \frac{\pi^2 EI}{L^2} = P_{cr}. \qquad (15.8.13)$$

Example 15.5: A simply supported rod of rectangular cross-section, with constant width b and varying depth $d(x)$, is subjected to an axial load P, as shown in Fig. (15.8.1). The second moment of the cross-sectional area, $I(x)$, is given as

$$I(x) = I_0 \left(1 + \gamma \sin \frac{\pi x}{L}\right),$$

where $I_0 = I(x = 0, L)$, $d_0 = d(x = 0, L)$ and γ is a constant. Determine an upper bound P_R for the critical load, using Rayleigh's method.

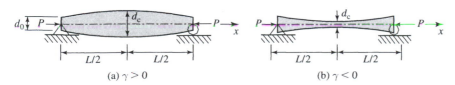

(a) $\gamma > 0$ (b) $\gamma < 0$ **Figure 15.8.1**

[†] We arrive at this conclusion by a qualitative reasoning. In Example 15.6, we shall verify this conclusion by comparing quantitatively the results for a series of assumed functions.

Solution: Using the Rayleigh method, let us assume that the buckling mode of this rod is the same as that of a rod of constant $I = I_0$; i.e., we assume an admissible function

$$v(x) = A \sin \frac{\pi x}{L}. \tag{15.8.14}$$

The numerator of the Rayleigh quotient is then

$$A^2 E I_0 \left(\frac{\pi}{L}\right)^4 \int_0^L \left(1 + \gamma \sin \frac{\pi x}{L}\right) \sin^2 \frac{\pi x}{L} \, dx = A^2 E I_0 L \left(\frac{\pi}{L}\right)^4 \left(\frac{1}{2} + \frac{4\gamma}{3\pi}\right) \tag{15.8.15}$$

by simple integration.

The denominator of the Rayleigh quotient, not being dependent on $I(x)$, is the same as for the prismatic rod [Eq. (15.8.12b)]. Therefore

$$P_R = \frac{A^2 E I_0 L (\pi/L)^4 (1/2 + 4\gamma/3\pi)}{A^2 \pi^2 / 2L} = \frac{\pi^2 E I_0}{L^2} \left(1 + \frac{8\gamma}{3\pi}\right). \tag{15.8.16}$$

We observe, by means of this very simple calculation, that we have easily obtained an approximation to the critical load. On the other hand, an exact value for the critical load of this non-prismatic beam can be found only from a solution of the equation $EI(x)v''(x) + Pv(x) = 0$ [cf. Eq. (11.4.4b)], i.e.,

$$EI_0 \left(1 + \gamma \sin \frac{\pi x}{L}\right) v''(x) + Pv(x) = 0,$$

where $v(x)$ satisfies the boundary conditions $v(0) = v(L) = 0$. We observe here that this is a second-order differential equation with variable coefficients which does not readily possess a simple analytic solution.

If, for example, γ is given as $\gamma = (d_c/d_0)^3 - 1$ [see Fig. (15.8.1)], the effect of the varying depth $d(x)$ is shown in the plot of $\frac{P_R}{P_E}$ vs. $\frac{d_c}{d_0}$ in Fig. (15.8.2), where P_E is the Euler buckling load, $P_E = \frac{\pi^2 E I_0}{L^2}$.

In the example below, we show a first means to obtain an improvement on the approximation for the Rayleigh load.

Example 15.6:   Determine an approximation for the critical load P_{cr} acting on a cantilever beam, as shown in Fig. (15.8.3a), using Rayleigh's method.

Solution: We assume an admissible function having the form [Fig. (15.8.3b)]

$$v(x) = A x^\zeta, \quad \zeta > 3/2, \tag{15.8.17a}$$

where A is a constant and ζ is a parameter. Then

$$v'(x) = \zeta A x^{\zeta - 1}, \tag{15.8.17c}$$

$$v''(x) = \zeta(\zeta - 1) A x^{\zeta - 2}. \tag{15.8.17d}$$

Substituting in the denominator of the Rayleigh quotient, and integrating, we obtain

$$\int_0^L [v'(x)]^2 \, dx = \zeta^2 A^2 \int_0^L x^{2(\zeta - 1)} \, dx = \frac{\zeta^2 A^2}{2\zeta - 1} L^{2\zeta - 1}. \tag{15.8.18a}$$

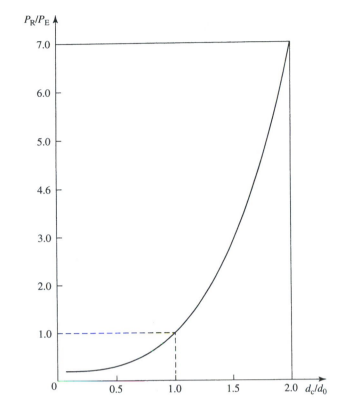

Figure 15.8.2

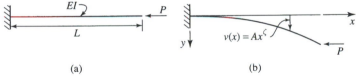

(a) (b)

Figure 15.8.3

Similarly, the integral of the numerator becomes

$$\int_0^L [v''(x)]^2 \, dx = \zeta^2(\zeta - 1)^2 A^2 \int_0^L x^{2(\zeta-2)} \, dx = \frac{\zeta^2(\zeta - 1)^2 A^2}{2\zeta - 3} L^{2\zeta-3}. \quad (15.8.18b)$$

Note that the condition $\zeta > 3/2$ is required in order that the preceding integral be bounded and positive.

The Rayleigh load is then

$$P_R = \frac{EI}{L^2}(\zeta - 1)^2 \left(1 + \frac{2}{2\zeta - 3}\right). \quad (15.8.19)$$

If, for example, $\zeta = 2$ then $P_R = \frac{3EI}{L^2}$. Similarly for $\zeta = 3$, $P_R = \frac{20}{3}\frac{EI}{L^2} = 6.667\frac{EI}{L^2}$. The exact critical load for this rod is known [see Eq. (11.9.9)], viz. $P_{cr} = \pi^2 EI/4L^2 = 2.467EI/L^2$. Therefore the percentage errors for the cases $\zeta = 2$ and 3 are 21.5 and 170%, respectively. We thus observe that, depending on the chosen admissible function $v(x)$, one may obtain a very poor approximation to P_{cr}. Clearly, we may suspect that the error increases sharply for an assumed function that differs greatly with the true buckling mode.

Now we note that the load P_R obtained in Eq. (15.8.19) is a function of the parameter ζ, i.e., $P_R = P_R(\zeta)$. Since P_R is an upper bound to P_{cr}, for $v(x)$ given by Eq. (15.8.17a), the value of ζ which will yield the best approximation to P_{cr} is that for which P_R is a minimum. To this end we set $\frac{dP_R}{d\zeta} = 0$. Upon differentiating, after some simple manipulations, we obtain

$$4\zeta^2 - 10\zeta + 5 = 0 \qquad (15.8.20a)$$

whose roots are

$$\zeta = \frac{1}{4}(5 \pm \sqrt{5}). \qquad (15.8.20b)$$

Since the lower root is less than unity, the only relevant root is $\zeta = \frac{1}{4}(5 + \sqrt{5}) = 1.8090$. Substituting this value in Eq. (15.8.19), we find

$$P_R = \frac{EI}{8L^2}(11 + 5\sqrt{5}) = 2.773\frac{EI}{L^2}, \qquad (15.8.21a)$$

which corresponds to an error of 12.4% with P_{cr}. Note that the assumed function is then

$$v(x) = Ax^{1.809}. \qquad (15.8.21b)$$

Thus, of all functions having the form given by Eq. (15.8.17a), that represented by Eq. (15.8.21b) yields the closest approximation to the true buckling mode.   □

Having obtained results for various values of ζ in this example, we are now in a position to relate the various shapes given by $v(x)$ to the calculated Rayleigh load. The shapes of the assumed buckling mode, Eq. (15.8.17a), are shown as dashed lines in Fig. (15.8.4) for three values of ζ ($\zeta = 2.0$, 3.0 and 1.809) as a function of x/L.[†]

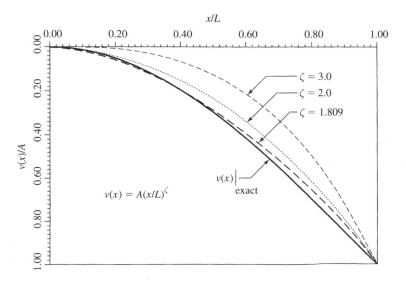

Figure 15.8.4

The exact buckling mode given in Chapter 11 [Eq. (11.5.11a)],

$$v(x) = A\left(1 - \cos\frac{\pi x}{2L}\right), \qquad (15.8.22)$$

is shown as a solid curve in the figure. We note that for $\zeta = 3$, the assumed shape differs greatly from the true mode. This explains the relatively large error when using this admissible function. For $\zeta = 2$, we note that the approximated shape is much closer to the true buckling mode, but still yields an error of 21.5% in the calculation for the critical load. For the value $\zeta = 1.809$ as obtained above, the assumed shape is indeed a very close approximation to the true mode and, as mentioned, is the best possible approximation for functions of the form given by Eq. (15.8.17a). The small deviation from the true buckling mode is reflected by the relatively small error of 12.4% for the buckling load.

From these results, we may conclude, as suggested previously in this section, that the closer the assumed shape $v(x)$ is to the true buckling mode, the closer the approximation of the Rayleigh load will be to the true critical load.

Finally, we note that the calculated Rayleigh load is often quite sensitive to the form of $v(x)$ and thus the method must be used with great care.

While this example illustrates a means to improve the first approximation, we observe that we started with a prescribed form of $v(x)$. By prescribing this general form, we are effectively imposing an unnatural constraint on the system. In finding the value of $\zeta = 1.809$, it was possible to reduce this imposed constraint. A more significant improvement of Rayleigh's method for stability analyses in which we further introduce a greater degree of flexibility for prescribed admissible functions, is developed in Section 9 of this chapter.

(b) Proof of the upper boundedness of the Rayleigh load (restricted proof)

In the discussion of the Rayleigh quotient, we concluded from the basic definition of the critical load that $P_R \geq P_{cr}$. We now prove this property mathematically for the case of the simply supported rod considered previously.

For a simply supported rod we require that the admissible function $v(x)$ satisfy the geometric boundary conditions $v(0) = v(L) = 0$. Now, *any* such admissible function can always be expanded in the region $0 \leq x \leq L$ as a Fourier sine series

$$v(x) = a_1 \sin\frac{\pi x}{L} + a_2 \sin\frac{2\pi x}{L} + a_3 \sin\frac{3\pi x}{L} + \cdots$$

$$= \sum_{n=1}^{\infty} a_n \sin\frac{n\pi x}{L}. \qquad (15.8.23)$$

We observe that the given geometric boundary conditions are automatically satisfied. The derivatives, expressed as a Fourier series, can be obtained by differentiating term by term; thus,

$$\frac{dv}{dx} = \sum_{n=1}^{\infty} a_n \frac{n\pi}{L} \cos\frac{n\pi}{L}x, \qquad (15.8.24a)$$

$$\frac{d^2v}{dx^2} = \sum_{n=1}^{\infty} a_n \left(\frac{n\pi}{L}\right)^2 \sin\frac{n\pi x}{L}. \qquad (15.8.24b)$$

Then

$$\left(\frac{dv}{dx}\right)^2 = \sum_{n=1}^{\infty}\sum_{m=1}^{\infty} a_n a_m \left(\frac{n\pi}{L}\right)\left(\frac{m\pi}{L}\right)\cos\frac{n\pi x}{L}\cos\frac{m\pi x}{L} \quad (15.8.24c)$$

and a similar expression exists for $(d^2v/dx^2)^2$. Substituting in Eq. (15.7.19) and interchanging the summation and integration processes,

$$P_R = \frac{EI\left(\frac{\pi}{L}\right)^4 \sum_{n=1}^{\infty}\sum_{m=1}^{\infty} n^2 m^2 a_n a_m \int_0^L \sin\frac{n\pi x}{L}\sin\frac{m\pi x}{L}\,dx}{\left(\frac{\pi}{L}\right)^2 \sum_{n=1}^{\infty}\sum_{m=1}^{\infty} nm\, a_n a_m \int_0^L \cos\frac{n\pi x}{L}\cos\frac{m\pi x}{L}\,dx}. \quad (15.8.25)$$

Recall now the orthogonality properties of trigonometric functions:

$$\int_0^L \sin\frac{n\pi x}{a}\sin\frac{m\pi x}{a}\,dx = \begin{cases} L/2, & m = n \\ 0, & m \neq n, \end{cases} \quad (15.8.26a)$$

$$\int_0^L \cos\frac{n\pi x}{a}\cos\frac{m\pi x}{a}\,dx = \begin{cases} L/2, & m = n \\ 0, & m \neq n. \end{cases} \quad (15.8.26b)$$

Hence, contributions to the sums appearing in Eq. (15.8.25), will occur only if $n = m$, while for $n \neq m$, the terms will vanish. Therefore, we have

$$P_R = \frac{\pi^2 EI}{L^2}\frac{\sum_{n=1}^{\infty} n^4 a_n^2}{\sum_{n=1}^{\infty} n^2 a_n^2}$$

or

$$P_R = \frac{\pi^2 EI}{L^2}\left(\frac{a_1^2 + 16a_2^2 + 81a_3^2 + \cdots}{a_1^2 + 4a_2^2 + 9a_3^2 + \cdots}\right). \quad (15.8.27)$$

Since all a_n are real, it is evident that if $a_n \neq 0$ for $n \geq 2$, the term within parentheses will be greater than unity, and thus $P_R > P_{cr}$. The condition $P_R = P_{cr}$ will be true only if $a_n = 0$, $n \geq 2$, i.e. if

$$v(x) = a_1 \sin\frac{\pi x}{L}, \quad (15.8.28)$$

which is the true buckling mode (eigenfunction). Hence P_R is an upper bound to P_{cr} and is equal to the critical load only if $v(x)$ is the actual buckling mode.

Having proved mathematically that P_R is always an upper bound to P_{cr}, it is worthwhile to interpret this result in physical terms in order to gain some physical insight into this class of problems.

We first remark that a member such as that shown in Fig. (15.7.1) has an *infinite number of degrees of freedom* since it may buckle freely into any shape $v(x)$. Describing the system anthropomorphically, we may say that the rod is free to 'choose' any buckling shape. The rod, in fact, thus chooses to buckle into a shape, called the *buckling mode*, which can be maintained by the smallest of all possible axial forces.

Now, when an admissible function $v(x)$ is chosen in Rayleigh's method, e.g. $v(x) = Ax(L - x)$ as in Eq. (15.8.1), the shape is artificially prescribed and the only remaining free variable that defines the displacement is the single constant A. Thus, in this example, the ∞-d.o.f. system is effectively transformed into 1-d.o.f.

system, with a generalised coordinate A. That is, as a result of the imposed constraint, the rod loses a degree of flexibility upon being 'forced' to assume a prescribed buckled shape. As a result, the load P_R, which is required to maintain the rod in this prescribed 'unnatural' shape, is greater than P_{cr}.

15.9   The Rayleigh–Ritz method for critical loads

We have observed that by choosing an admissible function such as that given by Eq. (15.8.1) for a simply supported rod, the ∞-d.o.f. system was transformed into a 1-d.o.f. system with a single generalised coordinate. The resulting system, having been made artificially less flexible, then requires a force P greater than P_{cr} to remain in a deflected state.

A modification of the Rayleigh method, which, in effect, increases the degrees-of-freedom of the system and thus reduces the calculated values of P_R, was proposed by Ritz. This improved method is known as the *Rayleigh–Ritz method*.

The improvement consists essentially of introducing additional free constants that are adjusted in such a way as to cause P_R to be the minimum for the type of admissible function prescribed. Thus, e.g., referring to the problem of the previous section, instead of choosing the parabola $v(x) = Ax(L - x)$, we let

$$v(x) = (a + bx + cx^2), \tag{15.9.1}$$

where a, b and c are arbitrary unknown constants. The only requirement on these constants is that the *resulting function $v(x)$ must be admissible*. When $v(x)$ and its derivatives are substituted into the Rayleigh quotient, the calculated Rayleigh load is a function of the constants, i.e., $P_R = P_R(a, b, c)$. Since we seek a minimum value for P_R, we take

$$\frac{\partial P_R}{\partial a} = 0, \qquad \frac{\partial P_R}{\partial b} = 0, \qquad \frac{\partial P_R}{\partial c} = 0, \tag{15.9.2}$$

thus obtaining a relation between the constants. It is to be emphasised, however, that the value of P_R will still be greater than P_{cr} since the resulting function $v(x)$ is not the true buckling mode.

To generalise our discussion, we assume an admissible function $v(x)$ to be of the form

$$v(x) = \sum_{i=0}^{n} a_i f_i(x), \tag{15.9.3}$$

where a_i are unknown constants. The functions $f_i(x)$ are called **coordinate functions**. For example, for $v(x)$ given in Eq. (15.9.1),

$$n = 2; \quad a_0 = a, \ f_0(x) = 1; \quad a_2 = b, \ f_2(x) = x; \quad a_2 = c, \ f_2(x) = x^2.$$

Then

$$v'(x) = \sum_{i=0}^{n} a_i f_i'(x), \tag{15.9.4a}$$

$$v''(x) = \sum_{i=0}^{n} a_i f_i''(x). \tag{15.9.4b}$$

Substituting in Eq. (15.7.19),

$$P_R = \frac{EI \int_0^L \left(\sum_{i=0}^n a_i f_i''(x) \right)^2 dx}{\int_0^L \left(\sum_{i=0}^n a_i f_i'(x) \right)^2 dx} \tag{15.9.5a}$$

or

$$P_R = \frac{N(a_1, a_2, \ldots, a_n)}{D(a_1, a_2, \ldots, a_n)}, \tag{15.9.5b}$$

where N and D denote the numerator and denominator respectively.

To obtain the minimum, we take

$$\frac{\partial P_R}{\partial a_i} = 0, \quad i = 0, 1, 2, \ldots, n. \tag{15.9.6a}$$

Hence

$$\frac{\partial P_R}{\partial a_i} = \frac{D \frac{\partial N}{\partial a_i} - N \frac{\partial D}{\partial a_i}}{D^2} = 0, \quad i = 0, 1, 2, \ldots, n. \tag{15.9.6b}$$

Since D is finite, Eq. (15.9.6b) implies

$$D \frac{\partial N}{\partial a_i} - N \frac{\partial D}{\partial a_i} = 0, \quad i = 0, 1, 2, \ldots, n \tag{15.9.7a}$$

and using Eq. (15.9.5b),

$$\frac{\partial N}{\partial a_i} - P_R \frac{\partial D}{\partial a_i} = 0, \quad i = 0, 1, 2, \ldots, n. \tag{15.9.7b}$$

Equation (15.9.7b) represents a set of homogeneous linear algebraic equations

$$\frac{\partial N}{\partial a_0} - P_R \frac{\partial D}{\partial a_0} = 0$$

$$\frac{\partial N}{\partial a_1} - P_R \frac{\partial D}{\partial a_1} = 0 \tag{15.9.7c}$$

$$\vdots$$

$$\frac{\partial N}{\partial a_n} - P_R \frac{\partial D}{\partial a_n} = 0.$$

Since $N = N(a_i)$, $D = D(a_i)$ [$i = 0, 1, 2, 3, \ldots, n$] we have a set of $(n + 1)$ equations in the $(n + 1)$ unknowns $a_0, a_1, a_2, \ldots, a_n$. For non-trivial solutions to exist, the determinant of the system of equations must vanish, i.e.,

$$\Delta = 0. \tag{15.9.8}$$

This last equation yields the discrete values for P_R from which the ratios existing among the constants a_i can be found.

The method is illustrated by means of the following example.

Example 15.7: For the cantilever rod shown in Fig. (15.8.3a), determine the Rayleigh load, assuming a buckled configuration of the form

$$v(x) = c_1(x/L)^2(x/L - 3) + c_2(x/L)^2. \tag{15.9.9a}$$

Solution: First, we note that the geometric boundary conditions $v(0) = v'(0) = 0$ are satisfied. Therefore $v(x)$ is an appropriate admissible function.

For mathematical simplicity, letting $\xi = x/L$, Eq. (15.9.9a) becomes

$$v(\xi) = \xi^2[c_1(\xi - 3) + c_2].\qquad(15.9.9b)$$

Then

$$\frac{dv(\xi)}{d\xi} = \xi[3c_1(\xi - 2) + 2c_2],\qquad(15.9.9c)$$

$$\frac{d^2v(\xi)}{d\xi^2} = 6c_1(\xi - 1) + 2c_2.\qquad(15.9.9d)$$

Substituting in the Rayleigh quotient, noting that

$$\int_0^L \left[\frac{dv(x)}{dx}\right]^2 dx \to L\int_0^1 \left[\frac{dv(\xi)}{d\xi}\right]^2 d\xi,$$

$$\int_0^L \left[\frac{d^2v(x)}{dx^2}\right]^2 dx \to \frac{1}{L}\int_0^1 \left[\frac{d^2v(\xi)}{d\xi}\right]^2 d\xi$$

and performing the integrations, from Eq. (15.9.5a) we obtain

$$P_R = \frac{4EI}{L^2}\frac{3c_1^2 - 3c_1c_2 + c_2^2}{24\,c_1^2/5 - 5c_1c_2 + 4c_2^2/3}.\qquad(15.9.10)$$

Equation (15.9.10) is observed to be of the form of Eq. (15.9.5b). Upon taking derivatives with respect to c_1 and c_2, substitution in Eqs. (15.9.7b) yields

$$(24EI/L^2 - 48P_R/5)c_1 + (-12EI/L^2 + 5P_R)c_2 = 0,$$
$$(-12EI/L^2 + 5P_R)c_1 + (8EI/L^2 - 8P_r/3)c_2 = 0.\qquad(15.9.11)$$

By setting the determinant of this system of equations to zero, we obtain the equation for the unknown P_R:

$$3P_R^2 - 104\left(\frac{EI}{L^2}\right)P_R + 240\left(\frac{EI}{L^2}\right)^2 = 0,\qquad(15.9.12)$$

whose roots are

$$P_R = \frac{1}{6}(104 \pm 89.084)\frac{EI}{L^2}.\qquad(15.9.13a)$$

Since we are interested in the lowest buckling load, the relevant root is

$$P_R = 2.48596\frac{EI}{L^2}.\qquad(15.9.13b)$$

The ratio c_2/c_1 is obtained by substituting P_R back in either of Eqs. (15.9.11). The assumed buckling mode, corresponding to this value of P_R, is then

$$v(\xi) = A\xi^2(\xi - 3.31355).\qquad(15.9.14)$$

We observe that the value of P_R, given by Eq. (15.9.13b), is the lowest possible value which can be obtained for an assumed admissible function of the general form given by Eq. (15.9.9b). Furthermore, for functions of this general form, the specific shape $v(\xi)$ given by Eq. (15.9.14) is the best approximation to the natural buckling mode.

It is noted that the Rayleigh load P_R obtained is still an upper bound for P_{cr} to the exact value $P_{cr} = \frac{\pi^2 EI}{4L^2} = 2.4674 \frac{EI}{L^2}$ [see Eq. (11.9.9)]. The relative error is

$$\frac{P_R - P_{cr}}{P_{cr}} = \frac{2.48596 - \pi^2/4}{\pi^2/4} = 0.75\%, \qquad (15.9.15)$$

which is very small.

PROBLEMS

Sections 3–5

15.1: A cylinder of weight W is attached to a hemisphere of radius R and weight W, as shown in Fig. (15P.1). The entire system, made of a rigid material, rests on a smooth surface. Determine the values of h/R for which the system is stable in the vertical position.

15.2: A top, made of a homogeneous rigid material with density ρ (N/m³), consist of two components: a lower component, which is in the shape of a hemisphere, and an upper component, which is in the shape of a cone, as shown in Fig. (15P.2). The top is made to spin about its vertical y-axis. Determine the ratio h/R for which the top will be in a vertical position when it ceases to spin.

15.3: Determine the value of the critical load for the rigid rod shown in Fig. (15P.3).

15.4: For what values of P is the system shown in Fig. (15P.4a) stable in the vertical configuration if the torsional spring at point A is governed by the nonlinear relation $M_A = C\theta^3$, where $C > 0$ is a constant [see Fig. (15P.4b)]. Explain this result in physical terms.

15.5: Two rigid rods, each of length L, connected by a hinge and a torsional spring with constant β (N-m/rad) at B, are supported at A and C, as shown in Fig. (15P.5). The rods are subjected to two axial forces P acting at B and C. Determine the value of the critical force P_{cr}.

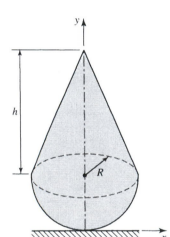

Figure 15P.1

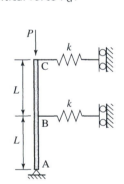

Figure 15P.2

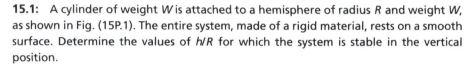

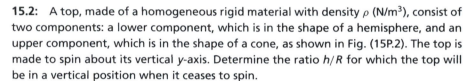

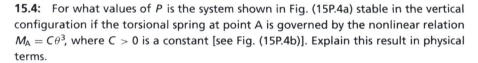

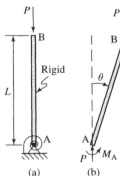

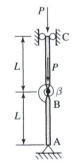

Figure 15P.3   **Figure 15P.4**   **Figure 15P.5**

Section 6

15.6: Given a system consisting of a rigid rod supported by linear springs with constants k_1 and k_2 and by a torsional spring with constant β, as shown in Fig. (15P.6). Determine (a) the critical load P_{cr} in terms of k_1, k_2, β and L and (b) the value of L that leads to a minimum value of P_{cr}. What is $(P_{cr})_{min}$?

15.7:* A system, consisting of two rigid rods, each of length L, is supported by a torsional spring having constant β_1 (N-m/rad) at A and a torsional spring with constant

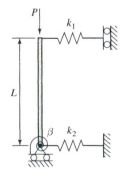

Figure 15P.6

β_2 at the hinge B, as shown in Fig. (15P.7a). (a) Determine the critical axial force P that causes instability. (b) Based on the answer to (a), and letting $\gamma \equiv \beta_2/\beta_1$, show analytically that the critical force P of the system of Fig. (15P.7b) is greater than that of Fig. (11P.7a) for any finite value of β_2.

15.8:* The system shown in Fig. (15P.8) consists of two rigid rods AC and CB, each of length L, which are hinged at C and which are supported by a linear spring with constant k at B and a torsional spring with constant β at C. (a) Determine the critical loads P_{cr} under which the vertical configuration becomes unstable. (b) If the values for the constants are given as $k = 5$ N/cm and $\beta = 500$ N-cm/rad, determine the critical load when (i) $L = 12$ cm and (ii) $L = 8$ cm. (c) Given the values k and β in (b). For what length L will the system can carry the greatest load P before instability occurs? What is this load?

15.9: A system, whose total length is L, consisting of three rods AB, BC and CD, is supported by springs with constant k (N/m) at the hinges B and C, as shown in Fig. (15P.9). An axial load P acts at A. (a) Determine the critical load P_{cr}. (b) What should be the ratio of b/a in order to maximise P_{cr}? What is $(P_{cr})_{max}$?

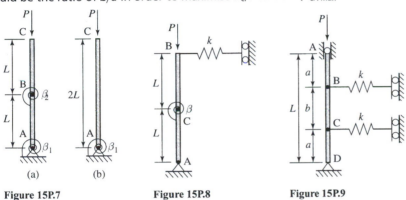

(a)	(b)	
Figure 15P.7	**Figure 15P.8**	**Figure 15P.9**

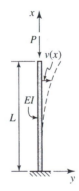

Figure 15P.10

Sections 7–9

15.10: (a) Determine the Rayleigh load P_R for the linear elastic column shown in Fig. (15P.10), using the expression for the internal strain energy U given by Eq. (15.7.3) together with the admissible function $v(x) = A(x/L)^2(1 + x/L)$. (b) Determine the percentage error with the true critical load.

15.11: (a) Using Rayleigh's method determine P_R, the approximation to the critical buckling load for a rod fixed at $x = 0$ and simply supported at $x = L$, which is subjected to an axial compressive load as shown in Fig. (15P.11). Use an admissible function of the form

$$v(x) = A\left[\left(\frac{x}{L}\right)^3 - \left(\frac{x}{L}\right)^2\right].$$

(b) Does this function satisfy *all* conditions, both geometric and natural (mechanical) conditions? (c) What is the percentage error with the known value of P_{cr}?

15.12: Repeat Problem 15.11, assuming the deflection is given by

$$v(x) = A\left[\frac{2}{3}\left(\frac{x}{L}\right)^4 - \frac{5}{3}\left(\frac{x}{L}\right)^3 + \left(\frac{x}{L}\right)^2\right].$$

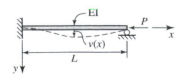

Figure 15P.11

15.13:* (a) Determine the Rayleigh buckling load for the elastic column shown in Fig. (15P.13), using as an assumed buckling mode $v = A(1 - \cos\frac{\pi x}{2L})$. (b) For what values

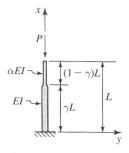

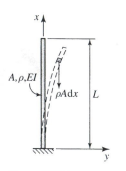

Figure 15P.13

Figure 15P.14

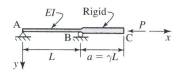

Figure 15P.15

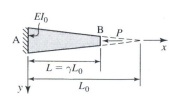

Figure 15P.16

of α or γ does $P_R = P_{cr}$? Why is this expression then 'exact'? (*Note*: See computer-related Problem 15.24.)

15.14:* A linear elastic column of constant cross-sectional area A and length L is fixed as shown in Fig. (15P.14). The material density of the column is given as ρ (N/m³). (a) Using Rayleigh's method, and assuming an admissible function of the form $v(x) = a(1 - \cos\frac{\pi x}{2L})$, determine an approximation for the critical weight W that causes instability of the column. (b) The exact critical weight (obtained from the solution of a Bessel equation) is $W = 7.837 EI/L^2$. Determine the percentage error obtained using the above admissible function.

15.15: A rod of length L and flexural rigidity EI is simply supported at A and B. A compressive axial load is applied at the end of a rigid strut BC having length $a = \gamma L$, which is rigidly connected to AB as shown in Fig. (15P.15). Determine an approximate value for the critical load using Rayleigh's method, assuming that the buckling mode of the segment AB is $v(x) = A\sin\frac{\pi x}{L}$. (*Note*: See computer-related Problem 15.25.)

15.16:* A rod AB of variable flexural stiffness $EI(x)$, in the form of a truncated cone, is fixed at one end and subjected to an axial load P, as shown in Fig. (15P.16). The moment of inertia $I(x)$ is given as $I(x) = I_0(1 - \frac{x}{L_0})^4$. (a) Assuming an admissible function in the shape of a parabola $v(x) = Ax^2$, determine the Rayleigh buckling load P_R. (b) The exact solution, obtained from the solution of a Bessel equation, for the case $L = L_0/2$ is $P_{cr} = 1.026 E I_0/L^2$. Determine the percentage error of P_R.

Review and comprehensive problems

15.17: A cylinder is attached to a hemisphere of radius R, as shown in Fig. (15P.1). The entire system, made of a rigid material whose density is ρ (N/m³), rests on a smooth surface. Determine the values of h/R for which the system is stable in the vertical position.

15.18: Given a rigid rod as shown in Fig. (15P.4), supported by a nonlinear torsional spring at A, which exerts a resisting moment on the rod according to the relation $M_A = C\tanh\theta$, $C > 0$. (a) Express the strain energy U of the system as a function of θ. (b) Determine the force P required to maintain equilibrium at any finite inclination θ of the rod with respect to the vertical position. (c) Determine the critical load that causes the vertical position to become unstable. (*Note*: See computer-related Problem 15.26.)

15.19:* Given a system consisting of four rigid rods of length L, connected by means of hinges and containing a spring, as shown in Fig. (15P.19). In the undeflected position, $\theta = \theta_0$. (a) Determine the range of θ for which the system is stable. (b) Determine the critical force P_{cr} at which the system becomes unstable. (c) Express the total potential of the system, Π, as a function of θ for any given θ_0. (*Note*: Here the force P can act either upwards or downwards. See computer-related Problem 15.27.)

15.20: A rigid rod is supported at A by a torsional spring β (N-m/rad) and by a linear spring having constant k (N/m), as shown in Fig. (15P.20). (a) Determine the vertical force P required to maintain the rod at any *finite* angle θ with respect to the vertical position. (b) What is the critical load P_{cr} for $\theta = 0$? For which values of the ratio $\alpha = \beta/kL^2$ is the bifurcation point stable or unstable? (*Note*: See computer-related Problem 15.28.)

15.21: Repeat Problem 15.14, using Eq. (15.9.9a) for the admissible function.

15.22:* A rigid rod AB, of length L, is attached to a linear elastic spring BC having constant k, as shown in Fig. (15P.22a). A vertical force P is applied at B. (a) Show that the rod is unstable in the position $\theta = 0$ under any downward load $P > 0$. Does the same hold true if an upward load P is applied? Justify the answer. (b) Show that if $0 < \theta < \pi$ as in Fig. (15P.22b), the rod is in a stable equilibrium position for all values of the equilibrium force

$$P = kL\,(1+\gamma)\Big\{1 - [\gamma^2 + 2\gamma + 2 - 2(1+\gamma)\cos\theta]^{-1/2}\Big\}.$$

(c) For what values of P can the rod be held in a position $\theta = \pi$? Express the answer in terms of $k,\ L$ and γ. (*Note*: See computer-related Problem 15.29.)

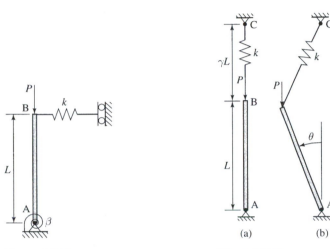

Figure 15P.19

Figure 15P.20

Figure 15P.22

(a)   (b)

15.23:* Two rigid bars AB and BC, each of length L, are connected by a hinge at B and supported at A and C, as shown in Fig. (15P.23a). The rods are initially inclined (when $P = 0$) at an angle θ_0 as shown in the figure. A vertical load P applied at B causes the system to displace to a position defined by θ ($\theta < \theta_0$), as shown in Fig. (15P.23b). (a) Determine the relation between P and θ for finite values of θ. (b) For what values of P will the system be in stable equilibrium? Evaluate the limiting value of P for stability if $\theta_0 = 30°$. (c) If $\theta_0 \ll 1$, show that $P_{cr} = \frac{2\sqrt{3}}{9} kL\,\theta_0^3$.

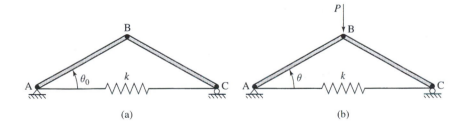

(a)   (b)   **Figure 15P.23**

The following problems require the use of a computer.

15.24:   Referring to Problem 15.13,
   (a) choose one or both of the following options: (i) Evaluate the Rayleigh load P_R for values $0 \le \gamma \le 1$ for several discrete values of α ($0 \le \alpha \le 1$), and, using a plotting routine, plot the resulting family of curves P_R/P_E vs. γ. (ii) Evaluate the Rayleigh load, P_R for values $0 \le \alpha \le 1$ for several discrete values of γ, ($0 \le \gamma \le 1$) and, using a plotting routine, plot the resulting family of curves P_R/P_E vs. α.

(b) (i) Using a computer, evaluate numerically the roots of the transcendental equation (see Problem 11.39),

$$\tan(\gamma \lambda L) = \cot\left[\frac{(1-\gamma)\lambda L}{c}\right], \quad c = \sqrt{\alpha},$$

which lead to the exact values of $P_{cr} = EI\lambda^2$. (ii) Using a plotting routine plot P_{cr}/P_E vs. γ.

(c) Determine the percentage error for several values of α and γ.

15.25: Referring to Problem 15.15,

(a) evaluate the Rayleigh load P_R for values $0 \le \gamma \le 10$, and, using a plotting routine, plot P_R/P_E vs. γ.

(b) (i) Using a computer, evaluate numerically the roots of the transcendental equation (see Problem 11.20)

$$\tan(\lambda L) = \frac{\gamma \lambda L}{\gamma - 1},$$

which lead to the exact values of $P_{cr} = EI\lambda^2$. (ii) Using a plotting routine, plot P_{cr}/P_E vs. γ.

(c) Determine the percentage error of the ratio P_R/P_E for values in the range $0 \le \gamma \le 10$.

15.26: Referring to Problem 15.18

(a) using a plotting routine, plot the ratio $\frac{P(\theta)}{C/L}$ as a function of θ.

(b) For $0 < \theta$, express the derivative $\frac{d^2\Pi(\theta)}{d\theta^2}$ at equilibrium as a function of θ only and evaluate numerically, by means of a computer, the value of θ where $\frac{d^2\Pi(\theta)}{d\theta^2}|_{equil.} = 0$.

(c) For what range of $0 < \theta \le 180°$ is the branch stable/unstable? Indicate these on the figure of (a) above.

(d) Assume that the rod is inclined under an equilibrium load at an angle of $20°$ with respect to the vertical. Based on the plot of (b) above, determine to what position the rod will 'jump' if given a perturbation $\theta = 20° + \epsilon$, ($\epsilon > 0$).

15.27: Referring to Problem 15.19, (a) using a plotting routine, plot a family of curves of the equilibrium force P as a function of θ for several discrete values of $\theta_0 : \theta_0 = 10°, 30°, 45°, 60°$. (b) Repeat (a) for the total potential $\Pi = \Pi(\theta; \theta_0)$.

15.28: Referring to Problem 15.20, (a) using a plotting routine, plot the relation $P(\theta)$ as a function of θ, using two values of α: $\alpha = 2$ and $\alpha = 4$. (b) For what range of $0 < \theta \le 180°$ is the branch stable/unstable when (i) $\alpha = 2$ and (ii) $\alpha = 4$.

15.29: Referring to Problem 15.22, (a) for the case $\gamma = 1$, evaluate numerically the required equilibrium force to maintain the rod in any finite position θ. (b) Using a plotting routine, plot the equilibrium force P as a function of θ for $0 \le \theta \le \pi$ and indicate the stable and unstable branches.

Appendix A

Properties of areas

A.1 General properties: centroids, first and second moments of areas

The definitions and properties given below relate to a general area A lying in the y–z plane of an (x, y, z) coordinate system, as shown in Fig. (A.1.1).

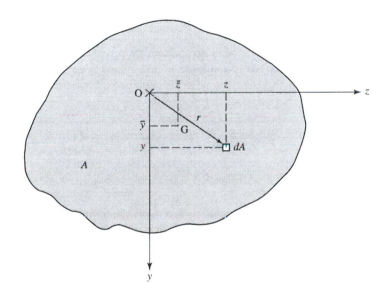

(a) Definitions
(i) Centroid of an area

The coordinates, y, z, of the centroid, G, of the area A are defined by

$$\bar{y} = \frac{\iint_A y \, dA}{\iint_A dA},$$
(A.1.1a)

$$\bar{z} = \frac{\iint_A z \, dA}{\iint_A dA}.$$
(A.1.1b)

Axes passing through the centroid are referred to as *centroidal axes*. If the origin of a y–z coordinate system coincides with the centroid G, then, *by definition*,

$$\iint_A y \, dA = 0, \tag{A.1.2a}$$

$$\iint_A z \, dA = 0. \tag{A.1.2b}$$

(ii) Second moments and mixed second moments of an area[†]

The *second moments* of the area A about the y- and z-axes, I_{yy} and I_{zz} respectively, are defined as

$$I_{yy} = \iint_A z^2 \, dA, \tag{A.1.3a}$$

$$I_{zz} = \iint_A y^2 \, dA. \tag{A.1.3b}$$

Note that second moments of an area are inherently positive. The *mixed second moment* ('product of inertia') of A about the y- and z-axes is defined as

$$I_{yz} = \iint_A yz \, dA. \tag{A.1.3c}$$

Note that, following its definition, the mixed second moment about an axis of symmetry is always zero.

(iii) Polar moment of an area

The *polar moment* of an area A (loosely called the *polar moment of inertia*) about an axis perpendicular to the y–z plane and passing through O, is defined as

$$I_{xx} = \iint_A r^2 \, dA. \tag{A.1.3d}$$

Noting that $r^2 = y^2 + z^2$, it follows that

$$I_{xx} = \iint_A y^2 \, dA + \iint_A z^2 \, dA = I_{zz} + I_{yy}. \tag{A.1.4}$$

(b) Properties of I_{yy}, I_{zz} and I_{yz}
(i) Parallel axis theorem

Consider two sets of orthogonal axes (y, z) and (y', z') that are parallel to each other, as shown in Fig. (A.1.2), such that

$$y' = y - a, \tag{A.1.5a}$$

$$z' = z - b. \tag{A.1.5b}$$

[†] Second moments and mixed second moments of an area are often loosely referred to as *moments of inertia* and *products of inertia*. This terminology is taken from the Dyanmics of Rigid Bodies where the mass of the body is considered.

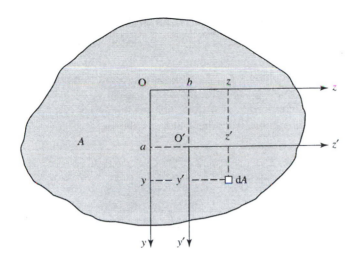

Figure A.1.2

The second moment of A about the z'-axis is then, by definition,

$$I_{z'z'} = \iint_A (y')^2 \, dA. \tag{A.1.6}$$

Substituting y' from Eq. (A.1.5a), we have

$$I_{z'z'} = \iint_A (y - a)^2 \, dA = \iint_A (y^2 - 2ay + a^2) \, dA$$

or

$$I_{z'z'} = \iint_A y^2 \, dA - 2a \iint_A y \, dA + a^2 \iint_A dA. \tag{A.1.7a}$$

From the definition given by Eq. (A.1.3b), we have

$$I_{z'z'} = I_{zz} - 2a \iint_A y \, dA + a^2 A. \tag{A.1.7b}$$

If the y- and z-axes are centroidal axes, then, by definition, $\iint_A y \, dA = 0$, and therefore

$$I_{z'z'} = I_{zz} + a^2 A. \tag{A.1.8a}$$

Similarly, for the moment of second moment about the y'-axis, we obtain

$$I_{y'y'} = I_{yy} + b^2 A. \tag{A.1.8b}$$

Equations (A.1.8) are known as the parallel axis theorem for second moments of an area. We observe, according to this theorem, that of all sets of axes that are parallel to each other, the second moment of an area about the centroidal axes will be a minimum. Similarly, using Eqs. (A.1.5), the parallel axis theorem for the mixed second moment is given by

$$I_{y'z'} = I_{yz} + ab \, A, \tag{A.1.8c}$$

where the y- and z-axes are centroidal axes.

(ii) Transformation laws for second moments and mixed second moments of an area

Given an orthogonal coordinate system (y, z) with origin at O, as shown in Fig. (A.1.3). The second moments I_{yy}, I_{zz} and mixed second moment I_{yz} are assumed known about these axes. A second orthogonal system (y', z') is constructed such that the (y', z') system is rotated with respect to the (y, z) system by an angle ϕ measured counter-clockwise from the y-axis, as shown.

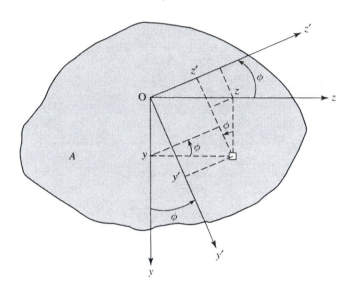

Figure A.1.3

The second moments $I_{y'y'}$, $I_{z'z'}$ and mixed second moment $I_{y'z'}$ can be determined in terms of the corresponding quantities about the original (y, z) system. From geometry,

$$y' = y \cos \phi + z \sin \phi, \qquad (A.1.9.a)$$

$$z' = -y \sin \phi + z \cos \phi. \qquad (A.1.9.b)$$

Therefore,

$$I_{y'y'} = \iint_A (z')^2 \, dA = \iint_A (-y \sin \phi + z \cos \phi)^2 \, dA$$

$$= \cos^2 \phi \iint_A z^2 \, dA - 2 \cos \phi \sin \phi \iint_A yz \, dA + \sin^2 \phi \iint_A y^2 \, dA$$

or

$$I_{y'y'} = \cos^2 \phi \, I_{yy} + \sin^2 \phi \, I_{zz} - 2 \cos \phi \sin \phi \, I_{yz}. \qquad (A.1.10a)$$

Similarly,

$$I_{z'z'} = \iint_A (y')^2 \, dA = \iint_A (y \cos \phi + z \sin \phi)^2 \, dA$$

from which we obtain

$$I_{z'z'} = \cos^2 \phi \, I_{zz} + \sin^2 \phi \, I_{yy} + 2 \cos \phi \sin \phi \, I_{yz}. \qquad (A.1.10b)$$

Following the same procedure, we obtain

$$I_{y'z'} = (\cos^2\phi - \sin^2\phi)\, I_{yz} + \sin\phi\,\cos\phi\,(I_{yy} - I_{zz}). \quad \text{(A.1.10c)}$$

Equations (A.1.10) are transformation laws that relate the second moments and mixed second moments about axes that are rotated by an angle ϕ with respect to each other.[†] Adding Eqs. (A.1.10a) and (A.1.10b), we find

$$I_{y'y'} + I_{z'z'} = I_{yy} + I_{zz}; \quad \text{(A.1.11)}$$

that is, the sum of the second moments of an area about any two orthogonal axes passing through a given point is constant. We refer to this property as an *invariant property*.

(iii) Stationary values of second moments of an area: principal axes[‡]

For given values of I_{yy}, I_{zz} and I_{yz}, we wish to find the stationary values of $I_{y'y'}$ and $I_{z'z'}$. Noting that $I_{y'y'}$ is a function of ϕ, we set

$$\frac{dI_{y'y}}{d\phi} = 0. \quad \text{(A.1.12)}$$

Performing this operation, we find

$$2\sin\phi\,\cos\phi\,(I_{zz} - I_{yy}) - 2(\cos^2\phi - \sin^2\phi)\,I_{yz} = 0$$

from which

$$\tan 2\phi = \frac{2I_{yz}}{I_{zz} - I_{yy}}. \quad \text{(A.1.13)}$$

Equation (A.1.13) has two relevant roots ϕ_1 and $\phi_2 = \phi_1 + \pi/2$, which define the orientation of two orthogonal axes about which the second moments of an area will be maximum and minimum. These axes are referred to as the *principal axes* of the area A passing through O. The two stationary values of the second moments of the area about these axes are called the *principal values of the second moments* and are denoted by I_1 and I_2, respectively, where by definition, $I_1 > I_2$. Using the proper trigonometric substitutions, I_1 and I_2 are given, respectively, by the expressions

$$I_{1,2} = \frac{I_{yy} + I_{zz}}{2} \pm \sqrt{\left(\frac{I_{yy} - I_{zz}}{2}\right)^2 + I_{yz}^2}. \quad \text{(A.1.14)}$$

From Eqs. (A.1.14), we have, at any given point,

$$I_1 + I_2 = I_{yy} + I_{zz} = \text{constant}, \quad \text{(A.1.15)}$$

[†] We observe the similarity of the transformation law for second moments and mixed second moments and that for the stress components, namely Eqs. (2.6.6a) and (2.6.8a). (Note that here the transformation law is in the y–z plane; the transformation law for plane stress is given in the x–y plane. Therefore, in order to compare properly, the present $y \to x$ and $z \to y$.) However, they are not identical: e.g., Eq. (2.6.6a) contains a plus ($+$) sign while a minus ($-$) sign appears in Eq. (A.1.10a). Thus the quantities

$$\begin{pmatrix} I_{yy} & I_{yz} \\ I_{yz} & I_{zz} \end{pmatrix}$$

are *not* components of a second-rank symmetric tensor since they do not conform with the definition of the transformation law for a two-dimensional second-rank tensor as given by Eqs. (2.6.6a) and (2.6.8a). However, if, instead of Eq. (A.1.3c), the mixed second moment is defined as $I_{xy} = -\iint yz\,dA$, then the resulting array is a tensor. Nevertheless, with the present definition of I_{yz}, the invariant property of the second moments of and area, Eq. (A.1.11), is the same as for the stress tensor.

[‡] The development is only outlined here since it follows closely the analogous development given in Chapter 2 for a two-dimensional stress state.

which conforms with the invariant property of Eq. (A.1.11). Note that the constant depends on the given point O. We observe from Eq. (A.1.13) that if $I_{yz} = 0$, then $\phi_1 = 0$ (and $\phi_2 = \pi/2$). Since $I_{yz} = 0$ about an axis of symmetry, it follows that if a given area A has an axis of symmetry, this axis is always a principal axis (and therefore an axis perpendicular to the axis of symmetry is also a principal axis). However, it should be remembered that the converse is not necessarily true; i.e., a principal axis need not be an axis of symmetry.

A.2   Properties of selected areas

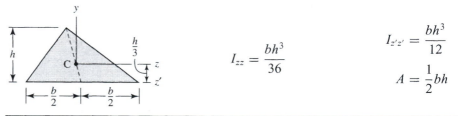

$$I_{zz} = \frac{bh^3}{12}$$

$$I_{z'z'} = \frac{bh^3}{3}$$

$$A = bh$$

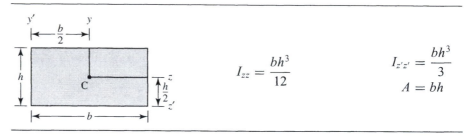

$$I_{zz} = \frac{bh^3}{36}$$

$$I_{z'z'} = \frac{bh^3}{12}$$

$$A = \frac{1}{2}bh$$

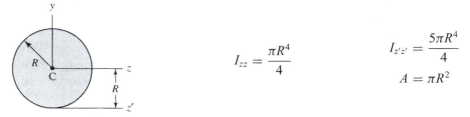

$$I_{zz} = \frac{\pi R^4}{4}$$

$$I_{z'z'} = \frac{5\pi R^4}{4}$$

$$A = \pi R^2$$

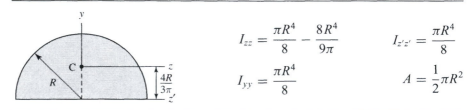

$$I_{zz} = \frac{\pi R^4}{8} - \frac{8R^4}{9\pi}$$

$$I_{yy} = \frac{\pi R^4}{8}$$

$$I_{z'z'} = \frac{\pi R^4}{8}$$

$$A = \frac{1}{2}\pi R^2$$

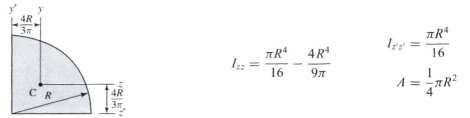

$$I_{zz} = \frac{\pi R^4}{16} - \frac{4R^4}{9\pi}$$

$$I_{z'z'} = \frac{\pi R^4}{16}$$

$$A = \frac{1}{4}\pi R^2$$

$$I_{zz} = \frac{R^4}{4}\left(\theta - \frac{1}{2}\sin 2\theta\right) \qquad z_c = \frac{2}{3}\frac{R\sin\theta}{\theta}$$

$$I_{yy} = \frac{R^4}{4}\left(\theta + \frac{1}{2}\sin 2\theta\right) \qquad A = \theta R^2$$

$$I_{zz} = \frac{\pi ab^3}{4}$$

$$A = \pi ab$$

$$I_{yy} = \frac{\pi a^3 b}{4}$$

Appendix B

Some mathematical relations

B.1 Curvature of a line $y = y(x)$

Consider two points P and Q located on a curve $y = y(x)$. Let these two points be defined by arc coordinates s and $s + \Delta s$, where s is measured from a fixed point A [Fig. (B.1.1)].

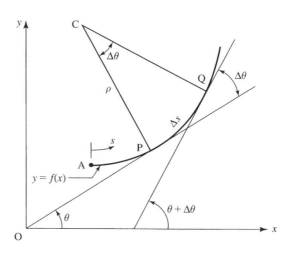

Figure B.1.1

Let θ and $\theta + \Delta\theta$ be the angles of inclination with respect to the x-axis of the tangents to $y(x)$ at the point P and Q, respectively.

Constructing the normals at P and Q, we note that they intersect at some point denoted by C. Furthermore, the angle $\angle PCQ = \Delta\theta$. Therefore

$$\lim_{\substack{\Delta s \to 0 \\ \Delta\theta \to 0}} \rho\,\Delta\theta = \Delta s \tag{B.1.1}$$

and thus

$$\frac{1}{\rho} = \lim_{\substack{\Delta s \to 0 \\ \Delta\theta \to 0}} \frac{\Delta\theta}{\Delta s} = \frac{d\theta}{ds}, \tag{B.1.2}$$

where ρ is the radius of curvature. Now, $\tan\theta = \frac{dy}{dx}$ and therefore $\theta = \tan^{-1}(\frac{dy}{dx}) \equiv \tan^{-1}[(y'(x)]$, where $y'(x) \equiv \frac{dy}{dx}$. Hence

$$\frac{d\theta}{dx} = \frac{d\theta}{dy'}\frac{dy'}{dx} = \frac{d}{dy'}[\tan^{-1}(y')] \cdot \frac{dy'}{dx}. \tag{B.1.3}$$

694

Furthermore, we recall that for any variable, say z,

$$\frac{\mathrm{d}}{\mathrm{d}z}[\tan^{-1}(z)] = \frac{1}{(1+z^2)}. \tag{B.1.4}$$

Therefore, substituting in Eq. (B.1.3), we have

$$\frac{\mathrm{d}\theta}{\mathrm{d}x} = \frac{y''}{[1+(y')^2]}$$

or

$$\mathrm{d}\theta = \frac{\mathrm{d}\theta}{\mathrm{d}x}\,\mathrm{d}x = \frac{y''}{[1+(y')^2]}\mathrm{d}x. \tag{B.1.5}$$

Now

$$\mathrm{d}s = \sqrt{\mathrm{d}x^2 + \mathrm{d}y^2} = \sqrt{[(\mathrm{d}x)^2 + (\mathrm{d}y/\mathrm{d}x)^2\mathrm{d}x^2]} = \mathrm{d}x\sqrt{[1+(y')^2]}. \tag{B.1.6}$$

Substituting Eqs. (B.1.5) and (B.1.6) in Eq. (B.1.2), we obtained

$$\frac{1}{\rho} = \frac{y''}{[1+(y')^2]^{3/2}}. \tag{B.1.7}$$

If $|y'| \ll 1$, then $\frac{1}{\rho} = y''(x)$. Since the curvature κ is defined as $\kappa \equiv 1/\rho$,

$$\kappa = \frac{y''}{[1+(y')^2]^{3/2}}. \tag{B.1.8}$$

B.2   Green's theorem

Let A be a simply connected domain lying in the y–z plane and bounded by a contour C_0 [Fig. (B.2.1a)] and let $P(y,z)$ and $Q(y,z)$ be functions that are continuous and have continuous partial derivatives in A. Then

$$\iint\limits_{A} \left(\frac{\partial Q}{\partial y} - \frac{\partial P}{\partial z}\right) \mathrm{d}y\,\mathrm{d}z = \oint\limits_{C_0} (P\,\mathrm{d}y + Q\,\mathrm{d}z), \tag{B.2.1}$$

where the integral along the closed contour C_0 is taken in the counter-clock-wise direction. (Note that the domain A then lies always to the left as the integration along C_0 proceeds counter-clockwise.)

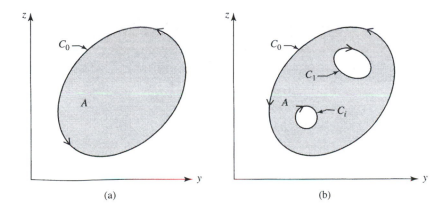

(a)   (b)   Figure B.2.1

For a multi-connected domain (i.e., one that contains n number of 'holes'), each defined by interior contours C_i, $i = 1, 2, \ldots, n$ [Fig. (B.2.1b)], Green's theorem becomes

$$\iint\limits_A \left(\frac{\partial Q}{\partial y} - \frac{\partial P}{\partial z} \right) \mathrm{d}y\,\mathrm{d}z = \oint\limits_{C_0} (P\,\mathrm{d}y + Q\,\mathrm{d}z) + \sum_{i=1}^{n} \oint\limits_{C_i} (P\,\mathrm{d}y + Q\,\mathrm{d}z), \quad \text{(B.2.2)}$$

where the integrals along the contours C_i are taken in the clock-wise direction. For this case, the domain A is the area bounded by the contours C_0 and C_i. (Note also that the domain A always lies to the left as integration proceeds along C_0 or any C_i.)

Alternatively, if we choose to integrate along the interior contours C_i ($i \geq 1$) in the counter-clock-wise direction, we have

$$\iint\limits_A \left(\frac{\partial Q}{\partial y} - \frac{\partial P}{\partial z} \right) \mathrm{d}y\,\mathrm{d}z = \oint\limits_{C_0} (P\,\mathrm{d}y + Q\,\mathrm{d}z) - \sum_{i=1}^{n} \oint\limits_{C_i} (P\,\mathrm{d}y + Q\,\mathrm{d}z). \quad \text{(B.2.3)}$$

B.3 The divergence theorem (Gauss' theorem)

Given a domain V in an (x, y, z) coordinate system, which is bounded by a surface S. We denote the unit (outward) normal to S at all points by $\boldsymbol{n}$ [(Fig. (B.3.1)]

$$\boldsymbol{n} = \ell_x \boldsymbol{i} + \ell_y \boldsymbol{j} + \ell_z \boldsymbol{k}, \quad \text{(B.3.1)}$$

where ℓ_x, ℓ_y and ℓ_z are its direction cosines. Furthermore, let

$$\boldsymbol{G}(x, y, z) = G_x \boldsymbol{i} + G_y \boldsymbol{j} + G_z \boldsymbol{k} \quad \text{(B.3.2)}$$

be a vector function that is continuous and has continuous first partial derivatives in V. Then,

$$\iiint\limits_V \nabla \cdot \boldsymbol{G} \, \mathrm{d}\Omega = \iint\limits_S \boldsymbol{n} \cdot \boldsymbol{G} \, \mathrm{d}s, \quad \text{(B.3.3)}$$

where $\mathrm{d}\Omega$ represents a volume element and $\mathrm{d}s$ represents a surface element on the surface S enclosing the volume V.

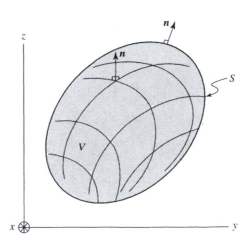

Figure B.3.1

In scalar components, the divergence theorem is written explicitly as

$$\iiint_V \left(\frac{\partial G_x}{\partial x} + \frac{\partial G_y}{\partial y} + \frac{\partial G_z}{\partial z} \right) d\Omega = \iint_S (\ell_x G_x + \ell_y G_y + \ell_z G_z)\, ds. \quad \text{(B.3.4a)}$$

In the two-dimensional case, i.e. where $\mathbf{G}(x, y) = G_x \mathbf{i} + G_y \mathbf{j}$ and $\mathbf{n} = \ell_x \mathbf{i} + \ell_y \mathbf{j}$, Eq. (B.3.4a) reduces to

$$\iiint_V \left(\frac{\partial G_x}{\partial x} + \frac{\partial G_y}{\partial y} \right) d\Omega = \iint_S (\ell_x G_x + \ell_y G_y)\, ds. \quad \text{(B.3.4b)}$$

Appendix C

The membrane equation

Consider a thin membrane of arbitrary area A, lying initially in the y–z plane. The membrane is subject to a uniform initial tensile force of intensity F per unit length along its perimeter and a constant uniform pressure p [Fig. (C.1)]. Assume that due to the pressure, the membrane undergoes small (infinitesimal) deflections w in the x-direction.

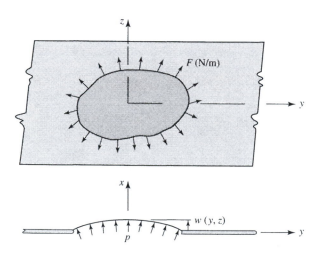

Figure C.1

We shall assume that the initial tensile force F is sufficiently large such that the increase in the tensile force due to the deflections is infinitesimal. Hence the tensile force can be assumed to *remain constant*. Furthermore, since the membrane is thin, no components of the internal force can exist at any point in a direction perpendicular to the deflected membrane and thus, at all points, F is tangential to the membrane surface.[†]

Consider now a small element $dA = dy\,dz$ of the deflected membrane as a free body [Fig. (C.2)]. Then, taking $\sum F_x = 0$,

$$(F\,dz)\,[\sin(\theta + d\theta)] - (F\,dz)\sin\theta + (F\,dy)\,[\sin(\psi + d\psi)]$$

$$- (F\,dy)\sin\psi + p\,dydz = 0, \tag{C.1}$$

---

† From Chapter 8, the relation between the curvature and moment, for the one-dimensional case, is given by $EI\kappa = M$. For a very thin membrane, the second moment of the cross-sectional area $I \to 0$. Since κ is finite, it follows that the moment $M \to 0$. Therefore, the shear force V (being the derivative of the moment M) will also tend to zero, i.e. $V \to 0$. Since V represents the force tangential to the deflected cross-section, it follows that the resultant F acting on the cross-section acts always tangential to the deflected membrane, as shown in Fig. (C.2). Consequently, the intensity of force at any point in the membrane has magnitude F (N/m) and acts, at all points, in a plane tangential to the deflected membrane.

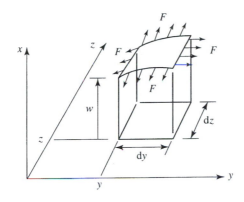

Figure C.2

where θ, $\theta + d\theta$, ψ and $\psi + d\psi$ are shown in Fig. (C.3).

Now, we restrict the analysis to membranes for which the displacements are sufficiently small such that the slopes θ and ψ are infinitesimal, i.e.

$$|\theta| \ll 1, \qquad |\psi| \ll 1.$$

Therefore

$$\sin \theta \simeq \theta \simeq \tan \theta = \frac{\partial w}{\partial y}; \qquad \sin \psi \simeq \psi \simeq \tan \psi = \frac{\partial w}{\partial z}. \qquad \text{(C.2a)}$$

Then

$$\sin(\theta + d\theta) \simeq (\theta + d\theta) \simeq \frac{\partial w}{\partial y} + d\left(\frac{\partial w}{\partial y}\right), \qquad \text{(C.2b)}$$

$$\sin(\psi + d\psi) = (\psi + d\psi) \simeq \frac{\partial w}{\partial z} + d\left(\frac{\partial w}{\partial z}\right). \qquad \text{(C.2c)}$$

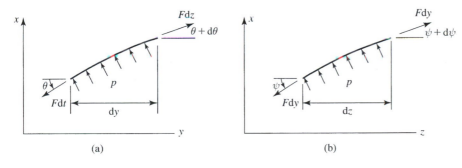

(a)   (b)   **Figure C.3**

Substituting in (C.1), we have

$$(F\,dz)\left[\frac{\partial w}{\partial y} + d\left(\frac{\partial w}{\partial y}\right)\right] - (F\,dz)\left(\frac{\partial w}{\partial y}\right) + (F\,dy)\left[\frac{\partial w}{\partial z} + d\left(\frac{\partial w}{\partial z}\right)\right]$$

$$- (F\,dy)\left(\frac{\partial \omega}{\partial y}\right) + p\,dy\,dz = 0,$$

or

$$(F\,dz)\,d\left(\frac{\partial w}{\partial y}\right) + (F\,dy)\,d\left(\frac{\partial w}{\partial z}\right) + p\,dy\,dz = 0. \qquad \text{(C.3)}$$

Now

$$d\left(\frac{\partial w}{\partial y}\right) = \frac{\partial}{\partial y}\left(\frac{\partial w}{\partial y}\right) dy = \frac{\partial^2 w}{\partial y^2} dy, \quad \text{along } z = \text{const.}$$

$$d\left(\frac{\partial w}{\partial z}\right) = \frac{\partial}{\partial z}\left(\frac{\partial w}{\partial z}\right) dz = \frac{\partial^2 w}{\partial z^2} dz, \quad \text{along } y = \text{const.} \tag{C.4}$$

Hence

$$(F\, dz)\frac{\partial^2 w}{\partial y^2}\, dy + (F\, dy)\frac{\partial^2 w}{\partial z^2}\, dz + p\, dy\, dz = 0,$$

or

$$F\left(\frac{\partial^2 w}{\partial y^2} + \frac{\partial^2 w}{\partial z^2}\right) dy\, dz + p\, dy\, dz = 0.$$

Therefore

$$\frac{\partial^2 w}{\partial y^2} + \frac{\partial^2 w}{\partial z^2} = -\frac{p}{F}, \tag{C.5}$$

which is the equation of equilibrium of a thin membrane, subjected to an initial tensile force F and lateral pressure p.

Recalling that the Laplacian operator ∇^2 in Cartesian coordinates is given by

$$\nabla^2 \equiv \frac{\partial^2}{\partial y^2} + \frac{\partial^2}{\partial z^2}, \tag{C.6}$$

we may rewrite Eq. (C.5) in the more concise notation

$$\nabla^2 w(y, z) = -\frac{p}{F}. \tag{C.7}$$

Appendix D

Material properties

Materials	Density (Mg/m³)	Elastic strength[a]			Ultimate strength			Endurance limit[c] (MPa)	Modulus of elasticity (GPa)	Modulus of rigidity (GPa)	Percentage elongation in 50 mm	Coefficient of thermal expansion (10⁻⁶ °C⁻¹)
		Tension (MPa)	Compression (MPa)	Shear (MPa)	Tension (MPa)	Compression (MPa)	Shear (MPa)					
Ferrous metals												
Wrought iron	7.70	210	b		330	b	170	160	190		30[d]	12.1
Structural steel	7.87	250	b		450	b		190	200	76	28[d]	11.9
Steel, 0.2% C hardened	7.87	430	b		620	b			210	80	22	11.9
Steel, 0.4% C hot-rolled	7.87	360	b		580	b		260	210	80	29	
Steel, 0.8% C hot-rolled	7.87	520	b		840	b			210	80	8	
Cast iron—gray	7.20				170	690		80	100		0.5	12.1
Cast iron—malleable	7.37	220	b		340	b			170		20	11.9
Cast iron—nodular	7.37	480			690				170		4	11.9
Stainless steel (18-8) annealed	7.92	250	b		590	b		270	190	86	55	17.3
Stainless steel (18-8) cold-rolled	7.92	1140	b		1310	b		620	190	86	8	17.3
Steel, SAE 4340, heat-treated	7.84	910	1000		1030	b	650	520	200	76	19	
Nonferrous metal alloys												
Aluminum, cast 195-T6	2.77	160	170		250		210	50	71	26	5	
Aluminum, wrought, 2014-T4	2.80	280	280	160	430	b	260	120	73	28	20	22.5
Aluminum, wrought, 2024-T4	2.77	330	330	190	470	b	280	120	73	28	19	22.5
Aluminum, wrought, 6061-T6	2.71	270	270	180	310	b	210	93	70	26	17	22.5
Magnesium, extrusion, AZ80X	1.83	240	180		340	b	140	130	45	16	12	25.9
Magnesium, sand cast, AZ63-HT	1.83	100	96		270	b	130	100	45	16	12	25.9
Monel, wrought, hot-rolled	8.84	340	b		620	b		270	180	65	35	14.0
Red brass, cold-rolled	8.75	410			520				100	39	4	17.6
Red brass, annealed	8.75	100	b		270	b			100	39	50	17.6
Bronze, cold-rolled	8.86	520			690				100	45	3	16.9
Bronze, annealed	8.86	140	b		340	b			100	45	50	16.9
Titanium alloy, annealed	4.63	930	b		1070	b			96	36	13	
Invar, annealed	8.09	290	b		480	b			140	56	41	1.1

(Continued)

(*Continued*)

Materials	Density (Mg/m³)	Elastic strength[a]			Ultimate strength			Endurance limit[c] (MPa)	Modulus of elasticity (GPa)	Modulus of rigidity (GPa)	Percentage elongation in 50 mm	Coefficient of thermal expansion (10⁻⁶ °C⁻¹)
		Tension (MPa)	Compression (MPa)	Shear (MPa)	Tension (MPa)	Compression (MPa)	Shear (MPa)					
Nonmetallic materials												
Douglas fir, green[e]	0.61	33	23			27	6.2		11			
Douglas, fir, air dry[e]	0.55	56	44			51	7.6		13			
Red oak, green[e]	1.02	30	18			24	8.3		10			3.4
Red oak, air dry[e]	0.69	58	32			48	12.4		12			
Concrete, medium strength	2.41		8			21			21			10.8
Concrete, fairly high strength	2.41		14			34			31			10.8

[a] Elastic strength may be represented by proportional limit, yield point or yield strength at a specified offset (usually 0.2% for ductile metals).

[b] For ductile metals (those with an appreciable ultimate elongation), it is customary to assume the properties in compression have the same values as those in tension.

[c] Rotating beam.

[d] Elongation in 200 mm.

[e] All timber properties are parallel to the grain.

Appendix E

Table of structural properties*

Wide-flange beams

Designation*	Area (mm²)	Depth (mm)	Flange Width (mm)	Flange Thickness (mm)	Web Thickness (mm)	Axis Z–Z I (10⁶ mm⁴)	Axis Z–Z S (10³ mm³)	Axis Z–Z r (mm)	Axis Y–Y I (10⁶ mm⁴)	Axis Y–Y S (10³ mm³)	Axis Y–Y r (mm)
W914 × 342	43610	912	418	32.0	19.3	6245	13715	378	391	1870	94.7
× 238	30325	915	305	25.9	16.5	4060	8880	366	123	805	63.5
W838 × 299	38130	855	400	29.2	18.2	4785	11210	356	312	1560	90.4
× 226	28850	851	294	26.8	16.1	3395	7980	343	114	775	62.7
× 193	24710	840	292	21.7	14.7	2795	6655	335	90.7	620	60.7
W762 × 196	25100	770	268	25.4	15.6	2400	6225	310	81.6	610	57.2
× 161	20450	758	266	19.3	13.8	1860	4900	302	60.8	457	54.6
W686 × 217	27675	695	355	24.8	15.4	2345	6735	290	184	1040	81.5
× 140	17870	684	254	18.9	12.4	1360	3980	277	51.6	406	53.8
W610 × 155	19740	611	324	19.1	12.7	1290	4230	257	108	667	73.9
× 125	15935	612	229	19.6	11.9	985	3210	249	39.3	342	49.5
× 92	11750	603	179	15.0	10.9	645	2145	234	14.4	161	35.1
W533 × 150	19225	543	312	20.3	12.7	1005	3720	229	103	660	73.4
× 124	15675	544	212	21.2	13.1	762	2800	220	33.9	320	46.5
× 92	11805	533	209	15.6	10.2	554	2080	217	23.9	228	45.0
W457 × 144	18365	472	283	22.1	13.6	728	3080	199	83.7	592	67.3
× 113	14385	463	280	17.3	10.8	554	2395	196	63.3	452	66.3
× 89	11355	463	192	17.7	10.5	410	1770	190	20.9	218	42.9
W406 × 149	18970	431	265	25.0	14.9	620	2870	180	77.4	585	64.0
× 100	12710	415	260	16.9	10.0	397	1915	177	49.5	380	62.5
× 60	7615	407	178	12.8	7.7	216	1060	168	12.0	135	39.9
× 39	4950	399	140	8.8	6.4	125	629	159	3.99	57.2	28.4
W356 × 179	22775	368	373	23.9	15.0	574	3115	158	206	1105	95.0
× 122	15550	363	257	21.7	13.0	367	2015	154	61.6	480	63.0
× 64	8130	347	203	13.5	7.7	178	1025	148	18.8	185	48.0
× 45	5710	352	171	9.8	6.9	121	688	146	8.16	95.4	37.8
W305 × 143	18195	323	309	22.9	14.0	347	2145	138	112	728	78.5
× 97	12325	308	305	15.4	9.9	222	1440	134	72.4	477	76.7
× 74	9485	310	205	16.3	9.4	164	1060	132	23.4	228	49.8
× 45	5670	313	166	11.2	6.6	99.1	633	132	8.45	102	38.6
W254 × 89	11355	260	256	17.3	10.7	142	1095	112	48.3	377	65.3
× 67	8580	257	204	15.7	8.9	103	805	110	22.2	218	51.1
× 45	5705	266	148	13.0	7.6	70.8	531	111	6.95	94.2	34.8
× 33	4185	258	146	9.1	6.1	49.1	380	108	4.75	65.1	33.8
W203 × 60	7550	210	205	14.2	9.1	60.8	582	89.7	20.4	200	51.8
× 46	5890	203	203	11.0	7.2	45.8	451	88.1	15.4	152	51.3
× 36	4570	201	165	10.2	6.2	34.5	342	86.7	7.61	92.3	40.9
× 22	2865	206	102	8.0	6.2	20.0	193	83.6	1.42	27.9	22.3
W152 × 37	4735	162	154	11.6	8.1	22.2	274	68.6	7.12	91.9	38.6
× 24	3060	160	102	10.3	6.6	13.4	167	66.0	1.84	36.1	24.6
W127 × 24	3020	127	127	9.1	6.1	8.87	139	54.1	3.13	49.2	32.3
W102 × 19	2470	106	103	8.8	7.1	4.70	89.5	43.7	1.61	31.1	25.4

* W means wide-flange beam, followed by the nominal depth in mm, then the mass in kg/m of length.

American standard beams

Designation*	Area (mm²)	Depth (mm)	Flange Width (mm)	Flange Thickness (mm)	Web thickness (mm)	Axis Z–Z I (10^6 mm⁴)	Axis Z–Z S (10^3 mm³)	Axis Z–Z r (mm)	Axis Y–Y I (10^6 mm⁴)	Axis Y–Y S (10^3 mm³)	Axis Y–Y r (mm)
S610 × 180	22970	622.3	204.5	27.7	20.3	1315	4225	240	34.7	339	38.9
× 158	20130	622.3	199.9	27.7	15.7	1225	3935	247	32.1	321	39.9
× 149	18900	609.6	184.0	22.1	18.9	995	3260	229	19.9	216	32.3
× 134	17100	609.6	181.0	22.1	15.9	937	3065	234	18.7	206	33.0
× 119	15160	609.6	177.8	22.1	12.7	874	2870	241	17.6	198	34.0
S508 × 143	18190	515.6	182.9	23.4	20.3	695	2705	196	20.9	228	33.8
× 128	16320	515.6	179.3	23.4	16.8	658	2540	200	19.5	218	34.5
× 112	14190	508.0	162.2	20.2	16.1	533	2100	194	12.4	153	29.5
× 98	12520	508.0	158.9	20.2	12.8	495	1950	199	11.5	145	30.2
S457 × 104	13290	457.2	158.8	17.6	18.1	358	1690	170	10.0	127	27.4
× 81	10390	457.2	152.4	17.6	11.7	335	1465	180	8.66	114	29.0
S381 × 74	9485	381.0	143.3	15.8	14.0	202	1060	146	6.53	91.3	26.2
× 64	8130	381.0	139.7	15.8	10.4	186	977	151	5.99	85.7	27.2
S305 × 74	9485	304.8	139.1	16.7	17.4	127	832	116	6.53	94.1	26.2
× 61	7740	304.8	133.4	16.7	11.7	113	744	121	5.66	84.6	26.9
× 52	6645	304.8	129.0	13.8	10.9	95.3	626	120	4.11	63.7	24.1
× 47	6030	304.8	127.0	13.8	8.9	90.7	596	123	3.90	61.3	25.4
S254 × 52	6645	254.0	125.6	12.5	15.1	61.2	482	96.0	3.48	55.4	22.9
× 38	4815	254.0	118.4	12.5	7.9	51.6	408	103	2.83	47.7	24.2
S203 × 34	4370	203.2	105.9	10.8	11.2	27.0	265	78.7	1.79	33.9	20.3
× 27	3490	203.2	101.6	10.8	6.9	24.0	236	82.8	1.55	30.5	21.1
S178 × 30	3795	177.8	98.0	10.0	11.4	17.6	198	68.3	1.32	26.9	18.6
× 23	2905	177.8	93.0	10.0	6.4	15.3	172	72.6	1.10	23.6	19.5
S152 × 26	3270	152.4	90.6	9.1	11.8	10.9	144	57.9	0.961	21.3	17.1
× 19	2370	152.4	84.6	9.1	5.9	9.20	121	62.2	0.758	17.9	17.9
S127 × 22	2800	127.0	83.4	8.3	12.5	6.33	99.8	47.5	0.695	16.6	15.7
× 15	1895	127.0	76.3	8.3	5.4	5.12	80.6	52.1	0.508	13.3	16.3
S102 × 14	1800	101.6	71.0	7.4	8.3	2.83	55.6	39.6	0.376	10.6	14.5
× 11	1460	101.6	67.6	7.4	4.9	2.53	49.8	41.7	0.318	9.41	14.8
S76 × 11	1425	76.2	63.7	6.6	8.9	1.22	32.0	29.2	0.244	7.67	13.1
× 8.5	1075	76.2	59.2	6.6	4.3	1.05	27.5	31.2	0.189	6.39	13.3

* S means standard beam, followed by the nominal depth in mm, then the mass in kg/m of length.

Standard channels

Designation*	Area (mm²)	Depth (mm)	Flange Width (mm)	Flange Thickness (mm)	Web thickness (mm)	Axis Z–Z I (10⁶ mm⁴)	Axis Z–Z S (10³ mm³)	Axis Z–Z r (mm)	Axis Y–Y I (10⁶ mm⁴)	Axis Y–Y S (10³ mm³)	Axis Y–Y r (mm)	z_c (mm)
C457 × 86	11030	457.2	106.7	15.9	17.8	281	1230	160	7.41	87.2	25.9	21.9
× 77	9870	457.2	104.1	15.9	15.2	261	1140	163	6.83	83.1	26.4	21.8
× 68	8710	457.2	101.6	15.9	12.7	241	1055	167	6.29	79.0	26.9	22.0
× 64	8130	457.2	100.3	15.9	11.4	231	1010	169	5.99	76.9	27.2	22.3
C381 × 74	9485	381.0	94.4	16.5	18.2	168	882	133	4.58	61.9	22.0	20.3
× 60	7615	381.0	89.4	16.5	13.2	145	762	138	3.84	55.2	22.5	19.7
× 50	6425	381.0	86.4	16.5	10.2	131	688	143	3.38	51.0	23.0	20.0
C305 × 45	5690	304.8	80.5	12.7	13.0	67.4	442	109	2.14	33.8	19.4	17.1
× 37	4740	304.8	77.4	12.7	9.8	59.9	395	113	1.86	30.8	19.8	17.1
× 31	3930	304.8	74.7	12.7	7.2	53.7	352	117	1.61	28.3	20.3	17.7
C254 × 45	5690	254.0	77.0	11.1	17.1	42.9	339	86.9	1.64	27.0	17.0	16.5
× 37	4740	254.0	73.3	11.1	13.4	38.0	298	89.4	1.40	24.3	17.2	15.7
× 30	3795	254.0	69.6	11.1	9.6	32.8	259	93.0	1.17	21.6	17.6	15.4
× 23	2895	254.0	66.0	11.1	6.1	28.1	221	98.3	0.949	19.0	18.1	16.1
C229 × 30	3795	228.6	67.3	10.5	11.4	25.3	221	81.8	1.01	19.2	16.3	14.8
× 22	2845	228.6	63.1	10.5	7.2	21.2	185	86.4	0.803	16.6	16.8	14.9
× 20	2540	228.6	61.8	10.5	5.9	19.9	174	88.4	0.733	15.7	17.0	15.3
C203 × 28	3555	203.2	64.2	9.9	12.4	18.3	180	71.6	0.824	16.6	15.2	14.4
× 20	2605	203.2	59.5	9.9	7.7	15.0	148	75.9	0.637	14.0	15.6	14.0
× 17	2180	203.2	57.4	9.9	5.6	13.6	133	79.0	0.549	12.8	15.9	14.5
C178 × 22	2795	177.8	58.4	9.3	10.6	11.3	127	63.8	0.574	12.8	14.3	13.5
× 18	2320	177.8	55.7	9.3	8.0	10.1	114	66.0	0.487	11.5	14.5	13.3
× 15	1850	177.8	53.1	9.3	5.3	8.87	99.6	69.1	0.403	10.2	14.8	13.7
C152 × 19	2470	152.4	54.8	8.7	11.1	7.24	95.0	54.1	0.437	10.5	13.3	13.1
× 16	1995	152.4	51.7	8.7	8.0	6.33	82.9	56.4	0.360	9.24	13.4	12.7
× 12	1550	152.4	48.8	8.7	5.1	5.45	71.8	59.4	0.288	8.06	13.6	13.0
C127 × 13	1705	127.0	47.9	8.0	8.3	3.70	58.3	46.5	0.263	7.37	12.4	12.1
× 10	1270	127.0	44.5	8.1	4.8	3.12	49.2	49.5	0.199	6.19	12.5	12.3
C102 × 11	1375	101.6	43.7	7.5	8.2	1.91	37.5	37.3	0.180	5.62	11.4	11.7
× 8	1025	101.6	40.2	7.5	4.7	1.60	31.6	39.6	0.133	4.64	11.4	11.6
C76 × 9	1135	76.2	40.5	6.9	9.0	0.862	22.6	27.4	0.127	4.39	10.6	11.6
× 7	948	76.2	38.0	6.9	6.6	0.770	20.3	28.4	0.103	3.82	10.4	11.1
× 6	781	76.2	35.8	6.9	4.6	0.691	18.0	29.7	0.082	3.31	10.3	11.1

* C means channel, followed by the nominal depth in mm, then the mass in kg/m of length.

Equal leg angles

Size and thickness (mm)	Mass (kg/m)	Area (mm²)	Axis Z–Z or Y–Y				Axis P–P
			I (10^6 mm⁴)	S (10^3 mm³)	r (mm)	z_C or y_C (mm)	r (mm)
L203 × 203 × 25.4	75.9	9675	37.0	259	62.0	60.2	39.6
× 22.2	67.0	8515	33.1	229	62.2	58.9	39.9
× 19.1	57.9	7355	29.0	200	62.7	57.9	40.1
× 15.9	48.7	6200	24.7	169	63.2	56.6	40.1
× 12.7	39.3	5000	20.2	137	63.5	55.6	40.4
L152 × 152 × 25.4	55.7	7095	14.8	140	45.7	47.2	29.7
× 22.2	49.3	6275	13.3	125	46.0	46.2	29.7
× 19.1	42.7	5445	11.7	109	46.5	45.2	29.7
× 15.9	36.0	4585	10.1	92.8	46.7	43.9	30.0
× 12.7	29.2	3710	8.28	75.5	47.2	42.7	30.0
× 9.5	22.2	2815	6.61	57.8	47.8	41.7	30.2
L127 × 127 × 22.2	40.5	5150	7.41	84.7	37.8	39.9	24.7
× 19.1	35.1	4475	6.53	74.2	38.4	38.6	24.8
× 15.9	29.8	3780	5.66	63.3	38.6	37.6	24.8
× 12.7	24.1	3065	4.70	51.8	39.1	36.3	25.0
× 9.5	18.3	2330	3.64	39.7	39.6	35.3	25.1
L102 × 102 × 19.1	27.5	3510	3.19	46.0	30.2	32.3	19.8
× 15.9	23.4	2975	2.77	39.3	30.5	31.2	19.8
× 12.7	19.0	2420	2.31	32.3	31.0	30.0	19.9
× 9.5	14.6	1845	1.81	24.9	31.2	29.0	20.0
× 6.4	9.8	1250	1.27	17.2	31.8	27.7	20.2
L89 × 89 × 12.7	16.5	2095	1.52	24.4	26.9	26.9	17.3
× 9.5	12.6	1600	1.19	18.8	27.2	25.7	17.4
× 6.4	8.6	1090	0.837	13.0	27.7	24.6	17.6
L76 × 76 × 12.7	14.0	1775	0.924	17.5	22.8	23.7	14.8
× 9.5	10.7	1360	0.732	13.7	23.2	22.6	14.9
× 6.4	7.3	929	0.516	9.46	23.6	21.4	15.0
L64 × 64 × 12.7	11.5	1450	0.512	11.9	18.8	20.5	12.4
× 9.5	8.8	1115	0.410	9.28	19.1	19.4	12.4
× 6.4	6.1	768	0.293	6.46	19.5	18.2	12.5
L51 × 51 × 9.5	7.0	877	0.199	5.75	15.1	16.2	9.88
× 6.4	4.75	605	0.145	4.05	15.5	15.0	9.93
× 3.2	2.46	312	0.079	2.15	15.9	13.9	10.1

* L means equal leg angles, followed by the nominal depth in mm, then the mass in kg/m of length.

Unequal leg angles

Size and Thickness (mm)	Mass (kg/m)	Area (mm²)	Axis Z–Z				Axis Y–Y				Axis P–P	
			$I\,(10^6$ mm⁴)	$S\,(10^3$ mm³)	r (mm)	y_c (mm)	$I\,(10^6$ mm⁴)	$S\,(10^3$ mm³)	r (mm)	z_c (mm)	r (mm)	$\tan\alpha$
L229 × 102 × 15.9	39.1	4985	27.0	188	73.7	85.3	3.46	43.4	26.4	21.8	21.5	0.210
× 12.7	31.7	4030	22.1	153	74.2	84.1	2.88	35.6	26.7	20.6	21.7	0.220
L203 × 152 × 25.4	65.8	8385	33.6	247	63.2	67.3	16.1	146	43.9	41.9	32.5	0.543
× 19.1	50.3	6415	26.4	192	64.3	65.0	12.8	113	44.7	39.6	32.8	0.551
× 12.7	34.2	4355	18.4	131	65.0	62.7	9.03	78.5	45.5	37.3	33.0	0.558
L203 × 102 × 25.4	55.7	7095	29.0	231	64.0	77.5	4.83	64.6	26.2	26.7	21.5	0.247
× 19.1	42.7	5445	22.9	179	64.8	74.9	3.90	50.3	26.7	24.2	21.6	0.258
× 12.7	29.2	3710	16.0	123	65.8	72.6	2.81	35.2	27.4	21.8	22.0	0.267
L178 × 102 × 19.1	39.0	4960	15.7	138	56.4	63.8	3.77	49.7	27.7	25.7	21.8	0.324
× 12.7	26.6	3385	11.1	95.2	57.2	61.5	2.72	34.7	28.2	23.3	22.1	0.335
× 9.5	20.2	2570	8.57	72.8	57.7	60.2	2.12	26.7	28.7	22.1	22.4	0.340
L152 × 102 × 19.1	35.1	4475	10.2	102	47.8	52.8	3.61	48.7	28.4	27.4	21.8	0.428
× 12.7	24.1	3065	7.24	71.0	48.5	50.5	2.61	34.1	29.2	25.1	22.1	0.440
× 9.5	18.3	3230	5.62	54.4	49.0	49.3	2.04	26.2	29.7	23.9	22.3	0.446
L152 × 89 × 12.7	22.8	2905	6.91	69.5	48.8	52.8	1.77	26.1	24.7	21.2	19.3	0.344
× 9.5	17.4	2205	5.37	53.1	49.3	51.8	1.39	20.2	25.1	20.0	19.5	0.350
L127 × 89 × 19.1	29.5	3750	5.79	70.1	39.4	44.5	2.31	36.4	24.8	25.3	19.0	0.464
× 12.7	20.2	2580	4.16	49.0	40.1	42.2	1.69	25.6	25.7	23.0	19.2	0.479
× 9.5	15.5	1970	3.24	37.5	40.6	40.9	1.32	19.8	25.9	21.9	19.4	0.486
× 6.4	10.4	1330	2.24	25.7	41.1	39.6	0.928	13.6	26.4	20.7	19.6	0.492
L127 × 76 × 12.7	19.0	2420	3.93	47.7	40.4	44.5	1.07	18.8	21.1	19.1	16.5	0.357
× 9.5	14.6	1845	3.07	36.7	40.9	43.2	0.849	14.6	21.5	17.9	16.6	0.364
× 6.4	9.82	1250	2.13	25.1	41.1	42.2	0.599	10.1	21.9	16.7	16.8	0.371
L102 × 89 × 12.7	17.7	2260	2.21	31.8	31.2	31.8	1.58	24.9	26.4	25.4	18.3	0.750
× 9.5	13.5	1725	1.74	24.4	31.8	30.7	1.23	19.2	26.9	24.3	18.5	0.755
× 6.4	9.22	1170	1.21	16.9	32.3	29.5	0.870	13.2	27.2	23.1	18.6	0.759
L102 × 76 × 12.7	16.5	2095	2.10	31.0	31.8	33.8	1.01	18.4	21.9	21.0	16.2	0.543
× 9.5	12.6	1600	1.65	23.9	32.0	32.5	0.799	14.2	22.3	19.9	16.4	0.551
× 6.4	8.63	1090	1.15	16.4	32.5	31.5	0.566	9.82	22.8	18.7	16.5	0.558
L89 × 76 × 12.7	15.2	1935	1.44	23.8	27.2	28.7	0.970	18.0	22.4	22.2	15.8	0.714
× 9.5	11.8	1485	1.13	18.5	27.7	27.4	0.770	13.9	22.8	21.1	15.9	0.721
× 6.4	8.04	1005	0.795	12.7	28.2	26.4	0.541	9.65	23.2	19.9	16.0	0.727
L89 × 64 × 12.7	14.0	1775	1.35	23.1	27.7	30.5	0.566	12.5	17.9	17.9	13.6	0.486
× 9.5	10.7	1360	1.07	17.9	27.9	29.5	0.454	8.70	18.3	16.8	13.6	0.496
× 6.4	7.29	929	0.749	12.4	28.4	28.2	0.323	6.75	18.7	15.6	13.8	0.506
L76 × 64 × 12.7	12.6	1615	0.866	17.0	23.2	25.4	0.541	12.2	18.3	19.1	13.2	0.667
× 9.5	9.82	1240	0.691	13.3	23.6	24.3	0.433	9.52	18.7	17.9	13.3	0.676
× 6.4	6.70	845	0.487	9.19	24.0	23.1	0.309	6.62	19.1	16.8	13.4	0.684
L76 × 51 × 12.7	11.5	1450	0.799	16.4	23.5	27.4	0.280	7.77	13.9	14.8	10.9	0.414
× 9.5	8.78	1115	0.637	12.8	23.9	26.4	0.226	6.08	14.2	13.7	10.9	0.428
× 6.4	6.10	768	0.454	8.88	24.3	25.2	0.163	4.26	14.6	12.5	11.0	0.440
L64 × 51 × 9.5	7.89	1000	0.380	8.96	19.5	20.7	0.214	5.95	14.7	14.8	10.7	0.614
× 6.4	5.39	684	0.272	6.24	19.9	20.0	0.155	4.16	15.0	13.6	10.8	0.626

* L means unequal leg angles, followed by the nominal depth in mm, then the mass in kg/m of length.

Structural tees

Designation*	Area (mm²)	Flange			Stem thick-ness (mm)	Axis Z–Z				Axis Y–Y		
		Depth of tee (mm)	Width (mm)	Thick-ness (mm)		I (10^6 mm⁴)	S (10^3 mm³)	r (mm)	y_c (mm)	I (10^6 mm⁴)	S (10^3 mm³)	r (mm)
WT457 × 171	21805	455.9	418.3	32.0	19.3	389	1098	133	102	196	936	94.7
× 119	15160	457.3	304.8	25.9	16.5	308	914	142	120	61.2	403	63.5
WT381 × 98	12515	384.9	267.8	25.4	15.6	175	613	118	99.1	40.8	305	57.2
× 80	10260	378.8	266.1	19.3	13.8	145	524	119	102	30.4	228	54.6
WT305 × 77	9870	305.6	323.9	19.1	12.7	78.7	328	89.2	65.8	54.1	333	73.9
× 70	8905	308.7	230.3	22.2	13.1	77.4	333	93.2	75.9	22.7	197	50.3
× 63	8000	306.1	229.1	19.6	11.9	69.1	300	93.2	75.4	19.6	172	49.5
× 46	5875	301.5	178.8	15.0	10.9	54.5	256	96.3	87.9	7.16	80.3	35.1
WT229 × 57	7225	231.3	280.3	17.3	10.8	29.9	161	64.5	45.7	31.7	226	66.3
× 45	5690	231.6	191.9	17.7	10.5	26.9	152	68.8	54.9	10.4	109	42.9
× 37	4730	228.5	190.4	14.5	9.0	22.3	128	68.6	53.8	8.32	87.7	41.9
× 30	3795	227.3	152.8	13.3	8.0	18.6	110	70.1	58.2	3.98	51.9	32.3
WT203 × 74	9485	215.5	264.8	25.0	14.9	32.0	187	57.9	44.7	38.8	293	63.8
× 37	4755	206.5	179.6	16.0	9.7	17.6	111	61.0	48.0	7.74	86.2	40.4
× 30	3800	203.3	177.7	12.8	7.7	13.8	87.7	60.2	46.0	5.99	67.5	39.9
× 19	2475	199.3	139.7	8.8	6.4	9.78	67.0	62.7	53.1	2.00	28.5	28.4
WT178 × 89	11420	183.9	372.6	23.9	15.0	21.5	141	43.4	31.5	103	552	95.0
× 61	7740	181.7	257.3	21.7	13.0	17.1	117	47.0	35.3	30.9	239	63.0
× 51	6445	178.3	254.9	18.3	10.5	13.6	93.2	46.0	32.8	25.3	198	62.5
× 36	4560	177.4	204.0	15.1	8.6	10.4	73.4	47.5	34.3	10.7	105	48.5
× 22	2850	175.8	170.9	9.8	6.9	7.91	58.2	52.6	40.1	4.07	47.7	37.8
× 16	2095	174.5	127.0	8.5	5.8	6.16	47.7	54.4	44.7	1.46	22.9	26.4
WT152 × 89	11355	166.6	312.9	28.1	18.0	18.1	135	39.9	32.5	71.6	459	79.5
× 71	9095	161.4	308.9	22.9	14.0	13.3	100	38.4	28.7	56.2	364	78.5
× 54	6840	155.6	305.8	17.0	10.9	9.66	74.4	37.6	25.9	40.6	265	77.2
× 37	4735	154.8	205.2	16.2	9.4	7.78	62.1	40.6	29.7	11.7	114	49.8
× 22	2840	156.7	165.6	11.2	6.6	5.62	45.1	44.5	32.3	4.25	51.1	38.6
× 12	1525	152.3	101.3	6.7	5.6	3.62	33.4	48.8	44.2	0.587	11.6	19.6
WT127 × 83	10645	144.3	264.5	31.8	19.2	11.9	105	33.5	30.7	49.1	370	68.1
× 65	8325	137.7	260.7	25.1	15.4	8.66	78.2	32.3	26.9	37.2	285	66.8
× 45	5690	129.8	256.0	17.3	10.7	5.37	49.8	30.7	22.5	24.2	188	65.3
× 22	2850	133.0	147.6	13.0	7.6	3.86	36.7	36.8	27.9	3.48	47.0	34.8
× 9	1140	125.3	100.6	5.3	4.8	1.81	20.0	39.9	34.5	0.454	9.03	19.9
WT102 × 43	5515	111.1	208.8	20.6	13.0	3.80	42.8	26.2	22.2	15.6	150	53.3
× 30	3785	104.8	205.0	14.2	9.1	2.39	27.7	25.1	18.7	10.2	99.6	51.8
× 18	2285	100.7	165.0	10.2	6.2	1.47	17.7	25.4	17.7	3.80	46.0	40.9
× 13	1695	103.4	133.4	8.4	5.8	1.42	17.2	29.0	21.2	1.66	24.9	31.2
× 7	955	100.2	100.1	5.2	4.3	0.895	11.7	30.5	24.2	0.437	8.72	21.4
WT176 × 15	1895	78.7	152.9	9.3	6.6	0.733	11.4	19.7	14.2	2.76	36.2	38.1
× 9	1150	76.6	101.6	7.1	5.8	0.549	9.24	21.9	17.2	0.624	12.3	23.3
WT51 × 10	1230	52.8	103.1	8.8	7.1	0.219	5.26	13.3	11.2	0.803	15.6	25.4

* WT means structural T-section (cut from a W-section), followed by the nominal dept in mm, then the mass in kg/m of length.

Reactions, deflections and slopes of selected beams

Load and support (Length L)	Equation of elastic curve	Maximum deflection (+ downward)	Slope at end ($+\curvearrowleft$)
	$v = -\dfrac{Px^2}{6EI}(x - 3L)$	$v_{max} = \dfrac{PL^3}{3EI}$ at $x = L$	$\theta = \dfrac{PL^2}{2EI}$ at $x = L$
	$v = \dfrac{wx^2}{24EI}(x^2 - 4Lx + 6L^2)$	$v_{max} = \dfrac{wL^4}{8EI}$ at $x = L$	$\theta = \dfrac{wL^3}{6EI}$ at $x = L$
	$v = \dfrac{w}{120EIL}[(L - x)^5 + 5L^4x - L^5]$	$v_{max} = \dfrac{wL^4}{30EI}$ at $x = L$	$\theta = \dfrac{wL^3}{24EI}$ at $x = L$
	$v = -\dfrac{M}{2EI}x^2$	$|v|_{max} = \dfrac{ML^2}{2EI}$ at $x = L$	$\theta = -\dfrac{ML}{EI}$ at $x = L$
	$x \le a:\ v = -\dfrac{Pbx}{6EIL}(x^2 - L^2 + b^2)$ $x = a:\ v = \dfrac{Pa^2b^2}{3EIL}$	$v_{max} = \dfrac{Pb(L^2 - b^2)^{3/2}}{9\sqrt{3}LEI}$ at $x = \sqrt{(L^2 - b^2)/3}$ $v_{\substack{center \\ not\ max}} = \dfrac{Pb(3L^2 - 4b^2)}{48EI}$	$\theta_1 = \dfrac{Pb(L^2 - b^2)}{6LEI}$ at $x = 0$ $\theta_2 = -\dfrac{Pa(L^2 - a^2)}{6LEI}$ at $x = L$
	$x \le L/2:\ v = -\dfrac{Px}{48EI}(4x^2 - 3L^2)$	$v_{max} = \dfrac{PL^3}{48EI}$ at $x = L/2$	$\theta_1 = \dfrac{PL^2}{16EI}$ at $x = 0$ $\theta_2 = -\dfrac{PL^2}{16EI}$ at $x = L$
	$v = \dfrac{wx}{24EI}(x^3 - 2Lx^2 + L^2)$	$v_{max} = \dfrac{5wL^4}{384EI}$ at $x = L/2$	$\theta_1 = \dfrac{wL^3}{24EI}$ at $x = 0$ $\theta_2 = -\dfrac{wL^3}{24EI}$ at $x = L$
	$v = -\dfrac{Mx}{6EIL}(x^2 - L^2)$	$v_{max} = \dfrac{ML^2}{9\sqrt{3}EI}$ at $x = L/\sqrt{3}$ $v_{\substack{center \\ not\ max}} = \dfrac{ML^2}{16EI}$	$\theta_1 = \dfrac{ML}{6EI}$ at $x = 0$ $\theta_2 = -\dfrac{ML}{3EI}$ at $x = L$

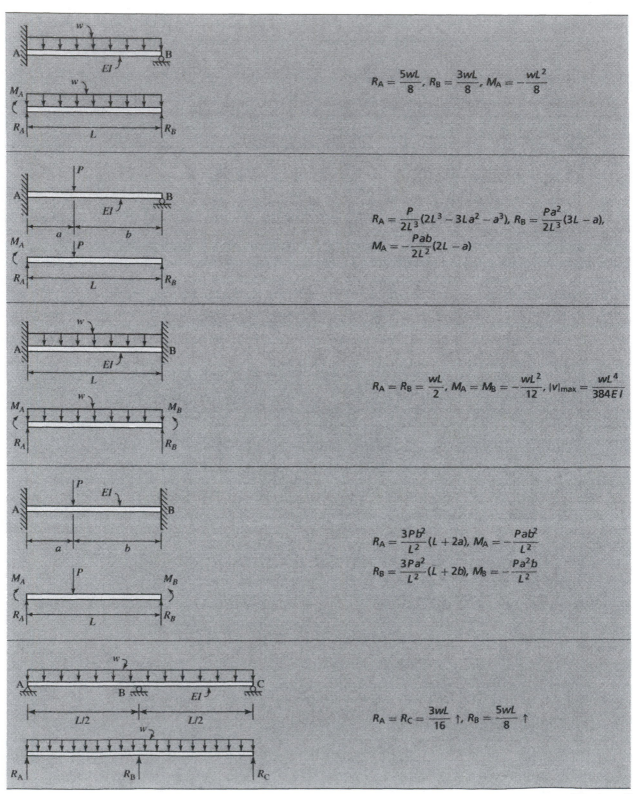

$$R_A = \frac{5wL}{8}, \; R_B = \frac{3wL}{8}, \; M_A = -\frac{wL^2}{8}$$

$$R_A = \frac{P}{2L^3}(2L^3 - 3La^2 - a^3), \; R_B = \frac{Pa^2}{2L^3}(3L - a),$$
$$M_A = -\frac{Pab}{2L^2}(2L - a)$$

$$R_A = R_B = \frac{wL}{2}, \; M_A = M_B = -\frac{wL^2}{12}, \; |v|_{max} = \frac{wL^4}{384EI}$$

$$R_A = \frac{3Pb^2}{L^2}(L + 2a), \; M_A = -\frac{Pab^2}{L^2}$$
$$R_B = \frac{3Pa^2}{L^2}(L + 2b), \; M_B = -\frac{Pa^2b}{L^2}$$

$$R_A = R_C = \frac{3wL}{16} \uparrow, \; R_B = \frac{5wL}{8} \uparrow$$

(Continued)

(*Continued*)

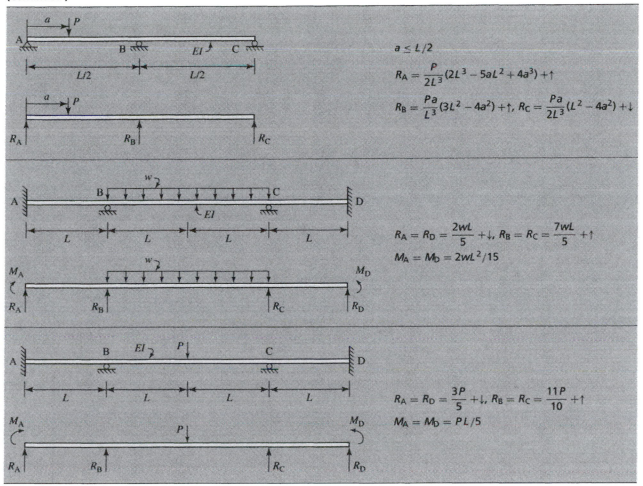

$a \leq L/2$

$$R_A = \frac{P}{2L^3}(2L^3 - 5aL^2 + 4a^3) + \uparrow$$

$$R_B = \frac{Pa}{L^3}(3L^2 - 4a^2) + \uparrow, \quad R_C = \frac{Pa}{2L^3}(L^2 - 4a^2) + \downarrow$$

$$R_A = R_D = \frac{2wL}{5} + \downarrow, \quad R_B = R_C = \frac{7wL}{5} + \uparrow$$

$$M_A = M_D = 2wL^2/15$$

$$R_A = R_D = \frac{3P}{5} + \downarrow, \quad R_B = R_C = \frac{11P}{10} + \uparrow$$

$$M_A = M_D = PL/5$$

Answers to selected problems

Chapter 1

1.1: (a) $P = 10\,\text{kN (compression)}$, (b) $P = 6\,\text{kN (tension)}$

1.2: (a) $P = 22{,}620\,\text{N}$, (b) $\bar{\sigma} = 72\,\text{MPa}$

1.3: $P = 120\,\text{kN}$

1.4: (a) $D = 5.04\,\text{cm}$, (b) $L = 7\,\text{m}$

1.5: $\sigma_{AB} = 0$, $\sigma_{AC} = 141.5\,\text{MPa}$

1.6: (b) $\beta = 54.7°$

1.8: (a) $\epsilon(x) = (2.34 \times 10^{-4}) \cdot x$, (c) $\Delta L = 0.468\,\text{mm}$

Chapter 2

2.1: (a) $F = 450\,\text{N}$, $V = 600\,\text{N}$, $M = 1200\,\text{N-m}$

2.3: (a) At A: $F = 600\,\text{N}$, $V = 0$, $M = 4200\,\text{N-m}$;
At C: $F = 0$, $V = 600\,\text{N}$, $M = 1200\,\text{N-m}$

2.5: $V = \dfrac{q_0 L}{\pi}$, $M = \dfrac{2q_0 L^2}{\pi}(1 - 2/\pi)$

2.7: $F = \dfrac{\pi \rho R^2}{h^2}(h^3 - y^3)$

2.8: (b) $F(x) = \dfrac{AcL}{\alpha^{1/2}}\left\{\ln\left[\alpha^{1/2}x/L + (1 + \alpha x^2/L^2)^{1/2}\right]\right\}$

2.9: (a) $V = -40x$, $M = -20x^2$ $(0 < x < 6)$
(c) $V = -5x^2/3$, $M = -x^3/9$
(f) $V = 10x^2 - 120x + 448$,
$M = 10x^3/3 - 60x^2 + 448x$, $(0 \le x < 6)$;
$V = -512$, $M = -512x + 4320$, $(6 < x < 10)$;
$V = 200$, $M = -200(14 - x)$, $(10 < x \le 14)$
(h) $V = 25x^2/3 - 100x + 700$,
$M = 25x^3/9 - 50x^2 + 700x - 4600$,
$(0 \le x \le 6)$;
$V = 400$, $M = 400(x - 10)$, $(6 \le x \le 10)$
(i) $V = 120$, $M = 120x - 720$, $(0 \le x \le 6)$;
$V = -60x + 480$,
$M = -30x^2 + 480x - 1800$, $(6 \le x \le 10)$

2.11: $V = -wL/2$, $M = -wLx/2$, $(0 \le x < L)$;
$V = w(L - x)$, $M = \dfrac{w}{2}(-x^2 + 4Lx - L^2)$,
$(L < x \le 2L)$;

$V = M = 0$, $(2L \le x < 3L)$; $V = -P$,
$M = -P(x - 3L)$, $(3L < x \le 4L)$

2.12: (a) $F = 1200$, $V = -400$, $M = -400x$ $(0 < x < 6)$;
$F = 1200$, $V = 0$, $M = -2400\,\text{N-m}$,
$(6 < x < 10)$
(c) $F = 600$, $V = 40\,\text{N}$, $M = 40x - 520\,\text{N-m}$,
$(0 \le x < 5)$;
$F = 0$, $V = 40\,\text{N}$, $M = 40x - 400\,\text{N-m}$,
$(5 < x \le 10)$
(e) $F = -1200\,\text{N}$, $V = 400\,\text{N}$, $M = 400x\,\text{N-m}$,
$(0 \le x < 8)$;
$F = 0$, $V = 5400\,\text{N}$, $M = 5400\,(x - 20)\,\text{N-m}$,
$(8 < x \le 20)$

2.13: $V_y = 0$, $V_z = 20\,\text{N}$, $T \equiv M_x = 600\,\text{N-cm}$,
$M_y = 20\,(100 - x)\,\text{N-cm}$, $M_z = 0$

2.15: (a) $F = P_x \cos\theta + P_y \sin\theta$,
$V = -P_x \sin\theta + P_y \cos\theta$,
$M = R[P_y \sin\theta - P_x(1 - \cos\theta)]$

2.17: $b/L = \sqrt{2}/2$

2.21: $F = a^3 A/4 + \sqrt{3}a^4 B/96$,
$M = 3\sqrt{3}Aa^4/64 + Ba^5/80$.

2.23: $F = V_y = V_z = 0$, $T \equiv M_x = 100\pi R^2$

2.24: (b) $M = \dfrac{b\sigma_0}{12}(3h^2 - 4c^2)$

2.28: $\sigma_x = 240$, $\tau_{nt}|_{\max} = 720\,\text{MPa}$

2.29: (a) $\sigma_n = 250 + 200\sqrt{3}$, $\sigma_t = 350 - 200\sqrt{3}$,
$\tau_{nt} = 200 + 50\sqrt{3}\,\text{MPa}$
(c) $\sigma_n = 150$, $\sigma_t = -150$, $\tau_{nt} = 0\,\text{MPa}$
(e) $\sigma_x = -87.5 - 50\sqrt{3}$, $\sigma_y = -62.5 + 50\sqrt{3}$,
$\tau_{xy} = 150 - 6.25\sqrt{3}\,\text{MPa}$
(g) $\sigma_x = 400$, $\tau_{xy} = -100\sqrt{3}$, $\sigma_t = 500\,\text{MPa}$

2.30: (a) $\tau_{nt} \equiv \tau_{xy} = \sigma_s - \dfrac{\sigma_n + \sigma_t}{2}$, (b) $\tau_{xy} = \dfrac{\sqrt{3}}{3}(\sigma_s - \sigma_t)$

2.31: (b) $\theta = -16.85°$, (c) $\theta = 22.5°$

2.32: $-1783 < \sigma_y < 716.7\,\text{kPa}$

2.35: (a) $\sigma_{1,2} = -20, 80$; $\tau_{\max} = 50$ MPa, $\theta_{1,2} = 16.85°$,
106.85°

(c) $\sigma_{1,2} = 947.2, 52.8$; $\tau_{\max} = 447.5$ MPa,
$\theta_{1,2} = 13.3°, 103.3°$

(e) $\sigma_{1,2} = 56.2, -356$; $\tau_{\max} = 206$ MPa, $\theta_{1,2} = 52°$,
$-38°$

(g) $\sigma_{1,2} = 42.5, -122.5$; $\tau_{\max} = 82.5$ MPa,
$\theta_{1,2} = -83°, 7°$

(i) $\sigma_{1,2} = 303.4, -403.4$; $\tau_{\max} = 359.4$ kPa,
$\theta_{1,2} = 57.6°, -32.4°$

2.36: (a) $\sigma_{1,2} = 100, 0$ MPa, $\theta_{1,2} = 26.6°, -63.4°$

(c) $\sigma_{1,2} = 220, 90$ MPa, $\theta_{1,2} = 56.3°, -33.7°$

(e) $\sigma_{1,2} = 110, 10$ MPa, $\theta_{1,2} = -61.8°, 28.2°$

2.37: (a) $\sigma_{1,2} = 101.2, 18.77$ MPa, $\theta_{1,2} = -7.0°, 83.0°$

(b) $\sigma_{1,2} = 117, -23.7$ MPa, $\theta_{1,2} = -11.7°, 78.3°$

2.39: (a) (i) $\sigma_2 = -260$ MPa, $\theta_{1,2} = 32.7°, -57.3°$

(ii) $\sigma_2 = -260$ MPa, $\theta_{1,2} = -32.7°, 57.3°$

(c) $\sigma_2 = 17.5$ MPa, $\theta_{1,2} = -36.9°, 53.1°$

2.41: (a) $C_2/C_1 = 1/3$, $C_3/C_1 = -1$

(b) AB: $X_n = Y_n = 0$;

BC: $X_n = -C_1 a^2 y$, $Y_n = -C_1 a y^2$;

CD: $X_n = -C_1 b^2 x$, $Y_n = C_1 b^3/3$

DA: $X_n = -\dfrac{C_1}{2}(\sqrt{3}x^2 y + xy^2)$,

$Y_n = -\dfrac{C_1}{2}(\sqrt{3}xy^2 + y^3/3)$

2.46: $\theta = 54.462°$

Chapter 3

3.1: (a) $\bar{\epsilon} = 5 \times 10^{-3}$, (b) $\epsilon_{\max} = 10^{-2}$,
(c) $\epsilon = 0.5 \times 10^{-2}$, (d) $u(x) = 0.25 \times 10^{-4} \cdot x^2$

3.2: (a) $\bar{\epsilon} = 1/a[2L^2(1 - \cos\theta) + a^2 + 2aL\sin\theta]^{1/2} - 1$

3.5: $\epsilon_n = \left(\dfrac{1 + 4b^2 x^2}{1 + 4a^2 x^2}\right)^{1/2} \cdot (b/a) - 1$

3.7: $\Delta L = \pi k R$

3.9: $\bar{\epsilon}_\theta = 4 \times 10^{-4}$

3.10: (b) $\bar{\epsilon}_{AB} = \dfrac{\sqrt{2}}{2L}(u + v)$, $\bar{\epsilon}_{BC} = \dfrac{\sqrt{2}}{2L}(v - u)$

3.13: $\gamma_{nt} = -1.32°$

3.14: $\gamma_{nt} = 2(\epsilon_y - \epsilon_x)\sin\theta\cos\theta$

3.16: (a) $\epsilon_x = a$, $\epsilon_y = c$, $\gamma_{xy} = 2b$; (c) $\epsilon_x = a + by$,
$\epsilon_y = 0$, $\gamma_{xy} = bx - 2a$

3.17: $\epsilon_x = \dfrac{\delta \cdot y}{L^2}$, $\epsilon_y = 0$, $\gamma_{xy} = \dfrac{\delta \cdot x}{L^2}$

3.19: $u(x, y) = (ax + e)(y/L)$, $v(x, y) = (by - e)(x/L)$

3.21: $\epsilon_x = \dfrac{1}{2}(3 \pm \sqrt{5}) \times 10^{-3}$

3.23: (a) $\epsilon_n = (250 + 200\sqrt{3})\mu$, $\epsilon_t = (350 - 200\sqrt{3})\mu$,
$\epsilon_{nt} = (200 + 50\sqrt{3})\mu$

(c) $\epsilon_n = 150\mu$, $\epsilon_t = -150\mu$, $\epsilon_{nt} = 0$

(e) $\epsilon_x = (-87.5 - 50\sqrt{3})\mu$, $\epsilon_y = (-62.5 + 50\sqrt{3})\mu$, $\epsilon_{xy} = (150 - 6.25\sqrt{3})\mu$

(g) $\epsilon_x = 400\mu$, $\epsilon_{xy} = -100\sqrt{3}\mu$, $\epsilon_t = 500\mu$

3.25: (a) $\theta = 18.43°$, (b) $\theta = -16.85°$, (c) $\theta = 22.5°$

3.26: (a) $\epsilon_{1,2} = -90.8, -30.8\mu$; $\theta_{1,2} = 49.7°, -40.3°$

(c) $\epsilon_{1,2} = 57.1, -77.1\mu$; $\theta_{1,2} = 31.7°, -58.3°$

(e) $\epsilon_{1,2} = 77.1, -57.1\mu$; $\theta_{1,2} = 58.3°, -31.7°$

(g) $\epsilon_{1,2} = 30.8, -90.8\mu$; $\theta_{1,2} = 40.3°, -49.7°$

3.27: (a) $\epsilon_{1,2} = -20, 80\mu$; $\theta_{1,2} = 16.85°, 106.85°$

(c) $\epsilon_{1,2} = 947.2, 52.8\mu$; $\theta_{1,2} = 13.3°, 103.3°$

(e) $\epsilon_{1,2} = 56.2, -356\mu$; $\theta_{1,2} = 52°, -38°$

(g) $\epsilon_{1,2} = 42.5, -122.5\mu$; $\theta_{1,2} = -83.7°, 7°$

(i) $\epsilon_{1,2} = 0.303, -0.403 \times 10^{-3}$;
$\theta_{1,2} = 57.6°, -32.4°$

3.28: (a) $\epsilon_{1,2} = 101.2, 18.77\mu$, $\theta_{1,2} = -7.0°, 83.0°$

3.29: (b) (i) $\epsilon_2 = -260\mu$, $\theta_{1,2} = 49.8°, -40.2°$

(ii) $\epsilon_2 = -20\mu$, $\theta_{1,2} = -49.8°, 40.2°$

(c) $\epsilon_2 = 17.5\mu$, $\theta_{1,2} = -36.9°, 53.1°$

3.30: (a) $\epsilon_y = -50\mu$, $\epsilon_{xy} = 275\mu$; (b) $\epsilon_y = -800\mu$,
$\epsilon_{xy} = -500\mu$;

(e) $\epsilon_y = 0\mu$, $\epsilon_{xy} = 200\sqrt{3}\mu$; (f) $\epsilon_y = -500\mu$,
$\epsilon_{xy} = -500\sqrt{3}/3\mu$

3.33: $|\epsilon_{nt}| \leq 10^{-3}$

3.34: $\delta A = ka^2 b^2/9$

3.36: $\delta A = 2\alpha ab(\delta T_0 + \delta T_1/\pi)$

3.37: $\gamma_{r\theta}(P) = \dfrac{ab\phi}{r(a - b)}$

3.41: (c) $a/c = -0.5$, $b/c = -0.5$,
(e) $\Delta A = 2aL^2|_{L=1} = 2a$

3.43: (a) $\Delta_{AC} = \sqrt{2}k/2$,

(b) $\Delta_{AEC} = \dfrac{k}{128}\left[\dfrac{16}{3}(5\sqrt{5} - 1)\right.$
$\left. + 18\sqrt{5} - \ln(2 + \sqrt{5})\right]$.

(c) $\bar{\epsilon}_{AC} = 0.5 k/L$, $\bar{\epsilon}_{AEC} = 0.4918 k/L$

3.45: (b) $\bar{\epsilon} = 3e/2$, (c) $\epsilon_n(D) = 11e/8L$,
(d) $\angle C^* D^* B^* = 88.57°$

3.46: (c), (d) $\Delta_{OB} = 3\sqrt{2}e/2$, (e) (i) $|u| = \sqrt{145}e/6$,
(ii) $|u_D|_{(OB)} = 17\sqrt{2}e/32$

3.50: $\theta = 54.462°$

3.51: $\theta = 19.548°$

3.52: $\theta = 42.213°$

Chapter 4

4.1: $\epsilon_{1,2} = 0.289, -0.324\mu$

4.6:   (d) (i) $\Delta = 0.256255 \times 10^{-4}$, (ii) $\delta\alpha = 0.028669°$,

(e) $\delta\alpha = \tan^{-1}\left[\dfrac{(1+\nu)(\sigma_0/E)(b/a)}{1 - \sigma_0/E + (1 + \nu\sigma_0/E)(b/a)^2}\right]$

4.10:   (b) $E = 44\,\text{GPa}$, $\sigma_p = 176\,\text{MPa}$

4.12:   (a) (i) $(E_t)_0 = 150\,\text{GPa}$, (ii) $E_t = 82.5\,\text{GPa}$,
$E_s = 141.7\,\text{GPa}$ (b) $U_0 = 111 \times 10^3\,\text{N-m/m}^3$

Chapter 6

6.1: $d = 13.8\,\text{mm}$

6.2: $A_{AB}/A_{BC} = 8E_{BC}/E_{AB}$

6.4: $\Delta L = \dfrac{k}{E}\left[\dfrac{L}{a} - \ln(1 + L/a)\right]$

6.6: $\Delta L = 4.2\,\text{mm}$

6.9: $a = 3.20d$

6.11: $v_B = 10.42\,\text{mm}$

6.12: $u_B = 4.57\,\text{mm}$, $v_B = 16.51\,\text{mm}$

6.14: $e = \dfrac{1}{2}\left(\dfrac{a^2 E_1 - c^2 E_2}{a E_1 + c E_2}\right)$

6.16:   (a) $e = \dfrac{2(2E_1 - 3E_2)}{3(\pi E_1 + 4E_2)}R$, (b) $\Delta L = \dfrac{2PL}{R^2(\pi E_1 + 4E_2)}$

6.19:   (a) $\beta = 62.2°$, (b) $P = 94.4\,\text{kN}$

6.20:   (a) $R_B = \dfrac{\ln[(a+b)/2a]}{\ln(b/a)} \cdot P$

6.21:   (a) $R_A = \dfrac{P}{1 + \frac{\ln(b/a)}{1-a/b}}$, $R_B = \dfrac{P}{1 + \frac{1-a/b}{\ln(b/a)}}$

6.25:   (a) $\sigma_s = 55.5\,\text{MPa}$, $\sigma_a = 222.2\,\text{MPa}$,
$\Delta L = 0.173\,\text{mm}$

6.27: $\sigma_{cu} = 4.65\,\text{MPa}$, $\sigma_s = 9.3\,\text{MPa}$

6.28:   (a) $F_{AD} = F_{CD} = -AE\alpha\Delta T\left(\dfrac{1 - \cos^2\beta}{1 + 2\cos^2\beta}\right)$,

$F_{BD} = 2AE\alpha\Delta T\left(\dfrac{(1 - \cos^2\beta)\cos\beta}{1 + 2\cos^2\beta}\right)$

6.30:   (a) $P_y = 5\sigma_0/6$, (b) $\Delta_F = 4\sigma_0 h/3E$, (c) $P_{ult} = \sigma_0 A$,
(d) $\Delta_F|_{ult} = 4\sigma_0 h/E$

6.31:   (a) $P_y = 3\sigma_0 A/2$, $\delta_c|_y = \sigma_0 L/3E$; $P_{ult} = 2\sigma_0 A$,
$\delta_c|_{ult} = 2\sigma_0 L/3E$

6.33:   (d) $\angle A^*B^*D^* = \pi/2 + \dfrac{\sqrt{3}aC}{4E}(1 + \nu)(e - \sqrt{e})$;

(e) $u_B = \dfrac{Ca^2}{2E}(2 + e - 2\sqrt{e})$, $v_B = -\dfrac{\nu Cab}{2E}(e - \sqrt{e})$

6.38:   (a) $\delta = \dfrac{P}{4\pi GL}\ln(D/d)$, (b) $\delta = 1.12\,\text{cm}$,

(c) $\Delta L = \dfrac{2PL}{\pi d^2 E_s}$, (d) $\Delta L = 0.06\,\text{mm}$

6.40:   (a) $\sigma_f/\sigma_m = E_f/E_m$, (e) $E_{eff} = 61.9\,\text{GPa}$,
$\overline{\sigma} = 15\,\text{MPa}$, $\Delta L = 0.97\,\text{mm}$, $\sigma_f = 72.3\,\text{MPa}$,
$\sigma_m = 580\,\text{kPa}$

Chapter 7

7.2: $t = 7.94\,\text{mm}$

7.3: $D = 1.427d$

7.4: $\Theta = \dfrac{49T}{8\pi^2 c^2 R^7}$

7.7: $d = 11.58\,\text{cm}$

7.10: $d = 5.65\,\text{cm}$

7.11: $\phi_a = \dfrac{7TL}{12\pi Gr_0^4}$

7.13: $\tau = 95\,\text{MPa}$

7.15: $N = 5699\,\text{rpm}$

7.16: $\sigma_{1,2} = 150.3, -208.8\,\text{MPa}$, $\theta_{1,2} = 60.9°$, $150.9°$

7.19:   (b) $\tau_{AB} = 53.92\,\text{MPa}$, $\tau_{CD} = 60.68\,\text{MPa}$

7.21:   (a) $T_A = 7t_0 L/12$, $T_C = 3t_0 L/4$,
(b) $\phi_B = 3t_0 L^2/\pi Gd^4$

7.23: $T_E = 14{,}577\,\text{N-m}$, (b) $b = 29.9\,\text{mm}$,
(c) $\Theta = 3.66°/\text{m}$

7.24:   (d) $\theta = \dfrac{2T_E R_0}{\pi R_i G\left(R_0^4 - R_i^4\right)}$

7.25:   (a) $T_y = 11{,}714\,\text{N-m}$, (b) $\phi = 4.65°$,
(c) $T = 13{,}619\,\text{N-m}$, (d) $T_p = 14{,}891\,\text{N-m}$
(e) $\Theta = 9.05°/\text{m}$

7.29: $\Theta = \dfrac{2T}{\pi}\left[(G_a - G_b)R_1^4 + G_b R_2^4\right]^{-1}$

7.29: $\phi_{B|A} = 3TL\left[16\pi Gt\overline{R}_a^3\right]^{-1}$

7.31:   (a) $\Theta_A = 5T/2\pi G_0 R^4$, $\Theta_B = 10T/\pi G_0 R^4$

7.33:   (b) $\tau = \left[\dfrac{(n+3)r^n}{2\pi R^{n+3}}\right]T$

7.34:   (a) $\tau_{x\theta} = \dfrac{5t_0}{2\pi R^5}r^2 x$, (b) $\phi = \dfrac{2}{3}\left(\dfrac{5t_0 L^3}{2\pi kR^5}\right)^{1/2}$

7.35:   (b) $\Theta = \dfrac{4T}{G_a\pi R_1^4}$, (c) $T_a/T = 2/3$,

(d) $\sigma = \dfrac{2\sqrt{2}E_s}{\pi G_a}T$

7.37:   (b) $\phi_C = \dfrac{T}{8\pi GR^4}(a + 16c)$, (c) $T_E/T_b = 1/8$,
(d) $b = 0.737R$

7.39:   (a) $T = 289.6\,\text{N-m}$, (b) $b = 6.79\,\text{mm}$

7.42:   (a) $T_y = 11{,}800\,\text{N-m}$, (b) $\Theta = 2.69°/\text{m}$,
(c) $T_p = 14{,}860\,\text{N-m}$

7.45:   (a) $D/d = [1 - (1 - 2/k)^4]^{-1/3}$

7.46:   (a) $D/d = 2^{1/4} = 1.1892$, (b) $D/d = 1.2207441$

Chapter 8

8.2: $e = a + c$

8.3:   (b) $V = -240$, $M = -240(x - 3)$,
$\quad\quad$ $(6 \leq x \leq 12)$
$\quad\quad$ (c) $V = -5x^2/3$, $M = -x^3/9$;
$\quad\quad$ (d) $V = 1200 - 300x$, $M = 1200x - 150x^2$
$\quad\quad$ (f) $V = 10x^2 - 120x + 448$,
$\quad\quad\quad$ $M = 10x^3/3 - 60x^2 + 448x$, $(0 \leq x < 6)$;
$\quad\quad\quad$ $V = -512$, $M = -512x + 4320$, $(6 < x < 10)$;
$\quad\quad\quad$ $V = 200$, $M = -200(14 - x)$, $(10 < x \leq 14)$
$\quad\quad$ (g) $V = 5x^2 - 200$, $M = 5x^3/3 - 200x$
$\quad\quad$ (h) $V = 25x^2/3 - 100x + 700$,
$\quad\quad\quad$ $M = 25x^3/9 - 50x^2 + 700x - 4600$,
$\quad\quad\quad$ $(0 \leq x \leq 6)$
$\quad\quad\quad$ $V = 400$, $M = 400(x - 10)$, $(6 \leq x \leq 10)$
$\quad\quad$ (j) $V = -80x + 160$, $M -40x^2 + 160x$, $(0 \leq x \leq 4)$;
$\quad\quad\quad$ $V = -160$, $M = -160x + 640$, $(4 \leq x < 8)$;
$\quad\quad\quad$ $V = 260$, $M = 260x - 2720$, $(8 < x < 10)$;
$\quad\quad\quad$ $V = 60$, $M = 60x - 720$, $(10 < x \leq 12)$

8.4: $V = P_1 + 2P_2(1 - a_2/L)$,
$\quad\quad$ $M = -P_1(x - a_1) + P_2[2(1 - a_2/L)x - (L - a_2)]$,
$\quad\quad$ $(0 \leq x < a_1)$; $V = -2P_2(a_2/L - 1/2)$,
$\quad\quad$ $M = 2P_2(a_2/L - 1/2)(L - x)$, $(a_2 < x \leq L)$

8.6:   (a) $F = 1200$, $V = -400$, $M = -400x$, $(0 < x < 6)$;
$\quad\quad\quad$ $F = 1200$, $V = 0$, $M = -2400$ N-m, $(6 < x < 10)$
$\quad\quad$ (b) $F = -400$, $V = -330$ N, $M = -300x$ N-m,
$\quad\quad\quad$ $(0 \leq x < 4)$; $F = -400$, $V = 1000 - 40x$ N,
$\quad\quad\quad$ $M = -20x^2 + 1000x - 5000$ N-m, $(4 < x \leq 10)$
$\quad\quad$ (d) $F = 0$, $V = -40x$ N, $M = -200x^2$ N-m,
$\quad\quad\quad$ $(0 \leq x < 8)$; $F = 0$, $V = 3200$ N,
$\quad\quad\quad$ $M = -3200x + 8000$ N-m, $(8 < x \leq 16)$

8.12:   (c) $R_B = \dfrac{q_0 L}{n + 1}$, $M_B = -\dfrac{q_0 L^2}{(n + 1)(n + 2)}$

8.13:   (a) $R_A = 2q_0 L(\pi - 2)/\pi^2$, $R_B = 4q_0 L/\pi^2$;
$\quad\quad$ (b) $V(x) = \dfrac{2q_0 L}{\pi^2}\left[\pi \cos\left(\dfrac{\pi x}{2L}\right) - 2\right]$,
$\quad\quad\quad$ $M(x) = \dfrac{4q_0 L}{\pi^2}\left[L \sin\left(\dfrac{\pi x}{2L}\right) - x\right]$

8.14:   (a) $(R_A)_a = 0.1321 q_0 L$, (b) $V(x) =$
$\quad\quad$ $\dfrac{q_0}{\alpha^2}\left[\left(1 + \alpha^2/2 - e^{-\alpha} - \alpha e^{-\alpha x/L}\right)L - \alpha^2 x\right]$,
$\quad\quad$ $M(x) = q_0\left\{\dfrac{1}{\alpha^2}\left[(\alpha^2/2 + 1 - \alpha e^{-\alpha})Lx - \left(1 - e^{-\alpha x/L}\right)L^2\right] - x^2/2\right\}$

8.15:   (a) $\overline{R} = \dfrac{d}{2}\left(\dfrac{2 + \epsilon_t + \epsilon_c}{\epsilon_t - \epsilon_c}\right)$

8.17:   (a) $d/b = \sqrt{3}$, (b) $d/b = \sqrt{2}$

8.19: $M = 2.462$ kN-m

8.24: $\overline{y} = \dfrac{1}{2}\dfrac{E_b d_b^2 - E_a d_a^2}{E_b d_b + E_a d_a}$

8.25:   (a) $\overline{y} = \dfrac{d}{3}\left[\dfrac{3 + 2\beta}{2 + \beta}\right]$, (b) $M = \dfrac{E_0 I}{R}\left[\dfrac{\beta^2 + 6\beta + 6}{3(\beta + 2)}\right]$,
$\quad\quad$ where $I = bd^3/12$

8.27: $M_E = 13\sigma_0 a d^2/30$

8.29:   (a) $\sigma_x = 27.5$ MPa, (b) $\sigma_x = -36.8$ MPa,
$\quad\quad$ (c) $\tau = 965$ kPa, (d) $\tau = 3.09$ MPa

8.31:   (a) $|\sigma_x| = 167.8$ MPa, (b) $\tau_{c-c} = 4.74$ MPa,
$\quad\quad$ (c) (i) $\tau = 21.1$ MPa, (d) $R = 181.4$ m

8.33:   (b) $\tau_{x\theta} = \dfrac{2P}{3\pi}\dfrac{(b^3 - a^3)\sin\theta}{(b^4 - a^4)(b - a)}$

8.35:   (a) $k = 9/8$

8.37:   (a) $(\tau)_y = \dfrac{3V_y}{2}\left[\dfrac{b_o(d_o^2 - d_i^2)}{(b_o - b_i)(b_o d_o^3 - b_i d_i^3)}\right]$

8.39:   (a) $a/L = (\sqrt{6} - \sqrt{2})/2 = 0.5176$,
$\quad\quad$ (c) $|\tau_{xy}|_{max} = 2wL(3 - \sqrt{3})/b^2$

8.41: $\tau = 83.97$ MPa

8.42:   (a) $h(x) = h_0\sqrt{1 - x/L}$, (b) $0 < x/L < 0.855$

8.43:   (a) $h_0 = \dfrac{L}{\pi}\sqrt{\dfrac{6q_0}{b\sigma_0}}$, (b) $h(x) = h_0\left(\sin\dfrac{\pi x}{L}\right)^{1/2}$,
$\quad\quad$ (c) $q_0 \leq \dfrac{2b\sigma_0}{3\pi^2}\tan 5°$, (d) $q_0 = 517$ N/m,
$\quad\quad\quad$ $h_0 = 16.7$ cm

8.46:   (a) $s = 28.3$ cm, (b) $s = 7.7$ cm

8.48:   (a) $L = 3.75$ m, (b) W356 × 45

8.53:   (a) $\sigma_{Al} = \sigma_B = 90$ MPa, (b) $\sigma_{Al} = 46.15$,
$\quad\quad$ $\sigma_B = 103.8$ MPa

8.55:   (a) $M = 83.06$ kN-m, (b) $R = 130$ m

8.58: $\sigma_c = -7.06$ MPa, (b) $\sigma_s = 120.5$ MPa

8.61: $\tau_{max} = \dfrac{2P}{\pi R^3}(L^2 + e^2)^{1/2}$

8.65:   B–C: $\sigma_x = \dfrac{P}{b^2}\left(\dfrac{60L}{b} - 8\right)$ [tension],
$\quad\quad$ $\sigma_x = -\dfrac{P}{b^2}\left(\dfrac{60L}{b} + 8\right)$ [compression], $\tau = \dfrac{9P}{b^2}$;
$\quad\quad$ C–D: $\sigma_x = \pm\dfrac{72PL}{b^3}$, $\tau = 0$

8.66:   At B: $\sigma = -69.1$, $\tau = 19.9$ MPa; At C: $\sigma = 72.7$,
$\quad\quad$ $\tau = 19.52$ MPa

8.68: $\theta = \tan^{-1}(d/4L)$

8.70:   (a) $M_E = \dfrac{\sigma_0 a d^2}{24}$, (b) $y_p = \dfrac{\sqrt{2}d}{2}$,
$\quad\quad$ (c) $M_P = \dfrac{\sigma_0 a d^2}{6}(2 - \sqrt{2})$

8.71: (a) $d_y = 3.75$ cm, (b) $M = -244.48\sigma_0$

8.73: $M_E = \dfrac{\sqrt{2}a^3\sigma_0}{12}$, $M_P = \dfrac{\sqrt{2}a^3\sigma_0}{6}$

8.74: $M_P = \dfrac{1}{3}(18 - 5\sqrt{10})ad^2\sigma_0$

8.75: (a) $M_{cr} = \alpha(\alpha + 1)bd^2\sigma_0$,

(b) $\kappa \equiv \dfrac{1}{R} = \dfrac{12}{Ebd^3}(M - M_{cr})$

8.78: (b) $\kappa = \dfrac{6AL}{Eh^2\pi}\left[1 - \cos\left(\dfrac{2\pi x}{L}\right)\right]$

8.80: $a = L/4$, $b = (3PL/8\sigma_{allow})^{1/2}$

8.82: (a) $h(x) = h_0(x/L)^{3/2}$, $h_0 = 20$ cm, (b) $x \leq 0.764L$,

(c) $L = 3.43$ m

8.83: $h(x) = h_0(x/L)^2$, $h_0 = 7.5$ cm

8.85: (a) $\sigma_{Al} = 36.4$ MPa, $\sigma_S = 121.3$ MPa,

(b) $R = 57.7$ m

8.86: (b) $(\sigma_x)_{max} = \dfrac{Md}{4I}(1 + \sqrt{E_t/E_c})$

8.89: (c) $R = \dfrac{4}{3}R_0$

8.92: (c) $M/M_P = 4(\ell/d)^3 - 4(\ell/d)^2 + 1$

8.97: (a) $[(1 - \pi\cos\theta)\sin^2\theta + 1]\tan\theta = \theta$,

(b) $\tilde{y} = 0.48823R$, $k = 1.4248$

8.98: (a) $M = \dfrac{\sigma a^3}{6}[(5 - 2\ell^2/a^2)\sqrt{1 - \ell^2/a^2}$

$+ 3(a/\ell)\sin^{-1}(\ell/a)]$

Chapter 9

9.1: (b) (i) $(0.4975L, -4.992L)$, (ii) $(0.5L, -4.950L)$

9.3: (a) $v(x) = \dfrac{wx^2}{24EI}(x^2 - 4Lx + 6L^2)$, $v_{max} = \dfrac{wL^4}{8EI}$

(c) $v(x) = \dfrac{w}{24EI}(x^4 - 4L^3x + 3L^4)$, $v_{max} = \dfrac{wL^4}{8EI}$

(e) $v(x) = \dfrac{q_0}{120EIL}(x^5 - 5L^4x + 4L^5)$,

$v_{max} = \dfrac{q_0L^4}{30EI}$

(g) $v(x) = \dfrac{q_0}{360EIL}[3(L - x)^5 - 10L^2(L - x)^3 - $

$7L^4x + 7L^5]$, $v_{max} = 6.522 \times 10^{-3}\dfrac{q_0L^4}{EI}$

at $\dfrac{x_{cr}}{L} = 1 - (1 - \sqrt{8/15})^{1/2} = 0.48067$

(i) $v(x) = \dfrac{M_0x}{6EIL}(2x^2 - 3Lx + L^2)$,

$|v|_{max} = \dfrac{\sqrt{3}M_0L^2}{108EI} = 0.01604\dfrac{M_0L^2}{EI}$ at

$\dfrac{x_{cr}}{L} = \dfrac{1}{2}\left(1 \pm \dfrac{1}{3}\sqrt{3}\right)$

9.5: (a) $v(x) = -\dfrac{q_0x}{90EIL^2}(x^5 - 3Lx^4 + 5L^3x^2 - 3L^5)$,

$v_{max} = \dfrac{23q_0L^4}{1920EI}$

(c) $v(x) = \dfrac{q_0L^4}{\pi^4 EI}\sin\dfrac{\pi x}{L}$, $v_{max} = \dfrac{q_0L^4}{\pi^4 EI}$

9.6: (a) $v(x) = \dfrac{q_0L^4}{6\alpha^4 EI}\{[-3(1 + \alpha)(\alpha x/L)^2 + $

$(\alpha x/L)^3 + 6]e^{-\alpha x/L} + 6(\alpha x/L) - 6\}$

9.8: (a) $v(x) = \dfrac{q_0L^4}{2\pi^2 EI_0}(x/L)(1 - x/L)$, $v(L/2) = $

$\dfrac{3q_0L^4}{8\pi^2 EI_0}$

9.11: (a) $v(x) = \dfrac{PL^3}{6EI}[-(x/L)^3 + 3(x/L)^2 + 6\gamma(x/L)]$

9.14: $\Delta_x/L = 1.19 \times 10^{-9}$

9.15: $v_s/v_f = \dfrac{1}{4}\left(\dfrac{d}{L}\right)^2\left(\dfrac{1}{1+\nu}\right)$

9.17: (b) $v(L) = \dfrac{4PL^3}{81EI} + \dfrac{163wL^4}{1944EI}$

9.19: (b) $v(L/2) = \dfrac{Pa}{24EI}(3L^2 - 4a^2)$

9.21: (b) $\dfrac{a}{L} = \dfrac{1}{6}(\sqrt{13} - 1)$

9.23: $v(L/2) = \dfrac{205wL^4}{31,104EI}$

9.25: (a) $R_A = 3M_0/2L\uparrow$, $M_A = -M_0/2$, $R_B = 3M_0/2L\downarrow$,

(c) $v_{max} = \dfrac{M_0L^2}{27EI}$

9.27: (a) $R_A = q_0L/24$, $R_B = 7q_0L/24$, $M_B = -q_0L^2/24$

9.29: (a) $R_A = 7wL/32\uparrow$, $R_B = 5wL/16\uparrow$,

$R_C = wL/32\downarrow$

9.31: (a) $R_A = R_B = wL/2\uparrow$, $M_A = M_B = -wL^2/12$,

(c) $v_{max} = v(L/2) = \dfrac{wL^4}{384EI}$

9.33: (a) $R_A = R_B = q_0L/4\uparrow$, $M_A = M_B = -5q_0L^2/96$,

(c) $v_{max} = v(L/2) = \dfrac{7q_0L^4}{3840EI}$

9.35: (a) $R_A = \dfrac{2wL(5 + 3\alpha)}{8 + 3\alpha}$, $M_A = -\dfrac{2wL^2(2 + 3\alpha)}{8 + 3\alpha}$,

(b) $v(x) = \dfrac{w}{24EI}\left[(x^4 - 8Lx^3 + 24L^2x^2)\right.$

$\left. + \dfrac{Lx^2(x - 6L)}{8 + 3\alpha}\right]$

9.37: $P = 2wL/3$

9.39: $M_0 = wL^2/3$

9.41: $R_A = 7wL/32\uparrow$, $R_B = 5wL/16\uparrow$, $R_C = wL/32\downarrow$

9.43: $R_C = \dfrac{5wL}{8}\left(\dfrac{\alpha}{1+\alpha}\right)$, $R_A = R_B = \dfrac{1}{2}(wL - R_C)$,

$\alpha = kL^3/48EI$

9.45: $R_A = R_B = P/2\uparrow$, $M_A = M_B = -PL/8$

9.48: $v_C = \dfrac{11PL^3}{96EI}$, $v_H = \dfrac{5PL^3}{48EI}$

9.49: $v_B = \dfrac{2q_0L^4}{3\pi^4EI}(\pi^3 - 24)$

9.53: (a) $v_C = \dfrac{5q_0L^4}{768EI}$, (b) $v_C = \dfrac{q_0L^4}{\pi^4}$

9.56: (a) $G_{V_C} = -\zeta/L$, $G_{M_C} = (L-x)\zeta/L$,

$(0 \le \zeta \le x_c)$;

$G_{V_C} = (L-\zeta)/L$, $G_{M_C} = (L-\zeta)x/L$,

$(x_c \le \zeta \le L)$,

(c) $M_C = wL^2/128$, (d) $M_C = 5wL^2/128$,

(e) $M_C = 21wL^2/512$ when placed in region $3L/16 \le$

$\zeta \le 7L/16$

9.58: (c) $|M_A|_{max} = 384.9$ N-m, $\zeta_{cr} = \dfrac{1}{3}(3 - \sqrt{3})L$

9.61: $v(x) = \dfrac{M_0 x}{6EIL}(2x^2 - 3Lx + L^2)$,

$|v|_{max} = \dfrac{\sqrt{3}M_0L^2}{108EI} = 0.01604\dfrac{M_0L^2}{EI}$ at

$\dfrac{x_{cr}}{L} = \dfrac{1}{2}\left(1 \pm \dfrac{1}{3}\sqrt{3}\right)$

9.63: $v(x) = \dfrac{q_0L^4}{3\pi^4EI}\left\{\pi^3[3(x/L)^2 - (x/L)^3] + \right.$

$\left. 48\left(\cos\dfrac{\pi x}{2L} - 1\right)\right\}$, $v_{max} = \dfrac{2(\pi^3 - 24)q_0L^4}{3\pi^4EI}$

9.65: $v_{max} = v(L/2) = \dfrac{Pa}{24EI}(3L^2 - 4a^2)$

9.67: $v_B = \dfrac{5PL^3}{384EI} + \dfrac{wL^4}{96EI}$, $v_D = \dfrac{5PL^3}{768EI} + \dfrac{37wL^4}{6144EI}$

9.69: (a) $v(x) = \dfrac{P}{48EI}(-4x^3 + 3Lx^2 + 8\langle x - L/2\rangle^3)$,

(b) $v(L/2) = \dfrac{PL^3}{192EI}$,

(d) $R_A = R_B = P/2$, $M_A = M_B = -PL/8$

9.71 $\theta_B = wL^3/6EI$, $v_B = wL^4/8EI$

9.73: (a) $\theta_A = \dfrac{L}{6EI}(2M_A + M_B)$, $\theta_B = -\dfrac{L}{6EI}$

$(2M_B + M_A)$, (b) $v(L/2) = \dfrac{L^2}{16EI}(M_A + M_B)$

9.75: (a) $v(a) = -\dfrac{M_0 a^2}{4EI}(1 - a/L)[-2 + 4(a/L) -$

$(a/L)^2]$, (d) down: $0 < a/L < 2 - \sqrt{2}$

9.77: (a) $v_C = \dfrac{Pa^2b^2(\alpha a + b)}{3\alpha EIL^2}$

9.79: (a) $v(x) = -\dfrac{q_0L^4}{5760EI}[40(x/L)^3 - 3\langle 2(x/L) - 1\rangle^5$

$- 37(x/L)]$, $v(L/2) = \dfrac{17q_0L^4}{11,520EI}$

$= 1.476 \times 10^{-3}\dfrac{q_0L^4}{EI}$

9.81: (a) $v(x) = \dfrac{q_0L^4}{16\pi^4}\sin(2\pi x/L)$

9.82: (b) $v_D = \dfrac{117wL^4}{384EI}$, (c) $k = 12EI/7L^3$

9.83: $R/L = \sqrt{3}/30$

9.84: $v(x) = \dfrac{q_0L^4}{24\pi^4EI}\{\cos(2\pi x/L) + \pi^2[\pi^2(x/L)^4$

$- 2\pi^2(x/L)^3 + 6(x/L)^2 + (\pi^2 - 6)x/L] - 3\}$

9.86: $R = \dfrac{5P}{16(1 + 3EI/kL^3)}$

9.88: (a) $R_A = 7P\downarrow$, $R_B = 8P\uparrow$, $M_A = 6PL$

9.90: (a) (i) $k = 192EI/L^3$, (ii) $k = 672EI/L^3$

9.91: (a) $v(x) = \dfrac{M_0 x^2}{4EI}(x/L - 1)$, (b) $R_A = 3M_0/2L\uparrow$,

$M_A = -M_0/2$

9.93: (a) $M_A = -\dfrac{wL^2}{12}\left(\dfrac{6 + \beta L/EI}{4 + \beta L/EI}\right)$,

$R_A = \dfrac{wL}{2}\left(\dfrac{5 + \beta L/EI}{4 + \beta L/EI}\right)$

(b) (ii) $M_A = M_B = -wL^2/12$,

$R_A = R_B = wL/2$

9.95: $R_B = R_D = \left(\dfrac{12AEI}{AL^2 + 24I}\right)(\alpha\Delta T - \delta/L)$

9.97: (b) $R_A = \dfrac{5P}{16}\left(\dfrac{\alpha}{3 + 2\alpha}\right)$, $R_D = \dfrac{P}{16}\left(\dfrac{48 + 27\alpha}{3 + 2\alpha}\right)$,

$M_D = -\dfrac{PL}{16}\left(\dfrac{24 + 11\alpha}{3 + 2\alpha}\right)$, $v_C = \dfrac{5PL^3}{48EI}\left(\dfrac{\alpha}{3 + 2\alpha}\right)$

9.98: $\delta = \dfrac{7wL^4}{1152EI}$

9.99: (b) $\gamma < 3EI/L^3$ stable, otherwise unstable

9.101: (b) $M_B = -wb^3/8L$, (c) $M_B = \dfrac{wL^2}{60} - \dfrac{PL}{10}$,

$M_C = \dfrac{PL}{40} - \dfrac{wL^2}{15}$

9.105: (b) $v_{max} = v(x = 0.236219L) = 5.7191 \times 10^{-4}\dfrac{wL^4}{EI}$

Chapter 10

10.1: $p = 4.8$ MPa

10.3: (a) $p_{cr} = \dfrac{2tE\Delta}{RL(1 - 2\nu)}$, (b) $p_{cr} = 14.7$ MPa

10.5: $h = 22.66$ m

10.6: (a) $(\sigma_\theta)_{(out)} = -(\sigma_\theta)_{in} = E\delta/2R$

10.8: $p = \dfrac{4\sigma_0 t}{R(3 + \cos 2\alpha)}$, σ_0 governs

10.9: $\Delta T = \dfrac{1}{\alpha}\left[\dfrac{\delta}{L} + \dfrac{pR(1 - 2\nu)}{2Et}\right]$

10.11: $\sigma = \dfrac{R}{6t}[3p + 2\rho R]$

10.12: (b) $(\sigma_x)_{A,C} = \dfrac{R}{8t}(4p - \sqrt{3}\rho L) \mp \dfrac{\rho L}{16t}(L - 4a +$

$2\sqrt{3}R)$, $(\sigma_x)_B = \dfrac{R}{8t}(4p - \sqrt{3}\rho L)$, $(\sigma_\theta)_{A,B,C} = pR/t$

10.15: (a) $(\sigma_\theta)_1 = \dfrac{pR_1}{t}$, $(\sigma_x)_1 = \dfrac{pR_1(R_1 + 2\nu R_2)}{2t(R_1 + R_2)}$,

$(\sigma_\theta)_2 = 0$, $(\sigma_x)_2 = \dfrac{p(1 - 2\nu)R_1^2}{2t(R_1 + R_2)}$,

(b) $\Delta_x = \left[\dfrac{(1 - 2\nu)R_1^2 L}{2Et(R_1 + R_2)}\right]p$

10.17: (a) $n_{\min} = 2\sqrt{2}$, (b) $\alpha = 35.26°$

10.19: (a) $\sigma_\phi = \dfrac{\rho R^2}{3t}\left[\cos\theta + \dfrac{(1 - \cos\theta)}{\sin^2\theta}\right]$,

(b) $\sigma_\phi = \dfrac{7}{18}\dfrac{\rho R^2}{t}$

Chapter 11

11.1: (a) $P_{cr} = 200$ N

11.3: $P_{cr} = \dfrac{\beta}{2L}(3 - \sqrt{5})$

11.5: (a) $P_{cr} = \dfrac{1}{2L}\left[\beta_1 + 2\beta_2 - [\beta_1^2 + 4\beta_2^2]^{1/2}\right]$

11.7: (b) (i) $P = 83.3$ N, (ii) $P = 80$ N, (c) $P_{cr} = 100$ N

11.9: (a) $P_{cr} = 32.9$ N, (b) $P_{cr} = 82.4$ N

11.11: $P_{cr} = 0.4601Ea^3 t/L^2$

11.13: (b) $P_{cr} = \pi^2 EI/4L^2$

11.17: $\tan(\lambda L) = \lambda L(1 + \gamma)$, $\lambda = \sqrt{P/EI}$

11.19: (a) $\tan(\lambda L) = -\gamma\lambda L$, $\lambda = \sqrt{P/EI}$

11.25: (b) $v(x) = \sin(\lambda x) + \cos(\lambda L) \cdot [(\lambda L)^3/\alpha - \lambda x]$,

$\lambda = \sqrt{P/EI}$, $\alpha = kL^3/EI$

11.27: (a) $v(x) = \dfrac{w}{P}\left\{\dfrac{1}{\lambda^2}\left[\left(\dfrac{1 - \cos\lambda L}{\sin\lambda L}\right)\sin\lambda x\right.\right.$

$\left.\left. + \cos\lambda x - 1\right] - Lx/2 + x^2/2\right\}$, $\lambda = \sqrt{P/EI}$

11.31: (a) $P_{cr} = \dfrac{(1 + c)^2\beta}{cL}$, (b) $(P_{cr})_{\min} = 4\beta/L$

11.34: $v(x) = e\left\{\cos(\lambda x) - \left[\dfrac{1 + \cos(\lambda L)}{\sin(\lambda L)}\right]\sin(\lambda x)\right.$

$\left. + (2x/L - 1)\right\}$

11.41: $W = \dfrac{3\pi^2 EI}{L^2}$

11.42: (a) $k = P_{cr}\left(L - \dfrac{\pi R^2}{4}\sqrt{\dfrac{\pi E}{P_{cr}}}\right)^{-1}$, (b) $k = 564$ N/cm

11.44: $k_{\max} = \dfrac{27\pi^2 EI}{16L^3}$

11.45: (b) $\phi = \tan^{-1}(\cot^2\alpha)$,

(c) $(P_{cr})_{\max} = \dfrac{2\sqrt{2}\pi^2 EI}{L^2}\left[\dfrac{(2 - \sin^2 2\alpha)^{1/2}}{\sin^2 2\alpha}\right]$

11.46: $k > 16\pi^2 EI/L^3$

Chapter 12

12.3: (c) (iii) $\tau_{\max} = 20\ T/a^3$

12.7: $\tau_{\max} = 17.63$ MPa

12.9 $GC = G\dfrac{bt_0^3}{12}$

12.10: (a) $\Delta_y = -\dfrac{3TL}{2Gt^3(1 + 2c/h)}$

12.13: $b/a = 4$

12.14: (a) $\dfrac{\tau_a}{\tau_b} = t/3R$, (b) $\dfrac{\Theta_a}{\Theta_b} = t^2/3R^2$

12.17: (a) Hexagon: $C = 41.784a^3 t$, (b) Hexagon:

$\tau = 0.04466T/a^2 t$, (c) $C = 43.77a^3 t$

12.18: (b) $\tau_{xy}|_B = -2\kappa/b$, $\tau_{xy}|_E = -2\kappa\alpha/b$,

$\tau_{xz}|_A = -2\kappa/a$, $\tau_{xz}|_D = -2\kappa\alpha/b$,

where $\kappa = \dfrac{T}{\pi ab(1 - \alpha^4)}$

(c) $\tau_{\max} = \dfrac{2T}{\pi ab^2(1 - \alpha^4)}$

12.20: (a) $C = \dfrac{na^3 t}{4}\cot^2(\pi/n)$, (c) $\tau = \dfrac{2T}{na^2 t}\tan(\pi/n)$

Chapter 13

13.5: (b) $\sigma_{xA} = -6.71$ MPa, $\sigma_{xB} = 0.872$ MPa

13.7: (b) $\beta = -\tan^{-1}\left(\dfrac{2Pab + \sigma_0 I_0}{Pab + 4\sigma_0 I_0}\right)$

13.9: $e = \dfrac{a}{2}\dfrac{3a + 2h}{3a + h}$

13.11: $e/d = 2/3$

13.13: (c) $e = 2R\left(\dfrac{\pi + 3}{\pi + 4}\right)$

13.15: $e = \left(\dfrac{h_1^3}{h_1^3 + h_2^3}\right)d$

13.17: $\alpha = \dfrac{1}{18}(\sqrt{157} - 7)$

13.18: $e = 3\pi R/8$

13.21: $e = a\sqrt{3}/6$

13.23:   (a) $e/a = 1 - \dfrac{4\sqrt{3}}{9} = 0.2302$

13.27:   (a) $\tau_A = \tau_C = \tau_D = \dfrac{W}{2\pi Rt}$, $\tau_B = \dfrac{3W}{2\pi Rt}$

13.29:   (a) $e = \sqrt{2}a/4$

13.31:   (a) $\alpha = 20.56°$, (b) $(\Delta_B)_z = \dfrac{9\sqrt{73}}{292} \dfrac{PaL}{t^3G}$

13.33: $\alpha = \dfrac{1}{2}(\pi + \sqrt{\pi^2 + 8})$

13.35: $b/a = (\sqrt{2}/3)^{1/2}$

13.36: $b/a = 3/4$

13.37:   (b) $\Delta_D = \dfrac{3TL}{2\pi t^3G}$

13.39:   (a) $e/d = \dfrac{1}{1 + \sin^2\theta}$

Chapter 14

14.5:   (a) $\Delta_B = 4PL^3/81EI\downarrow$, (b) $\Delta_A = 17PL^3/162EI\downarrow$,
(c) $\theta_A = PL^2/6EI$

14.7: $\Delta_G = \dfrac{7wL^4}{48EI}\downarrow$

14.8:   (a) $\theta_A = 33PL^3/2EI$

14.10: $M_0 = 7wL^2/64$ clock-wise

14.11:   (a) $v(x) = \dfrac{P}{6EI}(3Lx^2 - x^3), (0 \le x \le L)$

14.13: $\Delta_A = \dfrac{6PL^3}{Eb_0d^3}$

14.15: $\Delta L = \dfrac{4\nu P}{\pi DE}$

14.17:   (a) $M_A = 2wL^3/a$. (b) $R_C = \dfrac{wL}{2a}(5a + 4L)$,
(c) $M_C = -2wL^2$

14.21:   (b) (ii) Between D and F, $|M_G|_{\max} = 5wL^2/8$;
(c) (ii) between G and I, $(R_I)_{\max} = 7wL/4$

14.22:   (b) (i) Between E and F, $|M_D|_{\max} = 3wL^2/4$,
(ii) between D and E, $|V_C^-|_{\max} = 5wL/4$;
(c) (i) $4.667L \le \zeta \le 5.667L$, $|M_D|_{\max} = 5wL^2/6$,
(ii) $4.4L \le \zeta \le 5.4L$, $|V_C^-|_{\max} = 27wL/20$

14.27: $\Delta_B = \dfrac{3q_0L^4}{8EI}\uparrow$

14.29: $\Delta_A = \dfrac{6PL^3}{Eb_0d^3}$

14.31: $\Delta_B = \Delta_D = \dfrac{wL^4}{192EI}$

14.34: $\theta = 7.125°$

14.35:   (a) $\Delta_C = \dfrac{32\sqrt{3}(6 - \sqrt{3})}{33}\dfrac{PL}{AE}$; (b) $F_{BF} = F_{CE} = \dfrac{4\sqrt{3}(6 - \sqrt{3})}{33}P$

14.37:   (a) $v(x) = \dfrac{32wL^4}{\pi^4 EI}\displaystyle\sum_{n\,\text{odd}}^{\infty}\left[\dfrac{1 - 2\dfrac{(-1)^{\frac{n-1}{2}}}{n\pi}}{n^4}\right] \times \left(1 - \cos\dfrac{n\pi x}{2L}\right)$

14.38: $v(x) = \dfrac{2wL^4}{\pi^4 EI}\displaystyle\sum_{n=2,4,6}^{\infty}\dfrac{1}{n^4}\left(1 - \cos\dfrac{n\pi x}{L}\right)$

14.40: $\alpha = 10/9$

14.42:   (b) (i) $L \le \zeta \le 3L$, $|M_A|_{\max} = 19wL^2/6$,
(ii) $1.2L \le \zeta \le 3.2L$, $|M_A|_{\max} = 16wL^2/5$;
(c) (i) $7L \le \zeta \le 9L$, $(R_H)_{\max} = 2.5wL$;
(ii) $20L/3 \le \zeta \le 26L/3$, $(R_H)_{\max} = 8wL/3$

14.43:   (b) $R_A = q_0L/6$, $M_A = -q_0L^2/6$, $V_C = q_0L/24$

14.47: $\Delta_C = \dfrac{\pi PR^3}{4EI} + \dfrac{(3\pi - 8)PR^3}{4GC}$

14.49:   (a) $\alpha = 22.5°, 112.5°$

14.51: $v(x) = \dfrac{32PL^3}{\pi^4 EI}\displaystyle\sum_{n\,\text{odd}}^{\infty}\dfrac{1}{n^4}\left(1 - \cos\dfrac{n\pi x}{2L}\right)$;
$M(x) = \dfrac{8PL}{\pi^2}\displaystyle\sum_{n\,\text{odd}}^{\infty}\dfrac{1}{n^2}\cos\dfrac{n\pi x}{2L}$

Chapter 15

15.1: $h/R \le 3/4$

15.3: $P_{cr} = 5kL/2$

15.5: $P_{cr} = \dfrac{4\beta}{3L}$

15.7:   (a) $P_{cr} = \dfrac{1}{2L}\left[\beta_1 + 2\beta_2 - (\beta_1^2 + 4\beta_2^2)^{1/2}\right]$

15.8:   (b) (i) $P = 83.3$ N, (ii) $P = 80$ N, (c) $P_{cr} = 100$ N

15.9:   (a) $P_{cr} = \dfrac{kab}{b + 2a}$

15.11:   (a) $P_{cr} = 30EI/L^2$

15.12:   (a) $P_{cr} = 21.0EI/L^2$

15.13   (a) $P_{cr} = \dfrac{\pi^2 EI\alpha}{4L^2}\left\{1 + (\alpha - 1)\left[\gamma + \dfrac{1}{\pi}\sin(\pi\gamma)\right]\right\}^{-1}$

15.14:   (a) $W = 8.298EI/L^2$

15.15: $P_R = \dfrac{\pi^2 EI}{L^2}(1 + 2a/L)^{-1}$

15.16   (a) $P_R = \dfrac{3EI_0}{5L^2}\dfrac{[1 - (1 - \gamma)^5]}{\gamma}$,
(b) $P_R = 1.1625\,EI_0/L^2$

15.17: $h/R < \sqrt{2}R/2$

15.19:   (a) $\theta < \sin^{-1}\left[(\sin\theta_0)^{1/3}\right]$,
(b) $P = -kL\left[1 - (\sin\theta_0)^{2/3}\right]^{3/2}$

15.20:   (a) $P_{cr} = \dfrac{\beta}{L}\dfrac{\theta}{\sin\theta} + kL\cos\theta$

15.22:   (c) $P > \dfrac{kL(1 + \gamma)^2}{2 + \gamma}$

Index